Strickberger's
EVOLUTION
FIFTH EDITION

Strickberger's EVOLUTION

FIFTH EDITION

BRIAN K. HALL
Dalhousie University

BENEDIKT HALLGRIMSSON
University of Calgary

JONES & BARTLETT
LEARNING

WORLD HEADQUARTERS

Jones & Bartlett Learning
5 Wall Street
Burlington, MA 01803
978-443-5000
info@jblearning.com
www.jblearning.com

Jones & Bartlett Learning books and products are available through most bookstores and online booksellers. To contact Jones & Bartlett Learning directly, call 800-832-0034, fax 978-443-8000, or visit our website, www.jblearning.com.

Substantial discounts on bulk quantities of Jones & Bartlett Learning publications are available to corporations, professional associations, and other qualified organizations. For details and specific discount information, contact the special sales department at Jones & Bartlett Learning via the above contact information or send an email to specialsales@jblearning.com.

PRODUCTION CREDITS

Chief Executive Officer: Ty Field
President: James Homer
SVP, Chief Marketing Officer: Alison M. Pendergast
Executive Publisher: Kevin Sullivan
Senior Acquisitions Editor: Erin O'Connor
Editorial Assistant: Rachel Isaacs
Editorial Assistant: Michelle Bradbury
Production Editor: Daniel Stone
Associate Rights & Photo Research: Lauren Miller
Marketing Manager: Lindsay White
Composition: Circle Graphics
Cover/Title Page Image: © Galushko Sergey/ShutterStock, Inc.
Printing and Binding: Courier Kendallville
Cover Printing: Courier Kendallville

To order this product, use ISBN: 978-1-4496-9192-9

LIBRARY OF CONGRESS CATALOGING-IN-PUBLICATION DATA

Hall, Brian Keith, 1941-
  Strickberger's evolution / Brian K. Hall & Benedikt Hallgrímsson. — 5th ed.
    p. cm.
  Includes index.
  ISBN 978-1-4496-1484-3 (alk. paper)
  1. Evolution (Biology) I. Hallgrímsson, Benedikt. II. Strickberger, Monroe W. III. Title. IV. Title: Evolution.
  QH366.2.S78 2013
  576.8—dc23

                                            2012020619

6048

Printed in the United States of America
17 16 15 14 13   10 9 8 7 6 5 4 3 2 1

BRIEF CONTENTS

PART 7   *Populations, Speciation, and Extinctions*

PART 8   *Human Origins, Evolution, and Influence*

CONTENTS

PART 1

*Evolution
and Species*

PART **5**

*Natural
Selection
in Action*

© Photos.com

PART **7**

Populations, Speciation, and Extinctions

PART **8**

*Human
Origins,
Evolution,
and
Influence*

BOXED FEATURES

BOXED FEATURES AT END OF CHAPTER (END BOXES)

PREFACE

Available in four editions, the first in 1990, the fourth in 2008, *Strickberger's Evolution* has consistently remained abreast of the latest discoveries and approaches to the study of biological evolution and the science of evolutionary biology. The fourth edition incorporated the then fast developing fields of evolutionary developmental biology, paleobiology, and evolutionary ecology. In addition, insights from sequencing entire genomes and the increasing application of bioinformatic tools, cladistic analyses, and systems approaches to the study of organisms made their appearance. This, the fifth edition, continues the tradition of incorporating the latest discoveries concerning evolution. Additionally, this edition has all illustrations and figures in four colors, substantially enhancing the visual learning aspect of the book.

Many strands of evolutionary thought are old and have endured since 1858 when Charles Darwin was writing *On the Origin of Species*. Essential to our understanding of evolution is that groups of organisms are bound together by their common inheritance; that the past has been long enough for inherited changes to accumulate; and perhaps most essential of all, that discoverable biological processes and natural relationships among organisms explain the reality of evolution. Although each of these aspects had been studied and discussed at various times in human history, only after the mid-nineteenth century when Charles Darwin developed his theory of evolution by natural selection was biological evolution recognized as the science of life.

The central aim of the science of evolutionary biology is to explain the origins and diversity of life. The properties of different organisms—the organization and function of their component parts—are explained within the context of their organismal histories, which include adaptations to specific lifestyles at particular times. When historical conditions are repeated, and different organisms are subjected to similar selective evolutionary forces, some common features can be predicted; geographically widely separated populations and species, such as fish adapted to cave conditions, consistently show rudimentary eyes, enhanced development of chemosensory organs, and loss of pigment, among other common attributes. Other evolutionary changes are contingent on past history and environments, for example, the evolution of flight among reptiles and not among frogs or most mammals.

In this text we provide the evidence upon which general principles can be based and explore those aspects of organisms and biological processes that appear to be specific to lineages or stages of evolution. To this end, there is a logical arrangement of the subject matter, although the structure of the book allows you to begin with Part 8 or to read individual parts as units independent of the other parts:

- beginning with the history of evolutionary ideas, species and their relationships, and patterns of evolutionary change (Part 1);
- moving to the origins of the universe, Earth, rocks, continents, oceans and atmospheres (Part 2);
- before discussing the origin, molecules, cells, organisms and natural selection (Part 3).
- Part 4 considers how 19th century naturalists Charles Darwin and Alfred Russel Wallace came to almost the same theory of evolution by natural selection under the influence of social economic writings of Thomas Malthus; how Gregor Mendel's experiments provided basic principles of inheritance that were unknown to Darwin or to Wallace; and how later studies showed how environments and genes interact in inheritance (heredity).

- Operation of natural selection and how genotypes (which are inherited) and phenotypes (which are not) interact are the topics of Part 5.
- Natural selection is only effective if biological variation is present. Sources of variation in individuals and populations are the topics of Part 6.
- Interactions between individuals and populations and between both and the environment, as reflected in competition, predation, coevolution, and extinction, maintain species and are important components of speciation, the topics of Part 7.
- The book ends with how cultural, social, and biological evolution interact in our own populations, the impact humans have had on the evolution of other species, and how culture, evolution, and religion interact and coexist (Part 8).

Summaries are provided at the beginning of each chapter. We suggest that you read the summary both before and after you read each chapter—before to see whether what is coming is familiar or not; after to be sure you have captured the essence of the topic(s) of the chapter. Boxes within chapters are used to draw attention to particular topics. Important terms and concepts are highlighted in **bold** with summary text highlighted in blue.

Most chapters end with an End Box or Boxes (23 in all), each of which begins with a synopsis of the contents. Some End Boxes cover material that was distributed among several chapters in the fourth edition: Fossils as Evidence of Past Life (End Box 3.1 in Chapter 3) is an example. Other End Boxes are large and relevant to the content of more than one chapter. End Box 17.1 Animals Arise (4300 words, five figures) is an example. This End Box discusses the organisms present in the Precambrian and Early Cambrian, examines how stem taxa are distinguished from crown taxa, how limits on morphological variation are studied, and how "molecular clocks" are used to calibrate rates and times of evolution. Because knowledge of gene regulation obtained from living organisms tell us much about how animals arose and evolved, this End Box has been placed at the end of the chapter on gene regulation and the origin of variation.

Much more information and analysis is available on all topics discussed than we could possibly have included in the book. Many introductory textbooks provide no access to the literature upon which the science is based. We think this is a mistake for two major reasons:

(1) One is intrinsic to the nature of the subject matter. Evolution is a science, and as such, you as the reader of this scientific textbook should not take our representations and interpretations of scientific research for granted. Rejection or acceptance of a scientific hypothesis is based on whether data gathered to test the hypothesis refute it or not. As one example, the sequence of hominin, primate-like fossils extending from the far past to the present supports the hypothesis that humans have a primate origin (Chapter 25).

(2) The second reason is that whether this book is used for an introductory or a more advanced class—and it is, in our view, suitable for either—the primary literature cited provides an ideal basis for tutorials, discussions, essays and/or presentations.

As you should be able to check the primary literature for yourself, we have cited enough of that primary literature to enable you to enter it with ease. General references are highlighted in yellow in the text (e.g., Bowler, 2003) and grouped together at the end of each chapter. References to specific studies are provided in full as footnotes.

ORGANIZATION OF MATERIAL IN THE FIFTH AND FOURTH EDITIONS

Organization of some of the material in the fifth edition differs from the fourth. Specifically:

- Information on species and species concepts, found in chapters 2, 8, and 11 in the fourth edition is consolidated as Chapter 2 in this edition.

- Evidence of patterns of evolution in multicellular organisms (fungi, plants, animals) was discussed in six chapters (14–19) in the fourth edition. That information has been condensed into Chapter 3 and into End Boxes in this edition.
- Chapters 14 and 15 in the fifth edition bring together information on natural selection that was distributed throughout the fourth edition, especially in chapters 11, 12, and 14.
- The two major modes of speciation discussed in Chapter 24 in the fourth edition are each given separate chapters (22, 23) in the fifth edition.
- Importance of sources of variation in evolution is highlighted by placing the discussions of individual- and population-level variation into four chapters in Part 6.

SUPPLEMENTS TO THE TEXT

Jones & Bartlett Learning offers an array of ancillaries to assist instructors and students in teaching and mastering the concepts in this text. To request additional information and review copies of any of the following items please email **info@jblearning.com**.

FOR THE STUDENT

Developed exclusively for the fifth edition of *Strickberger's Evolution,* the Navigate Companion Website, **go.jblearning.com/Evolution5eCW**, offers a variety of resources to enhance understanding of evolution. The site contains quizzes and exercises to test comprehension and retention and an interactive glossary. This site also has links to other interesting and informative websites and seminal papers in the field of evolution and access comes free with each new printed edition of the text.

FOR THE INSTRUCTOR

Compatible with Windows and Macintosh platforms, the Instructor's Media CD provides instructors with the following traditional ancillaries:

- The PowerPoint™ Image Bank provides the illustrations, photographs, and tables (to which Jones & Bartlett Learning holds the copyright or has permission to reproduce digitally) inserted into PowerPoint slides. You can quickly and easily copy individual images or tables into your existing lecture slides.
- The PowerPoint Lecture Outline Presentation Package, created by Richard E. Strauss of Texas Tech University, provides lecture notes and images for each chapter of *Strickberger's Evolution*. Instructors with Microsoft PowerPoint software can customize the outlines, art, and order of presentation.

Additional resources are available for download online, including:

- The Test Bank of over 700 questions, created by DorothyBelle Poli of Roanoke College, is available as Rich Text Files and in several LMS compatible formats.
- Supplemental problems and exercises that were updated for the edition will enhance students' comprehension of and appreciation for the material in text. They are available as downloadable word documents at **go.jblearning.com/ Evolution5eCW**.

ACKNOWLEDGMENTS

The following individuals, each a specialist in evolutionary biology, in a particular area of evolutionary biology and/or who teach evolution, took time from their academic schedules to comment on drafts of chapters or boxes for this and previous editions. We appreciate your insights and forthright comments:

Ehab Abouheif, McGill University
Sina Adl, Dalhousie University
Christine Andrews, University of Chicago
Christopher C. Austin, Louisiana State University
Joseph Beilawski, Dalhousie University
Wouter Bleeker, Geological Survey of Canada
William A. Brindley, Utah State University
Michael A. Buratovich, Spring Arbor University
Albert Buckelew, Bethany College
Lisa Budney, Dalhousie University
Richard Burian, Virginia Tech
Nicholas J. Cheper, East Central University
Lee Christianson, University of the Pacific
M. Michael Cohen, Dalhousie University
David Deamer, University of California Santa Cruz
Leslie Dendy, University of New Mexico–Los Alamos
Tamara Franz-Odendaal, Mount Saint Vincent University
Barbara A. Frase, Bradley University
Ann Fraser, Kalamazoo College
Jonathan Frye, McPhearson College
Jennifer M. Gleason, University of Kansas
Kenneth Gobalet, California State University Bakersfield
Dalton R. Gossett, Lousiana State University
Rosemary Grant, Princeton University
Michael Gray, Dalhousie University
Robert Guralnick, University of Colorado, Boulder
William Hahn, Georgetown University
Jenna Hellack, University of Central Oklahoma
W. Wyatt Hoback, University of Nebraska at Kearney
John Hunt, University of Arkansas—at Monticello
John C. Jahoda, Bridgewater State University
Heather Jamniczky, University of Calgary
Megan Johnson, University of Calgary
Mark Johnston, Dalhousie University
Devon B. Keeney, Le Moyne College
Hollie Knoll, University of Calgary
Shigeru Kuratani, Riken Center for Developmental Biology, Japan
Robert Lee, Dalhousie University
Troy A. Ladine, East Texas Baptist University
Vicky M. Lentz, State University of New York College at Oneonta
Sally Leys, University of Alberta
Chris N. Lorentz, Thomas More College
Ian Mclaren, Dalhousie University
Gordon McOuat, University of Kings College
Jeff Meldrum, Idaho State University
Mary Murnik, Ferris State University
Audrey Napier, Alabama State University
Karl Niklas, Cornell University

Glen Northcutt, University of California, San Diego
John Olsen, Rhodes College
Wendy Olson, University of Northern Iowa
Leslie Orgel, The Salk Institute
Louise Page, University of Victoria
Keith W. Pecor, The College of New Jersey
Ray Pierotti, University of Arkansas
David Piper, Atlantic Geological Survey
DorothyBelle Poli, Roanoke College
David Polly, Indiana University, Bloomington
Mark Regan, Griffiths University, Brisbane
Andrew Roger, Dalhousie University
Harry Roy, Rensselaer Polytechnic Institute
Anthony Russell, University of Calgary
Gary D. Schnell, University of Oklahoma
David Scott, South Carolina State University
Kevin R. Siebenlist, Marquette University
Alastair Simpson, Dalhousie University
Shiva Singh, University of Western Ontario
Eric Sniverly, University of Calgary
George Spagna, Randolph-Macon College
Richard E. Strauss, Texas Tech University
John Stone, McMaster University
John Thomlinson, California State University, Dominguez Hills
James F. Thompson, Austin Peay State University
Michael E. Toliver, Eureka College
Matt Vickaryous, University of Calgary
Marvalee Wake, University of California Berkeley
Alan Weiner, University of Washington
Ken Weiss, Pennsylvania State University
Polly Winsor, University of Toronto
Pat Wise, University of Calgary
Nate Young, Stanford University
Stephen J. Zipko, Farleigh Dickinson University

The glossary is a vital part of any book and especially of a textbook. We have prepared a glossary of 900 terms to provide definitions of the key terms and processes discussed in the text. Along with the index, the glossary serves as a way to access similar or related terms and concepts. As one example, under the term *Adaptive radiation,* Biogeography, Innovation, Speciation, and Zoogeography are cross-referenced as topics related to adaptive radiation. We have also set up the glossary to draw attention to contrasting terms or concepts. As one key example, under the term *Individual,* Populations is cross-referenced as a fundamentally different level of biological organization others.

The index is also a vital part of any book and especially of a textbook. We have prepared an index of 2,870 terms to allow you to access the text with ease. As is the glossary, the index is cross-referenced extensively, enabling you to find related material on a topic of interest.

Brian K. Hall
Dalhousie University

Benedikt Hallgrímsson
University of Calgary

January, 2013

Brian Hall (left) and Benedikt Hallgrímsson (right) photographed outside Charles Darwin's home, Down House, in July 2006 after having participated in the symposium on Tinkering: The Micro-evolution of Development, held at the Novartis Foundation in London. Drs. Hall and Hallgrímsson have worked together for many years. They collaborated on the edited volume *Variation: A Central Concept in Biology* (2005), which addresses a concept of fundamental importance in evolutionary biology. The late Ernst Mayr concluded his foreword to the book with "In short, variation is an end-less source of challenging questions." Their latest edited volume is *Epigenetics: Linking Genotype and Phenotype in Development and Evolution* (2011), which addresses the fundamental question of how phenotypic traits and phenotypes are produced.

BRIAN HALL, born, raised, and educated in Australia, has been associated with Dalhou-sie University in Halifax, Nova Scotia since 1968, most recently as a University Research Professor and George S. Campbell Professor of Biology, and since July 2007 as University Research Professor Emeritus and Emeritus Professor of Biology. He was Killam Research Professor at Dalhousie University (1990–1995), Faculty of Science Killam Professor (1996–2001), and Canada Council for the Arts Killam Research Fellow (2003–2005).

Trained as an experimental embryologist, for the past 45 years he has undertaken research into vertebrate skeletal development and evolution and played a major role in integrating evolutionary and developmental biology into the discipline now known as Evolutionary Developmental Biology (evo-devo); he wrote the first evo-devo text book, published in 1990 and a second edition in 1999.

A fellow of the Royal Society of Canada and Foreign Honorary Member of the American Academy of Arts and Sciences, Dr. Hall has earned numerous awards for his research, teaching, and writing, including the 2005 Killam Prize in Natural Sciences, one of the top scientific awards in Canada.

BENEDIKT HALLGRIMSSON was born in Reykjavík, Iceland, and completed his studies at the University of Alberta and the University of Chicago. A biological anthropologist and evolutionary biologist, he combines developmental genetics and bioinformatics with morphometrics to address the developmental basis and evolutionary significance of phenotypic variation and variability. His work has focused on humans and other primates as well as mouse models and has employed both experimental and comparative approaches to study the evolutionary developmental biology of variation. He is the editor-in-chief of *Evolutionary Biology*, which is a journal dedicated to the synthesis of ideas in evolutionary biology and related disciplines. Based at the University of Calgary, Dr. Hallgrímsson teaches organismal biology and anatomy, for which he has received several Gold Star Teaching Awards, a Letter of Excellence Lecturer Award, and the McLeod Distinguished Achievement Award. From the American Association of Anatomists, he received the Basmajian/Williams and Wilkins Award for educational contributions in 2001. He is professor of Cell Biology and Anatomy and the Deputy Provost at the University of Calgary.

ANSWERING LIFE'S TIMELESS QUESTIONS

Where do we come from? What are we? Where are we going?

> —Translation of the title *D'où venons nous? Que sommes nous? Où allons nous?*, one
> of Paul Gaugin's most famous works, perhaps his ultimate masterpiece, in which he
> depicts his vision of life's great questions.

There is grandeur in this view of life, with its several powers, having been
originally breathed into a few forms or into one; and that, whilst this
planet has gone cycling on according to the fixed laws of gravity, from
so simple a beginning endless forms most beautiful and most wonderful
have been, and are being evolved.

> —The last sentence of *On the Origin of Species by Means of Natural Selection, or the
> Preservation of Favoured Races in the Struggle for Life*, by Charles Darwin, 1859.

Darwin's alienation of the inside from the outside was an absolutely
essential step in the development of modern biology. . . . The time has
come when further progress in our understanding of nature requires
that we reconsider the relationship between the outside and the inside,
between organism and environment.

> —From *The Triple Helix: Gene, Organism, and Environment,* by Richard Lewontin, 2000.

PART 1

Evolution and Species

© Photos.com

CHAPTER

1

Intellectual Origins of the Theory of Biological Evolution

CHAPTER SUMMARY

Many intellectual threads are woven into the fabric of the modern theory of evolution, a theory that requires recognition that Earth is ancient, that all organisms share a common inheritance, and that natural events can be explained by discoverable natural laws. It took a long time for the concept of change of species over time—evolution—to take hold. Idealistic philosophies that saw species as unchangeable and arranged in a hierarchical order from most imperfect to most perfect (expressed as the *Great Chain of Being*) discouraged any thought of species change. Widespread belief in the spontaneous generation of organisms hindered any thought that one organism could be transformed into another. Only in the late sixteenth and early seventeenth centuries did the recognition of inexplicable gaps in the chain of nature prompt European philosophers to propose that the universe might go through successive intermediate stages on the way to perfection. Not until the mid-nineteenth century, when Charles Darwin and Alfred Russel Wallace proposed a mechanism—evolution by natural selection—was species change tied to interactions between organisms and to interactions between organisms and their environment. It was then that the science of evolutionary biology was born.

Image © Jens Stolt/ ShutterStock, Inc.

NATURAL SELECTION *is the sum of the survival and fertility mechanisms that affect reproductive success, as measured by the differential survival and reproduction of individual organisms with particular features.*

UNIVERSAL PROPERTIES OF ORGANISMS

All organisms are bound together by three essential, shared universal properties:

1. A common mechanism of inheritance based on the genetic code carried in DNA.
2. An organization built upon cells.
3. The accumulation of inherited changes over billions of years of existence on Earth.

Only after Darwin and Wallace developed and published their theories of **species transformation in response to natural selection** did biological evolution become an acceptable scientific concept. Acceptance that organisms could change over time and that new types of organisms could arise from existing ones revolutionized the way we viewed the world and the way we understood and explained natural phenomena. Neither Darwin nor Wallace coined the term *evolution,* nor was either of them the first to propose that organisms changed over time. This chapter looks at the word and concept of '*evolution*' before Darwin and Wallace when the idea of evolution was applied to the **development of individual organisms** and not for transformation between generations.[1] We then turn to the concept **of evolution as transformation of species between generations**. A brief history of evolutionary views before Darwin and Wallace is followed by an outline of the evolutionary theory they proposed, and a discussion of how that theory has matured over the past 150 years. The chapter ends by discussing how the **theory of evolution is supported by the science of evolutionary biology**.

EVOLUTION AS DEVELOPMENT OF INDIVIDUALS

Evolution. **1** the process by which different living organisms are thought to have developed or diversified from earlier forms during the history of the earth. **2** the gradual development of an organism, especially from a simple to a more complex form: *the forms of written language undergo constant forms of evolution.* **3** *Chemistry* the giving off of a gaseous product, or heat. **4** a pattern of movements or maneuvers: *silk ribbons waving in fanciful evolutions.* **5** *Mathematics, dated* the extraction of a root from a given quantity. (*Oxford Dictionaries Online,* Copyright ©, 2012).

As this dictionary definition indicates, the word evolution has many different meanings. The concept of evolution can be applied to a wide variety of phenomena; the evolution of an argument, the evolution of the computer, the evolution of heart valves . . .

Using "evolution" to describe the development of an organism stems from the original seventeenth century use of the word. It began with the Latin term *evolutio*, which means unrolling, and was used to describe the unfolding of the parts and organs of an embryo to reveal a preformed body plan. An example would be the (mistaken) belief that a caterpillar "unfolds" into a butterfly as it emerges from the chrysalis (**FIGURE 1.1**), much as the shape of an umbrella is revealed when it is unfolded. Only in the nineteenth century did evolution come to mean transformation of a species or transformation of the features of organisms between generations.

Evolution as embryonic development can be traced to the Swiss botanist, physiologist, lawyer, and poet Albrecht von Haller (1708–1777). In Haller's time, it was presumed that the adult human was preformed in the egg. In 1774, Haller used evolution to describe the preformed development of the adult in the egg: "But the theory of evolution proposed by Swammerdam and Malpighi prevails almost everywhere . . . that there is in fact included in the egg a germ or perfect little human machine . . ."[2]

---

[1] Bowler (2003) and Hall and Olson (2003) are good source books for changing views of evolution.
[2] Haller (1774) cited from H. B. Adelmann, 1966, *Marcello Malpighi and the Evolution of Embryology.* Cornell University Press, Ithaca, NY, pp. 893–894.

FIGURE 1.1 The butterfly that appears to "unfold" from the chrysalis has, in fact, undergone a metamorphosis from a pupal stage.

Another Swiss naturalist and philosopher, Charles Bonnet (1720–1793), built on and developed this concept of **preformation**. Bonnet wrote that *all members of all future generations* are preformed within the egg: cotyledons within the seeds of plants; insect imagos inside pupae; future aphids in the bodies of parthenogenetic female aphids (FIGURE 1.2).[3] Some preformationists like Bonnet (now known as *ovists*) proposed that the miniature adult was contained within the egg. Others (now known as *spermists*

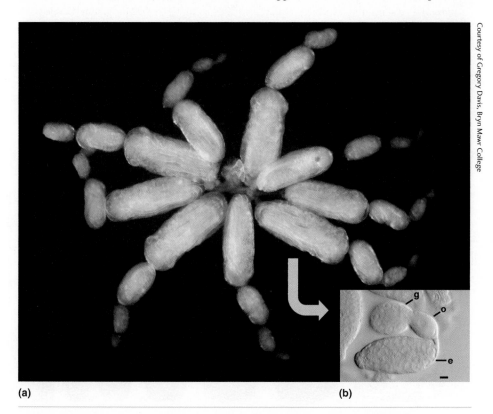

(a)   **(b)**

FIGURE 1.2 The ovaries of a female pea aphid (*Acyrthosiphon pisum*) spread out **(a)** to show embryos in various stages of development from the oldest (with orange eyespots) at the center to the youngest at the distal tips of each ovariole. **(b)** Dissection of one of these older embryos reveals that it possesses ovaries in which each ovariole contains an embryo (e), a developing oocyte (o) and an associated germarium (g) containing nurse cells and future oocytes.

[3] As you can see from Figure 1.2, "seeing" generations of embryos in female aphids in the eighteenth century would have been quite a challenge. My thanks to Greg Davis for the superb specimen shown in Figure 1.2.

FIGURE 1.3 Four views of adult humans as preformed within sperm **(a, b, c)** and within the body of the female **(d)**.

[(a) Adapted from Antoni van Leeuwenhoek. Philosophical Transactions, 1678; (b) Courtesy of National Library of Medicine; (c) Adapted from Nicholas Hartsoeker; (d) Adapted from Needham, J. *History of Embryology.* Cambridge University Press, 1959.]

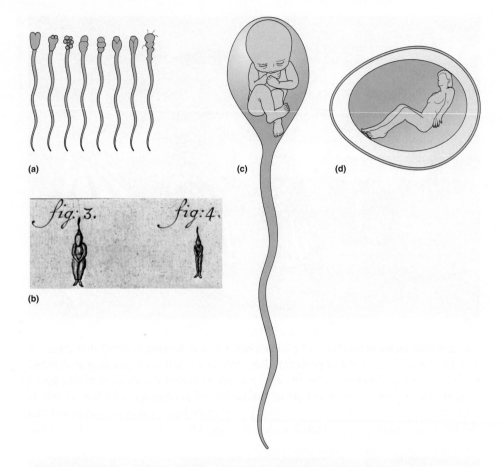

or animalculists) proposed that the adult in miniature was contained within the male seminal fluid. Indeed, they "saw" and drew adults within sperm (**FIGURE 1.3**). In its most highly developed form, preformationism held that the initial member of a species had within it the preformed "germs" of **all future generations**; Eve's ovaries contained the entire preformed human species nested like an infinite set of Russian dolls (**FIGURE 1.4**). Preformationism, therefore, reinforced the fixity of species and left the origin of species to an unknowable creation and/or creator.

FIGURE 1.4 Russian dolls illustrate nesting.

EVOLUTION AS SPONTANEOUS GENERATION

From the 1700s until the middle of the nineteenth century, it was a common belief that, although most large organisms reproduced by sexual means, smaller organisms arose spontaneously from mud or organic matter. According to some folklore, when larger organisms died, they decomposed into smaller ones. Over 400 years ago the Belgian physician and chemist Johann van Helmont (1577–1644) offered a classic expression of this belief of **spontaneous generation**:

> If you press a piece of underwear soiled with sweat together with some wheat in an open mouth jar, after about 21 days the odor changes and the ferment, coming out of the underwear and penetrating through the husks of wheat, changes the wheat into mice. But what is more remarkable is that mice of both sexes emerge, and these mice successfully reproduce with mice born naturally from parents . . . But what is even more remarkable is that the mice which come out of the wheat and underwear are not small mice, not even miniature adults or aborted mice, but adult mice emerge!

Two serious and somewhat contradictory obstacles to the development of evolutionary concepts therefore prevailed at this time:

- The concept of species constancy raised the question of the origin of species, but insistence on species fixity prevented any consideration of transformations between species.
- Acceptance of spontaneous generation seemed contrary to species fixity and cast doubt on any permanent continuity between organisms.

If species could arise *de novo* at any time or be capriciously changed into another species could there ever be a rational mechanism to explain the origin or sequence of appearance of species? In fact, experiments disproving spontaneous generation were published as early as during the seventeenth century as these contradictions began to be considered (BOX 1.1).

EVOLUTION AS TRANSFORMATION BETWEEN GENERATIONS

Not surprisingly (in hindsight), and given the nature of the fossil evidence discovered in the 1820 and 1830s, geologists were among the first to use the term evolution for the transformation of species and for progressive change through geological time.

One of the foremost naturalists, anatomists and geologists in Great Britain in the early nineteenth century was Robert Grant (1793–1874). Grant used the term evolution in 1826 to describe the gradual (and progressive) origin of invertebrate groups in successive strata of rock that "have evolved from a primitive model" by "external circumstances." Charles Lyell (1797–1875) used evolution in the second volume of *Principles of Geology* (1832) to illustrate gradual "improvement" of aquatic organisms to land-dwelling organisms recorded in the fossil record: "[shelled invertebrates] of the ocean existed first, until some of them, by gradual evolution, were *improved* into those inhabiting the land."[4] However, after Darwin published his theory of evolution, Lyell was one of the few geologists who did not embrace Darwin's theory but maintained that the distribution of organisms on Earth was explained by the geological history of the Earth.

---

[4] R. E. Grant, 1826. Observations on the nature and importance of geology. *Edinburgh New Philos. J.* **14**, 270–284; C. Lyell, 1830–1833. *Principles of Geology, Being an Attempt to Explain the Former Changes of the Earth's Surface by References to Causes Now in Operation.* J. Murray, London. (The quotation is from volume 2, page 11.) Both Grant and Lyell supported Lamarck's theory of evolution.

BOX 1.1	Early Experiments Disproving Spontaneous Generation of Insects

A number of seventeenth and eighteenth century experimentalists showed that spontaneous generation of insects does not occur.

In 1668, the Italian physician, Francesco Redi (1621–1697), who is credited with introducing controlled experiments to the study of spontaneous generation, demonstrated that maggots (larvae) arise only from eggs laid by flies, and that flies arise only from maggots.[a] Maggots and flies appeared within a few days after Redi placed meat in a glass jaw open to the air. If the jar was covered with gauze, adult flies did not enter the jar and lay their eggs, and maggots and flies did not appear (**FIGURE B1.1a, b**). A year later, the Dutch microscopist Jan Swammerdam (1637–1680) showed that the insect larvae found in the abnormal swellings (galls; **FIGURE B1.2**) of plants and trees arise from eggs laid by adult insects (**FIGURE B1.1c, d**).

Within a century, further experiments demonstrated that even the appearance of the microscopic "beasties" in decaying or fermenting solutions and broth observed by Antony van Leeuwenhoek (1632–1723)—another Dutch microscopist and the discoverer of microbes—could be explained as originating

Courtesy of National Library of Medicine

François Rédi (1626-1697). Galerie des Offices de Florence.

(a)

Open jar—maggots appear on meat

Covered jar— no maggots

(b)

© Hemera/Thinkstock

(c)

(d)

FIGURE B1.1 Spontaneous generation in the late seventeenth century. Francesco Redi **(a)** showed that maggots did not arise in jars covered with gauze **(b)**. Jan Swammerdam demonstrated that insect larvae in galls **(c)** arise from eggs laid by adult insect **(d)**. Antony van Leeuwenhoek **(e)** used his invention of a microscope **(f)** to demonstrate that microscopic organisms that appeared in broth originated from previously existing particles.

[(c) Courtesy of Dr. Tommi Nyman, Department of Biology, University of Eastern Finland.]

[a]F. Redi (1668). *Esperienze Intorno alla Generazione degl'Insetti (Experiments on the Generation of Insects)*, Florence.

| BOX 1.1 | Early Experiments Disproving Spontaneous Generation of Insects (Cont...) |

Courtesy of National Library of Medicine

ANTONIUS A LEEUWENHOEK.
Regiæ Societatis Londinensis membrum.

(e)

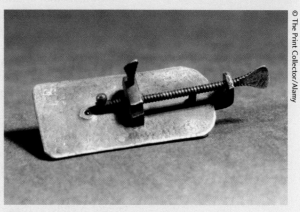

© The Print Collector/Alamy

(f)

from previously existing particles (**FIGURE B1.1e, f**). An Italian physiologist, Lazzaro Spallanzani (Abbé Spallanzani, 1729–1799) heated various types of broth in sealed containers and observed no growth of tiny organisms. Only when the containers were open to airborne particles did organisms appear and grow. Spontaneous generation as a hypothesis for the origin or transformation of life was on the way out, although it was not abandoned until the crucial nineteenth century experiments of the French chemist and microbiologist Louis Pasteur (1822–1895) and the English physician and man of science John Tyndall (1820–1893). As with many discoveries in science, Pasteur's evidence came incidentally, in his case, from experiments to understand the fermentation process used in making beer and wine. Nothing grew when Pasteur sealed broths of beer in airtight glass vessels, further disproving the idea of spontaneous generation.[b]

[b]See J. Farley, J., 1977. *The Spontaneous Generation Controversy from Descartes to Oparin.* The Johns Hopkins University Press, Baltimore, MD, for the definitive treatment of spontaneous generation.

© iStockphoto/Thinkstock

FIGURE B1.2 Image of a gall produced in response to an insect by a eucalyptus tree.

CHARLES DARWIN AND ALFRED RUSSEL WALLACE

The theory of evolution by natural selection transformed our understanding of the origins of and relationships between organisms into a single tree of life. The theory was developed independently by Charles Darwin and Alfred Russel Wallace, then revealed in a joint report to the Linnaean Society of London in July 1858, and published in a book length treatment, *On the Origin of Species by Means of Natural Selection,* by Darwin in 1859. With this 1859 publication of Darwin's, evolution became the study of:

The full title of Darwin's book is ON THE ORIGIN OF SPECIES BY MEANS OF NATURAL SELECTION OR THE PRESERVATION OF FAVOURED RACES IN THE STRUGGLE FOR LIFE.

- the origination and transformation of species (one species of horse gives rise to another species of horse);
- the transformation of major groups or lineages of organisms and the search for their ancestors (invertebrates as the ancestors of vertebrates; fish as the ancestors of amphibians); and
- the transformation of physical features such as jaws, limbs, kidneys, and nervous systems within lineages of organisms.

Acceptance that organisms could change over time and that new types of organisms could arise revolutionized our understanding of the origin of the natural world: As organisms interact with their environments, species adapt to those environments. The combination of limited resources and the production of more offspring than those resources can sustain results in competition. Individuals who possess heritable traits that make them better adapted to their environment are more likely to survive to produce more offspring (with the same beneficial heritable traits) than individuals who are less adapted to the environment. Because selection works in this way, allowing the better adapted to survive and reproduce, over time, species change. The combination of changing environments, the presence of hereditary variation, and differential reproduction results in the modification of existing characters, or the origin of new characters, that can spread throughout a population or species.

MENDEL AND MUTATIONS

In 1900, evolution entered a new phase of understanding following the rediscovery of Gregor Mendel's (1822–1884) breeding experiments with pea plants, and the ensuing rapid development of **Mendelian genetics**. Geneticists began to work with inbred lines of organisms with animals maintained in laboratories or plants in green houses, and with strains or cultivars that would have a hard time surviving in nature. The genes of these organisms could be manipulated in order to understand the genetics of inheritance.

The discovery of **mutations**—mostly those of large effect, resulting in changes in morphology that could be recognized and quantified—led to notions of large-scale evolution by jumps (saltations), rather than by gradual changes as proposed by Darwin. Evolution by jumps pitted geneticists against Darwinists, many of whom labeled geneticists as anti-Darwinian, as indeed many were. Two conceptual advances led to the reconciliation of Mendelism with Darwinism into what became known as **neo-Darwinism**: (1) The discovery in 1908 of what became known as the Hardy-Weinberg law for calculating gene frequencies in populations under natural selection; (2) the publication in 1918 by the English mathematician R. A. Fisher (1890–1962) of his paper, "The correlation between relatives on the supposition of Mendelian inheritance" (Fisher, 1918). The former showed that changes in a population's genetic structure (such as mutations or selection) can be used to study evolution. The latter proposes a genetic model in which variation in characters was based on Mendelian inheritance.

THE MODERN SYNTHESIS OF EVOLUTION

NEO-DARWINISM *is the theory that evolution occurs by natural selection and not through inheritance of acquired characters. The* MODERN SYNTHESIS *(also synthetic theory, evolutionary synthesis) is the integration of neo-Darwinism with Mendelian genetics.*

During the 1930s, the approaches put forth in the Hardy-Weinberg equation and by R. A. Fisher led to the rise of population genetics, the study of changes in gene frequency within a population under the influence of selection, mutation, genetic drift, and gene flow. In the 1940s, the synthesis of population genetics, systematics (the study of the diversification of life and the relationships of species, both past and present) and adaptive change forged what is known as the **Modern Synthesis of Evolution**. An overview of the evolutionary process that emerged from the Modern Synthesis and from the study of genetics and population biology is provided in **TABLE 1.1**.

Population genetics does not provide a complete theory of evolution, however. Why? Because evolution is **hierarchical**, operating on at least three levels:

- the *genetic level,* seen as changes in the genetic composition of individual organisms;
- the *organismal level,* seen as individual variation and differential survival through adaptation[5] (**FIGURE 1.5**) and the evolution of new structures, functions and/or behaviors; and

---

[5] For adaptation as a property of the phenotype that relates organisms to their environment through selection, see Bock (1980) and Morris and Lundberg (2011).

TABLE 1.1	Major Elements of the Modern Synthesis of Evolution

Organisms exist as individuals. Individual multicellular organisms develop, grow, mature, reproduce, and die.
Natural selection acts on individuals but individuals do not evolve. Individuals pass on their genes to individuals of the next generation.
Individuals exist in populations. Populations usually include different age classes of a single generation and individuals from other generations ('grandparents, grandchildren').
Populations of a sexually reproducing organism consist of individuals that are not identical to one another; they are not clones. Populations of asexually reproducing individuals may be clones.
Populations do not reproduce, individuals reproduce.
Resources are often limited.
Not all individuals in a population will **survive to reproduce** and contribute offspring to the next generation.
Variation is an essential prerequisite for evolution to act. **Natural selection** allows some variants to survive and others not.
Differential reproduction results in survival to the next generation of those individuals best suited to the conditions of their existence.
Because the **genetic background of individual sexually reproducing organisms differs**, those that are selected are more likely to pass their genes to the next generation.
Because of differential reproduction, mutations, and exchange of genes between populations, the **genetic composition of a population** will change gradually from generation to generation.
Populations may subdivide into smaller groups and so reinforce genetic differences that can provide the basis for **speciation**.
Populations or subsets or populations may "crash" or become extinct following environmental catastrophes.

[*Note*: Dates derived mostly from Gradstein et al. A Geological Time Scale. Cambridge University Press, 2004 and Geologic Time Scale, available from http://www.stratigraphy.org, Accessed January 2010.]

- changes in *populations* of organisms, seen as changes in gene flow between populations and the subsequent origin, radiation and adaptation of species (Table 1.1; Hall, 1999; Jablonski, 2007).[6]

Features that are heritable are evident because they reappear from generation to generation (Figure 1.5). Accumulation of heritable responses to selection of the phenotype, generation after generation, leads to evolution. Importantly, as an individual exists for only one generation, individuals do not evolve. Individuals within each generation, however, respond to **natural selection**, which is the sum of the survival and fertility mechanisms that affect reproductive success. **Genes** exist within individuals and are passed down from generation to generation in those individuals that reproduce. When natural selection on individuals is coupled with changes in the genotype in subsequent generations, populations of individuals evolve, evident in changes in the features shared by a group of organisms (Chanock et al., 2007).

GENOTYPE *is the term for all the genes of an individual. The* PHENOTYPE *is all the structural, functional, and behavioral characters of an organism.*

ONLY A THEORY AND NOT SCIENCE?

How our understanding of evolution originated and changed, and how evolution operates at the three levels of genes, organisms, and populations are the major topics of this text. All three levels have to be understood and integrated to paint a complete picture of evolution. That said, it is often claimed that because we cannot see evolution happening, evolution will always be a theory and not a proven body of knowledge.

---

[6] Authoritative essays on these three levels may be found in Pagel (2002) and in Ruse and Travis (2009).

Courtesy of Richard Borowsky, NYU

© Misulka/Dreamstime.com.

© Domen Lombergar/ShutterStock, Inc.

(a)

(b)

(c)

FIGURE 1.5 Heritable adaptations to life in the dark. **(a)** Surface-dwelling (eyed, pigmented) and cave-dwelling (blind, non-pigmented) forms of Mexican cave fish. Blind and non-pigmented (albino) cave-dwelling axolotls, *Proteus anguinus*, discovered in 1768, shown from the top **(b)** and from the front **(c)** to show the prominent gills.

The foundational theories of the physical sciences—the atomic theory, the theory of relativity, and the universal theory of gravitation—account for events we cannot see in every-day life. Nevertheless, we experience the effects of these forces of nature daily: The atomic theory explains chemical reactions, even though the atoms responsible for the reactions are not visible to the naked eye.[7] The universal theory of gravitation explains why objects remain "tethered" to Earth even though we cannot see the gravitational waves underlying the effect.

The facts of evolution are the anatomical similarities and differences among organisms, the places where they live, the metabolic pathways they use, the embryological stages through which they develop, the fossil forms they leave behind, and the genetic, chromosomal, and molecular features that connect them. The theory of evolution accounts for the historical sequence of organisms through time. It explains their existence through processes that cause changes in their genetic inheritance over time. The theory of evolution is a coherent explanation of the historical course of biology (facts) resulting from natural processes such as mutation, selection, genetic drift, migration, and alterations in how genes function. These explanations are consistent with all observations made so far.[8] For these reasons, evolution is a science.

---

[7] A scanning transmission electron microscope (STEM) unveiled in 2008 focuses a beam of electrons on a spot smaller than a single atom, scans the atoms, and photographs them in color.
[8] For further reading, see Bock (2007) and Committee on Defining and Advancing the Conceptual Basis of Biological Sciences in the 21st Century (2007).

EVOLUTION AND THE SCIENTIFIC METHOD

It has been claimed that evolution is not a science that can be studied and understood using the **scientific method**, which allows us to gather information that supports or rejects a hypothesis (**BOX 1.2**). The argument is that evolutionary hypotheses cannot be tested and supported (or falsified) in the same way that hypotheses in physics and chemistry are tested. One version of this argument states that because evolution deals with events that occurred in the past (**TABLE 1.2**)—events that are often impossible to repeat in a laboratory—evolutionary biology can never reach the status of the sciences of physics and chemistry. But we can explain historical events in evolution as rationally as we explain other past events, including the origin of Earth and the origin of continents through plate tectonics, both of which are discussed in other chapters. Indeed, just as data can be gathered to demonstrate the movement of continents across the globe over hundreds of millions of years (continental drift) and a theory developed to explain such movements (plate tectonics), so evolution that occurred in the past can be documented, studied, and tested.[9]

The fact of past evolutionary events can be tested scientifically and theories constructed. Note these important facts:

- evidence to test evolutionary hypotheses exists in the fossil record,
- evolution can be tested by comparing organisms and by constructing trees of life (phylogenetic trees), and
- evolution can be tested experimentally.

The sequence of primate-like hominin (humans and their closest relatives) fossils supports the hypothesis that humans have a primate origin. Correspondence in the basic chemical sequences of myoglobin and hemoglobin, two classes of iron-containing molecules that bind and transport oxygen, supports an evolutionary relationship between the two proteins. Because either hypothesis could be disproved by finding frog- or reptile-like hominid fossilized ancestors or by discovering a species that lacks chemical sequence similarities between myoglobin and hemoglobin, such hypotheses are scientific.

Furthermore, evolution can be demonstrated both in nature and experimentally. Indeed, early experimental tests of evolution were carried out during Darwin's lifetime, among the most convincing of which were studies undertaken by British scientist William Dallinger and published in 1878.[10] Dallinger cultivated flagellates (unicellular protozoa, each of which has a locomotory flagellum) in water in which the temperature was gradually increased. After several thousand generations the flagellates had evolved to survive and reproduce at much higher temperatures than those experienced by the initial population. Dallinger communicated with Darwin about his research and Darwin thought Dallinger's results explained how such organisms existed in hot springs.

A recent dramatic example of experimental evolution, documenting mutations, natural selection, and the effect of a beneficial mutation on the rate of evolution in bacterial populations, is outlined in **BOX 1.3**. The ability to undertake such experiments presupposes that we can identify and separate organisms into categories such as species.[11] How this is done, is the topic of another chapter.

---

[9] See Rice and Hostert (1993) for an overview of 40 years of laboratory experiments on the mechanisms of speciation, and see Colegrave and Collins (2008) for an overview of experimental studies on the evolution of mutation and genetic exchange.

[10] Rev. Dr. William Henry Dallinger (1839–1909), a Wesleyan Methodist minister, was among the first to undertake microscopic study of the life cycles of unicellular organisms.

[11] Further examples of the demonstration of evolution in experiments using microorganisms are discussed in a paper with the intriguing title "The *Beagle* in a bottle" (A. Buckling et al., 2009. The *Beagle* in a bottle. *Nature*, **457**, 824–829).

BOX 1.2	The Scientific Method

The essential nature of science is discovery using a method—the **scientific method**—that allows discoveries to be made. The discovery may be of a previously unknown object—tubeworms from deep-sea hydrothermal vents (**FIGURE B2.1**)—or a new explanation—how organisms survive and evolved at depth.

The scientific method as we know it developed in Europe in the late seventeenth century. With the publication of *De Motu Cordis* (On the Motion of the Heart and Blood) in 1628, the English physician William Harvey (1578–1657) demonstrated with a brilliant synthesis of theory, observation and experimentation that the circulation of the blood could be explained in terms of pumps and valves. In the words of another famous physician Sir William Osler (1849–1919) two hundred and fifty years after Harvey's death:

> ... *De Motu Cordis* marks the break of the modern spirit with the old traditions. No longer were men to rest content with careful observation and with accurate description; no longer were men to be content with finely spun theories and dreams, which "serve as a common subterfuge of ignorance"; but here for the first time a great physiological problem was approached from the experimental side by a man with a modern scientific mind, who could weigh evidence and not go beyond it, and who had the sense to let the conclusions emerge naturally but firmly from the observations.*

Applying the scientific method involves four steps:

1. Thinking up a hypothesis.
2. Designing and performing controlled experiments or making observations that allow information (data) relevant to the hypothesis to be collected.
3. Analyzing the data in an objective way against the background of existing knowledge.
4. Drawing conclusions that support or refute the hypothesis (**FIGURE B2.2**).

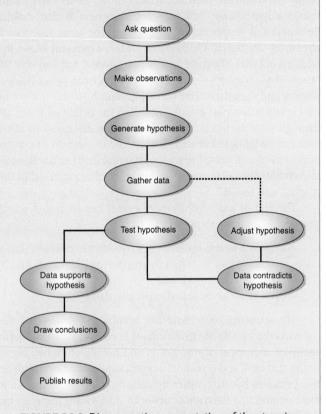

FIGURE B2.2 Diagrammatic representation of the steps in the scientific method.

Through the repeated application of this method, science progresses by accumulating evidence consistent with one interpretation of an event or process and inconsistent with others. When possible, experimentation is an important way to test hypotheses. However, when experimentation is not possible, data can be collected and hypotheses accepted or rejected without experimental verification. Consequently, the scientific method can be applied to astronomy, geology, and past evolutionary events; the scientific method is sufficiently precise to allow explanations of past events. If the explanations contradict present events, new hypotheses are generated and new data obtained.

Systems that are alike in many respects but differ in others can be investigated using the scientific method. Ants and termites can be studied as two types of social insects. Chimpanzees and modern humans can be studied as closely related primates. Here, the hypotheses relate to evolution within related organisms. The scientific method can take us further, however. We can compare ants and humans as two groups of social organisms. Here the hypotheses are that the evolution of social

Courtesy of Monika Bright - University of Vienna, Austria/NOAA

FIGURE B2.1 Deep-sea red tube worms.

*From the Harveian oration delivered by William Osler at the Royal College of Physicians London, 18 October 1906 and reprinted in the *Brit. Med. J*, 1957, **8**, 1257–1263.

BOX 1.2 **The Scientific Method (Cont...)**

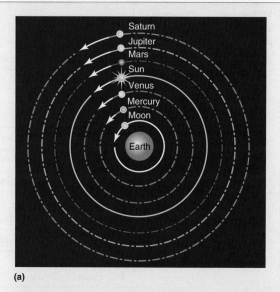

(a)

FIGURE B2.3 Two hypotheses of the orbits of the planets, one in which the planets orbit Earth **(a)**, the other in which the planets orbit the Sun **(b)**.

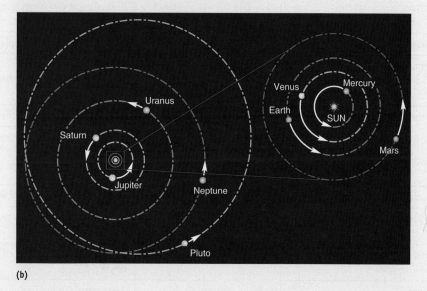

(b)

organization has elements in common throughout the animal kingdom.

Through systematic application of the scientific method, unsupported hypotheses are eliminated and a single interpretation emerges as the one best explaining the data. This process may take years or centuries, as it did for the discovery of the relationship between the orbits of Earth and the Sun (**FIGURE B2.3**), and for the discovery of deoxyribonucleic acid (DNA) as the molecule of inheritance that carries the genetic code from generation to generation. Other discoveries may be, or appear to be, instantaneous: The rising level of the water as he stepped into his bath led Archimedes to shout "Eureka" as he "saw" the principle of buoyancy. Even such "Eureka moments," however, cannot be isolated from past

knowledge or current ways of thinking. This is so even when the discovery totally changes how we view natural phenomena, as when Albert Einstein discovered that matter and energy are not separate but inter-convertible, expressed in the formula $E = mc^2$ (energy equals mass times the speed of light squared). Even though the formula revolutionized our thinking, earlier theories of matter and energy existed, just as earlier theories of evolution existed before Darwin and Wallace revolutionized our thinking. Similarly, as you progress through this text you will see that evolutionary biology has a history in which earlier hypotheses, types of data, and conclusions have been replaced with later hypotheses, different types of data, and different conclusions. This is why evolution is a science that is understood and explained through the application of the scientific method.

TABLE 1.2 — Major Events in Organismal Evolution in Relation to the Geological Time Scale

Eon	Era	Period	Epoch	Millions of Years Before Present (approx.)	Duration in Millions of Years (approx.)	Some Major Organic Events
Phanerozoic	Cenozoic	Quaternary	Recent (last 5,000 years)	0.01	1.8	Appearance of humans
			Pleistocene	1.8		
		Tertiary	Pliocene	5.3	3.5	Dominance of mammals and birds
			Miocene	23.8	18.5	Proliferation of bony fishes (teleosts)
			Oligocene	34	10.2	Rise of modern groups of mammals and invertebrates
			Eocene	55	21	Dominance of flowering plants
			Paleocene	65	10	Radiation of primitive mammals
	Mesozoic	Cretaceous		142	77	First flowering plants / Extinction of dinosaurs
		Jurassic		206	64	Rise of giant dinosaurs / Appearance of first birds
		Triassic		248	42	Development of conifer plants
	Paleozoic	Permian		290 / 320	42	Proliferation of reptiles / Extinction of many early forms (invertebrates)
		Carboniferous	Pennsylvanian	354	30	Appearance of early reptiles
			Mississippian	417	34	Development of amphibians and insects
		Devonian		443	63	Rise of fishes / First land vertebrates
		Silurian		495	26	First land plants and land invertebrates
		Ordovician		545	52	Dominance of invertebrates / First vertebrates
		Cambrian		900	40	Sharp increase in fossils of invertebrate phyla
Precambrian	Proterozoic	Upper		1,600	355	Appearance of multicellular organisms
		Middle		2,500	700	Appearance of eukaryotic cells
		Lower		4,000–4,400	900	Appearance of planktonic prokaryotes
	Archean			4,560	1,400	Appearance of sedimentary rocks, stromatolites, and benthic prokaryotes
	Hadean				160–560	From the formation of Earth until first appearance of sedimentary rocks; no observable fossil organisms

[*Note*: Dates derived mostly from Gradstein et al. *A Geological Time Scale.* Cambridge University Press, 2004, and from *Geologic Time Scale,* available from http://www.stratigraphy.org, accessed August 2012.]

BOX 1.3	Experimental Evidence for Evolution

A dramatic illustration of our ability to follow evolution as it occurs—to conduct an experiment in evolution—is provided by a study using the bacterium *Escherichia coli*, which, as of February 2010, had been running for nearly 25 years and had more than 50,000 generations (Lenski and Travisano, 1994 and Lenski, 2010). The design of the study is deceptively simple, considering the wealth of knowledge it gives us. The novelty comes from the tenacity to measure mutation rates and assess the magnitude of their effects in generation after generation, for over two decades.

In February 1988, 12 genetically identical populations of *E. coli* were established in colonies under a selection pressure (glucose in the environment was limited) and were monitored after 2, 5, 10, 15, 20, 40, and 50 generations. Any change (evolution) at the nucleotide level was detected by determining the complete nucleotide sequence of the bacteria at each of these six times and comparing the sequence with earlier time periods and with the genome of the initial population. The **rate of evolution** after 20,000 generations, measured by the accumulation of mutations in nucleotide sequences, was 2 nucleotide changes/1,000 generations. Reproductive success, which was used as a measure of **adaptation,** (Bock, 1980) was 1.5 times higher at 20,000 generations than in the starting populations and continued to rise at a slower rate in subsequent populations. The authors concluded that beneficial mutations were arising and accumulating in the populations. This simple metric provides evidence for the operation of natural selection in these populations.

Interestingly, the trends of rate of evolution, accumulation of mutations, and increased reproductive success seen in the first 20,000 generations did not continue in subsequent generations. Why? Because evolution is not predictable and because some mutations have much greater effects than others. Sometime between generations 20,000 and 40,000 a mutation arose in the gene *mutT* that codes for an enzyme involved in repairing damaged DNA, making the process much more efficient. Because damaged DNA was now being repaired more efficiently, the rate of accumulation of mutations increased from 45 mutations in the first 20,000 generations to 600 mutations between generations 20,000 and 40,000. The rate of evolution of nucleotide sequences changed because a single mutation enabled subsequent mutations to accumulate.

■ KEY TERMS

adaptation
genes
genotype
hierarchical
Mendelian genetics
modern synthesis
mutations

natural selection
neo-Darwinism
phenotype
preformation
scientific method
spontaneous generation

■ EVOLUTION ON THE WEB

Explore evolution on the Internet! Visit the accompanying website for *Strickberger's Evolution, Fifth Edition,* at **go.jblearning.com/Evolution5eCW** for exercises and links relating to topics covered in this chapter.

■ REFERENCES

Appleman, P. (ed.), 2001. *A Norton Critical Edition. Darwin: Texts, Commentary,* 3d ed. Philip Appleman (ed.). W. W. Norton & Company, New York. [Contains reprints of eight essays on scientific thought before Darwin, an evaluation of Darwin's life, selections from Darwin's writing, and almost 100 commentaries on Darwin's influence on science, social thought, philosophy, ethics, religion, and literature.]

Barrick, J. E., D. S. Yu, S. H. Yoon, et al., 2009. Genome evolution and adaptation in a long-term experiment with *Escherichia coli. Nature,* **461,** 1243–1247.

Bock, W. J., 1980. The definition and recognition of biological adaptation. *Am. Zool.,* **20,** 217–227.

Bock, W. J., 2007. Explanations in evolutionary theory. *J. Zool. Syst. Evol. Res.,* **45,** 89–103.

Bowler, P. J., 2003. *Evolution: The History of an Idea,* 3d ed. University of California Press, Berkeley, CA.

Chanock, S. J., T. Manolio, M. Boehnke, et al., 2007. Replicating genotype-phenotype associations. *Nature*, **447**, 655–660.

Cobb, M., 2006. Heredity before genetics: a history. *Nature Rev. Genet.*, **7**, 953–958.

Colegrave, N., and S. Collins, 2008. Experimental evolution: experimental evolution and evolvability. *Heredity*, **100**, 464–470.

Committee on Defining and Advancing the Conceptual Basis of Biological Sciences in the 21st Century, 2007. *The Role of Theory in Advancing 21st Century Biology: Catalyzing Transformative Research*. The National Academies Press, Washington, DC.

Dallinger, W., 1878. On the life history of a minute septic organism: with an account of experiments made to determine its thermal death point. *Proc. R. Soc. Lond.*, **27**, 332–350.

Fisher, R. A., 1918. The correlation between relatives on the supposition of Mendelian inheritance. *Phil. Trans R. Soc. Edinburgh*, **52**, 399–433

Gould, S. J., 1977. *Ontogeny and Phylogeny*. Harvard University Press, Cambridge, MA. [The book that launched the current reintegration of development and evolution as evolutionary developmental biology.]

Hall, B. K., 1999. *Evolutionary Developmental Biology*, 2nd ed. Kluwer Academic Publishers, Dordrecht, The Netherlands.

Hall, B. K., and W. M. Olson (eds.), 2003. *Keywords & Concepts in Evolutionary Developmental Biology*. Harvard University Press, Cambridge MA.

Jablonski, D., 2007. Scale and hierarchy in macroevolution. *Palaeontology*, **50**, 87–109.

Lenski, R. E., 2010. Chance and necessity in the evolution of a bacterial pathogen. *Nat Genet*, **43**, 1174–1176.

Lenski, R. E., and M. Travisano, 1994. Dynamics of adaptation and diversification—a 10,000 generation experiment with bacterial populations. *Proc. Natl Acad. Sci. USA*, **91**, 6898–6814.

Mayr, E., 1982. *The Growth of Biological Thought: Diversity, Evolution, and Inheritance*. Harvard University Press, Cambridge, MA. [A magisterial treatment by one of the founders of the modern synthesis of evolution.]

Mayr, E., 2001. *What Evolution Is*. With a Foreword by Jared Diamond. Basic Books, New York. [A treatment for those with less background in the sciences.]

Morris, D. W., and P. Lundberg. 2011. *Pillars of Evolution. Fundamental Principles of the Eco-Evolutionary Process*. Oxford University Press, Oxford, U.K.

Pagel, M. (ed. in chief), 2002. *Encyclopedia of Evolution*, 2 vol. Oxford University Press, New York.

Rice, W. R., and E. E. Hostert, 1993. Perspective: Laboratory experiments on speciation: What have we learned in forty years? *Evolution*, **47**, 1637–1653.

Ruse, M., and Travis, J., 2009. *Evolution: The First Four Billion Years*. Harvard University Press, Cambridge, MA. [Contains authoritative essays and entries.]

Species and Their Relationships

CHAPTER SUMMARY

Understanding the nature and reality of species has occupied humans for centuries. By the seventeenth and eighteenth centuries, explorations to previously unknown regions of the globe opened Europeans' eyes to the wonders of exotic animals and plants, which led to an increased interest in classifying organisms. In the mid-eighteenth century, the Swedish founder of taxonomy, Carl von Linné (Carolus Linnaeus), revolutionized our view of nature by proposing the species as the basic unit of biological classification, and by building a hierarchical classification system based on the species. A French naturalist, Georges Buffon, went much farther. Reasoning that the species is much more than a category in classification, Buffon proposed that species are real, unchanging, and the only natural grouping of historical and interbreeding entities. For Jean-Baptiste Lamarck, on the other hand, species were arbitrary; he believed forms intermediate between species must exist. The species imagined by Lamarck would never become extinct but could transform into another species. Charles Darwin's theory of evolution included *both* species origin *and* species extinction, with the fundamentally important addition that ancestral species could *coexist* with descendant species—new species could arise without the loss of the parent species.

WHAT IS A SPECIES AND HOW ARE THEY RELEATED?

The aim of this chapter is to answer two deceptively simple questions: "What is a Species?" and "How are Species Related?" The answers to these questions are not easy, even though hypotheses to name and relate Earth's organisms can be traced all the way back to Greek philosophers of the fourth and fifth centuries BCE. Of these, Aristotle (384–322 BCE) and Plato (428–348 BCE) had the greatest impact on Western thought through their philosophies of idealism and teleology. Because of the importance of understanding how we came to the present day concepts of species and relationships this chapter includes some historical information on both topics.

While the precise definition of a species remains a constant source of debate, we continue to recognize species as biological units, which we classify as morphological species using morphological data. Biological and evolutionary species concepts define species as interbreeding or evolutionarily isolated units, respectively. Any comparison between any two species is an implicit statement about their relationships; two species in the same genus—*Homo sapiens* and *Homo erectus,* for example—are more closely related than are species in different genera. Understanding relationships between species therefore is critical.

SPECIES

SPECIES AS IDEALS

Plato's writing was founded in his belief that the physical world around us is not reality, but "shadows" thrown by an unseen but ideal reality. As an illustration, consider a circle. A circle that you draw will never be perfectly round, yet in your mind you can visualize a perfect circle. And because you can visualize such a thing, Plato argues, it must exist in the ideal world. The concept of **idealism** is easily applied to diverse religions, in which the unseen ideal world is that of a supreme being who transcends human understanding and sensory capabilities.

Plato's student, Aristotle, whom many regard as the founder of biology and other sciences, disagreed with Plato on this point. He believed that the ideal form of an object exists within that object, and not within the ideal realm of forms. The presence of the ideal in any object, including living creatures, informed his philosophy that "final causes" direct all natural processes. Aristotle believed that the last stage of embryonic development—the adult form—explains the changes that occur in the immature forms. This type of explanation is called *teleological* (goal-oriented), with the adult representing the "telos," or final goal, of the embryo.[1] Consequently, for many later thinkers, **teleology** became associated with processes by which advanced stages influence and affect earlier stages. As idealism had come to imply conscious creation, it seemed as though organs and organisms were **designed** for some special purpose, that each species was created as an **ideal** in anticipation of its future use—an **idealist species concept**. In this way, natural science became interwoven with religion. The following quotation from Carl von Linné (1697–1778) is typical of the situation in the 1700s:

> If the Maker has furnished this globe, like a museum, with the most admirable proofs of his wisdom and power; if this splendid theater would be adorned in vain without a spectator; and if man the most perfect of all his works is alone capable of considering the wonderful economy of the whole; it follows that man is made for the purpose of studying the Creator's work that he may observe in them the evident marks of divine wisdom.[2]

CARL VON LINNÉ, *who took the Latinized form of his name, Carolus Linnaeus, when he was knighted in 1761, is most often referred to as Carl Linnaeus.*

---

[1] For how teleology remains hard to avoid even today, see K. Weiss, 2002, Biology's theoretical kudzu: The irrepressible illusion of teleology. *Evol. Anthropol.,* **11**, 4–8.

[2] Linnaeus, C., 1754. *Reflections on the Study of Nature.*

Despite the prevalence of these views, however, evidence indicates that individuals were coming up with new species concepts and had moved well beyond teleology and the **Great Chain of Being** (BOX 2.1) when Linnaeus came along (Hull, 1965). Recent analysis by Wilkins (2009) goes a long way toward filling in the two-thousand-year gap between Aristotle and Linneaus. Especially important is Wilkins' thesis that "there was a single species 'concept' from antiquity to the arrival of genetics, the *generative conception . . .* the marriage of reproduction or generation, with form" (pp. xi, 10). Thus Wilkins sees a direct connection between the eighteenth century and the biological species concept of the mid-twentieth century. A brief discussion of the elements of these later concepts follows.

SPECIES ARE REAL BUT DO NOT CHANGE

Although Linnaeus placed special emphasis on species as the practical unit of classification, another naturalist, Georges-Louis Leclerc, Comte de Buffon (1707–1788), known in English as Georges Buffon, codified the concept that species are the only biological units that have a natural existence ("*Les espéces sont les seuls êtres de la nature*").[3]

In 1749 Buffon introduced the fundamentally biological concept that species distinctions should be made on the basis of whether there are reproductive barriers to crossbreeding between groups (whether there is "reproductive isolation"). Reproductive isolation was determined by whether fertile or sterile hybrids were produced:

> We should regard two animals as belonging to the same species if, by means of copulation, they can perpetuate themselves and preserve the likeness of the species; and we should regard them as belonging to different species if they are incapable of producing progeny by the same means.[4]

For Buffon, considerable variation could occur between individual members of a species, eventually even producing completely new varieties as seen in different breeds of dogs. Despite such variation, a species itself remained permanently distinguished from other species if viable progeny were produced. At times, Buffon seemed to propose an evolutionary position, implying that a species could change significantly (albeit through degeneration), as the last lines of the quotation below shows,

> . . . all the families, among plants as well as animals, have come from a single stock, and that all animals are descended from a single animal, from which have sprung in the course of time, as a result of progress or of degeneration, all the other races of animals (*Natural History,* 4th volume, 1753).

Ultimately, however, Buffon rejected species transformation.

The eighteenth century barrier to the acceptance of evolution rested mostly on the reality of species. If species were real, they seemed inevitably fixed. And if they were fixed, how then could new species arise?[5] Buffon, who had proposed evolutionary events on cosmological and geological scales, established three basic arguments *against* biological evolution, arguments that were used by antievolutionists well into the nineteenth century:

- New species have not appeared during recorded history.
- Mating between different species either fails to produce offspring or produces sterile hybrids; mating between individuals of the same species produces viable

---

[3] For an accessible and comprehensive history of classification/taxonomy see the book *Naming Nature* by Yoon (2009).

[4] As translated by A. O. Lovejoy (1968) in Buffon and the problem of species. In *Forerunners of Darwin: 1745–1859* (eds Glass B., Temkin O., Straus W. L.), pp. 84–113. Johns Hopkins University Press, Baltimore, MD.

[5] For discussion of the species concept from biological, historical, and philosophical points of view, see Mayr (1982), Wilson (1999, 2005), Lee (2003), Levy (2010), and Wilkins (2009, 2010).

The Great Chain of Being and Progress to Perfection

Plato defined species as representing the initial mold for all later replicates of that species: "The Deity wishing to make this world like the fairest and most perfect of intelligible beings, framed one visible living being containing within itself all other living beings of like nature."[a] Aristotle expanded this worldview to a chain-like series of forms called the **Scale of Nature** (*Scala naturae*), each form representing a link in the progression from least perfect to most perfect (**FIGURE B1.1**). This concept continued long into the history of European thought, merging with other ideas into the **Ladder of Nature** and the **Great Chain of Being**, introduced in the mid-1750s by the Swiss naturalist and preformationist Charles Bonnet.[b]

Indeed, the Great Chain of Being was accepted well into the eighteenth century, most notably by Johann von Goethe (1749–1832) and other members of the German school of Natural Philosophy (*Naturphilosophie*). According to Goethe the formation of each level of organisms was based on a fundamental idealized plan, often referred to as an **archetype** or *Bauplan* (pl. *Baupläne*). Plant morphology was founded on an *Urpflanze* (ancestral plant) that had only one main organ, the leaf, from which the stem, root, and flower parts derived as variations (**FIGURE B1.2a**).[c] Bones of the skull were modifications of the vertebrae of an *Urskeleton* (animal archetype);

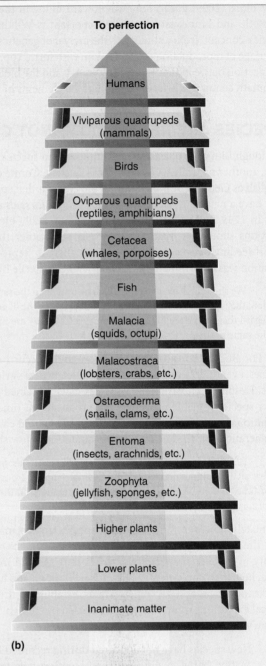

To perfection

Humans

Viviparous quadrupeds (mammals)

Birds

Oviparous quadrupeds (reptiles, amphibians)

Cetacea (whales, porpoises)

Fish

Malacia (squids, octupi)

Malacostraca (lobsters, crabs, etc.)

Ostracoderma (snails, clams, etc.)

Entoma (insects, arachnids, etc.)

Zoophyta (jellyfish, sponges, etc.)

Higher plants

Lower plants

Inanimate matter

(a)   **(b)**

FIGURE B1.1 **(a)** Aristotle (above) and **(b)** his Scale of Nature.

[(a) Courtesy National Library of Medicine; (b) Adapted from descriptions in Guyénot, E., 1941. *Les Sciences de la Vie: L'Idee d'Evolution.* Albin Michel, Paris.]

[a]Plato, *Timaeus*, circa 360 BCE.
[b]The most comprehensive discussion of the great Chain of Being remains the book by Lovejoy (1936). *Species: A History of the Idea* by Wilkins (2009), contains an informative discussion in the context of the nature of species, a reproduction of Bonnet's 1764 "Chain of Natural Beings" as Figure 4.
[c]It is now known that leaves evolved independently at least three times: in ferns, club mosses, and in flowering plants.

BOX 2.1   **The Great Chain of Being and Progress to Perfection (Cont...)**

ribs were modifications of vertebral processes (**FIGURE B1.2b**; see Hall, 1999).

Among the few who disputed this laddered concept of nature was François-Marie Arouet (1694–1778), a French philosopher and writer who wrote under the pen name Voltaire. Voltaire addressed the problem of the many observed gaps between species, gaps that did not seem to be in accord with the expected innumerable steps in the continuous progression

from imperfect to perfect. Voltaire proposed that although there were no living species to fill these gaps, such gaps were real and might result from the extinction of species.[d] As with so many changes in thought during the eighteenth century in Europe, these evolutionary rumblings were associated with the *Age of Enlightenment* (the ascent of philosophy and culture based on reason), the rise of *empiricism* (knowledge arisen from our experiences), and challenges to Papal authority.

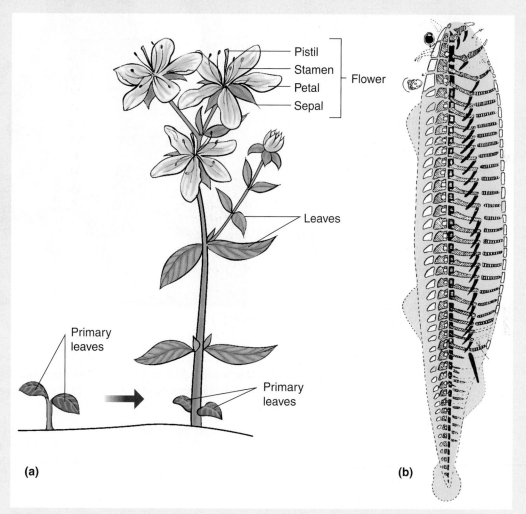

FIGURE B1.2 Archetypes of plants and vertebrates. The idealized plant **(a)** shows Goethe's concept of the derivation of all plant parts from the leaf. The segments in the vertebrate skeleton pictured by Owen **(b)** are alike from head to tail.

[(a) Adapted from Wardlaw, C. W. *Organization and Evolution in Plants*. Longmans, 1965; (b) Adapted from Owen R. *On the Archetype and Homologies of the Vertebrate Skeleton*. Voorst, 1848.]

---

[d]Voltaire was echoing the writings of the German philosopher and mathematician, Gottfried Leibniz (1646–1716), for whom the evolution of species was part of the perfection toward which the universe was continually progressing (Ruse, 1996).

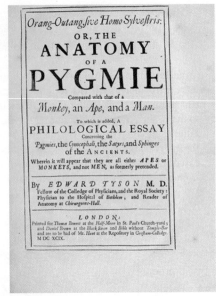

FIGURE 2.1 **(a)** Title page of Tyson's book. **(b)** Orangutan.

[Reproduced from Edward Tyson. *The anatomy of a Pygmy compared with that of a monkey, an ape and a man.* Printed for T. Bennet and D. Brown, and are to be had of Mr. Hunt (1699).]

offspring. How could individuals of a single species be separated from others of the same kind and be able to reproduce to transform into a new species?

- Where are all the missing links between existing species if transformation from one to the other has taken place?

Numerous missing links had been imagined or postulated in previous centuries. Buffon, who maintained that none had been found, seemed unaware that an English physician and comparative anatomist, Edward Tyson (1650–1708), had published dissections and comparisons of monkeys, orangutans (FIGURE 2.1), apes, and humans, which he interpreted as variations on a single type (Tyson, 1699).

SPECIES ARE REAL AND THEY EVOLVE

Recognizing the species problem, the French proponent of biological evolution, Jean-Baptiste de Lamarck (1744–1829), proposed that one must do away with the concept of species fixity in order to establish the possibility of evolution.

Lamarck (1809) shared the concept from the Great Chain of Being that species do not become extinct and proposed that species evolved from other species rather than being created separately. Lamarck's branching classification of animals (FIGURE 2.2) introduced a direct challenge to the venerable doctrine of a Scale of Nature that goes only in one direction, from imperfect to perfect. The mechanism Lamarck offered to account

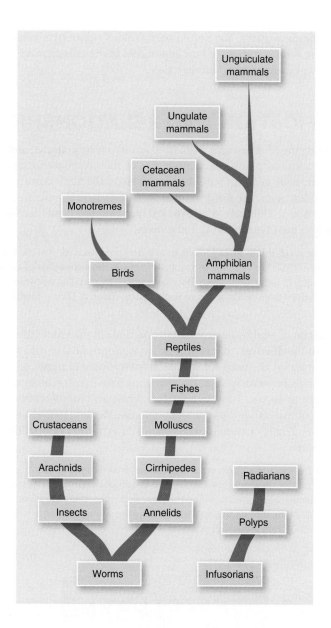

for these evolutionary changes—the inheritance of acquired characters—was ingenious, although flawed.[6] Lamarck claimed that features used frequently by an organism became stronger, and features not often used became weaker and eventually disappeared. He also maintained that physical modifications occurring within an organism's lifetime could be passed along to their offspring in the next generation. Through these two mechanisms, Lamarck proposed that organisms worked their way up the Scale of Nature towards perfection.

Despite its idealistic nature, the Great Chain of Being led almost directly to the idea that the perfection of organisms may demand multiple intermediate stages. By the eighteenth century, the concept of **evolution** as the transformation of one species into another reflected the acceptance of change between **entities** in the Great Chain of Being. Thus arose the fundamental question of how these entities should be named, classified, and related to one another. Recall from the close of the section on *Species as Ideals* above that Wilkins (2009) documented a consistent generative conception of species running

---

[6] See Tassy (2010) for depiction of relationships between organisms before and after Darwin.

through pre-Linnean writings. You should keep Wilkins' perspective in mind when reading the following short summary of how approaches to the classification of and relationships between organisms changed through time.

CLASSIFICATION AND RELATIONSHIPS

In Europe during the Middle-Ages, plants and animals were collected and described on the basis of their culinary or medical properties or their response to seasonal changes. In his six volume *Book of Plants* published in Arabic in the ninth century and consulted through the Middle Ages, the Kurdish botanist, historian, geographer, astronomer, and mathematician Abu al-Dinawari (828–896) grouped over 630 plants according to at least two different systems: over-wintering and growth.

> Plants are divided into three groups: in one, root and stem survive the winter; in the second the winter kills the stem, but the root survives and the plant develops anew from this surviving rootstock; in the third group both root and stem are killed by the winter, and the new plant develops from seeds scattered in the earth. (*Book of Plants,* English translation in 1953).

The expansion of worldwide exploration and trade in the sixteenth and seventeenth centuries led to the discovery of many new lands, floras, faunas, and species. This, in turn, greatly increased the need for species classification and began to raise questions about relationships between species. Comparisons based on the observable features of organisms began to appear. For example, *Insectorum sive minimorum animalium theatrum* (Theater of Insects), (**FIGURE 2.3**), by Thomas Moufet (1553–1604), a prominent sixteenth century English naturalist, physician, and expert on insects, described grasshoppers and locusts:

> Some are green, some black, some blue. Some fly with one pair of wings, others with more; those that have no wings they leap, those that cannot either fly or leap, they walk; some have longer shanks, some shorter. Some there are that sing, others are silent. And as there are many kinds of them in nature, so their names were almost infinite, which through the neglect of naturalists are grown out of use.

FIGURE 2.3 Title page of *Insectorum sive minimorum animalium theatrum* (Theater of Insects) by Thomas Moufet (1553–1604), published in London in 1634.

[Reproduced from Thomas Moffet, *Insectorum sive minimorum animalium theatrum*, Londres, Thom. Cotes, 1634.]

Evolutionary relationships between species of plants or insects could not be established easily using such a classification. As a consequence, classification and evolution became inextricably linked.

CARL LINNAEUS AND CLASSIFICATION

Before the **biological importance of species** could be recognized, it had to be determined how species should be defined, classified, distinguished one from another, and placed into groups that reflect their most significant features. Early classification schemes usually followed Aristotle in postulating a broad category (for example, "substance") and then creating subsidiary categories, each with its distinguishing elements (for example, body, animal), until an individual species could be placed into a particular subdivision.

Consequently, the method of classification devised by Linnaeus, the founder of modern systematics, revolutionized the way we view species and their relationships[7] (Williams and Forey, 2004). Beginning with a precise description of each species, Linnaeus grouped species closely related by their morphology into a category called a **genus** (pl. **genera**). Thereby, he established a system of **binomial nomenclature** in which each **species** name defines its membership in a genus and provides it with a unique name and identity: For example, *Homo sapiens* (humans), with *Homo* as the genus name, and *sapiens* as the species name (or specific epithet).[8] Linnaeus then grouped related genera into **orders**, orders into **classes**, and so forth.

In Linnaeus's scheme, species were identified and grouped on the basis of fundamental structural and morphological features, in contrast to the schemes of many predecessors. For instance, one pre-Linnaean classification separated animals into those that can fly and those that cannot. Consequently, flying fish were recognized as hybrids between birds and fish. By ignoring categories based on lifestyle and confining his attention to a detailed description of the species itself, Linnaeus showed that a basic relationship of flying fish to other fish (and not to birds) underlay the change in the fins that enable flying fish to "fly." Except for those features shared by all vertebrates, there are no special bird-like structures in flying fish at all. Although his system was idealistic—it treated species as ideal forms—Linnaeus's classification was a major advance, one that was necessary before natural evolutionary relationships between organisms could be revealed.

Although later in his life Linnaeus toyed with the concept of transitions between species, for much of his career he regarded species as fixed entities that could not change, albeit with a proviso: **varieties** within a species could show considerable non-heritable differences among themselves. Two examples from the work of Linnaeus show his belief that varieties can exist and still be members of the same species:

- subdivision in 1758 of the humans species, *Homo sapiens* into four races (Asiatic, American, European, African), and
- designation in 1753 of the species *Beta vulgaris* for beets, whose cultivated varieties (spinach beet, chard, beetroot, fodder beet, and sugar beet) were given varietal names, for example, *Beta vulgaris perennis* for sea beet.

FLYING FISH do not fly. Rather, they glide above the surface.

A **VARIETY** *is a natural or artificially bred population of a species with only minor differences to other populations of the species. Charles Darwin treated species and varieties equally; varieties, in his view, were species in transformation.*

---

[7] Much more than a classifier of species, Linnaeus spent much of his energy striving to organize the economy of Sweden according to scientific principles, to adapt crops such as rice and tea to grow in the Arctic tundra, and to domesticate elk, buffalo, and guinea pigs as farm animals.

[8] The convention when referring to two species in the same genus is to use only a capitalized letter for the second reference to the genus. For example, *Homo sapiens* and *H. erectus*. Generic and species names are *always* in italics, with the genus name capitalized and the specific epithet in lower case. The specific epithet is *never* used alone. It (e.g. *erectus*) and the genus name (*Homo*) or abbreviation (*H.*) is the official scientific binomial name of the species.

Sometimes you can find an indication of "subspecies" (*Homo sapiens neanderthalensis* and *H. s. sapiens*) or the name of the taxonomist who conferred the name as *Musca domestica* L. (the house fly named by Linnaeus).

Classification developed rapidly under Linnaeus. Many species were described, mainly on the basis of their reproductive parts, and classified into groupings that are still valid today. Classification was almost always based on appearance and not on ancestry or other factors like behavior or origin; many descriptions were based on preserved specimens whose natural behavior and origins were often unknown. This **typological** concept of species was reinforced by taxonomists who were required to deposit "type" specimens in museums or herbaria as the standard (type) against which other specimens can be compared.

SPECIES CONCEPTS

As you now can see, the problem of the nature of species has a history that is thousands of years old. Before Darwin, concepts of species reflected philosophies of idealism, teleology, creation, and fixity, all of which were inconsistent with evolutionary thinking; species were fixed entities separated by gaps, either morphological or spatial. Today, we have many species concepts. Four, summarized in **TABLE 2.1**, are discussed below.

MORPHOLOGICAL SPECIES

Morphological characters of adults and other life history stages of multicellular organisms have long been used to define species. The **morphological species** concept (sometimes referred to as morphospecies) is based on uniting individuals who share more characters (features) with one another than they do with any other organisms. Both adult anatomy and features of embryos provide important classes of information from which relationships between morphospecies can be determined. Specialists in particular organisms developed branching (dichotomous) keys based on morphological features as aids to species identification. A sample and simplified key is shown in **FIGURE 2.4**.

For some decades after Darwin, comparative anatomy and especially **comparative (evolutionary) embryology** was the most successful way to study animal evolution. Comparative anatomists followed the logic that organisms with similar structures shared a common ancestor. Organisms with dissimilar structures represented divergent evolutionary pathways. Knowing evolutionary relationships made it possible to trace many stepwise changes in tissues and organs (Hall, 1994, 1999).

Morphological classification remains in common usage and is generally reliable, especially when supplemented by molecular criteria. Classification of whales and

TABLE 2.1	Species Concepts

Definitions of the four species concepts discussed in this chapter
Morphological species (sometimes morphospecies): unites individuals that share more characters (features) with one another than they do with any other organisms. **Paleontological species** are a subset of morphological species.
Biological species: based on the inability of individuals in populations to interbreed with individuals in other populations to produce viable offspring. The biological species concept cannot be used for fossils or for many unicellular organisms. Because their hybrids often survive and reproduce, the concept can only rarely be applied to plants or fungi.
Evolutionary species: defined on the basis of their evolutionary isolation from each other using morphological, genetic, molecular, behavioral, and/or ecological features.
Phylogenetic species (also phylospecies, cladospecies): ancestor-descendant relationships define lineages of organisms.

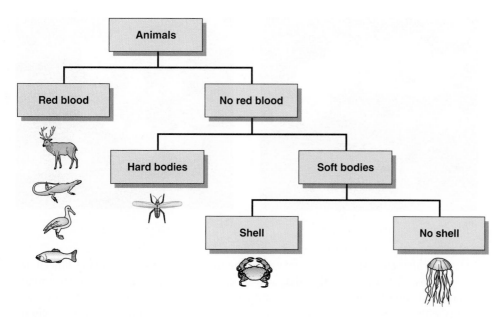

FIGURE 2.4 Dichotomous key.

their allies provides a good example. For a long time cetaceans—whales, dolphins, porpoises—were classified as a separate mammalian group. More recent analyses of fossil, morphological, and molecular data support the hypothesis that cetaceans are closely related to extant artiodactyls (even-toed ungulates), a group that includes camels, pigs, hippos, and ruminants. Early whale fossils morphologically similar to artiodactyls support this phylogeny. Further, evidence from mitochondrial, nuclear, and chromosomal DNA shows that cetaceans are not only related to artiodactyls but arose deep within artiodactyls.

Another class of comparative anatomists were (and are) paleontologists, for whom anatomy is often the only evidence on which species can be erected, lineages recognized, and evolutionary trends identified. Consequently, **paleontological species** are sometimes recognized as a subset of morphological species (Table 2.1). Because reproductive isolation cannot be determined for fossils, paleontological species are not biological species. However, if a set of paleontological species can be identified as a series of ancestral and descendant populations, paleontological species may be recognized as evolutionary species.

BIOLOGICAL SPECIES

The morphological species concept says nothing about whether members of a species interbreed or have the potential to interbreed. Many twentieth century biologists therefore adopted the **biological species concept** propounded by Ernst Mayr in 1942.[9] A biological species is a sexually interbreeding or potentially interbreeding group of individuals separated from other species by the absence of genetic exchange because of reproductive and other barriers. The concept of a species as an interbreeding group of individuals distinct from other such groups is based on the knowledge that such groups exist in nature and are separated by gaps across which interbreeding does not occur, as first articulated by Buffon (discussed earlier).

---

[9] A volume celebrating Mayr's major contributions to speciation and systematics was published in 2005 (Hey et al., 2005).

FIGURE 2.5 **(a)** Bird's eye gilia *G. tricolor* and **(b)** angel's gilia *G. angelensis* are an example of sibling species from the plant kingdom.

(a)   (b)

*A **CLONE** is an organism descended from and genetically identical to another organism. Under this definition, all offspring produced by asexual means—a method of reproduction used by bacteria and many eukaryotes, especially plants—are clones. Clones provide genetic uniformity across the generations, an advantage when organisms face the same conditions for a long time; however, clones lack genetic diversity, which can be a handicap in the face of environmental challenges such as disease.*

The biological species concept cannot be used for fossils or for many unicellular organisms. The concept only rarely can be applied to plants or fungi because many plant and fungal hybrids are viable and can reproduce. Nor does it apply to organisms with asexual reproduction in which the progeny of an individual form a **clone**. Although the individuals in each clone are genetically identical to one another and are isolated genetically from other clones, few biologists would classify each clone as a separate species.

Whether a population fits the definition of a biological species can be tested by whether individuals in populations in the same locality normally interbreed. Should interbreeding occur in nature, we can ask whether embryos develop and/or whether the progeny are viable and fertile. If the answer to these questions is no, we consider the organisms as biological species separated by (a) reproductive barrier(s).

Applying the biological species concept also allows taxonomists to *separate* into different species groups that had been *regarded as a single species* on the basis of morphological and/or geographical criteria. Such pairs of species are often known as **sibling species**. Two leafy-stemmed sibling species in a genus of the phlox family, the bird's eye gilia, *Gilia tricolor*, and angel's gilia, *G. angelensis*, are an example from the plant kingdom (**FIGURE 2.5**). Although they are closely related species and co-exist in the same populations there is a strong reproductive incompatibility barrier, which allows for minimal hybridization. European short-toed and common (Eurasian) tree creepers (*Certhia brachydactyla* and *C. familiaris*)—an example from the animal kingdom—differ morphologically only in the size of the third toe and in the patterning of the feathers on the wing (**FIGURE 2.6a,b**) but their distribution, behavior and ecology are sufficiently distinct to prevent interbreeding (Figure *2.6b*). The fruit fly species *Drosophila pseudoobscura* and *D. persimilis* are almost identical in appearance and cannot be distinguished on the basis of their morphology. They do interbreed but only the female offspring are fertile.

Species that have long been isolated geographically may still be able to interbreed if brought into contact, usually in a botanical or zoological garden. Examples include two widely separate populations of trees occupying similar habitats: the Chinese catalpa (*Catalpa ovata*) from China and the Indian bean tree (*Catalpa bignonioides*) found in the eastern United States. Although they can be crossed in a nursery and produce hybrids that are as viable and fertile as the parents, these populations have been geographically separate for many millions of years, and so are identified as separate species.

Applying the biological species concept also allows taxonomists to *unify* different groups into single species that had been *classified as distinct species* on the basis of morphological and/or geographical criteria. One example is the union of several species of North American sparrows into a single **polytypic** species, the song sparrow. This polytypic species, *Passarella melodia*, consists of multiple geographic subspecies. Similarly, there are several populations of yarrow in the genus *Achillea*, each with distinct ecological adaptations restricting their growth to particular, different environments. However, because

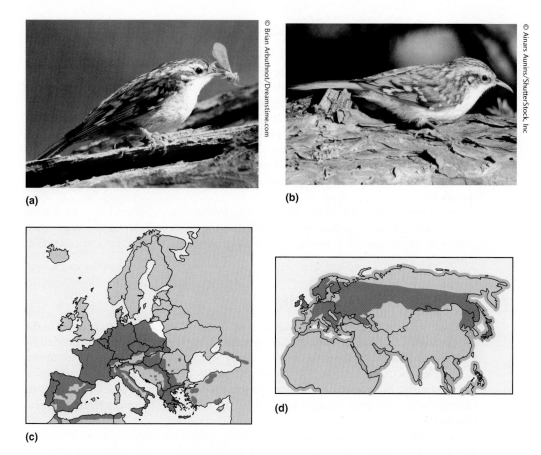

FIGURE 2.6 A sibling species pair, the European short-toed and common (Eurasian) tree creepers
(a) *Certhia brachydactyla* and *C. familiaris* **(b)** are almost indistinguishable morphologically. Distinctive
geographical distributions **(c)** and **(d)** and behaviors reinforce their inability to interbreed.

these populations can exchange genes if they came into contact, they are generally identified as subspecies of the same species, rather than distinct species.

In a subtle but potentially important extension of the biological species concept, Baker and Bradley (2006) proposed the **genetic species concept**: "a group of
genetically compatible natural populations that is genetically isolated from other such
groups." They document many instances in mammals in which species are genetically,
but not reproductively, isolated. They trace the genetic species concept back to Bateson,
Dobzhansky, and Muller's models of speciation, which describe how ancestrally identical
but geographically separated populations become genetically incompatible through gene
mutations.

In reality, a genetic species concept is applied whenever species are separated on
the basis of molecular differences. For example, forest- and savannah-dwelling African
elephants have long been recognized by taxonomists as different populations of the
same species. Through sequencing a large amount of DNA, Rohland et al. (2010) not only
demonstrated that these are distinct species—*Loxodonta cyclotis* and *L. africana*—but that
they diverged from one another as long ago (between 6.6 and 8.8 million years ago) as the
Asian elephant diverged from the wooly mammoth.[10]

---

[10] For how whole genome sequences open up new approaches to the genetic basis of speciation
and ecological adaptation, see J. F. Storz and H. E. Hoekstra, 2007. The study of adaptation and
speciation in the genomic era. *J. Mammal.*, **88**, 1–4.

EVOLUTIONARY SPECIES

To take evolution into account explicitly, various specialists proposed an **evolutionary species concept**, which considers the species as an evolutionary entity. Here, species are defined by differences that are dependent on their "evolutionary" isolation rather than on their reproductive isolation. American vertebrate paleontologist George Gaylord Simpson (1902–1984) wrote in 1961, "an evolutionary species is a lineage (an ancestor-descendant sequence of populations) evolving separately from others and with its own unitary evolutionary role and tendencies."[11] Here, for the first time, was a species concept incorporating change over time (evolution) rather than static features, and which allowed consideration of changes resulting from competition and interaction among species.

Of course, as with all species concepts, the evolutionary concept has its problems. In particular, because speciation is an evolutionary process, defining the stage when groups of organisms have reached complete separation (that is, have speciated) is difficult.

PHYLOGENETIC SPECIES

Defining taxa strictly according to cladistic relationships (see End Box 2.1) has led to the development of a **phylogenetic species concept**. A phylogenetic species is defined by ancestor-descendant relationships rather than by reproductive isolation. In these terms, E. O. Wiley modified Simpson's evolutionary species definition to read: "A species is a single lineage of ancestral descendant populations of organisms which maintains its identity from other such lineages and which has its own evolutionary tendencies and historical fate."[12]

Cracraft (1983) proposed that a phylogenetic species is a monophyletic group composed of "the smallest diagnosable cluster of individual organisms within which there is a parental pattern of ancestry and descent." The "diagnosable" element in this definition is based on similar, if not identical, characters and measurements used to classify organisms in previous systematic approaches. Emphasis on the "smallest diagnosable cluster" can result in (perhaps false) designation of varieties or smaller genetically distinct groups as phylogenetic species.

Despite devoting several decades of thought to the problem, Darwin never resolved the nature of species, although taxonomy was fundamental to his theory (Winsor, 2009). The definitions of species described above were not developed until well after Darwin died. If we apply them to Darwin's work we see that, although he recognized and delineated species using their morphology—the morphological species concept—his theory is about biological or evolutionary species. Some historians maintain that Darwin's theory would have received far less attention and acceptance had he defined species as biological units in transition. How, they ask, could those trying to describe, separate, and name species deal with such a definition?

CLASSIFICATION REFLECTS EVOLUTIONARY RELATIONSHIPS

Ideally, to obtain the most informative phylogenetic picture of a particular population of organisms, we would look at a portion of a multi-limbed phylogenetic tree whose branches connect all extant and ancestral populations of that organism. Such connections would indicate the degree of each population's relationship to all other

---

[11] G. G. Simpson (1961). *Principles of Animal Taxonomy.* Columbia University Press, New York (p. 153).
[12] E. O. Wiley. (1978). The evolutionary species concept reconsidered. *Syst. Zool.,* **27**, 17–26 (quotation from p. 18).

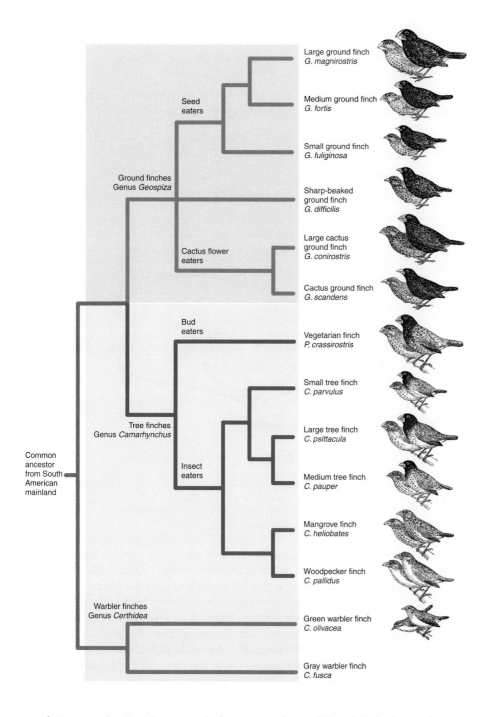

FIGURE 2.7 Evolutionary tree of Darwin's finches provided in 1947 showing adaptations of the beaks of different species. A phylogeny based on mtDNA produced over 50 years later supports some aspects of this phylogeny, distinguishing between tree finches and ground finches and showing that distinctions among members within each group are not well established. This molecular study by Sato et al. (1999) also indicates that these finches are all descended from a single species, now identified as a warbler-type, "dull-colored grassquit" (genus *Tiaris*).

[Adapted from Lack, D. Darwin's Finches: An Essay on the General Biological Theory of Evolution. Cambridge University Press, 1947. Illustration by Lt. Col. William Percival Cosnahan Tenison.]

populations on the tree. An example from an analysis of Darwin's finches is shown in **FIGURE 2.7**.

However, 10 to 30 million species of organisms may exist today. Some are identified and named as species but many are not. Additionally, hundreds of millions of species may have existed in the past. Creating a Universal Tree of Life that includes even the named species, let alone the unnamed ones, appears an impossible task. But then, two decades ago sequencing the genome of an organism seemed an impossible task. However, nearly 200 whole genomes were sequenced between 1995 and 2012.

Each unit of classification, whether a species, genus, family, order, or so on, is a **taxon** (plural, **taxa**). Each taxon is given a distinctive name; *Homo sapiens* for humans, Insecta for insects, and so forth. Taxa are arranged in nested categories so that a taxon in a "higher" category, such as an order, includes one or more taxa in "lower" categories, such

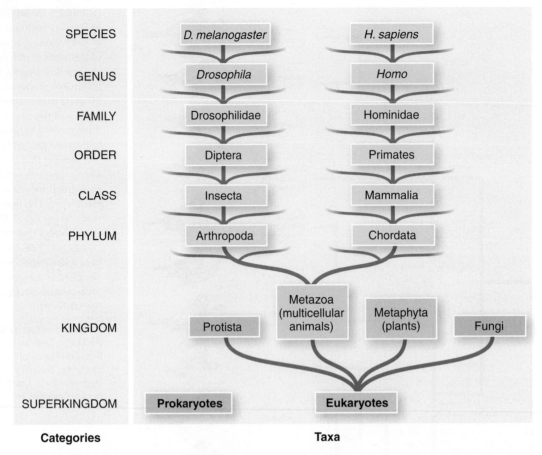

FIGURE 2.8 The nested taxonomic categories traditionally used to classify fruit flies and humans. As indicated in the text, the arrangement is not a phylogenetic tree but a nested hierarchical classification.

as families. Some of the categories and taxa used in classifying modern humans (*Homo sapiens*) and fruit flies (*Drosophila melanogaster*) are shown in **FIGURE 2.8**. The arrangement in the figure is not a phylogenetic tree but shows how classification produces nested categories.

The approximately two million named species of animals are arranged into some 37 phyla.[13] Some phyla have enormous numbers of species. The phylum *Arthropoda* (arthropods) has more than a million species, *Nematoda* (roundworms) half a million. At the other extreme, the phylum *Placozoa* (from the Latin for flat animal) consists of a single marine species, *Trichoplax adhaerens,* individuals of which are 300-μm "long" (perhaps we should say "short" or "across" as it has no obvious symmetry). The phylum *Cycliophora* (from the Latin for small wheel-bearing) also consists of a single microscopic marine species, *Symbion pandora,* which spends its life attached to the mouth parts and appendages of the Norway lobster *Nephrops norvegicus.*[14] These are "small" phyla in more ways than one!

This mode of classification offers a simple scheme for identifying and cataloging large numbers of species. For example, we can use phyla such as Arthropoda and

A PHYLUM *(division, in plants), the taxonomic category immediately below the kingdom in traditional classification schemes, is a group of organisms that share a common plan and have a closer evolutionary relationship to one another than they do to organisms in another phylum/division.*

---

[13] For criteria to distinguish phyla and for descriptions of phyla, see Hall (1999), Budd (2003), and Valentine (2004).

[14] At least one laboratory, that of Roger Croll in Halifax, NS, has successfully established colonies of *T. adhaerens* and developed protocols to analyze its cellular organization and behavior (Heyland et al., 2011).

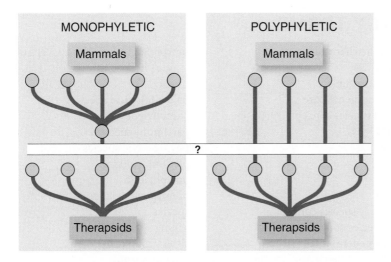

FIGURE 2.9 Monophyletic and polyphyletic schemes used to explain the evolution of mammals from therapsid reptiles. In monophyletic evolution, only a single reptilian group served as ancestor to the mammalian radiation. Two or more groups gave rise to mammals in the polyphyletic scheme.

Chordata to delineate two major groups of animals from all other animals (Figure 2.8), and use mammalian orders such as Primates and Rodentia to distinguish two groups of mammals from all other mammals. Because each traditionally recognized taxon in such a classification is not always **monophyletic** (derived from a single common ancestor), these classification schemes do not necessarily reflect evolutionary history. As one example, mammals, a long-recognized group of animals, may have had a **polyphyletic** origin (**FIGURE 2.9**), arising from two or more groups of reptiles. Amphibians almost certainly are polyphyletic. Many recent studies show that fungi are more closely related to animals than to plants, weakening any correlation between traditional classification and phylogeny of the relationship between plants and fungi.

The difficulty of classifying organisms above the species level has led to a variety of proposals to make classification more objective. **Cladistics (phylogenetic systematics)** is the latest departure from previous approaches and has had a major impact on classification, and therefore on evolutionary biology. For example, cladistic analysis of variation in the Y chromosome in humans informs our understanding of human origins, showing that humans originated in Africa and later expanded throughout the "old world," and finally into the "new world." There is also evidence of human range expansion from Asia back into Africa, indicating bidirectional gene flow.[15] Cladistics and the methods of phylogenetic systematics are outlined in some detail in **END BOX 2.1**.

In summary, the goals of classification include: (1) the arrangement of groups into a pattern that accurately reflects their evolutionary relationships and/or (2) the placement of groups into a reference system so that their major features are easily and efficiently described and identified. No single classification system fully accomplishes either of these goals. Traditional morphological classification simplifies the placement of organisms into a classification scheme but runs the risk of ignoring their evolutionary relationships. Cladistic classification, discussed in End Box 2.1, offers hypotheses of relationships that can be tested as new data arise. Morphological classification remains in common usage and is generally reliable, especially when supplemented by molecular criteria.

---

[15] M. F. Hammer et al. (1998). Out of Africa and back again: nested cladistic analysis of human Y chromosome variation. *Mol. Biol. Evol.*, **15**, 427–441.

■ KEY TERMS

apomorphic character
archetype
Bayesian inference
binomial nomenclature
biological species concept
cladistics
cladogram
Class
clone
evolution
evolutionary species concept
genetic species concept
genus (pl. genera)
Great Chain of Being
idealism
idealist species concept
Ladder of Nature
maximum likelihood estimates
monophyletic

morphological species concept (also
  morphospecies)
Order
paleontological species
paraphyletic
parsimony method
phylogenetic species concept
plesiomorphic
polyphyletic
polytypic species concept
species
sibling species
sister group (axon)
synapomorphy
taxon (plural, taxa)
teleology
Typology
variation

■ EVOLUTION ON THE WEB

Explore evolution on the Internet! Visit the accompanying website for *Strickberger's Evolution, Fifth Edition,* at **go.jblearning.com/Evolution5eCW** for exercises and links relating to topics covered in this chapter.

■ REFERENCES

Baker, R. J., and R. D. Bradley, 2006. Speciation in mammals and the genetic species concept. *J. Mammal.* **87,** 643–662.

Budd, G. E., 2003. Animal phyla. In *Keywords and Concepts in Evolutionary Developmental Biology,* B. K. Hall and W. M. Olson (eds.). Harvard University Press, Cambridge, MA, pp. 1–10.

Cracraft, J., 1983. Species concepts and speciation analysis. *Curr. Ornithol.,* **1,** 159–187.

Hall, B. K. (ed.), 1994. *Homology: The Hierarchical Basis of Comparative Biology.* Academic Press, San Diego, CA [paperback issued 2001].

Hall, B. K., 1999. *Evolutionary Developmental Biology,* 2nd ed. Kluwer Academic Publishers, Dordrecht, The Netherlands.

Hey, J., W. M. Fitch, and F. J. Ayala (eds.), 2005. *Systematics and the Origin of Species on Ernst Mayr's 100th Anniversary.* The National Academies Press, Washington, DC.

Heyland, A., R. P. Croll, S. Goodall, et al., 2011. *Trichoplax adhaerens,* an enigmatic basal metazoan with potential. In: *Developmental Biology of the Sea Urchin and Other Marine Invertebrate Model Systems.* D. J. Carroll and S. A. Stricker (eds.). Humana Press, New York (in press).

Hull, D. L., 1965. The effect of essentialism on taxonomy: two thousand years of stasis. *Brit. J. Philos. Sci.,* **15,** 314–326, **16,** 1–18.

Lamarck, J. B., 1809. *Philosophie zoologique, ou Exposition des considérations relatives à l'histoire naturelle des animaux . . .* Dentu, Paris.

Lee, M. S. Y., 2003. Species concepts and species reality: salvaging a Linnean rank. *J. Evol. Biol.,* **16,** 179–188.

Levy, A., 2010. Pattern, process and the evolution of meaning: species and units of selection. *Theory Biosci.* **129,** 159–166.

Lovejoy, A. O., 1936. *The Great Chain of Being. A Study of the History of an Idea.* Harvard University Press, Cambridge, MA. (Reprinted in 2005 by Harper & Row).

Mayr, E., 1942. *Systematics and the Origin of Species, from the Viewpoint of a Zoologist.* Harvard University Press, Cambridge, MA.

Mayr, E., 1982. *The Growth of Biological Thought: Diversity, Evolution, and Inheritance.* Harvard University Press, Cambridge, MA. [A magisterial treatment by one of the founders of the modern synthesis of evolution.]

Rohland, N., D. Reich, S. Mallick, et al., 2010. Genomic DNA sequences from mastodon and woolly mammoth reveal deep speciation of forest and savanna elephants. *PLoS Biol.,* **8**(12): e1000564. doi:10.1371/journal.pbio.1000564.

Ruse, M., 1996. *From Monad to Man: The Concept of Progress in Evolutionary Biology.* Harvard University Press, Cambridge, MA.

Sato, A., C. O'Huigin, F. Figueroa, et al., 1999. Phylogeny of Darwin's finches as revealed by mtDNA sequences. *Proc. Natl Acad. Sci. USA,* **96**, 5101–5106.

Sereno, P. C., 2007. Logical basis for morphological characters in phylogenetics. *Cladistics,* **23**, 565–587.

Tassy, P., 2010. Trees before and after Darwin. *J. Zool. Syst. Evol. Res.,* **49**, 89–101.

Tyson, E. 1699. *Orang-Outang, Sive Homo Sylvestris: or, the Anatomy of a Pygmie Compared With That of a Monkey, an Ape, and a Man.* Thomas Bennet & Daniel Brown, London. [Facsimile edition published in 1966 with an introduction by M. F. Montague.]

Valentine, J. W., 2004. *On the Origin of Phyla.* The University of Chicago Press, Chicago.

Williams, D. M., and Forey, P. L. (eds.), 2004. *Milestones in Systematics.* CRC Press, Boca Raton, Florida.

Wilkins, J. S., 2009. *Species: A History of the Idea.* University of California Press, Berkeley, CA.

Wilkins, J. S., 2010. What is a species? Essences and generation. *Theory Biosci.* **129**, 141–148.

Wilson, R. A. (ed.), 1999. *Species. New Interdisciplinary Essays.* MIT Press, Cambridge, MA.

Wilson, R. A., 2005. *Genes and the Agents of Life. The Individual in the Fragile Sciences,* Cambridge University Press, Cambridge, UK.

Winsor, M. P., 2009. Taxonomy was the foundation of Darwin's evolution. *Taxon,* **58**, 43–49.

Yoon, C. K., 2009. *Naming Nature: The Clash Between Instinct and Science.* W. W. Norton New York, NY.

END BOX 2.1

Cladistics

© Photos.com

SYNOPSIS: This end box summarizes the branching approach to classification introduced into biology 45 years ago. Features of organisms—*characters*—are classified as ancestral (*plesiomorphic*) or derived (*apomorphic*). Shared-derived characters (*synapomorphies*) are used to produce phylogenetic trees known as *cladograms*. Cladistics only recognized groups (clades) that share a common ancestor—are *monophyletic*. Cladistics usually produces a number of cladograms for any given group of organisms being considered. Each cladogram represents a hypothesis about relationships between the organisms. Statistical methods are used to determine which one of multiple trees is the most likely (which is the most *parsimonious* in cladistic terminology).

Cladistics is an approach to classification that is based on branching. The original proponent of this method was German entomologist and systematist Willi Hennig (1913–1976), whose pivotal role in founding cladistics is recognized in the name of the professional association of cladisticians, the Willi Hennig Society (Hennig, 1966, 1975). Cladistics informs us about the relationships of taxa, not about their ancestry. In cladistic terminology, a parental taxon that produces a new taxon becomes the sister group of its offspring, essentially annulling its ancestral relationship.

The Willi Hennig Society (http://www.cladistics.org/) publishes the journal CLADISTICS, *which keeps alive Hennig's name and his pivotal role in founding cladistics.*

CHARACTERS

Cladistics stresses the separation of ancestral (**plesiomorphic**) characters from newly derived (**apomorphic**) characters, and emphasizes the latter to establish phylogenies. The sharing of derived characters (**synapomorphies**) between organisms dictates the phylogeny (Forey et al., 1992; Sereno, 2007). For example, how species A, B and C are related can be determined by noting which share the newly derived character X (**FIGURE EB1.1**). Character X shared by species A and C indicates a closer relationship of these two species to one another than to species B. In traditional taxonomy, the conclusion would be that A and C branched off together from a common ancestor, and that species B branched off separately (Figure EB1.1**b**). Emphasis on shared-derived characters, however, contains no suggestion as to which species is ancestral (cf. panels **a** and **b** in Figure EB1.1). Indeed, as evolutionary patterns can be networks—more like a shrub with many branches from a single node than like a tree with only two branches from each node—dichotomous branching may oversimplify the evolutionary process, especially for lineages with many species or with rapid evolution. It is important to remember that each phylogenetic tree (**cladogram**) is a hypothesis about relationships, which, like any hypothesis, requires constant testing. How those tests are performed is discussed in the last section.

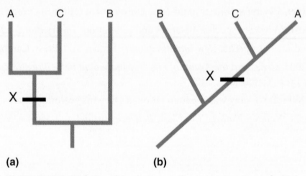

FIGURE EB1.1 Relationships between three species, two of which share a newly derived character (X), as depicted using a cladistic approach **(a)**, which does not make any statement of ancestry, and as depicted in a more traditional tree **(b)** in which species B is ancestral to both species A and species C. Note that in both trees species A and C are more closely related to one another than either is to species B.

PATTERNS OF BRANCHING

In cladistics, every significant evolutionary step marks a dichotomous branching event that produced two genetically separated **sister taxa** equal to each other in rank. For example, as birds and crocodiles derive from a common reptilian stem ancestor, cladistics consider them to be sister groups of equal rank within the group archosaurs ("first lizards"), which in turn ranks lower than the group Reptilia, within which it nests.

As noted above, lineages can be "reticulate," analogous to the branching patterns seen in a shrub versus the smaller number of branches in a tree. Such patterns emerge because a lineage can connect genetically with related lineages through hybridization, convergent evolution, or horizontal gene transfer from even more distant lineages. Such phylogenies are networks rather than simple dichotomous branching trees. Consequently, many of the traditional groups of organisms are not monophyletic lineages and so do not accurately reflect evolutionary lineage relationships. Arthropods are one such group. Crustaceans (lobsters, crabs, hermit crabs, barnacles), long regarded as a single class of arthropods, turns out to contain three monophyletic lineages, each nested separately within arthropods (Andrew, 2011). Perhaps the most difficult change for us to accept relates to the term *birds*.

BIRDS AS NON-AVIAN REPTILES

We all (think we) know what birds are. We know their identifying features (feathers, wings, flight) and can see how they must be related. Cladistics, however, does not follow the traditional classification of ranking birds (Aves) separate from reptiles (Reptilia). In cladistic analysis, groups that do not include all the descendants of a common ancestor—for example, the class Reptilia does not include their mammalian or avian offshoots—are **paraphyletic,** and so are not valid taxa (Padian and Chiappe, 1998). Similarly, cladistic schemes exclude polyphyletic groups such as arthropods and others that consist of convergent or parallel lineages. In cladistic analysis, a group of taxa (a clade) can only achieve taxonomic status if it is **monophyletic** (Figure EB1.1). Furthermore, the group must include its common ancestor and all of the common ancestor's descendants. Therefore, birds, which are related to dinosaurs and are descended from dinosaurs, are dinosaurs (and reptiles). To distinguish birds as a monophyletic group we have to speak of avian-reptiles and non-avian reptiles, meaning those reptiles that gave rise to birds (and include birds) and those that did not.

SELECTING BETWEEN COMPETING HYPOTHESES OF PHYLOGENY

When the approach outlined in Figure EB1.1 is extended to many taxa using many characters, complex phylogenies are possible. Selecting the most likely hypothesis of relationships from the set of possible hypotheses requires sophisticated statistical methodology (Felsenstein, 2004). One approach—the **parsimony method**—is based upon selecting the phylogenetic tree that minimizes the number of changes necessary to explain its evolutionary history.

To use the example in **FIGURE EB1.2**, if a particular character needs to pass through more steps in one phylogenetic tree than it would have passed through in another tree, the latter tree is preferred. Cladogram (a) in Figure EB1.2 is selected over cladogram (b), because (a) requires two evolutionary events to explain the data, whereas (b) requires three. Cladogram (a) is said to be more parsimonious than cladogram (b). This preference exists because evolution is assumed to occur by simple binary branching.

Parsimony, however, is not a statistical method. **Maximum likelihood estimation**, a statistical method based on probability distribution, is extensively used. Maximum likelihood, developed by mathematician R. A. Fisher in the early 1920s, maximizes the most likely distribution for a set or sets of data. It can handle very large data sets, which is important in today's world of mechanized genome sequencing. Because of its formal terminology, precise rules, and strict genealogical consistency in assigning branching points and patterns, cladistic classification combined with maximum likelihood analysis are now established methods for determining relationships.

A third approach, **Bayesian inference**, takes its name from the Reverend Thomas Bayes (1702–1761). Bayes' theorem is a way of considering how the probability that a theory is correct is influenced by a new piece of evidence. More specifically, the probability that a theory is correct can be evaluated by considering some prior probability then revising it in light of new data. "Spam" filters use Bayesian analysis to assess the probability that an incoming message is junk based on the presence or absence of "token" words, thus determining whether or not the message should be routed to the spam folder or eliminated entirely. Because the validity of the prior distribution cannot be assessed statistically, Bayesian analysis requires sophisticated methods to determine levels of significance (Gelman et al., 2003).

END BOX 2.1

Cladistics (Cont...)

© Photos.com

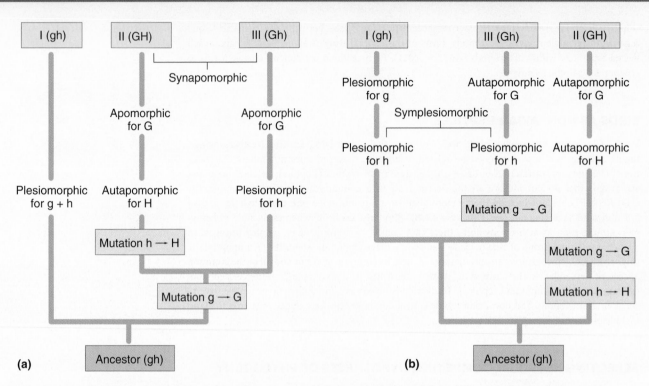

FIGURE EB1.2 Cladograms for variations of two characters g and h in hypothetical taxa I, II, and III and their presumed common ancestor. Taxon I has the ancestral (plesiomorphic) characters g and h, taxon II has derived (apomorphic) characters G and H, and taxon III is apomorphic for G and plesiomorphic for h. As shown, two alternative cladograms are possible. In **(a)** the assumption is made that character G shared by taxa II and III evolved only once by mutation of g → G, and character H evolved in the lineage of taxon II also by a single mutation of h → H. Thus, taxa II and III share the derived G character (synapomorphy) and are sister taxa, while taxon I, in turn, is the sister group of (II + III). In **(b)** the assumption is made that I and III are sister taxa, because they share ancestral character h (symplesiomorphy), and taxon II is the sister group of (I + III). Note, however, for the hypothesis in (b) to be accepted two mutations of g → G are necessary (as well as one of h → H). Thus, by the principle of parsimony, the (a) cladogram requiring a total of only two mutations is accepted over the (b) cladogram that requires three mutations. Derived character states that are unique rather than shared with other lineages are called autapomorphies, and do not indicate relationship.

REFERENCES

Andrew, D. R., 2011. A new view of insect-crustacean relationships II. Inferences from expressed sequence tags and comparisons with neural cladistics. *Arthropod Struct. Dev.,* **40**, 289–302.

Felsenstein, J., 2004. *Inferring Phylogenies.* Sinauer Associates, Sunderland, MA.

Forey, P. L., C. J. Humphries, I. J. Kitching, et al. (eds.), 1992. *Cladistics: A Practical Course in Systematics,* 10. Systematics Association Publications, Oxford University Press, Oxford, UK.

Gelman, A., J. B. Carlin, H. S. Stern, and D. B. Rubin, 2003. *Bayesian Data Analysis,* 2nd ed. Chapman and Hall/CRC, Boca Raton, FL.

Hennig, W., 1966. *Phylogenetic Systematics.* University of Illinois Press, Urbana, IL.

Hennig, W., 1975. "Cladistic analysis or cladistic classification?" A reply to Ernst Mayr. *Syst. Zool.,* **24**, 244–256.

Padian, K., and L. M. Chiappe, 1998. The origin of birds and their flight. *Sci. Am.,* **278**, 38–47.

Sereno, P. C., 2007. Logical basis for morphological characters in phylogenetics. *Cladistics,* **23**, 565–587.

© Photos.com

Similarity and Patterns of Evolution

CHAPTER SUMMARY

Structures with an underlying similarity found in different species share their essential similarity (homology) because the species share a common ancestor. Structures not related through a common ancestor arise by independent evolution and so are not homologous. Essentially, three patterns of evolution exist: (1) common descent because of similarity based on homology; (2) parallelism, the evolution of similar features in related lineages; and (3) convergence, the evolution of similar features in independent lineages. Many evolutionary biologists are actively involved in identifying and separating these different patterns of evolution in order to determine evolutionary origins and relationships.

Careful anatomical dissections and comparisons of adults and of embryos provided the basis upon which late nineteenth and twentieth century biologists identified species and constructed detailed evolutionary trees. Similarities among vertebrate embryos during early developmental stages provided evidence for their common evolutionary past.

Because in almost every instance fossils differ from present-day forms, fossils provide the hard physical evidence for evolution (END BOX 3.1). Horses comprise one of the most complete and continuous fossil records of the evolution of an animal lineage. From the fossil record, we know that horses evolved from a small, four-toed leaf-browsing animal to a large,

Image © skilpad/ ShutterStock, Inc.

single-toed animal with continuously growing teeth, adapted to chewing tough grasses. In some instances, the fossil record contains forms with features of two major extant groups of organisms. Perhaps the most famous of these is *Archaeopteryx* ("ancient wing"), which has features of both reptiles and of birds. Other ancient lineages, often described as "living fossils," persist with minimal morphological changes to the present day. The coelacanth (*Latimeria chalumnae*) is a notable example of a living fossil fish, the Ginko (*Ginkgo biloba*) of a living fossil tree.

INTRODUCTION

At least four (and in fact many more) species concepts are in use, reflecting different types of evidence used to identify a species. These species concepts also reflect speciation as an ongoing process; species are classified at various stages during speciation or when speciation is complete. Because evolutionary change differs in intensity and duration, temporally and spatially (especially when environments are changing rapidly), some organisms are difficult to classify.

Members of a species are identified by their similarity derived from a shared history. Although a species name indicates a distinct group with shared characters, the individual members of a species display variation. Because classification and evolution inevitably emphasize different aspects of organisms, the basis on which similarity is determined is a central issue. Two aspects of this issue are discussed below in the context of evolutionary patterns seen when we compare organisms with different degrees of shared relatedness and evolutionary history. The first aspect considers how similarity of features is determined. The second deals with two classes of evidence—comparative embryology and the fossil record—used to compare organisms.

SIMILARITY: KNOWING WHEN CHARACTERS ARE THE SAME OR DIFFERENT

In general, the more similar features shared by a group, the more likely the group descended from a common ancestor. Once again, a classic example lies in the evolution of horses and the reduction in their number of toes over the past 60 million years (My). (MacFadden, 1992; Vila et al., 2001). We can readily recognize and equate the parts of the horse skeleton at the ends of the feet as toes, even when the number decreased over time from four to one (FIGURE 3.1). The fossil record is detailed, enabling us to reconstruct evolutionary changes, recognize now extinct lineages, and identify the lineage that led to the modern horse. The evolutionary record is rarely as complete as it is for horse evolution, however.

Greater difficulties in interpretation occur when similar characters arise in different lineages, as seen in organisms that evolve to mimic another species in their environment. Examples include palatable insects that mimic a poisonous insect species, moths that mimic leaves, and seahorses that mimic seaweeds or corals (FIGURE 3.2). Organisms in far-flung parts of the globe may evolve in parallel or by convergence, even though they do not share gene flow or a recent common ancestor. Examples include placental and marsupial "tigers" or "wolves" (FIGURE 3.3), marsupial, placental, and monotreme "anteaters," and African euphorbs and American cacti (FIGURE 3.4). These examples bring us face to face with the "**apples and oranges**" problem (FIGURE 3.5). How can we tell whether similarity reflects evolutionary origin from a common ancestor or independent evolution? When does similarity mean sameness and when does similarity mean close resemblance?

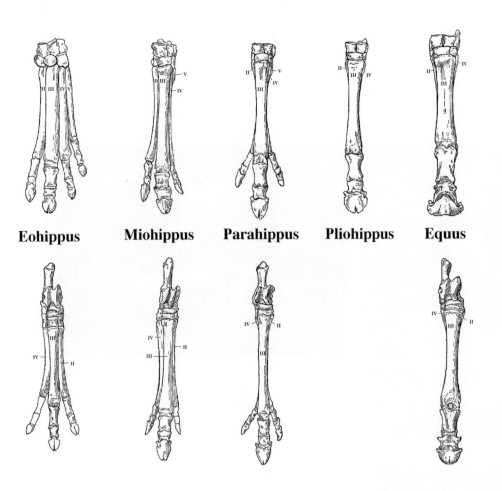

Eohippus **Miohippus** **Parahippus** **Pliohippus** **Equus**

FIGURE 3.1 Reduction of the toes from four to one in both forelimbs (top row) and hind limbs (bottom row) in horses from the Eocene "dawn horse," *Eohippus* (*Hyracotherium*), to the modern horse genus *Equus*, which appeared in the Pleistocene and has persisted to today. Digit III is retained in all, digits II and IV are reduced to splint bones in *Parahippus* and *Equus*, while digits I and V (the outer digits) were lost as early as *Miohippus*. (See also Figure B3.1.)

[Modified from Gregory, W. K., 1951. *Evolution Emerging. A Survey of Changing Patterns from Primeval Life to Man.* Two Volumes. The Macmillan Company. New York.]

(a) (b)

FIGURE 3.2 Mimicry. **(a)** The pygmy seahorse *Hippocampus bargibanti*, which is no more than 2.5 cm in length, mimics the sea fan coral (*Muricella* sp.) in which it resides. **(b)** Many moths mimic leaves as shown in this example of the lappet moth (*Gastropacha quercifolia*) that mimics dried oak leaves. *(continues)*

FIGURE 3.2 *(continued)* **(c)** Many flowers mimic insects, In this example the upper petals resemble antennae, the iridescent blue mimics the blue luster of the wings of a fly, while the two glistening patches mimic eyes. **(d)** The leafy (Glauerts) sea dragon (*Phycodurus eques*) mimics seaweed. **(e)** The giant Devil's Flower Mantis (*Idolomantis diabolica*) mimics flowers.

(a) Borhyaenid marsupial (Miocene, Argentina)

(b) Marsupial Tasmanian wolf (Tasmania, Australia)

(c) Placental wolf (North America)

FIGURE 3.3 Striking examples of parallel evolution involving independent evolution of the "wolf" phenotype on three continents. **(a)** *Prothylacynus patagonicus*, a marsupial from the early Miocene in southern Argentina. **(b)** *Thylacinus cynocephalus*, the recently extinct marsupial Tasmanian wolf. **(c)** *Canis lupus*, the placental North American wolf.

[Adapted from Marshall, L. Q. Marsupial paleogeography. In L. L. Jacob (ed.), *Aspects of Vertebrate History.* Museum of Northern Arizona Press, 1980.]

(a)

(b)

(c)

(d)

FIGURE 3.4 Convergent evolution between representative desert species of American Cactaceae, illustrated by the American saguaro **(a, b)** and African Euphorbiaceae, illustrated by two species of African euphorbs **(c, d)**.

FIGURE 3.5 When weighed on the balance of homology an apple is an apple and an orange is an orange.

© Raymond Gregory/ShutterStock, Inc.

SIMILARITY AND PATTERNS OF EVOLUTION

Phenotypes may be similar (1) because of recent shared ancestry; (2) because similar characters arose in groups with a more distant shared ancestor; or (3) because similar evolutionary patterns arose independently in different lineages.[1] Three concepts and terms were proposed in the mid-nineteenth century to deal with these situations.

1. **Homology: similarity resulting from shared ancestry:** when similarity of a feature arises because organisms in two species or lineages share a recent common ancestor that possessed the feature, the features are **homologous** (**FIGURE 3.6**).
2. **Parallelism: similarity based on shared genes or developmental pathways:** when similar features arise in related lineages whose common ancestor lacked them, the features are considered to have evolved independently and in **parallel**. Because of the shared earlier evolutionary history of the two lineages, parallel features normally develop using similar genetic or developmental pathways (Figure 3.6).
3. **Convergence: similarity resulting from evolution in independent lineages:** when a feature arises independently in unrelated organisms because of similar responses to the same selective pressures, we regard the features as **convergent**. Convergent evolution also is known as homoplasy. Because the lineages have independent evolutionary histories, convergent features usually develop using different genetic or developmental pathways (Figure 3.6). Similarity or dissimilarity of development is therefore a key criterion separating parallelism from convergence, although distinguishing parallelism from convergence can be subjective. There are no rules that specify how far in the past one should search for a common ancestor in parallel evolution, and even convergent lineages have common, albeit very distant, ancestors (Hall et al., 2003, Hall, 2006; Wake et al., 2011).

Each of these patterns of evolution is now treated in more detail.

HOMOLOGY

Derived from terminology introduced in the 1840s by the English comparative anatomist Richard Owen (1804–1892), organs identified as the same, even if serving different functions, are considered homologous. The humerus in the upper arm of a climbing monkey is homologous with the humerus in the forelimb of a digging mole, even though put to different uses by the two animals. Studies of similar bones in a wide range of vertebrates—the

---

[1] See Hall (1994, 2003, 2006) and Scholtz (2010) for discussions.

(a) Homology: two species bearing the same phenotype caused by common ancestry for the same genotype

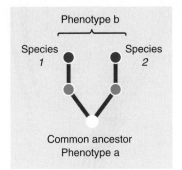

(b) Parallelism: two species with the same phenotype descended from a common ancestor with a different phenotype and genotype

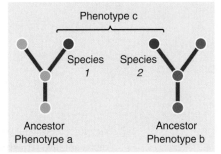

(c) Convergence: two species with the same phenotype whose common ancestor is very far in the distant past

FIGURE 3.6 Homology, parallelism, and convergence diagrammed for two species (*1, 2*) that share a similar phenotypic character (phenotype a, b, c).

humerus in the upper arm, radius and ulna in the lower arm, digits—demonstrated that the forelimbs of widely different vertebrates are homologous, albeit with such different functions as walking, flying, and swimming (**FIGURE 3.7**). Homology can be applied at different levels in the biological hierarchy; for example, to individual bones (the humerus) and to parts of the body (the forelimbs). Organs that perform the same function in different groups but do not share a similarity of structure are **analogous**. Wings of bats or birds, which are built around a bony skeleton, and the wings of insects, which are based on an exoskeleton associated with a network of veins, do not show a common underlying structural plan and so are analogous, not homologous (**FIGURE 3.8**). Because analogues are found in organisms that do not share a recent common ancestor[2], analogous features develop by convergence.

COMPARATIVE ANATOMY *is introduced elsewhere in this text as a major means of identifying morphological species.*

HOMOLOGY STATEMENTS

Toes of a 60-My-old horse are recognizable and easily identified as toes (Figure 3.1). Toes are homologous throughout the lineages of horse evolution because they arose from an ancestor that had the same features in the same position.

[2] Of course, if we trace evolutionary history far enough back and construct evolutionary trees we find that all organisms share a common ancestor, and that some genetic and developmental processes are very ancient indeed and have persisted.

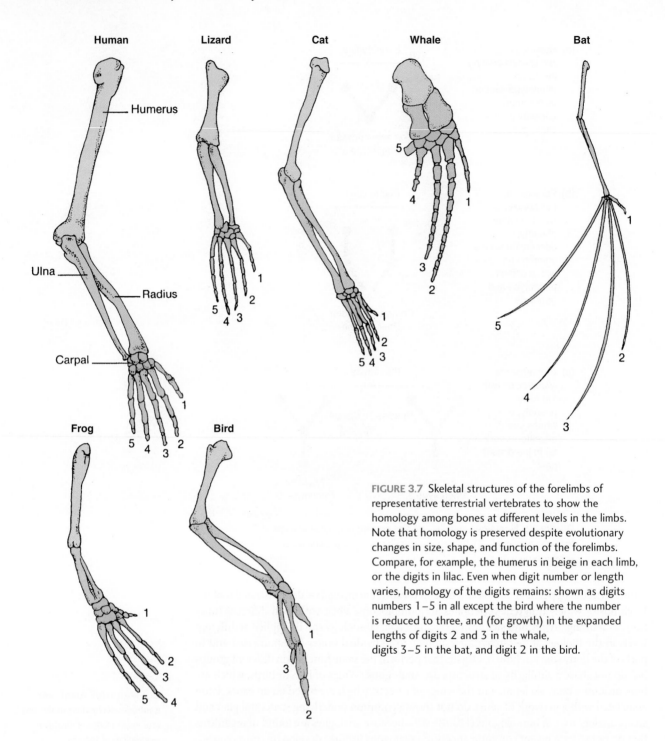

FIGURE 3.7 Skeletal structures of the forelimbs of representative terrestrial vertebrates to show the homology among bones at different levels in the limbs. Note that homology is preserved despite evolutionary changes in size, shape, and function of the forelimbs. Compare, for example, the humerus in beige in each limb, or the digits in lilac. Even when digit number or length varies, homology of the digits remains: shown as digits numbers 1–5 in all except the bird where the number is reduced to three, and (for growth) in the expanded lengths of digits 2 and 3 in the whale, digits 3–5 in the bat, and digit 2 in the bird.

Statements of homology make no comment about features having to be identical or even to look the same to be homologues. The similar features in the forelimb skeleton of the different vertebrates shown in Figure 3.7 are homologous because they arose from the same feature in a common ancestor. They are homologues even though they have changed in appearance with the evolution of wings in birds, flippers in seals, and so forth. As Charles Darwin wrote in Chapter 14 of *The Origin of Species*:

What can be more curious than that the hand of man formed for grasping, that of a mole, for digging, the leg of a horse, the paddle of a porpoise and the wing of a bat, should

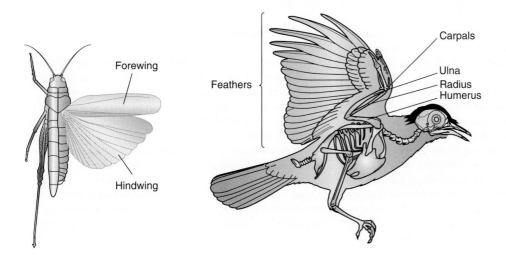

FIGURE 3.8 Comparison of insect and bird wings to show their analogy.

Carpals
Ulna
Radius
Humerus
Feathers
Forewing
Hindwing

all be constructed on the same pattern and should include similar bones and in the same relative positions?

Just as it is easy to see homology in the toes of ancestral and descendant horses, it is more difficult to see wings and flippers as homologues. Compare Figures 3.7 with **FIGURE 3.9** to see the essential similarities.

An additional source of evidence for homology, one used to considerable extent by Darwin, is the presence of vestigial features in descendant species that are more fully developed in their ancestors. Examples of vestigial feature or **vestiges** may be found in **BOX 3.1**.

*The **LARGEST KNOWN SNAKE**, a Paleocene relation of the boa constrictors, 13 m in length and with an estimated weight of 1,135 kg, was described in 2009.*

LEVELS OF HOMOLOGY

Whenever a statement about homology is made, *the level at which the comparison is being made should always be specified*: forelimbs are homologous as forelimbs or as the anterior set of paired appendages; humeri are homologous as the single bone located in the

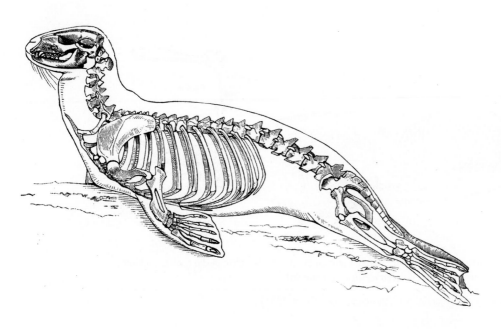

FIGURE 3.9 Skeleton of a harbor seal, *Phoca vitulina*, showing how the homology of the bones in the flippers relates to the limb bones of the other tetrapods shown in Figure 3.7.

[Adapted from Romanes, G. J. *Darwin, and After Darwin.* Open Court, 1910.]

BOX 3.1   **Vestigial Organs**

Comparative anatomy and embryology come together in various ways, one of which is the study of **vestiges**: homologous structures that seem to have lost some or all of their ancestral functions (FIGURE B1.1).

From an evolutionary viewpoint, rudimentary or vestigial organs occur when an organism adapts to a new environment without losing some previously evolved structures (Hall, 2003). As evolution continues, structures that are no longer used tend to diminish, showing only traces of their former size and function. Examples are the rudiments of hind limb and pelvic girdle bones in some species of whales and snakes (FIGURE B1.2), even though hind limbs and pelvic girdles were lost in both groups when they diverged from limbed ancestors. (Although to speak of snakes with legs may seem paradoxical, the direct ancestors of modern-day snakes had legs. At least four genera of limbed fossil snakes are now known: *Haasiophis, Pachyrhachis, Eupodophis,* and *Najash.*)

Adult whales and dolphins have forelimbs (flippers) but no hind limbs. Flippers develop from flipper buds in early embryos. The existence of vestigial of hind limb skeletal elements indicates that hind limb buds must occur in whale and dolphin embryos, which they do (FIG. B1.3; Bejder and Hall, 2002; Hall, 2007).

Organisms that have evolved in dark environments such as caves also provide evidence that obsolete structures gradually become rudimentary. Cave-dwelling crustaceans possess only reduced eyestalks, and some Mexican cavefish have eyes so reduced that the fish are blind. In such cases, other sensory organs assume a greater role to compensate for the lack of vision.

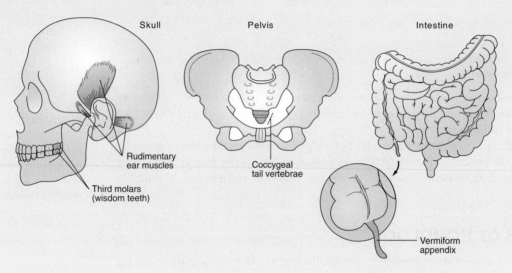

FIGURE B1.1 Vestigial (non-functioning) structures found in humans include the third set of molar (wisdom) teeth, muscles that move the ears in other mammals, ear muscles, tail vertebrae, and the appendix.

[Adapted from Romanes, G. J. *Darwin, and After Darwin.* Open Court, 1910.]

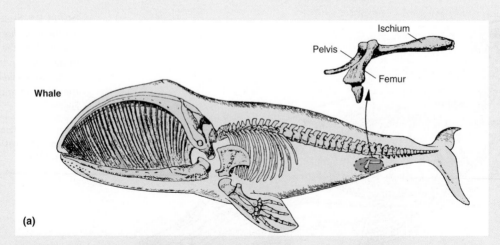

FIGURE B1.2 **(a)** Rudimentary hind limb (femur) and elements of the pelvic girdle (pelvis, ischium) in a bowhead (Greenland) whale, *Balaena mysticus. (continues)*

[Adapted from Romanes, G. J. *Darwin, and After Darwin.* Open Court, 1910.]

BOX 3.1   Vestigial Organs (Cont...)

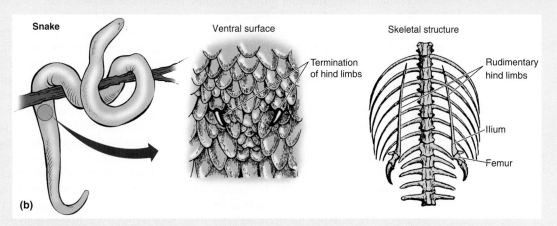

FIGURE B1.2 *(continued)* **(b)** Spurs at the termination of the hind limbs, represented by a vestigial femur and pelvic girdle elements (ilium) in a python.

[Adapted from Romanes, G. J. *Darwin, and After Darwin*. Open Court, 1910.]

Courtesy of Erin Green, Crown Copyright, Dept. of Conservation, NZ

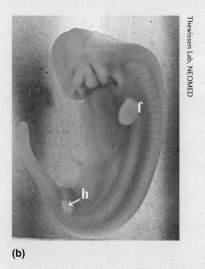

Thewissen Lab, NEOMED

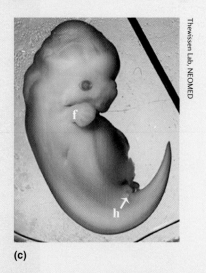

Thewissen Lab, NEOMED

FIGURE B1.3 **(a)** Hector's dolphin, *Cephalorhynchus hectori*, showing the well-developed flippers and absence of hind limbs. Embryos of the spotted dolphin, *Stenella attenuata*, at 24 **(b)** and 48 **(c)** days of gestation to show the well-developed flipper buds (f) and the rudimentary hind limb buds (h).

forelimb between shoulder girdle and elbow. If in one species two bones were found in the position occupied by the humerus, one of the bones would not be a homolog of the single bone in other species. Indeed, such an occurrence (no such specimen has ever been found) would lead us to suspect that the organism containing the two bones had an independent evolutionary history from those organisms containing a single humerus. The two bones would be given different names and not be considered homologous. Furthermore, features of organisms that are not homologous as structures (limbs and genitalia in terrestrial vertebrates, for example) may share genes or gene pathways that are homologous. Homology is a hierarchical concept that takes into account the fact that evolutionary change at different levels (genes, development, structures) need not, and often does not, occur in tandem.

Features are not only the morphological or structural aspects of organisms, but the physiological, developmental, behavioral, molecular, or genetic aspects as well. Behavioral characters, for example, patterns of grooming in rodents, stand as homologous features in their own right. When the heritable basis of behavior is conserved, homologous behaviors are based on homologous features (forepaws and whiskers used by rodents in grooming) that, in turn, may be based on homologous developmental processes. However, as the developmental basis of behavior is both heritable and evolvable (Stamps, 1991), the structural and developmental basis of homologous behavior need not be the same.

Importantly, dissimilarities between organisms do not render behaviors non-homologous; homology at one level does not require homology at the levels upon which the homology is based. For example, consider the homologous behavior of sharks that produce an electric shock to stun their prey. Although the electric organs of all sharks produce an electric shock, electric organs may be modified muscles or modified nerves depending on the taxon, as shown in **FIGURE 3.10**. Homology at one level (stunning prey with electric organs) does not imply homology at another level (the developmental origin of the electric organs).

FIGURE 3.10 Independent evolution of electric organs (and ventrally curved jaws) in electric fish from South American (*Sternarchorhynchus mormyrus*) and African waters (*Campylomormyrus phantasticus*) results in different patterns of electrical discharge.

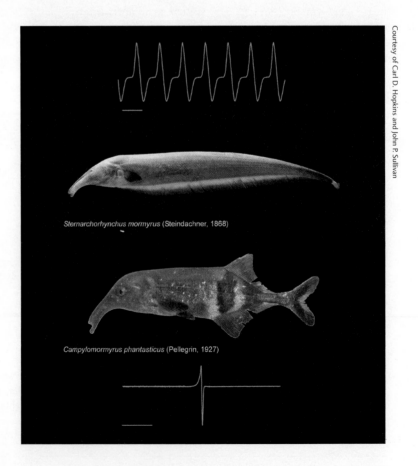

Sternarchorhynchus mormyrus (Steindachner, 1868)

Campylomormyrus phantasticus (Pellegrin, 1927)

HOMOLOGY OF MOLECULES

With the discovery and comparison of the molecular sequences underlying proteins and nucleic acids, the term homology was extended from features of the phenotype to features of the genotype (Hillis, 1994). Genes *shared between species* because of shared species ancestry are called **orthologous genes** (orthologues). **Paralogous genes** (paralogues) are genes *duplicated within a species*; that is, they are extra copies of a gene, rather like the extra vertebrae discussed below as an example of serial homology (Fitch and Margoliash, 1967).

SERIAL HOMOLOGY

Parts repeated in an individual are liable to vary in number, structure, and/or function in response to natural selection. The term **serial** (iterative) **homology** is used for similarities among parts of the same individual; for example, similarities between neck and tail vertebrae, as shown in Figure 3.9, or variants of hemoglobin molecules (α, β, γ chains). Serial homology often reflects the duplication of a gene responsible for producing or affecting a particular structure. Duplication of globin genes led to the large number of hemoglobin variants present in organisms today. Serial homology also can reflect duplication of a particular structure, such as the duplication of vertebrae. Such duplicates may have originally had similar features, but subsequently evolved independently of each other, as illustrated by the independent evolution of neck (cervical) and tail vertebrae in mammals.

PARALLELISM

Parallelism is the evolution of similar features in lineages that are related but do not share a most recent common ancestor. Examples cited above include marsupial and placental "tigers" and "wolves" (Figure 3.3) and the anteater-like features found in several lines of mammals, each of which descended from a non-anteater mammalian group[3] (**FIGURE 3.11**). As with homology and convergence, parallelism can occur in both the phenotype and the genotype.

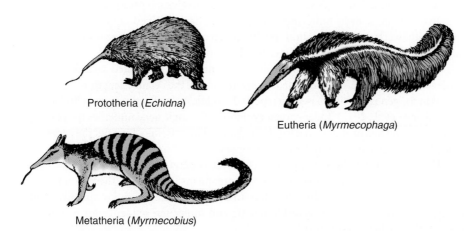

Prototheria (*Echidna*)

Eutheria (*Myrmecophaga*)

Metatheria (*Myrmecobius*)

FIGURE 3.11 Similar phenotypic features—long snout, long tongue, powerful claws—among anteaters that evolved independently within each of the three major groups of extant mammals.

---

[3] For an assessment of parallel evolution in early mammals, see Z.-X. Luo (2007), Transformation and diversification in early mammal evolution. *Nature*, **450**, 1011–1019.

As discussed above, organisms are organized hierarchically, with information building upon the level(s) below. New emergent properties arise at each level and, importantly, cannot be predicted from the properties of the level below. If evolution is constrained—for example, by processes that limit variation—we might expect to find parallel features based on similar genetic or developmental processes.

CONVERGENCE

As the examples of parallel evolution of the "wolf" phenotype on three continents (Figure 3.3) illustrate, why we need to understand evolutionary relationships in order to separate parallelism from convergence. In parallelism, similar features evolve in related lineages, based on similar genetic or developmental processes. In **convergence** (convergent evolution), similar features evolve in independent lineages, based on different genetic or developmental processes. Euphorbs and cacti are a good example of convergence (Figure 3.4). The two types of plants look a great deal alike, and share a number of physiological and metabolic features; however, cacti evolved in South America, and euphorbs evolved in Africa. Cacti and euphorbs converged on a number of features through adaptation to their respective environments, but the features are based on different underlying processes.

The number of examples of convergence in evolution is evidence that responses to similar environmental conditions can, and often do, lead to functionally similar anatomical structures in different evolutionary lineages (Figure 3.3). Evolution of wings in insects and in vertebrates is an example of convergence, as two independent lineages of animals are responding to selection for flight through modification of existing but different appendages. Wings of birds and bats are homologous as limbs with digits; they share an ancestor that possessed limbs with digits, built using similar regulatory processes. However, neither bat nor bird wings are homologous to insect wings, because no common ancestor has a feature from which both types of wings could have been derived.

Likewise, the structural similarity of squid and vertebrate eyes does not come from an ancestral visual structure in a recent common ancestor of mollusks and vertebrates, but from convergent evolution; similar selective pressures led to similar organs that enhance visual acuity. What is shared deep in metazoan ancestry is the ability to form light-gathering cells or organs.[4] From this ability, such convergences arose independently in numerous animal lineages subject to similar selective visual pressures (because of such selective pressures, even butterflies and primates have evolved color vision photopigments with overlapping absorption spectra, based on similar amino acid sites, despite the large separation between the two lineages [Frentiu et al., 2007]).

Modification of shared genetic or developmental, long postulated as underlying, convergence is now being demonstrated. Convergence in the relative length of the limbs in lizards in the genus *Anolis* is based on repeated modification of early stages of limb development (Sanger et al., 2012). Independent evolution of a single gene underlies the convergent evolution of the loss of abdominal legs in spiders and insects; knocking out the genes results in spiders with an extra pair of legs—a 10-legged spider (Khadjeh et al., 2012).

Having discussed how similarity of features is determined, we turn to two classes of evidence that have been used to compare organisms (animals in these examples) and assess evolutionary relationships for close to 200 years. The first is comparative embryology, the second the fossil record. (Genetic and molecular evidence for evolution are discussed elsewhere in this text.)

---

[4] Interestingly, genetic studies indicate that a similar inherited factor (the *Pax-6* gene) regulates the development of anterior sense organ patterns in invertebrates and vertebrates. Nevertheless, despite some common regulatory features, specific cellular pathways in embryonic eye development differ substantially between squid and vertebrates. Squid photoreceptor cells derive from the epidermis; vertebrate retinae derive from the central nervous system.

COMPARATIVE EMBRYOLOGY AS EVIDENCE FOR SHARED SIMILARITY AND EVOLUTION

Comparative embryology is the study of relationships among anatomical structures in the embryos of different species. Animal species have long been the targets of active comparative embryology research, though similar knowledge informs our understanding of plant evolution.[5]

Early in the nineteenth century the Estonian comparative embryologist Karl von Baer (1792–1876) discovered remarkable similarities among the embryos of vertebrates whose adult forms were quite different from each other. The earlier in development the comparisons were made the more similar the embryos were found to be. von Baer generalized his findings into a "law": *early embryos of related species bear more common features than do later, more specialized developmental stages.* Throughout his life and in his publications (even after Darwin's *The Origin of Species* was published), von Baer's views remained comparative and taxonomic, not evolutionary; he used categories of embryos to erect a scheme of classification, not evolutionary lineages.

One of Darwin's major insights was to use comparative embryology as evidence for evolutionary change. As he stated in *The Origin of Species*:

> In two groups of animals, however much they may at present differ from each other in structure and habits, if they pass through the same or similar embryonic stages, we may feel assured that they have both descended from the same or nearly similar parents, and are therefore in that degree closely related. Thus, community in embryonic structure reveals community of descent (p. 481).

In 1861 (after *The Origin of Species* had been published) and in a lifetime of publications and lectures, German embryologist, naturalist, philosopher, and artist Ernst Haeckel (1834–1919) used von Baer's research on comparative embryology to propose that during their development animal embryos repeat the evolutionary history of the groups to which they belong. For Haeckel, developing embryos were stages of evolution, and evolution could be studied in embryos (**FIGURE 3.12**). In this way, Haeckel integrated comparative embryology with evolution in what became known as the **biogenetic law**.

> Ontogeny [development of the individual] is a short rapid recapitulation of phylogeny [the ancestral sequence]… The organic individual repeats during the swift brief course of its individual development the most important of those changes in form that its ancestors traversed during the slow protracted course of their paleontological evolution according to the laws of heredity and adaptation.[6]

To Haeckel, this meant that the tadpole developmental stage of an extant frog reflected, or recapitulated, a *tailed frog ancestor.*

We now understand, however, that early stages of embryonic development recapitulate only *early ancestral developmental stages,* not ancestral adults (Hall, 1999, 2002). Juvenile stages of ancestral organisms can be retained in the adult forms of their descendants, as, for example, in the preservation of juvenile ape features in adult humans. This observation directly contradicts the Haeckelian notion that descendants retain ancestral adult features. Rather, organisms that share common descent make use of common underlying embryological patterns. Further, related organisms use shared genes and gene networks to produce characteristic developmental stages that have persisted for tens of millions of years. Evidence from genetics, molecular and developmental biology, and from the integration of evolutionary and developmental biology (evo-devo) provide strong support for this view.

---

[5] See Niklas (1977), Hall (1999), and Hall and Olson (2003) for further information.
[6] E. Haeckel (1866). *Naturliche Schöpfungsgeschichte.* Reimer, Berlin. See Richards (2008) for the authoritative scientific biography of Haeckel.

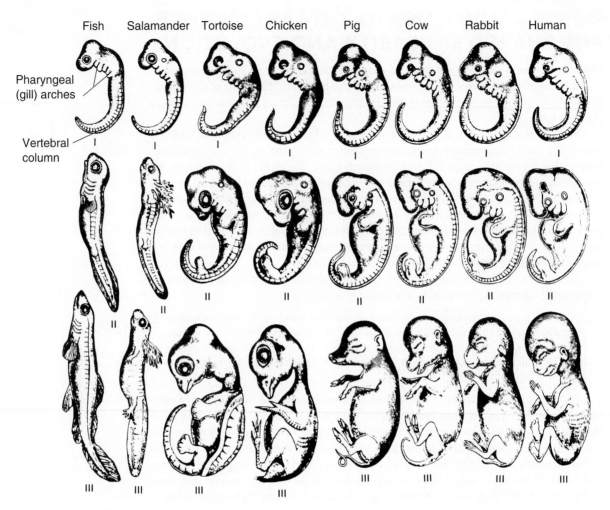

Fish Salamander Tortoise Chicken Pig Cow Rabbit Human

Pharyngeal
(gill) arches

Vertebral
column

FIGURE 3.12 One of Haeckel's classic nineteenth century illustrations of different vertebrate embryos at comparable stages of development. Although Haeckel took some liberties in drawing these figures, the earlier stages are more similar to one another than later stages are to one another. Embryos in the different groups have been scaled to the same approximate size so that comparisons can be made among them.

[Adapted from Romanes, G. J. *Darwin, and After Darwin.* Open Court, 1910.]

FOSSILS AS THE PHYSICAL EVIDENCE OF EVOLUTION

Discovery of the fossil record of life gave us a rich source of data with which to understand relationships among organisms. It led to the hypothesis that Earth's surface and the organisms on it had existed for a long time, and to the conclusion that organisms succeeded one another through time (Fortey, 2002; Rudwick, 2005).

End Box 3.1 contains a brief history of the recognition that fossils are the remains of past life, as well as a discussion of how fossils form.

Charles Lyell (1797–1875), a contemporary and close friend of Charles Darwin, broke with the popular theories that catastrophic or miraculous events (**catastrophism**) were responsible for Earth's geological structure. In the mid-nineteenth century, Lyell and others began to consider that species may have changed in concert with changes in Earth's geology. He developed the principle of **uniformitarianism** (**BOX 3.2**), that the natural laws and processes functioning in the universe today, and the rate at which they

BOX 3.2	How Rocks Are Deposited: Catastrophism and Uniformitarianism

Georges Cuvier (1769–1832) was one of the most gifted French comparative anatomists and the founder of modern paleontology. In the late 1700s and early 1800s, his theory of catastrophism was commonly believed to explain the formation of geological structure. Transition from catastrophism to an alternate theory—uniformitarianism—had profound effects on our understanding of the natural world. It helped liberate scientific thinking from the concept of a static universe powered by unexplainable changes to one that is dynamic and understandable in natural terms.

CATASTROPHISM

According to catastrophism, sharp discontinuities in the geological record—stratifications of rocks, layering of fossils, transition from marine to freshwater fossils—were evidence of sudden upheavals caused by glaciations, floods, and other catastrophes. Fossils were recognized as extinct species whose place has been taken by species alive today. Swiss paleontologist, geologist, naturalist, and founder of the Museum of Comparative Zoology at Harvard University Louis Agassiz (1807–1873) proposed that there may have been as many as 100 successive special divine creations.

UNIFORMITARIANISM

Jean-Baptiste Lamarck was introduced elsewhere in this text when discussing his theory of the inheritance of acquired characters. In contrast to Cuvier's catastrophist position, Lamarck proposed that geological discontinuities represented gradual changes in the environment and climate to which species were exposed. Through environmental effects on organisms, these changes led to species transformation.

This uniformitarian concept, that the steady, uniform action of the forces of nature could account for Earth's features (foreshadowed by Buffon and others), was strongly developed in the work of the Scottish geologist James Hutton (1726–1797). Later, Charles Lyell offered the *uniformitarian reply to catastrophism* with the following hypotheses:

1. Sharp, catastrophic discontinuities are absent if geological strata are examined over widespread geographical areas. Most often, a widely distributed stratum shows regularity in its structure and composition. Only in specific localities do rapid shifts seem to appear, as a response to local changes.
2. Changes in the geological record arise from the action of erosive natural forces such as rain, wind, volcanic activity, and flood deposits. The laws of motion and gravity that govern natural events are constant through time. Therefore, past events were caused by the same forces that produce phenomena today (although the extent of phenomena, such as volcanism, might have fluctuated in the past). Consequently, all natural explanations for phenomena should be investigated before supernatural causes are used to explain them.
3. Earth must be very old for so many geological changes to have taken place by such gradual processes.

Thus, uniformitarianism did not exclude sudden geological changes such as floods, volcanic eruptions, and meteorite impacts—events that were of common or recorded knowledge. Instead, it led to the position that even such "catastrophes" could be natural and rationally explained.

operate, have not changed since the beginning of time. However, Lyell did not explicitly identify the changes in Earth's geology and climate as a selective pressure that could drive evolution of life. Indeed, he struggled with the ideas put forth by Darwin and Wallace, which connected species change with (a) interactions between organisms and (b) interactions between organisms and their environment, and which proposed a mechanism for species change—evolution by natural selection.

In the mid-nineteenth century the known fossil record was sporadic, the result of serendipitous collecting and random finds. Nevertheless, during Darwin's lifetime a few paleontological findings came to light that strongly supported his theory of descent with modification. One was the discovery in 1861 of what had been proposed as a true "missing link," in this case, an animal that was interpreted as intermediate between reptiles and birds. As shown in **FIGURE 3.13**, this fossil, *Archaeopteryx*, had a number of *reptilian features,* including teeth and a tail of 21 vertebrae. However, it also had a number of *bird-like features,* such as a wishbone and feathers. English biologist and Darwin-proponent Thomas Henry Huxley argued convincingly that *Archaeopteryx* was a "cousin" to the lineage running from reptiles (dinosaurs) to birds. He believed that such "primitive" forms were predictable consequences of evolution that helped prove the theory. An even more

Analysis of mtDNA from 22 fossil horses has revealed TWO NEW SPECIES OF HORSES *and revised the patterns of relationships known previously only from the fossil record (Orlando et al., 2009).*

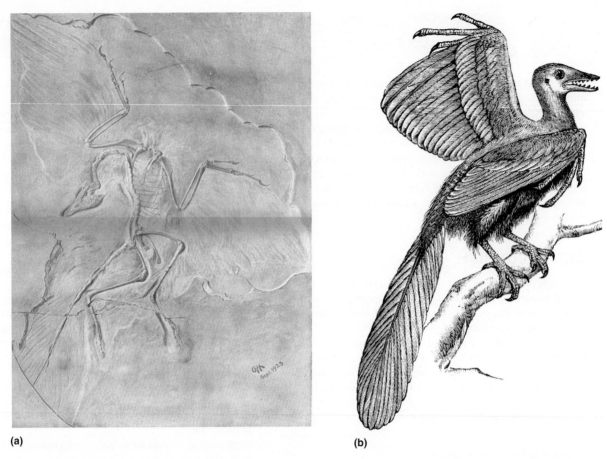

(a)   (b)

FIGURE 3.13 **(a)** The Berlin specimen of *Archaeopteryx*. **(b)** (Gerhard) Heilmann's reconstruction of what *Archaeopteryx* may have looked like in real life.

[(a) and (b) Reproduced from Heilmann, G. *The Origin of Birds.* Appleton, 1927 (Reprinted Dover Publication, 1972).]

complete history of the evolution of a lineage exists and, once again, we return to the evolution of horses, which was discerned through the fossil record. The finely detailed phylogeny of horses is one of the best illustrations of some of the realities and complexities of evolution.

EVOLUTION OF HORSES

EOHIPPUS is an alternate but later genus name for Hyracotherium. *Priority goes to* Hyracotherium.

One year after the publication of *On the Origin of Species,* Richard Owen (1804–1892) described the earliest known horse-like fossil, first called *Hyracotherium* but often referred to as *Eohippus* (the dawn horse). *Hyracotherium* was some 50-cm high (about 20 inches, the size of an average Border Collie) and weighed about 23 kg (50 pounds). It had four toes on its front legs and three on its hind legs, adapted to walking on soft, moist forest floors, and simple teeth adapted for browsing on soft vegetation (**FIGURES** 3.1 and 3.14). Later fossil finds revealed that *Hyracotherium* was actually a number of herbivorous species present from North America to Europe, some no larger than an average-sized modern-day fox; this is an excellent example of parallel evolution in the horse lineage.

In the approximately 60 My after *Hyracotherium* arose, horses changed radically. Today, horses have only a single toe on each foot (Figure 3.1). They show special adaptations for running on hard ground. Their elongated legs are built for speed, bearing most of the limb muscles in the upper part of the legs, enabling a powerful, rapid swing. This arrangement, coupled with a special set of ligaments, provides them with a pogo-stick-like

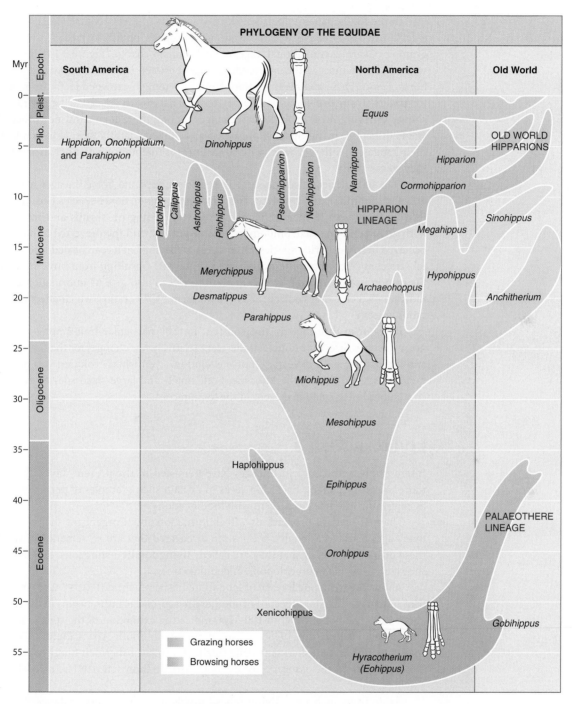

FIGURE 3.14 Evolutionary relationships among various lineages of horses, with emphasis on North American and Old World groups. Sample reconstruction of the digits ("toes") of the hind feet of some fossil horses and of the extant horse *Equus* are shown. The number of digits declined from four to one during evolution of the lineage. These horse lineages show both branching and non-branching patterns of evolution.

[Adapted from MacFadden, B. J. *Fossil Horses: Systematics, Paleobiology and Evolution of the Family Equidae.* Cambridge University Press, 1992.]

springing action while running on hard ground. Horses also show adaptations for chewing tough, silica-containing grasses; their teeth are much longer than the teeth of other grazing animals.

By the 1870s, paleontologists such as America's Othniel C. Marsh (1831–1899) were able to use fossils of North American and European horses to present a now classic example of various transitional stages of evolutionary change (Figure 3.14). Remarkably, we now know almost all the intermediate stages between *Hyracotherium* and the modern horse, *Equus*. These include transitions from low- to high-crowned teeth, from browsers to grazers, from pad-footed to spring-footed, and from small- to large-brained. As shown in Figure 3.14, evolutionary changes among these forms did not proceed in a single direction, being better represented as a "bushy" family tree. Horses adapted to their habitats in different ways, with some lineages maintaining distinct structures until they went extinct.

Although all occupied the same general area, separate horse lineages made use of different environmental resources (*resource partitioning*). Some species became grazers, feeding on grasses. Others remained browsers, feeding on shrubs and trees. Others both grazed and browsed. Still others became grazers, and then reverted to browsing, as occurred in some Florida species. Differences in feeding habits can be deduced from dental scratches (resulting from grazing) and dental pits (resulting from browsing). Testing the carbon isotope ratios ($^{12}C/^{13}C$) in fossil teeth can also give us information about the individual's diet: grasses and shrubs have different $^{12}C/^{13}C$ ratios, which affect the $^{12}C/^{13}C$ ratios in teeth.[7]

Rates of evolution were not constant for any particular trait among the various horse lineages. Size, for example, underwent relatively few changes for the first 30 million and the last few million years of horse evolution. Even when evolution was proceeding rapidly, as it did during the Miocene, both small- and large-sized species evolved. No continuous linear trend is present in the fossil record.

LIVING FOSSILS

Interestingly, some ancient lineages have persisted to the present day with minimal morphological changes. Evolution seeks to explain such examples of persistence as well as explaining examples of descent with modification.

The fossil record provides us with information about organisms that went extinct many ages ago. Occasionally, species are discovered that are so remarkably similar to these extinct organisms that they are called "**living fossils**": sturgeons, lungfish, horseshoe crabs, *Lingula* (a brachiopod), and ginkgo trees.

About 200 Mya, one lineage of lobe-finned fishes evolved (FIGURE 3.15) into terrestrial vertebrates. The fossil record of another lineage, coelacanths, begins in the Devonian about 380 Mya and ends 80 to 100 Mya, indicating extinction of the species in the late Cretaceous. However, in 1938 a museum curator in South Africa found a coelacanth specimen amongst a fisherman's daily catch. Still today, fishermen find live coelacanths in deep waters off the eastern coast of South Africa (Thomson, 1991). The coelacanth is a living fossil.

Aside from such rare "living relics," fossils in almost every instance differ from present-day forms, often in proportion to their age. More recent geological strata contain forms more like the present than those in older strata. Fossils provide the hard evidence of evolution.

---

[7] B. J. MacFadden, N. Solounias, and T. E. Cerling, 1999. Ancient diets, ecology and extinction of 5-million-year-old horses from Florida. *Science,* **283**, 824–827; J. T. Eronen et al., 2009. The impact of regional climate on the evolution of mammals: a case study using fossil horse. *Evolution,* **64**, 398–408.

(a)

(b)

(c)

FIGURE 3.15 Representative "living fossils." **(a)** The coelacanth (*Latimeria chamulnae*). **(b)** Leaves of the ginkgo tree (*Ginkgo biloba*). **(c)** The Atlantic horseshoe crab (*Limulus polyphemus*).

■ KEY TERMS

analogy

biogenetic law

catastrophism

convergent

convergence

homology

living fossils

orthologous genes

parallelism

paralogous genes

serial (iterative) homology

uniformitarianism

vestigial organs

vestiges

■ EVOLUTION ON THE WEB

Explore evolution on the Internet! Visit the accompanying website for *Strickberger's Evolution, Fifth Edition,* at **go.jblearning.com/Evolution5eCW** for exercises and links relating to topics covered in this chapter.

■ REFERENCES

Bejder, L., and B. K. Hall, 2002. Limbs in whales and limblessness in other vertebrates: mechanisms of evolutionary and developmental transformation and loss. *Evol. Dev., 4*, 445–458.

Fitch, W. M., and E. Margoliash, 1967. Construction of phylogenetic trees. *Science, 155*, 279–284.

Fortey, R., 2002. *Fossils: The Key to the Past.* Natural History Museum, London, UK.

Frentiu, F. D., G. D. Bernard, C. I. Cuevas, et al., 2007. Adaptive evolution of color vision as seen through the eyes of butterflies. *Proc. Natl Acad. Sci. USA, 104* (Suppl 1), 8634–8640.

Hall, B. K. (ed.), 1994. *Homology: The Hierarchical Basis of Comparative Biology.* Academic Press, San Diego, CA [paperback issued 2001].

Hall, B. K., 1999. *Evolutionary Developmental Biology,* 2nd ed. Kluwer Academic Publishers, Dordrecht, The Netherlands.

Hall, B. K., 2003. Descent with modification: the unity underlying homology and homoplasy as seen through an analysis of development and evolution. *Biol. Rev. Camb. Philos. Soc., 78*, 409–433.

Hall, B. K., 2002. Palaeontology and evolutionary developmental biology: a science of the 19th and 21st centuries. *Palaeontology, 45*, 647–669.

Hall, B. K., 2006. Homology and homoplasy. In: *Handbook of the Philosophy of Biology,* D. Gabbay, P. Thagard, and J. Woods (eds.), 2006, Volume *3* (*Philosophy of Biology*), M. Matthen and C. Stephens (eds.). Elsevier BV, Amsterdam, pp. 441–465.

Hall, B. K., 2007. *Fins into Limbs. Development, Transformation, and Evolution.* The University of Chicago Press, Chicago.

Hall, B. K., and W. M. Olson (eds.), 2003. *Keywords & Concepts in Evolutionary Developmental Biology.* Harvard University Press, Cambridge, MA.

Hillis, D. M., 1994. Homology in molecular biology. In *Homology: The Hierarchical Basis of Comparative Biology,* B. K. Hall (ed.). Academic Press, San Diego, CA, pp. 339–368.

Khadjeh, S., N. Turetzek, M. Pechmann, et al. 2012. Divergent role of the *Hox* gene *Antennapedia* in spiders is responsible for the convergent evolution of abdominal limb repression. *Proc. Natl Acad. Sci. USA., 109*, 4921–4926.

MacFadden, B. J., 1992. *Fossil Horses: Systematics, Paleobiology, and Evolution of the Family Equidae.* Cambridge University Press, Cambridge, UK.

Niklas, K. J., 1997. *The Evolutionary Biology of Plants.* The University of Chicago Press, Chicago.

Orlando, L., J. L. Metcalf, M. T. Alberdi, et al., 2009. Revising the recent evolutionary history of equids using ancient DNA. *Proc. Natl Acad. Sci. USA, 196*, 21754–21759.

Richards, R. J., 2008. *The Tragic Sense of Life: Ernst Haeckel and the Struggle over Evolutionary Thought.* University of Chicago Press, Chicago.

Rudwick, M. J. S., 2005. *Bursting the Limits of Time. The Reconstruction of Geohistory in the Age of Revolution.* The University of Chicago Press, Chicago. [The best history of geology available.]

Sanger, T. J., L. J. Revell, J. J. Gibson-Brown, J. B. Losos. 2012. Repeated modification of early limb morphogenesis programmes underlies the convergence of relative limb length in *Anolis* lizards. *Proc. R. Soc. Lond. (B), 279*, 739–748.

Scholtz, G., 2010. Deconstructing morphology. *Acta Zool. (Stockh.), 91*, 44–63.

Stamps, J. A., 1991. Why evolutionary issues are reviving interest in proximate behavioral mechanisms. *Am. Zool., 31*, 338–348.

Thomson, K. S., 1991. *Living Fossil; The Story of the Coelacanth.* W. W. Norton & Co., New York, NY.

Vila, C., J. A. Leonard, A. Götherström, et al., 2001. Widespread origins of domestic horse lineages. *Science, 291*, 474–477.

Wake, D. B., M. H. Wake, and C. D. Specht, 2011. Homoplasy: from detecting pattern to determining process and mechanism of evolution. *Science, 331*, 1032–1035.

END BOX 3.1

Fossils as Evidence of Past Life

SYNOPSIS: This is a brief history of our understanding of the nature of fossil evidence that explains *succession*—the replacement of one form of organism by another—as an explanation of why complete sequences of fossils are rarely found.

The discovery and study of fossils provides an essential basis for understanding evolutionary relationships between past organisms and for appreciating their lengthy history. During the sixteenth and seventeenth centuries, fossils were regarded as "naturally formed" images of God's creation, placed on Earth for man's admiration (Rudwick, 2005; Ruse and Travis, 2009).

It had long been known that the fossilized bones or shells in exposed riverbanks, in mines, and on eroded surfaces did not resemble extant species (**FIGURE EB1.1**), and that seashells could be found in most unlikely places, such as mountaintops. Ancient Greeks were aware of such fossils, and a number of ancient writers, including Herodotus (484–425 BC), hypothesized that they could be explained by changes in the positions of sea and land. Studies by English physician and naturalist Robert Hooke (1635–1703) and Danish anatomist and geologist Nicolaus Steno (1638–1686) concerning the reality of fossil species led to naturalistic proposals to understand fossil origins. The frontispiece of Charles Lyell's 1830 *Principles of Geology* is a portrait of the three remaining columns of the ruined Temple of Serapis in Pozzuoli, Italy, which show evidence of historical rise and fall in sea level. A three-meter section of these columns contains holes bored by gastropods, indicating that the bases of the columns were once submerged (**FIGURE EB1.2**).[a]

When arranged by stratigraphic age, with deeper strata signifying older ages than superimposed strata, older fossils show greater morphological differences from extant species than do later fossils. This evidence of change over time provided the foundation for a "law of succession" in which one form replaced another.

FOSSIL FINDS

Fossil remains are predominantly found in sedimentary rocks, which originated as a succession of deposits in seas, lakes, riverbeds, or deserts (**FIGURE EB1.3**). Even in appropriate sedimentary environments, many dead organisms decompose before they fossilize or, if they have fossilized, are destroyed by erosion.

Because isolation of populations encourages and sustains their differences, we rarely find intermediate forms in the same place as the original forms. Consequently, a complete evolutionary progression of fossils from most ancient to most recent has never been found in a single locality. Nevertheless, fossils provide the hard evidence for evolution. One of the most complete fossil sequences is the evolution of horse lineages. Many of the fossils in this sequence were discovered soon after Darwin published *The Origin of Species*.[b]

Fossils are not always the result of organisms' remains; other forms such as footprints can be enormously informative. The earliest land-dwelling vertebrate (tetrapod) fossils are dated to the Late Devonian Period, whereas fossil trackways left by tetrapods date to the early Middle Devonian, some 18 My earlier (Niedzwiedzki et al., 2010). Further, the trackways attest to life in coral reef lagoons whereas the first fossilized remains imply that the first tetrapods lived in a river delta or lake environment (Markey and Marshall, 2007). Burrows and disturbed sediments provide us with the evidence for Precambrian adult animal life, while multicellular Precambrian embryos demonstrate the antiquity of animal development and of animals themselves (Chen et al., 2009). Even single-celled organisms have left traces of their existence as far back as 1.8 Bya (Matz et al., 2008). Finally, molecular signatures reveal the presence of life billions of years ago.

---

[a]Lyell used such an image through 12 editions of his book as an example of gradual geological change.
[b]See Fortey (2002), Hall (2002), and Rudwick (2005) for three perspectives on the fossil record.

END BOX 3.1

Fossils as Evidence of Past Life (Cont...)

FIGURE EB1.1 Succession of fossils as revealed when geological strata are exposed in a quarry or canal.

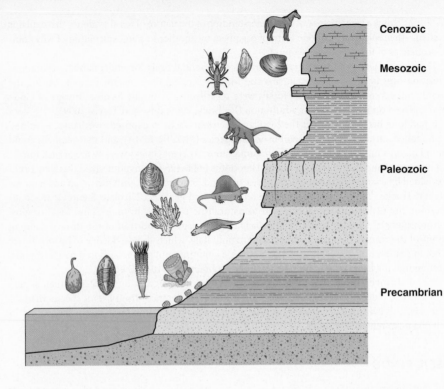

Cenozoic

Mesozoic

Paleozoic

Precambrian

FIGURE EB1.2 A contemporary photograph of the three remaining columns of the ruined "Temple of Serapis" in Pozzuoli, Italy, showing that they had been historically subjected to rise and fall in sea level. A dark, three-meter section of the columns is filled with holes bored by marine organisms, evidence that the columns were once partly submerged.

FIGURE EB1.3 The process of fossilization in which an organism (in this case, an animal) **(a)** dies in a watery environment that protects it from scavengers. Reduced oxygen levels in deeper water further resist deterioration **(b)**. The remains are gradually silted over **(c)** and eventually covered by successive layers of soil that compact into sedimentary rock **(d)**. In time, because of erosion, the fossil surface may become exposed **(e)**.

[From Kardong, K. V., 2006. Vertebrates: Comparative Anatomy, Function, Evolution, 4th ed. McGraw-Hill, New York. © The McGraw-Hill Companies, Inc.]

(a) (b) (c) (d) (e) (f)

END BOX 3.1

Fossils as Evidence of Past Life (Cont...)   © Photos.com

REFERENCES

Chen, J-Y., D. J. Bittjer, M. G. Hadfield, F. Geo., et al., 2009. Complex embryos displaying bilaterian characters from Precambrian Doushantou phosphate deposits, Wend'an, Guizhou, China. *Proc. Natl Acad. Sci. USA*, **106**, 19056–19060.

Fortey, R., 2002. *Fossils: The Key to the Past.* Natural History Museum, London, UK.

Hall, B. K., 2002. Palaeontology and evolutionary developmental biology: a science of the 19th and 21st centuries. *Palaeontology*, **45**, 647–669.

Kardong, K. V., 2006. *Vertebrates: Comparative Anatomy, Function, Evolution,* 4th ed. McGraw-Hill, New York.

Markey, M. J., and C. R. Marshall, 2007. Terrestrial-style feeding in a very early aquatic tetrapod is supported by evidence from experimental analysis of suture morphology. *Proc. Natl Acad. Sci. USA*, **104**, 7134–7138.

Matz, M. V., T. M. Frank, N. J. Marshall, E. A. Widder, and S. Johnsen. 2008. Giant deep-sea protist produces Bilaterian-like traces. *Curr. Biol.*, **18**, 1849–1854.

Niedzwiedzki, G., P. Szrek, K. Narkiewicz, M. Narkiewicz, and P. R. Ahlberg, 2010. Tetrapod trackways from the early Middle Devonian period of Poland. *Nature*, **463**, 43–48.

Rudwick, M. J. S., 2005. *Bursting the Limits of Time. The Reconstruction of Geohistory in the Age of Revolution.* The University of Chicago Press, Chicago. [The best history of geology available.]

Ruse, M., and Travis, J., 2009. *Evolution: The First Four Billion Years.* Harvard University Press, Cambridge, MA. [Contains authoritative essays and entries.]

PART 2

The First Ten Billion Years: 13.7–3.7 Bya

CHAPTER

4

Universe and Earth Arise

CHAPTER SUMMARY

The universe is about 13.73-billion-years old, may be infinitely large, and has been expanding since its origin. Within two hundred million years of its origin (i.e., by 13.6 Bya), the first stars composed of hydrogen and helium had formed. Nuclear reactions in stars, particularly stars more massive than our own Sun, eventually fused a small percentage of the hydrogen and helium into all other chemical elements, including biologically important ones such as carbon, oxygen, nitrogen, phosphorus, and iron. In less than another billion years (by 12.6 Bya) hundreds of galaxies existed. It took more than another 8 By, however, for the solar system to arise, by condensation from a rotating mass of gas and dust. The central mass became the Sun; peripheral masses became the planets. Earth and its moon formed between 4.56 and 4.53 Bya. Earth's primary atmosphere was composed mostly of hydrogen and helium, but these elements were later replaced by a secondary atmosphere composed of gases such as carbon dioxide, water vapor, ammonia, methane, and nitrogen, forming an atmosphere with little or no oxygen.

INTRODUCTION

This chapter provides an overview of the first 10 billion years of our universe, from its almost incomprehensible origin in an instant, 13.73 Bya, to the origin of Earth 4.6 Bya, and the formation of Earth's secondary atmosphere 3.8 to 4.1 Bya (**TABLE 4.1**).[1] Without that most fleeting of beginnings, and the events that have since occurred, there would be no atoms, no molecules, no solar system, no Earth, and thus no life, or no life as we know it (Davies, 2006). Evidence briefly outlined here comes from **cosmology** (the study of the universe), **astronomy** (the study of that part of the universe we can see in the night sky), and **geology** and **geophysics** (the earth sciences). Data were collected with space- and ground-based telescopes, satellites, and space probes, by exploring mountaintops and ocean depths, and with scanners and probes of ever-increasing sophistication, capable of extracting more and more refined and precise information.[2] The chapter ends by asking the question: What features have made life possible here on Earth?

THE FIRST 380,000 YEARS

Humans have long speculated about the origin of the universe, but only in recent decades have we had the tools to investigate our theories. Today, however, cosmology—the study of the universe—is an exciting whirlwind of research and discovery. Aided by the Hubble Space Telescope, satellites and space probes, as well as research on the ground, we are "seeing" further and further into the past, discovering planets orbiting far-distant stars, and much more. Yet the more we learn, the more we realize how little we know. For instance, is ours the only universe that has ever existed (or indeed, exists now), and is there life on other planets?

Although debate continues, the available evidence leads most cosmologists to support the conclusion that the universe began 13.73 Bya in a sequence of events known as the **Big Bang**, a name that came into our vocabulary as a term of derision for the hypothesis known as **the evolutionary universe**. Despite its name, the Big Bang did *not* involve an explosion, nor did it begin at a particular place or expand *into* anything. Amazingly, space appeared everywhere, all at once. From this unimaginable beginning

*The term **Big Bang** was coined by one of the evolutionary universe theory's greatest opponents, the British astronomer and mathematician, Sir Fred Hoyle (1915–2001), whose own theory of the* STEADY–STATE UNIVERSE *has not survived the evidence gathered since he proposed it in 1948.*[3]

[3]See Mitton (2005) for a history of these theories.

TABLE 4.1	The First Ten Billion Years
Time (Bya)	**Event**
13.7	Origin of the universe
13.6	First stars form (composed of hydrogen and helium)
12.8	Hundreds of galaxies exist
4.6	Origin of solar system and sun
4.56	Origin of Earth
4.53	Origin of Earth's moon
4.2–3.8	Secondary atmosphere forms around Earth

[1]Throughout the text we follow the convention for deep time used in geology and paleontology: By, billion years; Bya, billion years ago; My, million years; Mya, million years ago. Cosmologists use Ga (gigaannum) for a billion (10^9) years. Thus, 10 By = 1 Ga.
[2]For cosmology see Hawking (2001), Longair (2006), and Weintraub (2010), or visit http://map.gsfc.nasa.gov/m_uni.html, a NASA website devoted to the Wilkinson Microwave Anisotropy Probe (WMAP). For Earth see Condie and Sloan (1998), Redfern (2003), or visit http://nasascience.nasa.gov/earth-science, a NASA website devoted to the nature of Earth.

TABLE 4.2	The First 100 Seconds of the Life of the Universe	
Time (seconds)	**Event**	**Temperature (˚C)**
0	Birth of the universe	
10^{-43}	Era of quantum gravity and exotic physics	10^{32}
10^{-35}	Universe expands exponentially	10^{28}
10^{-11}	Electromagnetic and weak forces differentiate; quarks and gluons emerge	10 quadrillion
0.1 microseconds		20 trillion
1 microsecond		6 trillion
10 microseconds	Quarks bound into protons and neutrons	2 trillion
100 seconds	Beginning of formation of the nuclei of helium and a few other elements from hydrogen	1 billion

[*Source:* Riordan and Zajc (2006).]

arose everything that exists today: hundreds of billions of stars in more than 100 billion galaxies—every atom of which owes its origin to the Universe's early days—and much else besides.[4]

TABLE 4.2 provides a recent view of the first 100 seconds of our universe. Look particularly at the left column to gain a sense of the speed of the processes involved and at the right column for the extraordinarily high temperatures (up to 100,000 times hotter than the Sun's core) and rapid rate of cooling. Then factor in size: the universe is thought to have increased from virtually nothing to about 10^{24} meters or one hundred million light years, all in an instant at around 10^{-35} of a second. Since then, the universe has continued to expand, albeit at a far, far slower pace. **Inflation**, as this expansion is known, has been called one of the most important scientific discoveries of the 20th century (see below for more on this topic).

Within the hot, dense brew of these first seconds, a great deal of complicated physics was going on, physics that is well beyond the scope of this book. Simplifying greatly, once the initial inflation ended, the universe entered a phase during which a variety of elementary particles first appeared, followed by electrons, protons, and neutrons, and finally the atomic nuclei of helium and a few other low-molecular-weight elements (Table 4.2). The latter stage involved nuclear fusion and ended about 20 minutes after the Big Bang, by which time it was too cool (though still extremely hot) for fusion to continue. Among other important early events, the four fundamental forces of the universe, including gravity and the electromagnetic force, came into being. Matter, energy, space, and time all now existed (Cline, 2004).

Experiments carried out at the Brookhaven National Laboratory in Long Island, New York have given us a glimpse of what the universe may have looked like early in this process, 10 microseconds into the Big Bang:

> During those early moments, matter was an ultra hot, super dense brew of particles called quarks and gluons rushing hither and yon and crashing willy-nilly into one another. A sprinkling of electrons, photons and other light elementary particles seasoned the soup. This mixture had a temperature of trillions of degrees, more than 100,000 times hotter than the sun's core (Riordan and Zajc, 2007).

But still no atoms, nor would there be for a long time. It was just too hot for nuclei to combine with electrons to form atoms. Instead, **photons**—the elementary particles

*The National Aeronautics and Space Administration (*NASA*) provides a fascinating introduction to the history of the universe at http://map.gsfc.nasa.gov/universe/.*

10^{-35} SECONDS (SEC) is an unimaginably short period of time. Recollect that 10^{-1} sec = one tenth of a second, 10^{-2} = a hundredth of a second, and so on. At the other extreme, 10^{24} meters is 10 followed by 23 zeroes or a trillion trillion (or septillion) meters, according to U.S. terminology.

The nucleus of the most common form (isotope) of HYDROGEN consists of a single proton. Hydrogen has dominated the universe from the time it first formed.

---

[4]Lineweaver and Davis (2005) and see NASA, 2006. Foundations of Big Bang Cosmology. Available at http://map.gsfc.nasa.gov/m_uni/uni_101bb2.html.

responsible for light and electromagnetic radiation—dominated the energy of the universe. Because free electrons scatter photons quite efficiently, an observer at the time (if such were possible) would have been immersed in a dense fog, like the inside of a cloud. Once it was cool enough, electrons (which are negatively charged) could combine with positively charged protons to form neutral hydrogen atoms, and combine with helium nuclei to form neutral helium atoms. Finally—around 380,000 years after the Big Bang—the universe became transparent, and the photons streamed forth.

Today, the universe is bathed in a remarkably uniform sea of those same photons—the afterglow of radiation from the initial Big Bang known as the **cosmic microwave background** (CMB). In 1992, data from the Cosmic Background Explorer (COBE), the first NASA satellite devoted to cosmology, revealed minute fluctuations—parts per hundred thousand—in the temperature of the CMB from place to place, which now averages 2.725 K. Discovery of these minute fluctuations in the CMB has proved invaluable in providing an exact snapshot of the universe as it was back then. Most scientists regard these fluctuations (**anisotropies**) as reflecting reflect minute variations in the distribution of matter in the early universe, the seeds of future structure.

A second satellite, the Wilkinson Microwave Anisotropy Probe (WMAP), which was launched in 2001 and continued transmitting data until October 2010 (though data analysis is ongoing at the time of this writing), has produced even more accurate data than COBE. Among other findings, it has:

- pegged the age of the universe at 13.7 billion years, with only a one percent margin of error;
- produced an intricate map or "fingerprint" of the CMB fluctuations;
- pinned down the time when the CMB streamed forth; and
- revealed that the first stars appeared about 200 My after the Big Bang.

THE UNIVERSE EVOLVES

FIGURE 4.1 provides a time line of the evolution of the universe, shown as a bell to simulate change in three dimensions over time (the universe is actually flat, not bell-shaped; FIGURE 4.2a). On the left, you can see the instantaneous inflation of the Big Bang and the afterglow of the CMB. Next comes a band labeled Dark Ages (Loeb, 2006). For 200 million years, the universe would have appeared dark—there were no stars to emit visible

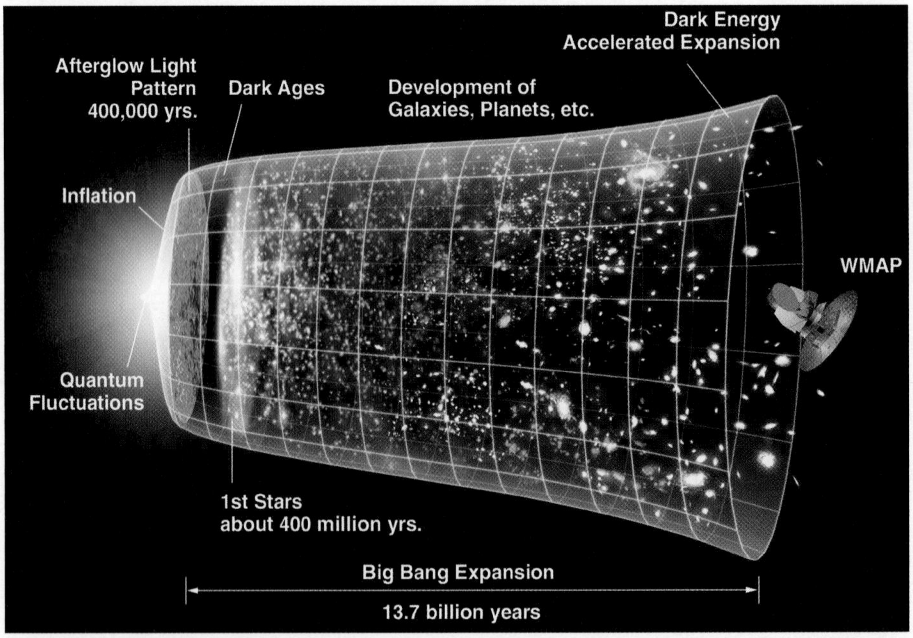

The word EVOLVED is used in a different sense in astronomy and cosmology than in biology. In astronomy, evolution means changes with time of individual objects and systems, changes that can be explained by the laws of physics. In biological evolution, contingency (chance) plays a major role; in cosmology, it almost certainly does not.

FIGURE 4.1 Time line of the universe, from the Big Bang to now, shown as a bell to simulate change in three dimensions over time. Initial inflation at 10^{-35} of the first second (left) was replaced by a far smaller rate of inflation that is currently accelerating as a result of dark energy (hence the bell's flare), implying that the universe will expand forever. WMAP—the Wilkinson Microwave Anisotropy Probe, a probe operated by the National Aeronautics and Space Administration (NASA)—here represents the present.

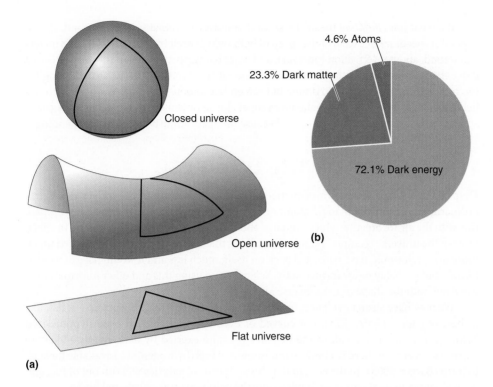

4.6% Atoms

23.3% Dark matter

72.1% Dark energy

Closed universe

Open universe **(b)**

Flat universe

(a)

FIGURE 4.2 The universe, **(a)** which is flat rather than spherical (closed) or saddle-shaped (open), is made up of atoms, dark matter, and dark energy in the proportions shown in **(b)**, although it is important to realize that most of the universe is virtually empty.

light. Gradually, however, gravity had its way, and matter began to clump together, making larger and larger bodies, until finally the first generation of stars blinked on. Indeed, by 900 My after the Big Bang, there already existed many hundreds of galaxies bright enough for their light to reach the location of modern Earth, meaning that the total number of galaxies at the time must have been even larger (Bouwens and Illingworth, 2006).

Stars are crucial to this story because they are the nuclear fusion reactors where chemical elements are manufactured, and from which they are dispersed in a star's dying days. It is an enormously complicated story involving several types of stars and a great deal of time. Those first-generation stars—massive, hot, and short-lived—produced a few heavy elements, especially oxygen, neon, silicon, magnesium, and sulfur (Abel et al., 2002).[5] As a result, all later stars, including those we see today, contain heavy elements. Eventually, through a process of accumulation, all the naturally occurring elements of the periodic table were formed. Our own solar system has the full quota of elements, and thus owes its origin to the debris of many earlier generations of stars.

Today the universe is highly structured, with planets and moons within solar systems, and solar systems within galaxies, some of which form clusters and super clusters that appear as occasional bumps (more properly, nodes) in sheets and filaments of galaxies, separating immense, bubble-like voids. Within this unimaginable amount of material, hydrogen still rules. In our own galaxy, hydrogen accounts for 74 percent of material by mass, helium for a further 24 percent. Add in oxygen at 1 percent and carbon at 0.46 percent, and not much is left for the other elements.

Amazingly, ordinary atoms make up only 4.6 percent of the contents of the universe (**FIGURE 4.2b**). Mysterious (and hypothetical) **dark matter** and **dark energy**, neither of which emits or absorbs light, account for the rest. We can sense dark matter through its gravitational effects on regular matter. Dark energy brings us back to Figure 4.1, and the topic of inflation (Cline, 2004). Notice that the bell continues to expand (inflate) as time passes, and that it is flared near its mouth, reflecting the fact that the rate of expansion of the universe is picking up speed—another recent discovery. Dark energy is thought to mediate this increase (Davies, 2006).

MISSING *from this account are all sorts of things: black holes, quasars, supernovae, and much, much more.*

[5] The most massive star known, R136a1 has a mass 265 times the mass of the Sun. At its birth R136a1 may have been 320 times the mass of the Sun. (Crowther, et al., 2010)

But what does *inflation* mean? How was the universe, in that first split second, able to expand at speeds far in excess of the speed of light (which itself travels at 299,792,458 meters per second, about 9.46 trillion km/year), and what has happened since? The answer is that space–time itself was inflating; any matter inside it was (and is) merely carried along for the ride. A difficult concept, no doubt; but remember, gravity continues to operate. Things held together by gravity—say, a galaxy or a super cluster or you—are not affected. If a galaxy appears to be moving away from us, it's because the space–time between us is expanding.

THE FUTURE OF THE UNIVERSE

Despite the enormous amount of matter involved, the density of the universe is extra-ordinarily small, equivalent to six atoms of hydrogen per cubic meter on average. Compare this with the air we breathe, which contains about 10^{25} molecules per cubic meter. In effect, most of the universe is empty. Yet research tells us that if the universe had contained much more matter, it would have collapsed back on itself; much less and it would have expanded forever, but probably never formed stars. Scientists are using this and other information to work out both the shape of the universe and its future.

We now have strong evidence, from a variety of sources, that the universe is flat, like a sheet of paper (Figure 4.2**a**), not curved or even spherical, as had been hypothesized before. It is also—as a result of the negative pressure exerted by dark energy—likely to expand forever.[6] But there is a limit to the amount of fuel (hydrogen, etc.) available for stars to burn (Barger, 2005). In the end, under this scenario, all galaxies will run out of fuel and we will return to total darkness, leaving only the cold remnants of celestial bodies.

ORIGIN OF THE SOLAR SYSTEM

On February 14, 1990, from a distance of 6.4 billion km, cameras onboard the spacecraft *Voyager I* gave us the first view of our solar system. Had they been present 5 to 5.6 Bya, they could have witnessed its very beginnings, for it was then that the huge cloud of dust and gas (the **solar nebula**), out of which our system developed, began to collapse under its own weight.

Astronomers have proposed two main theories for the origin of planets in our solar system. According to the **collision theory**, proposed in 1749 by Georges Buffon and pro-moted in various forms by others until the early twentieth century, a star or comet either collided with our sun, throwing out debris that formed the planets and other bodies, or passed close enough to pull out, through gravity, the material that became the planets. The most serious of many difficulties with this theory is the extreme rarity of such events.[7]

Most astronomers today accept an alternative theory, the **condensation theory** (nebu-lar hypothesis), first proposed by Immanuel Kant (1724–1804) in 1755 and later elaborated by Pierre-Simon Laplace (1749–1827). In its modern form, the condensation theory posits that a huge mass of dust and gas began collapsing under its own gravity, perhaps as the result of a nearby disturbance such as a supernova—the spectacular explosion of a massive, dying star. The large condensing mass at the center of the cloud began heating up, eventu-ally reaching thermonuclear reaction temperatures about 4.6 Bya. The resulting star was our sun (Kasting and Catling, 2003). The remaining material formed a whirling "accretion disk" around the sun in which smaller condensations grew as the material cooled (FIGURE 4.3). At first, dust particles merely adhered to each other, but once the resulting bodies reached around a kilometer in diameter, gravity kicked in. This boosted the process of planet forma-tion as vast numbers of bodies known as **planetesimals** collided and coalesced, eventually forming **protoplanets** and then **planets**, a violent process occupying some 100 million years.

[6]NASA, 2005. Is the Universe Infinite? Available at http://map.gsfc.nasa.gov/universe/uni_shape.html. Accessed October May 22, 2012.
[7]See the NASA website, http://nasascience.nasa.gov/planetary-science for up-to-date information, and see Rudwick (2005) for an excellent history.

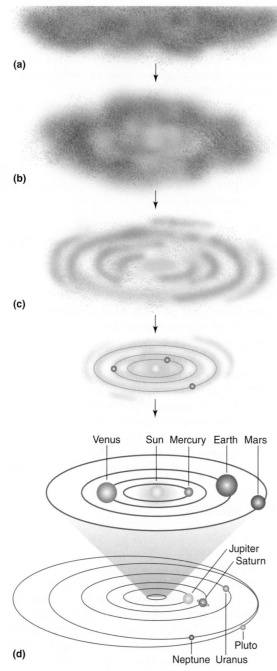

FIGURE 4.3 Stages during the condensation of the solar nebula into the solar planetary system. **(a)** Fragmentation of an interstellar cloud. **(b)** Contraction and flattening of the solar nebula. **(c)** Condensation of nebular material into protoplanets and smaller bodies such as asteroids. **(d)** Solidification of planets, with an indication of present orbits. The four inner planets are shown at a larger scale than are the outer planets.

(a)

(b)

(c)

Venus Sun Mercury Earth Mars

Jupiter
Saturn

Pluto
(d) Neptune Uranus

The dust and gas out of which our solar system developed contained the full range of elements. But as the planets formed the distribution of those elements changed dramatically. The result? Four small, terrestrial inner planets (Mercury, Venus, Earth, and Mars) that are composed mainly of rock and metal and have only three satellites (moons) between them, and four giant, gaseous outer planets (Jupiter, Saturn, Uranus, and Neptune) that are made primarily of helium and hydrogen and are surrounded by rings and many moons. Completing the picture are asteroids, comets, interplanetary dust, and **dwarf planets** such as Pluto (long considered a planet but voted out at a meeting of the International Astronomical Union on August 24, 2006—an extraordinary way to come to a scientific conclusion).[8]

[8]Although Pluto is a celestial object that orbits the Sun and is large enough for the force of its gravity to have formed it into a sphere—two of the criteria for recognizing a planet—it does not meet the third criterion: that it dominate the neighborhood around its orbit, sweeping up and incorporating unto itself asteroids, comets, and other small celestial bodies.

This distribution of the planets reflects the distance of each one from the Sun. Unlike the materials comprising the inner planets, hydrogen and helium solidify only at extremely low temperatures and were unable to form condensations in such small planets so close to the Sun. They were therefore unable to withstand the *solar wind,* a powerful, fast-moving stream of particles that swept clean the warmer inner areas of these gases and any remaining dust following the initial ignition of the sun.

The Sun itself has changed, both on astronomical time scales (it has become progressively warmer since its origin 4.6 Bya) and on more human time scales (it has undergone periods of warming and cooling).[9] The Little Ice Age, for example, was a period of cooling that lasted from about 1300 till the late nineteenth century in Europe and North America, though the exact dates vary from location to location.[10] The Sun has undergone a slight cooling over the past 35 years, its surface temperature declining by 0.05% to 1336 kilowatts/square meter (W/m^2).

EARTH'S FIRST 700 MILLION YEARS: THE HADEAN ERA

The HEAT released by relatively short-lived radioactive elements already present would have helped raise the temperature of Earth at this time.

Earth owes its origin to a series of cataclysmic collisions that generated an enormous amount of heat, causing the fledgling planet to become molten and allowing material to move around. Influenced by gravity, the heavier elements (especially iron) sank inwards and lighter material rose towards the surface, where a crust eventually formed. This process was far from smooth, as sections re-melted during periods of bombardment by meteorites and other bodies. As a result, our planet today consists of concentric layers nested within one another. And, as a consequence of its iron core, Earth is surrounded by an enormous magnetic bubble, the **magnetosphere**. This, together with our atmosphere, protects us from incoming radiation, including the electrically charged particles carried by solar winds, which pass Earth at speeds of 400 km/second. According to a recent study, however, this crucial development did not occur until 3.45 Bya, by which time both solar activity and the heat of Earth's core had declined sufficiently (Tarduno et al., 2010).

ORIGIN OF THE MOON

Development of Earth's layers must have been well underway around 4.53 billion years ago when, according to the currently favored hypothesis, a Mars-sized object crashed into Earth at an oblique angle, blasting vast quantities of material from both bodies into orbit. So catastrophic was the impact (often called the **Big Whack**) that much of both Earth and the intruder from space would have melted. Some of the debris cast into space clumped together to form the **Moon**, which remained tied to Earth's orbit. The bulk of the intruder would have merged with Earth. This momentous event happened a mere 30 million or so years after Earth had reached a size approximating that of today.

However it arose,[11] formation of the Moon helped to stabilize our planet's axial tilt (obliquity) in respect to our orbit around the Sun, thus producing regular seasons. This

---

[9] For an up-to-date analysis of time, see S. Carroll (2010), *From Here to Eternity. The Quest for the Ultimate Theory of Time.* Dutton, New York.

[10] In 1645, the inhabitants of the village of Chamonix, located at the foot of Mont Blanc in Switzerland, were concerned that an advancing glacier would overtake and destroy their villages and farms. They called in the Bishop of Geneva, who performed an exorcism at the front of the advancing glacier. The glacier retreated only to advance again despite repeated exorcisms. These villages were facing the same Little Ice Age that iced up the North Atlantic and killed centuries-old orange groves in China.

[11] Prompted to explain a persistent problem—why one "half" of the Moon consists of high mountains and the other of basins composed of volcanic rock rich in potassium and uranium—Jutzi and Asphaug (2011) proposed a model in which a "moonlet" that was in stable orbit around Earth for tens of millions of years floated out of that orbit, collided with the Moon, and in the space of only a few hours fused with the Moon, giving the Moon its distinctive features.

stabilization is not absolute: slight variations in the tilt of our planet (between 22.1 and 24.5 degrees) continue to affect climate in dramatic ways through a series of complicated long-term cycles, yet still allow life to persist. Today Earth is tilted at 23.44 degrees, roughly the midpoint of the possible extremes. In addition, the Moon has acted as a drag on the planet's speed of rotation, progressively slowing it.

Not surprisingly, no rocks remain from Earth's earliest days—the oldest rock discovered thus far is 4.03 billion years old—but the Moon can tell us much about the timing and nature of the event that created it. Analysis of rock samples brought back to Earth by the American (manned) and Soviet Union (unmanned) missions of the late 1960s and early-mid 1970s reveals that the Moon (and thus Earth) is at least 4.5 billion years old. And because the Moon has far less iron than does Earth, the impact must have happened *after* much of Earth's iron had sunk into its middle.

HADEAN ERA

Little is known about Earth in the years following the Moon's birth. Until recently it was assumed that the planet was mostly a place of roiling magma for the next 500 million or so years, a time that would have richly deserved its name, the **Hadean Era** (from *Hades*, hell). But since the early 2000s, when zircon crystals 4.4 By old were discovered in Western Australia, scientists have had to rethink this assumption. Zircon crystals require liquid water and low temperatures to form, so the planet may well have cooled, and landmasses formed, far earlier than previously thought (Valley, 2005).[12] But because intense bombardment between 4.1 and 3.8 Bya is likely to have sterilized the planet's surface, life could not have originated any earlier than the end of the Hadean. Again, evidence for the bombardment is indirect: large impact craters formed on the Moon and Mercury around this time. Not surprisingly, almost everything to do with the Hadean remains controversial.

The first fragments of crust are likely to have been highly unstable, parts of it re-melting and folding into the magma beneath during times of bombardment (Hawkesworth and Kemp, 2006). Those few rocks identified so far from the end of this period, however, already show some evidence that plate tectonics—the movements of areas of crust across Earth's surface—had already begun. The earliest crust, however, differed in composition from today's, and possibly has no analogues with what exists now.

KNOWING WHEN THESE EVENTS OCCURRED

All of this raises the question of how we determine the age of such ancient things. Without the ability to date Earth and its past inhabitants we could not construct a timeline of life on Earth. Before the mid-twentieth century, scientists used a variety of methods to arrive at their own conclusions, none of which even vaguely approached the real situation. Development of **radiometric dating**, which relies on the decay of radioactive isotopes, has made accurate dating possible (BOX 4.1). When it comes to dating the solar system, and Earth itself, we apply the method to meteorites, which are kilometer-sized fragments of asteroids that have fallen to Earth. Meteorites are some of the oldest and most primitive objects in our system (*primitive* meaning little altered since the origin of the solar system).

ATMOSPHERES FORM

Over time, Earth's atmosphere has changed radically from the hydrogen and helium briefly present at the outset (a smaller Earth lacked gravity to hold onto any gases, particularly such light ones), through a second, more lasting atmosphere that accumulated

The MOON ROCKS *sampled are between 3.16 and 4.5 By old. The youngest parts of the Moon, the result of volcanic activity, are calculated to be about 1.2 By old, but no rocks of that age were brought back to Earth.*

We know this because 50 IMPACT CRATERS *each more than 300 km across were formed on the moon at this time, and thus by inference on Earth and the other inner planets (Canup, 2004).*

---

[12]Zircon (zirconium silicate, $ZrSiO_4$), the oldest mineral in Earth's crust, incorporated small amounts of uranium when it formed, and so is an excellent mineral for determining the timing of events in Earth's history.

BOX 4.1 **Radiometric Dating**

The atoms of a particular element all have the same number of protons, but the number of neutrons (and so the atomic weight of the atom) can vary. Different forms of an element are known as **isotopes,** some of which decay naturally over time, producing new isotopes and releasing energy. Decay of radioactive isotopes is orderly, allowing tiny fragments of rocks even billions of years old to be dated with great accuracy. All methods of radiometric dating rely on knowing the stable products into which the atoms of an element disintegrate, and the rate at which the disintegration occurs (see the examples in **TABLE B1.1**).

At one extreme, samarium[147] (Sm^{147}) disintegrates into neodymium[147] (Nd^{147}) with a half-life of 110 billion years (the half-life is the time it takes for one-half of the original amount of a radioactive isotope to decay into another isotope). Decay of Sm^{147} into Nd^{147} can be used to date objects that formed more than a billion years ago. Indeed, the oldest rocks identified on Earth, located on Hudson Bay in northern Quebec, were dated as having formed 4.03 Bya using the decay of Sm^{147} into Nd^{147}.[*]

At the other extreme, carbon[14] (C^{14}) disintegrates into nitrogen[14] (N^{14}) with a half-life of 5,730 years, and so can be used to date objects such as organic remains that existed no more than some 50,000 years ago. Produced continuously in the upper atmosphere, C^{14} eventually finds its way via photosynthesis into all living plants, and thence into animals in a fixed ratio to non-radioactive carbon. After the organism dies the amount of C^{14} slowly decreases, allowing time of death to be calculated. In this way, Egyptian mummies, mammoths, ancient trees, and more have been dated.

TABLE B1.1	**Radioactive Isotopes Used in Dating**		
Parent Isotope → Daughter Product	**Half-Life (By)**	**Usable Range**	**Use for Dating**
Samarium[147] → Neodymium[147]	110	> 1 By	Basalt, ancient meteorites
Rubidium[87] → Strontium[87]	49	> 100 My	Igneous and metamorphic rocks
Thorium[232] → Lead[208]	14	> 300 My	Mineral crystals in crustal rocks
Uranium[238] → Lead[206]	4.5	> 100 My	Mineral crystals in crustal rocks
Potassium[40] → Argon[40]	1.3	> 0.1 My	Volcanic rocks
Uranium[235] → Lead[207]	0.7	> 100 My	Intrusions and mineral grains
Uranium[234] → Thorium[230]	0.25 My	> 1 My	Animal bones and teeth, corals
Carbon[14] → Nitrogen[14]	5,730 years	< 60,000 years	Organic remains

[*] J. O'Neil et al. (2008). Neodymium-142 evidence for Hadean mafic crust. *Science,* 321, 1828–1831.

GREENHOUSE GASES *such as methane and CO_2 are much in the news today. By trapping radiant energy from the sun, thereby preventing it from returning to space, they keep the planet warm, and therefore suitable for life. But through the burning of fossil fuels, land clearing, and other means, humans are raising the atmospheric levels of these gases, the result being global warming.*

between 4.2 and 3.8 Bya, to a third and very different atmosphere dominated by nitrogen and oxygen, the one we breathe today.

The second atmosphere owed its origin to massive out-gassing from volcanoes and to material imported from meteorites and comets. It consisted primarily of water vapor and carbon dioxide (CO_2); other gases present in smaller quantities probably included hydrogen (H_2), nitrogen (N_2), carbon monoxide (CO), ammonia (NH_3), hydrochloric acid (HCl), and hydrogen sulfide (H_2S). Several of the gases present, and especially CO_2 and methane (CH_4), are greenhouse gases, and would have acted as a warming blanket even as Earth cooled, given that the Sun then gave off only about 70% of its current heat. The much dimmer light reaching Earth would have been made even dimmer by the heavy cloud cover and volcanic debris above; it would have been hot and dark on Earth's surface.

As Earth and its atmosphere cooled, water vapor condensed to form rain (a lot of it!) and eventually rivers, lakes, and oceans, allowing the water cycle (from water vapor in the atmosphere to liquid water or ice on Earth's surface and back again) to become established. At first, though, rain would have vaporized as it hit the hot surface. As mentioned above, we now have evidence that liquid water existed on Earth well before 4 Bya, though it may not have remained liquid at all times during the heavy bombardment period.[13] It is important to note, however, that atmospheric pressure under this regime of gases would have been greater than at present, allowing liquid water to exist at higher temperatures than it does now.

Although the source of all this water remains a topic of some controversy, likely sources are:

- release of liquid water from the material making up planetesimals as early Earth cooled, and then from volcanoes;
- from meteorites and asteroids, especially between 4.5 and 3.8 Mya (see Table 4.1); and
- from the release of water from hydrated minerals and from the action of the Sun.

The first indirect evidence of life dates from 3.7 Bya, not long in geological terms after the end of the Hadean. This first life had a profound influence on the planet but certainly did not mark the end of major changes on Earth. In particular, the atmosphere would continue to change for a very long time to come.

WHAT MAKES EARTH SO SPECIAL?

We conclude this chapter by asking the question: What makes Earth so special? What features have allowed life to evolve and flourish over the last 3.7 billion years? (Here we define life as carbon-based and dependent on water, though other systems may be possible.) A standard list would include the following:

- a sun of moderate size providing even radiant energy over hundreds of millions of years;
- an orbit that is just the right distance from the sun, providing just the right amount of heat and light, and thus a moderately stable climate;
- the right mix of atomic elements, especially carbon, hydrogen, nitrogen, oxygen, phosphorus, and sulfur;
- presence of liquid water; and
- an iron core that casts a gigantic magnetic shield far into space, protecting us from harmful solar and cosmic radiation.

By the end of the Hadean, almost everything was in place. Only the **ozone layer** that protects life from harmful ultraviolet rays was missing; it would accompany the rise of free oxygen in the air beginning around 2.3 Bya. Free oxygen results from the metabolic activities of living organisms, and is one of the many ways that life has influenced conditions here on Earth.

So . . . is Earth unique? Over the last 20 or so years, scientists have located a large number of planets orbiting stars other than our own—760 as of February 2012, according to the Habitable Exoplanets Catalog. Almost all, however, are gigantic balls of gas; only four of those discovered so far appear to resemble Earth in any way or to be potentially habitable. Given that there are billions of galaxies in the universe, we can only guess at the likelihood that there is life elsewhere. Closer to home with the NASA Mars Exploration Rover mission (http://marsrovers.nasa.gov/home/), we continue to search for clues that life might once have existed on Mars.

---

[13] Water also is present in the Moon's interior (A. E. Saal et al., 2008. Volatile content of lunar volcanic glasses and the presence of water in the Moon's interior. *Nature* 454, 192–195), and ice has been detected just below the surface near the Moon's North Pole (http://www.universetoday.com/58410/water-ice-found-on-moons-north-pole/).

■ KEY TERMS

astronomy
Big Bang theory
collision theory
condensation theory
cosmic microwave
   background (CMB)
cosmology
dark energy

dark matter
dwarf planets
Earth
evolutionary universe
inflation
isotopes
magnetosphere
Moon

ozone layer
photons
planetesimals
planets
protoplanets
radiometric dating
solar nebula
Sun

■ EVOLUTION ON THE WEB

Explore evolution on the Internet! Visit the accompanying website for *Strickberger's Evolution, Fifth Edition,* at **go.jblearning.com/Evolution5eCW** for exercises and links relating to topics covered in this chapter.

■ REFERENCES

Abel, T., G. L. Bryan, and M. L. Norman, 2002. The formation of the first star in the universe. *Science,* **295**, 93–98.

Barger, A., 2005, The midlife crisis of the cosmos. *Sci. Am.* **292**, 46–53.

Bouwens, R. J., and G. D. Illingworth, 2006. Rapid evolution of the most luminous galaxies during the first 900 million years. *Nature,* **443**, 189–192.

Canup, R. M., 2004. Dynamics of lunar formation. *Annu. Rev. Astron. Astrophys.,* **42**, 441–475.

Cline, J. M., 2004. The origin of matter. *Am. Sci.,* **92**, 148–157.

Condie, K. C., and R. E. Sloan, 1998. *Origin and Evolution of Earth: Principles of Historical Geology.* Prentice Hall, Upper Saddle River, NJ.

Crowther, P. A., O. Schnurr, R. Hirschi, et al., 2010. The R136 star cluster hosts several stars whose individual masses greatly exceed the accepted 150 Msun stellar mass limit. *Monthly Not. R. Astron. Soc.,* **308**, 731–751.

Davies, P., 2006. *The Goldilocks Enigma: Why is the Universe Just Right for Life?* Allen Lane, London.

Hawkesworth, C. J., and A. I. A. Kemp, 2006. Evolution of the continental crust. *Nature,* **443**, 811–817.

Hawking, S. W., 2001. *The Universe in a Nutshell.* Bantam Books, Toronto.

Jutzi, M., and E. Asphaug, 2011. Forming the lunar farside highlands by accretion of a companion moon. *Nature,* **476**, 69–72.

Kasting, J. F., and D. Catling, 2003. Evolution of a heritable planet. *Annu. Rev. Astron. Astrophys.,* **41**, 429–463.

Lineweaver, C. H., and T. M. Davis, 2005. Misconceptions about the Big Bang. *Sci. Am.,* **292**, 24–33.

Loeb, A., 2006. The dark ages of the universe. *Sci. Am.,* **295**(5), 46–53.

Longair, M. S., 2006. *The Cosmic Century.* Cambridge University Press, Cambridge, UK.

Mitton, S., 2005. *Fred Hoyle: A Life in Science.* Aurum Press, London, UK.

National Aeronautics and Space Administration (NASA), 2008. *Cosmology; The Study of the Universe* (website devoted to the Wilkinson Microwave Anisotropy Probe). Accessed May 22, 2012 from http://map.gsfc.nasa.gov/m_uni.html.

Redfern, M., 2003. *The Earth: A Very Short Introduction.* Oxford University Press, New York.

Riordan, M., and Zajc, W. A., 2006. The first few microseconds. *Sci. Am.,* **294**(5), 24–31.

Rudwick, M. J. S., 2005. *Bursting the Limits of Time. The Reconstruction of Geohistory in the Age of Revolution.* The University of Chicago Press, Chicago.

Tarduno, J. A., R. D. Cottrell, M. A. Watkeys, et al., 2010. Geodynamo, Solar Wind, and Magnetopause 3.4 to 3.45 Billion Years Ago. *Science,* **327**, 1238–1240.

Valley, J. W., 2005. A cool early Earth? *Sci. Am.,* **291**, 58–65.

Weintraub, D. A., 2010. *How Old is the Universe?* Princeton University Press, Princeton, NJ.

CHAPTER

5

The Atmosphere, Rocks, and Continents

CHAPTER SUMMARY

Mediated first by methanogens and then by photosnthetic organisms, Earth's atmosphere has transformed from one dominated by carbon dioxide and water vapor to one now dominated by oxygen and nitrogen. The second atmosphere, which developed between 3.8 and 4.2 Bya, began acquiring oxygen many hundreds of millions of years after the first oxygen-producing organisms arose. Much as the atmosphere of the earth has changed over time, so has the planet itself. Earth today is made up of concentric layers differing in both composition and physical properties. The core consists of iron and nickel and is surrounded by a thick, partially ductile mantle. Blanketing the mantle is a thin crust of igneous, sedimentary, and metamorphic rocks, distributed in strata (layers). Different strata of sedimentary rocks can be dated relative to one another by the fossils they contain. The history of the continents and many of Earth's features can be explained by the movement of giant plates. These plates can separate from, slide past or converge on each other. We call such activity *plate tectonics,* and explain the distribution of many organisms on the basis of the historical fusion and separation of these areas of Earth's crust.

INTRODUCTION

By the end of the Hadean Era, some 4 billion years ago, most of the conditions necessary for the initiation of life were present. Since that time, Earth has gone through astounding changes. Continents and oceans have come and gone, while the atmosphere we breathe today bears little resemblance to what was present back then. In this chapter we continue our discussion of Earth's evolution by considering the structure of the planet, the way in which continents have formed and moved around, and the impacts of this latter phenomenon on living organisms. We begin, however, by describing some significant developments in the evolution of Earth's atmosphere.

EARTH'S CHANGING ATMOSPHERE

Earth's second **atmosphere**, which consisted mostly of water vapor and carbon dioxide (CO_2), is hypothesized to have developed between 4.2 and 3.8 Bya. Of the many other gases present, methane is especially interesting. Although organisms were not involved in the production of gases in the early phases of this atmosphere, recent studies provide evidence of methane derived from biotic sources, and of single-celled microbes known as **methanogens**, in rocks dated at 3.7 Bya[1] (**TABLE 5.1**). By adding to atmospheric levels of methane, a greenhouse gas, these organisms may well have helped keep Earth warm after its initial cooling. The gas would have survived far longer in an atmosphere devoid of oxygen than it does today.

The present composition of Earth's atmosphere is outlined in **TABLE 5.2**. Importantly, oxygen makes up 21% of the atmosphere by volume. Earth's first atmosphere lacked oxygen, and the second atmosphere did not begin to accumulate significant levels of the gas until more than a billion years had passed (Table 5.1). Geochemists generally agree that the proportion of free oxygen in the atmosphere began to increase about 2.3 Bya. Driving this increase were aquatic **cyanobacteria**, the first fossils of which are more than 3.5 By old. Known to some as "the architects of the atmosphere," cyanobacteria are photosynthetic; they use the energy of sunlight to synthesize organic compounds from carbon dioxide and water, releasing oxygen in the process. Electron transfer in the photosynthetic process removes hydrogen ions from water molecules, producing free oxygen, which diffuses to the atmosphere. Once oxygen levels began to rise, there were disastrous consequences for life as it then existed; many of Earth's anaerobic inhabitants would have been wiped out.

TABLE 5.1	Major Events that Occurred from 4.1 to 2.3 Billion Years Ago
Time (Bya)	**Events**
4.0–4.1	Oldest rocks on Earth
3.8	Meteoric bombardment of Earth ends; oldest possible signs of life (based on carbon-isotope ratio)
3.7	Oldest signs of methane-producing organisms (methanogens) as molecular fossils
3.5	Photosynthetic organisms (cyanobacteria)
2.3	Oxygen starts to accumulate in the atmosphere

[1]J. F. Kasting and D. Catling, 2003. Evolution of a heritable planet. *Annu. Rev. Astron. Astrophys.*, **41**, 429–463.

Gas	Percent by Volume
Nitrogen	78.09
Oxygen	20.95
Argon	0.93
Water vapor	0.40*
Carbon dioxide	0.039
Neon	0.002
Carbon monoxide, helium, hydrogen, krypton, methane, nitrous oxide, ozone, xenon	< 0.001

TABLE 5.2 Composition of Earth's Atmosphere Today

*This is the average over the entire atmosphere. Percent water vapor varies between typically one and four percent at the surface.

EARTH'S STRUCTURE

Our knowledge of Earth's structure comes from many sources. Earthquakes, in particular, have helped unravel the mysteries of Earth's interior. Sensitive seismographs detect seismic waves, the paths and velocities of which depend on the composition, fluidity, and thickness of the materials through which they travel. Combined with studies of Earth's magnetic, electric, and gravitational fields, seismic information has revealed a complex picture.[2]

At Earth's center is the extremely hot **inner core**, a solid iron mass (with some nickel) 1,287 km in radius, surrounded by a 2,090 km thick **outer core**, a molten iron envelope mixed with sulfur or silicon (**FIGURE 5.1**). Surrounding the iron core is the **mantle**, a layer of repeatedly melted and re-crystallized rock about 2,900 km thick that comprises approximately four-fifths of Earth's volume. Floating on the surface of the mantle is the **crust,** consisting of a thicker (and mostly older) **continental crust**, and a thinner, younger **oceanic crust**. The crust plus the uppermost portion of the mantle are collectively known as the **lithosphere**. Crust and mantle are separated by a discontinuity known as the Moho (Mohorovičić) discontinuity, named after Croatian seismologist Andrija Mohorovičić (1857–1936), who observed a sharp change in the velocity of earthquakes when they move through this layer (Figure 5.1). The crust is composed of *three fundamental types of rocks:*

- **Igneous rocks** form when molten rock (magma) cools and solidifies. If magma cools slowly, which it does deep within Earth, minerals crystallize to form coarse-grained rocks such as *granite*. Magma also may be deposited rapidly and directly on the surface in the form of *lava,* which cools quickly to form fine-grained rocks such as *basalt*.
- **Sedimentary rocks.** Weathering of existing rocks by water, wind, glaciers, and chemical reactions produces particles that are transported and reformed into new arrangements. Ash and rocks thrown up by volcanic activity also contribute. Thus, a stream may deposit its sediments at the bottom of a lake and blowing sand may form dunes. With time, layers of considerable depth form. As these layers are compressed and harden they form sedimentary rocks. *Sandstone* (sand origin) and

Two papers published in the 22 May, 2010 issue of Science *provide evidence that Earth's* INNER CORE IS ASYMMETRICAL *(actually, hemispherical). The solid inner core, which is some 1,220 km in radius, grows at a rate of 0.5 mm/year, the result of iron crystallization in the liquid outer core. Crystallization in the Western Hemisphere and melting in the Eastern results in an eastward drift of the inner core (Monera et al., 2010; Dues et al., 2010; see Bergman, 2010 for even more recent studies).*

The largest EARTHQUAKE *recoded so far, which occurred off the coast of Chile in 1960, measured 9.5 on the Richter scale (an open-ended logarithmic scale).*

LIMESTONE *also forms from the deposition of calcium carbonate by organisms that live and die in marine reefs and shallow seas.*

[2]See Condie and Sloan (1998), Redfern (2003), Rudwick (2005), and Tarbuck and Lutgens (2006) for analyses from different perspectives.

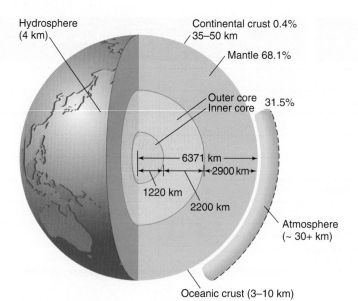

FIGURE 5.1 A section through Earth. The lithosphere consists of relatively rigid plates composed of the rocklike crust plus a portion of the underlying mantle, reaching a depth of about 96 km at the oceanic basins and 96 to 145 km at the continents. Below the lithosphere is a more plastic, deformable layer that allows the lithospheric plates to move about.

shale (mud origin) are familiar examples. *Limestone* (calcium carbonate) is formed chiefly from the remains of marine organisms, such as corals and mollusks, that incorporate calcium carbonate into their skeletons.

- **Metamorphic rocks** are igneous or sedimentary rocks that have undergone significant changes in response to heat, pressure, and/or chemical interactions. *Marble* is a metamorphic rock that was originally limestone; *slate* is a metamorphic rock that was once shale.

Today, Earth's crust consists of 65% igneous, 8% sedimentary, and 27% metamorphic rocks (by volume). A layer of sedimentary rocks covers most of the surfaces of continental landmasses and much of the ocean floor. Over time, however, these three major rock types are transformed from one type to another in a **rock cycle** (FIGURE 5.2). They also undergo substantial movement, especially through the process of plate tectonics (Hawkesworth and Kemp, 2006), which is discussed below.

Fossils, the physical evidence of once-living organisms that provide the hard evidence for evolution, are usually found in sedimentary rocks, occasionally (although often distorted) in metamorphic rocks such as marble but never in igneous rocks.

The age of sedimentary rocks can be estimated quite accurately by dating layers of volcanic rock and ash that lie between the sedimentary strata. Using this method, we date the oldest sedimentary rocks yet found on Earth (in Greenland) to 3.9 Bya, early in the Archean era.[3] The oldest animal fossils are found in sediments from the Cambrian Era that coincide with igneous rocks that are 545 My old. Later sedimentary rocks can be dated fairly precisely up to the Holocene Epoch, which began 11,700 years ago.

Introduction of the terms *Cambrian*, *Holocene*, and *Archean* leads us into a discussion of the naming and classification of geological strata.

[3] If confirmed, 4.03-By-old minerals in 4.0-By-old rocks in Hudson Bay Canada are the oldest minerals and rocks discovered so far (J. O'Neil et al., 2008. Neodymium-142 Evidence for Hadean Mafic Crust. *Science* **321**, 1828–1831.)

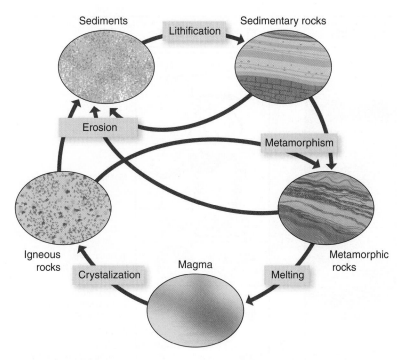

FIGURE 5.2 Diagrammatic representation of the rock cycle. Some crustal rocks have been recycled many times; others have persisted with little change from their initial formation. Geologists estimate that about half of all crustal rocks have formed during the last 600 My.

[Adapted from Hawkesworth, C. J., *and* A. I. A. Kemp, (2006).]

GEOLOGICAL STRATA

From the seventeenth century on, the positions of different rocks relative to one another have been used to determine their respective ages, although not the amount of time it took for them to form. Nicholas Steno (1638–1686), a Danish founder of geology and early proponent of the validity of fossils, was one of the first to establish a basic law of rock formation, the **law of superposition:** If a series of sedimentary rocks has not been overturned, the oldest **strata** (layers) will be at the bottom of the series and the youngest strata will be at the top.

Rocks do not form nor are sediments deposited in identical thicknesses or at equal rates from time to time or place to place. Furthermore, large sections of the geological record in all localities have been worn away by weathering and erosion or have been destroyed by new rock formations and Earth movements; nowhere does the geological record offer a complete sequence that can be traced continuously to the present time. In the early nineteenth century, English geologist William Smith (1769–1839) discovered a way to identify different strata by the unique kinds of fossils within them, and produced the first geological map, published in 1815.

Because they represent only a partial sampling of organisms, fossils are infrequent in all geological strata. Soft-bodied organisms are especially rare in the fossil record. Furthermore, because they may have lived only in restricted habitats or areas, the same fossils are not always present in all the locations where a stratum is found. Nevertheless, a particular stratum can be identified because it contains at least some fossils characteristic of that period (**FIGURE 5.3**). Fossils therefore became a primary means by which a particular geological stratum (layer) or group of strata (a system) could be traced from one region to another.

For an insightful analysis of the events leading to the production of this FIRST GEOLOGICAL MAP, *see Winchester (2001).*

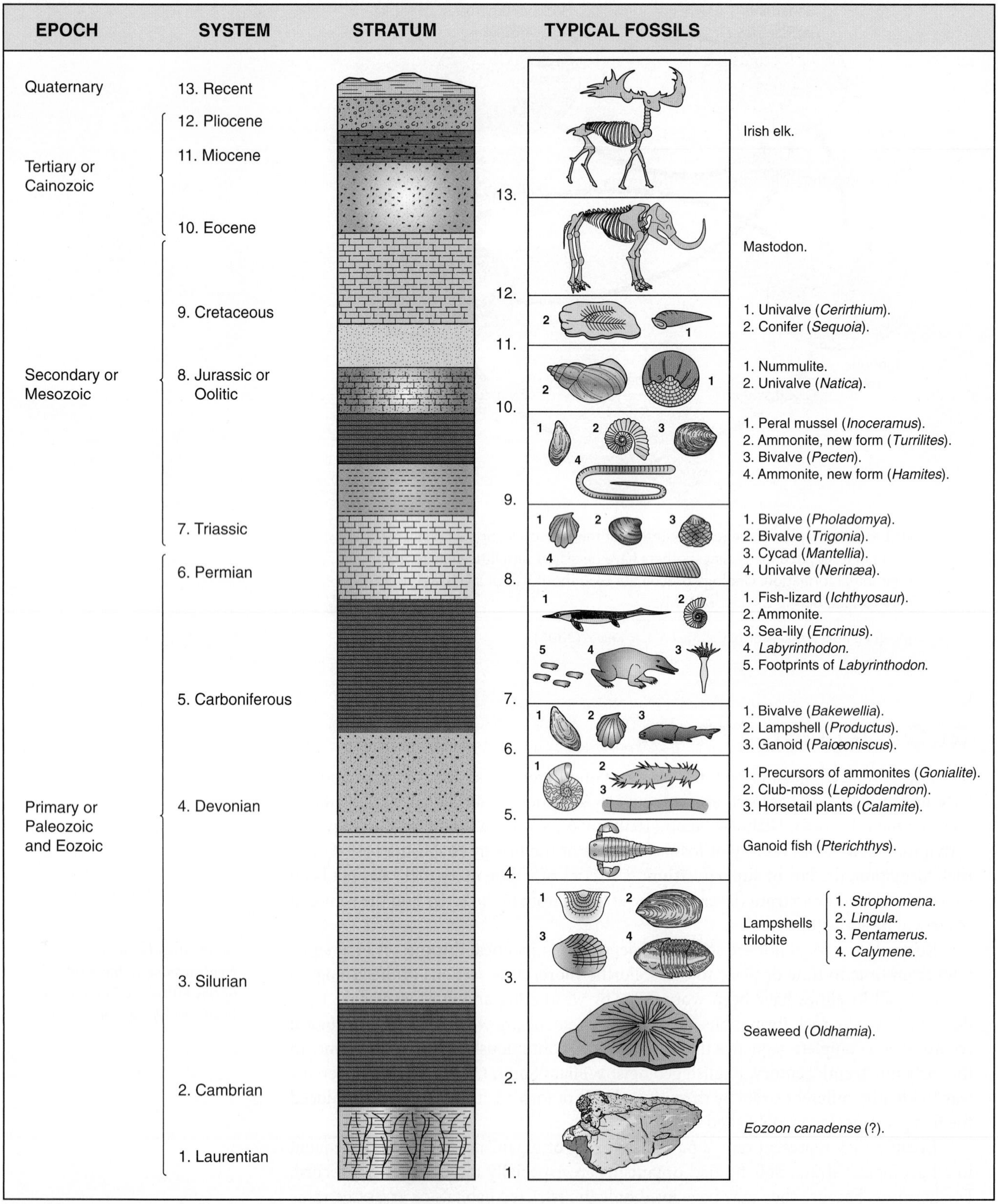

EPOCH	SYSTEM	STRATUM	TYPICAL FOSSILS	
Quaternary	13. Recent			Irish elk.
Tertiary or Cainozoic	12. Pliocene			
	11. Miocene			Mastodon.
	10. Eocene			1. Univalve (*Cerirthium*). 2. Conifer (*Sequoia*).
Secondary or Mesozoic	9. Cretaceous			1. Nummulite. 2. Univalve (*Natica*).
	8. Jurassic or Oolitic			1. Peral mussel (*Inoceramus*). 2. Ammonite, new form (*Turrilites*). 3. Bivalve (*Pecten*). 4. Ammonite, new form (*Hamites*).
	7. Triassic			1. Bivalve (*Pholadomya*). 2. Bivalve (*Trigonia*). 3. Cycad (*Mantellia*). 4. Univalve (*Nerinæa*).
Primary or Paleozoic and Eozoic	6. Permian			1. Fish-lizard (*Ichthyosaur*). 2. Ammonite. 3. Sea-lily (*Encrinus*). 4. *Labyrinthodon*. 5. Footprints of *Labyrinthodon*.
	5. Carboniferous			1. Bivalve (*Bakewellia*). 2. Lampshell (*Productus*). 3. Ganoid (*Paiœoniscus*).
	4. Devonian			1. Precursors of ammonites (*Gonialite*). 2. Club-moss (*Lepidodendron*). 3. Horsetail plants (*Calamite*).
				Ganoid fish (*Pterichthys*).
	3. Silurian			Lampshells trilobite {1. *Strophomena*. 2. *Lingula*. 3. *Pentamerus*. 4. *Calymene*.
	2. Cambrian			Seaweed (*Oldhamia*).
	1. Laurentian			*Eozoon canadense* (?).

FIGURE 5.3 Nineteenth century illustration of a table of stratified rocks that classifies geological strata according to their relative age and shows some of the fossils associated with each period.

[Adapted from Clodd, E., 1888. *The Story of Creation.* Longmans Green, London.]

As Georges Cuvier and others showed, the relative ages of fossils correspond closely to the relative ages of the strata in which the fossils are found; fossils from the uppermost strata seemed more like extant organisms than fossils from lower strata. The Cambrian system—named by Adam Sedgwick in 1835 after Cambria, the Latinized version of *Cymry* ("the Welsh people")—contains strata in which many marine invertebrate skeletons, such as trilobites, brachiopods, and mollusks, first appear (Figure 5.3). Cambrian strata are found on all continents and occupy the same relative positions wherever they occur, lying above Precambrian strata (identified, in part, by the absence of fossil shells) and below Ordovician and Silurian strata,[4] which contain corals, echinoderms, small early fishes, and so on (Figure 5.3).

By these means, geologists have determined that significant numbers of hard-bodied organisms existed more than half a billion years ago. The time that has elapsed since then (545 My) is known as the **Phanerozoic Eon**, from *phanero,* visible, and *zoon,* life. The Phanerozoic Eon consists of three major **eras** of geological strata, beginning with the Paleozoic. As shown in **TABLE 5.3**, each era contains a number of subsidiary systems or **periods**, often further subdivided into **epochs**. William Smith's nephew, John Phillips, who provided the first estimate of Earth's age based on measuring rates of sedimentation,[5] used changes in fossil assemblages to identify and name the three major eras of post-Cambrian geological time:

- Paleozoic—"old life," dubbed the Age of Fishes;
- Mesozoic—"middle life," the Age of Reptiles; and
- Cenozoic—"new life," the Age of Mammals.

The billions of years before the Phanerozoic are known as the Precambrian Eon, which dates from the onset of the Hadean Era to the end of the Proterozoic Era 545 Mya (Table 5.3).

ORIGIN AND MOVEMENT OF CONTINENTS

Like the oceans, continents on Earth have come and gone. Those that survived have changed their shapes and locations in the most amazing ways. Continents can be traced back to the origin of stable regions of Earth's crust (sometimes called cratons) that arose between 3.8 and 1.5 Bya.

Earth today contains seven major continents, which are defined by convention rather than strict criteria: Africa, Antarctica, Asia, Australia, Europe, North America, and South America. Go back in time, however, and the global map of Earth was very different from what it is today; the continents were not where we would expect them to be.

The concept that continents have been lost is an ancient one. English essayist and master of inductive reasoning Francis Bacon (1561–1626) proposed that a continent named **Atlantis** once existed in the middle of the Atlantic Ocean, but later sank beneath the surface (**FIGURE 5.4**). Although Atlantis remains a mythical continent, later scientists were able to demonstrate that the existing continents had moved, expanded, and shrunk during their history. Even further, they proposed that continents had subdivided, spawning new continents.

Between 1912 and 1930, German geophysicist and meteorologist Alfred Wegener (1880–1930) developed the concept that all the continents were at one time a single landmass, **Pangaea** (named from the Greek for entire [*pan*] and Earth [*Gaia; Gaea* in Latin]). Wegener hypothesized that fissures occurred within Pangaea, resulting in its fragmentation and separation into smaller continents (**FIGURE 5.5**). He collected considerable evidence for his hypothesis, which he called **continental drift**, but failed to come up with a convincing mechanism of how it worked. Indeed, it is hard to imagine a force that could

---

[4] Named after the Silures, a Celtic tribe who inhabited the Welsh Borderlands in Roman times.
[5] J. Phillips, 1860. *Life on the Earth: Its Origin and Succession.* Macmillan, London. (Reprinted 1980, Arno Press, New York.)

TABLE 5.3	Geological Ages and Associated Organic Events

Time Scale				Millions of Years Before Present (approx.)	Duration in Millions of Years (approx.)	Some Major Organic Events
Eon	Era	Period	Epoch			
Phanerozoic	Cenozoic	Quaternary	Holocene (last 11,700 years)	0.01	1.8	Appearance of humans
			Pleistocene	1.8		
		Tertiary	Pliocene	5.3	3.5	Dominance of mammals and birds
			Miocene	23.8	18.5	Proliferation of bony fishes (teleosts)
			Oligocene	34	10.2	Rise of modern groups of mammals and invertebrates
			Eocene	55	21	Dominance of flowering plants
			Paleocene	65	10	Radiation of primitive mammals
	Mesozoic	Cretaceous		142	77	First flowering plants / Extinction of dinosaurs
		Jurassic		206	64	Rise of giant dinosaurs / Appearance of first birds
		Triassic		248	42	Development of conifer plants
	Paleozoic	Permian		290 / 320	42	Proliferation of reptiles / Extinction of many early forms (invertebrates)
		Carboniferous	Pennsylvanian	354	30	Appearance of early reptiles
			Mississippian	417	34	Development of amphibians and insects
		Devonian			63	Rise of fishes / First land vertebrates
		Silurian		443 / 495	26	First land plants and land invertebrates
		Ordovician			52	Dominance of invertebrates / First vertebrates
		Cambrian		545	40	Sharp increase in fossils of invertebrate phyla
Precambrian	Proterozoic	Upper		900 / 1,600	355	Appearance of multicellular organisms
		Middle		2,500	700	Appearance of eukaryotic cells
		Lower		4,000–4,400	900	Appearance of planktonic prokaryotes
	Archean				1,400	Appearance of sedimentary rocks, stromatolites and benthic prokaryotes
	Hadean			4,560	160–560	From the formation of Earth until first appearance of sedimentary rocks; no observable fossil organisms

[*Note*: Dates derived mostly from Gradstein et al. *A Geological Time Scale*. Cambridge University Press, 2004, and from *Geologic Time Scale*, available from http://www.stratigraphy.org, accessed August 2012.]

FIGURE 5.4 Hypothetical image of the lost city of Atlantis.

move continents. Such a force would have to operate on a global scale. Until a mechanism was discovered, continental drift remained a hypothesis, not a theory.

Through the 1930s, 40s and 50s, most geologists considered Wegener's hypothesis of drifting continents as little more than an imaginative fantasy. During this time, however, evidence was being gathered, such as the fit of the profiles of adjacent (but widely separated) continents and the similarity of rocks, fossils, and ancient glaciations on these continents, both discussed below. Two further lines of evidence, paleomagnetism and ocean-floor spreading, are discussed in **BOX 5.1**. Eventually evidence for continental drift became so overwhelming that it could no longer be ignored.[6]

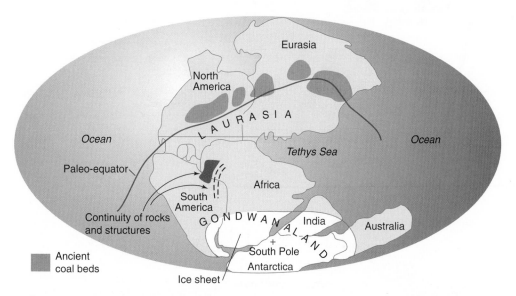

FIGURE 5.5 Fit of the continental shelf of Pangaea at a depth of 500 fathoms (915 meters) with the position of the present-day continental masses indicated.

[Adapted from Eichler, D. L. and A. L. McAlester. *History of the Earth.* Prentice Hall, 1980.]

---

[6] For an analysis of the role of continental drift in establishing conditions reflected in the distribution of animals and plants today, see Cox and Moore (2005).

| BOX 5.1 | Keys to Continental Drift: Paleomagnetism and Ocean-Floor Spreading |

PALEOMAGNETISM

As new rocks arise from the cooling of magma, ferrous material within them (for example, magnetite, Fe_3O_4) magnetizes in a direction that depends on the location and strength of Earth's magnetic field at the time. Should this magnetic field change, the orientation of magnetic particles in newly forming rocks also would be expected to change. Thus, we can study igneous rocks from all continents and eras for their fossilized magnetism (**paleomagnetism**), and deduce the direction and distance of Earth's magnetic poles relative to these rocks.

Geologists were amazed to discover that the magnetic poles had apparently drifted thousands of kilometers during past ages and that the *magnetic poles of different continents did not coincide for long periods of time* (FIGURE B1.1). The magnetic poles derived by analyzing continental rocks that date between the Silurian and Permian (290–443 Mya) are different for each continent. Because several magnetic poles could not exist simultaneously, these different polar locations are best explained by the movement of continents relative to each other and to the poles.

Figure 5.5 shows that the continents were united during the Paleozoic. The magnetic orientation of iron in rocks that formed during the Paleozoic all point to the same geographic position for the South magnetic pole. As the continents separated in the Mesozoic, the "fossilized" magnetic orientations of these deposits pointed to different South magnetic pole positions, giving rise to the anomalies shown in Figure B1.1. What changed was not the position of the Paleozoic magnetic poles, but the geographic location of the Paleozoic continents. (The fact that the magnetic poles occasionally reverse their orientation—see below—does not affect this outcome.)

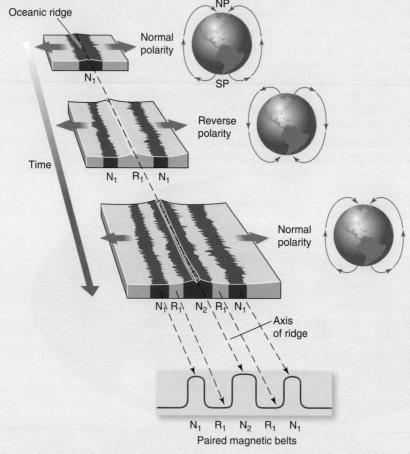

FIGURE B1.1 Diagrammatic sections through an oceanic ridge showing how sea-floor spreading produces differently magnetized belts, with normal (N) or reversed (R) polarity. Hot molten material adds to the ridge from the mantle, falls away on both sides and, as it cools, magnetizes in the orientation of the prevailing magnetic field. With time, the magnetic field changes in strength and/or direction, and new material added to the ridge forms a pair of belts that differ from adjacent belts.

| BOX 5.1 | Keys to Continental Drift: Paleomagnetism and Ocean-Floor Spreading (Cont...) |

OCEAN-FLOOR SPREADING

A further line of evidence for continental drift comes from ocean-floor geology. We know that the oceans are ancient, yet soon after sampling began, scientists were surprised to discover that the ocean floor is geologically young, with sediments no older than 100 to 200 My. Also, in contrast to the sedimentary rocks that comprise most continental mountains, oceanic mountains consist almost exclusively of igneous basalts. The relative youth of the present ocean basins indicates that they replaced older ocean floors.

Another unusual oceanic feature is the presence of magnetized belts that parallel the long **mid-oceanic ridges** found in almost all ocean basins (Figure B1.1). While each belt is paired with a belt of approximately equal width and of the same magnetic orientation on the other side of the ridge, belts adjacent to each other on the same slope magnetize differently. Radiometric dating shows that the youngest belts are closest to the crest of the ridge and the older belts farther away

(Figure B1.1). Changes in magnetic orientation between adjacent belts, known as **reversals**, are caused by a 180° reversal in the polarity of Earth's magnetic pole. This is because the degree of magnetism weakens over time, until it reverses so that the south-pointing needle on a compass now points north. The timing of such reversals is irregular.

All these observations can best be explained if the mid-oceanic ridges are fissures out of which new ocean floor emerges and spreads to either side, something that is now well documented. Molten rock oozing from the oceanic ridge magnetizes upon cooling and is displaced from the ridge by later-emerging material. Just as trees record their lives in growth rings, the ocean floor retains its history in a series of parallel rock belts marked by the magnetic fields prevailing at their time of origin (Figure B1.1). **Sea-floor spreading** is not uniform, however. Its annual rates vary from about 1 cm at the North Atlantic Ridge and 3 cm at the South Atlantic Ridge—the latter being the rate of annual growth of our fingernails—to as much as 9 cm in some portions of the Eastern Pacific Ridge.

EVIDENCE FOR CONTINENTAL DRIFT

As shown in Figure 5.5, one of the most striking geographic correlations is the almost exact match between the profiles of the east coast of South America and the west coast of Africa. Not quite so obvious, but observable, is the match between the east coast of North America and the northwest coast of Africa. These and other geographical juxtapositions are consistent with the idea that the continents were joined or were extremely close together in the past (Cox and Moore, 2005).

A group of rock strata in India, the **Gondwana system**, dates from the late Carboniferous to the early Cretaceous, 290 to 142 Mya. Formations of an extremely similar nature and composition exist in South Africa, South America, Antarctica, the Falkland Islands, and Madagascar. As shown in FIGURE 5.6, associated with these Gondwana formations are distinctive types of fossil plants (*Glossopteris*) and animals (*Mesosaurus, Lystrosaurus, Cynognathus*). Sedimentary evidence also suggests that all of the areas bearing Gondwana formations (along with Australia) were covered by the same glaciation event during a Paleozoic ice age. To account for these observations, geologists proposed the existence of a massive southern continent, **Gondwana**, which included the areas that now carry the Gondwana formations and Australia.

The interpretation that emerges from this evidence for continental drift (and from the evidence outlined in BOX 5.1) is shown in FIGURE 5.7. It begins 200 Mya with giant landmass Pangaea, which began to break up about 225 Mya, during the Triassic. By 135 Mya Pangaea had broken up into a Gondwana group of continents and a North American–Eurasian group called **Laurasia**. Sea-floor spreading (Box 5.1) separated North America from Africa (shown in Figure 5.7 as it would have appeared 35 Mya). During the Cretaceous continued continental drift separated North America from Greenland and South America from Africa, producing essentially the distribution of continents found today as shown in Figure 5.7. In the *Western Hemisphere* the rapid drift

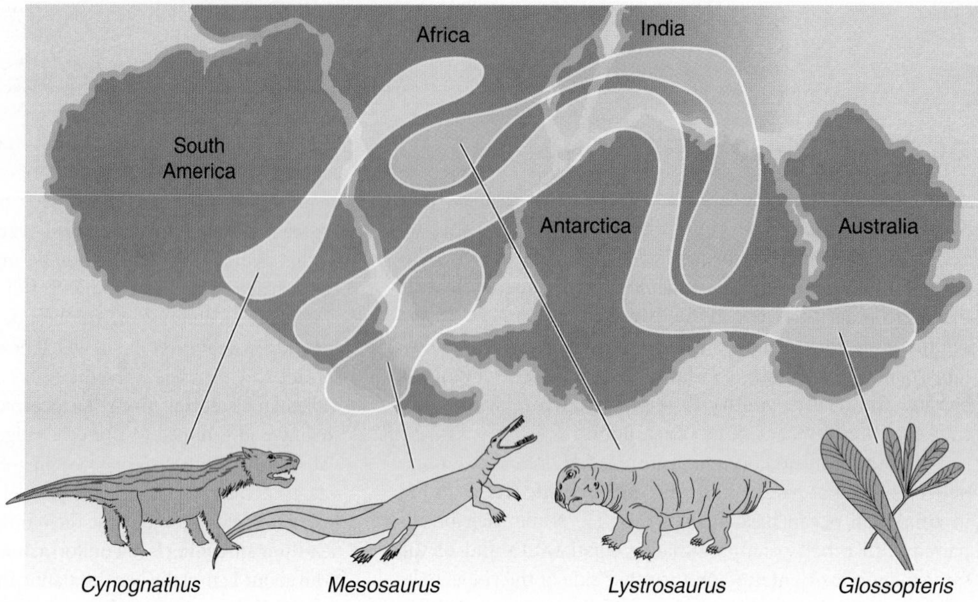

FIGURE 5.6 Distribution of several fossil animals on and between Gondwanan continents. The presumed fit of the continental margins during the Permian-Triassic periods is also shown. Fossil leaves of *Glossopteris*, a group of plants that lived in the early Permian, are found on all Gondwana formations.

[Adapted from Colbert, C. H. *Wandering Lands and Animals.* Hutchinson, 1973.]

of South America away from Africa, which began about 100 Mya, led to a reunion with North America some 4 or 5 Mya. In the *Southern Hemisphere,* New Zealand drifted away from the Australian–Antarctic–South American landmass before the end of the Cretaceous. The mechanism by which continents drifted over so many hundreds of millions of years—**plate tectonics**—is outlined below.

PLATE TECTONICS

The dramatic and long-lasting movements of continents are based on movements (**tectonics**) of gigantic **plates** lying atop the Earth's mantle. Today we recognize eight major plates and several minor ones, identified primarily using evidence from earthquake belts that accompany movements at plate edges (**FIGURE 5.8**). Three major types of events at plate boundaries have been described:

1. Plates **separate from each other** by the addition of new lava at their *adjoining boundary.* Separation of plates in the oceanic ridges account for sea-floor spreading and the size increase of some oceanic basins (Figure 5.8).[7]
2. Adjoining plates **slide past each other** *at a common boundary or fault,* without any significant change in the sizes of the plates. Earthquake activity at the San Andreas Fault is caused by the movement of the Pacific Plate, carrying a section of western California, past the North American Plate. The speed at which this is occurring—6 cm per year—will bring Los Angeles to the same latitude as San Francisco in about 10 My and to the Aleutian Islands near Alaska in about 60 My.
3. One plate **moves toward another** causing a *convergent boundary.* When one plate carries oceanic crust, the convergent event is often marked by the loss of plate material as the ocean crustal mass plunges (is **subducted**) beneath the continental crust into the mantle. The descending plate changes or deforms the mantle, which produces volcanic activity and mountain formation in the crustal region above. For

---

[7] Volcanic eruptions and accompanying release of magma on Earth today allow ongoing accumulation of evidence for the mechanisms of plate divergence (F. Sigmundsson, 2006. Magma does the splits. *Nature,* **442**, 251–252).

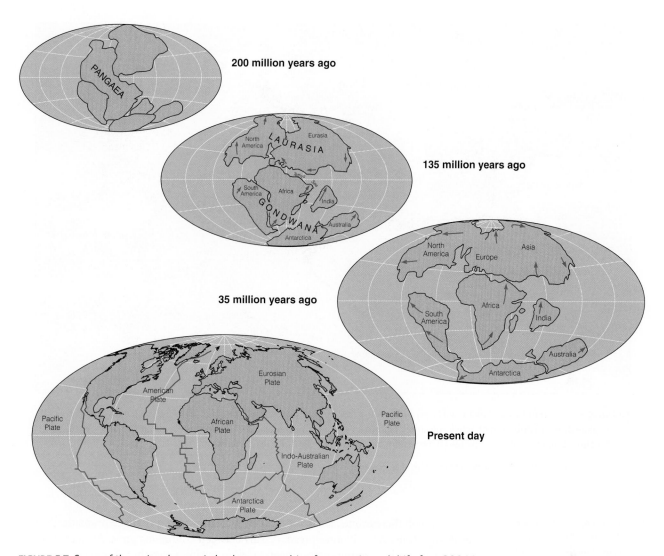

FIGURE 5.7 Some of the major changes in landmasses resulting from continental drift, from 200 Mya to now.

example, in its motion northward, the Pacific Plate descends under a border of the North American Plate, resulting in the Aleutian Islands. Charles Darwin envisioned this process as the source of volcanic and mountain building activity as early as 1838.

"The contemplation of volcanic phaenomena in South America has induced the author to infer, that the crust of the globe in Chile rests on a lake of molten stone, undergoing some slow but great change . . . that mountain building and volcanos are due to the same cause, and may be considered as mere subsidiary phaenomena, attendant on continental elevations; that continental elevations, and the action of volcanos, are phaenomena now in progress, caused by some slow but great change in Earth's interior; and, therefore, that it might be anticipated that the formation of mountain-chains is likewise in progress; and at a rate which may be judged of, by either actions, but most clearly by the growth of volcanoes."[9]

Whether the PACIFIC PLATE *is being pushed or pulled remains unclear; hence the use of the more neutral term* descends. *Knowledge acquired from recent volcanic eruptions is providing new evidence on mechanisms of plate tectonics.*[8]

---

[8] F. Sigmundsson, 2006. Magma does the splits. *Nature*, **442**, 251–252.
[9] Charles Darwin, 1838 p. 656. On the connexion of certain volcanic phænomena, and on the formation of mountain-chains and volcanos, as the effects of continental elevations. *Proc. Geol. Soc. London*, **2**, 654–660.

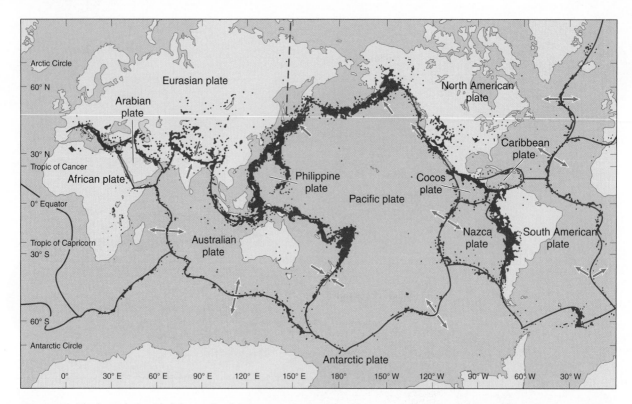

FIGURE 5.8 The boundaries (dark brown lines) between tectonic plates (for example between the Australian and Pacific plates), the distribution of active volcanoes (shown by the density and amount of dark brown along the boundaries), and the direction of movement of the plates (shown as double red arrows) illustrate the dynamics of plate tectonics.

[Adapted from Cloud, P. *Cosmos, Earth and Man: A Short History of the Universe.* Yale University Press, 1978.]

Fifty percent of the OCEAN FLOOR *is no older than the beginning of the Tertiary, which was 65 Mya.*

Through these events, Earth's continents have moved and changed dramatically over geological time. As seafloors spread, oceans between continents grow. As oceanic crust plunges beneath a landmass, continents may grow closer together. These changes have profound effects on the organisms living on these continents, determining their distribution and so influencing their evolution.

BIOLOGICAL CONSEQUENCES OF CONTINENTAL DRIFT

Plate tectonics subjects moving continents and their living inhabitants to new climatic conditions and geographical relationships. Because of the breakup of landmasses, continental drift can separate groups of organisms that were formerly associated, setting each isolated population or even species on its own evolutionary pathway. On the other hand, the joining of landmasses can lead to competition among previously separated groups of plants and animals that had evolved specialized adaptations to their habitats. These organisms can then interact, leading to increased complexity for some groups, and extinction for others (Benton, 2003; Rudwick, 2005; Tarbuck and Lutgens, 2006).

MAMMALIAN RADIATIONS

Mammals are introduced here because they illustrate the biological consequences of continental drift so well. The **mammals** found in Australia and South America demonstrate clearly the effect of continental drift on the distribution and evolution of organisms.

Mammals, all of which possess hair, mammary glands, and a specialized middle ear, are classified into three groups:

- **Monotremes** (prototherians) such as the platypus and echidna, which lay eggs;
- **Marsupials** (metatherians) such as kangaroos and the koala, which carry their young in a pouch; and
- **Placental mammals** (eutherians) such as cats, dogs, and humans, which have a placenta.

Only placental mammals are found in most parts of the world today. Indeed, by the Pliocene (5.3 Mya), placentals had replaced marsupials in all areas except Australia and South America.

Australia's fauna includes representatives of two families of monotremes and 13 families of marsupials. With the exception of bats—Australia has two indigenous species—and rodents, no placental mammals were present on the continent until humans introduced dogs, rabbits, sheep, and so on.

Mammals in South America followed a different evolutionary path from that experienced in the rest of the world (minus Australia). Until the Mid-Tertiary, some 30 Mya, South American mammals included a number of placental families alongside five families of marsupials.

The picture of mammalian evolution that emerges begins when early monotremes and marsupials entered southern parts of Pangaea by the Late Jurassic and Early Cretaceous (**FIGURE 5.9**). Subsequent rifting of Australia isolated its mammalian fauna, where marsupials diversified. Meanwhile, placental mammals evolved in western Pangaea during the Late Cretaceous and Early Tertiary, 60 to 70 Mya (Figure 5.9**b**). In South America, marsupials and early placental mammals had replaced monotremes by the Early and Mid-Tertiary, by

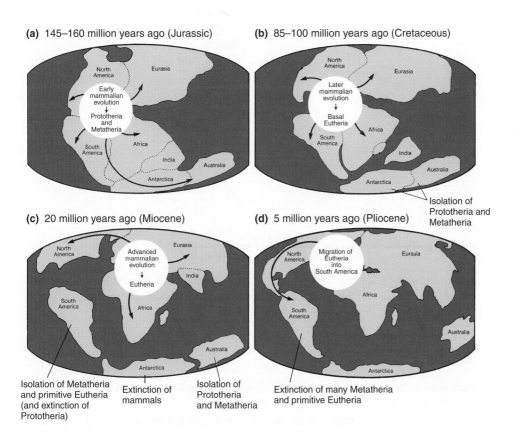

FIGURE 5.9 Effect of continental drift on the dispersion and isolation of major lineages of mammals over the 150 My from the Jurassic Period to the Pliocene Epoch.

which time the South American continent had drifted away from Africa and separated from North America (Figure 5.9**c**). Thereafter, until South America rejoined North America via the Panama Isthmus during the Pliocene, the evolution of mammals on South America was largely independent of mammalian evolution elsewhere (Figure 5.9**d**). During the Oligocene, 34–24 Mya, however, some island hopping, combined with transport on floating debris (*rafting*) allowed various monkeys and caviomorph rodents (capybaras, chinchillas, and the like) to make their way from Africa or North America to South America.

When the Pleistocene began, massive invasions of northern eutherians south across the Central American land bridge resulted in the rapid extinction of many South American mammalian families, which proved unable to withstand the competition. Only rarely, as with opossums, have South American mammals managed to successfully invade North America.

An important lesson emerges: biological evolution and historical circumstances of all kinds are closely connected. Evolutionary changes are allied with geological changes. Geological events can produce various environmental effects, which affect organismal interactions and have evolutionary consequences. Charles Darwin advanced our understanding enormously when he built upon Charles Lyell's analysis of life as influenced by geology to demonstrate that interactions among organisms play a major role in their distribution and evolution.

■ KEY TERMS

Atlantis	metamorphic rocks
atmosphere	methanogens
continental drift	mid-oceanic ridges
core	monotremes
crust	paleomagnetism
cyanobacteria	Pangaea
Epoch	Period
Era	Phanerozoic Eon
fossils	placentals
Gondwana	plate tectonics
igneous rock	rock cycle
Laurasia	sea-floor spreading
law of superposition	sedimentary rock
lithosphere	strata
mammals	subduction
mantle	tectonic plates
marsupials	

■ EVOLUTION ON THE WEB

Explore evolution on the Internet! Visit the accompanying website for *Strickberger's Evolution, Fifth Edition,* at **go.jblearning.com/Evolution5eCW** for exercises and links relating to topics covered in this chapter.

■ REFERENCES

Benton, M. J., 2003. *When Life Nearly Died: The Greatest Mass Extinction of All Time.* Thames and Hudson, London.

Bergman, M. I., 2010. An inner core slip-sliding away. *Nature,* **466**, 697–698.

Condie, K. C., and R. E. Sloan, 1998. *Origin and Evolution of Earth: Principles of Historical Geology.* Prentice Hall, Upper Saddle River, NJ.

Cox, C. B., and P. D. Moore, 2005. *Biogeography: An Ecological and Evolutionary Approach,* 7th ed. Blackwell Publishing, Oxford, UK.

Dues, A., Irving, J. C. E., and J. H. Woodhouse, 2010. Regional variation of inner core anisotropy from seismic normal mode observations. *Science* **328**, 1018–1020.

Hawkesworth, C. J., and A. I. S. Kemp, 2006. Evolution of the continental crust. *Nature,* **443**, 811–817.

Monera, M., Calved, M., Margerie, L., and A. Saurian, 2010. Lopsided growth of Earth's inner core. *Science* **328**, 1014–1017.

Redfern, M., 2003. *The Earth: A Very Short Introduction.* Oxford University Press, New York.

Rudwick, M. J. S., 2005. *Bursting the Limits of Time. The Reconstruction of Geohistory in the Age of Revolution.* The University of Chicago Press, Chicago.

Tarbuck, E. J., and F. K. Lutgens, 2006. *Earth Science,* 9th ed. Prentice Hall, Englewood Cliffs, NJ.

Winchester, S, 2001, *The Map that Changed the World,* Harper Collins, London, UK.

PART 3

From Molecules to Organisms: 3.7–1.5 Bya

CHAPTER 6

Origin of the Molecules of Life

CHAPTER SUMMARY

The fundamental molecules of life are nucleic acids (informational and hereditary molecules) and amino acids (peptides and proteins). In all life forms, production and storage of information are carried out by distinct nucleic acids: DNA for information storage, and several types of RNA as messengers, regulators, and translators. Transfer RNA carries the genetically coded message from the information storage molecule, DNA, to the protein synthesizing machinery in the form of three-nucleotide units known as codons. Elements from which these molecules originated were present early in Earth's history. So too was molecular selection. Once life arose, molecular selection became natural selection. Part III discusses how these molecules arose and become organized into biological structures such as membranes and cells subject to natural selection. We will see that compounds necessary for the origin of life were present before and during the formation of our solar system. Their origin and synthesis derived from abiotic chemical interactions, independent of living systems. This chapter summarizes the nature of amino and nucleic acids, discusses how the double helical structure responsible for DNA replication was discovered, and examines three possible sites for the origin of molecules on Earth (volcanoes, deep sea vents, and clays).

INTRODUCTION

Compounds necessary for the origin of life were present before and during the formation of our solar system. Their origin and synthesis derived from **abiotic** chemical interactions, *independent of living systems.*[2]

As a consequence of the high cosmic abundance of hydrogen and hydrogen-based compounds, hydrogen-containing gases existed early in Earth's history. All hypotheses about the composition of the early atmosphere agree that the gases emitted from Earth's interior included hydrogen (H_2) and methane (CH_4). Sparked by energy from solar and ultraviolet radiation, the origin of a variety of hydrogen-containing and organic[1] molecules provided the fundamental molecular structures from which life arose. These molecules included ammonia (HN_3), carbon monoxide (CO), and hydrogen sulfide (H_2S). Spectroscopy demonstrates that these and many other "organic" molecules were present in the dense interstellar clouds that gave rise to the stars and planets (**FIGURE 6.1**). Presence of critical elements—H, O, C, N, S, P, and Ca—provided chemical diversity enabling reactions that formed organic molecules involving carbon to occur. The terrestrial presence of such molecules is not unique (a class of meteorites known as carbonaceous chondrites contains them; **BOX 6.1**).

Large amounts of water, an excellent solvent, which is stable in liquid form over a wide range of temperatures, and which enables acids and bases to ionize and react, were present on early Earth. Geochemical evidence shows that water was present in the cold planetesimal condensations and appeared in liquid form as soon as the lithosphere reached temperatures that allowed water to condense. Additional water has been continually emitted into the atmosphere through volcanic activity, which was greater in the past than it is now. Energy sources, chemicals, temperature, and water laid the

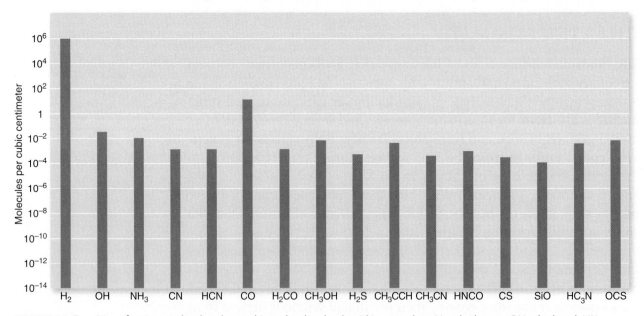

FIGURE 6.1 Densities of various molecules observed in molecular clouds within our galaxy. H_2 = hydrogen, OH = hydroxyl, NH_3 = ammonia, CN = cyanogen, HCN = hydrogen cyanide, CO = carbon monoxide, H_2CO = formaldehyde, CH_3OH = methyl alcohol, H_2S = hydrogen sulfide, CH_3CCH = methylacetylene, CH_3CN = methyl cyanide, HNCO = isocyanic acid, CS = carbon monosulfide, SiO = silicon monoxide, HC_3N = cyanoacetylene, and OCS = carbonyl sulfide.

[Adapted from Buhl, D., *Origins of Life,* **5**, 1974:29–40.]

[1]"Organic" is used here for molecules based on carbon, not molecules produced by organisms. Such carbon-based molecules are abiotic in origin.
[2]The term *abiotic* refers to processes not associated with living organisms.

BOX 6.1   Molecules in Meteorites

Chondrites are meteorites that formed in the condensed interstellar dust of the solar nebula. Among the most complex (and the rarest) are **carbonaceous chondrites.** Fewer than 100 are known, all containing carbon-based molecules.

As an example of the molecules found in carbonaceous chondrites, the **Murchison meteorite**, which fell to Earth in Australia in 1969, contains more than 80 kg of carbonaceous material. About one percent of this material is organic carbon in the form of amino acids (**FIGURE B1.1**).[a] Because a number of different laboratories analyzed the meteorite almost immediately after it landed and the results were consistent overall, researchers concluded that the amino acids did not result from terrestrial contamination. Recent research is consistent with the conclusion that the early solar system contained a greater diversity of molecules than currently found on Earth. The Murchison meteorite formed before the origin of the Sun and picked up molecules while passing through the early solar system; more than 14,000 different chemicals have been identified in the Murchison meteorite including purines, pyrimidines, glycerol, and many amino acids such as glycine, alanine, valine, proline, and aspartic acid (Schmitt-Kopplin et al., 2010). Remarkably, practically all of the amino acids found in the meteorite (whether assembled into proteins or not), are similar to the amino acids produced in the laboratory experiments discussed in Box 6.3, by applying an electric spark to mixtures. However, a number of the amino acids found in meteorites do not appear in proteins in Earth's organisms (Scott and Krot, 2007; Herd et al., 2011).

Interior portions of the Murchison meteorite contain fatty acids up to eight carbons long. Thus, we can conclude

FIGURE B1.1 Individual components in a test tube, isolated from the Murchison meteorite (black fragment) are unchanged since their condensation from material ejected by a star.

Courtesy of Argonne National Laboratory

that fatty acids, which are important constituents of cell membranes, formed early in the life of the solar system and did not originate with life. Moreover, research shows that samples of uncontaminated Murchison meteorite compounds can produce fatty acid-like structures *and* bounded vesicles that resemble membranes.[b]

---

[a]Three percent by weight of the Tagish Lake meteorite is organic material (Herd et al., 2011).

[b]For an overview of these studies, see D. W. Deamer, 2003. A giant step towards artificial life? *Trends Biotechnol.,* **23**, 336–338. Also see the chapters in Chela-Flores et al., 2002.

---

foundations of a "universal organic chemistry" billions of years ago (Morowitz, 2002; Knoll, 2003).

MOLECULES OF LIFE

Criteria allowing us to answer the question "What is life?" are outlined in **BOX 6.2**. The most fundamental features of "life" are present in the most basic organisms: the presence of those molecules and reactions necessary for survival and reproduction. The two most fundamental classes of molecules of life, those performing the functions outlined in Box 6.2, are **amino acids** from which proteins are constructed, and **nucleotides** from which deoxyribonucleic acid (DNA) and ribonucleic acid (RNA) are constructed. Although life can be defined in a number of ways, the continuity of life is based on the presence, function, and replication of DNA within cells (Box 6.2). DNA, RNA, and cells are required in any definition of life or living.

We will now examine briefly the nature of amino and nucleic acids.

BOX 6.2   What Is Life?

Any search for an answer to the question "What is Life?" over the past 69 years takes as its starting point the 1994 book of the same name by the physicist, theoretical biologist, developer of quantum mechanics, and Nobel Laureate, Erwin Schrödinger (1887–1961). The subtitle of the book, *The Physical Aspect of the Living Cell* (1944), reveals Schrödinger's position, which is that life can be understood in physical and material terms. Many physicists turned to biology after reading *What Is Life?* Indeed, the foundation of modern molecular biology and the accelerated search for the nature of genes from the mid-1940s on have been ascribed to the influence of Schrödinger's book, which has never been out of print.

Five attributes of a **definition of life** that allow us to **distinguish living from nonliving** are summarized below.

1. Living things perform the reactions necessary for survival and replication.
2. Life can be defined as the ability to transform external sources of material and energy into such processes as metabolism, growth, and development.
3. Living things exist as individuals and collections of individuals (populations).
4. Cells are the units of life and so life is based on membranous structures.
5. Life is based upon three classes of molecules: DNA, RNA, and proteins.

Although life can be defined using any or all of these attributes, the **continuity of life** from generation to genera-

tion is based on the presence, function, and replication of DNA within cells. Life defined by metabolic activity requires the presence of RNA. Therefore, DNA, RNA, proteins, *and* cells are required in any definition of life or living. Why? Because:

- DNA does not enable a cell to function (to be alive) unless the DNA is transcribed into RNA.
- RNA does not enable a cell to function (to be alive) unless it in turn is translated into a protein or triggers the activity of other molecules of DNA, RNA, or protein.

Higher levels of organization such as DNA in chromosomes, RNA in ribosomes, proteins derived from the translation of RNA, or cells organized into tissues and organs are not included in this minimal definition of life. Philosopher Robert A. Wilson (2005) included such higher-level organization and phenomena when he concluded that living agents:

- have parts that are heterogeneous and specialized,
- include a variety of internal mechanisms,
- contain diverse organic molecules, including nucleic acids and proteins,
- grow and develop,
- reproduce,
- repair themselves when damaged,
- have a metabolism,
- bear environmental adaptation, and
- construct the niches they occupy.

AMINO ACIDS

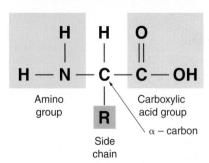

Amino Acid Structure

An ENZYME *is a protein that catalyzes (allows or regulates) a chemical reaction.*

Foremost among the metabolic agents that enable cells to function are many different **proteins** that catalyze and regulate practically all the chemical reactions within an organism. Amino acids form the basic structure of proteins. These amino acids consist of a central α-carbon atom (C) to which are attached the following:

- an amino group with a potential positive charge (NH_2^+),
- a carboxylic acid group (COOH) with a potential negative charge,
- a hydrogen atom (H), and
- a side chain (R) that varies in structure among the different amino acids.

Amino acids link by **peptide linkages** into linear **polypeptide chains** (typically 10 to 100 amino acids long), which are the building blocks of proteins. The highly specific structure of any protein molecule, whether it functions as an **enzyme** (catalyst) or in another role, derives from the exact linear placement of the amino acids. Specific amino acid sequences enable polypeptide chains to fold into specific three-dimensional forms that confer individual properties on proteins. Most activities associated with life derive from the enzymatic and regulatory activities of amino acid sequences.

NUCLEIC ACIDS

The sequences of amino acids in proteins are determined by the **nucleotide sequences in nucleic acids** (END BOX 6.1). Each functional unit of DNA or RNA consists of three linked nucleotides (bases) known as a **codon**, a different codon coding for each of the 20 amino acids. Nucleotides are the building blocks of the nucleic acids DNA and RNA, each long-chain nucleic acid consisting of a pentose (5-carbon) sugar (**ribose** or **deoxyribose** in **RNA** and **DNA**, respectively), a monophosphate group, and a nitrogenous base (FIGURE 6.2). Connected to the 1′ carbon of each sugar is one of four kinds of nitrogenous bases, two of which in DNA *and* RNA are purines (*adenine* [A] and *guanine* [G]), and two of which are pyrimidines —*cytosine* (C) and *thymine* (T) in DNA; cytosine (C) and *uracil* (U) in RNA—FIGURE 6.3.

You might think that the restriction of nucleic acid composition to only four different kinds of bases would limit the message-bearing capacity of these molecules to only four kinds of messages. It doesn't. Nucleic acid molecules may be many thousands or millions of nucleotides long, with each message encoded by a unique linear sequence of nucleotides, endowing these molecules with the capacity to carry an immense variety of highly complex messages (End Box 6.1 and see Lynch, 2007 and Darnell, 2011). For any one nucleotide position four different messages are possible (A, G, C, or T); for two nucleotides in tandem 4^2 or 16 different messages are possible (AA, AG, AC, AT, GG, GC, and so on). The rule simply is that for a linear sequence of *n* nucleotides, 4^n different possible messages can be encoded. Thus, a linear sequence of only 10 nucleotides can discriminate among billions (4^{10}) of potentially different messages.

THE DOUBLE HELIX AND REPLICATION OF DNA

The biochemical information discussed above and in End Box 6.1 helps explain the information-carrying role of nucleic acids, but it does not explain how nucleic acids *replicate and transmit this information*. The model for nucleic acid replication derives from the familiar **double helix** structure deduced by James Watson and Francis Crick in 1953. The DNA double helix consists of two antiparallel strands coiled around each other in the form of a right-hand helix (FIGURE 6.4), with complementary pairing between strands, linking purine and pyrimidine bases (A–T, G–C).

An important discovery published in 1950 by the biochemist Erwin Chargraff (1905–2002), now known as Chargraff's rule, is that the number of adenine residues always equals the number of thymine residues and that the number of guanine residues always equals the number of cytosine residues, irrespective of the organism providing the DNA. Chargraff told Watson and Crick of his finding, providing them with information essential to devising their model for DNA replication; if a DNA molecule could dissociate and reassociate, only complementary bases would form bonds in the new DNA.

DNA takes various structural forms depending on relative humidity, salt concentration, and other factors. In the B (hydrated) form diagrammed in Figure 6.3, the 10 base pairs for each complete turn of the helix are stacked almost perpendicularly to the helical axis. In the A (dry) form, each turn of the helix has 11 bases tilted 20° relative to the helical axis.[3]

---

[3] Rosalind Franklin (1920–1958), an English x-ray crystallographer whose photographs provided Watson and Crick with essential information in devising the double helix model, gave the two DNA forms the names A and B, respectively.

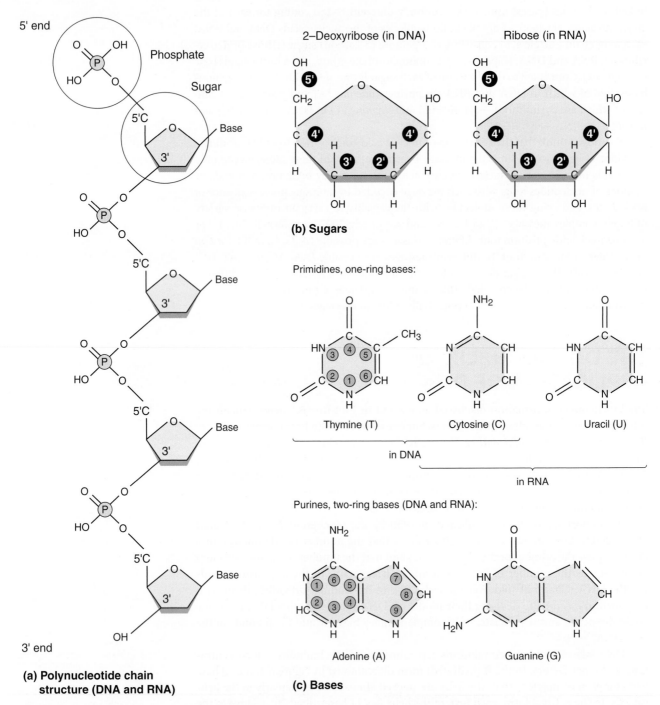

FIGURE 6.2 **(a)** General structure for DNA and RNA chains, which are composed of a linear sequence of nucleotides. Each nucleotide consists of a phosphate group, a sugar, and a nitrogenous base, linked as shown. **(b)** Differences between the sugars found in DNA (deoxyribose) and RNA (ribose). **(c)** The basic kinds of nitrogenous bases found in DNA (T, C, A, G) and RNA (C, U, A, G).

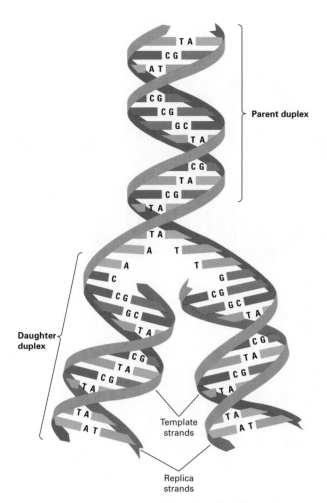

Parent duplex

Daughter duplex

Template strands

Replica strands

FIGURE 6.3 Examples of hydrogen bond pairing between bases (Adenine–Thymine (A–T) and Guanine–Cytosine (G–C)). The two strands on the double helix are bridged by parallel rows of such paired nucleotide bases stacked at regular 3.4 Å intervals, where 1 Å = 0.0000001 mm.

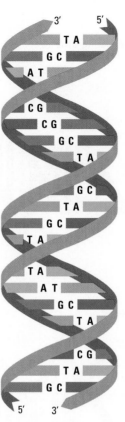

FIGURE 6.4 Helical structure of DNA as determined by Watson and Crick with complementary pairing of purine and pyrimidine bases (for which see Figure 6.3).

The replicative power of the DNA double helix derives from each of the two strands serving as a template for a newly complementary strand. The two new double helices that form bear nucleotide sequences identical to each other *and* to the parent molecule (**FIGURE 6.5**). This quality of exact molecular replication enables similar messages to be transmitted from generation to generation and confers on nucleic acids their function as the *genetic (hereditary) material.* Hereditary information is contained in the triplet genetic code outlined in End Box 6.1.

This code was proposed in 1961 in a paper that epitomizes the scientific method at its best (Crick et al., 1961[4]). Using what today looks to us to be a minimal amount of molecular information, Crick and his colleagues deduced (a) that the code was based on nonoverlapping triplets of amino acids, (b) that each triplet was read from a specific

[4]For a discussion of this seminal paper, see C. Yanofsky, 1997. Establishing the triplet nature of the genetic code. *Cell,* **128**, 815–817.

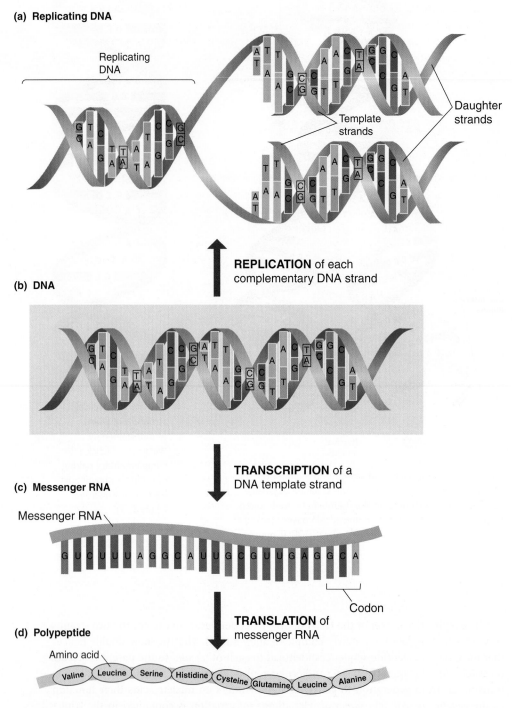

FIGURE 6.5 Illustration of how DNA replicates and how information is transferred from DNA to RNA to protein. In DNA replication **(b** to **a)** proteins break the hydrogen bonds between paired bases, allowing the two template strands to unwind. Each unwound daughter strand then acts as a template producing a new complementary strand. As a result, two double-stranded DNA molecules form, each an exact replica of the original parental double helix. In transcription **(b** to **c)**, one of the two DNA strands serves as a template upon which a molecule of messenger RNA is transcribed (green). This messenger then serves in turn as a template on which a molecule of protein is translated **(c** to **d)**. The messenger RNA is exactly complementary to its DNA template. A sequence of three nucleotides (a triplet codon) on the messenger specifies one amino acid **(d)**.

starting amino acid, and (c) that the code was degenerate (more than one codon codes for each amino acid) and comma-less (there are no spacing nucleotides between codons).[5]

Now that we understand the basic molecules of life, we turn to the most likely places on Earth in which these molecules evolved, then turn to a brief discussion of experimental generation of peptides in the laboratory to help elucidate our understanding of molecular origins.

POSSIBLE SITES FOR THE ORIGIN OF EARLY MOLECULES

Heat-loving (thermophilic) organisms have been found at temperatures as high as 320°C, some living a kilometer and a half below the ocean surface in deep sea hydrothermal vents, others thriving in hot springs (**FIGURE 6.6**). The astounding living conditions of these organisms demonstrate that life can be maintained (and may have arisen) under the most stringent conditions. Shallow marine hydrothermal plumes are possible sites for early life on Earth (see below). Two further sites, also discussed below, are near volcanic regions or on the surfaces of layered clays.

Organisms that live and reproduce at the high temperatures (45° to 80°C) found in deep sea vents, hot springs or near volcanoes are known as THERMOPHILES. *Organisms that live under any form of extreme conditions are known as* EXTREMOPHILES.

VOLCANOES

Formation of minerals in association with volcanic activity over geological time demonstrates that "a common chemistry apparently links the living entities on the Earth with the cosmos."[6] Formation of minerals and the transition from non-living to living was driven by changing environments. Indeed, the appearance and modification of minerals

(a)

(b)

FIGURE 6.6 Heat-loving (thermophilic) organisms thrive in hot springs **(a)** and in hydrothermal vents 1.5 km below the ocean surface **(b)**.

© Pichugin Dmitry/ShutterStock, Inc.

Courtesy of OAR/National Undersea Research Program (NURP)/NOAA

---

[5] See End Box 6.1 for more detailed discussion of the genetic code, and see Lynch (2007) for the evolution of the architecture of the genome.
[6] Burns, 2010, p. 581.

during Earth's history can be viewed as mineral evolution in the same way that we understand molecular and biological evolution:

- Thousands of mineral species exist—4,714 according to Hazen et al. (2008);
- Minerals originated in sequence in ten evolutionary stages, five of which are: in the cosmic dust associated with the Big Bang; in the magma from which the first rocks formed; in the deep crust; on Earth's surface; through the activity of organisms.
- As with biological evolution, the stages of mineral evolution were initiated by changes in Earth's environment: concentration of elements after origination of the pre-solar nebula; changes in physical parameters such as pressure, temperature, and atmospheric levels of H_2O, CO_2, and O_2; and by the origin and evolution of organisms (Hazen et al., 2008; Johnson et al., 2008).

Much of Earth's early landmass was composed of volcanic islands the sediments of which produced sponge-like minerals (*zeolites*) that retain and catalyze organic compounds available as molecules for life to arise. Furthermore, high global "greenhouse" temperatures persisted for more than a billion years because of high CO_2 atmospheric pressures. Greenhouse gases such as carbon dioxide (CO_2) and methane (CH_4) absorb infrared radiation, reduce the loss of heat from Earth's surface, and raise global temperature, all of which have been implicated in the origin of the first molecules of life.

Without the current GREENHOUSE EFFECT, *the average temperature on Earth would be of the order of 15°C lower than it is, an eventuality that, had it occurred, would have taken the average global temperature in 2001 from 14.5°C to just below 0°C; not a happy prospect.*

HYDROTHERMAL DEEP SEA VENTS

Hydrothermal deep sea vents were first discovered in 1977 on the Galapagos Rift off the coast of Ecuador, at a depth of 2.5 km. Since then, the theory that life may have originated in deep-sea vents (Figure 6.6**b**) has gained considerable support. Lipids isolated from sediments deposited 2.7 Bya in what is now Ontario, Canada have been interpreted as evidence of unicellular life in a "subsurface hydrothermal biosphere . . . believed to have a near-global extent" (Ventura et al., 2007). However, because many organic compounds are degraded under the conditions found in hydrothermal vents (Miller and Lazcano, 1995), much more research needs to be conducted to identify the range of conditions under which the molecules of life arose.

The superheated plumes arising from hydrothermal vents today produce large amounts of ammonia and can produce amino acids such as alanine. It has been hypothesized that as oceans formed billions of years ago, hydrothermal deep sea vents, with water temperatures of 350°C, produced large amounts of ammonia by combining nitrogen or its oxides with overheated water in the presence of iron and other mineral catalysts.[7] In this scenario, metallic ions played the catalytic role now played by enzymes. Inorganic catalysts function by lowering the energy level necessary for a reaction, thereby increasing its frequency. By providing specific sites at which potentially reacting molecules can be localized, organic catalysts (enzymes) speed up this process even more. For example, inorganic ferric ion (Fe^{3+}) shows some catalytic activity in a variety of reactions (FIGURE 6.7a). When such ions are incorporated into porphyrin molecules to form **heme** (Figure 6.7**b**), the molecules are about a thousand times more effective than Fe^{3+} alone. If the protein component of the enzyme catalase then adds to the heme unit, catalytic efficiency increases by a further factor of one billion (Figure 6.7**c**).

Iron pyrite is abundant in rocks of the ocean bottom. Some researchers believe that these molecules were instrumental in the origin of life molecules in deep sea vents. They hypothesize that metallic ions combined with water to produce acetic acid, which is one of the simplest acids and a primary initiator of many metabolic pathways in extant organisms. This acetic acid, combined with CO_2, would have produced pyruvic

---

[7] J. A. Brandes et al., 1998. Abiotic nitrogen reduction on early Earth. *Nature,* **395**, 365–367.

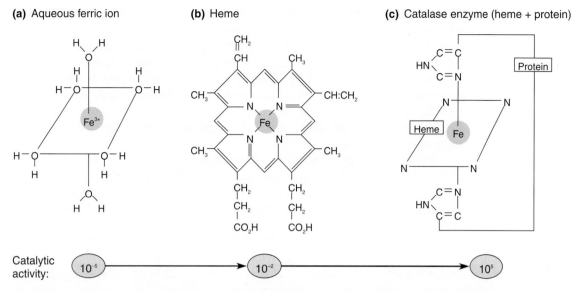

(a) Aqueous ferric ion

(b) Heme

(c) Catalase enzyme (heme + protein)

Catalytic activity: 10^{-5} → 10^{-2} → 10^{5}

FIGURE 6.7 Change in catalytic activity for the reaction $2H_2O_2 \rightarrow 2H_2O + O_2$ when the iron atom is used by itself **(a)**, or in different molecular combinations **(b, c)**. The catalase protein in **(c)** provides an enzymatic enhancement to the reaction because it binds rapidly to hydrogen peroxide molecules and distorts them so their decomposition proceeds at a lower "activation energy" than without the enzyme.

[Adapted from Calvin, M., 1969. *Chemical Evolution*. Oxford University Press, Oxford, UK.]

acid, which, in turn, interacted with ammonia to produce alanine. Experiments by Huber and Wächtershäuser, published in the late 1990s,[8] provided laboratory-based evidence for the production of acetic acid through this pathway at 100°C. Tests at a wider range of temperatures and at the pressures found at ocean depths produced ammonia that was stable to temperatures as high as 800°C. When the ammonia combined with pyruvic acid, the reaction produced alanine, which then formed short peptides (Hazen, 2005).

CLAYS

Crystals, which grow by adding subunits into their highly ordered structures, have some *self-replicatory powers*. It has been proposed that the earliest self-replicating structures may have consisted of organic, clay-like silicate crystals or layered clays such as *montmorillonite*. Several pieces of evidence have been used to support this theory: that when activated by phosphate, they can condense to form high yields of polypeptide chains in layers of clay; nucleotides can be joined into chains 55 nucleotides long on montmorillonite clays; and amino acids can polymerize into peptides on mineral surfaces such as hydroxyapatite or illite.[9] If nucleic acids such as RNA became incorporated into such an assembly, formation of peptides and the evolution of a protein-synthesizing system could have followed (Darnell, 2011).

MONTMORILLONITE *is an aluminum silicate named after the 12th century town of Montmorillon in France. Montmorillonite forms as one µm crystals that produce a clay used in agriculture to hold soil moisture, in drilling for oil to keep the drilling mud viscous, and in earthenware dams and foundry sand.*

[8]C. Huber and G. Wächtershäuser, 1998. Peptides by activation of amino acids with CO on (Ni, Fe)S surfaces: Implications for the origin of life. *Science*, **281**, 670.

[9]J. P. Ferris, 2005. Mineral catalysis and prebiotic synthesis: Montmorillonite-catalyzed formation of RNA. *Elements*, **1**, 145–149.

Although mineral surfaces are gaining attention as a locale for biosynthetic reactions, their ability to develop a highly complex metabolic sequence extending from self-replicating RNA to protein synthesis is difficult to visualize. Some biologists have therefore emphasized the possibility that *proteins themselves* or *protein-nucleic acid combinations* were the first self-replicating systems. Early peptides could have been templates for the aggregation of nucleotides, which, when bonded, would form a precise mold for the replication of these same peptides. Many experiments (some of which are discussed in BOX 6.3) have demonstrated the origin of such molecules in the laboratory under abiotic conditions.

Therefore, molecules associated with life can be traced to abiotic processes. Amino acids can be synthesized from hydrogen, ammonia, methane, and water when exposed to an external source of energy, and mixtures of amino acid subjected to high temperatures can polymerize to form chains with peptide-like properties.

BOX 6.3 Generating Peptides in the Laboratory

In the 1920s, Aleksandr Oparin (1894–1980), a Russian biochemist, and J. B. S. Haldane (1892–1964), an English physiologist and geneticist, independently hypothesized that organic compounds found in Earth's early atmosphere might be similar to those found in organisms. Almost 30 years elapsed, however, before this hypothesis was tested experimentally.

In 1953, Stanley Miller (1930–2007), an American chemist and biologist, placed methane, ammonia, and hydrogen gases in a glass apparatus, generated an electric spark in a large five-liter flask, and boiled water in a smaller flask to provide vapor to the spark and to circulate the gases[a] (FIGURE B3.1). Any compounds formed in the apparatus were condensed (or recirculated if they were volatile). After a week of continuous electrical discharge, Miller used chromatography to analyze products that may have accumulated in the aqueous phase. Remarkably, compounds previously only known from organisms, including amino acids and other substances, such as urea, had formed in the flasks. All of the molecules had simple structures, but all were compounds known to be essential for life.

On the early Earth, in addition to electrical discharges, other energy sources such as β–, γ– or x-rays from cosmic radiation, ultraviolet light, and elevated temperatures facilitated the production of amino acids and other organic compounds from simple gases, including purine and pyrimidine nucleotide bases, that the Miller reactions did not produce. Regardless, these and subsequent experiments are consistent with the hypothesis that the early atmosphere of Earth had a significant number of simple organic (but abiotic) molecules that arose as a response to cosmic radiation or high temperatures following impacts of extraterrestrial material.[b]

Also in the 1950s, American biochemist Sidney Fox (1912–1968) and colleagues developed a technique in which heat could be used to produce peptides from dry mixtures of amino acids.[c] Depending on the kinds of amino acids in the mixture, Fox obtained as much as a 40 percent yield of peptide-like products with molecular weights between 4,000 and 10,000 Daltons. Fox called these polymers **proteinoids** (also *thermal proteins*), and demonstrated that they had protein-like features: they reacted positively to the same reagents as did proteins, their solubilities resembled proteins, they were precipitable with similar reagents, and had other protein-like traits. According to these analyses, the proteinoids possess nonrandom proportions of amino acids; they all show similar properties when tested by sedimentation rates, electrophoretic techniques, column fractionation, and other measurements. Preferential interaction between amino acids seems to dictate their position and frequency, leading to some degree of uniformity in the kinds of molecules produced.

---

[a]S. L. Miller, 1953. A production of amino acids under possible primitive Earth conditions. *Science*, **117**, 528–529.
[b]S. Miyakawa et al., 2000. Abiotic synthesis of guanine with high-temperature plasma. *Origins of Life and Evolution of the Biosphere*, **30**, 557–566. Extracts from Miller's original experiments were reanalyzed by Johnson et al. (2008) with results consistent with volcanic origins of organic molecules.
[c]For an overview of his work, see S. W. Fox, 1984. Proteinoid experiments and evolutionary theory. In *Beyond Neo-Darwinism*, M.-W. Ho and P. T. Saunders (eds.). Academic Press, London, pp. 15–60.

BOX 6.3 Generating Peptides in the Laboratory (Cont...)

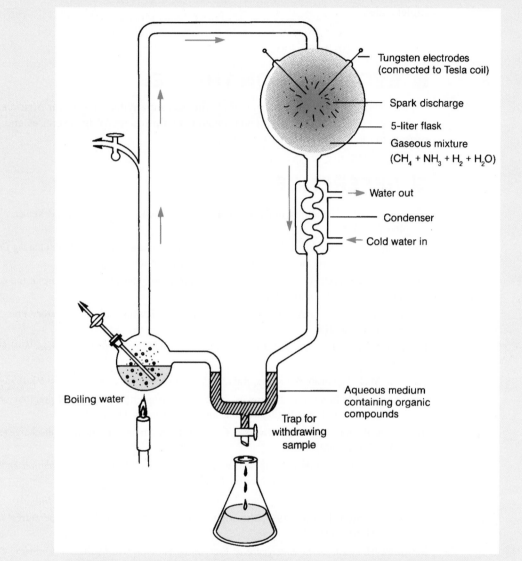

FIGURE B3.1 Diagrammatic representation of the apparatus Stanley Miller used to demonstrate the synthesis of organic compounds by electrical discharge in a reducing atmosphere.

■ KEY TERMS

abiotic

amino acids

anticodon

carbonaceous meteorites

chondrites

codon

degenerate (redundant) code

DNA (deoxyribonucleic acid)

enzyme

extremophiles

frozen accidents

life (definition)

Murchison meteorite

nucleic acid

nucleotide

peptide

polypeptide

protein

proteinoids

RNA (ribonucleic acid)

stop codon
thermophilic organisms (thermophiles)
third codon position
translation
triplet code

universal genetic code
universality
wobble hypothesis
wobble pairing

■ EVOLUTION ON THE WEB

Explore evolution on the Internet! Visit the accompanying website for *Strickberger's Evolution, Fifth Edition,* at **go.jblearning.com/Evolution5eCW** for exercises and links relating to topics covered in this chapter.

■ REFERENCES

Burns, J. A., 2010. The four hundred years of planetary science since Galileo and Kepler. *Nature,* **466**, 575–584.

Chela-Flores, J., G. Lemarchand, and J. Oró (eds.), 2002. *Astrobiology: Origins from the Big-Bang to Civilization.* Kluwer Academic Press, Dordrecht, The Netherlands.

Crick, F. H. C., Barnett, L., Brenner, S., and R. Watts-Tobin, 1961. General nature of the genetic code for proteins. *Nature,* **192**, 1227–1232.

Darnell, J. 2011. *RNA. Life's Indispensable Molecule.* Cold Spring Harbor Laboratory Press, Cold Spring Harbor, NY.

Hazen, R. M., 2005. *Genesis. The Scientific Quest for Life's Origins.* Joseph Henry Press, Washington, DC.

Hazen, R. M., D. Papineau, W. Bleeker, et al., 2008. Mineral evolution. *Amer. Mineral.* **93**, 1693–1720.

Herd, C. D. K., A. Blinova, D. N. Simkus, et al., 2011. Origin and evolution of prebiotic organic matter as inferred from the Tagish Lake meteorite. *Science,* **332**, 1302–1307.

Johnson, A. P., H. J. Cleaves, J. P. Dworkin, et al., 2008. The Miller volcanic spark discharge experiment. *Science,* **322**, 404.

Knoll, A. H., 2003. *Life on a Young Planet: The First Three Billion Years of Evolution on Earth.* Princeton University Press, Princeton, NJ.

Lynch, M., 2007. *The Origins of Genome Architecture.* Sinauer Associates, Sunderland, MA.

Miller, S. L., and A. Lazcano, 1995. The origin of life—did it occur at high temperatures? *J. Mol. Evol.,* **41**, 689–692.

Morowitz, H. J., 2002. *The Emergence of Everything: How the World Became Complex.* Oxford University Press, New York.

Schmitt-Kopplin, P., Z. Gabelica, R. D. Gougeon, et al., 2010. High molecular diversity of extraterrestrial organic matter in Murchison meteorite revealed 40 years after its fall. *Proc. Natl Acad. Sci. USA,* **107**, 2763–2768.

Schrödinger, E. 1944. *What Is Life? The Physical Aspect of the Living Cell.* Cambridge University Press, Cambridge, UK.

Scott, E. R. D., and A. N. Krot, 2007. Chondrites and their components In *Meteorites, Comets and Planets* (ed. A. M. Davis), Chapter 1.07, Treatise on Geochemistry Update 1, 1–72. Elsevier Pergamon, Oxford, UK.

Time Magazine, 2006. *Nature's Extremes. Inside the Great Natural Disasters that Shape Life on Earth.* Time, Inc., New York.

Ventura, G. T., F. Kenig, C. M. Reddy, et al., 2007. Molecular evidence of Late Archean archaea and the presence of a subsurface hydrothermal biosphere. *Proc. Natl Acad. Sci. USA,* **104**, 14260–14265,

Watson, J. D., and F. H. C. Crick, 1953. A structure for deoxyribose nucleic acid. *Nature,* **171**, 737–738.

Wilson, R. A., 2005. *Genes and the Agents of Life. The Individual in the Fragile Sciences,* Cambridge University Press, Cambridge, UK.

END BOX 6.1

The Genetic Code

SYNOPSIS: Here, 10 fundamental features of the *genetic code* are listed that determine the sequences of three nucleotides (codons) in each mRNA molecule, explain how the same products can be coded for by different codons (*redundancy*), and suggest a hypothesis for the evolution of the code and the likelihood that it has always been triplet.

Information transfer between nucleic acids and proteins follows a genetic code that uses the placement of a specific trinucleotide sequence in mRNA to determine the placement of a particular amino acid within a protein. **TABLE EB1.1** outlines the essential features of the genetic code; **TABLE EB1.2** gives the code itself.

Ten fundamental features characterize the genetic code:

1. Messenger RNA molecules consist of only *four kinds of nucleotide bases,* adenine (A), guanine (G), uracil (U), and cytosine (C). Combinations of these four nucleotides form chains of varying lengths and varying sequence.
2. An mRNA codon that specifies a particular amino acid consists of a chain of *three nucleotides.*
3. The code is nonoverlapping (comma-less; Table EB1.1). Each codon *translates* in a continuous, uninterrupted sequence, three successive nucleotides at a time, from one end of an mRNA reading frame to the other.
4. *Reading frames* in messenger RNA generally begin with the codon AUG, and terminate at the stop codons UAA, UAG, or UGA.
5. The codon sequence complements an *anticodon* sequence on the transfer (tRNA) molecule that carries a particular amino acid to the mRNA codon (Table EB1.1).
6. All organisms share the *universal* coding dictionary outlined in Table EB1.2, with some codon differences in mycoplasmas (bacteria lacking polysaccharide cell walls) and ciliate protists. Mitochondrial organelles show a few codon differences from those used by nuclear genes.
7. Ambiguities have not been found in the code; that is, *the same codon does not specify two or more different amino acids.*
8. With the exception of methionine and tryptophan, *more than one codon designates each amino acid.*
9. The pattern of code *redundancy is mostly in the third codon position* (see following section). For example, eight amino acids (including valine, threonine, and alanine) are coded for by four separate codons, called a quartet. Each of these codons in the quartet varies only at the third position.
10. When an amino acid uses only a *duet* (two) of the codons in a quartet, the third codon positions in the duet are both pyrimidines (U and C) or both purines (A and G), never one pyrimidine and one purine.

REDUNDANCY AND THE GENETIC CODE

From an evolutionary perspective, the genetic code's most essential features are its **universality** and its **redundancy** or "**degeneracy**," terms that refer to the use of several codons for the same amino acid. Relative nonspecificity (redundancy) in the third nucleotide in each codon minimizes errors in replicating the master code or in translating it into protein.

Explanations for some features of code redundancy have not been difficult to find. In part, redundancy derives from the presence of more than one kind of tRNA for a single amino acid. tRNAs used for leucine, for example, may have the anticodons AAU, AAC, and GAG. Francis Crick explained that the redundancy pattern mostly attaches to the third codon position because of **wobble pairing** between certain bases of the tRNA anticodon and the mRNA codon in this position.

In Crick's **wobble hypothesis**, published in 1966 and since proved correct, anticodons bearing inosine (I) at the **third codon position** can pair with either U, A, or C. Anticodons bearing guanine (G) can pair with either of the pyrimidines uracil (U) and cytosine (C); anticodons bearing U can pair with either of the purines adenine (A) and guanine (**FIGURE EB1.1**). **Third codon position redundancy** therefore points to the importance of the first two codon positions in specifying amino acids. These two positions suffice to code for the eight amino acids that use codon quartets.

The next most error-prone translational event occurs at the **first codon position** where the code again shows some redundancy—for example, UUA and CUA both code for leucine—or is

END BOX 6.1

The Genetic Code (Cont...)

© Photos.com

TABLE EB1.1	Essential Features of the Genetic Code

Term	Meaning
Code letter	Nucleotide, for example, A, U, G, C (in mRNA) or A, T, G, C (in DNA). Note that there are four code letters in each nucleic acid "alphabet," forming two kinds of complementary base pairs: A–U and G–C in RNA, and A–T and G–C in DNA.
Codon or code word	Sequence of nucleotides specifying an amino acid, e.g., the RNA codon for leucine = CUG (or GAC in DNA).
Anticodon	Sequence of nucleotides on transfer RNA that complements the codon, e.g., GAC = anticodon for leucine.
Genetic code or coding dictionary	Table of all the codons, each designating the specific amino acid into which it translates (Table EB1.2)
Codon length or word size	Number of letters in a codon, e.g., three letters in a **triplet code** (these are the same as coding ratio in a nonoverlapping code).
Nonoverlapping code	Code in which only as many amino acids are coded as there are codons in end-to-end sequence, e.g., for a triplet code, UUUCCC = phenylalanine (UUU) + proline (CCC).
Redundant or degenerate code	Presence of more than one codon for a particular amino acid, e.g., UUU, UUC = Code = phenylalanine. Twenty different amino acids are therefore coded by a total of more than 20 codons.
Synonymous codons	Different codons that specify the same amino acid in the redundant code, e.g., UUU = UUC = phenylalanine.
Ambiguous code	Circumstances when one codon can code for more than one amino acid, e.g., GGA = glycine, glutamic acid. No ambiguities exist in the present code although such ambiguities may have existed in the past.
Comma-less code	Absence of nucleotides (spacers) between codons, e.g., UUUCCC = two amino acids in triplet non-overlapping code.
Reading frame	Particular nucleotide sequence coding for a polypeptide that starts at a specific point and partitions into codons until it reaches the final codon of that sequence.
Frame shift mutation	Change in the reading frame because of the insertion or deletion of nucleotides in numbers, other than multiples of the codon length. This changes the previous partitioning of codons in the reading frame and causes a new sequence of codons to be read.
Sense word	Codon that specifies an amino acid normally present at that position in a protein.
Replacement mutation	Change in nucleotide sequence, either by deletion, insertion or substitution, resulting in the appearance of a codon that produces a different amino acid (**missense mutation**) in a particular protein, e.g., UUU (phenylalanine) mutates to UGU (cysteine).
Stop mutation	Mutation that results in a codon that does not produce an amino acid, e.g., UAG (also called a chain-terminating codon or **nonsense codon**).
Universal code	Use of the same genetic code in all organisms, e.g., UUU = phenylalanine in bacteria, mice, humans, and tobacco (with some exceptions, e.g., mitochondria, see Table EB1.2).

[Adapted from Strickberger, M. W., 1985. *Genetics, Third Edition.* Macmillan, New York.]

TABLE EB1.2	The Genetic Code of Nucleotide Sequences in mRNA Codons Specifying Individual Amino Acids[a]		

U U U } phe U U C U U A } leu U U G	U C U U C C U C A } ser U C G	U A U } tyr U A C U A A } STOP[b] U A G	U G U } cys U G C U G A STOP[b] U G G trp
C U U C U C } leu C U A C U G	C C U C C C } pro C C A C C G	C A U } his C A C C A A } gln C A G	C G U C G C } arg C G A C G G
A U U A U C } ile A U A A U G[c] met	A C U A C C } thr A C A A C G	A A U } asn A A C A A A } lys A A G	A G U } ser A G C A G A } arg A G G
G U U G U C } val G U A G U G	G C U G C C } ala G C A G C G	G A U } asp G A C G A A } glu G A G	G G U G G C } gly G G A G G G

[a]Rare exceptions to this code occur in various animal mitochondria in which AUA also specifies methionine, and AGA specifies serine or functions as a stop codon. Such mitochondrial changes seem to be in the direction of economizing in the kinds of transfer RNA produced in a small organelle that makes few polypeptides. Some other exceptions are found in several organisms displayed in Figure EB 1.3.
[b]These codons are also called *chain-terminating codons* or, in the past, *nonsense codons*.
[c]This is the common codon used to initiate protein synthesis.

constructed so that an error may occasionally enable the substitution of an amino acid with related function, for example, if G is substituted for U, UUA (leucine) becomes GUA (valine).

The **second codon position** is the least error prone during translation. It is hypothesized that this position at one time may have separated entire classes of amino acids with unique functions. Such a coding system would offer a selective advantage by ensuring that amino acids are rarely substituted between classes. Possible remnants of such a grouping may be:

- assignment of U to the second codon position for leucine, isoleucine, and valine, all of which are **hydrophobic amino acids** that exist mostly in the interior of proteins, and
- assignment of A to the second codon position for glutamic acid, histidine, aspartic acid, and other **hydrophilic amino acids** that commonly exist on the protein surface (**TABLE EB1.3**).

REDUNDANCY AND CODE EVOLUTION

Although no ancestral codes persist in any extant organisms, as with any other biological trait, the genetic code is hypothesized to have evolved from an earlier and presumably simpler form. In evolutionary terms, an early genetic code, even one composed of three nucleotides, would have provided information for fewer amino acids than now, and/or for amino acids with broader functions than now.

As the code evolved, the first codon position came into use for amino acid positioning as a consequence of modifications that could be made to the immediately adjacent transfer RNA

END BOX 6.1

The Genetic Code (Cont...)

FIGURE EB1.1 Wobble base pairing for the amino acid inosine (I), which can pair with cytosine (C), uridine (U), or adenine (A).

(tRNA) nucleotide. With the advent of two accurately translated codon positions, the genetic code could expand its repertoire of amino acids. Retention of third codon position redundancy or wobble may have economized on the number of tRNA molecules needed to translate amino acids such as valine, threonine, and alanine.

Following this logic, Jukes (1983) proposed an early stage in the evolution of the genetic code somewhat like that shown in **FIGURE EB1.2a.**[a] Each **codon quartet** or **codon family** of four codons specified perhaps one amino acid. Each amino acid was represented by only one type of

[a]For an evaluation of this proposal, see T. H. Jukes and S. Osawa, 1991. Recent evidence for evolution of the genetic code. In *Evolution of Life: Fossils, Molecules, and Culture*, S. Osawa and T. Honjo (eds.). Springer, Tokyo, pp. 79–95.

TABLE EB1.3	Classification of 18 Amino Acids According to the Nucleotide Found at their Second Codon Positions and the Hydration Potentials of Their Side Chains[*]		
Amino Acid	**Second Codon Letter**	**Hydration Potential (k/cal/mole)**	
Gly	G	+2.39	**Most hydrophobic**
Leu	U	+2.28	
Ile	U	+2.15	
Val	U	+1.99	
Ala	C	+1.94	
Phe	U	−0.76	
Cys	G	−1.24	
Met	U	−1.48	
Thr	C	−4.88	
Ser	C(G)	−5.06	
Trp	G	−5.89	
Tyr	A	−6.11	
Gln	A	−9.38	
Lys	A	−9.52	
Asn	A	−9.68	
Glu	A	−10.19	
His	A	−10.23	
Asp	A	−19.92	**Most hydrophilic**

[*]In other proposals, the first codon position specifies amino acids made through similar biosynthetic pathways, or distinguishes between small and large amino acids (see Maynard Smith and Szathmáry, 1995). [*Source:* Data from R. V. Wolfenden et al., 1979. Water, protein folding and the genetic code. *Science, 206,* 575–577.]

tRNA molecule whose anticodon could pair with all four codons in the family; such a code is common in present-day mitochondrial organelles that, as a consequence of their small size and limited function, economize in the kinds of tRNA *and* proteins they produce. Such extreme wobble limited this ancient code to no more than 15 or 16 amino acids. Genetic code exceptions found in mitochondria and in the nuclei of a few organisms (**FIGURE EB1.3**) indicate that some proteins in these entities tolerated some variation in amino acid composition or chain termination without ill effect. These exceptions may not be ancient relics but of recent origin. Some codon changes may still be evolving in such organisms.

END BOX 6.1

The Genetic Code (Cont...)

© Photos.com

(a) An early code: 16 anticodons for perhaps 15 amino acids

UUU UUC UUA UUG **AAU** phe?	UCU UCC UCA UCG **AGU** ser	UAU UAC **AUG** tyr / UAA UAG [STOP]	UGU UGC UGA UGG **ACU** cys?
CUU CUC CUA CUG **GAU** leu	CCU CCC CCA CCG **GGU** pro	CAU CAC CAA CAG **GUU** his?	CGU CGC CGA CGG **GCU** arg
AUU AUC AUA AUG **UAU** ile	ACU ACC ACA ACG **UGU** thr	AAU AAC AAA AAG **UUU** asn?	AGU AGC AGA AGG **UCU** ser?
GUU GUC GUA GUG **CAU** val	GCU GCC GCA GCG **CGU** ala	GAU GAC GAA GAG **CUU** asp?	GGU GGC GGA GGG **CCU** gly

tRNA gene
duplications
and
mutations

— — — — →

Evolution of
new
anticodons

(b) A later code: 31 anticodons for perhaps 18 amino acids

UUU UUC **AAG** phe / UUA UUG **AAU** leu	UCU UCC **AGG** ser / UCA UCG **AGU** ser	UAU UAC **AUG** tyr / UAA UAG [STOP]	UGU UGC **ACG** cys / UGA UGG **ACU** trp
CUU CUC **GAG** leu / CUA CUG **GAU** leu	CCU CCC **GGG** pro / CCA CCG **GGU** pro	CAU CAC **GUG** his / CAA CAG **GUU** glu	CGU CGC **GCG** arg / CGA CGG **GCU** arg
AUU AUC **UAG** ile / AUA AUG **UAU** ile	ACU ACC **UGG** thr / ACA ACG **UGU** thr	AAU AAC **UUG** asn / AAA AAG **UUU** lys	AGU AGC **UCG** ser / AGA AGG **UCU** arg
GUU GUC **CAG** val / GUA GUG **CAU** val	GCU GCC **CGG** ala / GCA GCG **CGU** ala	GAU GAC **CUG** asp / GAA GAG **CUU** glu	GGU GGC **CCG** gly / GGA GGG **CCU** gly

(c) The modern code: 43 known anticodons for 20 amino acids

UUU UUC	**AAG** phe / phe	UCU UCC	**AGI** ser / ser	UAU UAC	**AUG** tyr / tyr	UGU UGC	**ACG** cys / cys
UUA	**AAU** leu	UCA	**AGU** ser	UAA UAG	[STOP]	UGA	[STOP]
UUG	**AAC** leu	UCG	**AGC** ser			UGG	**ACC** trp
CUU CUC	**GAG** leu / leu	CCU CCC	**GGI** pro / pro	CAU CAC	**GUG** his / his	CGU CGC	**GCI** arg / **GCG** arg
CUA	**GAU** leu	CCA	**GGU** pro	CAA	**GUU** gln	CGA	arg
CUG	**GAC** leu	CCG	pro	CAG	**GUC** gln	CGG	arg
AUU	**UAI** ile	ACU	**UGI** thr / **UGG** thr / **UGU**	AAU AAC	**UUG** asn / asn	AGU AGC	**UCG** ser / ser
AUC	**UAG** ile	ACC	thr				
AUA	**UAC*** ile	ACA	thr	AAA	**UUU** lys	AGA	**UCU** arg / arg
AUG	**UAC** met	ACG	thr	AAG	**UUC** lys	AGG	arg
GUU	**CAI** val	GCU	**CGI** ala / **CGU** ala	GAU GAC	**CUG** asp / asp	GGU GGC	**CCG** gly / gly
GUC	**CAG** val	GCC	ala				
GUA	**CAU** val	GCA	ala	GAA	**CUU** glu	GGA	**CCU** gly
GUG	**CAC** val	GCG	ala	GAG	**CUC** glu	GGG	**CCC** gly

Further
tRNA gene
duplications
and mutations

— — — — →

Evolution of
new anticodons
(and deletion
of ACU
anticodon to
produce a UGA
stop codon)

FIGURE EB1.2 Possible stages in the evolution of the modern genetic code based on the scheme proposed by Jukes in 1983. mRNA codons are at the left of each box. tRNA anticodons are in shaded capital letters to their right. Evolution of tRNA genes and anticodons are proposed to have transformed an early code for some 15 amino acids **(a)** to codes producing 18 **(b)** and finally 20 **(c)** amino acids.

[*Source:* Strickberger, M. W., 1985. *Genetics, Third Edition.* Reprinted by permission of Pearson Education, Inc., Upper Saddle River, NJ.]

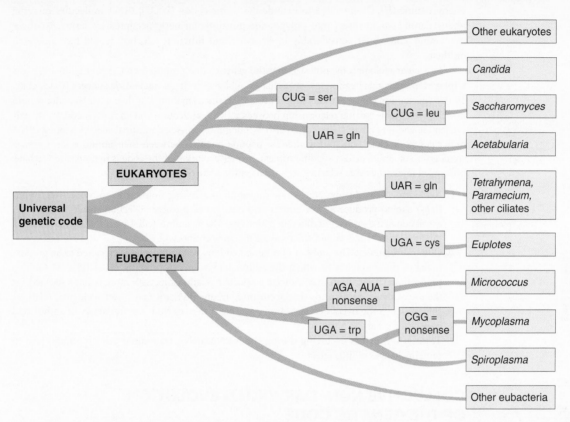

FIGURE EB1.3 Changes in the universal genetic code (Table EB1.2) found in nuclei of some basal eukaryotes and eubacteria. R designates a purine nucleotide, for example, UAR = UAA or UAG.

[Adapted from Osawa, S., 1995. *Evolution of the Genetic Code.* Oxford University Press, Oxford, UK.]

In subsequent evolution (**FIGURE EB1.2b, c**), tRNA gene duplications enabled new anticodons to evolve, some of which mutated so that new amino acids activated them. By such means the types of tRNA molecules increased and new amino acids were added to the code. The hypothesis is that when 20 different amino acids had incorporated into the code, ancestral organisms were producing a large enough number of proteins that codon changes necessary to include any further amino acids would have led to widespread protein malfunction and lethality. At that point the code "**froze**," limiting its codon assignments to the prevailing amino acids. Universality of the genetic code (with a few rare codon exceptions) indicates that only the ancestral bearers of this particular code survived.

FROZEN ACCIDENTS

All proteins in extant organisms consist of L-amino acids. All nucleic acids consist of nucleotides with D-sugars. DNA double helices coil in a right-handed rather than a left-handed direction. Phenomena known as **frozen accidents** could explain these universal features. Further evolution of the code was restricted because protein synthesis, in its mature form, precisely positioned (and still positions) each particular amino acid in every polypeptide in which it was found. Any change in the genetic code for a widely used amino acid would significantly change many different proteins carrying that amino acid (Szathmáry, 1991). As one example, if the code for phenylalanine (UUU) extended to include the serine codon (UCU), tRNA molecules carrying phenylalanine would insert into polypeptide positions formerly occupied by serine. As these two amino acids differ considerably in structure and function selection would have operated on them.

Molecular biologists hypothesize that the genetic code proposed by Crick et al. in 1961 was **triplet** even during its beginnings, or perhaps doublet with single **nucleotide spacers** (Crick et al., 1961).[b] Mechanical considerations support this view, since anything less than a triplet codon would not provide a stable pairing relationship between a tRNA anticodon and an mRNA codon. The trinucleotide width of a triplet anticodon also is thought to provide a minimal space, enabling tRNA molecules to lie close enough together for peptide bonding between their amino acids. Given a small group of amino acids coded by triplets, further evolution of the code is hypothesized to have proceeded under the three selective conditions outlined below.

1. **Nucleotide Substitution.** Nucleotide substitutions resulting from errors in replication (mutations) should produce as few amino acid changes as possible. Selection would result in an expansion of the number of different codons used by an amino acid. Under these circumstances random base changes would still produce the same amino acid.
2. **Codon Frequency.** The number of different codons per amino acid is expected to be proportional to the frequency in which the amino acid occurs in proteins. (**FIGURE EB1.4**). This proportional correlation may indicate a selective relationship, with more codons selected for the use of those amino acids that occur more frequently; or it may be an accidental relationship in which the overall composition of proteins derives from the frequency of amino acid codons; or it may be both.
3. **Translation.** Errors occurring during the mRNA–tRNA translation process should lead to protein changes of little effect.

COLLECTIVE NON-DARWINIAN EVOLUTION OF THE GENETIC CODE

In an overview of the evolution of cells, Carl Woese, one of the three authors of the 2006 study, termed the emergence of vertical evolution the DARWINIAN TRANSITION *(Woese, 2002).*

A novel theory to account for the universality and optimality of the genetic code was proposed by Vetsigian et al. (2006). The theory is *collective* because it proposes that the code was stabilized when there was extensive horizontal gene exchange between organisms, before the evolution of vertical gene transfer—that is, the transfer of genetic information from generation to generation.[c]

These researchers (a physicist, a microbiologist, and a molecular geneticist) emphasize that the code is not important in and of itself, but because it provides the link between gene replication and gene expression. An important and potentially controversial aspect of this theory is that the evolution of life would not require a single common ancestor to explain the origin of the code or of the cellular basis of life. Organisms would have sorted into communities that shared related genetic codes in what Vetsigian and colleagues hypothesize to have been a three-stage process:

- Several or many genetic codes, which, through horizontal gene transfer, led to the consolidation of a single genetic code.
- Increasing complexity of genes in a community of organisms.
- Vertical transmission and the isolation of groups of organisms.

---

[b]For a recent discussion of the elegant scientific reasoning that led to the discovery of the triplet nature of the genetic code, see C. Yanofsky (2007). Establishing the triplet nature of the genetic code. *Cell,* **128**, 815–818.

[c]As discussed elsewhere in this text, extensive horizontal gene transfer would have promoted gene sharing between evolving organisms before vertical transfer began to impose hereditary barriers between groups of organisms.

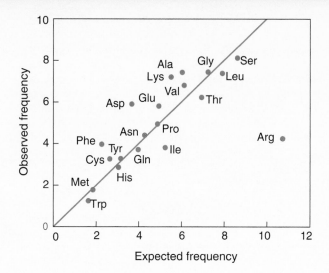

FIGURE EB1.4 Comparison between observed and expected frequencies of amino acids at 5,492 positions in 53 different vertebrate polypeptides. Excluding arginine (Arg), the correlation coefficient for these data is 0.89. (The expected frequency of each amino acid derives from calculating the nucleotide composition of the mRNA from the frequency of bases in the first two codon positions used by the various amino acids that compose the proteins: U = 0.220, A = 0.303, C = 0.217, G = 0.261. Randomized triplet codons for this nucleotide composition furnish the expected amino acids.)

[Reproduced from King, J. L, et al. *Science* 164 (1969):788–798. Reprinted with permission from AAAS.]

EVOLUTION ON THE WEB

Explore evolution on the Internet! Visit the accompanying website for *Strickberger's Evolution, Fifth Edition,* at **go.jblearning.com/Evolution5eCW** for exercises and links relating to topics covered in this chapter.

REFERENCES

Crick, F., 1966. Codon–anticodon pairing: the wobble hypothesis. *J. Mol. Biol.* **19**, 548–55.

Crick, F. H. C., Barnett, L., Brenner, S., and R. Watts-Tobin, 1961. General nature of the genetic code for proteins. *Nature,* **192**, 1227–1232.

Jukes, T. H., 1983 Evolution of the amino acid code. In *Evolution of Genes and Protein,* M. Nei and R. K. Koehn (eds.), Sinauer Associates, Sunderland, MA, pp. 191–207.

Loomis, W. F., 1988. *Four Billion Years: An Essay on the Evolution of Genes and Organisms.* Sinauer Associates, Sunderland, MA.

Lynch, M., 2007. *The Origins of Genome Architecture.* Sinauer Associates, Sunderland, MA.

Maynard Smith, J., and E. Szathmáry, 1995. *The Major Transitions in Evolution.* W.H. Freeman and Company, Oxford, UK.

Osawa, S., 1995. *Evolution of the Genetic Code.* Oxford University Press, Oxford, UK.

Pagel, M. (ed. in chief), 2002. *Encyclopedia of Evolution.* 2 Volumes, Oxford University Press, New York.

Szathmáry, E., 1991. Four letters in the genetic alphabet: A frozen evolutionary optimum? *Proc. Roy. Soc. Lond., (B),* **245**, 91–99.

Vetsigian, K., C. Woese, and N. Goldenfeld, 2006. Collective evolution and the genetic code. *Proc. Natl Acad. Sci. USA,* **103**, 10696–10701.

Woese, C., 2002. On the evolution of cells. *Proc. Natl Acad. Sci. USA,* **99**, 8742–8747.

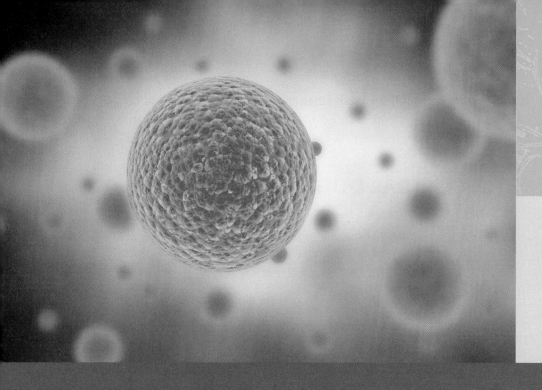

© Photos.com

Molecules, Membranes, and Protocells

CHAPTER SUMMARY

DNA as the hereditary genetic material, transcription of DNA into RNA, and translation of RNA into protein form the chemical basis of life on Earth today. The weight of current evidence, however, is that RNA evolved before DNA or protein. This scenario is known as the RNA World, during which RNA was both the hereditary nucleic acid and the molecule from which proteins were translated. Once DNA evolved and acquired the function of hereditary transmission, three separate functional classes of RNA nucleotide sequences arose: storage, messenger, and translational (both ribosomal and transfer RNA), with the functions of transcribing and translating information stored in DNA into amino acids and proteins.

Sometime before 3.5 Bya, linear structural molecules became organized into membranes, vesicles, protocells, and then cells. Although remarkable, this sequence of events is not miraculous; the basic elements of a cell—a membrane-bounded, three-dimensional arrangement of molecules—can readily be produced experimentally in the laboratory. These human-engineered cells exhibit features of living systems such as organization, selective permeability, energy use, and response to selection, all of which argue for these structures as a reasonable replication of the origin of protocells/cells on Earth. Those early protocells that could best maintain themselves perpetuated; those that could not, failed to survive—

Image © Jezper/ ShutterStock, Inc.

125

the origins of selection. With further change, protocells that could reproduce would have been favored by selection, successively enhancing the probability of new molecules, reactions, and structures arising.

THE CENTRAL DOGMA OF TRANSCRIPTION AND TRANSLATION

In organisms alive today the amino acid sequences of catalytic enzymes derive entirely from the nucleotide sequences in RNA (BOX 7.1), which in turn derive from the nucleotide sequences of DNA. The following statements are fundamental to our understanding of the relationship between genetic material and proteins:

As outlined in Box 7.1, RNA exists in several forms in multicellular organisms, different types of RNA functioning in different organelles.

- the three-dimensional structure of a protein—its form, shape, and subsequent function—is primarily determined by the linear sequence of amino acids, which in turn
- derives from the linear sequence of bases in nucleic acids by means of a protein-synthesizing apparatus involving three different kinds of RNA.

The discovery in the 1960s that DNA → RNA → protein is a one-way and nonreversible pathway became known as the **central dogma of molecular biology**, although, as discussed below, we now know that some RNA exerts functional roles without being translated into protein.

The essential functions and nature of the interactions between DNA, RNA, and proteins are threefold:

Classes of RIBOSOMAL RNA (rRNA) include 16S rRNA molecules in prokaryotes and 18S rRNA in eukaryotes, both of which are essential for ribosomal protein synthesis. S is the Svedberg unit of sedimentation of a molecule when centrifuged for 10^{-13} seconds. It is named for a Swedish chemist and Nobel Laureate Theodor Sverberg (1884–1971).

- **Replication,** the process of duplication by which DNA makes another copy of itself, a process we can represent as **DNA → DNA.**
- **Transcription,** the process by which the message coded in DNA is transferred to RNA, a process we can represent as **DNA → RNA.**
- **Translation,** the process by which the RNA message is translated into the peptides from which proteins are made, a process we can represent as **RNA → protein.**

The chain of information transfer—*the central dogma*—therefore comprises:

replication of DNA → DNA, transcription of DNA → RNA, and translation of RNA → protein.

DNA provides the **genotype,** or **genetic endowment,** of an organism, except for some **viruses** in which RNA is the genetic material (BOX 7.2) and **prions,** which have neither DNA nor RNA, but infectious agents responsible for some diseases, and as recently discovered, molecular units of inheritance (BOX 7.3). Expression of this nucleic acid information, via transcription and/or translation, provides the features of an organism,

BOX 7.1	Types of RNA

RNA exists in several forms in multicellular organisms with different types of RNA functioning in different organelles.

- **Structural** or storage **RNA**, known by the acronym **sRNA**, stores information in the nucleus.
- **Messenger RNA**—discovered in 1960 and known by the acronym **mRNA**—transfers information from DNA, specifying the sequences into which each

amino acid is to be placed in the protein synthesizing machinery, which is found on small cytoplasmic vesicles (ribosomes) composed of **ribosomal RNA (rRNA).**
- Transfer RNA (tRNA) adapts the positional information from mRNA by transferring amino acids to ribosomes to produce polypeptides.

BOX 7.2 Viruses

Viruses are intracellular infectious agents composed of DNA *or* RNA in a protein coat. In those viruses where the nucleic acid is RNA and not DNA, RNA acts as the genetic material. Interestingly, the mutation rates for RNA viruses are usually higher (often much higher) than for DNA viruses.

Although one could consider that viruses show some of the properties of living entities, all viruses depend on host cells to be able to replicate the viral genetic material. Named from the Latin *virus* for a toxin or poison, viruses are agents associated with organisms and not organisms themselves. Viruses are not independent living entities.* A typical definition for a virus is "a sub-microscopic infectious agent that can neither grow nor replicate (reproduce) outside a host cell," or "an acellular entity that can only replicate inside the cells of a host organism." The first virus—tobacco mosaic virus (TMV)—was discovered in 1899 (Figure 7.7). Since then, it has been joined by more than other 5,000 viruses isolated from all groups of organisms.

Viruses infect the cells of virtually all living organisms, producing their protein coats and replicating their genetic material within the host cells. **Bacteriophages** infect and later destroy bacteria. Production of the protein coat of the T4 bacteriophage, a DNA virus that infects the common bacterium *Escherichia coli* and that has been especially well studied is now discussed.

ASSEMBLY OF THE T4 BACTERIOPHAGE

A completed T4 bacteriophage particle is composed of 30 to 40 different components produced in a sequence of coordinated steps, each of which is under genetic control (**FIGURE B2.1**). Phage DNA uses bacterial host DNA-dependent RNA polymerase to synthesize phage mRNA, which translates on host ribosomes. This produces a number of early-stage proteins, many of which are necessary for the subsequent synthesis of phage DNA. Within 5 to 7 minutes after infection, these early enzymes catalyze the formation of a pool of phage DNA fibrils.

*For an insightful analysis that concludes that viruses can be regarded as living, see Dupré and O'Malley (2009).

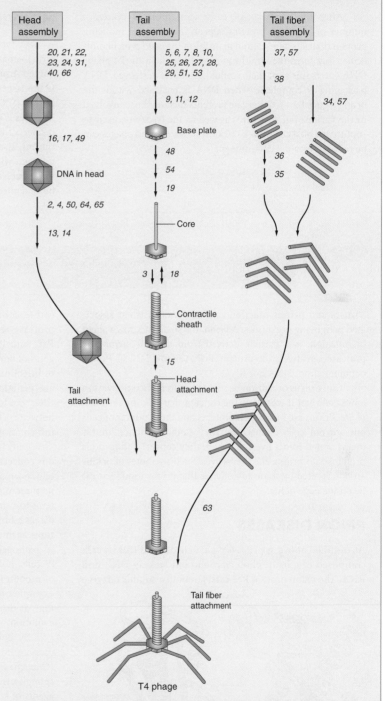

FIGURE B2.1 Sequence of T4 phage development and the genes involved in the major stages. Many of the genes (indicated by numbers) produce proteins or polypeptides directly incorporated into the T4 assembly. The products of other genes (for example, *38, 57, 63*) are not incorporated as structural elements but are necessary for the assembly.

[Modified from Wood, W. B., 1980, Bacteriophage T4 morphogenesis as a model for assembly of subcellular structure. *Q. Rev. Biol.* **55**, 353–367.]

After the synthesis of early viral proteins ceases, a number of late-stage proteins appear, including an inducing protein that acts as a scaffold to position the *head protein* molecules that form the viral head capsule or shell. As T4 phage DNA enters this viral shell, scaffold protein is destroyed. DNA packaging is completed when DNA is enclosed within the head (Figure B2.1). Other late proteins include those involved in the various tail structures as well as the lysozyme used to rupture the host cell walls so that the newly formed phages can be released into the environment.

MIMIVIRUSES

A new type of virus—a **mimivirus** (<u>mi</u>micking <u>mi</u>crobe <u>virus</u>)—that infects amoebae was discovered in 2003 and named *Acanthamoeba polyphaga* mimivirus (APMV). APMV is the largest virus known and has the largest viral genome known (900 protein-coding genes). A second small virus with 21 genes is associated with the mimivirus in the amoebae, but it infects the mimivirus, not the amoeba. Although APMV is the only such virus known (yet), a virus infecting a virus opens up a new window on the link between life and non-life. Viruses have much to tell us about the level of sophistication that can be generated with seemingly small amounts of information.

Viruses and **prions** function as living entities when associated with living organisms. A prion, the smallest known agent of infection, is a particle derived from the cell membrane protease-sensitive prion protein (PrPᶜ) (**FIGURE B3.1**). Prions contain neither DNA nor RNA.

Can a virus or a prion be said to be alive when not within a live cell? Not if we follow the concept that *cells are the units of life*. Prions are misshapen proteins and the products of cells but are not cells themselves. They cannot function unless incorporated into a live cell, where they convert endogenous proteins into prions like themselves. Such attributes of prions are interpreted by some as evidence that prions (and viruses) lie on the edge of life.

PRION DISEASES

As outlined above, a prion is a particle derived from a cell, composed of a hydrophobic protein but lacking DNA and RNA. The prion protein PrP exists in both a soluble enzyme (protease)-sensitive form (PrPᶜ) and a highly insoluble protease-resistant form, PrPˢᶜ (Figure B3.1) Conversion of PrPᶜ into PrPˢᶜ is associated with the deposition of aggregates of amyloid fibers in the brain and acquisition by PrPˢᶜ of infectious properties. Although prion protein sequences are genetically determined, the interactions that change their conformation can be transmitted and reproduced between cells and individuals without further genetic information.

Only functional when incorporated into a living cell, prions convert cellular proteins into new prions. The prion protein, normally harmless, can undergo a pathogenic change in its three-dimensional shape, which converts other such proteins to a similar form. Through such "domino effect," a **prion disease**, functioning remarkably like a non-nucleic acid infectious agent, progressively develops as prion proteins change to pathogenic form.[a] Prions therefore alter the phenotype of cells, not by changes in their DNA but as a consequence of protein folding. This phenomenon provides a remarkable example of proteins functioning as units of heredity, a role that, as discussed below, they may have played early in life's evolution.

By entering brain cells and taking over protein synthesis, prions initiate neurological degeneration in humans (*Creutzfeldt-Jakob disease*) and "mad cow" disease (bovine spongiform encephalopathy) in cattle. Prions also are the agents of kuru, or "laughing death," in the Fore tribe of New Guinea, whose custom was to eat the brains of dead relatives. Kuru, which is a form of transmissible spongiform encephalopathy, is characterized by loss of coordination, dementia

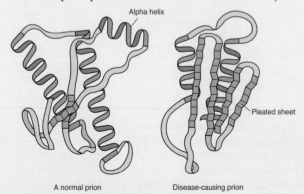

FIGURE B3.1 A normal prion (left), compared to an aberrant, disease-causing prion (right).

[a]S. B. Prusiner, 1995. The prion diseases. *Sci. Am.,* **272**, 48–57; G. Miller. Could they all be prion diseases? *Science,* **326**, 1337–1339.

BOX 7.3 | **Prions and Prion Diseases (Cont...)**

and, paradoxically, outbursts of laughter as the disease progresses; hence the name laughing death.

PRIONS AS MOLECULAR UNITS OF INHERITANCE

The patterns by which prion protein fold and self-propagating, at least in the yeast *Saccharomyces cervisiae*, in which nine proteins are known to form prions. When the best known prior protein (Sup35) switches to the prion state, translation is modified, resulting in altered cell-to-cell adhesion, uptake of nutrients, and resistance of the yeast to toxins and antibiotics in patterns that are yeast-strain specific. Modification in proteins that form prions in environmentally controlled—the greater the environmental stress, the more prions that form—providing a class of those epigenetic mechanisms of inheritance discussed elsewhere in the text.[b]

[b]R. Halfmann and S. Lindquist, 2010. Epigenetics in the extreme: Prions and the inheritance of environmentally acquired traits. *Science*, **330**, 629–632.

its **phenotype**. This information transfer and processing depends on the universal existence of enzymes that mediate each step (**FIGURE 7.1**).

In brief, and as diagrammed in Figure 7.1, the genetic material, through the process of transcription, produces a molecule of **messenger RNA (mRNA)**, that is, base for base, a complement to one of the DNA strands. Through the mediation of cellular organelles known as ribosomes, which consist **of ribosomal RNA (rRNA)** and protein, a sequence of bases in mRNA translates into a sequence of amino acids. This translation follows the **triplet rule**; a sequence of three mRNA bases (a codon) designates one of the 20 different kinds of amino acids used in protein synthesis. No physical material is inserted by the

The bacterium ESCHERICHIA COLI *is almost universally referred to in the literature as* E. COLI.

PRIONS *containing the same sequence of the prion protein (PrP^{Sc}) can give rise to different disease states, an aspect of prion function that is not understood.*

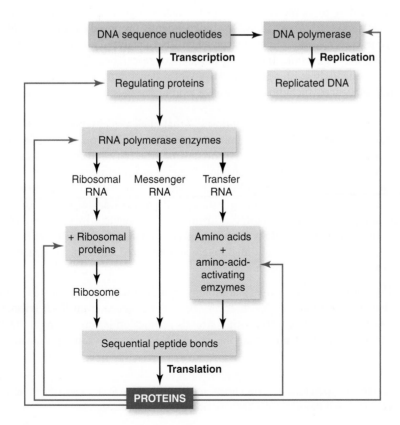

FIGURE 7.1 Schematic diagram showing the mutual dependence of information carried by nucleotide sequences and function governed by proteins. *Solid lines* indicate the general directions of information transfer, and *dashed lines* point to proteins that this process synthesizes. Clearly, the nucleotide sequence information determines the amino acid sequences of proteins, and proteins in turn regulate and catalyze the transfer of nucleotide information. One process could not have developed without the other.

mRNA into the protein during the *translation* of mRNA into protein; *only information is transferred.*

Organismal function depends on transforming external sources of material and energy into processes that take place within organisms. To this end, cells depend totally on protein catalysts (enzymes) for the production of proteins and nucleic acid templates. The development of chemical energy providers such as adenosine triphosphate (ATP) was critical to the evolution of life at the cellular level. These molecules release small but significant amounts of phosphate-bond energy that the enzyme apparatus of the cell controls and localizes to specific reactions.

With the aid of ribosomes and various enzymes, amino acids connect in sequence through peptide linkages. A polypeptide chain forms in which the precise position of each component amino acid is designated by the genetic code for those amino acids. Because 20 different amino acid alternatives exist for each position in a polypeptide, a sequence of five amino acids has 20^5 (more than three million) possible arrangements. By contrast, a sequence of five nucleotides in a nucleic acid has only 4^5 or 1,024 possibilities. Moreover, many amino acids are quite different in structure, enabling them to interact with each other, with molecules such as water, with metal ions, and with various monomers and polymers. These differences confer an astronomical variety of possible three-dimensional configurations on a protein, in contrast to the more rigid shapes assumed by many nucleic acids.

The universal interdependence of the three processes of replication, transcription, and translation, and the disparity in numbers of molecules that can form using nucleic or amino acids, has led molecular biologists to ask whether these functional and informational systems could have evolved independently of one other. The weight of current evidence is that RNA evolved before DNA or protein in what is now known as the **RNA World** (Yarus, 2010; Darnell, 2011).

RNA FIRST

In the early 1980s, three molecular biologists—Leslie Orgel (1927–2007), Francis Crick (1916–2004), and Carl Woese (b.1928)—independently proposed the *RNA World* as the first stage in the evolution of life, a stage when RNA catalyzed all molecules necessary for survival and replication. Increasing numbers of researchers have provided further evidence for the early existence of an RNA World dependent on self-replicating RNA nucleotide sequences, including synthesis of ribonucleotides under simulated prebiotic conditions.[1]

Supporters of the "RNA first hypothesis"—that nucleic acid replication arose first—argue that only a self-replicating system can provide the basis on which selection can build a cooperative functional unit (an organism). The self-replicating power of RNA would have enabled responses to selection to develop protein systems that supported further self-replication. Protein synthesis then would have evolved either through specific amino acids directly interacting with specific nucleotide sequences, or perhaps through indirect placement of amino acids into such sequences by transfer molecules. The hypothesis that an early RNA World antedated cellular enzymatic proteins is based on several types of evidence, outlined below and analyzed in depth by Hazen (2005), Yarus (2010), and Darnell (2011).

- The central dogma of molecular biology includes the fundamental statement that enzymes are proteins. The discovery of enzymes made from RNA (**ribozymes**, which cut and rejoin preexisting strands of nucleotides and so replicate RNA) has been used to demonstrate that short RNA sequences can replicate and form

---

[1] See Orgel (2004), Hazen (2005), Yarus (2010), and Darnell (2011) for reviews, Chen et al. (2005) for RNA catalysis, and Powner et al. (2009) for ribonucleotide synthesis.

templates for complementary RNA sequences. The discovery that ribozymes possess catalytic properties *in the absence of proteins* supports the feasibility of autocatalytic nucleic acid replication in an RNA World.

- Some cellular RNA molecules catalyze reactions that include binding to ATP, which is the most widespread energy transfer molecule in organismal metabolic pathways. A catalyst that functioned to join oligonucleotides together was isolated from random oligonucleotides and shown to obtain energy for the interactions from triphosphates. Therefore, an RNA molecule (a ribozyme) can function as protein catalysts do today. The existence of such RNAs increases the likelihood that RNA could have replicated itself in the RNA World. These experiments are important in providing evidence that enzymatic pathways required for RNA replication could have been present in a pre-DNA and/or pre-protein world.[2]

- Selection experiments in the laboratory led to the evolution of RNAs into new kinds of molecules with catalytic activities many orders of magnitude greater than in the initial mixture. Some selected ribozymes can combine a ribose sugar with a thiouracil base to make nucleotides *at a rate more than ten million times greater* than the uncatalyzed reaction.

- The protein translation system used by all cells depends on messenger, ribosomal, and transfer RNAs. However, peptide formation catalyzed by ribosomal enzymatic action, called "peptidyl transferase," depends more on ribosomal RNA than on ribosomal proteins. And, as noted above, a selected noncellular ribozyme can perform this same amino acid-binding function,[3] although finding ribozymes that can synthesize long chains of RNAs has proven difficult (Yarus, 2010). Nevertheless, the RNA in ribosomes, rather than ribosomal proteins, may function to catalyze peptide bonds.

- Synthetic RNA molecules can be designed to perform precise catalytic reactions and, when replicated with ribonuclease, to become resistant to this enzyme—which normally degrades RNA—thereby ensuring their survival.[4]

- Some RNA molecules function as gene regulators by binding to nucleic acids and affecting gene expression. Some viruses use RNA as the genetic material (Box 7.2).

- RNA "fragments" are found in coenzymes in various metabolic reactions, for example, in coenzyme A, nicotinamide adenine dinucleotide (NAD), and flavin adenine dinucleotide (FAD). Because other chemicals could have assumed the function of these RNA fragments, researchers have interpreted their presence as "historical," a remnant of an earlier RNA World.

Ribose is the sugar that forms the backbone of RNA. Because ribose is extremely unstable, a consistent pre-biological supply of ribose needed to form the RNA backbone has been questioned. To address such difficulties, G. F. Joyce and colleagues proposed that early nucleic acid genetic material was not based on ribose-containing nucleotides but rather on *ribose-like analogues*, in which pairing difficulties were minimized, and which were more easily synthesized than ribose. Such a system has been demonstrated experimentally.[5]

---

[2] M. Sassanfar, and J. W. Szostak, 1993. An RNA motif that binds ATP. *Nature*, **364**, 550–553. For the latest synthesis of a ribozyme that catalyzes transcription from an RNA template, see Wochner et al., 2011.

[3] B. Zhang, and T. R. Cech, 1997. Peptide bond formation by in vitro selected ribozymes. *Nature*, **390**, 96–100. The equivalent to an "intron" portion of transcribed RNA in the protozoan ciliate *Tetrahymena* splices itself out of the RNA molecule and helps form a chemical bond between the protein coding RNA sections ("exons") on either side.

[4] For an overview, see M. M. Hanczyc et al., 2003. Experimental models of primitive cellular compartments: encapsulation, growth, and division. *Science*, **302**, 618–622.

[5] G. F. Joyce et al., 1987. The case for an ancestral genetic system involving simple analogues of the nucleotides. *Proc. Natl Acad. Sci. USA*, **84**, 4398–4402; R. Larralde et al., 1995. Rates of decomposition of ribose and other sugars: implications for chemical evolution. *Proc. Natl Acad. Sci. USA*, **92**, 8158–8160.

RNA DIVERSIFIES

Three separate functional classes of RNA nucleotide sequences eventually arose: **storage, messenger,** and **translational** (*ribosomal* and *transfer*).

RNA still fills messenger and transfer functions in all organisms and fills a genetic storage functions in some viruses (Box 7.2). The evolution of ribosomes not committed to the production of single proteins would have enabled the *same ribosome to translate different mRNAs.* Thus, the transcriptional process became responsible for regulating which proteins to make from ribosomes.

Particular proteins could be selected by regulating which mRNA molecules to transcribe from the stored genetic material, a process that eventually gave rise to sophisticated regulatory systems. Selection would have favored dependence on only few enzymes and proteins, each with multiple functions and lower coupling specificities, over employing greater numbers with restricted functions and higher binding specificity. Such **compartmentalization** is a hallmark of evolution at all levels.

RNA lost its role of independent self-replication when DNA evolved as the replicating and hereditary material. This simple statement fails to capture how little we understand the transition from an RNA-dominated world to a world in which RNA and DNA cooperated to produce the first organic structures. We can hypothesize that those first structures would have been linear arrays of molecules forming simple monolayers. Sometime before 3.5 Bya, linear structural molecules became organized into membranes, then vesicles, protocells, and finally cells.

MEMBRANES: THE FIRST STRUCTURES

The presence of appropriate peptides and proteins was only the beginning of the origin of life. Processes of metabolism and function arose as organismal organization arose. How did such organization come about? At its earliest, interactions among molecules must have led them to assume relative positions based on forces such as hydrogen bonding, ionization, solubility, adhesion, and surface tension.

Phospholipids are the basic layered molecules of cell membranes. They are organic molecules with a phosphorus-containing polar group at one end and non-polar fatty acid groups at the other (FIGURE 7.2a). In water, which is a polar solvent, the polar ends of these molecules are oriented toward water (they are *hydrophilic*), while their non-polar ends are oriented toward each other; that is, away from water (they are *hydrophobic*). As a result, phospholipid membranous structures can form quickly, yielding vesicles composed of bimolecular layers in which the non-polar surfaces of each of the two layers "dissolve" in each other (Figure 7.2b). Such vesicles would have trapped other molecules.[6] *Membranous droplets* or vesicles composed of lipids, polypeptides, or other molecules, produced by the mechanical agitation of molecular films on liquid surfaces such as tide pools (FIGURE 7.3) would have accumulated.[7] Such membranous droplets are **protocells**.

PROTOCELLS

Attainment of the **droplet level of organization** was an important step in the origin of life, not least because, depending on its structure and permeability, the membrane surrounding the droplet can exhibit **selective permeability**; it can permit some compounds to enter from the environment and others to exit from the droplet. These droplets function as **protocells**: a **self-assembled membrane-bounded system containing molecules**.

---

[6] P.-A. Monnard et al., 2002. Influence of ionic solutes on self-assembly and polymerization processes related to early forms of life: implications for a prebiotic aqueous medium. *Astrobiology* **2**, 213–219.

[7] M. M. Hanczyc et al., 2003. Experimental models of primitive cellular compartments: encapsulation, growth, and division. *Science*, **302**, 618–622.

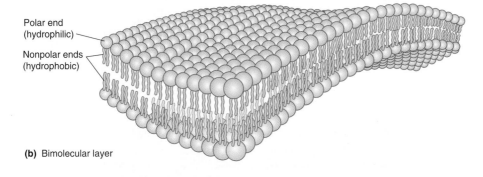

(a) Lecithin

Polar end
(hydrophilic)

Nonpolar ends
(hydrophobic)

(b) Bimolecular layer

FIGURE 7.2 **(a)** A phospholipid molecule, lecithin. **(b)** A diagrammatic view of a bimolecular sheet like double layer of phospholipid molecules that have self-assembled with their hydrophilic phosphate heads (colored circles) facing the water solvent, and their hydrophobic hydrocarbon tails facing each other. Such polar–non-polar molecules characterize cell membranes.

This organization and the small size of protocells allow the establishment of concentration gradients, facilitating reactions that could not have taken place at lower concentrations.

Protocell membranes can incorporate peptides that span phospholipid bilayers, forming channels for ionic and molecular transfers. Because of their semipermeable membranes, protocells are not isolated from their environment. They can acquire external energy and matter to retain, and even enhance, their organizational and informational structures, as do living things. For early protocells to have shown the most essential "living" attributes, they had to maintain their individuality, to grow and to divide. Some authors ascribe such properties to bimolecular vesicles.[8]

Two types of fairly simple structures produced in the laboratory in the last century—Oparin's coacervates and Fox's microspheres—possess some of the basic prerequisites of protocells. Although formed artificially, these structures point to the likelihood that nonbiological membrane enclosures (protocells) could have sustained reactive systems of the type required for cellular-based life to arise (see below). More recently, membrane-bound, cell-like vesicles were produced in the laboratory from phospholipids and from

[8] See H. J. Morowitz, 2002. *The Emergence of Everything: How the World Became Complex.* Oxford University Press, New York.

FIGURE 7.3 Effect of wave action on a surface film containing **(a)** molecules oriented with one end pointed away from water (hydrophobic) and the other toward the water (hydrophilic). Wave action **(b)** and **(c)** results in the formation of droplets and the bilayered vesicles shown in **(d)**. Double-membrane structures may have formed through incorporation of one bilayered vesicle within another.

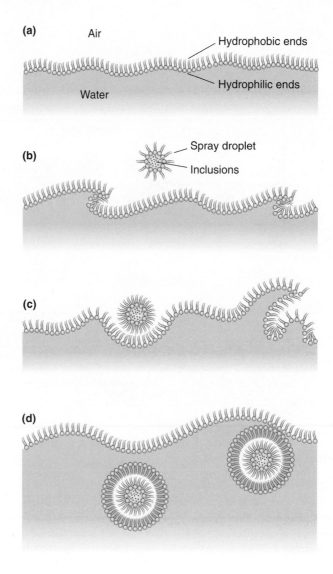

J. W. Szostak shared the 2009 Nobel Prize in Physiology or Medicine for the discovery of how chromosomes are protected by telomeres and the enzyme telomerase. For Szostak "the origin of life and the origin of Darwinian evolution are essentially the same thing" (cited in Science, *2009, **323**, 199).*

mixtures of fatty acids, fatty alcohols, and esters of glycerol. These vesicles extend tubular processes into the surrounding medium, and absorb water across the membrane. Synthetic DNA inserted into these protocells underwent elongation, with nucleotides added one by one (Mansy et al., 2008).

Working from the hypothesis that the first polymers were RNA-based, Jack Szostak and his colleagues are working to create protocells from fatty acid membranes, and RNA-like molecules that can both store information and replicate (Mansy and Szostak, 2009). Studies from this group involving the production of model protocells are consistent with hypotheses that first life was membrane-based (cellular). Indeed, competition between protocells has been demonstrated: those protocells with the most efficient RNA replication expand and assimilate "weaker" protocells with less robust RNA synthesis. Competition and selection would have been essential elements of the prebiotic world.

PRODUCTION OF COACERVATES IN THE LABORATORY

A **coacervate** is a spherical aggregation of colloidal particles in liquid suspension. Coacervates form in response to special conditions of acidity, temperature, and so on,

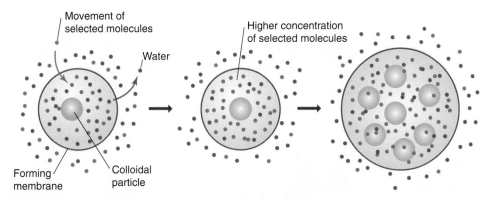

FIGURE 7.4 Formation of coacervates by the exclusion of water molecules (blue dots) from associated colloidal particles (tan circles). The intervening water molecules can be removed through dehydration (for example, increased salt concentration), or when colloidal particles are attracted to each other because they have opposite charges or because some colloids are basic (for example, histones) and others are acidic (for example, nucleic acids).

[Adapted from Kenyon, D. H. and G. Steinman. *Biochemical Predestination.* McGraw-Hill, 1969.]

dispersed colloidal particles separating spontaneously out of solution into droplets (**FIGURE 7.4**). If there is more than one type of macromolecule in the colloid, complex coacervates form and show a number of interesting properties:

- a simple but persistent organization;
- remaining in solution for extended periods; and
- increasing in size.

Aleksandr Oparin, the first to draw serious attention to these droplets, developed artificial coacervate systems that could incorporate enzymes able to perform functions, such as the synthesis and hydrolysis of starch (**FIGURE 7.5**) and the synthesis of poly-nucleotides (Oparin, 1924).[9]

A **COLLOID** *is a mixture in which tiny particles are evenly dispersed throughout a liquid medium.*

PRODUCTION OF MICROSPHERES AND VIRUSES IN THE LABORATORY

Sidney Fox showed that 10^8 to 10^9 **microspheres**/gm of polymer formed when thermally produced polymers were boiled in water and allowed to cool. The microspheres—essentially protocells—are uniform in size, stable, bounded by double membranes (as cells are) and can undergo **fission** and **budding** (**FIGURE 7.6**).[10]

These microspheres possess several qualities that indicate active internal processes, such as **selective absorption**, **growth** in size and mass, as well as **osmosis**, **movement**, and **rotation** (Figure 7.6). Moreover, microspheres show the potential for transferring and/or exchanging information; polymer particles pass through junctions between them. Although it is now clear that the quantities of amino acids required in these experiments were much higher than likely to have been present on Earth under prebiotic conditions, Fox's studies (like Oparin's) are important in demonstrating that the origin of life can be investigated experimentally.

[9] This is the original Russian edition of Oparin's theory; a revised edition was published in English 1938, and reprinted 1953 by Dover Publications.

[10] See S. W. Fox, 1984. Proteinoid experiments and evolutionary theory. In *Beyond Neo-Darwinism*, M.-W. Ho and P. T. Saunders (eds.). Academic Press, London, pp. 15–60.

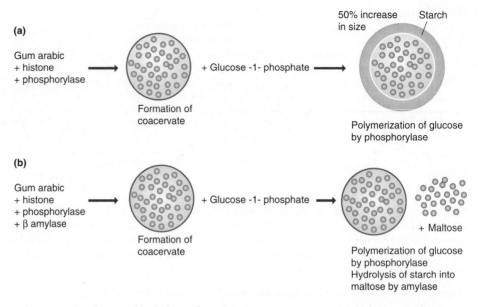

FIGURE 7.5 Synthesis and hydrolysis of starch in coacervate systems in which enzymes have been included in the droplets. In **(a)** the enzyme acts to polymerize phosphorylated glucose into starch, while in **(b)** the starch formed this way is hydrolyzed into maltose by the enzyme amylase.

The spontaneous **self-assembly** of macromolecules into vesicles and protocells indicates that the occurrence of similar entities early in Earth's history would not have been an unusual event. Such entities are not cells, of course, and much time may have elapsed before more elegant structures with more complex metabolic capabilities developed. However, their self-assembly and potential to undergo selection justifies the term and concept protocell.

Substantial evidence shows that the component materials of even more complex structures can self-assemble without the requirement of a prior pattern. One important line of evidence is the formation from DNA in the laboratory of an "artificial virus"

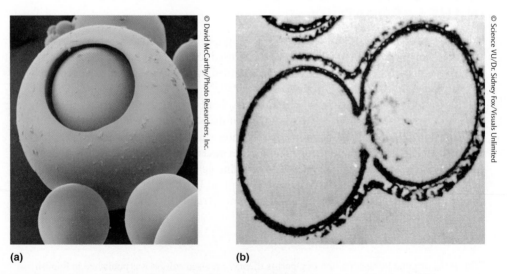

FIGURE 7.6 Microspheres produced in the laboratory. **(a)** Polystyrene microspheres of different sizes. **(b)** Protenoid microspheres form a double membrane and undergo budding, as visualized with transmission electron microscopy.

TABLE 7.1 — What Is a Gene?

It may surprise you to find that there are several definitions for the base unit of inheritance, the *gene*. In large part, this is because different specialists study the gene in different ways and at different levels. Reflecting these different approaches, a gene can be defined as one or more of the following (an example of each is provided):

- a region of DNA, the activation of which leads to the formation of a feature or character (the gene for blue eyes);
- a region of DNA, the activation of which leads to the formation of a protein or RNA (the gene for the protein collagen);
- a region of DNA encompassing coding and non-coding segments (exons and introns);
- a unit of inheritance located on a chromosome (the gene for muscular dystrophy in humans mapped to chromosome 9);
- a set of nucleotides reliably copied and transferred from generation to generation (the unit of heredity).

The first definition is based on the concept that a gene "makes" a feature: "the gene for blue eyes." However, genes do not make features. This definition arises from the conception that a blueprint for the organism—a genetic program—is written in the sequences of nucleotide bases.* A gene as "a unit of inheritance" can give the same impression.

*The metaphor of a genetic or developmental program was raised independently in 1961 by the evolutionary biologist Ernst Mayr and the molecular biologists Jacques Monod and François Jacob; see elsewhere in this text for the contributions of Monod and Jacob to our understanding of gene regulation.

(Box 7.2). Using readily available DNA, a "virus" containing 5,386 base pairs and capable of infecting and killing bacterial cells was made in the laboratory.[11] Of course, it was produced using the DNA-protein system that characterizes modern life. Even this simple form must be much more complex that the first life form on which selection would have acted; the simplest form of cellular life that has been produced artificially had around 300 genes. (The nontrivial issue of defining "What is a Gene" is discussed in **TABLE 7.1**.)

A second line of evidence is the self-assembly of viruses such as the **tobacco mosaic virus**. The protein and RNA components of tobacco mosaic virus spontaneously aggregate into the precise configuration needed to produce an active virus (**FIGURE 7.7**). Cellular organelles such as ribosomes also can self-assemble.

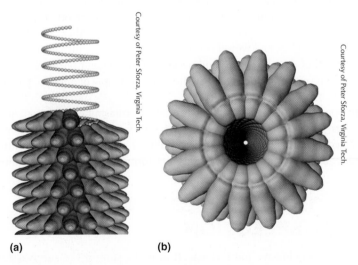

Courtesy of Peter Sforza, Virginia Tech.

Courtesy of Peter Sforza, Virginia Tech.

(a)　　　　(b)

FIGURE 7.7 A model of the tobacco mosaic virus (TMV) as seen from the side **(a)** and the top **(b)** to show the relationship between the RNA (yellow) and the protein coat (blue).

[11] To equate 5,386 base pairs with 300 genes, the Phage φ-X174, with a genome of 5,386 base pairs, has 11 genes.

WHEN DID LIFE ARISE?

The answer to this question depends on how life is defined.

Life can be considered to have arisen with the first molecules around 4.5 Bya; with the first cells 3.5 to 3.8 Bya; or with the first organisms. Because molecules produced by organisms (biotic organic molecules) can be formed abiotically in the absence of life, and because such molecules arose before the origin of organismal life, we recognize a continuum between the abiotic production of molecules and first life. Life requires cellular organization and so can be considered to have arisen with the first protocells. Selection arose equally early.

EVOLUTION OF SELECTION

Selection bridged the gap between chemical evolution (changes in the composition of non-reproductive or poorly reproductive molecules, protocells, and so on) and biological evolution (changes in inherited differences among reproductive organisms). If protocells initially formed only through the operation of the then prevailing environmental chemistry, selection would not have been very efficient. Once protocell numbers increased by self-replication, selection would have become more efficient, with advantageous traits transmitted to succeeding generations during replication.

Selection is the only natural mechanism that can account for changes among non-reproductive individuals that could have resulted in the origination of organisms. Although the events may be complex, the device is simple: organisms that react to their environment with features that enhance their survival in that environment replace those that lack such features. Researchers have found that a specific enzyme that breaks down RNA (a ribozyme) can be selected from among trillions of random RNA molecules with only few selective steps. This evidence demonstrates that selection operates on molecules in the absence of any higher levels of organization.[12]

Because of its historical continuity, selection enables a succession of adaptations that permit new features to accumulate. Selection has allowed biological organizations that would otherwise have been highly improbable. Once the game of life began, the evolutionary replacement of players bearing information—whether those players were coacervates, microspheres, protocells, genes, organisms, species, or other entities—became inextricably bound to the players' ability to play the game further. Selection had previously molded this ability, and would continue to measure it as life continued. Some random replacement of the players certainly occurred by accident rather than by selection, but because the resources of life are always limited in one way or another, participation in life is selective by its very nature.

Given reproductive expansion, limited resources, transmitted variation, and environmental change—the organismal condition—selection characterizes living systems. This set of attributes led to a coupling between two very different life processes—**function** (*proteins*) and **reproduction** (*nucleic acids*)—providing living forms with their rapid evolutionary rates.

ORIGIN OF POPULATIONS AND NATURAL SELECTION

When applied to organisms, **natural selection** is the sum of the survival and fertility mechanisms that affect reproductive success. Selection characterizes living systems because of (i) population expansion through reproduction, (ii) resource limitation, (iii) variation among individuals in a population, and (iv) environmental change over time.

---

[12] B. Zhang and T. R. Cech, 1997. Peptide bond formation by in vitro selected ribozymes. *Nature*, **390**, 96–100; and see I. A. Chen et al., 2004. The emergence of competition between model protocells. *Science*, **305**, 1474–1476.

The earliest structures, whether protocells or other localized organizations, had one important evolutionary feature: they were the first distinctive, multi-molecular **individuals** that could interact as units with their environment. Together with their various neighbors and progenies, such individuals would form a group or population on which selection could act.

Selection would have arisen when the following conditions were fulfilled:

- A population of individuals existed.
- The properties of these individuals were governed by reactions in which they absorbed and transformed environmental material into their own material.
- Individuals differed in the efficiency with which these processes took place.
- Availability of materials and energy was limited so that not all types of individuals could form or survive.

Protocells incorporating such organization and metabolic properties allowing them to grow and divide—and hence to perpetuate themselves—would have increased most, either in relative frequency or in area occupied. Early selection would have been confined to the survival of non-reproductive individuals that could wrest the most material from their environment and transform it for their own benefit with the least expenditure of energy. Although differences among such individuals could not be transmitted precisely—they lacked a hereditary system—the fact that some individuals survived and others did not would have affected the composition and further interactions of succeeding populations. The important consequence is that selection evolved from chemical non-reproductive selection to biological natural selection. The mechanism that enabled the formation of protocells is therefore a crucial issue in understanding both the origin and the nature of selection.

■ KEY TERMS

bacteriophage	replication
central dogma (molecular biology)	reproduction
coacervate	ribosomal RNA (rRNA)
genotype	ribozymes
individuals	RNA World
messenger RNA (mRNA)	selection
microspheres	self-assembly
mimivirus	structural (storage) RNA (sRNA)
natural selection	tobacco mosaic virus
phenotype	transfer RNA (tRNA)
phospholipid	translation
population	translational
prion	triplet code
prion disease	vesicles
protocells	virus

■ EVOLUTION ON THE WEB

Explore evolution on the Internet! Visit the accompanying website for *Strickberger's Evolution, Fifth Edition*, at **go.jblearning.com/Evolution5eCW** for exercises and links relating to topics covered in this chapter.

REFERENCES

Chen, I. A., K. Salehi-Ashtiani, and J. W. Szostak, 2005. RNA catalysis in model protocell vesicles. *J. Am. Chem. Soc.,* **127,** 13213–13219.

Darnell, J. 2011. *RNA. Life's Indispensable Molecule.* Cold Spring Harbor Laboratory Press, Cold Spring Harbor, NY.

Dupré, J., and M. A. O'Malley, 2009. Varieties of living things: life at the intersection of lineages and metabolism. *Philos. Theor. Biol., 1:e003,* 1–25.

Hazen, R. M. 2005. *Genesis. The Scientific Quest for Life's Origins.* Joseph Henry Press, Washington, DC.

Loomis, W. F., 1988. *Four Billion Years: An Essay on the Evolution of Genes and Organisms.* Sinauer Associates, Sunderland, MA.

Mansy, S. S., and J. W. Szostak, 2009. Reconstructing the emergence of cellular life through the synthesis of model protocells. *Cold Spring Harb. Symp. Quant. Biol.,* **74,** 47–54.

Mansy, S. S., J. P. Schrum, M. Krishnamurthy, et al., 2008. Template-directed synthesis of a genetic polymer in a model protocell. *Nature,* **454,** 122–125

Maynard Smith, J., and E. Szathmáry, 1995. *The Major Transitions in Evolution.* WH Freeman and Company, Oxford, UK.

Oparin, A. I., 1924. *Proiskhozhdenie Zhizny* (The Origin of Life). Moscovsky Robotschii, Moscow.

Orgel, L. E. 2004. Prebiotic chemistry and the origin of the RNA world. *Crit. Rev. Biochem. Mol. Biol.,* **39,** 99–123.

Osawa, S., 1995. *Evolution of the Genetic Code.* Oxford University Press, Oxford, UK.

Pagel, M. (ed. in chief), 2002. *Encyclopedia of Evolution.* 2 Volumes, Oxford University Press, New York.

Powner, M. W., B. Gerland, and J. D. Sutherland 2009. Synthesis of activated pyrimidine ribonucleotides in prebiotically plausible conditions. *Nature,* **459,** 239–242.

Szathmáry, E., 1991. Four letters in the genetic alphabet: a frozen evolutionary optimum? *Proc. Roy. Soc. Lond. (B),* **245,** 91–99.

Wochner, A., J. Attwater, A. Coulson, and P. Holliger, 2011. Ribozyme-catalyzed transcription of an active ribozyme. *Science,* 332, 209–212.

Yarus, M. 2010. *Life from an RNA World: The Ancestor Within.* Harvard University Press, Cambridge, MA.

CHAPTER

8

The First Cells and Organisms Arose 3.5 Bya

CHAPTER SUMMARY

Earth's early atmosphere supported single-celled organisms known as methanogens (methane generators) that only survive in the absence of oxygen. Methanogens produce methane as a byproduct of their metabolism and in so doing, modified Earth's early atmosphere. Because of their sequential nature—in many cases, one pathway is integral to another—such metabolic pathways could not have arisen randomly. Many pathways—and the enzymes controlling individual reactions within them—are found in extant organisms, having persisted because of their selective advantages. A fundamental revolution occurred when some organisms began to reduce carbon and release O_2 as a byproduct, using light as an energy source: this is the process we now know as photosynthesis. Evolution of new metabolic pathways led to the evolution of aerobic metabolism and aerobic cells/organisms, the first of which—cyanobacteria—were producing stromatolite reefs 3 Bya.

Methanogens and cyanobacteria are prokaryotic cells. They are small (about 1 μm), contain no nuclear membrane, cytoskeleton or complex organelles, and divide by binary fission. Extant organisms based on a prokaryotic cellular organization are classified into two domains: Eubacteria and Archaebacteria. Eukaryotic cells, which are classified into five superkingdoms, arose from one or more lineages of prokaryotic cells. They contain organelles such as mitochondria, possess a nuclear membrane, have organelle DNA, and arose from one or more lineages of prokaryotic cells to produce multicellular organisms.

EARTH'S ATMOSPHERE(S) AND THE FIRST CELLS/ORGANISMS

Two different atmospheres have enveloped Earth during its history. From radiometric dating and analyses of samples of cosmic material we know that the **first or primary atmosphere**, composed of hydrogen and helium, persisted for less than half a billion years after the origin of Earth 4.6 Bya. Possible reasons for loss of the primary atmosphere are the low gravitational force of Earth itself, extremely strong winds from the Sun, and/or heat generated by Earth and by the Sun. Organisms had nothing to do with the origin or loss of this primary atmosphere, which existed before life arose.

Four billion years ago, volcanoes and hot springs were spewing gases into the atmosphere. Comets colliding with Earth also may have contributed gases, especially water vapor. These combined events resulted in the formation of a **secondary atmosphere**, composed largely of water vapor and carbon dioxide, with smaller amounts of hydrogen, carbon monoxide, nitrogen, ammonia, methane, hydrochloric acid, and hydrogen sulfide. This atmosphere, which arose between 4.2 and 3.8 Bya (**TABLE 8.1**), contained no or very little oxygen. Like the first, the second atmosphere arose before the origin of organisms. However, organisms did contribute to changes in the second atmosphere, especially to a steady increase in the concentration of O_2, thereby establishing conditions suitable for the evolution of new types of cells and organisms (Berner et al., 2007; Heinz and van Holde, 2011).

METHANOGENS: MOLECULAR FOSSILS

By 3.5 Bya, methane (CH_4) levels in the atmosphere began to increase in response to release of methane by the first organisms, which may have originated as early as 3.8 Bya. These were single-celled **methanogens** (methane generators), organisms that only survive in the absence of oxygen, producing methane as a byproduct of their metabolism.

Methanogens are not known from fossilized cells but from fossilized molecules: **molecular fossils**. Researchers detected the presence of methane of biological origin in

TABLE 8.1	Major Events that Occurred from 4.3 to 1.5 Bya
Bya	**Event**
4.3	Volcanoes, comet impacts, water forms
4.2–3.8	Second atmosphere with increasing amounts of O_2
3.8	$^{13}C/^{12}C$ ratios indicative of biological activity
3.76	Banded iron formation (may be biologically produced)
3.5	Methanogens as molecular fossils; levels of methane in atmosphere increase
3.5–3.0	H_2O increases and CO_2 decreases
3.0–3.5	Cyanobacteria as stromatolite fossil reefs
2.5	Oldest terrestrial bacterial fossils
2.0–2.3	O_2 progressively increases in the second atmosphere as a consequence of organismal evolution
1.5	First eukaryotic organisms as single-celled fossils

rocks dated at 3.5 Mya. Whether all the methane in these rocks is the result of methanogen activity (that is, is biotic), or whether some was deposited by geological-chemical (abiotic) processes is notoriously difficult to determine. However, *methane of microbial origin* within minute fluid inclusions was detected through analysis of silica dykes in cherts of the Dresser Formation in Western Australia's Pilbara craton, which are more than 3.5 By old (Ueno and colleagues, 2001, 2006). Given the careful distinction between abiotic and biotic sources of the methane, and the coexistence of one of the oldest microfossils—filamentous single cells—within the same geological unit, this study provides strong evidence for the origin of methanogens more than 3.5 Bya.[1]

Ueno and colleagues further argued that methane produced by methanogens would have played an important role in regulating the climate 3.5 Bya.[2] This is important because methane is now recognized as a **greenhouse gas**. In the early oxygen-free atmosphere, accumulating methane formed an insulating layer that raised the surface temperature, preventing ice from building up. Between 3.5 and 3.0 Bya, Earth's surface temperature cooled, liquid water formed, and the level of CO_2 in the atmosphere fell (Table 8.1). The presence of sand composed of sulfides of iron, lead, and zinc in 3-By-old river deposits is consistent with an absence of atmospheric oxygen 3 Bya; sulfides do not form in the presence of oxygen.

Along with RUMINANTS *such as cows, we produce methane as the result of the activity of methanogens in our intestines. A cow releases 250–500 liters of methane day, a human 500–1500 ml.*

OXYGEN AND PHOTOSYNTHESIS

Before organisms requiring oxygen for metabolism arose, any O_2 released would have reacted with other elements in the atmosphere and with minerals on the surfaces of rocks. These interactions initiated the first weathering of rocks whose inexorable breakdown produced the soils of today.

The current level of oxygen in the atmosphere is 21% (**FIGURE 8.1**). Oxygen concentration remained at 1% of the present level until about 2 Bya. Geochemists have

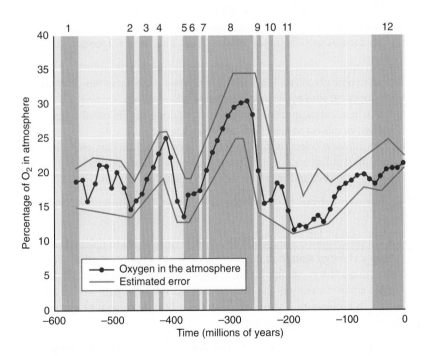

FIGURE 8.1 Changes in the percentage of oxygen in the atmosphere over the past 600 million years along with upper and lower limits of the estimated error in predicting oxygen levels. Numbers 1 to 12 correspond to significant evolutionary events, for example: origin of animal body plans (1) and increase in body size of mammals (12) correlate with high concentrations of oxygen; major extinction events (5, 9, 11) correlate with low levels of oxygen.

[Reproduced from Berner, R. A. et al., *Science* **316** [2007]: 557 [http://www.sciencemag.org]. Reprinted with permission from AAAS.]

[1] For an earlier study of these molecular fossils see J. J. Brocks et al., 1999. Archean molecular fossils and the early rise of Eukaryotes. *Science*, **285**, 1033–1036. For a different (and minority) interpretation of such features, see J. M. Garcia-Ruiz et al., 2003. Self-assembled silica-carbonate structures and detection of ancient microfossils. *Science*, **302**, 1194–1197.

[2] For influences on Earth's early climate, see J. F. Kasting and D. Catling, 2003. Evolution of a heritable planet. *Annu. Rev. Astron. Astrophys.*, **41**, 429–463.

The major PHOTOSYNTHETIC ORGANISMS *today are flowering plants on land and unicells in the plankton. Net primary productivity—mostly from photosynthesis—is 1.5 tonnes/hectare/year in the oceans and 7.8 tonnes/hectare/year on land.*

Often labeled the energy molecule of life, ADENOSINE TRIPHOSPHATE (ATP) *is a coenzyme derived from adenosine, involved in reactions that provide energy by splitting high-energy bonds between phosphate groups.*

determined that a gradual accumulation of oxygen in the secondary atmosphere began 2.0 to 2.3 Bya and gradually increased to its present 21% concentration as photosynthetic organisms diversified and spread. Much of this O_2 accumulation is considered to be the consequence of metabolic activity of newly evolving life. Organisms released oxygen during **photosynthesis**, the synthesis and production of organic compounds from carbon sources and water (END BOX 8.1), during which free O_2 is produced as hydrogen ions are removed from water molecules.[3]

METABOLISM

Cellular protein and nucleic acid synthesis became possible only after the origin and diversification of biochemical pathways in which components of such polymers could be produced. We see the results of such biochemical selection everywhere; cellular metabolism is highly organized in time and space so that each metabolic step in a sequence occurs in a highly repeatable order. Moreover, different metabolic sequences are often precisely coupled and regulated so that the products of one sequence—for example, adenosine triphosphate (ATP)—are used in other sequences/pathways such as the Embden-Meyerhof pathway shown in Figure EB 2.1. Finally, metabolic pathways evolved into hierarchical networks with modular properties (Ravasz et al., 2002).

As there are no existing relics of ancient metabolic pathways, we cannot obtain direct evidence of precellular or early cellular metabolism, or of the evolution of metabolic pathways. Two types of indirect evidence are available, however: comparisons of metabolic pathways shared between lineages of organisms, and the phylogenetic history of DNA sequences, including entire genes and genomes.

The molecular and cellular record of early life discussed above indicates that first life was anaerobic. Aerobic metabolism (including photosynthesis; END BOX 8.2) arose perhaps a billion years later (3–3.5 Bya; Margulis et al., 2006). Anaerobic and aerobic metabolisms are outlined in End Box 8.2 and END BOX 8.3 respectively.

Essential conclusions from the evidence in End Box 8.2 concerning anaerobic metabolism are:

- that anaerobic glycolysis is an almost universal pathway in which energy released as glucose degrades to pyruvic acid;
- that biochemical pathways did not arise randomly, but followed rules of chemistry in pathways subject to selection; and
- that continuous selection for the same enzymatic function essential in glycolysis in different lineages of organisms preserved sequence similarity of the shared enzymes involved, and may illuminate its evolutionary past.

Essential conclusions from the evidence in End Box 8.3 concerning aerobic metabolism are:

- that a revolution occurred when some organisms began to reduce carbon by using light as an energy source in the process we know as photosynthesis; and
- that perhaps the most significant change in metabolism to accompany the new aerobic environment was the evolution of a respiratory pathway by which oxygen is used to produce much more energy from the breakdown of glucose than produced from anaerobic glycolysis.

We relied on molecular fossils for evidence of the origin of methanogens. We can now turn to the origin and evolution of the first cells and organismal structures with a fossil record.

---

[3] R. Buick, 2008. When did oxygenic photosynthesis evolve?. *Philos. Trans. R. Soc. Lond. (B),* **363,** 2731–2743.

THE EARLIEST FOSSIL CELLS: CYANOBACTERIA

From what we can discern so far, the evolution of metabolic pathways in early cellular life evolved from:

- simple **anaerobic** systems dependent on energy sources in the primeval "soup" and functioning in the absence of oxygen (End Box 8.2); to
- **autotrophic** systems capable of generating organic compounds using atmospheric CO_2 as the energy source; to
- **photoautotrophic** systems capable of generating organic compounds using light as the source of energy (End Box 8.1); to
- **aerobic** systems that derive energy from the transfer of electrons to oxygen (End Box 8.3).

Analysis of 15 complete cyanobacterial genomes is consistent with the origin of photosynthesis in anaerobic protocyanobacteria in response to selection pressure from UV light and depletion of electron receptors.[4]

The consequences of using water as an electron and hydrogen donor in photosynthesis were profound. Liberation of molecular oxygen began to produce an oxidizing, aerobic environment whose chemical effects were quite different from the previous, relatively more reducing, environment. The speed at which oxygen accumulated by photosynthesis (Figure 8.1) is reflected in the presence of oxygen-generating cyanobacteria in the 3-By–old South African Bulawayan limestone and gunflint strata and in the 3.5-By–old filamentous cells at Warrawoona, Australia (FIGURE 8.2). Accumulation of oxygen in the atmosphere facilitated an increase in the number and kinds of organisms capable of utilizing aerobic metabolic pathways[5] (End Box 8.3; Berner et al., 2007).

The geological strata in which the oldest cyanobacteria have been found are mostly unmetamorphosed rocks called **cherts**, which are dark or black, reflecting their high carbon content. They also contain considerable silicon deposits. The oldest cyanobacteria specimens, located in the Warrawoona group of Western Australia, are structurally similar to cyanobacteria isolated from 850-My-old cherts (Figure 8.2). Like many other Archaean cherts, those in the Warrawoona group are associated with layered organic deposits called **stromatolites**. A fascinating find, published in 2006, documented evidence for seven types of stromatolites from a 10-km–long exposure of 3.4-By-old shallow water marine reef in the Strelley Pool Chert in Western Australia (Allwood et al., 2006). The conclusion is that these stromatolite reefs formed an extensive and structured biological ecosystem.[6] Furthermore, by contributing oxygen to the atmosphere, stromatolites played a major role in the evolution of the atmosphere. So, what are stromatolites?

STROMATOLITES

On Earth today, stromatolites consist of mats of microorganisms that trap various aqueous sediments, which they cement together to form characteristic laminated

Autotrophic organisms (AUTOTROPHS) can synthesize food from simple organic compounds. By contrast, heterotrophic organisms (HETEROTROPHS) depend on obtaining complex molecules from their prey.

By the Cambrian or somewhat earlier, oxygen levels had risen sufficiently to permit rapid evolution of large MULTICELLULAR ORGANISMS that derived their O_2 from the atmosphere.

The ARCHAEAN ERA, which began 3.8 Bya and lasted until 2.5 Bya, was characterized by much volcanic activity, the formation of rift valleys, an atmosphere without O_2, but with water, and temperatures not dissimilar to those on Earth today.

---

[4] A. Y. Mulkidjanian et al., 2006. The cyanobacterial genome core and the origin of photosynthesis. *Proc. Natl Acad. Sci. USA*, **103**, 13126–13131.

[5] J. M. Olson, 2006. Photosynthesis in the Archean era. *Photosyn. Res.* **88**, 109–117, and see J. F. Allen and W. Martin, 2007. Out of thin air. *Nature* **445**, 610–612 for ways of determining when oxygen levels began to increase.

[6] For an accessible analysis of the evolution of ecosystems, see P. Seldon and J. Nudds (2005), *Evolution of Fossil Ecosystems*, University of Chicago Press, Chicago.

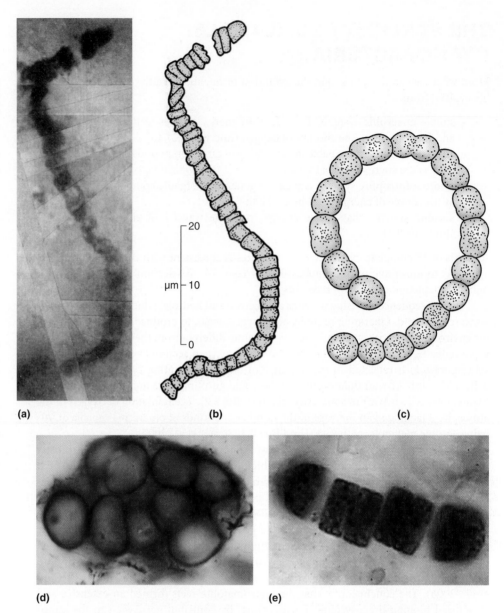

(a) (b) (c)

(d) (e)

FIGURE 8.2 (a) Thin section of a filamentous unicellular fossil found in stromatolite chert in the 3.5-By-old Warrawoona formation in Western Australia. **(b)** Reconstruction of this fossil. **(c)** Diagram of a phase-contrast microphotograph of a filament of cells of an extant cyanobacterium. Ancient fossil bacteria. **(d)** A colonial and **(e)** the filamentous genus *Palaeolyngbya* from the Bitter Springs chert of central Australia, a site dating to the Late Proterozoic, about 850 million years old. Among extant prokaryotes that appear similar to the microfossil in (a) are colorless sulfur-gliding and green sulfur bacteria.

[(a) through (c) Reproduced from Schopf, J.W., *Science* 260 (1993):640–646. Reprinted with permission from AAAS. Courtesy of J. William Schopf, Professor of Paleobiology & Director of IGPP CSEOL; (d) and (e) Courtesy of J. William Schopf, Professor of Paleobiology & Director of IGPP CSEOL.]

structures shaped like giant knobs. Extant and ancient stromatolite reefs differ in global distribution. Extant stromatolites occur only in extremely inhospitable environments—salinities twice that of normal seawater and temperatures greater than 65°C—where they are protected from grazing animals such as snails and sea urchins. Ancient stromatolites were distributed more widely, presumably because such herbivores were absent.

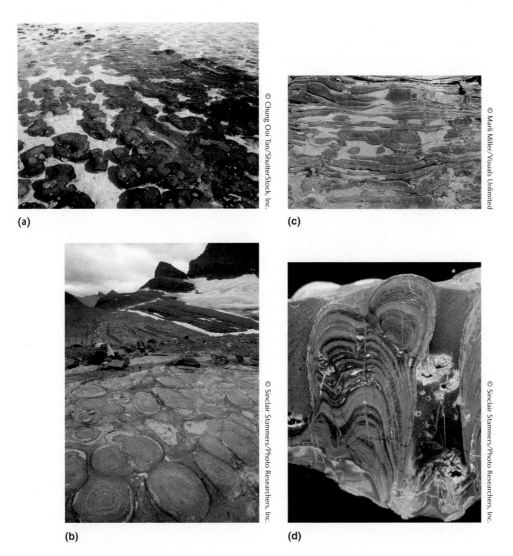

FIGURE 8.3 Stromatolites composed of calcium carbonate secreted by cyanobacteria are the oldest organisms known (3.5 By old) and the longest lasting. Compare the living stromatolites in Namibia **(a)** with the 2-By-old fossils from the Helena Formation in Glacier National Park, Montana **(b)**. Cross sections of the Namibian **(c)** and fossil **(d)** stromatolites show the similar organization of the internal layers of calcium carbonate.

As shown in **FIGURE 8.3**, these extant structures are remarkably similar to ancient stromatolites, which therefore are hypothesized to have been deposited by single-celled organisms. According to Golubic and Knoll,

> Stromatolites. . . are initiated by the establishment of a thin mat of microbes on a sediment surface. As sediment particles accumulate on top of mats, they are trapped and bound into a coherent layer by the microorganisms. Microbially mediated precipitation of calcium carbonate can also contribute to sediment accumulation and stabilization. . . Through time, commonly, a laminated structure accretes, each lamina marking a former position of the living mat community. . . On the present-day Earth, filamentous cyanobacteria are predominant mat-builders, but coccoid cyanobacteria, other types of bacteria, and a variety of eukaryotic algae produce well-defined mats.[7]

[7] S. Golubic and A. H. Knoll, 1993. Prokaryotes. *In Fossil Prokaryotes and Protists*, J. H. Lipps (ed.), Blackwell Scientific, Boston, pp. 51–76.

Indeed, many of the fossil organisms found in stromatolite deposits are remarkably similar to unicellular organisms alive today (Figure 8.2c). Schopf (1996) points out that aerobic cyanobacteria and anaerobic bacteria can coexist in stromatolites by using different light-gathering pigments that enable them to occupy different habitats:

> The oxygen-producing cyanobacteria live in the uppermost layers of stromatolites, with the non-oxygen-producing photosynthesizers just beneath. Much of the light energy is absorbed by the cyanobacteria. . . but this does not snuff out the anoxygenic photosynthesis of the green sulfur and purple bacteria that live below because these more primitive photosynthetic anaerobes are literally able to see through the cyanobacterial layer—their pigments absorb light unused by the cyanobacteria above.

These are complex communities of organisms able to utilize different aspects of their environment.

^{13}CARBON/^{12}CARBON RATIOS

A completely different type of evidence—^{13}C/^{12}C ratios—is consistent with the conclusion that life existed when ancient stromatolite reefs were deposited.[8]

The two isotopes ^{13}C and ^{12}C differ in respect to their participation in cellular metabolism. Organisms contain more ^{12}C than ^{13}C because of differences in how several key metabolic enzymes function. Thus, biological materials have different ^{13}C/^{12}C ratios than nonbiological material.[9] Through analyses, we know that almost all the stromatolite deposits dating from 3.5 Bya and later have carbon isotope ratios similar to rocks from the Carboniferous and other strata in which living forms appear. Such biologically-derived isotope carbon ratios also have been found in the oldest sedimentary rocks on Earth, the Isua Greenstone Belt, which is an outcrop of greenstone rocks in Greenland that formed on Earth's surface 3.8 Bya. These carbon ratios support the conclusion that life on Earth existed 3.8 Bya.[10]

Lest you take the ages of the first appearance of organisms as settled, we note a study in which ^{13}C enrichment is interpreted as arising from petroleum and not organismal activity and the earliest origin of cyanobacteria and eukaryotes dated at 2.15 and 1.78–1.68 Bya, respectively.[11]

BANDED IRON FORMATIONS

Banded iron formations are sedimentary rocks that are deposited as a result of the precipitation of chemicals. Deposited in layers that vary from millimeters to centimeters in thickness, the alternating colours (bands) reflect differing mineral content between layers (**FIGURE 8.4**). The largest deposits of iron ore are found in banded iron formation, the oldest of which date to 3.76 Bya. Levels of oxygen as much as 20 times that found in Earth's atmosphere today are locked within banded iron formations, which may have formed when photosynthesizing cyanobacteria living in ocean sediments split water molecules and released O_2 into the surrounding seawater. The released O_2 combined with

---

[8] Methods for using isotopes in radiometric dating are outlined elsewhere in this text.

[9] K. M. Brindle et al., 1982. Observation of carbon labelling in cell metabolites using proton spin echo NMR. *Biochem. Biophys. Res. Commun.*, **109**, 864–891; G. Tcherkez et al., 2010. On the ^{13}C/^{12}C isotopic signal of day and night respiration at the mesocosm level. *Plant Cell Environ.*, **33**, 900–913.

[10] H. D. Holland, 1997. Evidence for life on Earth more than 3850 million years ago. *Science*, **275**, 38–39.

[11] B. Rasmussen et al., 2008. Reassessing the first appearance of eukaryotes and cyanobacteria. *Nature*, **455**, 1101–1104.

FIGURE 8.4 Banded iron formations composed of layers (varves) of iron oxides (both hematite [Fe_2O_3] and magnetite [Fe_3O_4] interposed with layers of shale and chert).

ferrous ions, which were available only episodically, to form insoluble iron oxides (hematite and magnetite).

Now that we have discussed the origins of cellular life on Earth, we will turn our attention to the two important types of cellular life: **prokaryotic** and **eukaryotic cells**. Of what were these early cells composed? What types of organisms did they form? How were these organisms related? We answer these three fundamental questions by examining the nature of prokaryotic and eukaryotic cells, and discuss how classifications of life reflect advances in our understanding of the relationships between organisms composed of these cell types.

PROKARYOTIC AND EUKARYOTIC CELLS

Before our awareness of the early RNA World, the problem of how life could have originated by ordinary chemical means seemed insuperable. This is hardly surprising given the **complexity** of even the simplest cell, surrounded by a highly selective permeable membrane composed of lipids and proteins that regulate the kinds of substances that pass through. Within the cell, the cytoplasm consists of a multitude of structures and substructures involved in the synthesis, storage and breakdown of a large variety of chemical compounds.

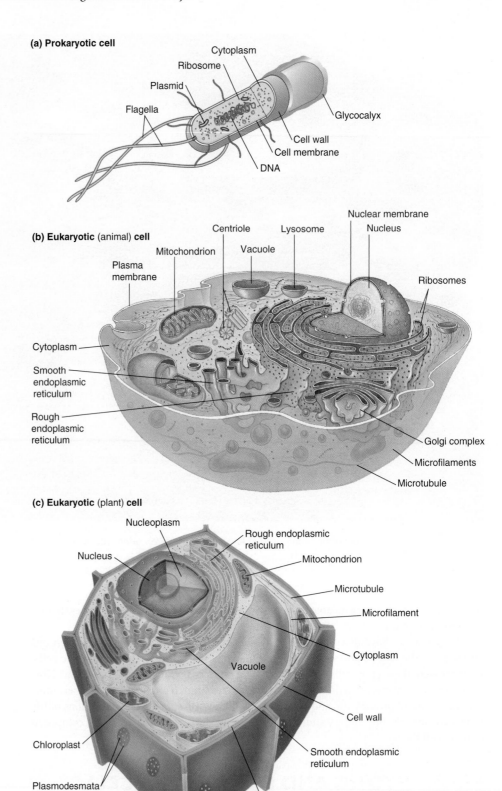

FIGURE 8.5 Diagrammatic representations of generalized prokaryotic **(a)** and eukaryotic animal **(b)** cells, and eukaryotic plant **(c)** cell showing cross sections through various important cellular organelles.

FIGURE 8.5 shows representative prokaryotic and eukaryotic cells, the latter as seen in a typical plant and a typical animal. The most obvious differences between the two cell types shown in Figure 8.5 is the absence in the generally smaller prokaryotic cells (0.5–10 μm) of any internal membranous network, such as a nuclear membrane or cytoskeleton, and the presence of these features in the generally larger (10–100 μm) eukaryotic cells.

In addition, prokaryotic cells reproduce by **binary fission**, which does not involve the **mitotic cell division** found in almost all eukaryotic cells. Consequently, prokaryotic cells lack **organelles** such as mitotic spindles and centrioles and do not have the histone proteins that structurally organize the relatively larger and more numerous eukaryotic chromosomes.[12]

Comparing the cells in Figure 8.5, the *complexity* of the eukaryote cells when compared with the prokaryote cell is evident. But look more closely. There is an *amazing similarity* between the three cell types, reflecting **fundamental (deep) homologies**, some of which (such as the cell membrane and genes) are billions of years old (Loomis, 1988). Structures such as the nucleus and organelles seen in the plant and animal cells (ribosomes, Golgi apparatus, endoplasmic reticulum, cytoskeleton) had their origin in the earliest single-celled eukaryotes. Others (mitochondria, chloroplasts) have only a slightly more recent origin, but, in their essential similarity, reflect common solutions to common problems as plants and animals originated. Deep homologies both at the cellular/structural and genetic/molecular levels underlie life. Such comparisons and discussion of differences in cellular complexity between prokaryotic and eukaryotic cells raise two obvious and absolutely central questions: can we measure or assess complexity, and has complexity increased during evolution? Both questions are addressed in **END BOX 8.4**.

Existence of prokaryotic and eukaryotic cells, and organized life composed of one cell or the other, raises the obvious question of how these organisms are related. As the summary below shows, our views have changed over the years in parallel with our enhanced understanding of cell structure, the origin of cellular components, and the nature of the genetic material in prokaryotic and eukaryotic cells.

PROKARYOTES AND EUKARYOTES: KINGDOMS OF LIFE

FIGURE 8.6 summarizes some of the major biological and geological evolutionary events from the lowermost Hadean division 4.6 Bya to the Phanerozoic (current) Eon. As the many transitions in the history of life became understood more completely, the way in which forms of life were classified changed. Each change reflects accumulated knowledge of the differences and similarities between forms of life and our increased understanding of the single tree of life.

For over two thousand years, life was classified into **two kingdoms**, **Plantae** (L. *planta*, plant) and **Animalia** (L. *anima*, breath, life). Assignment to one kingdom or the other was based on structure, function, metabolism—plants use photosynthesis, animals do not—and locomotion (animals move from place to place, plants do not, other than during seed or spore dispersal).

With the origination of photosynthesis, the most significant biological change was the transformation of one or more lineages of prokaryotic cells into eukaryotic cells. Evidence of major differences in cellular organization between prokaryotic and eukaryotic cells was

---

[12] In a major technical breakthrough published in 2010, researchers at the J. Craig Venter Institute in Maryland synthesized the chromosome of the bacterium *Mycoplasma mycoides* in the laboratory and used this "prosthetic genome" to replace the bacterium's normal DNA (D. G. Wilson et al., 2010). Whether this can be regarded as the first synthetic cell has been debated as, for example, in the journal *Nature* (2010, **465**, 422–424).

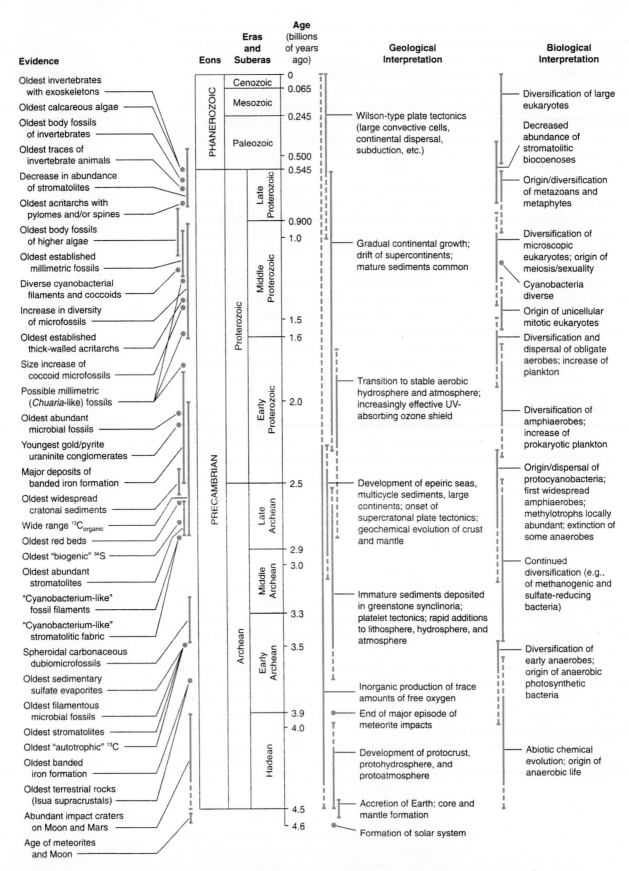

FIGURE 8.6 A summary of the types of evidence for evolution of Earth, rocks, and organisms from 4.6 By ago to the present.

[Adapted from Sogin, M.L., *Current Opinion Genet. Devel.,* 1 (1991): 457–463 and Wheelis, M.L., et al., *Proc. Natl Acad. Sci* USA 89 (1992): 2930–2934.]

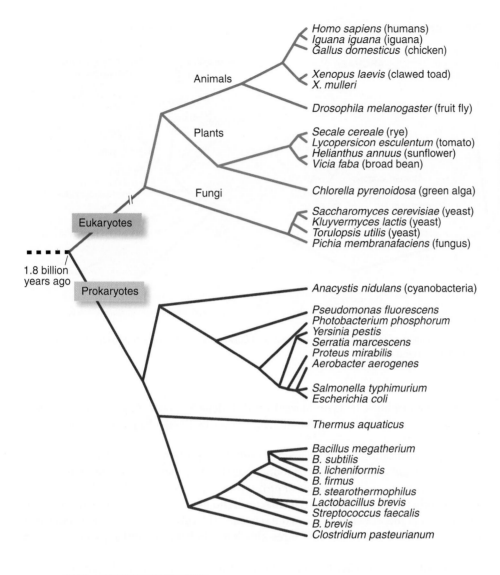

Homo sapiens (humans)
Iguana iguana (iguana)
Gallus domesticus (chicken)

Animals

Xenopus laevis (clawed toad)
X. mulleri

Drosophila melanogaster (fruit fly)

Plants

Secale cereale (rye)
Lycopersicon esculentum (tomato)
Helianthus annuus (sunflower)
Vicia faba (broad bean)

Fungi

Chlorella pyrenoidosa (green alga)

Saccharomyces cerevisiae (yeast)
Kluyvermyces lactis (yeast)
Torulopsis utilis (yeast)
Pichia membranafaciens (fungus)

Eukaryotes

1.8 billion years ago

Prokaryotes

Anacystis nidulans (cyanobacteria)

Pseudomonas fluorescens
Photobacterium phosphorum
Yersinia pestis
Serratia marcescens
Proteus mirabilis
Aerobacter aerogenes

Salmonella typhimurium
Escherichia coli

Thermus aquaticus

Bacillus megatherium
B. subtilis
B. licheniformis
B. firmus
B. stearothermophilus
Lactobacillus brevis
Streptococcus faecalis
B. brevis
Clostridium pasteurianum

FIGURE 8.7 A phylogenetic tree based on early data gathered from comparing 5S rRNA nucleotide sequences among over 30 species. The separation between eukaryotes and prokaryotes is estimated at 1.8 Bya. Other studies comparing amino acid sequences from 57 prokaryotic and eukaryotic enzymes also indicated that the two groups shared a common ancestor about 2 Bya.

[*Source:* Hori, H., and S. Osawa, *Proc. Natl. Acad. Sci. USA,* **76**, 1979: 381–385.]

used in 1941 by Stanier and van Neil to separate life into two broad domains: prokaryotes, which arose 3.8 Bya, and eukaryotes, which arose about 1.8 Bya[13] (**FIGURE 8.7**). So major were the differences between the domains considered to be that Stanier and van Neil designated prokaryotic and eukaryotic cellular organization as comprising two major divisions of life—two super kingdoms:

- **Prokaryotes** for all bacteria based on the prokaryotic cell type, and
- **Eukaryotes** for all uni- or multicellular organisms based on eukaryotic cells.

An immediate note of caution is required concerning the use of the names prokaryotes and eukaryotes for two major divisions of the tree of life. Species within each group are united by whether they are based on prokaryotic or eukaryotic cells, but neither group is natural (if by natural we mean having arisen from a single ancestor).

Recognizing the unnatural nature of these superkingdoms, Whittaker (1959) organized prokaryotes, the 96 phyla of animals and divisions of plants, and other eukaryotes into **five kingdoms**, one prokaryote and four eukaryotes. *Five Kingdoms*—the title

The technical term for a group with a single common ancestor is MONOPHYLETIC *(one origin). Groups that include lineages of organisms with separate origins are* POLYPHYLETIC *(many origins). Both prokaryotes and eukaryotes are polyphyletic assemblages.*

---

[13] By 1962, Stanier and van Neil had established the characters listed as distinguishing prokaryotes from eukaryotes (Stanier, R. Y., and C. B. van Neil, 1962. The concept of a bacterium. *Arch. Mikrobiol.,* **42**, 17–35).

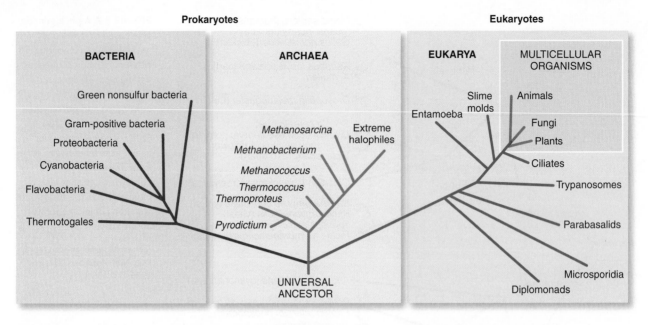

FIGURE 8.8 A simplified version of the three domains of life, Eubacteria, Archaea, and Eukarya, showing some representative groups within the Eukarya. See Woese et al. (1990) and Friend (2007).

of an influential book by Lynn Margulis and Karlene Schwartz, first published in 1982 and twice in updated editions (Margulis and Schwartz, 1998)—helped widely popularize this scheme.[14]

This classification scheme remained until the early 1980s when Woese (1981) and later Woese and colleagues (1990) found that rRNA sequences from an archaebacterium (a representative of a lineage of prokaryotic cells) were sufficiently different from those of other prokaryotes that **archaebacteria** should be placed into a separate domain or division of life, which they termed **Archaea**.[15] The current most natural subdivision of single celled organisms is into three kingdoms: **Eubacteria**, **Eukarya** and **Archaea** (**FIGURE 8.8**).

Eubacteria encompass the major forms of bacteria, as well as the cyanobacteria. Practically all of these possess unique peptidoglycan or murein cell walls consisting of chains of sugars cross-linked with short peptides, some of which contain D-amino acids. Archaea use other materials for their cell walls and often live under more rigorous environmental conditions than Eubacteria, such as hot sulfur springs and extreme salt concentrations.[16] These three kingdoms cut across the old prokaryote–eukaryote division; Archaea share several core similarities with eukaryotic cells, while the eukaryote assemblage includes many single-celled organisms, including photosynthetic and non-photosynthetic protozoans and algae.

---

[14] For the fourth edition, published in 2009, the title of this book changed to *Kingdoms and Domains* to reflect the changing views on how life is organized (L. Margulis and M. J. Chapman, 2009. *Kingdoms and Domains: An Illustrated Guide to the Phyla of Life on Earth*. Academic Press, New York).

[15] The books by Tim Friend (2007) and Jan Sapp (2009) contain fascinating accounts of this, the third domain of life. For the first genome isolated from Archaea, see R. A. Clayton et al., 1997. The first genome from the third domain of life. *Nature*, **387**, 459–462.

[16] Some 10% of the unicells in the photic zone of the open ocean are archaebacteria, but because they are difficult to culture they are little understood (Alastair Simpson, personal communication).

MULTICELLULAR EUKARYOTES

Five groups of multicellular organisms—green algae/plants, brown algae, red algae, animals, fungi—reside within the eukaryote kingdom. Four eukaryotic branches of the tree of life—protists, fungi, plants, and animals—have been recognized for some time. Over the last two decades, however, as molecular and evolutionary biologists have discovered new types of data and more and more organisms to analyze, increasing concern has been voiced that the four eukaryotic kingdoms do not accurately reflect evolutionary relationships among eukaryotes. For example, the Kingdom Protista includes organisms that are closely related to animals (choanoflagellates), organisms closely related to plants (e.g., green and red algae) and other major lineages that are extremely distantly related (ciliates and slime molds, both classified as protists, are more distantly related to each other than animals are to fungi; **FIGURES 8.8** to **8.10**).

The term MULTICELLULAR *can be tricky. Fungi are no more multicellular than filamentous algae; some brown algae are multicellular (Cock et al., 2010); many so-called fungi are actually (unicellular) yeasts; and multicellularity has been evoked experimentally in the unicellular yeast,* Saccharomyces cerevisiae *(Ratcliff et al., 2012).*

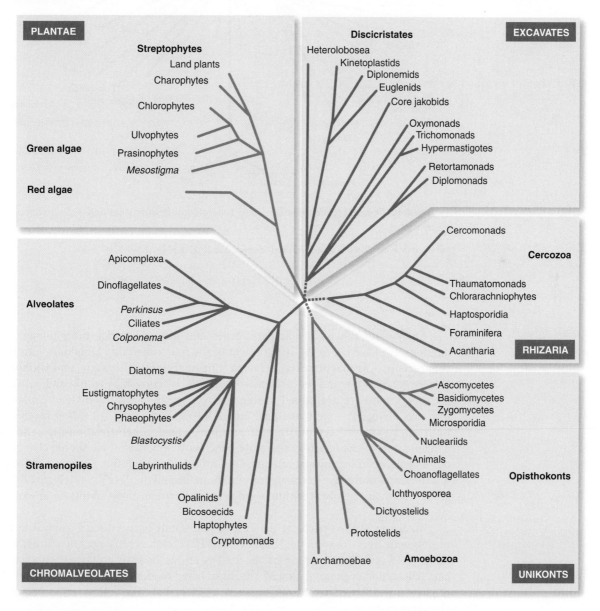

FIGURE 8.9 The eukaryotic tree of life as five supergroups. Relationships between the five supergroups are unresolved.

[Modified from Keeling, P. J., *Trends Ecol. Evol.*, **20**, 2005: 670–676.]

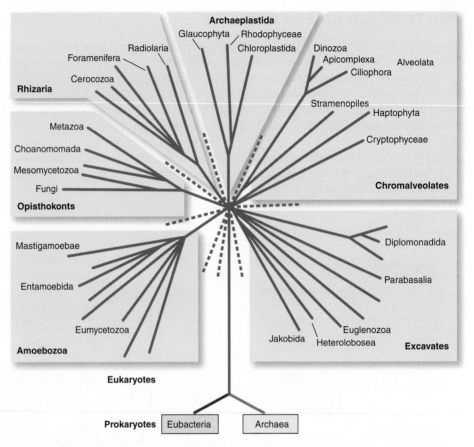

FIGURE 8.10 The eukaryotic tree of life as six supergroups. Relationships between supergroups are unresolved.

[Adapted from Adl, S. M., et al. *J. Eukaryot. Microbiol.,* **52,** 2005: 399–451.]

Analyses of the molecular data, genes and gene networks, morphology and development have led to the replacement of the four eukaryotic kingdoms with **five supergroups—Archaeplastida, Excavata, Chromalveolata, Rhizaria, and Unikonta** (Figure 8.9, BOX 8.1, and see Keeling et al., 2005). Relationships within and between the supergroups are not carved in stone; a large group of experts divided Unikonta into two, giving six superkingdoms (Figure 8.10, Box 8.1, and see Adl et al., 2005). Both schemes reflect accumulating information on phylogenetic relationships, and the dynamic state of classification and of nomenclature, which, after all, should change as our knowledge of organismal relationships, origins, and evolution changes. Although it is unclear how the five to six supergroups branched off the Tree of Life (Figure 8.10), affinity between members within a group is greater than their affinity to any other supergroup.

No aspect of the search for the Tree of Life stands in more stark contrast to the phylogenetic trees drawn in the 1880s by Ernst Haeckel than placement of animals with fungi (Jones et al., 2011), some parasitic protists, and as the sister group to choanoflagellates. Discussion of the evolution of organisms based on eukaryotic cells, especially the origin of organelles, of nuclear and organelle DNA, and of eukaryotic genes, is discussed elsewhere in this text.

BOX 8.1 **Five Eukaryote Supergroups**

This supergroup classification recognizes the origin of the multicellular eukaryotes—animals, plants, fungi—from monophyletic lineages of protists but leaves relationships between the supergroups unresolved; Figure 8.10 shows only broad ancestral relationships to Eubacteria and Archaea. In part this is a consequence of uncertainty associated with (a) the phylogenetic methods employed, (b) the statistical approaches used to determine significant results, and (c) horizontal gene transfer, which clouds ancestral relationships.[a]

ARCHAEPLASTIDA (ALSO KNOWN AS PLANTAE)

A supergroup that includes red and green algae, land plants and Charophyta (the ancestors of plants), characterized by chloroplastids and whose ancestors are thought to have been the first photosynthetic organisms. Erection of the supergroup

recognizes several monophyletic protist lineages as close allies of plants and of red and green algae.

EXCAVATA

This supergroup contains organisms that previously were members of the **Protista**. Commonly known excavates include *Giardia* (**FIGURE B1.1a**), which is responsible for the intestinal illness giardiasis (beaver fever, traveler's tummy), *Trypanosoma brucei*, which results in sleeping sickness (Figure B1.1b) and *Trichomonas*, which causes trichomoniasis. Excavates are not united by any single morphological or molecular feature. Rather ultrastructural or molecular features unite overlapping subsets of the 10 groups within this supergroup (Figure 8.9). Consequently, there is considerable controversy about this supergroup and relationships are in flux, although a recent analysis by Hampl and colleagues supports Excavata as monophyletic.[b]

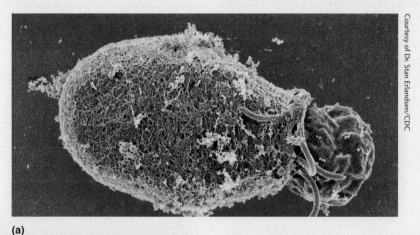

(a)

Courtesy of Dr. Stan Erlandsen/CDC

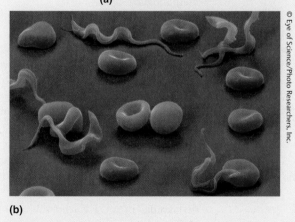

(b)

© Eye of Science/Photo Researchers, Inc.

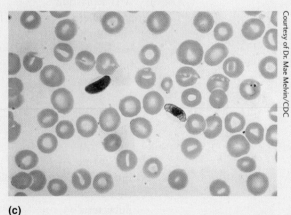

(c)

Courtesy of Dr. Mae Melvin/CDC

FIGURE B.1 Two excavates, **(a)** *Giardia* and **(b)** the trypanosome *Trypanosoma brucei* (purple), amongst red blood cells (red), and **(c)** a chromalveolate, *Plasmodium falciparum*, among red blood cells.

[a]Horizontal gene transfer is taken up in another chapter in the context of the origination of the organelles in eukaryotic cells.
[b]V. Hampl et al., 2009. Phylogenomic analyses support the monophyly of Excavata and resolve relationships among eukaryotic "supergroups." *Proc. Natl Acad. Sci. USA*, **106**, 3859–3864.

BOX 8.1	Five Eukaryote Supergroups (Cont...)

CHROMALVEOLATA

A supergroup of some 23 previous groups, including various types of algae (kelp, dinoflagellates, diatoms) that possess chloroplasts., as well as some important non-photosynthetic groups, notably ciliates and parasites such as *Plasmodium falciparum*, (Figure B1.1**b**), which is responsible for one form of malaria. One subgroup, the **alveolates** (ciliates, dinoflagellates, and others in Figure 8.9) is especially well supported through phylogenies of nuclear genes, but the monophyly of Chromalveolates as a whole is controversial.

RHIZARIA

A group of eukaryotic organisms recognized by Tom Cavalier-Smith (one of the world's experts) on the basis of molecular data alone (Cavalier-Smith, 2004). For some, this demonstrates the power of molecular approaches to the Tree of Life. Others, for whom groups of organisms should share some aspect(s) of their morphology, find it disturbing. Many, but not all, rhizarians are heterotrophic cells that capture and digest prey such as prokaryotes and other eukaryotes. Perhaps the most well-known are the 4,000+ species of foraminiferans and radiolarians, examples of which Ernst Haeckel drew in such wonderful and glorious detail.

UNIKONTA

This may be the most surprising group, for it unites organisms that were placed within three of Whittaker's five kingdoms—some parasitic protists, choanoflagellates, fungi, and animals—into a single group, the **opisthokonts** (Figure 8.10). The second group of unikonts, **amoebozoans**, includes slime molds and many types of more typical amoebae. (Analysis by Adl et al. [2005] identified sufficient differences between opisthokonts and amoebozoans to place each into a separate supergroup: **Opisthokonta**—animals, fungi, sponges, choanoflagellates, parasitic protists, Mesozoa and *Trichoplax*—and **Amoebozoa**—most amoebae, slime molds, some amoeboflagellates and several species lacking mitochondria [Figure 8.10]).

■ KEY TERMS

alveolates
anaerobic metabolism
Archaea
Archaebacteria
Archaeplastida
atmosphere
autotroph
banded iron formations
binary fission
Calvin cycle
chert
chlorophyll
Chromalveolata
complexity
Cope's rule
Embden-Meyerhof glycolytic pathway
environment
Eubacteria
Eukarya

eukaryotic cells
Excavata
greenhouse gas
Krebs cycle
methanogens
mitosis
molecular fossils
monophyletic
organelles
parasitism
photosynthesis
polyphyletic
prokaryotic cells
Protista
Rhizaria
secondary atmosphere
stromatolites
thylakoids
Unikonta

■ EVOLUTION ON THE WEB

Explore evolution on the Internet! Visit the accompanying website for *Strickberger's Evolution, Fifth Edition*, at **go.jblearning.com/Evolution5eCW** for exercises and links relating to topics covered in this chapter.

■ REFERENCES

Adl, S. M., A. G. B. Simpson, M. A. Farmer, et al., 2005. The new higher level classification of Eukaryotes with emphasis on the taxonomy of protists. *J. Eukaryot. Microbiol.*, **52**, 399–451.

Allwood, A. C., M. R. Walker, B. S. Kamber, et al., 2006. Stromatolite reef from the Early Archaean era of Australia. *Nature*, **441**, 714–718.

Berner, R. A., VandenBrooks, J. M., and Ward, P. D., 2007. Oxygen and evolution. *Science*, **316**, 557–558.

Cavalier-Smith, T., 2002. The phagotrophic origin of eukaryotes and phylogenetic classification of Protozoa. *Int. J. Syst. Evol. Microbiol.*, **52**, 297–354.

Cavalier-Smith, T., 2003. Protist phylogeny and the high-level classification of Protozoa. *Eur. J. Protistol.*, **39**, 338–348.

Cavalier-Smith, T., 2004. Only six kingdoms of life. *Proc. R. Soc. Lond. B*, **271**, 1251–1262.

Cock, J. M., L. Sterk, P. Rouzé, et al., 2010. The *Ectocarpus* genome and the independent evolution of multicellularity in brown algae. *Nature*, **465**, 617–621.

Friend, T., 2007. *The Third Domain: the Untold Story of Archaea and the Future of Biotechnology.* Joseph Henry Press, Washington, DC.

Garcia-Ruiz, J. M., S. T. Hyde, A. M. Carnerup, A. G. Christy, M. J. Van Krankendonk, and N. J. Welham, 2003. Self-assembled silica-carbonate structures and detection of ancient microfossils. *Science*, **302**, 1194–1197.

Heinz, D., and K. E. van Holde (eds.), 2011. *Oxygen and the Evolution of Life.* Springer, New York.

Jones, M. D. M., I. Forn, C. Gadelha, et al., 2011. Discovery of novel intermediate forms redefines the fungal tree of life. *Nature*, **474**, 200–203.

Keeling, P. J., G. Burger, D. G. Durnford, B. F. Lang, R. W. Lee, R. E. Pearlman, A. J. Roger, and M. W. Gray, 2005. The tree of eukaryotes. *Trends Ecol. Evol.*, **20**, 670–676.

Knoll, A. H., 2003. *Life on a Young Planet: The First Three Billion Years of Evolution on Earth.* Princeton University Press, Princeton, NJ.

Margulis, L., M. Chapman, R. Guerrero, and J. Hall, 2006. The last eukaryotic common ancestor (LECA): acquisition of cytoskeletal motility from aerotolerant spirochetes in the Proterozoic Eon. *Proc. Natl Acad. Sci. USA*, **103**, 13080–13085.

Margulis, L., and K. V. Schwartz, 1998. *Five Kingdoms. An Illustrated Guide to the Phyla of Life on Earth*, 3rd ed. W. H. Freeman and Co., New York. [1st edition, 1982].

Maynard Smith, J., and E. Szathmáry, 1995. *The Major Transitions in Evolution.* W. H. Freeman and Co., Oxford, UK.

Morowitz, H. J., 2002. *The Emergence of Everything: How the World Became Complex.* Oxford University Press, New York.

Pagel, M. (ed. in chief), 2002. *Encyclopedia of Evolution.* 2 Volumes, Oxford University Press, New York.

Ratcliff, W. C., R. F. Denison, M. Borrello, and M. Travisano. 2012. Experimental evolution of multicellularity. *Proc. Natl Acad. Sci. USA*, **109**, 1595–1600.

Ravasz, E., A. L. Somera, D. A. Mongru, Z. N. Oltvai, and A.-L. Barabási, 2002. Hierarchical organization of modularity in metabolic networks. *Science*, **297**, 1551–1555.

Sapp, J., 2009. *The New Foundation of Evolution: On the Tree of Life.* Oxford University Press, Oxford, UK.

Schopf, J. W., 1996. Metabolic memories of Earth's earliest biosphere. In *Evolution and the Molecular Revolution*, C. R. Marshall and J. W. Schopf (eds.). Jones and Bartlett, Sudbury, MA, pp. 73–107.

Stanier, R. Y., and C. B. van Neil, 1941. The main outlines of bacterial classification. *J. Bacteriol.*, **42**, 437–466.

Szathmáry, E., and J. Maynard Smith, 1995. The major evolutionary transition. *Nature*, **374**, 227–232.

Ueno, Y., Y. Isozaki, H. Yurimoto, and S. Maruyama, 2001. Carbon isotope signatures of individual Archean microfossils (?) from Western Australia. *Int. Geol. Rev.*, **43**, 186–212.

Ueno, Y., K. Yamada, N. Yoshida, et al., 2006. Evidence from fluid inclusions for microbial methanogenesis in the early Archaean era. *Nature*, **440**, 516–519.

Whittaker, R. H., 1959. On the broad classification of organisms. *Quart. Rev. Biol.*, **34**, 210–226.

Wilson, D. G., J. I. Glass, C. Lartigue, et al., 2010. Creation of a bacterial cell controlled by a chemically synthesized genome. *Science*, **329**, 52–56.

Woese, C. R., 1981. Archaebacteria. *Sci. Am.*, **244**, 92–122.

Woese, C. R., O. Kandler, and M. L. Whellis, 1990. Toward a natural system of organisms: proposal for the domains Archaea, Bacteria and Eucarya. *Proc. Natl Acad. Sci. USA.*, **87**, 4576–4579.

END BOX 8.1

Photosynthesis

© Photos.com

SYNOPSIS: How light-absorbing pigments move hydrogen ions across cell membranes to generate the co-enzyme adenosine triphosphate as a source of energy is discussed. This process of **photosynthesis**, which greatly expanded the environments available to organisms during Earth's evolution, is based on a metabolic pathway in which CO_2 provides the source of carbon to produce sugars and other organic compounds upon which much of life on Earth depends.

Despite advantages offered by electron transport systems in the early stages of evolution, reliance on chemical energy sources restricted organisms to specific localities or conditions containing organic and inorganic compounds. Perhaps the most important step toward **environmental independence** occurred when a mechanism evolved allowing light-absorbing (photosensitive) pigments cofactor to move H^+ ions (protons) across membranes to generate adenosine triphosphate (ATP) through the process of **photophosphorylation**. Most photosynthetic mechanisms now depend on **chlorophyll**, although photosynthesis in various prokaryotic cells uses the transmembrane protein bacteriorhodopsin as a light-driven proton pump.

Cyclic photosynthesis is an ancient photosynthetic pathway in which solar energy acting on light-sensitive chlorophyll excites the molecule to a high-energy state, allowing an electron to pass on to other electron transfer agents (**FIGURE EB1.1**). At an early evolutionary stage, this system must have bound to the membrane of protocells such as those discussed elsewhere in this text, allowing photoactive energy to couple to existing membrane-contained systems that could phosphorylate ADP to ATP. ATP generated by this new, coupled system would be available for metabolic needs and would enable growth to occur independently of environmental chemical energy sources such as glucose (End Box 8.2).*

One system originated as a membrane complex able to deacidify the cell interior by transporting protons out of the cell using energy provided from ATP → ADP breakdown. Incorporation of anaerobic oxidation-reduction enzymes and electron transfer components into the cell membrane allowed for the evolution of a system that could also reverse the ATP → ADP reaction to produce an ATP-synthesizing system energized by re-entry of protons into the cell. Association of chlorophyll with such membrane components offered a pathway for powering proton gradient formation—the **proton pump**—by photoactivity.

CARBON SOURCE

Some bacterial photophosphorylators, such as purple nonsulfur bacteria in the genus *Rhodospirillum*, depend in part on complex organic molecules for their carbon sources. They also use organic substrates as electron donors, for example, the in the oxidation of succinic acid to fumaric acid. By contrast, purple sulfur bacteria in the genus *Chromatium* use CO_2 exclusively and can obtain electrons from inorganic material such as H_2S. Although it is unclear which condition occurred first, the ability to use CO_2 as a source of carbon must have offered early photosynthesizers an important opportunity to expand their ecological distribution. To this end, the **Calvin cycle**, the most common pathway for the reduction of CO_2, is present in practically all photosynthetic organisms. As shown in **FIGURE EB1.2**, one CO_2 molecule is incorporated for each turn of this cycle, and one molecule of ribulose bisphosphate is regenerated for each CO_2 molecule incorporated. Six turns of the cycle are necessary to produce one glucose molecule. The overall reaction is:

$$6 \ CO_2 + 18 \ ATP + 12 \ NADPH + 12 \ H^+ \rightarrow glucose + 18 \ ADP + 18 \ P_i + 12 \ NADP^+.$$

Despite the advantage of using readily available CO_2 and easily obtainable photosynthetic energy, the distribution of early photosynthesizers would have been restricted because they depended on compounds such as H_2S for hydrogen sources. Such dependence is found today

---

*For overviews of photosynthesis in the context of the evolution of metabolic pathways, see Blankenship (2002), Knoll (2003), Heinz and van Holde (2011), and West-Eberhard et al. (2011).

(a) Semi-isolated double bond in nucleus III; Mg bound to nuclei I and II.

(b) Semi-isolated double bond in nucleus II; Mg bound to nuclei I and III.

(c) Semi-isolated double bond in nucleus I; Mg bound to nuclei II and III.

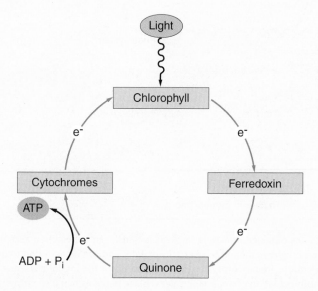

FIGURE EB1.1 Three resonance forms of chlorophyll *a*, showing stability as these double and single bonds shift around the ring system (C^1–C^8) in various ways (*colored lines*). This ability to resonate enables chlorophyll to temporarily retain a high electron energy level resulting either from the excitation of electrons exposed to appropriate wavelengths of light or from electrons transferred to chlorophyll from other pigments. Chlorophyll also can transmit such energy to other molecules used in photosynthetic reactions.

[Adapted from Wald, G., 1974. Fitness in the universe: choices and necessities. In *Cosmochemical Evolution and the Origins of Life*, J. Oró et al. (eds.). D. Reidel, Dordrecht, Netherlands, pp. 7–27.]

FIGURE EB1.2 Simplified diagram of the Calvin cycle, which is the main metabolic pathway for carbon dioxide fixation in photosynthetic organisms, using light and chlorophyll to produce energy and transfer electrons (e-) transfer. The Calvin cycle relies on reducing power contributed formed during photosynthetic reactions and on energy provided by ATP. Further details are provided in the text.

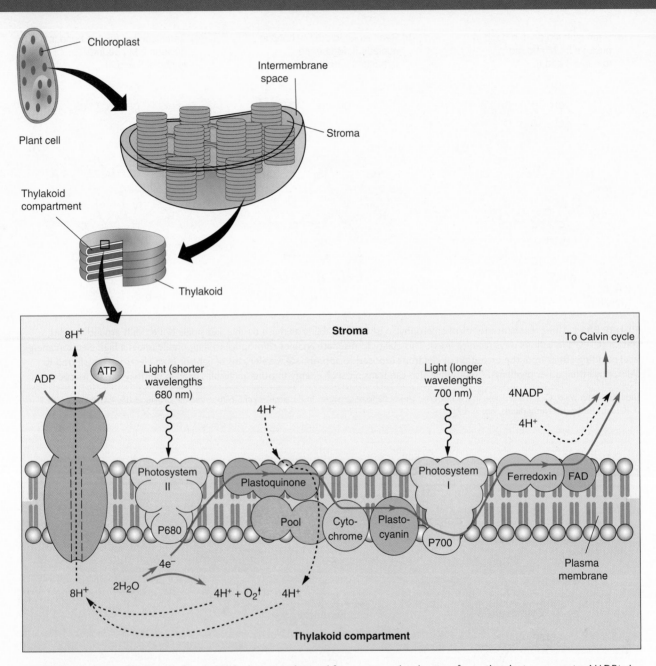

FIGURE EB1.3 Noncyclic electron flow in which electrons obtained from water molecules transfer to the electron acceptor NADP+ via two photosystems (I and II) each of which is sensitive to slightly different wavelengths of light, and can be activated to high energy levels to pass electrons along the chain. Transfer of electrons from photosystem II compensates for the loss of electrons to NADP+ by photosystem I. In turn, electrons derived from the photo-oxidation of water molecules replace photosystem II electrons. As in cyclic photophosphorylation, the flow of electrons produces a proton gradient that can supply energy for the phosphorylation of ADP to ATP. This appears more clearly in the *thylakoid membrane* structure where the ATP-synthesizing enzyme, ATPase, acts as a proton pump generating ATP by tapping energy from the gradient of H+ ions flowing across the membrane into the thylakoid compartment. Also shown are the general locations of the two components required for photosynthesis and some of the electron transport chain proteins. An *unbroken line* indicates the presumed flow of electrons. *Dashed lines* indicate the flow of H+ ions.

[Membrane sequence from S. L. Wolfe, 1981. *Biology of the Cell*, 2nd ed. Wadsworth, Belmont, CA, and also G. Zubay, 1988. *Biochemistry*, 2nd ed. Macmillan, New York, NY.]

among some bacterial photosynthesizers (purple and green sulfur bacteria) in which hydrogen sulfide provides the electrons for the hydrogenation of carbon:

$$2\,H_2S + CO_2 \xrightarrow{\text{light}} (CH_2O) + H_2O + 2\,S.$$
$$\text{carbohydrate}$$

A primary revolution in organismal distribution must have occurred when photosynthetic mechanisms evolved that could derive their electrons from readily available water molecules. This process now involves two chlorophyll systems (**noncyclic photosynthesis**), resulting from horizontal gene transfer between two kinds of photosynthetic bacteria, each possessing a somewhat different protein reaction center for transferring electrons. As shown in **FIGURE EB1.3**, the source of electrons for the photosystem II chlorophyll component is the oxidation (dissociation) of water into electrons and protons and the release of molecular oxygen:

$$2\,H_2O \rightarrow 4e^- + 4H^+ + O_2\uparrow.$$

The photosystems are localized within distinctively specialized photosynthetic membranes (**thylakoids**) found today in cyanobacteria and in the chloroplasts of eukaryotic algae and plants (Figure EB1.3). Union of these two photosystems (I and II) would have aided CO_2 reduction, especially with the advent of an aerobic atmosphere.

REFERENCES

Blankenship R. E., 2002. *Molecular Mechanisms of Photosynthesis*. Blackwell Science, Oxford, UK.

Heinz, D., and K. E. van Holde (eds.), 2011. *Oxygen and the Evolution of Life*. Springer, New York.

Knoll, A. H., 2003. *Life on a Young Planet: The First Three Billion Years of Evolution on Earth.* Princeton University Press, Princeton, NJ.

West-Eberhard, M. J., J. A. C. Smith, and K. Winter, 2011. Photosynthesis reorganized. *Science*, **332**, 311–312.

END BOX 8.2

Anaerobic Metabolism

SYNOPSIS: The basic features of metabolism are discussed that take place in the absence of oxygen—*anaerobic metabolism*—in which breakdown products of sugars such as glucose are made available as a source of energy. Anaerobic metabolism characterized the first cells that evolved on Earth.

The number of metabolic pathways shared by most organisms provides a window onto the origins and evolution of metabolism. Because of their sequential nature, metabolic pathways could not have arisen randomly but were channeled by available preexisting compounds. Many pathways, and the enzymes controlling individual reactions, survived because of their selective advantages.

ANAEROBIC GLYCOLYSIS

Anaerobic glycolysis is an almost universal pathway in which energy is released from the breakdown of glucose to pyruvic acid in the absence of oxygen. Reactions comprising anaerobic glycolysis illuminate its evolutionary past.

Anaerobic glycolysis is the most elemental metabolic pathway; all living organisms share various sections of this pathway. This universality exists because all organisms derive their free energy from the chemical breakdown of monosaccharides, which are the simplest and most biologically important carbohydrates, each consisting of a single sugar molecule. Glucose and fructose are monosaccharides. Sugars, starches, and celluloses are carbohydrates.

Organisms obtain monosaccharides either by making them (by reducing carbon dioxide) or by ingesting organic materials and converting them to monosaccharides. In either situation, glycolytic pathways may begin directly with glucose or with almost any organic material—sugars, fats, or amino acids—that can be converted into glucose. The **Embden-Meyerhof glycolytic pathway** leads from glucose to pyruvic acid (pyruvate), providing a net yield of two high-energy phosphate bonds in ATP, the which is basic currency for cellular chemical energy (**FIGURES EB2.1** and **EB2.2**):

$$C_6H_{12}O_6 + 2 \text{ ADP} + 2 P_i + 2 \text{ NAD}^+ \rightarrow$$
$$2 C_3H_4O_3 + 2 \text{ ATP} + 2 \text{ NADH} + 2 H^+ + 2 H_2O.$$

Only two of the reactions shown here furnish ATP; other steps in the pathway are preparatory to these primary reactions. Thus, organisms would need only to add or modify one or a few enzymes for each additional step beyond the primary reactions and not continually elaborate entirely new metabolic pathways (Margulis et al., 2006).

RULES OF CHANGE

Biochemical pathways did not arise randomly, but followed rules of chemistry in pathways subject to selection. Survival of such pathways depended on their ability to cope with frequently encountered chemical problems. For example, although glucose may not have been the first energy-yielding compound, it is now a common sugar whose stability and ready availability in plants and animals led to the importance of glycolysis in virtually all organisms; each enzyme in the glycolytic pathway is found in all multicellular and in most single-celled organisms.

As an interesting testimonial to evolutionary economy, seven of the glycolytic enzymes also function in glucose biosynthesis, **exactly reversing the direction of glycolysis**. Furthermore, the amino acid sequences in many of these enzymes are remarkably conserved in organisms that have been evolving separately for at least a billion years; the sequence of 10 to 11 amino acids in the catalytically active site in the enzyme triosephosphate isomerase is found in organisms ranging from bacteria (*Escherichia coli*) to corn (*Zea mays*). The prediction that a single ancestral sequence existed for this catalytic purpose in the common progenitor of all these organisms has been borne out by comparisons across an even wider range of organisms (**FIGURE EB 2.3**). Continuous selection for the same enzymatic function essential in glycolysis in these various lineages preserved sequence similarity (Canback et al., 2002).

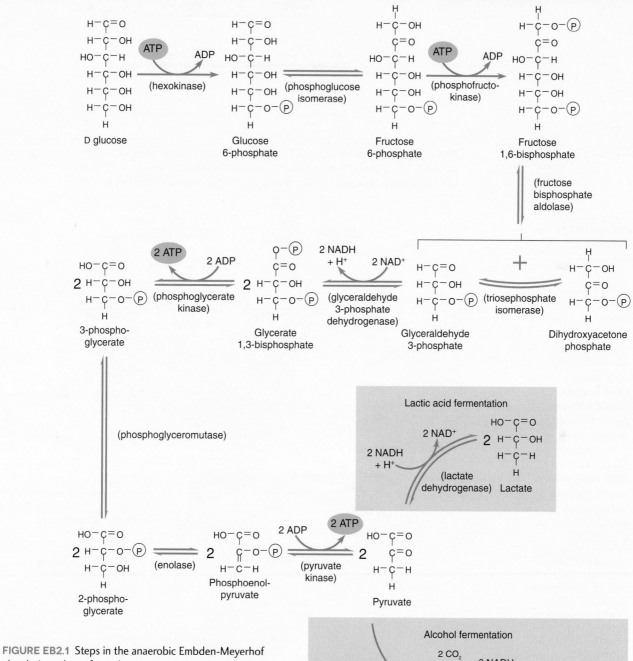

FIGURE EB2.1 Steps in the anaerobic Embden-Meyerhof glycolytic pathway from glucose to pyruvate. Beginning with one molecule of glucose, the pathway degrades two ATP molecules to ADP but phosphorylates four ADP molecules to ATP. The overall advantage of this pathway to the cell derives from the net formation of two high-energy phosphate bonds. Also indicated is the reduction of the pyridine nucleotide coenzyme, NAD^+. The reduced form of this compound (NADH) can be oxidized to regenerate NAD^+ by reactions that donate hydrogens and electrons to form either lactic acid or ethanol.

END BOX 8.2

Anaerobic Metabolism (Cont...)

© Photos.com

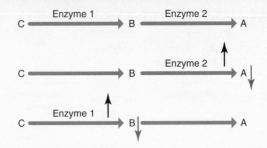

FIGURE EB2.2 Evolution of a new metabolic pathway by modification of an existing pathway. Depletion of product A (red downward arrow) favors selection for enzyme 2. Depletion of product B (green downward arrow) favors selection of enzyme 1. See text for details.

FIGURE EB2.3
Phylogenetic distribution of triosephosphate isomerase based on nucleotide sequences. Eukaryotes, brown; bacteria, blue; α-proteobacteria, blue background; green plants, green background; unresolved nodes, gray background. Numbers for green plants are bootstrap values (maximum of 100), using minimum evolution, parsimony, and maximum likelihood estimates, respectively.

[Reproduced from B. Canback, S. G., Andersson, and C. G. Kurland, 2002. The global phylogeny of glycolytic enzymes. *Proc. Natl Acad. Sci. USA*, **99**, 6097–6102.]

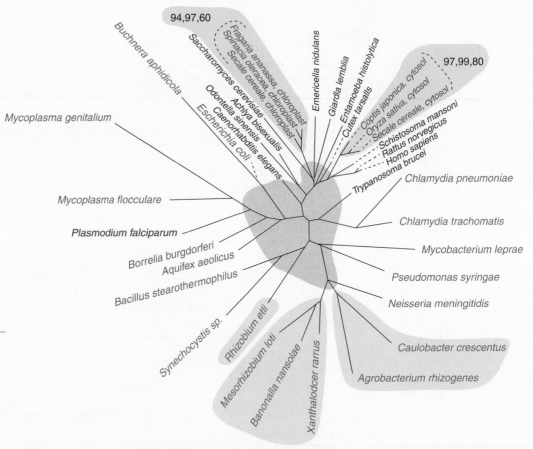

REFERENCES

Canback, B., S. G., Andersson, and C. G. Kurland, 2002. The global phylogeny of glycolytic enzymes. *Proc. Natl Acad. Sci. USA*, **99**, 6097–6102.

Margulis, L., M. Chapman, R. Guerrero, and J. Hall, 2006. The last eukaryotic common ancestor (LECA): acquisition of cytoskeletal motility from aerotolerant spirochetes in the Proterozoic Eon. *Proc. Natl Acad. Sci. USA*, **103**, 13080–13085.

END BOX 8.3

Aerobic Metabolism

SYNOPSIS: The basic features of metabolism are covered that take place in the presence of oxygen—*aerobic metabolism*—in which energy in the form of adenosine triphosphate (ATP) is generated from sugars such as glucose in pathways that depend on the presence of oxygen.

As discussed in the context of anaerobic metabolism in End Box 8.2, many pathways and enzymes controlling individual reactions survive in many living systems as a result of their selective advantages. Eventually, high-energy carbon compounds became less plentiful and some organisms evolved the ability to reduce more abundant carbon compounds such as CO_2, using a membrane-bound oxidation-reduction system.

A revolution occurred when some organisms began to reduce carbon by using light as an energy source in the process we know as **photosynthesis** (End Box 8.1). Such reduction requires a source of electrons, for which substances such as H_2S were used. It later became advantageous to split water, which was plentiful, to gain electrons. Oxygen released as a byproduct of splitting water molecules increased to the extent that it was exploited to enhance energy yields from glucose breakdown. A new pathway—the **Krebs cycle**—arose. Electrons removed from oxidized compounds in the cycle transfer to a **membrane-bound electron transport system** for which oxygen acts as the ultimate electron acceptor molecule.

KREBS CYCLE

Although tied to aerobic metabolism, the Krebs cycle (also known as the citric acid cycle or the tricarboxylic acid cycle) does not immediately depend on molecular oxygen, and so may have originated from anaerobic pathways (Margulis et al., 2006). Regardless, this process releases much more energy than anaerobic glycolysis.

Although the Krebs cycle itself does not use molecular oxygen, its evolution and adoption by aerobic organisms is based on a membrane-bound electron transport system where oxygen serves as the final electron acceptor in the chain: another stage in the evolution of membrane-bound systems. In this electron transport process, the pyridine nucleotide coenzyme NAD^+ picks up electrons and associated protons (e^- and H^+, respectively) and transfers them into a respiratory chain consisting of various electron carriers. As diagrammed in **FIGURE EB3.1**, such electron transfers occur along an electrical potential gradient that provides sufficient energy exchange at three coupling sites to allow an ADP molecule to be phosphorylated.

In summary, complete aerobic oxidation of a molecule of glucose, including **oxidative phosphorylation**, produces maximally about 38 molecules of adenosine triphosphate (ATP), compared to only two molecules of ATP formed by anaerobic glycolysis:

$$C_6H_{12}O_6 + 6\ H_2O + 6\ O_2 + 38\ ADP + 38\ P_i \rightarrow$$
$$6\ CO_2 + 12\ H_2O + 38\ ATP.$$

From Krebs cycle reactions:

Malate   Succinate   α-ketoglutarate   Isocitrate

Cell (eukaryote)

Oxaloacetate   Fumarate   Succinate   α-ketoglutarate

$e^- + H^+$ $e^- + H^+$ $e^- + H^+$ $e^- + H^+$

NAD   ADP + P$_i$ FAD

ATP

Matrix

Flavoprotein

Intermembrane space

Coenzyme Q

Cytochrome *b*

ADP + P$_i$

Mitochondrion   **Respiratory transport chain**

ATP

Cytochrome *c*

Cytochrome *a*

ADP + P$_i$

ATP

Oxygen

$O_2 + 4\,e^- + 4\,H^+ \longrightarrow 2H_2O$

FIGURE EB3.1 A simplified diagram of a respiratory pathway for the transfer of electrons and hydrogen protons occurring on the inner membrane of mitochondria. Electrons and protons from coenzymes NAD$^+$ and FAD pass down the respiratory transport chain, leading ultimately to oxygen and the production of water. Although different microorganisms use different electron carriers, the general sequence of oxidative phosphorylation is the same: as electrons transfer down the chain, hydrogen ions pump across the membrane, and their return flow drives ATP synthesis. For simplicity, the diagram provides coupling sites at which ATP generates, but these sites have not been precisely localized. Depending on which set of reactions bring electrons from NADH to the respiratory chain, as many as 38 ATPs can be formed during complete oxidation of glucose.

REFERENCE

Margulis, L., M. Chapman, R. Guerrero, and J. Hall, 2006. The last eukaryotic common ancestor (LECA): acquisition of cytoskeletal motility from aerotolerant spirochetes in the Proterozoic Eon. *Proc. Natl Acad. Sci. USA*, **103**, 13080–13085.

END BOX 8.4

Complexity

SYNOPSIS: Here the issue of whether some organisms are more complex than others is covered. We discuss and evaluate how to define *complexity*, whether complexity has increased over evolutionary time (are humans more complex than worms?), and how to measure complexity (numbers of cell types, the size of organisms, number of life history stages?).

The evolution of multicellular from unicellular organisms raises the difficult issue of **complexity**, a topic introduced in this chapter when comparing prokaryotic and eukaryotic cells. Extinction and replacement of species or biota also raises the issue of whether complexity has increased during organismal evolution. Almost all evolutionary biologists would accept that complexity is part of life. For example:

> "Biological complexity is displayed at many hierarchical levels, from molecular and cellular operations within an organism to species interactions in ecological communities" (Avise and Ayala, 2007, p. 8564).

That said, we can ask the question: are some organisms more complex than others? Are mammals more complex than the lineages of reptiles—mostly large dinosaurs—they replaced? It seems self-evident that multicellular organisms such as animals and plants are more complex than unicellular organisms. But are some multicellular organisms more complex than others? Are we more complex than plants? Are flies more complex than fleas? Some researchers view the appearance of many different animal body plans in the Early Cambrian as evidence for a burst in the evolution of complexity in the late Precambrian.[a] Analysis of the nature of limbs shows that their intricacy increased during the early Cambrian within many lineages of crustaceans and that intricacy of limb types in extant crustaceans correlates with species diversity (Adamowicz et al., 2008).

Nevertheless, revolutionary biologists shy away from the concept that evolution results in increasing complexity almost as much as they shy away from the concept of progress. To quote Szathmáry and Maynard Smith (1995):

> There is no theoretical reason to expect evolutionary lineages to increase in complexity with time, and no empirical evidence that they do so. Nevertheless, eukaryotic cells are more complex than prokaryotic cells, animals and plants are more complex than protists, and so on.

As an explanation, Szathmáry and Maynard Smith propose that "this increase in complexity may have been achieved as a result of a series of major evolutionary transitions. These involved changes in the way information is stored and transmitted." Nonetheless, no agreed-upon criteria exist to define the "information" used to measure complexity, or to identify increasing complexity. Nor do all agree that complexity *has* increased throughout the evolution of life; the issue is as much philosophical as it is scientific.

Criteria used to measure complexity are listed in **TABLE EB4.1**. Several of these criteria, notably genome size, gene number, gene networks, and compartmentalization, are discussed in the text. The question is, which of these attributes of organisms allow us to answer "yes" to whether complexity has increased, yes being the answer most biologists and laypersons would give if asked whether some organisms are more complex than others.

The numbers of interactions between proteins (the "interactome") in the genomes of humans (*Homo sapiens*), a fruit fly (*Drosophila melanogaster*), and a nematode (*Caenorhabditis elegans*) have been calculated and compared (Stumpf et al., 2008). The estimated 650,000 interactions between human proteins is an order of magnitude higher than *Drosophila* and three times higher than for *C. elegans*. These fascinating data notwithstanding, what they tell us about complexity remains enigmatic.

Perhaps the most commonly used criterion—**number of cell types**—is discussed below, as are two criteria—increase in organismal size and life history stages that are not on the list, and for good reasons.

---

[a]D. W. McShea, 2005. A universal generative tendency toward increased organismal complexity. In *Variation. A Central Concept in Biology*, B. Hallgrímsson and B. K. Hall (eds.). Elsevier/Academic Press, New York, pp. 435–453.

END BOX 8.4

Complexity (Cont...)

TABLE EB 4.1 **Criteria Used to Measure Complexity**	
Criterion	**Source**
Genome size or the total number of genes in an organism	Cavalier-Smith, 1985
Number of genes that encode proteins Number of parts or units in an organism (where parts might be segments, organs, tissues, and so forth)	Szathmáry and Maynard Smith, 1995; Adamowicz et al., 2008
Number of cell types possessed by an organism	Valentine et al., 1994; Valentine, 2003; Vickaryous and Hall, 2006
Increased compartmentalization, specialization, or subdivision of function between organisms being compared	Hall, 1999; Adamowicz et al., 2008
Number of genes, gene networks, or cell-to-cell interactions required to form the parts of an organism	Larsen, 1992; Stumpf et al., 2008
Number of interactions between the parts of an organism, reflecting functional complexity or a high degree of integration	Hall, 1999; Stumpf et al., 2008
Length and complexity of the minimal statement required to describe the organism, an approach that effectively combines the other eight.	Wilson, 2005

INCREASE IN NUMBERS OF TYPES OF CELLS

Increase in complexity is perhaps easiest to see during the embryonic development of an animal, when the individual transforms from a single-celled zygote to a multicellular organism. Adult humans possess as many as several hundred different cell types.

Reflecting the parallel between development and evolution discussed elsewhere in this text, the most commonly used metric of complexity is the number of different types of cells possessed by an organism. As shown in **FIGURE EB4.1**, this metric has increased over the course of animal evolution. Estimates of numbers of cell types range from 6 to 12 in sponges and cnidarians, 20 to 30 in flatworms, 50 to 55 in mollusks, arthropods, annelids, and echinoderms to as high as 200 to 400 in humans. James Valentine and his colleagues (1994) plotted the time of origin of animal body plans against number of cell types and found that the upper limits of complexity increased from the earliest forms, such as sponges, to the vertebrates, and that early rates of increase were high in comparison to later changes within the vertebrates (Figure EB4.1). Similar conclusions come from studies across kingdoms.[b]

It is difficult to escape the conclusion that, when measured by numbers of cell types, complexity has increased during at least part of the evolutionary history of life. What about increase in size through evolution as an indicator of increasing complexity?

INCREASE IN ORGANISMAL SIZE

Embryonic development (and so increasing complexity) is accompanied by increasing size. Early stages of the evolution of life were accompanied by an increase in organismal size.

---

[b]G. Bell and Mooers, A. O. (1997) Size and complexity among multicellular organisms. *Biol. J. Linn. Soc.*, **60**, 345–363.

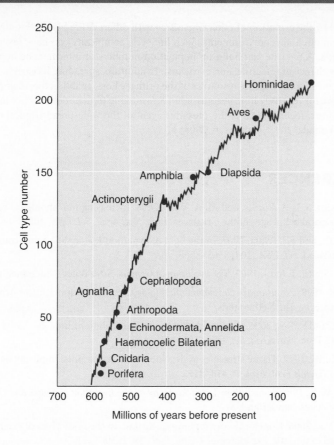

FIGURE EB4.1 Time of origin of various animal groups (My before present) with the estimated numbers of cell types in each. The marked increase from low to high is considered by many evolutionary biologists to reflect an increase in complexity.

[Adapted from Valentine, J. W., A. G. Collins, and C. P. Meyer. Morphological complexity increases in metazoans. *Paleobiology*, **20**, 1994: 131–142.]

So, why do we not use increase in organismal size over evolutionary time, as a criterion of complexity? Evolutionary size increase has occurred—it is known as **Cope's rule.**[c] Evolution is often for decrease in size, however, as seen in parasites in many different groups of animals, or in organisms that live between the sand grains at high tide levels on beaches. Different groups of organisms of the same size show different levels of complexity, when complexity is measured by criteria other than organism size. Related animals of vastly different sizes—mice and elephants, Chihuahuas and Great Danes, hummingbirds and ostriches—show comparable complexity.

LIFE HISTORY STAGES

Once relatively large potential host organisms evolved, **parasitism** became a successful way of life for many invertebrates. As parasites evolve many of their organs become simplified or disappear altogether. In contrast, the **life cycle of a parasite** may become enormously complex and involve multiple intermediate hosts. Some tapeworms, for example, pass through a number of intermediate hosts ranging from arthropods to fish before the adult stage develops in the primary mammalian host. Such developmental and life history networks take advantage of opportunities

---

[c]Moen (2006) compared body size evolution across six clades of extant turtles *and* through the turtle phylogenetic tree, without finding any evidence for Cope's rule. Indeed, Moen argues that no analysis of extant taxa has provided evidence for Cope's rule. Similarly, Hunt and Roy (2006) found no support for Cope's rule in their analysis of evolution of body size in a deep-sea ostracod genus during the Cenozoic.

END BOX 8.4

Complexity (Cont...)

© Photos.com

provided by the evolution of other organisms with different life histories and which exploit different and often new environments. Such life cycle complexity can have several other adaptive advantages. A parasite can build up population numbers in intermediate hosts, enhancing the parasite's chance of infecting a primary host. In addition, spreading its early stages among intermediate hosts conserves the resources of the primary host, enabling the adult parasite to remain productive for relatively long periods; adult *Schistosoma* trematodes may live for 30 years in their human hosts. (*Schistosoma* is the causative agent of the widespread tropical disease schistosomiasis [Yamada, 2003; Thompson, 2005].)

REFERENCES

Adamowicz, S. J., A. Purvis, and M. A. Wills, 2008. Increasing morphological complexity in multiple parallel lineages of the Crustacea. *Proc. Natl Acad. Sci. USA*, **105**, 4786–4791.

Avise, J. C., and F. J. Ayala, 2007. In the light of evolution I: adaptation and complex design. *Proc. Natl Acad. Sci. USA*, **104**, 8563–8566.

Cavalier-Smith, T. (ed.). 1985. *The Evolution of Genome Size*. Wiley, Chichester, UK.

Hall, B. K, 1999. *Evolutionary Developmental Biology*, Second edition. Kluwer Academic Publishers, Dordrecht, The Netherlands.

Hunt, G., and K. Roy, 2006. Climate change, body size evolution, and Cope's rule in deep-sea ostracods. *Proc. Natl Acad. Sci. USA*, **103**, 1347–1352.

Larsen, E. W., 1992. Tissue strategies as developmental constraints: implications for animal evolution. *Trends Ecol. Evol.*, **7**, 414–417.

Loomis, W. F., 1988. *Four Billion Years: An Essay on the Evolution of Genes and Organisms*. Sinauer Associates, Sunderland, MA.

Moen, D. S., 2006. Cope's rule in cryptodiran turtles: do the body sizes of extant species reflect a trend of phyletic size increase? *J. Evol. Biol.*, **19**, 1210–1221.

Stumpf, M. P. H., T. Thorne, E. de Silva, et al., 2008. Estimating the size of the human interactome. *Proc. Natl Acad. Sci. USA*, **105**, 6959–6964.

Szathmáry, E., and J. Maynard Smith, 1995. The major evolutionary transition. *Nature*, **374**, 227–232.

Thompson, J. N., 2005. *The Geographic Mosaic of Coevolution*. The University of Chicago Press, Chicago.

Valentine, J. W., 2003. Cell types, numbers, and body plan complexity. In *Keywords and Concepts in Evolutionary Developmental Biology*, B. K. Hall and W. M. Olson (eds.). Harvard University Press, Cambridge, MA, pp. 35–43.

Valentine, J. W., Collins, A. G., and Meyer, C. P., 1994. Morphological complexity increase in metazoans. *Paleobiology*, **20**, 131–142.

Vickaryous, M., and Hall, B. K., 2006. Human cell type diversity, evolution development classification with special reference to cells derived from the neural crest. *Biol. Rev. Camb. Philos. Soc.*, **81**: 425–455.

Wilson, R. A., 2005. *Genes and the Agents of Life. The Individual in the Fragile Sciences*, Cambridge University Press, Cambridge, UK.

Yamada, T. (ed.), 2003. *Textbook of Gastroenterology*, Fourth edition. Lippincott Williams & Wilkins Philadelphia, PA.

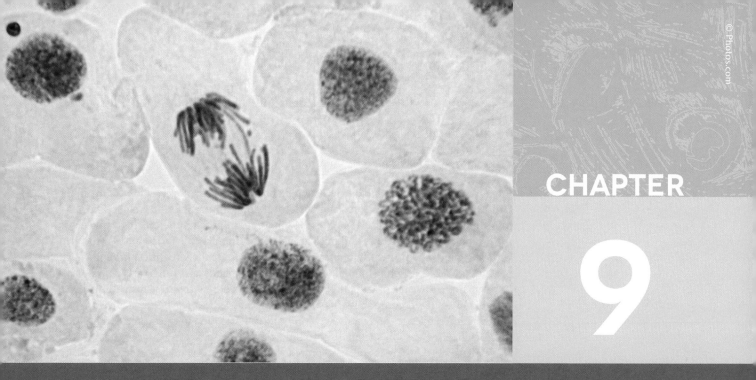

© Photos.com

Eukaryotic Organelles, Genes, and Organisms Arose 1.8 Bya

CHAPTER SUMMARY

Prokaryotic cells arose 3.5 or more Bya as methane-producing, anaerobic methanogens. About two billion years later (1.8 Bya), eukaryotic cells had arisen from one or more prokaryotic lineages, initially as unicellular organisms. A billion years later, around 750 Mya, multicellular eukaryotes appeared and underwent several explosive radiations. Three groups of multicellular organisms are recognized: plants, animals, and fungi. Eukaryote assemblages include many single-celled organisms, among which are photosynthetic and non-photosynthetic protozoans and algae. While similar in organismal complexity to eubacteria, archaeal (archaebacterial) cells share several core similarities with eukaryotes, sufficient that all three lineages may have arisen from a single common ancestor.

Eukaryotic cells are generally aerobic and more complex than prokaryotic cells, although complexity is hard to determine. Eukaryotic cells have many organelles, internal membranes, a cytoskeleton, and a microtubular apparatus for mitotic cell division. The physical and genetic features of mitochondria and chloroplasts show that these eukaryotic cellular organelles are remnants of ancient symbiotic relationships acquired by the engulfment of one unicellular organism by another, through endosymbiosis. Genes in eukaryotic cells, unlike genes in prokaryotic cells, often have nucleotide sequences—introns—within them that do not translate into peptide sequences.

We begin with a discussion of the intracellular organelles found in eukaryotic, but entirely absent from prokaryotic, cells.

ORGANELLES

Membranes were the first biological structures to arise. Eukaryotic cells possess membranous structures (**organelles**) such as a nucleus, endoplasmic reticulum, Golgi bodies, and mitochondria or chloroplasts (**FIGURE 9.1**) not found in prokaryotic cells. The nucleus is the location of chromosomes and nuclear DNA, the endoplasmic reticulum is associated with ribosomes and protein synthesis, the Golgi bodies are the structures through which proteins are secreted, while mitochondria (in animals, plants, and fungi) and chloroplasts (in plants) are the sites of oxidative phosphorylation and energy transformation. These membrane-enclosed **compartments** allow eukaryotic cells to isolate molecules for specific reaction sequences and pathways, confining transcription to the nucleus, translation to the cytoplasm, aerobic metabolism to mitochondria, and so forth.

All eukaryotes share mitotic and protein-synthesizing mechanisms and almost all are aerobic. Exceptions include some forms such as microsporidians (single-celled parasites of arthropods, fish, and humans) and yeast that can function anaerobically, and some single-celled species such as *Giardia, Trichomonas,* and *Pelomyxa* (**FIGURE 9.2**), which have anaerobic energy metabolism systems but arose from aerobic ancestors. Even yeast can adapt to aerobic conditions, as demonstrated by the discovery of two mutations in a member of the flocullin gene family, *Flo11,* in the yeast *Saccharomyces cerevisiae.*[1] The *Flo11* gene increases

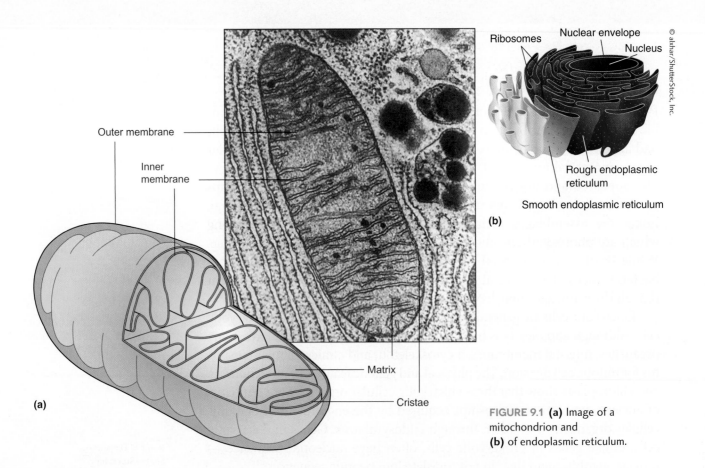

Outer membrane

Inner membrane

Matrix

Cristae

(a)

Ribosomes

Nuclear envelope

Nucleus

Rough endoplasmic reticulum

Smooth endoplasmic reticulum

(b)

© akhbar/ShutterStock, Inc.

FIGURE 9.1 **(a)** Image of a mitochondrion and **(b)** of endoplasmic reticulum.

---

[1] Flocculation is the clumping of yeast in the presence of sugar, as occurs in brewing beer.

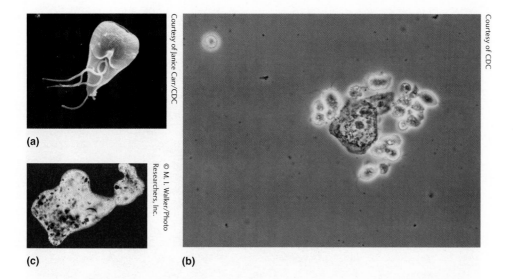

FIGURE 9.2 Anaerobic protists. **(a)** A scanning electron micrograph of *Giardia lamblia*. **(b)** A phase contrast image of *Trichomonas vaginalis*. **(c)** A light micrograph of *Pelomyxa palustris*.

the ability of the cell surface to repel water in those individuals that float on the surface, providing them with access to oxygen. Individuals without a key mutation in *Flo11*[2] are restricted to the oxygen-poor lower layers of the medium, demonstrating a mutational basis for adaptation to aerobic versus anaerobic conditions.

EVOLUTION OF EUKARYOTIC ORGANELLES

Information has been sought for several decades concerning the origin of the mitochondria and chloroplasts found within animal and plant cells. The most well-supported hypothesis is that eukaryotic cells evolved by physically incorporating entire prokaryotic organisms into their cytoplasm. Over time, these engulfed organisms became mitochondria or chloroplasts. Sufficient data have accumulated that we now have a mechanism for the endosymbiotic hypothesis and so can speak of the **endosymbiotic theory** of organelle origins.

ENDOSYMBIOTIC THEORY

According to the endosymbiotic theory, ancestral anaerobic eukaryotic cells acquired the ability to engulf and ingest supramolecular particles after acquiring new surface membrane properties. An internal cytoskeleton composed in part of actin filaments and microtubules (**FIGURE 9.3**) helped maintain cell shape and allowed these cells to move about and change their shape. Provided with these mechanisms, some early eukaryotic cells became active predators able to engulf (endocytose) "large" objects, such as prokaryotic cells.[3]

The hypothesis is that selection would have led to increased predatory abilities, including enlarged predator cell size, as well as to other innovations that affect movement, capture of prey, and digestion. Among the various unicells on which these cells fed were aerobic bacteria capable of oxidative phosphorylation and cyanobacteria capable of photosynthesis. These ingested aerobic and photosynthetic prokaryotic cells assumed mutually advantageous **symbiotic** relationships with their hosts, similar to those seen

---

[2] The basis for the mutation is a 111-nucleotide deletion in a repression region that increases gene expression, and a rearrangement in the central tandem repeat domain of the coding region that yields a more hydrophobic gene product; M. Fidalgo et al., 2006. Adaptive evolution by mutations in the *FLO11* gene. *Proc. Natl Acad. Sci. USA*, **103**, 11228–11233.

[3] For discussions of the endosymbiotic origin of eukaryote organelles, see Gillham (1994), Gray et al. (1999, 2001), Cavalier-Smith (2002), Knoll (2003), and Lane (2005).

FIGURE 9.3 The cytoskeleton of eukaryotic cells includes both microfilaments composed of actin (red) and microtubules composed of tubulin (green). Nuclei are stained purple.

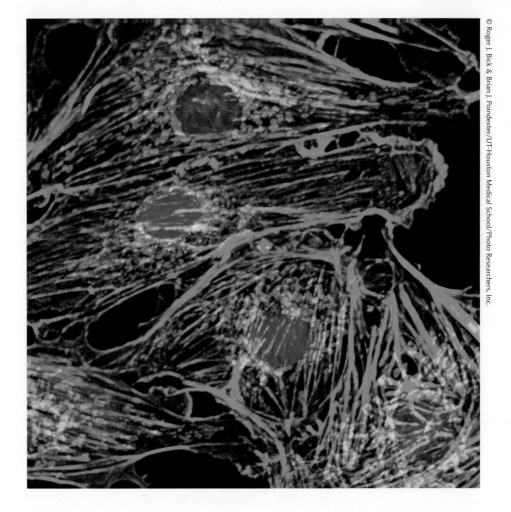

GRAM–POSITIVE BACTERIA *are so named because they stain with the crystal violet stain used in the Gram-staining technique, which is named after the Danish bacteriologist J. M. C. Gram (1853–1938).*

in living taxa such as *Paramecium*, as illustrated in **FIGURE 9.4**. Cyanobacterial viruses (cyanophages) express cyanobacterial-like photosynthesis genes, which they transfer horizontally (below) to other cyanophages and to cyanobacteria, a series of exchanges that would have facilitated the origin and spread of photosynthesis.[4]

According to one scheme, an early eukaryote lineage *first* established a symbiotic relationship with mitochondrion-like aerobic bacteria that enhanced eukaryotic metabolism and broadened eukaryotic predatory activity. Twenty-five years ago Woese and his coworkers pointed out that five of the major eubacterial groups—Gram-positive, purple, green sulfur, green nonsulfur, and cyanobacteria—possess photosynthetic species, and suggest that most, perhaps all, extant eubacterial groups may have derived from a common photosynthetic ancestor.[5] Later, one or more lines of these new aerobic eukaryotic cells began similar symbiotic relationships with photosynthetic cyanobacteria, eventually evolving into the various eukaryotic algae (brown, red, and green) and plants (**FIGURE 9.5**).

Evidence for the origins of mitochondria and chloroplasts from prokaryotic endosymbionts (**primary symbiosis**) is now strong (Figure 9.5). Evidence also now indicates that some eukaryotic algae acquired chloroplasts through **secondary endosymbiosis**—*from a eukaryotic rather than from a prokaryotic cell.* Such secondary invasions by chloroplast-carrying eukaryotes may account for the retention of an algal-type cell within some

[4] M. B. Sullivan et al., 2006. Prevalence and evolution of core photosystem II genes in marine cyanobacterial viruses and their hosts. *PLoS Biol* **4**(8): e234. doi:10.1371/journal.pbio.0040234.
[5] C. R. Woese et al., 1985. Gram-positive bacteria: possible photosynthetic ancestry. *Science*, **229**, 762–765.

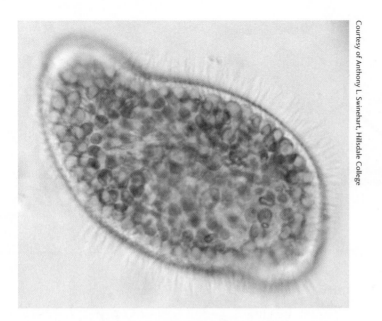

Courtesy of Anthony L. Swinehart, Hillsdale College

FIGURE 9.4 Symbiotic relationships between a eukaryote and its photosynthetic organelles. The protozoan ciliate *Paramecium bursaria* harbors hundreds of symbiotic algae (green) that may be released from the cell and cultured independently.

protozoan parasites (Burger et al., 2004). Both primary and secondary endosymbiosis would have conferred substantial selective benefits to their eukaryotic hosts.[6] A further process—**chloroplast capture**—has been shown to occur between living plant species following the grafting of cultivated tobacco (*Nicotiana tabacum*) into tobaccos of two other species. Chloroplasts from *N. tabacum* traveled across the graft and replaced chloroplasts of the species receiving the graft (Stegemann et al., 2012).

Further supporting a symbiotic origin was the discovery that ribosomes of mitochondria and chloroplasts are more similar in size, biochemistry, sensitivity to antibiotics, and nucleotide sequences of ribosomal RNA to ribosomes of prokaryotic cells than they are to host ribosomes. Such similarities are obvious in the mitochondrion of a freshwater protozoan, *Reclinomonas*, which bears the largest collection of mitochondrial genes so far discovered, and is therefore interpreted by some as more basal than other organisms, which lost genes as they evolved in concert with more derived eukaryotes. The mitochondrial genome of *Reclinomonas americana* strain ATCC 50394 is 69,034 base pairs and codes for 97 genes. Most mitochondria are only a quarter this size and code for 18 to 30 genes. Many genes remaining in *Reclinomonas* mitochondria are strikingly similar to the obligate intracellular bacterial parasite *Rickettsia prowazekii*, which is responsible for epidemic typhus, a finding that is consistent with the proposal that an aerobic rickettsial ancestor, adapted to parasitizing eukaryotic cells, gave rise to mitochondria.

MITOCHONDRIAL DNA

Further support for the symbiotic origin of mitochondria and chloroplasts comes from the discovery that both organelles have their own genetic material (DNA), inherited independently from the nuclear DNA/genes, usually from the female parent. Here, we concentrate on **mitochondrial DNA (mtDNA)**.

---

[6] Intriguingly, one eubacterium (*Gemmata obscuriglobus*) has evolved endocytosis (Forterre and Gribaldo, 2010).

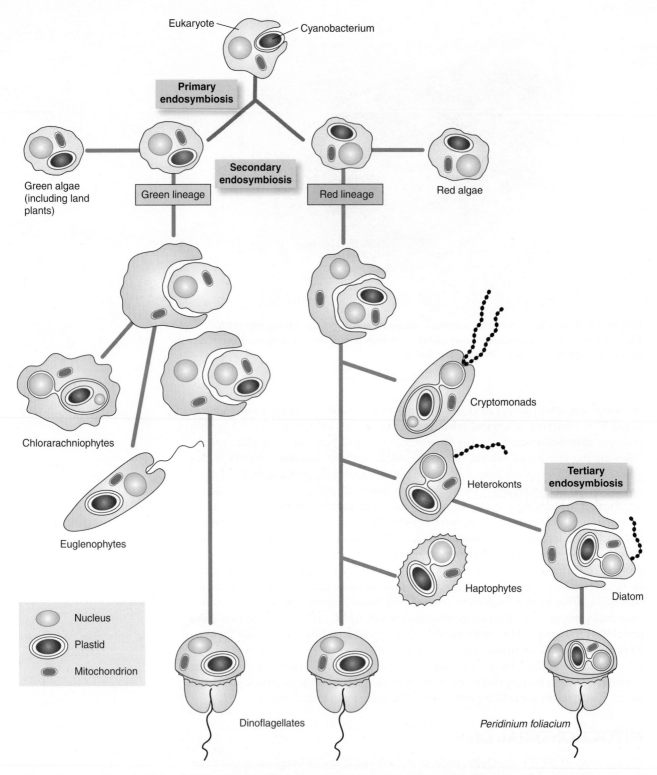

FIGURE 9.5 Primary, secondary, and tertiary endosymbiosis and the origin of organelles in various lineages of eukaryotes including algae, land plants, dinoflagellates, and cryptomonads, all of which are chimeras. As discussed in the text, horizontal gene transfer between lineages complicates reconstruction of a phylogenetic tree of organelle evolution.

[Adapted from Cracraft, J., and M. J. Donoghue [eds.], 2004. *Assembling the Tree of Life*. Oxford University Press, Oxford, UK.]

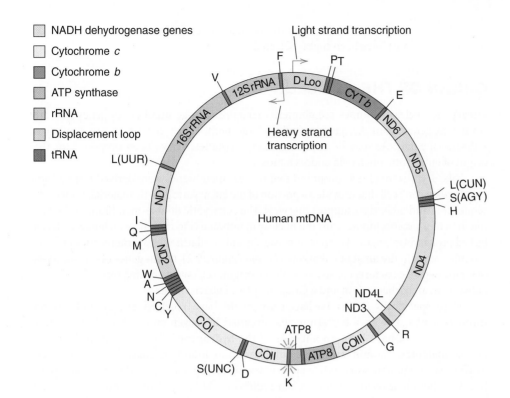

NADH dehydrogenase genes
Cytochrome *c*
Cytochrome *b*
ATP synthase
rRNA
Displacement loop
tRNA

Light strand transcription

F
PT
V
12SrRNA
D-Loo
CYT b
E
16SrRNA
ND6
ND5
Heavy strand transcription
L(UUR)
Human mtDNA
L(CUN)
S(AGY)
H
ND1
I
Q
M
ND4
ND2
W
A
N
C
Y
ND4L
ND3
COI
R
ATP8
G
COII
ATP8
COIII
S(UNC)
D
K

FIGURE 9.6 Genetic organization of the human mitochondrial genome showing the location of the genes for seven products as colored areas in the circular DNA. Most genes are transcribed clockwise along the light strand, but some are transcribed counterclockwise along the heavy strand (shown by the red arrows at the top).

[Reprinted with permission from the *Annual Review of Genetics*, Volume 29 © 1995 by Annual Reviews [www.Annual Reviews.org]. Illustration courtesy of David A. Clayton, Howard Hughes Medical Institute.]

In most animals, mitochondrial DNA consists of a single circular molecule, typically around 16,000 base pairs; human mtDNA is 16,569 base-pairs long (FIGURE 9.6). Although two copies of nuclear DNA are inherited, one from each parent, only a single copy of mtDNA in each mitochondrion is inherited from a single parent, in almost all cases in the egg and therefore from the female parent. The mtDNA in a single mitochondrion neither recombines with nuclear DNA nor with the DNA in other mitochondria. Mitochondria do divide, however; thus each cell may have thousands of copies of mtDNA.

Changes in nuclear DNA result from genetic recombination at meiosis *and* from random mutations. As a result of the uniparental inheritance of mitochondria, mutations, rather than recombination, are the only source of difference among copies of mtDNA. Because mtDNA does not undergo genetic recombination, and because mitochondria cannot repair damage to their DNA, more changes accumulate over time in mitochondrial than in nuclear DNA. Recently, a case has been made for a major role of mtDNA in the evolution of eukaryotes, the 200,000-fold increase in eukaryotic genomes over prokaryotic being attributed to the oxidative phosphorylation across cell membranes facilitated by mtDNA (Lane and Martin, 2010).

ORGANELLE GENES MOVE TO THE NUCLEUS

Nuclear genes that code for symbiotic organelle proteins have been identified widely in eukaryotes. The proteins are made by the common cellular cytoplasmic translation process and are transported to the correct organelle position using special "signal" and "transit" peptides (coupled with signal peptides in those chloroplasts of secondary endosymbiotic origin).

Once incorporated into the eukaryotic cytoplasm, symbiotic organelles transferred many of their genes to the eukaryotic nucleus (Gillham, 1994). Other organelle genes were lost; genes such as those for anaerobic glycolysis and amino acid biosynthesis could be provided by host nuclei. Such transfers and deletions maintained and replicated only two copies of a symbiotic gene in a host nucleus instead of sustaining a separate gene copy within each of many cellular organelles. Because some cells carry enormous numbers of

More than 8,000 COPIES *of the mitochondrial genome exist in some human cells and even more copies of chloroplast genomes in some plant cells.*

organelle genomes (above), reductions in organelle gene number by deletion or nuclear incorporation must have been highly selected.

ORIGIN OF THE NUCLEUS

Eukaryotic nuclear genomes are chimeric, containing large numbers of genes that are similar to genes from Archaea as well as large numbers of genes that are similar to Eubacterial genes. As may be expected, many hypotheses have been proposed for the origin of eukaryotic nuclei via endosymbiosis.

Lake and Rivera (1994) proposed that nuclear membranes were derived from a captured prokaryotic cell that provided a portion of the eukaryotic genetic material. Molecular sequencing and other data support this view. The eukaryotic double-membrane endoplasmic reticulum is continuous with the nuclear membrane (Figure 9.1), which may reflect a linked prokaryotic origin.[7] Furthermore, significant similarities exist among endoplasmic reticular genes in the ancient eukaryote *Giardia* (Figure 9.2) and in genes of certain types of Gram-negative bacteria consistent with the original eukaryote having formed by fusion between an archaebacterium and a Gram-negative eubacterium.

More specifically, an archaebacterium could have received energy and carbon through symbiosis with a bacterium that excreted hydrogen and carbon dioxide. This bacterium would have been an ancestor of the **hydrogenosome**, a 1 μm in diameter, double-membrane bound modified mitochondrion found in anaerobic ciliates such as *Giardia*, fungi, and some other eukaryotes, including species in the enigmatic phylum Loricifera that reside deep in ocean sediments. Although hydrogenosomes share a common ancestry with mitochondria, they have a radically different, anaerobic energy metabolism. As discussed in a subsequent section, other microbiologists explain the presence of both "eubacterial-like" and "archaeal-like" genes in the eukaryotic nucleus as the product of numerous horizontal gene transfer events involving endosymbionts and prey organisms, rather than one dramatic cell fusion event.[8]

GENE STRUCTURE

Split genes are among the features of eukaryotic genes that are almost entirely absent from prokaryotes. A split gene is a gene in which amino acid–coding nucleotide sequences that may be separated by hundreds of bases in DNA are combined in the final mRNA to translate into a single polypeptide product. The nucleotide sequences that code for amino acids in such polypeptides are called **exons** (*expressed sequences*). The intermediate noncoding nucleotide sequences are called **introns** (*intervening sequences*). Split genes transcribe their nucleotide sequences from DNA to RNA, but the RNA is processed so that the introns are removed and the exons spliced together (**FIGURE 9.7a**). Processed and spliced mature mRNA molecules are transported through pores of the nuclear envelope to the cytoplasm where they are translated into polypeptides.

Split genes occur in almost all vertebrate protein-coding genes examined thus far, and in many similar genes in other eukaryotes. Originally each exon may have coded for a single polypeptide domain with a specific function. Because exon arrangement and intron removal are flexible, the exons coding for these polypeptide subunits act as **modules** (sometimes called *domains*), combining in various ways to form new genes. Single genes can produce different functional proteins by arranging their exons in several different ways

GRAM–NEGATIVE BACTERIA *are so named because they do not stain with crystal violet in the Gram staining technique.*

---

[7] The eubacterium *Gemmata obscuriglobus*, which displays endocytosis, also has a double membrane around its DNA (Forterre and Gribaldo, 2010).
[8] For hydrogenosomes, see W. Martin and M. Müller, 1998. The hydrogen hypothesis for the first eukaryote. *Nature*, **392**, 37–41. For lateral gene transfer, see W. F. Doolittle, 1998. You are what you eat: a gene transfer ratchet could account for bacterial genes in eukaryotic nuclear genomes. *Trends Genet.*, **14**, 307–311.

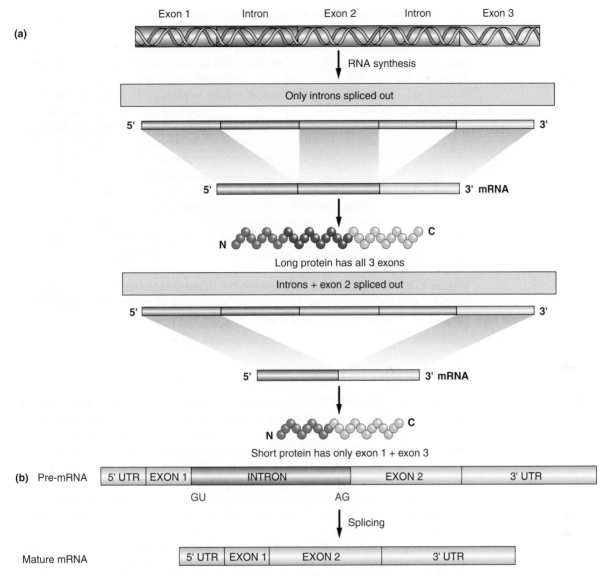

FIGURE 9.7 **(a)** Alternate splicing allows a single gene to produce more than one mRNA and so more than one protein. **(b)** Further details of introns splicing. The regions preceding and following the exons are transcribed but not translated 5′ and 3′ untranslated regions (UTRs). The intron, which lies between two exons, contains splicing signals (GU and AG dinucleotides) that are recognized by a nuclear assembly known as the spliceosome that splices the intron from the pre-mRNA.

through **alternate splicing** patterns (Figure 9.7**a**), a mechanism now known to play a major role in generating both protein and functional diversity in animals (Blencowe, 2006).[9]

INTRONS EARLY OR INTRONS LATE

Because genes of prokaryotic cells lack introns but genes of eukaryotic cells possess introns, it is logical to assume that, as with organelles, introns arose with eukaryotic genes. However, molecular geneticists have hypothesized that introns could have existed *before* eukaryotes arose.[10] Doolittle (1978) and others, notably Fedorov et al. (2001), proposed

[9] In alternate splicing, introns are removed from pre-mRNA and exons are spliced together to generate a code that can then be translated.
[10] See Catania and Lynch (2008) for an overview of the theories for the origin of introns.

an "**introns early**" hypothesis in which cellular organisms ancestral to both prokaryotic and eukaryotic cells had introns. As prokaryotic genes *lack* introns, a consequence of this hypothesis is that introns were lost as prokaryotic cell lineages evolved, perhaps as selection streamlined DNA replication and improved transcription efficiency. Furthermore, as prokaryotic cells lack nuclei, and because ribosome attachment and protein synthesis occur simultaneously with transcription of mRNA in prokaryotic cells, any introns in a prokaryotic cell would translate as part of mRNA, resulting in nonfunctional proteins.

An "**introns late**" hypothesis suggests that introns first inserted into full-length eukaryotic genes as transposable elements *after* eukaryotes arose (Sverdlov et al., 2007); gene assembly did not depend on the initial use of introns to connect and shuffle modular exons. Evidence for this hypothesis includes tests showing lack of correspondence between exons and functional protein units, and the demonstration that some introns are of recent origin in eukaryotes such as diatoms and *Drosophila*.

A decision between the two hypotheses is yet to come. Indeed, both may be correct, for it has been proposed that as many as 40% of introns were "early" and the remainder "late."[11] In part, a multiple origin hypothesis reflects that not all introns are the same: spliceosomal introns found only in eukaryotes are removed during the formation of mRNA (Figure 9.7**b**); Group-I, -II and -III introns splice themselves (self-splicing) from the pre-mRNA.[12] Given their ability to modulate the acquisition of transposable elements and introns, spliceosomes may be important mediators of variation underlying complexity (Gray et al., 2010).

ACQUIRING INTRONS?

If introns arose late, how did introns enter eukaryotic genes? Why did they become so prevalent? Why are they maintained? Answers vary, but introns are mobile DNA sequences that can splice themselves out of, as well as into, specific "target sites," acting like **mobile transposon-like elements** that mediate transfer of genes between organisms—**horizontal (lateral) gene transfer** (below). Insertion of such elements adds excess nucleotides to mRNA and so could disrupt normal gene expression. Consequently, their survival relies on splicing themselves out of RNA transcripts before translation, that is, by acting as introns. Indeed, a recently published method allows the importance of alternate splicing of exons to be determined on a tissue by tissue basis by assembling a "splicing code." Features of RNA from mouse exons and their neighboring and intervening introns were gathered and used to predict whether an exon is alternatively spliced and whether patterns of splicing are tissue specific. The complexity of the code was reflected by the number of tissue-specific features per exon (Barash et al., 2010)[13].

Using knowledge that the protein product of one gene can regulate the expression of other genes in prokaryotes, some researchers have proposed that nucleotide sequences in eukaryotic introns play regulatory roles. Unlike prokaryotic cells, RNA processing takes place in the nucleus in eukaryotic cells. As intron-less mRNA translates in the cytoplasm, nuclear introns could function in a regulatory role without affecting mRNA translation, providing a mechanism that would have enabled the evolution of the more complex gene

---

[11] J. M. Logsdon Jr. et al. 1998. Molecular evolution: recent cases of spliceosomal intron gain? *Curr. Biol.*, **8**, R560–R563; S. J. De Souza et al., 1998. Towards a resolution of the introns early/late debate: only phase zero introns are correlated with the structure of ancient proteins. *Proc. Natl Acad. Sci. USA*, **95**, 5094–5099; J. Coulombe-Huntington and Majewski, 2007. Intron loss and gain in *Drosophila*. *Mol Biol Evol.*, **24**, 2842–2850.

[12] A spliceosome is made up of five small nuclear ribonucleoproteins (snRNPs) and a variable number (often hundreds) of other proteins.

[13] Barash and colleagues estimate that 15–20% of mutations resulting in human diseases affect selection of splice sites. For the evolution of exons in the human genome, see X. H.-F Zhang and L. Chasin, 2006. Comparison of multiple vertebrate genomes reveals the birth and evolution of human exons. *Proc. Natl Acad. Sci. USA*, **103**, 13427–13432.

systems found in eukaryotes. The increase in the number of genes from a few thousand in prokaryotic cells to tens of thousands in many eukaryotic cells may then have depended on adding introns to the gene regulatory system.

HORIZONTAL GENE TRANSFER AND THE TREE(S) OF LIFE

Over the past decade, a major change has emerged in our thinking concerning the origin of genes in individual lineages of organisms The change results from the discovery of widespread horizontal gene transfer (HGT), the relocation and successful incorporation of one organism's or species' DNA into the nuclear DNA of another organism or species. HGT between individuals contrasts dramatically with the vertical transmission of genes from generation to generation. HGT obscures phylogenetic relationships when otherwise distantly related organisms share a gene or sequence obtained through HGT rather than through common ancestry.

Horizontal gene transfer occurs between lineages of prokaryotic cells, from prokaryotic to eukaryotic cells, and more rarely, between eukaryotic cells. Once thought impossible, then possible but insignificant, then possible but only between prokaryotic cells, HGT is now hypothesized to have played a major role in the evolution of life on Earth. Furthermore, HGT continues today. Indeed, fluorescent tracking enabled Babic et al. (2008) to visualize and record HGT (including segregation of the gene into different chromosomes during replication) in *E. coli* in real time. At least 10% of the genomes of unicellular organisms arose by HGT. In some lineages the value is as high as 90%. Overall, as much as one third of the genome of some prokaryotic organisms has been acquired through HGT. Disproportionate numbers of genes have transferred to prokaryotic cells with large genomes, preferentially enriching gene classes that are present in high levels in those genomes. These data suggest that gene transfer is not random. Evidence for HGT in eukaryotic cells is based on the mosaic nature of eukaryotic nuclear genomes, which contain large numbers of genes that are similar to genes from Archaea and from Eubacteria.

HGT is carried out via a virus or a small, circular DNA particle known as a plasmid containing a foreign gene that can be transferred. Antibiotic resistance genes are transmitted between bacterial strains by such a mechanism. So too are transposons, which are nucleotide sequences that promote their own movement (transposition) among different genetic loci. Movement of transposons has been so extensive that, to take only one example, more than one third of the genome of the western clawed frog *Xenopus tropicalis* consists of transposable elements.

A second mode of HGT in eukaryotes (below) is via intracellular symbionts (which live in harmony with their host) or parasites (which harm their host). A survey by Choi and Kim (2007) of some 8,000 protein domain families between and across kingdoms estimated that greater than 50% of members of the Archaea, 30% to 50% of bacteria, but less than 10% of eukaryotes, acquired one or more protein families by HGT.

TRANSFER BETWEEN UNICELLULAR ORGANISMS

Successful gene transfer has been most common between closely related unicellular organisms. Many bacterial clones share similar genes and gene sequences, indicative of HGT. So too do lineages with multiple species within a single genus, including such bacteria as *Escherichia* spp., *Salmonella* spp., and *Shigella* spp. Once established in their new hosts, mutations in horizontally transferred genes provide a source of new genetic information in addition to that provided by nuclear and organelle genes. Many pathogenic bacteria can widen their spectrum of resistance to antibiotic drugs in a decade or less as a consequence of the rapid adaptive evolution of genes obtained by HGT.

Through the analysis of enormously large data sets, increasingly of completely sequenced genomes (Wessler, 2006; Dunning Hotopp et al., 2007), we can estimate the

Data are available from 5,287 species ranging from viruses to eukaryotes. They are accessible from GenBank (http://www.ncbi.nlm.nih. gov/Genbank/). Complete GENOME SEQUENCES, *now known for over 860 organisms (83 eukaryotes, 55 Archaea, and 723 bacteria) are accessible in GenBank.*

amount and rate of HGT between prokaryotic cells over longer evolutionary time. As one example, *E. coli* diverged from the *Salmonella* lineage 100 Mya. More than 200 horizontal gene transfers occurred during the subsequent 100 My so that today almost 20% of the *E. coli* genome can be traced to HGT.[14]

In a major study published in 2005 by Beiko and colleagues, over 220,000 proteins from the completely sequenced genomes of 144 species were compared. The analyses revealed patterns of gene sharing so prevalent that the researchers referred to them as *"highways of gene sharing."*[15] As part of the analyses, closely related taxa were compared with one another and with distantly related taxa that inhabited similar environments, the latter to check for parallel evolution. Genes that code for 16S rRNA, cell wall, cell division, and ribosomal proteins were found to have rarely been subject to horizontal transfer. By contrast, other genes with more specialized functions were found to have been more subject to transfer. More recently, Dagan and Martin (2007) concluded that among 57,670 gene families from 190 genomes, at least two thirds, and possibly all, of the gene families have been subjected to HGT at sometime in their evolutionary history. HGT, the origin of organelles (and their genes) by endosymbiosis, and mutation are three major means by which lineages of organisms acquire new genetic information.

TRANSFER TO OR BETWEEN MULTICELLULAR ORGANISMS

Among eukaryotes investigated to date, HGT appears most prevalent among bdelloid rotifers, a group of asexually-reproducing freshwater invertebrates. The rotifer species *Adineta vaga* has inherited genes from bacteria, fungi, and plants. Because the species lacks sexual reproduction, HGT provides a means of enhancing its genetic diversity. The marine green alga *Ostreococcus tauri*, which is the smallest known free-living eukaryote, *may have* acquired all of one of its 19 chromosomes by HGT. Even more dramatic, the fruit fly *Drosophila ananassae* has acquired the entire genome of a bacterial parasite (*Wolbachia*) by HGT. The first demonstration (from one of the authors' laboratories) of an endosymbiotic intracellular association between a green alga and all life history stages of the spotted salamander (*Ambystoma maculatum*) suggests that such endosymbioses may be more common than previously imagined (Kerney et al., 2011).

EVOLUTION OF EUKARYOTES

Unlike eukaryotes, extant prokaryotic unicellular organisms—which include Eubacteria and Archaea (see below)—are small, contain no nuclear membrane or complex organelles, and divide by binary fission (Moseley and Nurse, 2010). Eukaryotic genes contain introns, which prokaryote genes do not. Eukaryotes are generally aerobic and contain many organelles and a microtubular apparatus for mitotic cell division (below).

Whatever the origin of their membranes, organelles, genes, and metabolism, eukaryotic cells appeared as fossils at least 1.8 Bya (Parfrey et al., 2011). The 1.7 By interval between the age of fossils[16] and the earliest prokaryotic cells at 3.5 Bya may reflect the length of time needed to incorporate and coordinate the many profound changes in cell structure and function associated with the evolution of eukaryotic cells. Gaining a foothold in a world dominated by prokaryotic cells would have taken hundreds of millions of years. As

---

[14] J. G. Lawrence, and H. Ochman, 1998. Molecular archaeology of the *Escherichia coli* genome. *Proc. Natl Acad. Sci. USA*, **95**, 9413–9417.

[15] R. G. Beiko et al., 2005. Highways of gene sharing in prokaryotes. *Proc. Natl Acad. Sci. USA*, **102**, 14332–14337.

[16] Molecular fossils/biomarkers attributed to eukaryotes have been found in 2.7 By old rocks, so the gap between prokaryote and eukaryote origins may be much much less than 1.7 By and closer to 0.8 By (Parfrey et al., 2011).

one example, the near-universal presence of cytoskeletal microtubules in eukaryotic cells (Figure 9.3) reflects their association with nuclear chromosomal division, which contrasts with the prokaryotic method that depends on separating dividing chromosomes by their individual attachment to a lengthening cell membrane.[17] Each of the relatively larger, more complex, and often more numerous eukaryotic chromosomes has a **centromere** (or **kinetochore**) that attaches to a microtubular network of sliding spindle fibers during division, allowing the daughter chromosomes to move to opposite poles of the cell (FIGURE 9.8b).

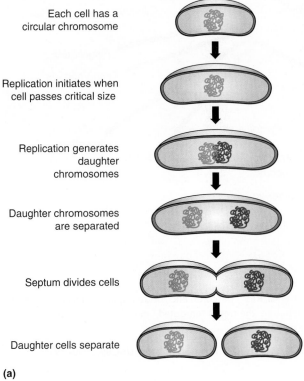

Each cell has a circular chromosome

Replication initiates when cell passes critical size

Replication generates daughter chromosomes

Daughter chromosomes are separated

Septum divides cells

Daughter cells separate

(a)

Subunits

Microtubule

CENTROMERE

Microtubule

Shortens

Sister kinetochores

Elongates

(b)

FIGURE 9.8 Prokaryote binary cell division **(a)** and centromere and kinetochores **(b)**.

[17] The association between chromosome type and nuclear membranes as a better dichotomy than prokaryote-eukaryote is elaborated by A. J. Bendish and K. Drlica, 2000, Prokaryotic and eukaryotic chromosomes: what's the difference? *BioEssays*, **22**, 481–486.

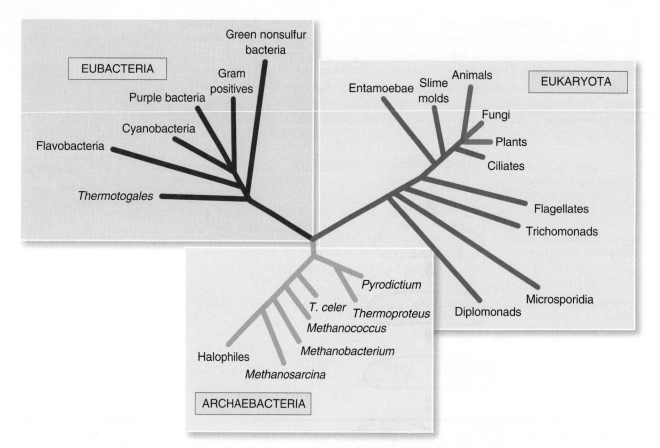

FIGURE 9.9 One possible phylogenetic tree for all cellular life ("The Universal Tree") based on nucleotide sequence similarities among some ribosomal components (small-subunit RNA, 16S rRNA). In this illustration, the tree is "unrooted," meaning that no source ("root") is indicated for the origin of the common ancestor. See text for details.

[Adapted from Sogin, M. L., 1991. *Curr. Opin. Genet. Dev.* **1**, 457–463, and Wheelis, M. L., et al., 1992. *Proc. Natl. Acad. Sci. USA*, **89**, 2930–2934.]

One of the first organisms to have its complete genome sequenced was the methane-producing marine "bacterium," *Methanococcus jannaschii*, found at depths of 2,600 m where the pressure is over 200 atmospheres.[18] Although Eubacteria do not contain histone proteins, *M. jannaschii* has genes coding for histones. Furthermore, 56% of the genome of *M. jannaschii* is not found in other organisms. These data are sufficient to place *M. jannaschii* and other such organisms into a separate domain of life, the Archaea. With this and subsequent genetic analyses, **Eubacteria**, **Archaea**, and **Eukarya** were recognized as three domains of life that cut across the traditional five kingdom and two division (prokaryotes and eukaryotes) scheme. This scheme divided Prokaryotes into two groups, and increases them in rank and equivalence to eukaryotes. The status of other organisms also changed. As one example, slime molds and ciliates, long regarded as prokaryotes, now nest with plants and animals within the Eukarya[19] (**FIGURE 9.9**). This scheme of three domains has worked well and fits well with the current consensus that Archaea are more closely related to Eukaryotes than they are to Eubacteria.

All early eukaryotes were single-celled organisms or simple filaments (**FIGURE 9.10**). Larger, visible **multicellular fossils** with differentiation into several cell types or tissues

Organisms that live under such extreme conditions are often called EXTREMOPHILES.

[18] C. J. Bult et al., 1996. Complete genome sequence of the methanogenic archaeon, *Methanococcus jannaschii. Science*, **273**, 1058–1073.
[19] The first complete genome of a species from the Eukarya domain—the budding yeast, *Saccharomyces cerevisiae*—was published in 1997 (R. A. Clayton et al., 1997. The first genome from the third domain of life. *Nature*, **387**, 459–462).

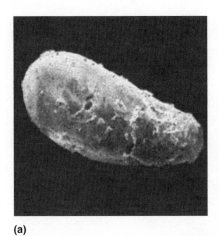
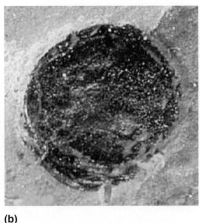
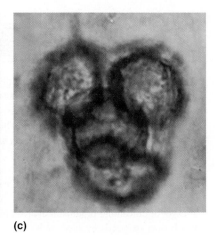

(a) (b) (c)

FIGURE 9.10 Microfossils of probable eukaryotic cells that date back to the Proterozoic. The cells in **(a)** and **(b)** are many times larger than any known prokaryotic cells and are considerably more complex. The group of cells in **(c)** is in a characteristic tetrahedral arrangement, suggesting they formed through either mitotic or meiotic eukaryotic cell division mechanisms.

[Reproduced from *The Evolution of the Earliest Cells* by J. W. Schopf, 1978. *Sci. Am. 239.* Reprinted by permission. Images courtesy of J. William Schopf, Professor of Paleobiology & Director of IGPP CSEOL.]

are first found just before the beginning of the Phanerozoic. When the Cambrian, the first of the Phanerozoic periods, began about 545 Mya, numerous forms of multicellular animals that could fossilize made a sudden marked appearance. In an interval of perhaps 10 My or less before then, an explosive radiation of eukaryotes occurred in which a large number of animal lineages (phyla) appeared.

Evolutionary biologists have long focused on understanding the derivations and relationships of these animals, as well as on the **origin of multicellularity**, a state that evolved many times in the history of life (**BOX 9.1**). The advantages of multicellularity over the unicellular condition are many and include:

- Potential for increase in size beyond the limits set by the surface-to-volume of single cells;
- Specialization into distinct cell types, each (or each compartment within the organism) with its own function, such as food gathering, reproduction, or protection from predators; and
- Enhanced dispersal, especially of immature stages.

Once distinct cell lineages arise, selection acts on those cell lineages, which are major parts of the phenotype of each organism. A key innovation of multicellularity was the ability to regulate where and when transcription occurs within multicellular embryos or organisms (Box 9.1). Within distinctive cell lineages genes became linked into networks or cascades, a different cascade for each cell lineage, thus furthering diversification. Multicellularity facilitated increasing complexity. Ravasz and colleagues provided support for tightly organized and coherent metabolic networks organized hierarchically across 43 organisms at very different levels of organization—a systems-level feature whose basic elements would have responded strongly to selection associated with the evolution of multicellularity.[20]

CONSTRUCTING TREES OF LIFE

Determining relationships between Archaea and Eubacteria on the basis of molecular data and then generating a tree based on those relationships may be an impossible task if much of the DNA and so many of the genes in these organisms were acquired by

[20] E. Ravasz et al., 2002. Hierarchical organization of modularity in metabolic networks. *Science,* **297**, 1551–1555; and see elsewhere in this text.

BOX 9.1 Multicellularity

Increase in organismal size can, and usually does, have a considerable effect on relative fitness. One way to achieve increase in size is to become **multicellular**; that is, to consist of more than one cell. Plants (metaphytes) and their allies the green algae, animals (metazoans), brown algae, red algae, and fungi all contain species that are multicellular.

Whether we examine genomic and cellular organization or phylogenetic relationships, the evidence indicates that multicellularity has evolved numerous times. At least 25 different lineages of unicellular organisms evolved multicellularity. Red and brown seaweeds evolved multicellularity independently of true plants; multicellularity evolved on six occasions in the green algae Volvocales (*Volvox*)[a] and independently in other flagellated green algae; fungi evolved multicellularity more than once; and so on (Grosberg and Strathmann, 2007).

The origin of multicellularity would have required single cells that divided into two cells (individuals) to develop mechanisms either to:

- **prevent the two cells from separating**, producing an organism whose cells would have had the *same genetic constitution*, or
- **facilitate aggregation with (a) cell(s) from another individual**, potentially producing an organism whose cells would have *different genetic constitutions*.

Some multicellular organisms such as *Volvox* arise by cells staying together. Others, such as slime molds and some ciliated protozoa, arise by aggregation. Although aggregation and failure to separate are different cellular mechanisms, both reflect properties of cell membranes, such as the presence of cell adhesion molecules or ionic coupling.

A detailed comparison of the genome of *Volvox carteri* with that of the single-celled green alga, *Chlamydomonas reinhardtii*, was published in 2010. The researchers expected to find many differences that would shed light on one of the many origins of multicellularity.[b] One expectation was a major increase in numbers of genes with multicellularity. Although each individual *V. carteri* consists of some 2,000 flagellated cells and 16 germ cells embedded in an extracellular matrix, while each *C. reinhardtii* is a single cell (**FIGURE B1.1**), the genomes each contain around 14,500 genes. (At 138×10^6 base pairs, the *Volvox* genome is 17% larger, but this is a consequence of more repetitive DNA, not more genes; Prochnik et al., 2010.) *If* these genomes are representative of ancestral forms of

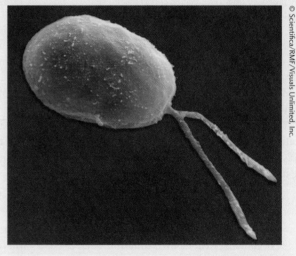

(a)

20μm

(b)

FIGURE B1.1 **(a)** *Volvox carteri.* **(b)** *Chlamydomonas reinhardtii.*

uni- and multicellular green algae, the implication is that the acquisition of multicellularity among algae was not associated with major increases in numbers of genes. The origin of multicellularity must have explanations other than increase in the number of genes in the genome.

Choanoflagellates, the closest known relatives to animals, are a group of 125 species of unicellular, aquatic eukaryotes. Their phylogenetic position makes choanoflagellates candidate species to provide insights into the nature of the genome in the last common unicellular ancestor of choanoflagellates and animals, as well as into genomic changes associated with this particular evolutionary origin of multicellularity (Carr et al., 2008). The 41.6 Mb genome of a choanoflagellate, *Monosiga brevicollis*, has 9,200 intron-rich genes, including genes that encode cell adhesion, cell signaling, and extracellular matrix molecules such as cadherins and tyrosine kinases otherwise restricted to animals. Researcher Nicole King and

---

[a] The smaller multicellular Volvocales are comprised of between four and 64 cells embedded in a common extracellular matrix. The largest, such as *Volvox*, consist of thousands of cells (Bonner, 2006).

[b] The genome of *Chlamydomonas reinhardtii*, published in 2007, revealed previously unknown genes and protein products that could shed light on the evolution of chloroplast and flagella function (S. S. Merchant et al., 2007. The *Chlamydomonas* genome reveals the evolution of key animal and plant functions. *Science* **318**, 245–250).

BOX 9.1   **Multicellularity (Cont...)**

her colleagues conclude that introns were lost and genes for proteins involved in cell signaling modified substantially with the origin of multicellularity.[c]

Brown algae (Phaeophyceae) are photosynthetic seaweeds distantly related to green plants. The 214 Mbp genome of *Ectocarpus siliculosus* contains many signal transduction genes, especially receptor kinases not found in green plants. The 16,256 protein-coding genes in *E. siliculosus* are intron-rich (seven introns per gene) with almost a quarter of the genome consisting of transposons, many of which produce small RNAs that silence the transposons.[d] New insights into the origin of multicellularity continue to accumulate. A further study from King's laboratory demonstrated that the transition from single cells to multicellular colonies in the choanoflagellate *Salpingoeca rosetta* occurs by cell division rather than by cell aggregation. These researchers conclude that multicellularity may have arisen by cell division (rather than by cell aggregation) in the last common ancestor of animals and choanoflagellates (Fairclough et al., 2010).

PLURICELLULARITY

Conditions that resemble multicellularity, named **pluricellularity**,[e] evolved independently in several lineages of unicellular organisms. Pluricellularity is used to embrace such multi-individual types of organization as colony formation by corals, filament formation in bacteria, and aggregation in slime molds (**FIGURE B1.2**). Different forms of pluricellularity evolved within bacteria; some bacteria aggregate, some are colonial, while others are filamentous. The earliest organisms interpreted as colonial have been isolated from 2.1-By-old formation of black shale in Gabon. The deposition, in what is interpreted as an oxygenated water column, of these 7–120 mm long, 5–70 mm wide, and 1–10 mm thick sediment-surface dwellers could indicate that they used aerobic metabolism. So too does the age of their occurrence—2.1 Bya—which is after the increases in global oxygen concentrations 2.4–2.3 Bya.[f]

Why distinguish between multi- and pluricellularity? Because the former produces a single, many-celled organism, and the latter is the coming together of single-celled individuals, as in colony formation. Growth regulation, cell-cell recognition systems, and modes of cooperation differ fundamentally in the two conditions.

(a)

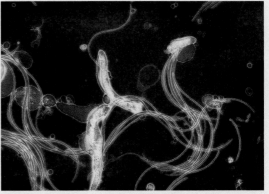

(b)

(c)

FIGURE B1.2 Examples of pluricellularity: Coral colonies **(a)**, bacterial filaments **(b)**, and aggregates of slime molds **(c)**.

[c] N. King et al., 2008. The genome of the choanoflagellate *Monosiga brevicollis* and the origin of metazoans. *Nature*, **451**, 783–788; M. Abedin and N. King, 2008. The premetazoan ancestry of cadherins. *Science*, **319**, 946–948.
[d] J. M. Cock et al., 2010. The *Ectocarpus* genome and the independent evolution of multicellularity in brown algae. *Nature*, **465**, 617–621.
[e] Sina Adl (Dalhousie University) drew the fundamental differences between multi- and pluricellularity to our attention (*pers comm.*).
[f] A. E. Albani et al., 2010. Large colonial organisms with coordinated growth in oxygenated environments 2.1 Gyr ago. *Nature*, **466**, 100–104.

FIGURE 9.11 The "Tree of Life," depicted as a reticulate web by Doolittle (2009).

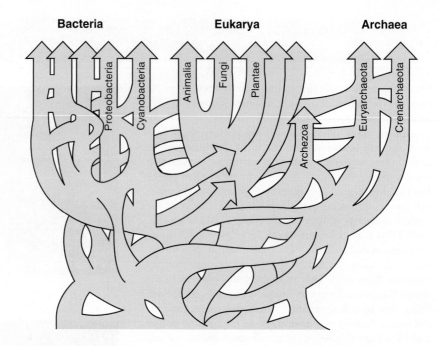

horizontal transfer rather than through vertical inheritance from a common ancestor. Even depicting the Tree of Life as "unrooted" (Figure 9.9), a depiction that makes no claims about which lineage was ancestral, may not be possible.

Because HGT is not a major factor in eukaryote evolution, recreating a eukaryote tree of life is a more realistic but still daunting project. Still, reconstructing a Universal Tree of Life (UToL) is severely compromised by HGT among prokaryotic cells. Indeed, according to Doolittle (2009), because of rampant HGT, the very notion of a Tree of Life may be an entirely inappropriate way to represent prokaryote evolution (**FIGURE 9.11**).[21] Such an interpretation is countered by studies such as that by Theobald (2010) in which phylogenetic analyses based on ubiquitously conserved proteins are interpreted as providing strong support for a UToL. Despite the practical problems it presents, the idea of a UToL has long been attractive. Indeed, Charles Darwin and Alfred Russel Wallace were the first to offer an acceptable explanation for organismal relationships and the UtoL when they tied all organisms together by community of descent through evolution by natural selection.

■ KEY TERMS

alternate splicing	hydrogenosome
Archaea	intron
chloroplast	mitochondrial DNA (mtDNA)
choanoflagellates	multicellularity
compartments	organelles
endosymbiosis	pluricellularity
Eubacteria	primary endosymbiosis
Eukarya	secondary endosymbiosis
exon	symbiosis
horizontal (lateral) gene transfer	Universal Tree of Life (UToL)

---

[21] For the development of an alternative to the Universal Tree of Life, see Doolittle (1999, 2009) and Doolittle and Bapteste (2007).

■ EVOLUTION ON THE WEB

Explore evolution on the Internet! Visit the accompanying website for *Strickberger's Evolution, Fifth Edition,* at **go.jblearning.com/Evolution5eCW** for exercises and links relating to topics covered in this chapter.

■ REFERENCES

Adl, S. M., A. G. B. Simpson, M. A. Farmer, et al., 2005. The new higher level classification of Eukaryotes with emphasis on the taxonomy of protists. *J. Eukaryot. Microbiol.*, **52**, 399–451.

Babic, A., A. B. Lindner, M. Vulic, E. J. Stewart, and M. Radman, 2008. Direct visualization of horizontal gene transfer. *Science*, **319**, 1533–1536.

Barash, Y., J. A. Calarco, W. Gao, et al., 2010. Deciphering the splicing code. *Nature*, **465**, 53–59.

Blencowe, B. J., 2006. Alternative splicing: new insights from global analyses. *Cell*, **126**, 37–47.

Bonner, J. T., 2006. *Why Size Matters: From Bacteria to Blue Whales*. Princeton University Press, Princeton, NJ.

Burger, G., M. W. Gray, and B. F. Lang, 2004. Mitochondrial genomes: anything goes. *Trends Genet.*, **19**, 709–716.

Carr, M., B. S. C. Leadbeater, R. Hassan, M. Nelson, and S. L. Baldauf, 2008. Molecular phylogeny of choanoflagellates, the sister group to Metazoa. *Proc. Natl Acad. Sci. USA*, **105**, 16641–16646.

Catania, F., and M. Lynch, 2008. Where do introns come from? *PloS Biol.* **6**(11), e283. doi:10.1371/journal.pbio.0060283.

Cavalier-Smith, T., 2002. The phagotrophic origin of eukaryotes and phylogenetic classification of Protozoa. *Int. J. Syst. Evol. Microbiol.*, **52**, 297–354.

Cavalier-Smith, T., 2003. Protist phylogeny and the high-level classification of Protozoa. *Eur. J. Protistol.*, **39**, 338–348.

Choi, I.-G., and S.-H. Kim, 2007. Global extent of horizontal gene transfer. *Proc. Natl Acad. Sci. USA*, **104**, 4489–4494.

Dagan, T., and W. Martin, 2007. Ancestral genome sizes specify the minimum rate of lateral gene transfer during prokaryote evolution. *Proc. Natl Acad. Sci. USA*, **104**, 870–875.

Diao, X., M. Freeling, and D. Lisch, 2006. Horizontal transfer of a plant transposon. *PLoS Biol.*, **4**, 119–128.

Doolittle, W. F., 1978. Genes-in-pieces: were they ever together? *Nature*, **272**, 581.

Doolittle, W. F., 1999. Lateral genomics. *Trends Cell Biol.*, **9**, M5–M8.

Doolittle, W. F., 2009. The practice of classification and the theory of evolution, and what the demise of Charles Darwin's tree of life hypothesis means for both of them. *Phil. Trans. Roy. Soc. Lond., Series B*, **364**, 2221–2228.

Doolittle, W. F., and E. Bapteste, 2007. Pattern pluralism and the tree of life hypothesis. *Proc. Natl Acad. Sci. U.S.A.*, **104**, 2043–2049.

Dunning Hotopp, J. C., M. E. Clark, D. C. Oliveira, et al., 2007. Widespread lateral gene transfer from intracellular bacteria to multicellular eukaryotes. *Science*, **317**, 1753–1756.

Fairclough, S. R., M. J. Dayel, and N. King, 2010. Multicellular development in a choanoflagellate. *Curr. Biol.*, **20**, R875–R876.

Fedorov, A., X. Cao, S. Saxonov, et al., 2001. Intron distribution difference for 276 ancient and 131 modern genes suggests the existence of ancient introns. *Proc. Natl Acad. Sci. USA*, **98**, 13177–13182.

Forterre, P., and S. Gribaldo, 2010. Bacteria with a eukaryotic touch: a glimpse of ancient evolution? *Proc. Natl. Acad. Sci. USA*, **107**, 12739–12740.

Gillham, N. W., 1994. *Organelle Genes and Genomes*. Oxford University Press, Oxford, UK.

Gray, M. W., G. Burger, and B. F. Lang, 1999. Mitochondrial evolution. *Science*, **283**, 1476–1481.

Gray, M. W., G. Burger, and B. F. Lang, 2001. The origin and early evolution of mitochondria. *Genome Biol.*, **2**, 1–5.

Gray, M. W., J. Lukes, J. M. Archibald, et al., 2010. Irremediable complexity? *Science*, **330,** 920–921.

Grosberg, R. K., and R. R. Strathmann, 2007. The evolution of multicellularity: a minor major transition? *Annu. Rev. Ecol. Evol. Syst.* **38**, 621–654.

Keeling, P. J., G. Burger, D. G. Durnford, et al., 2005. The tree of eukaryotes. *Trends Ecol. Evol.*, **20**, 670–676.

Kerney, R., E. Kim, R. P. Hangarter, et al., 2011. Intracellular invasion of green algae. *Proc. Natl Acad. Sci. USA*, **108**, 6497–6502.

Knoll, A. H., 2003. *Life on a Young Planet: The First Three Billion Years of Evolution on Earth.* Princeton University Press, Princeton, NJ.

Lake, J. A., and M. C. Rivera, 1994. Was the nucleus the first endosymbiont? *Proc. Natl Acad. Sci. USA*, **91**, 2880–2881.

Lane, N., 2005. *Power, Sex, Suicide. Mitochondria and the Meaning of Life.* Oxford University Press, Oxford, UK.

Lane, N., and W. Martin., 2010. The energetics of genome complexity. *Nature*, **467**, 929–934.

Moseley, J. B., and P. Nurse, 2010. Cell division intersects with cell geometry. *Cell*, **142**, 189–193.

Pagel, M. (ed. in chief), 2002. *Encyclopedia of Evolution.* 2 Volumes, Oxford University Press, New York.

Parfrey, L. W., D. J. G. Lahr, A. H. Knoll, and L. A. Katz. 2011. Estimating the timing of early eukaryotic diversification with multigene molecular clocks. *Proc. Natl Acad. Sci. USA*, **108**, 13624–13629.

Prochnik, S. E, J. Umen, A. M., Nedelcu, et al., 2010. Genomic analysis of organismal complexity in the multicellular green alga *Volvox carteri. Science* **329**, 223–226.

Simpson, A. G., 2003. Cytoskeletal organization, phylogenetic affinities and systematics in the contentious taxon Excavata (Eukaryota). *Int. J. Syst. Evol. Microbiol.*, **53**, 1759–1777.

Stegemann, S., M. Keuthe, S. Greiner, and R. Bock. 2012. Horizontal transfer of chloroplast genomes between plant species. *Proc. Natl Acad. Sci. USA*, **109**, 2434–2438.

Sverdlov, A. V., M. Csuros, I. B., Rogozin, and E. V. Koonin, 2007. A glimpse of a putative pre-intron phase of eukaryotic evolution. *Trends in Genet.*, **23**, 105–108.

Theobald, D. L., 2010. A formal test of the theory of universal common ancestry. *Nature*, **465**, 219–222.

Wessler, S. R., 2006. Transposable elements and the evolution of eukaryotic genomes. *Proc. Natl Acad. Sci. USA*, **103**, 17600–17601.

Whittaker, R. H., 1959. On the broad classification of organisms. *Quart. Rev. Biol.*, **34**, 210–226.

Woese, C. R., O. Kandler, and M. L. Whellis, 1990. Toward a natural system of organisms: proposal for the domains Archaea, Bacteria and Eucarya. *Proc. Natl Acad. Sci. USA*, **87**, 4576–4579.

PART 4

Theories of Evolution and Heredity

Image © Photos.com

CHAPTER

10

Voyages of Discovery, Natural Selection, and Evolution

CHAPTER SUMMARY

Charles Darwin developed a theory of evolution based on the mechanism of natural selection. Amazingly, around the same time Alfred Russel Wallace independently developed essentially the same theory. The two theories were communicated to a meeting of the Linnaean Society of London on July 1, 1858. In this chapter, we outline evolutionary thinking before the mid-1850s (including ideas on natural selection) in order to place Darwin and Wallace's theory of a *mechanism* of evolution in context and to explain why these two British naturalists were searching for evidence that organisms change over time. Although their social, educational, and financial backgrounds differed enormously, voyages of discovery and collecting exposed both men to the diverse yet related faunas and floras of the tropics; Darwin's voyage around the world on *H.M.S. Beagle* and Wallace's trips to Brazil, Malaysia, and Indonesia. Today, Wallace is considered as having begun the study of biogeography, which explains the distribution and abundance of species around the globe. The lives of the two men are outlined, both because they are fascinating and as background to the discussion of the theory of evolution by natural selection.

INTRODUCTION

Accumulating knowledge of fossils in the first half of the nineteenth century led many to accept the concept that species change through time. Indeed, Lamarck proposed a mechanism—use and disuse of organs during an individual's lifetime—to explain how

Image © SidEcuador/ ShutterStock, Inc.

195

such change came about. While Charles Darwin considered that Lamarckian inheritance played a role in evolution, he and Alfred Wallace independently conceived of **natural selection** as the primary mechanism responsible for evolution. Although our current knowledge of genetics has refined our understanding of the mechanism of evolution, natural selection acting on natural variation remains the primary basis for evolutionary change of organisms' features.

EVOLUTION BY NATURAL SELECTION BEFORE DARWIN AND WALLACE

Lamarck, Darwin, and Wallace were not the first to think about change in organisms over time. Such thinking goes back to ancient Greece and the *Scala Naturae,* or Great Chain of Being, as a fixed progression of steps from lowest to highest stages.[1]

In the eighteenth century, Buffon saw natural selection—and the artificial selection of domesticated animals and plants by humans—as *agents responsible for the extinction of species* but not for the *generation of new species.* Buffon maintained that new species arose by spontaneous generation and that differences in the conditions under which spontaneous generation occurred resulted in differences between species. Later in the eighteenth century, Lamarck sought a systematic, comprehensive, and materialistic explanation for the diversity of organisms on Earth. In Lamarck's theory, environmental effects stimulated the production of new varieties in an adaptive direction, resulting in organisms that were better adapted to their environments. According to this theory, "imperfect" or "defective" species did not go extinct; organisms always could adapt their way out of extinction.

Histories of evolutionary thinking before Darwin and Wallace are usually written as a progression from Lamarck's theory of the inheritance of acquired characters to Darwin's theory of natural selection with a passing nod to Wallace. Early in the nineteenth century, however, a number of natural historians separated the origin of variations from the forces responsible for preserving variation by using a **principle of natural selection** to explain changes within species. Among these was Scottish/American physician, medical researcher, printer, and bookseller, William Charles Wells (1757–1817) and Scottish landowner, horticulturalist, and self-proclaimed evolutionary theorist, Patrick Matthew (1790–1874). Matthew, who remained active after *The Origin of Species* was published in 1859, claimed he had introduced the concept of natural selection in 1831. To emphasize the point, Matthew had "Discoverer of the Principle of Natural Selection," printed on his calling cards. In later editions of *The Origin of Species* Darwin acknowledged these prior developments of a theory of evolution by natural selection, a number of which are outlined briefly in **BOX 10.1**, along with Darwin's comments on them.

All of DARWIN'S PUBLISHED WORK *can be accessed at http://darwin-online.org.uk/. Extracts from his writings may be found in* Appleman (2001).

CHARLES DARWIN AND ALFRED WALLACE

Almost contemporaries and both active in research, Charles Robert Darwin (1809–1882) was 14 years older than Alfred Russel Wallace (1823–1913) who outlived Darwin by 31 years (**FIGURES 10.1** and **10.2**).[2] Wallace esteemed Darwin as his senior and leader in the study of evolution, an attitude that explains much of the dynamic of their interaction and shared reputation.

---

[1] See Oldroyd (1983) and Bowler (2003) for excellent histories of the study of evolution.
[2] For biographical details of the lives of Darwin and Wallace, see F. Darwin (1887), C. Darwin (1958), Raby (2001), Shermer (2002), Browne (2003), Slotten (2004), Quammen (2006), and Secord (2010). A website on Wallace is maintained at http://wallacefund.info/, and a website on Darwin at The Darwin Project (http://www.thedarwinproject.com/). The unusual spelling of Russel (usually Russell) appears to have arisen from a mistake on Wallace's birth certificate.

| BOX 10.1 | Proposals of Evolution by Natural Selection Before Darwin and Wallace |

As noted in the text, neither Darwin nor Wallace was the first to propose that organisms evolved. Darwin's grandfather published a theory of evolution in 1794 and 1796 (E. Darwin, 1796; Elliott, 2003) and Robert Chambers provided considerable evidence for evolution in a book published anonymously in 1844 (Chambers, 1844).

It is generally concluded, however, that *a theory of evolution by natural selection* became known with Darwin and Wallace's publication in 1858 and Darwin's book in 1859. Yet, as introduced in the text, a number of early 19th century natural historians, notably **William Wells** and **Patrick Matthew**, foreshadowed theories of evolution by natural selection. As neither provided sufficient evidence, their ideas seemed highly speculative. Furthermore, their papers appeared in obscure publications and so did not attract attention (this was also true for the publication by Gregor Mendel on a mechanism of inheritance in 1866). Darwin acknowledged these prior studies in those editions of *The Origin of Species* that included a historical sketch of prior work on evolutionary theory.[a]

WILLIAM WELLS (1757–1817) AND THE ORIGIN OF HUMAN SKIN COLOR

In addition to being a physician, printer, and bookseller, Wells undertook medical research. Among Wells' considerable medical investigations were studies on the origin of skin color in humans. Wells (1818) advocated that change in skin color was a response to natural selection and thought that such a mechanism could operate on other features as well.[b] He recognized that individual animals vary from one another, that domesticated animals have been "improved" by selection, and proposed that human races (which Wells defined by skin color) arose by selection on varieties with different skin colors. A relevant section from Wells' study is:

> [What was done for animals artificially] seems to be done with equal efficiency, though more slowly, by nature, in the formation of varieties of mankind, fitted for the country which they inhabit. Of the accidental varieties of man, which would occur among the first scattered inhabitants, some one would be better fitted than the others to bear the diseases of the country. This race would multiply while the others would decrease, and as the darkest would be the best fitted for the

[African] climate, at length [they would] become the most prevalent, if not the only race.[c]

In his historical sketch included in later editions of *The Origin of Species,* Darwin stated that, "[Wells] distinctly recognizes the principle of natural selection, and this is the first recognition which has been indicated. . . but he applies it only to man, and to certain characters alone."

PATRICK MATTHEW (1790–1874) AND NATURAL SELECTION OF TREES AND TIMBER QUALITY

Matthew, a Scottish landholder and fruit farmer, became sufficiently informed on the uses of timber, especially by the Navy, that he wrote a book on the topic (Matthew, 1831, and see Dempster, 1983). Matthew was aware that culling only the tallest and strongest trees (artificial selection) resulted, over time, in a decline of timber quality. He argued that culling inferior trees would improve the quality of the timber produced in subsequent generations and might even result in the formation of new varieties. Unlike Wells, Matthews discussed the broader impact of his idea:

> There is a law universal in nature, tending to render every reproductive being the best possible suited to its condition that its kind, or organized matter, is susceptible of. . . . This law sustains the lion in his strength, the hare in her swiftness, and the fox in his wiles. As nature, in all her modifications of life, has a power of increase far beyond what is needed to supply the place of what falls by Time's decay, those individuals who possess not the requisite strength, swiftness, hardihood, or cunning, fall prematurely without reproducing—either a prey to their natural devourers, or sinking under disease, generally induced by want of nourishment, their place being occupied by the more perfect of their own kind, who are pressing on the means of subsistence . . . [The] progeny of the same parents, under great differences of circumstance, might, in several generations, even become distinct species, incapable of co-reproduction.

In 1860 Matthew communicated his theory to the publishers of Darwin's *The Origin of Species*. Darwin's response in the April 21, 1860 issue of the *Gardener's Chronicle* was to:

> freely acknowledge that Mr. Matthew has anticipated by many years the explanation which I have offered of the origin of species, under the name of natural selection. I think that no one will feel surprised that neither I, nor apparently any other naturalist, has heard of Mr. Matthew's views, considering how briefly they are given, and that they appeared in the Appendix

---

[a] For an analysis of the origin of this historical sketch, see C. N. Johnson (2007), The preface to Darwin's *Origin of Species*; the curious history of the "historical Sketch." *J. Hist. Biol.,* **40**, 529–556.
[b] Wells had presented his ideas in a presentation to the Royal Society of London in 1813. The 1818 book was published after his death (see Green, 1957). The most complete discussion of Wells and the others introduced in this box may be found in Bowler (1988).

---

[c] The quotation is from an appendix to Wells (1818), which is entitled "*An account of a female of the white race of mankind, part of whose skin resembles that of a negro, with some observations on the cause of the differences in colour and form between the white and negro races of man.*"

to a work on Naval Timber and Arboriculture. I can do no more than offer my apologies to Mr. Matthew for my entire ignorance of his publication.

Darwin wrote that Matthew "clearly saw . . . the full force of the principle of natural selection."[d]

At least two others explored variation and selection.

WILLIAM LAWRENCE (1783-1867): HUMAN EVOLUTION AND HEREDITY

Sir William Lawrence, an English surgeon, rose to become President of the Royal College of Surgeons of London and Sergeant Surgeon to Queen Victoria. Lawrence published two books of his lectures in which he outlined his ideas on human evolution (Lawrence, 1816, 1819).

As Africans produced children with dark skin color even when born in climates such as found in England, Lawrence concluded that human skin color must be inherited and not a response to local climate: "The offspring inherit only [their parents'] connate peculiarities and not any of the acquired qualities" (Lawrence, 1816, p. 196). Lawrence went further than Wells in proposing that mental attributes were inherited, that human races arose through hereditary changes (what we would now regard as mutations), and that, as the breeder has improved domestic breeds, selection has improved races of humans:

> These signal diversities [in humans] which constitute differences of race in animals . . . can only be explained by

two principles . . . namely, the occasional production of an offspring with different characters from those of the parents, as a native or congenital variety; and the propagation of such varieties by generation. (Lawrence, 1819, pp. 343, 348)[e]

EDWARD BLYTH (1810-1873): SELECTION TO MAINTAIN SPECIES

English zoologist and pharmacist Edward Blyth was curator of the museum at the Royal Asiatic Society of Bengal for 21 years, during which time he established our basic knowledge of the fauna of India. Although Blyth did not use the term "natural selection," he discussed variation and artificial selection in three papers published between 1835 and 1837 (Blyth, 1835). Blyth thought that selection in nature returned organisms to an archetypal form by eliminating variants; selection maintained species, it did not create them. Although not evolutionary in current terms, the emphasis on selection does recognize an evolutionary mechanism.

Darwin corresponded with Blyth early in 1855, seeking information on variation in domestic animals, and acknowledged in the first chapter of *The Origin of Species* the great value of Blyth's knowledge. Indeed, Blyth drew Darwin's attention to the 1855 paper by Wallace on the origin of new species.

---

[d] Cited from the 6th edition (1866) of Darwin, C., *On the Origin of Species. . .* , p. xv.

[e] Such considerations of human race could be used as a basis for discrimination and racism, and so, fortunately, are not accepted today.

(a)   **(b)**

FIGURE 10.1 Portraits of Charles Darwin at 40 **(a)**, and in later life **(b)**.

(a)   (b)

FIGURE 10.2 Portraits of Alfred Wallace in middle life **(a)**, and in later life **(b)**.

[(a) Reproduced from Marchant, James (1916) *Alfred Russel Wallace — Letters and Reminiscences* (Vol. 1 ed.), London, New York, Toronto and Melbourne: Cassell and Company; (b) Reproduced from Alfred Russel Wallace (1889) *Darwinism*, London and New York: Macmillan and Co.]

Wallace, who died in 1913, witnessed the impact of the theory of evolution by natural selection for over half a century after its publication in 1859. His last word on evolution was in an interview published two weeks after his death[3]. His last book on evolution was published in March 1913, eight months before he died (Wallace, 1913).

In 1908, Wallace received the highest honor the English monarch can bestow—appointment to the Order of Merit—an exclusive order restricted to 24 living recipients. In the same year, Wallace (and Darwin posthumously) received the first (and only) Darwin–Wallace Gold Medal (**FIGURE 10.3**), struck by the Linnaean Society of London to commemorate the 50th anniversary of the publication of their statements on the theory of evolution by natural selection.[4]

Given this coupling of their careers, the early lives of these two English men were vastly different. Darwin was born to a wealthy, upper-middle-class British family whose fortunes derived largely from his father, Robert Darwin (1766–1848), and paternal grandfather, Erasmus Darwin (1731–1802), both of whom were prosperous physicians. Darwin was, in Victorian parlance, a *gentleman*; a man of good social position in possession of wealth and leisure.

By this definition, Wallace was not a gentleman. Born in Wales to a respectable but poor middle-class family, Wallace was forced to leave his grammar school at age 13 because of his father's financial difficulties, becoming apprenticed as a surveyor to one of his older brothers. Self-educated by attending lectures at the London Mechanic's Institute, Wallace worked as a surveyor until age 20.

In contrast, at age 16, Charles left grammar school for Edinburgh University to study medicine. Surgical procedures at that time were dreadfully brutal. Having witnessed two operations—one of them on a child—Darwin realized he could never become a surgeon. He transferred to Cambridge University with the intention of becoming a minister of the Church of England; being a "country parson" was considered a

FIGURE 10.3 The Darwin–Wallace Gold Medal struck by the Linnaean Society of London in 1908 to commemorate the 50th anniversary of the first report of their theories of evolution by natural selection.

[Reproduced from Linaean Society.]

---

[3] Alfred Russel Wallace interview by W. B. Northrop (1913). *The Outlook* (New York) **105**, 618–622 (also available online at http://people.wku.edu/charles.smith/wallace/S753.htm; accessed June 7, 2012).
[4] Silver medals also were awarded in 1908 to Joseph Dalton Hooker, August Weismann, Ernst Haeckel, Francis Galton, E. Ray Lankester, and Eduard Strasburger. Fifty years later (1958), 20 silver medals were awarded. One hundred years later (February 2009; the 200th anniversary of Darwin's birth) 13 silver medals were awarded (http://en.wikipedia.org/wiki/Darwin%E2%80%93Wallace_Medal). Silver Medalists from 2008 whose research is discussed in this book, include Joseph Felsenstein (phylogenetic systematics), Peter and Rosemary Grant (Darwin's finches, speciation), Lynn Margulis (endosymbiotic origin of organelles), John Maynard Smith (evolutionary theory), and H. Allen Orr (speciation).

FIGURE 10.4 Some of the species discovered by Wallace in his expedition to Malaysia and Indonesia. **(a)** The Little Paradise-Kingfisher, *Tanysiptera hydrocharis*, collected in the Aru Islands in 1858. **(b)** Wallace's golden birdwing butterfly, *Ornithoptera croesus*, collected by Wallace in Indonesia and named by him in 1859.[5] **(c)** Wallace's flying frog, *Rhacophorus nigropalmatus*, discovered in Sarawak and painted by him.

[(a) Reproduced from Richard Bowdler Sharpe Family of Kingfishers, published from 1868 to 1871; (c) Reproduced from Alfred Russel Wallace (1869) *The Malay Archipelago*, London and New York: Macmillan and Co.]

(a)

(b)

(c)

© jps/ShutterStock, Inc.

Wallace (1869) wrote: "This is, I believe, the first instance known of a 'FLYING FROG,' *and it is very interesting to Darwinians as showing that the variability of the toes which have been already modified for purposes of swimming and adhesive climbing, have been taken advantage of to enable an allied species to pass through the air like the flying lizard."*

A career naval officer, ROBERT FITZROY *rose to become Vice-Admiral. An extraordinarily able navigator, Fitzroy pioneered meteorological forecasting and the development of barometers to measure atmospheric pressure. He was Governor of New Zealand for two years.*

respectable position for a person drawn to the study of the natural world. However, his interests were neither in academic nor in clerical pursuits but in hunting, collecting, natural history, botany, and geology. Darwin despised the formal and classical education at Cambridge and was no more than a mediocre student. His father, who believed Charles had betrayed the family trust of industrious professionalism, castigated him: "You care for nothing but shooting, dogs, and rat-catching, and you will be a disgrace to yourself and all your family."

VOYAGES OF DISCOVERY

CHARLES DARWIN

In 1831 Darwin was able to put off further study for the ministry and set off on his now famous voyage around the world on *H.M.S. Beagle* (**FIGURE 10.4** and **BOX 10.2**).

The voyage of the *Beagle* lasted almost five years, from December 27, 1831 to October 2, 1836. The traditional view has been that during this time Darwin transformed himself from a casual amateur into a dedicated geologist and naturalist, although the historian of science James Secord maintains that: "By the summer after graduating from Cambridge in 1831, Darwin was at 22 probably the best educated naturalist of his age in Britain" (Secord, 2010, p. xi).

Darwin's letters recounting many of the observations made during the voyage, along with his collections of plants, animals, fossils, and minerals, excited considerable scientific interest, even before the *Beagle* returned to England in 1836. Darwin's account of the voyage was published as *Journal of Researches into the Natural History and Geology of the Countries Visited During the Voyage of H.M.S. Beagle Round the World, Under the Command of Capt. FitzRoy* (Darwin, 1838). An abridged popular version, *Voyage of the Beagle*, published in 1845, remains one of the most perceptive chronicles of exploration

---

[5] Wallace was ecstatic at capturing this specimen, recording in his field notes: "On taking it out of my net and opening the glorious wings, my heart began to beat violently, the blood rushed to my head, and I felt much more like fainting than I have done when in apprehension of immediate death. I had a headache the rest of the day, so great was the excitement produced by what will appear to most people a very inadequate cause."

BOX 10.2 Charles Darwin and the Voyage of the *Beagle*

H.M.S. Beagle was a 10-gun brigantine (a sailing ship with two masts and square-rigged sails) (FIGURE B2.1), 27 meters (about 88.5 feet) long, weighing 240 tons, the weight of a large train engine. The maiden voyage of the *Beagle* (1826–1830) was to undertake a hydrological survey of Patagonia and Tierra del Fuego. Two years into the voyage, the enormous difficulty of the task led the captain to commit suicide. He was replaced by Robert FitzRoy (1805–1865), who was then a 23-year-old Flag Lieutenant. FitzRoy completed the survey with considerable success, returning to London in October 1830 with four natives from Tierra del Fuego, whom he hoped to train as missionaries (FIGURE B2.2). In 1831 (the year Darwin began his voyage), *Beagle* had been refitted for circumnavigation in order to fix world longitudinal markings and to complete the charting of the coast of South America.

Darwin was a naturalist on board but not the official naturalist, which was a special unpaid position created by the British Admiralty on naval ships making broad geographical surveys. The Navy had appointed the ship's surgeon, Robert McCormick (1800–1890), as the official naturalist. Darwin's primary role was to serve as a dining companion to the Captain, Robert FitzRoy, to whose service his duties as naturalist were secondary. For this reason, it was of no importance to the British Admiralty that Darwin had not distinguished himself academically or even finished his studies at Cambridge. His qualifications as a gentleman, good shot, and sportsman were quite sufficient.[a]

This voyage of the *Beagle* was no pleasure cruise. In an account of the voyage published in 1995, Keith Thomson noted, "To say that the *Beagle* was extremely cramped, even given

FIGURE B2.2 A native from Tierra del Fuego.

FIGURE B2.1 *H.M.S.* (His Majesty's Ship) *Beagle* in the Strait of Magellan at the southern tip of South America.

[a] Upstaged by Darwin—who was given first choice of any natural history specimens found—McCormick left the *Beagle* in South America. Later he was the surgeon and naturalist on James Ross' 1839–1843 exploration of the Antarctic.

the expectations of the time, would be a supreme understatement. The ship was, after all, no longer than the distance between two bases on a baseball field," that distance being 27.4 m (90 ft). Darwin shared the poop cabin at the stern of the ship, a 3 × 3.5 m (about 10 × 11 ft) space with the mast in the middle, with two officers. This cabin also held a 3 × 1.8 m (10 × 6 ft) chart table and various chart lockers, as well as drawers for equipment and specimens. Darwin wrote, "I have just room to turn around and that is all."

The five-year voyage on the *Beagle* (FIGURE B2.3) enabled Darwin to observe and think about a wide range of organisms and geological formations. Darwin collected in the Brazilian tropical forests. At Punta Alta on the coast of Argentina, Darwin unearthed fossil bones of a 6 meter (19.7 ft)-tall giant sloth (*Megatherium*), and other fossil mammals resembling extant species, yet recognizably different. The primitiveness and wildness of the Tierra del Fuego Indians at the southern tip of South America

BOX 10.2 Charles Darwin and the Voyage of the *Beagle* (Cont...)

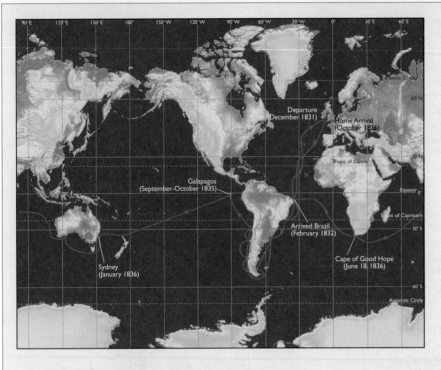

FIGURE B2.3 Route of the five-year voyage of the *Beagle,* beginning at Plymouth, England, in December 1831, and ending in Falmouth, England, in October 1836. Almost four years were spent in South America, including one month (September–October 1835) among the Galápagos Islands.

impressed Darwin, as did the severity of their struggle to subsist in a meager and unrelenting environment (Figure B2.2).

On the voyage, Darwin carried with him the first—and, at the time, the only—published volume of Charles Lyell's *Principles of Geology,* a gift from FitzRoy. Darwin had also the multivolume set of Humboldt's *Personal Narrative of Travels . . .* (Humboldt, 1814), which inspired him as it did Wallace. *H.M.S. Beagle* also had a considerable library.[b]

Darwin noted assiduously the geological features of the many terrains he covered. To explain some of the geological uplifting processes that shaped the South American landscape, Darwin gathered evidence showing the distribution of marine shells at various locations above sea level (recall the frontispiece to Lyell's book, a recent photograph of which is shown in Figure EB2.2 and an enlargement in **FIGURE B2.4**), and the terracing of land by erosion as the land was elevated. While Darwin was ashore near the Bay of Concepción on the coast of Chile, a severe earthquake raised the level of the land in some places from 1 to as much as 3 meters (from about 3 ft to 10 ft) above sea level. This experience had a deep effect on Darwin, although it did not shake his belief in uniformitarianism or gradualism. As he wrote in his account of the voyage:

> A bad earthquake at once destroys our oldest associations: the earth, the very emblem of solidity, has moved beneath our

feet like a thin crust over a fluid; one second of time has created in the mind a strange idea of insecurity, which hours of reflection would not have produced.

Darwin's account of his voyage on the *Beagle* was published in 1838, in an abridged "popular" version some years later (Darwin, 1845), and in 1988 in an edition edited by his great grandson Richard Darwin Keynes. *Voyage of the Beagle* remains one of the most perceptive chronicles of exploration published in the 19th century.

FIGURE B2.4 A contemporary photograph of the three remaining columns of the ruined "Temple of Serapis" in Pozzuoli, Italy, showing the dark, 3-meter sections is filled with holes bored by marine organisms.

[b] For the books on board *H.M.S. Beagle,* see F. Burkhardt and S. Smith, 1985. Appendix IV. Books on board the *Beagle.* Volume 1, pp. 553–566 in *C. R. Darwin, 1885–1888. The Correspondence. Volumes 1–3.* F. Burkhardt and S. Smith, (eds.), Cambridge University Press, Cambridge, UK.

(a)   **(b)**

FIGURE 10.5 (a) A map showing Wallace line separating two major zoogeographical regions in Asia. **(b)** A small fraction (0.1%) of the beetles collected by Alfred Russel Wallace in the Malay Archipelago.

published in the nineteenth century. For the famous English anatomist Richard Owen, it was "the most delightful book in my collection . . . as full of good original wholesome food as an egg . . ." For the German explorer and naturalist Alexander von Humboldt (1769–1859), whose record of his own travels inspired Darwin and Wallace, it was an "excellent and admirable book . . . which enlarged and corrected my views."[6]

ALFRED RUSSEL WALLACE

Wallace was so inspired by reading Darwins' *Journal of Researches* and other journals of naturalists/explorers/collectors, that he and a close friend, naturalist Henry Walter Bates (1825–1892),[7] set off for Brazil in 1848. They intended to collect insects to sell to museums and collectors back in England, and to seek evidence for species transformation. Their entire collection, including all their notes and diaries, was lost when the ship transporting them back to England in 1852 caught fire and sank. Darwin, on the other hand, returned to England with collections that he distributed to the experts of the day. These, along with the publication of his *Journal of Researches,* established his reputation as a naturalist.

Wallace spent a further 8 years (1854–1862) exploring and collecting in what is now Malaysia and Indonesia, this time collecting 125,000 specimens—1,000 of which were species new to science (**FIGURE 10.5**)—and laying the foundation for his ground-breaking studies in the distribution of animals (**zoogeography**), now extended as biogeography (Cox and Moore, 2005).[8] A biogeographical barrier in the Indonesian region identified by Wallace as part of his explorations and research, known today as Wallace's Line, separates species with links to Australian from species with East Asian connections (Figure 10.5).

---

[6] Letter from Owen to Darwin, 11 June, 1839; letter from Humboldt to Darwin, 18 September, 1839.
[7] See Bates (1863) for the record of his voyages along the Amazon. A form of mimicry that became known as Batesian mimicry (discussed elsewhere in this text) was first outlined by Henry Bates. For engaging accounts of how voyages of discovery influenced Darwin, Wallace, Thomas Henry Huxley (the Stanley survey of the Great Barrier Reef on board *HMS Rattlesnake*), and Joseph Hooker (the Ross Antarctic expedition on board *HMS Erebus* in 1839), see McCalman, 2009)
[8] Over the next 14 years Wallace wrote two books on zoogeography, one on his research in Malaysia and Indonesia, and a broadly based treatment, *The Geographical Distribution of Animals* (Wallace, 1869, 1876).

FIGURE 10.6 A contemporary photograph of Down House where Charles Darwin and his family lived from 1842 to 1882.

In an interview conducted just before his death in November 1913, Wallace stated that during this 8-year period in Malaysia and Indonesia:

> I did a great deal of work on the natural selection theory, and my paper came before Darwin in 1858. It seems that Darwin had been working along the same lines, and shortly after reading my paper he published his 'Origin of Species.'[9]

DARWIN RETURNS TO ENGLAND

After Darwin's return to England in 1836, he was able to dedicate himself to natural history through a substantial income and inheritance, supplemented during his lifetime by shrewd financial management and investment. He married his cousin, Emma Wedgwood (1808–1896), and in 1842 settled in Down House near the town of Downe in Kent, 23 km (about 14 miles) from London (FIGURE 10.6). There he and Emma began to raise a large family. Darwin participated on the 12-man committee that drafted the 'Strickland Code' in 1842. This code developed the standard *Rules of Zoological Nomenclature,* including how species should be named, a code that stands to this day (Strickland et al., 1843).[10]

For the last 40 years of his life, Darwin lived at home, mostly as a semi-invalid subject to heart palpitations, rashes, and gastric discomfort. "He could bear neither heat nor cold; half an hour of conversation beyond his habitual time was sufficient to cause insomnia and hinder his work on the following day . . . He suffered also from dyspepsia, from spinal anaemia and giddiness . . . and he could not work more than three hours a day."[11] The cause(s) of Darwin's disability (or disabilities) is not known. Hypotheses range from parasitic infection—trypanosomes that cause trypanosomiasis (Chagas disease), which affects heart and intestines—to heavy metal (arsenic) poisoning from some of the so-called cures of his time, to psychosomatic illness involving severe symptoms of panic disorder. Whatever the cause, his illnesses isolated him—or were used by Darwin to isolate himself—from direct contact with much of the world around him. Nevertheless, Darwin

[9] W. B. Northrop, 1913. Alfred Russel Wallace interview. *The Outlook* (New York) **105**, 618–622 (quotation is from p. 621). See elsewhere in this text for further discussion.

[10] Darwin's position on the nature of species as developed in *The Origin of Species* was introduced elsewhere in this text.

[11] C. Lombroso (1891), *The Man of Genius.* Walter Scott, London, UK, pp. 356–357.

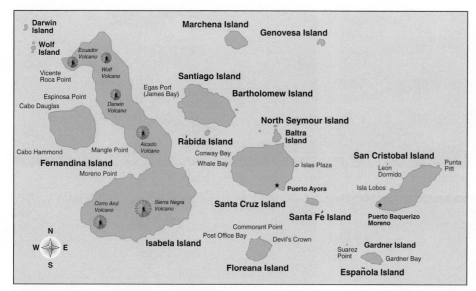

(a)

FIGURE 10.7 A map of the Galápagos Islands **(a)** and a photograph of the lava-dominated landscape of the islands **(b)**.

(b)

maintained enormously important contacts through letters, publications, and meetings with colleagues at Down House. He remained extraordinarily well-connected socially and academically, corresponding extensively with naturalists, scholars, and breeders. When *The Origin of Species* was published on 24 November 1859, Darwin was completing a 9-week stay in a remote Yorkshire village where he was taking the "water cure." When not soaking in cold water or wrapped in wet sheets, Darwin spent his time corresponding with fellow naturalists, seeking their support for his theory (Dixon and Radick, 2009).

GALÁPAGOS ISLANDS

One experience that, years later, had great impact on Darwin's thinking about evolution was the month he spent in the bleak, lava-covered Galápagos Islands off the coast of Ecuador (FIGURE 10.7). Lying 800 km (nearly 500 miles) from the mainland, these islands contain an exotic collection of land and marine animals and plants, including giant tortoises and meter-long marine and land iguanas (FIGURE 10.8).[12] As Darwin had noted on the mainland, different islands with environmentally similar habitats were not always occupied

---

[12] See Stewart (2007) for an engaging analysis of the importance of the Galápagos Islands to Darwin's work.

FIGURE 10.8 Among the many organisms Darwin saw on the Galápagos islands were: **(a)** marine iguanas (*Amblyrhynchus cristatus*); **(b)** giant tortoises such as the saddleback tortoise *Geochelone nigra;* **(c)** colorful crabs such as the common red rock crab, *Grapsus grapsus;* and **(d)** the Galápagos lava cactus, *Brachycereus nesioticus,* growing from the lava.

by similar species. He was particularly struck by the situation in the Galápagos where different species of tortoises and mockingbirds were found on each island. Insect-eating warblers and woodpeckers were absent but others species of seed- and insect-eating birds were present. Later, these were shown to be related species of finches (FIGURE 10.9).

Ornithologist David Lack and various Darwin historians have pointed out that although Darwin's *Journal of Researches* expresses evolutionary forethoughts, the significance of his observations on the Galápagos Islands and elsewhere did not become apparent to Darwin until after his return to England. This was especially true for the various Galápagos finches, which were first classified in England by John Gould (1804–1881), a leading British ornithologist. Gould pointed out to Darwin that (1) these were related species of finches with differing morphologies (FIGURE 10.10) and (2) different combinations of the 12 to 13 species occurred on different islands. Darwin had not kept notes as to which island was home to which species. Realization that each island appeared to have its own constellation of species raised the important question: What could account for this distribution of organisms? In Darwin's words,

> It is the circumstance that several of the islands possess their own species of the tortoise, mocking-thrush, finches, and numerous plants, these species having the same general habits, occupying analogous situations, and obviously filling the same place in the natural economy of this archipelago, that strikes me with wonder.[13]

---

[13] We see here Darwin's appreciation of what are now important elements of the field of ecology, such as the concept of the ecological niche discussed elsewhere in this text.

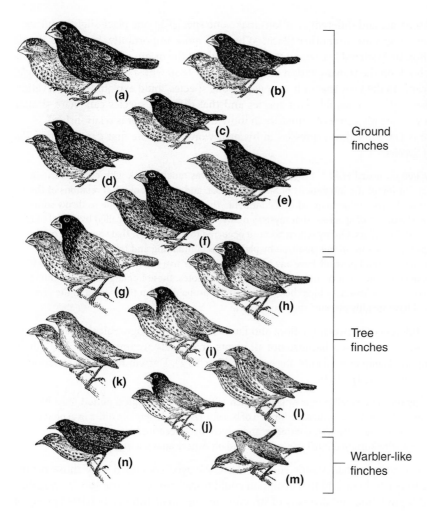

FIGURE 10.9 Species of finches (male on left, female on right; about 20 percent of actual size) observed by Darwin on the Galápagos Islands. **(a–f)** Species in the genus *Geospiza*. **(g–l)** Species in the genus *Camarhynchus*. **(m)** *Certhidea olivacea*, the warbler finch. **(n)** *Pinaroloxias inornata*, the Cocos finch.

(Reproduced from Lack, D. *Darwin's Finches: An Essay on the General Biological Theory of Evolution.* Cambridge University Press, 1947. Illustration by Lt Col William Percival Cosnahan Tenison.)

Ground finches

Tree finches

Warbler-like finches

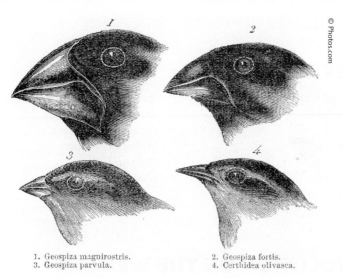

1. Geospiza magnirostris.
2. Geospiza fortis.
3. Geospiza parvula.
4. Certhidea olivasca.

© Photos.com

FIGURE 10.10 Drawings by Charles Darwin of four species of Darwin's finches showing variation in size and beak morphology.

Did separate and different creations make one species in one place slightly different from another species in another place? Why, and more importantly, how? The *Beagle* voyage (Box 10.2) stirred Darwin to think deeply about the origin of species; he began his first notebook on the transmutation of species late in 1836 whilst on board HMS *Beagle*. Darwin came to the view that only changes among species could explain the observation that present species resemble past species and that different species can share similar structures: "The only cause of similarity in individuals we know of is relationship."

Here is Darwin's aim, expressed in his own words from the first paragraph of *The Origin of Species*:

> When on board *H.M.S. Beagle*, as naturalist, I was much struck with certain facts in the distribution of the inhabitants of South America, and in the geological relations of the present to the past inhabitants of that continent. These facts seemed to me to throw some light on the origin of species—that mystery of mysteries, as it has been called by one of our greatest philosophers. On my return home, it occurred to me, in 1837, that something might perhaps be made out on this question by patiently accumulating and reflecting on all sorts of facts which could possibly have any bearing on it. After five years' work I allowed myself to speculate on the subject, and drew up some short notes; these I enlarged in 1844 into a sketch of the conclusions, which then seemed to me probable: from that period to the present day I have steadily pursued the same object.

The differences between the flora and fauna of different geographical areas, Darwin thought, must have arisen because not all plants or animals are universally distributed. Regarding the Galápagos Islands, for example (Box 10.2), Darwin raised the question in *The Origin of Species* (p. 457):

> Why on these small points of land, which within a late geological period must have been covered with ocean, which are formed of basaltic lava, and therefore differ in geological character from the American continent, and which are placed under a peculiar climate—why were their aboriginal inhabitants . . . created on American types of organization?

Darwin realized that islands such as the Galápagos contained only those organisms able to reach them, and that evolution could transform only those species available. "Seeing this gradation and diversity of structure in one small, intimately related group of birds, one might really fancy that from an original paucity of birds in this archipelago one species had been taken and modified for different ends."

Uncovering the mechanism for the transformation of species, however, was nowhere near as obvious as was the realization that such transformation had occurred.

■ KEY TERMS

Alfred Russel Wallace
Charles Darwin
Edward Blyth
Galápagos Islands
natural selection
Patrick Matthew
natural selection

theories of natural selection
voyage of *HMS Beagle*
voyages of discovery
William Lawrence
William Wells
zoogeography

■ EVOLUTION ON THE WEB

Explore evolution on the Internet! Visit the accompanying website for *Strickberger's Evolution, Fifth Edition*, at **go.jblearning.com/Evolution5eCW** for exercises and links relating to topics covered in this chapter.

■ REFERENCES

Appleman, O. (ed.), 2001. *A Norton Critical Edition. Darwin: Texts, Commentary,* 3rd ed., W. W. Norton & Company, New York.

Bates, H. W., 1863. *The Naturalist on the River Amazons.* 2 Volumes. Murray, London.

Blyth, E., 1835. An attempt to classify the "varieties" of animals, with observations on the marked seasonal and other changes which naturally take place in various British species, and which do not constitute varieties. *Mag. Nat. Hist.* **8**, 40–53.

Bowler, P. J., 1988. *The Non-Darwinian Revolution: Reinterpretation of a Historical Myth.* The Johns Hopkins University Press, Baltimore. MD.

Bowler, P. J., 2003. *Evolution: The History of an Idea,* 3rd ed. University of California Press, Berkeley, CA.

Browne, J., 2003. *Charles Darwin: The Power of Place.* Alfred A. Knopf, New York.

Chambers, R., 1844. *Vestiges of the Natural History of Creation.* John Churchill, London.

Cox, C. B., and P. D. Moore, 2005. *Biogeography: An Ecological and Evolutionary Approach,* 7th ed. Blackwell Publishing, Oxford, UK.

Darwin, C., 1838. *Journal of Researches into the Natural History and Geology of the Countries Visited During the Voyage of H.M.S. Beagle Round the World, Under the Command of Capt. FitzRoy.* John Murray, London.

Darwin, C., 1845. *The Voyage of the Beagle.* John Murray, London.

Darwin, C., 1859. *On the Origin of Species by Means of Natural Selection or the Preservation of Favoured Races in the Struggle for Life.* John Murray, London.

Darwin, C., 1958. *The Autobiography of Charles Darwin 1809–1882.* (ed. N. Barlow.) The Norton Library, New York.

Darwin, E., 1796. *Zoonomia; or the Laws of Organic Life. Part 1–III.* J. Johnson, London.

Darwin, F. (ed.), 1887. *The Life and Letters of Charles Darwin, including an Autobiographical Chapter.* John Murray, London.

Darwin Keynes, R. (ed.), 1988. *Charles Darwin's Beagle Diary.* Cambridge University Press, New York.

Dempster, W. J., 1983. *Patrick Matthew and Natural Selection: Nineteenth Century Gentleman-Farmer, Naturalist and Writer.* Harris, Edinburgh, UK.

Dixon, M., and G. Radick, 2009. *Darwin in Ilkley.* The History Press, Stroud, UK.

Elliott, P., 2003. Erasmus Darwin, Herbert Spencer, and the origins of the evolutionary worldview in British provincial scientific culture, 1770–1850. *Isis,* **94**, 1–29.

Green, J. H. S., 1957. William Charles Wells FRS (1757–1817). *Nature,* **179**, 997–999.

Humboldt, A. von, 1814. *Personal Narrative of Travels to the Equinoctial Regions of America, During the Years 1799–1804.* Longman, Hurst, Rees, Orme, and Brown, London.

Lawrence, W, 1816. *An Introduction To The Comparative Anatomy And Physiology, Being The Two Introductory Lectures Delivered At The Royal College Of Surgeons On The 21st And 25th Of March 1916.* J. Callow, London.

Lawrence, W., 1819. *Lectures On Physiology, Zoology And The Natural History Of Man.* J. Callow, London.

Matthew, P., 1831. *On Naval Timber and Arboriculture: With Critical Notes on Authors Who Have Recently Treated the Subject of Planting.* Black, Edinburgh, and London.

McCalman, I., 2009. *Darwin's Armada: Four Voyages and the Battle for the Theory of Evolution.* W. W. Norton/Simon & Schuster, New York.

Oldroyd, D. R., 1983. *Darwinian Impacts. An Introduction to the Darwinian Revolution.* 2nd revised edition. New South Wales University Press, Kensington, Australia.

Quammen, D., 2006. *The Reluctant Mr. Darwin.* W. W. Norton, New York.

Raby, P., 2001. *Alfred Russel Wallace, A Life.* Princeton University Press, Princeton, NJ.

Secord, J. (ed.) 2010. *Charles Darwin. Evolutionary Writings: Including the Autobiographies.* Oxford University Press, Oxford, UK.

Shermer, M., 2002. *In Darwin's Shadow: The Life and Science of Alfred Russel Wallace.* Oxford University Press, New York.

Slotten, R. A., 2004. *The Heretic in Darwin's Court: The Life of Alfred Russel Wallace.* Columbia University Press, New York.

Stewart, P. D., 2007. *Galápagos: The Islands that Changed the World.* Yale University Press, New Haven, CT.

Strickland, H. E., J. Phillips, J. Richardson, et al., 1843. Report of a committee appointed "to consider of the rules by which the nomenclature of zoology may be established on a uniform and permanent basis. *Rep. British Assoc. Adv. Sci.,* **1842**, 105–121.

Thomson, K. S., 1995. *HMS Beagle: The Story of Darwin's Ship.* W. W. Norton, New York.

Wallace, A. R., 1869. *The Malay Archipelago.* Harper, New York.

Wallace, A. R., 1876. *The Geographical Distribution of Animals.* Harper and Brothers, New York.

Wallace, A. R., 1913. *Social Environment and Moral Progress.* Cassell & Co., Ltd., London, New York, Toronto, and Melbourne.

Wells, W. C., 1818. *Two Essays: Upon a Single Vision with Two Eyes, the Other on Dew.* Constable, London.

CHAPTER

11

The Theory of Evolution by Natural Selection

CHAPTER SUMMARY

The essence of the theory of evolution by natural selection proposed independently by Charles Darwin and Alfred Russel Wallace is that the combination of limited resources and the production of more offspring than those resources can sustain results in competition. Competition results in some individuals leaving more offspring than other individuals. Writings of the Rev. Thomas Malthus (1766–1834) showed Darwin and Wallace (independently) how limited resources could lead to competition among species for these resources. Features of those individuals who survive to produce the most offspring are represented at higher frequencies in the next generation(s) than are those characters that are less well adapted to the environment. Natural selection increases the numbers of individuals with features best adapted to the resources and to the environment; the environment includes other species. The combination of changing environments, hereditary variation, differential reproduction, and natural selection results in modification of existing characters or the origin of new characters that become established and spread throughout a population or species. Should part of the population(s) become reproductively isolated a new species can arise.

THE DARWIN-WALLACE THEORY OF EVOLUTION BY NATURAL SELECTION

The evolutionary theories derived by Darwin and Wallace are outlined in FIGURE 11.1. The evolutionary process the two men independently proposed is twofold.

1. Excess reproduction + limited resources leads to competition, which, because of natural variation and the action of natural selection, allows those best adapted to pass their characters to the next generation(s).
2. Changing environments + hereditary variation + natural selection can result in the modification of existing characters or the origin of new characters that become established, and spread throughout a population or species.

Darwin and Wallace saw that differential reproduction in generation after generation of those individuals with increased chances of survival (because of enhanced adaptation for a particular environment) could change the composition of the population. The evolutionary process is a continual one; enhanced reproductive ability of some individuals is followed by further competition for the limited resources and further natural selection in subsequent generations. Of course, random events play their part; an individual killed by a falling tree is prevented from reproducing whether well adjusted to the environment or not.

REV. THOMAS MALTHUS, DARWIN, AND WALLACE

Both Darwin and Wallace were aware that many more individuals of each species are born than can possibly survive. They came to the conclusion that there must be a recurring struggle for existence; any individual that varies, however slightly, in a way that is advantageous will have a better chance of surviving (of being selected) and of leaving offspring. Behind this simple explanation is a complex set of explanations, which Darwin and Wallace spent most of their lives investigating.

You should not be left with the impression, however, that Wallace and Darwin came up with identical theories. They did not. They differed over the use of Herbert Spencer's term *struggle for existence*, but more fundamentally, Wallace was unconvinced of the utility of **artificial selection** as a means to understand natural selection. Darwin used artificial selection to demonstrate the fact of selection, a position borne out by current knowledge of the speed with which selection can act (below and see Kutschera, 2003 and Elliott, 2003). Wallace placed more importance on geographical isolation than did Darwin (Wallace, 1913). Wallace, who adopted spiritualism later in life, did not believe that natural selection applied to the human spirit or mind, or to mathematical, musical, or artistic abilities.

> Evolution is extremely interesting, and men fastened on it as the only explanation for all the manifold mysteries with which they are confronted. Evolution is true in part, but it does not account by any means for all the facts. . . My argument has always been that the mind and the spirit, while being influenced by the struggle for existence, have not originated through natural selection.[1]

Both attributed the origin of their theory that species tend to produce more members than resources can sustain (which is the primary population pressure that leads to competition and selection), not to the biological literature of the time but to the literature of political economy: "Darwin [and Wallace] revolutionized the study of nature by taking the actual variation among actual things as central to the reality, not as an annoying

---

[1] W. B. Northrop (1913). Alfred Russel Wallace interview. *The Outlook* (New York), **105**, 618–622.

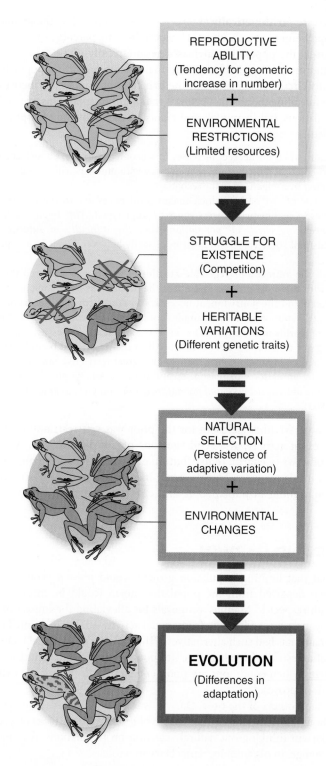

FIGURE 11.1 Schematic of the main conceptual arguments for evolution by natural selection given by Charles Darwin and Alfred Russel Wallace.

[Adapted from Wallace, A. R., 1889. *Darwinism: An Exposition of the Theory of Natural Selection with Some of Its Applications.* Macmillan, London.]

and irrelevant disturbance to be wished away" (Lewontin, 2000). Remarkably, Darwin and Wallace obtained their inspiration for their theories from the same source, the writings of the Rev. Thomas Malthus. If alive today, Malthus would be classified as a political economist, an expert in population structure, and a specialist in the statistical analysis of populations (demography).

One consequence of the rising tide of poverty in Victorian England, famously expressed by Malthus, was that the fate of the poor is inescapable; their reproductive powers will always exhaust their means of subsistence. Food supplies, Malthus pointed out, at best increases **arithmetically** $(1 \rightarrow 2 \rightarrow 3 \rightarrow 4 \rightarrow 5 \ldots)$ by the gradual accretion of land and improvement of agriculture. However, because parents usually produce more than two children, the population increases **geometrically** $(2 \rightarrow 4 \rightarrow 8 \rightarrow 16)$. Thus, famine, war, and disease inevitably become major factors limiting population growth. In 1798 Malthus circulated an essay on the principle of population increase.[2] What is impressive when reading the abstract below is that Malthus applies his reasoning not only to human populations, but to animals and plants, as the following excerpt demonstrates (emphasis added).

> Population, when unchecked, increases in a geometrical ratio. Subsistence increases only in an arithmetical ratio. A slight acquaintance with numbers will shew the immensity of the first power in comparison of the second.
>
> By that law of our nature which makes food necessary to the life of man, the effects of these two unequal powers must be kept equal. *This implies a strong and constantly operating check on population from the difficulty of subsistence.* This difficulty must fall somewhere and must necessarily be severely felt by a large portion of mankind.
>
> Through the animal and vegetable kingdoms, nature has scattered the seeds of life abroad with the most profuse and liberal hand. She has been comparatively sparing in the room and the nourishment necessary to rear them. The germs of existence contained in this spot of earth, with ample food, and ample room to expand in, would fill millions of worlds in the course of a few thousand years. Necessity, that imperious all pervading law of nature, restrains them within the prescribed bounds. The race of plants and the race of animals shrink under this great restrictive law. And the race of man cannot, by any efforts of reason, escape from it. *Among plants and animals its effects are waste of seed, sickness, and premature death. Among mankind, misery and vice* (Malthus, 1803).

The only hope Malthus held out for the poor was self-restraint: delay marriage, refrain from sexual activity, and send the sons of the poor to war. All other solutions—the Poor Laws (welfare), redistribution of wealth, improvements in living conditions—were, in his view, inadequate. Such measures would only stimulate further increase in the number of poor people and continue the cycle of famine, war, and disease.

MALTHUS AND DARWIN

Malthus argued that limiting population growth would *prevent* evolutionary change; individuals who departed from the population norm would be more susceptible to extinction. With respect to Malthus' proposals for alleviating the impact of population increase—delay marriage and refrain from sexual activity—Darwin pointed out that neither plants nor animals had such alternatives. "There can be no artificial increase in food, and no prudential restraint from marriage" (although animals and plants both can refrain from reproducing when conditions are poor, enhancing survival of the existing adults until conditions improve).

As were many others of the time, Darwin was deeply impressed by the Malthusian argument, although Malthus did not espouse evolution. Malthus' ideas played an important role in Darwin thinking out natural selection on populations as a mechanism of evolutionary change. In his autobiography Darwin wrote:

> I soon perceived that selection was the keystone of man's success in making useful races of animals and plants. But how selection could be applied to organisms living in a state of nature remained for some time a mystery to me.

---

[2] Malthus' 1798 *An Essay on the Principle of Population* is reprinted in Appleman (2001). Malthus was not the first to see the link between numbers of individuals and resources. On the first page of his treatment Malthus paraphrases an identical proposal made by Benjamin Franklin.

In October 1838, that is, fifteen months after I had begun my systematic enquiry, I happened to read for amusement "Malthus on Population" and being well prepared to appreciate the struggle for existence which everywhere goes on from long-continued observation of the habits of animals and plants, it at once struck me that under these circumstances favourable variations would tend to be preserved, and unfavourable ones to be destroyed. The result of this would be the formation of new species. Here then I had at last got a theory by which to work; but I was so anxious to avoid prejudice, that I determined not for some time to write even the briefest sketch of it. In June 1842 I first allowed myself the satisfaction of writing a very brief abstract of my theory in pencil in 35 pages; and this was enlarged during the summer of 1844 into one of 230 pages, which I had fairly copied out and still possess.[3]

MALTHUS AND WALLACE

Wallace read Malthus' essay before leaving on his voyage to Brazil in April 1848. Before Wallace and Bates set off for the Amazon, they had discussed the possibility of the transformation of species; Wallace wrote to Bates in 1847 of his aim to "gather facts toward solving the problem of the origin of species." In an interview just before his death in 1913, Wallace confirmed that during his eight years in Malaysia and Indonesia he had done "a great deal of work on the natural selection theory."[4]

Darwin had the basic idea of natural selection almost a decade, and had his long outline written about four years, before Wallace read Malthus. In 1855 while still in Malaysia, Wallace published a paper on the distribution of species in relation to geographical isolation, a paper of which Darwin was aware. Edward Blyth immediately recognized the significance of Wallace's 1855 paper and brought it to Darwin's attention in a letter written of December 8 of that year: "What think you of Wallace's paper in the Ann. M. N. H.? Good! Upon the whole! . . . Wallace has, I think, put the matter well; and according to his theory, the various domestic races of animals have been fairly developed into species. . . . A trump of a fact for friend Wallace to have hit upon!" In his paper Wallace concluded, "every species has come into existence coincident both in space and time with a closely allied species" (p. 51), indicating that he was seeking a mechanism to explain species change.

Three years later, Wallace was still collecting birds, insects, and mammals in the islands of Southeast Asia, and continuing his search for a mechanism for evolution (FIGURE 11.2), when, remarkably, Malthus performed the same function for Wallace as he had for Darwin 20 years earlier (emphasis added).

> At that time [February 1858] I was suffering from a rather severe attack of intermittent fever at Ternate in the Moluccas . . . and something led me to think of the positive checks described by Malthus in his "Essay on Population," a work I had read several years before, and which had made a deep and permanent impression on my mind. These checks—war, disease, famine, and the like—must, it occurred to me, act on animals as well as on man. Then I thought of the enormously rapid multiplication of animals, causing these checks to be much more effective in them than in the case of man; and, while pondering vaguely on this fact there suddenly flashed upon me the idea of the survival of the fittest—that the individuals removed by these checks must be on the whole inferior to those that survived (Wallace, 1895. Introductory note to Chapter II).

It took Darwin some time to accept and use Herbert Spencer's term and metaphor "SURVIVAL OF THE FITTEST." In several letters, Wallace tried to convince Darwin to use the phrase, which he (Wallace) had adopted. Only after he had completed several editions of Origin of Species *did Darwin adopt the term for the processes that lead to the differential survival of organisms from one generation to the next. Darwin wrote to Thomas Huxley that it might have been preferable had he used* struggle for reproduction *rather than* struggle for existence.

---

[3] Darwin wrote but did not publish an autobiography (*Recollections of the Development of My Mind and Character*) in 1876, intending it for his family. After his death, his son Francis edited the autobiography and had it published along with his own biography of his father, and with letters to and from Charles Darwin (F. Darwin, 1887).

[4] W. B. Northrop (1913). Alfred Russel Wallace interview. *The Outlook* (New York), **105**, 618–622 (quotation is from p. 621).

FIGURE 11.2 A small fraction (0.1%) of the beetles collected by Wallace in the Malay Archipelago.

PUBLICATION OF THE DARWIN-WALLACE THEORIES ON EVOLUTION BY NATURAL SELECTION

Despite having read Malthus in 1838, having written out a 35-page abstract of his theory in 1842, and a 230-page extended abstract in 1844, Darwin delayed publishing his theory for 21 years. Why? Why did he not publish the 1842 sketch or the 1844 manuscript? In part these delays reflect Darwin's desire to amass a wealth of evidence that would overcome any opposition to the generality of the theory. But, only in part; Darwin had other important reasons for delaying publication.

The social structure and religious attitudes in England made Darwin fearful of the reception his ideas would receive. The publication in 1844 of the book *Vestiges of the Natural History of Creation* made Darwin all too aware of the likelihood of a hostile public reception to his theory. Published anonymously—the author shared Darwin's concerns about public reaction—*Vestiges* was shown after the author's death to have been written by Robert Chambers (1802–1871), a member of a prominent Scottish publishing house (Chambers, 1844). *Vestiges* elaborated the theory that all matter, whether inorganic or organic, evolved out of (transformed from) preexisting matter by the accumulation of accidental changes Although enormously popular for a time—there was much public discussion of the 14 editions in Britain—considerable religious and scientific denunciation focused on this work (Chambers, 1994). This made Darwin fearful of exposing his own ideas to ridicule. Reading *Vestiges* and the responses it elicited convinced Darwin that he needed as much evidence as he could accumulate before exposing his own theory to public scrutiny.

Darwin devoted the next 14 years (1844–1858) to documenting variation in nature, differences between varieties and species, the geographical distribution of organisms, and accumulating and analyzing evidence for artificial selection. During this time, Darwin did not publish a single word on his theory of evolution. Some historians maintain that the "delay" in publishing was because Darwin's theory that natural selection was the mechanism of speciation was incomplete until 1852, perhaps even later.[5] Only in the early 1850s did Darwin devise a mechanism for speciation, the **principle of divergence**, which he represented in the only figure in *The Origin of Species* (FIGURE 11.3). Evolution as a branching tree has been the metaphor used ever since (FIGURE 11.4; see Winsor, 2009).

---

[5] Much has been written on why Darwin delayed publishing his theory. Browne (2003) and Van Wyhe (2007) challenges whether Darwin really did delay publication, setting *The Origin of Species* in a timetable they claim Darwin set out and adhered to.

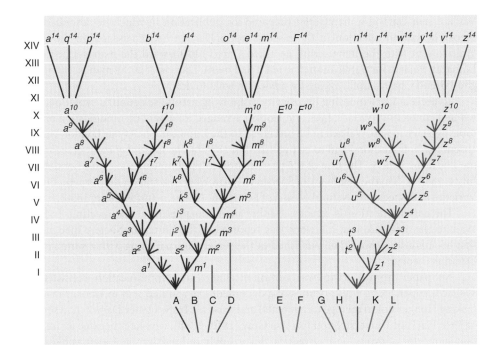

FIGURE 11.3 A representation of the illustration Charles Darwin used in his book, *On the Origin of Species* to represent progressive divergence within individual species (a^1–a^{10}; m^1–m^{10}; z^1–z^{10}) and the splitting of species into multiple lineages (S^2, i^2, i^3 from m^1 and m^2, for example).

WALLACE COMMUNICATES WITH DARWIN

As introduced above, in 1855 Wallace had published a paper on the consequences of geographical isolation on species distribution and exchanged letters with Darwin on the topic, without any discussion of mechanisms of evolution. Three years later (June 1858) Darwin discovered that Wallace had independently developed the same mechanism for evolution as Darwin had worked out. As Wallace outlined in February 1858, in his recollection of recovering from a fever:

> In the two hours that elapsed before my ague fit was over I had thought out almost the whole of the theory, and the same evening I sketched the draft of my paper, and in the two succeeding evenings wrote it out in full, and sent it by the next post to Mr. Darwin (Wallace, 1895. Introductory note to Chapter II).

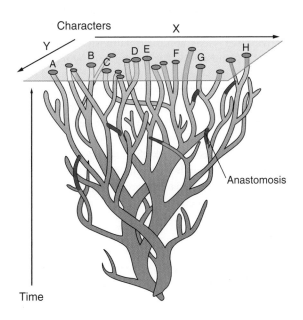

FIGURE 11.4 A phylogenetic tree shown as continuous branches undergoing evolutionary changes through time. Time is shown as the vertical axis. Some populations have become extinct. Others have merged or diverged to produce new and different forms (compare with Darwin's tree in Figure 11.3). *X* and *Y* axes represent different genetic traits or phenotypes. The differences between A and H may be sufficient to warrant separate taxonomic designations.

[Adapted from Levin, 1979.]

Unfortunately, the LETTERS *to and from Darwin and Wallace during the ensuing several weeks have all disappeared.*

These COMMUNICATIONS *can be cited as Darwin and Wallace (1858) using the common title under which the material was communicated to the Linnaean Society by Lyell and Hooker, or as separate publications (Darwin, 1858; Wallace 1858) using the titles of each part (and with separate pagination). All three are listed in the references for this chapter.*

Wallace sent Darwin a manuscript prepared for publication in which he described the theory of natural selection in the essential form Darwin had envisaged. In the letter Wallace sent with the manuscript, he asked Darwin to forward the manuscript to the Linnaean Society for publication if he felt it had merit. Darwin sent the manuscript to his good friend Charles Lyell, asking for advice on what to do.

To prevent Darwin losing his priority in applying natural selection to evolution, and without contacting Wallace, Lyell and Hooker arranged for a joint communication on the topic by Darwin and Wallace to be presented to the Linnaean Society in London (in the absence of both authors) and to be published in 1858 in *The Journal of the Proceedings of the Linnaean Society*. The title of the presentation to the Linnean Society of London on July 1, 1858 is: "On the Tendency of Species to Form Varieties; and on the Perpetuation of Varieties and Species by Natural Means of Selection" (Darwin and Wallace, 1858).

The communication is strange. Neither man was present at the meeting. Indeed, Wallace was unaware that his work was reported at the Linnaean Society meeting or that his manuscript would be published in the society journal along with a summary of Darwin's theory.

No paper from Darwin was communicated. The communication consisted of Wallace's paper, along with portions of a document Darwin had sent to Hooker in 1847 (asking Hooker not to divulge its contents) and excerpts from a letter Darwin had sent in 1857 to Harvard University botanist Asa Gray (1810–1888), who later became an important supporter of evolution in America. Surprisingly in hindsight, the joint publication evoked little response from either the scientific or the nonscientific communities, indicating that the theory itself, without supporting evidence and without enlisting large-scale evolutionary phenomena, neither invited serious interest nor threatened established views. Only with the publication in November 1859 of Darwin's expanded work with its huge weight of supporting evidence did the world take notice. Wallace's reputation gained much from his association with Darwin.

ARTIFICIAL AND NATURAL SELECTION

Wallace and Darwin arrived independently at a theory of the transformation of species by natural selection following extensive firsthand analysis of variation in nature. Both were influenced by Malthus in recognizing the importance of competition for resources, but they differed in their emphasis on the role of selection. Darwin also made more use of the available fossil record as evidence for evolution (Hall, 2010).

Darwin's principle of divergence asserts that as long as competition between sub-populations exists, co-existing subpopulations can begin to specialize and diverge to the point of speciation. With this principle, Darwin maintained that competition was the most powerful selective force. Geographical isolation featured much less prominently in Darwin's thinking than it did for Wallace. For Darwin, artificial selection involving human choice was the key to understanding natural selection and so he discussed many examples of artificial selection. Wallace thought artificial selection an inadequate way to understand natural selection and placed his emphasis on geographical isolation. So, although both derived a theory of evolution by natural selection, the theories were not identical.

In artificial selection, the breeder—whether of dogs, cats, pigeons, cattle, horses, peas, or wheat—selects the parents with the features deemed most desirable to propagate in the next generation. As the selected parents produce a variety of offspring, the breeder can continue to select in a particular direction until the features of interest are consistently present.[6]

---

[6] Artificial selection is now a much more complex topic that it was in Darwin's day, often a by product of human exploitation of forests, oceans, and fisheries or of the use of pesticides or drugs.

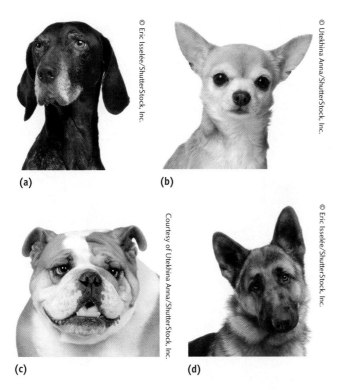

© Eric Isselée/ShutterStock, Inc.

© Utekhina Anna/ShutterStock, Inc.

Courtesy of Utekhina Anna/ShutterStock, Inc.

© Eric Isselée/ShutterStock, Inc.

(a) (b)

(c) (d)

FIGURE 11.5 Heads of four breeds of dogs produced by artificial breeding: **(a)** German shorthair pointer, **(b)** Chihuahua, **(c)** English bulldog, **(d)** German Shepard. Molecular evidence from mitochondrial DNA sequencing shows that the domestic dog is more closely related to the gray wolf than to any other species of the canid family and that domestication started independently on several occasions 10,000 to 15,000 years ago.

For thousands of years humans have selected for various types of dogs, all of which are variants of the single species *Canis lupus familiaris*. Dogs now range in size from the St. Bernard to the Chihuahua, in body forms from greyhound to bulldog (**FIGURE 11.5**) and express a wide variation in coat color, length, and curl.[7] Fanciers have long bred pigeons, which now show a wide variety of beak and body shapes and feather patterns (**FIGURE 11.6**). The same is true for sheep, cattle, and all the many agricultural species of plants and animals, including the hybrid corn and hybrid wheat many of us consume as food (**FIGURE 11.7**).

Darwin was amazed at the ease with which domesticated plants and animals could be changed and by the small number of generations of selection it took to bring change about. He knew that in a few generations a pigeon breeder could produce a pigeon with a head so big that the bird could not fly. Such "sports" and variations often were so different from their stock a few generations before that they would have been considered different species if found in the wild or as fossils. Artificial selection demonstrated to Darwin and others that continued selection was powerful enough to produce observable changes in almost any species (Wilner, 2006).

Darwin also had gathered data to produce his own estimate of geological time. His estimate—300 My—allowed more than enough time for life to have evolved into the many known forms and varieties. (**BOX 11.1**). The claim that natural selection in particular environments could accomplish even greater changes than artificial selection and lead to speciation therefore seemed reasonable, given the much longer periods of time in evolutionary history and the "unrelenting vigilance" of natural selection.

---

[7] Variation in coat color and type in over 80 breeds of dogs is now known to result from the function of only three major genes, short-haired dogs carrying the ancestral alleles (E. Cadieu et al., 2009. Coat variation in the domestic dog is governed by variants in three genes. *Science*, **326**, 150–153.)

FIGURE 11.6 A sample of the range of varieties of pigeons known to Charles Darwin, as depicted in the *London Illustrated News.*

Darwin's theory therefore was a revolutionary combination reflecting:

- the ease of effecting change via artificial selection,
- the vastness of geological time,
- the application of the principle that present day processes could be extrapolated back in time, and
- identification of a natural selector, the pressure of continuously limited resources.

SELECTION ON SMALL CONTINUOUS VARIATION

The RED NOTEBOOK *(our name, not Darwin's) was the first of a series of notebooks in which Darwin recorded his ideas on evolution. Begun while on the Beagle in 1836, Darwin had filled his first notebook by early 1837.*

Darwin had to deal with the difficulty of how selection recognizes each small modification. In a notebook started in 1836 (the red notebook), he recorded that he had concluded that, in some instances, differences between species were so great that they could only be achieved *per saltum*—by huge leaps.

Courtesy of the Plant and Soil Science eLibrary

FIGURE 11.7 Breeding from plants with small corn cobs (shown on the left and right) produces hybrid corn with much larger and more nutrient cobs.

BOX 11.1 Darwin's Calculations of Geological Time

Essential to Darwin's theory of evolution by natural selection was the assumption that Earth's age extended beyond anything ever proposed before. As Darwin pointed out in the final chapter of *The Origin of Species*:

> The mind cannot possibly grasp the full meaning of the term of a hundred million years; it cannot add up and perceive the full effects of many slight variations, accumulated during an almost infinite number of generations.

This emphasis on evolution taking a long time ran counter to the prevailing estimates of Earth's age. The heliocentric theory, in which the Sun was the center of the universe, tied the origin of Earth to the Sun. Newton had calculated that a sphere the size of Earth would take about 50,000 years to cool down. As even such a relatively short period contradicted the 5,000 or so years of history allowed in the Judeo-Christian Bible, Newton piously rejected these calculations. Buffon, in contrast, calculated that Earth was approximately 75,000 years old (74,832 to be precise).[a]

In Darwin's time, British physicist and engineer William Thomson (Lord Kelvin, 1824–1907), after whom the Kelvin scale of temperature is named, reassessed the temperature gradient observed in mine shafts, the conductivity of rocks, and the presumed temperature and cooling rate of the sun to calculate the total age of Earth's crust at about 100 My. Kelvin thought that only the last 20 to 40 My could have been sufficiently cool for life to exist. Geologist, paleontologist, and

Reader in Geology at Oxford, John Phillips (1800–1874) calculated a similar number, 96 My, based on sedimentation rates in the Ganges River, though in Kelvin's opinion this number was still too small to account for the Darwinian evolution of organisms.

Tradition has it that Darwin did not consider the long geological periods of time needed for life to evolve, but did calculations using his own observations. Darwin included his estimate of geological time—300 My—in the first and second editions of *The Origin* but removed them from later editions, which is unfortunate as his estimate was closer to reality than any other. Darwin used rates of erosion of the Weald (**FIGURE B1.1**) to demonstrate the enormity of geological time. The appropriate section from page 295 of the second edition of *The Origin of Species* (with emphases added) reads:

> If, then, we knew the rate at which the sea commonly wears away a line of cliff of any given height, we could measure the time requisite to have denuded the Weald. This, of course, cannot be done; but we may, in order to form some crude notion on the subject, assume that the sea would eat into cliffs 500 feet in height at the rate of one inch in a century . . . At this rate, on the above data, the denudation of the Weald

[a] For an excellent discussion of these early estimates see pp. 93–121 in M. Gorst (2001). *Measuring Eternity: The Search for the Beginning of Time*. Broadway Books, New York.

BOX 11.1	Darwin's Calculations of Geological Time (Cont...)

must have required 306,662,400 years; or say three hundred million years. But perhaps it would be safer to allow two or three inches per century, and this would reduce the number of years to 150 or 100 million years. So that it is not improbable that a longer period than 300 million years has elapsed since the latter part of the Secondary period.

Darwin's estimates of geological time caused him much anguish. How could a piece of the English countryside be three times older than Kelvin's estimate of the age of Earth itself, especially when Kelvin (a member of the British nobility) based his calculations on physical rates while Darwin's were "back of the envelope" calculations? On November 20, 1860, in a footnote to a letter to Charles Lyell concerning changes for the next edition of his book, Darwin wrote, "The confounded Wealden Calculation to be struck out, and a note to be inserted to the effect that I am convinced of its inaccuracy from a review in the *Saturday Review*, and from Phillips, as I see in his Table of Contents[b] that he alludes to it."

Darwin never referred to the age of Earth again!

The Weald is the rolling countryside between the South and North Downs in Kent and adjacent counties in southern England.

© Ian Goodrick/Alamy

FIGURE B1.1 The Weald between the South and North Downs in Kent and adjacent counties as it is today.

[b] The Table of Contents referred to by Darwin is the printed version of the 1860 presidential address to the Geological Society of London by John Phillips (Phillips, 1860).

Darwin's concept of variation in relation to evolution focused on individuals rather than POPULATIONS. Today, however, we recognize variation as a property of populations rather than of individuals. The current view of variation—populations contains reservoirs of individual-level variation on which selection can act—finally emerged with the modern synthesis in the 1930s and 1940s.

By 1859 in *The Origin of Species*, Darwin explicitly confined the origination of species by natural selection to small, continuous variations. In the earliest of the six editions of *The Origin of Species* written by Darwin[8] he excluded larger variations as not being useful, adopting the dictum that "nature makes no leaps." If much of variation involves small changes that do not depart from the species pattern, how can new species arise? Darwin replied that no limits apply to variation; each stage in the evolution of a species entails further variation upon which selection acts. A succession of changes through time, rather than a single simultaneous set of changes, leads to species differences.

[8] Several editions of *The Origin of Species* were produced after Darwin's death.

We now know that many traits stem from small heritable changes ascribed to many different genes (POLYGENES), each with small effect, from plasticity of development, and from what are known as epigenetic factors.

Darwin's search for small modifications led him to place less emphasis on the many traits involving distinct steps and differences, large variations such as different colors, presence and absence of structures, and different numbers of structures. The rediscovery in 1900 of Mendel's breeding experiments, in which he used large observable differences in plant traits to develop the basic laws that explain inheritance, and the subsequent development of the concept of the gene added considerable support for gradual change, although not for small modifications.

■ KEY TERMS

artificial selection	geological time
Alfred Russel Wallace	natural selection
artificial selection	principle of divergence
Charles Darwin	theories of evolution by natural selection
continuous variation	Thomas Malthus
Darwin's calculation of age of Earth	variation

■ EVOLUTION ON THE WEB

Explore evolution on the Internet! Visit the accompanying website for *Strickberger's Evolution, Fifth Edition,* at **go.jblearning.com/Evolution5eCW** for exercises and links relating to topics covered in this chapter.

■ REFERENCES

Appleman, O. (ed.), 2001. *A Norton Critical Edition. Darwin: Texts, Commentary,* 3rd ed., W. W. Norton & Company, New York.

Browne, J., 2003. *Charles Darwin: The Power of Place.* Alfred A. Knopf, New York.

Chambers, R., 1844. *Vestiges of the Natural History of Creation.* John Churchill, London.

Chambers, R., 1994. *Vestiges of the Natural History of Creation and Other Evolutionary Writings,* with a new Introduction by James A. Secord and a new Index by Brian W. Ogilvie. University of Chicago Press, Chicago.

Darwin, C., 1858. I. Extract from an unpublished Work on Species, by C. DARWIN, Esq., consisting of a portion of a Chapter entitled, "On the Variation of Organic Beings in a state of Nature; on the Natural Means of Selection; on the Comparison of Domestic Races and true Species." II. Abstract of a Letter from C. DARWIN, Esq., to Prof. ASA GRAY, Boston, U.S., dated Down, September 5th, 1857. *J. Proc. Linn. Soc. (Zool.),* **3**, 45–53.

Darwin, C., 1859. *On the Origin of Species by Means of Natural Selection or the Preservation of Favoured Races in the Struggle for Life.* John Murray, London.

Darwin, F. (ed.). 1887. *The Life and Letters of Charles Darwin,* including an Autobiographical Chapter. John Murray, London.

Darwin, C., and Wallace, A. R., 1858. On the Tendency of Species to Form Varieties; and on the Perpetuation of Varieties and Species by Natural Means of Selection. *J. Proc. Linn. Soc. (Zool.),* **3**, 45–62.

Elliott, P., 2003. Erasmus Darwin, Herbert Spencer, and the origins of the evolutionary worldview in British provincial scientific culture, 1770–1850. *Isis,* **94**, 1–29.

Hall, B. K., 2010. Charles Darwin, embryology, evolution and skeletal plasticity. *J. Appl. Ichthyol,* **26**, 148–161.

Kutschera, U., 2003. A comparative analysis of the Darwin-Wallace papers and the development of the concept of natural selection. *Theory Biosci.,* **122**, 343–359.

Levin, D. A. (ed.), 1979. *Hybridization: An Evolutionary Perspective.* Dowden, Hutchinson and Ross, Stroudsburg, PA.

Lewontin, R. C., 2000. *The Triple Helix: Gene, Organism, and Environment.* Harvard University Press, Cambridge, MA.

Malthus, T., 1803. *An Essay on the Principle of Population*, 2nd ed., Cambridge University Press, Cambridge, UK. (First edition, 1798.)

Phillips, J., 1860. *Life on the Earth: Its Origin and Succession.* MacMillan and Co., Cambridge and London. [Reprint edition, 1980. Arno Press, New York.]

Van Wyhe, J., 2007. Mind the gap: did Darwin avoid publishing his theory for many years? *Notes Rec. R. Soc.,* **61**, 177–205.

Wallace, A. R., 1855. On the law which has regulated the introduction of new species. *Ann. Nag. Nat. Hist.,* **16** (2nd Series), 184–196.

Wallace, A. R., 1858. III. On the tendency of varieties to depart indefinitely from the original type. *J. Proc. Linn. Soc. (Zool.),* **3**, 53–62.

Wallace, A. R., 1889. *Darwinism: An Exposition of the Theory of Natural Selection, with Some of its Applications.* Macmillan, London.

Wallace, A. R., 1895. *Natural Selection and Tropical Nature. Essays on Descriptive and Theoretical Biology.* New Edition With Corrections And Additions, MacMillan and Co., London.

Wallace, A. R., 1913. *Social Environment and Moral Progress.* Cassell & Co., Ltd., London, New York, Toronto & Melbourne.

Wilner, E., 2006. Darwin's artificial selection as an experiment. *Stud. Hist. Philos. Sci. (C) Stud. Hist. Philos. Biol. Biomed. Sci.,* **37**, 26–40.

Winsor, M. P., 2009. Taxonomy was the foundation of Darwin's evolution. *Taxon,* **58**, 43–49.

CHAPTER

12

Mendel, Inheritance, and a Theory of Heredity

CHAPTER SUMMARY

Neither Darwin nor Wallace knew how variation could be produced in nature. The two accepted theories of the time were the blending of parental traits in each generation (blending inheritance) and the inheritance of acquired characters. In seeking an alternate theory of inheritance, Darwin revived the ancient theory of pangenesis in which particles (gemmules) were thought to travel from all the organs to accumulate in the gonads of each individual, from where they contributed to the next generation via the eggs or sperm (gametes). Discovery of the separation of germ plasm (from which gametes arose) and soma or body plasm (from which the rest of the body formed) disproved both the inheritance of acquired characters and pangenesis. Gregor Mendel's breeding experiments with pea plants conducted in the mid-nineteenth century provided a mechanism of inheritance based in hereditary factors, subsequently shown to be genes. Mendel proposed two laws for the behavior of these hereditary factors: the principle of segregation and the principle of the independent assortment of hereditary units (genes).

INTRODUCTION

To explain how species arise and change, Darwin and Wallace proposed a mechanism—natural selection—that depended on heritable variations on which selection could act. No one knew, though, how such variation could arise. Most nineteenth century biologists advocated blending inheritance, by which offspring were thought to represent blended copies of their parents' features. No one, however, except for Gregor Mendel, had a mechanism to explain inheritance. Not until the early twentieth century was the source of inherited variation shown to reside in genes. But discovery of genes and their functions still did not explain all the issues surrounding transformation or origination of species.

Darwin was aware that blending inheritance created a major problem for his theory: If blending inheritance operated, natural selection would be incapable of maintaining a trait for more than a few generations (unless both parents possessed the same trait). Blending inheritance therefore confronted evolutionary theory with a serious problem. If characters are blended out when individuals mate with other members of the population, how can beneficial variations be preserved by natural selection? Darwin spent many years searching for an alternative to blending inheritance, settling finally on the **theory of pangenesis**, a theory disproved by the discovery of the separation of germ cells from body cells in many animals (below).

In the mid-1860s—after *The Origin of Species* had been published—Gregor Mendel published experiments that led him to propose two principles of inheritance. One is the **principle of segregation**, by which discrete entities segregate from each other in the gametes. The second is the **principle of independent assortment**—the autonomous segregation of these discrete entities during gamete formation—made evident because Mendel chose traits that we now know reside on different chromosomes. Both of these principles contradicted blending inheritance; phenotypic effects of some genes may blend with those of other genes but genes do not blend with each other.

We know now that these entities are ALLELES *of a single gene.*

SEEKING A MECHANISM OF HEREDITY

Darwin's most fundamental insight was that small differences among individuals within populations are the raw material for evolution. Before and during his time, most scholars—even Darwin's most staunch supporter Thomas Huxley—held the view that **saltations** (jumps or large discontinuous changes in organisms or parts of organisms) were the changes most relevant to evolution. However, because the mechanisms of inheritance were not known, resolving the important issue of small versus large change was not possible.

At times, Darwin proposed that (a) environmental change, (b) a large increase in numbers in a population, or (c) some disturbance of the reproductive organs might enhance variation, all of which are plausible agents of change. At other times, Darwin adopted a version of Lamarckian inheritance (Howard, 2009).

We can gain an appreciation of the similarities and differences in Darwin and Lamarck's approaches to variation and heredity by comparing their positions on several key issues:

- Both Darwin and Lamarck held that species descended from a common ancestor; that is, both held evolutionary theories.
- As to why some features persisted while others disappeared, Lamarck invoked use and disuse and the inheritance of acquired characters. Darwin (and Wallace) invoked natural selection.
- As to how new variations arose, Lamarck invoked use and disuse. So did Darwin, although he came to place more emphasis on environmental change than use and disuse. Darwin (and Wallace) proposed natural selection as the guiding principle in evolution.

BLENDING INHERITANCE

The fundamental problem for Darwin, as for all of his contemporaries, was that offspring were thought to represent perfect or *blended copies* of their parents. (Mendel's studies brought him to a different conclusion, but for many years his studies were unknown.) Because blended parts will tend to produce features that average out those of the parents—much as mixing red and blue produces purple (**FIGURE 12.1**)—**blending inheritance** sets variation and inheritance in opposition; new adaptations are diluted with each generation of interbreeding and natural selection is incapable of maintaining a trait for more than a few generations. In defense of his theory, Darwin laid out numerous arguments for the compatibility of blending inheritance and natural selection. Four are listed here.

1. A beneficial character could maintain itself if the individuals with the character were isolated from the rest of the population, just as breeders isolated newly appearing "sports" and their offspring to develop new stocks of domesticated animals or plants. We now know this as **mate choice** and **reproductive isolation**.
2. Some characters are prepotent and so appear undiluted in later generations. This is the **dominance effect** discovered by Mendel.
3. A new trait does not appear only once in a population but must arise fairly often to produce the large amount of variation present in most populations, a proposition confirmed by modern studies. Common variants would not dilute out as easily as would rare variants. Although he had no mechanism, Darwin maintained that some forms carrying a particular variation **pass on to future generations the tendency for the same variation to arise again**.
4. **Natural selection** both enhances the reproductive success of favorable variants and diminishes the reproductive success of unfavorable ones. Thus, the frequency of favorable variations increases when unfavorable ones die out, reducing the likelihood of diluting out favorable variations.

PANGENESIS

Twelve years after the publication of *The Origin* and after much investigation and experimentation (and with no mechanism having been proposed by others), Darwin (1868)

FIGURE 12.1 An example of blending in which mixing red and blue gives purple.

© Chatan Vekariya/ShutterStock, Inc.

(a) Pangenesis theory (all body parts contribute genetic material to sex cell)

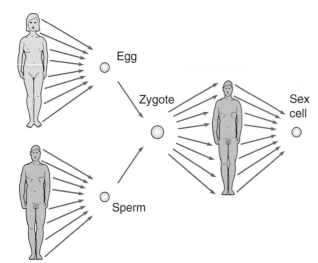

(b) Germ plasm theory (only gonads contribute genetic material to sex cell)

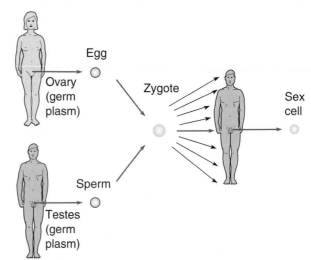

FIGURE 12.2 Comparison between **(a)** pangenesis and **(b)** germ plasm theories in humans. In pangenesis, all structures and organs throughout the body contribute particles (pangenes, gemmules) to the gametes. In the germ plasm theory, only the germ cells contribute hereditary units (genes) to the next generation.

[Adapted from *Genetics, Third Edition,* by Monroe W. Strickberger. © 1985 by Monroe W. Strickberger. Reprinted by permission of Pearson Education, Upper Saddle River, NJ.]

revived an old theory, pangenesis—meaning whole birth or whole origin—to provide a mechanism of inheritance that would counter blending inheritance (**FIGURE 12.2**).

According to pangenesis, small particulate known as **gemmules** or pangenes are produced by all the tissues of the parents and incorporated into developing eggs and sperm in the gonads. After fertilization of an egg by a sperm, the gemmules disperse throughout the developing embryo to form the various tissues and organs of the offspring. Among Darwin's presumptions was the existence of many gemmules for each trait. Darwin further proposed that the number of gemmules varies during passage from one generation to another—providing a mechanism for the origin of variation—and that gemmules could be lost but were not changed by "blending." Thus, pangenesis helped account for the presumed effects of use and disuse, for observations that not all traits become blended, and for the origin of variation.

Despite the many experiments conducted in search of evidence for pangenesis by Darwin's cousin, Francis Galton (1822–1911), no convincing evidence was obtained. Galton, who spent much of his life investigating heredity, especially eugenics, transfused blood between rabbits with different coat colors, which he then allowed to breed. Galton saw no evidence in their offspring that would support pangenesis; coat colors from blood donors did not show up in the offspring. Darwin response to his cousin was that he had not claimed that gemmules circulated in the blood.[1] German biologist and evolutionary theorist August Weismann (1834–1914) was convinced he had disproved pangenesis when he cut off the tails of mice over 22 generations and showed that tail length was not affected by the presumed loss of tail gemmules in each generation. This proof against pangenesis was enough to kill the theory and to demonstrate lack of inheritance of acquired characters.

[1] See M. Bulmer (2003), *Francis Galton: Pioneer of Heredity and Biometry,* Johns Hopkins University Press, Baltimore, MD. Historians of biology continue to examine Darwin's pangenesis theory and Galton's influence on Darwin, Liu (2008) being a particularly insightful example.

Nineteenth-century Lamarckians knew that animals and plants developed from eggs or seeds fertilized by sperm or pollen. They understood "acquired characters" to be body features (somatic traits) that were reproduced in each generation without instructions from germ-line tissue. Weismann developed the **germ plasm theory** of inheritance, in which only the gametes in the reproductive tissues (testes and ovaries) transmit the heredity factors for the entire organism. Because any changes that occur in non-reproductive body tissues are not transmitted, changes in heredity can neither be explained by the inheritance of acquired characters nor by use and disuse (Churchill, 1968).[2] So, the search for the hereditary basis of constancy and variation continued.

CONSTANCY AND VARIATION

Organismal evolution relies on two fundamental aspects of biological inheritance: **constancy** and **variation**. Constancy resides in the observation that **like produces like**. Constancy has the evolutionary significance that all life processes depend on the transmission of information from previous generations. In contrast, variation resides in the observation that **like can produce unlike**.[3]

The search for the basis of constancy and variation—the search for a basis for heredity—is usually taken to begin with Mendel's experiments in the mid-nineteenth century and his demonstration that hereditary material must be distinct from the physical features of organisms; features not expressed in one generation can appear in the next (see below). Many consider that the search culminated a century later with the discovery that DNA is the molecule of hereditary (**TABLE 12.1**)[4]. But demonstrating the nature of the hereditary material still leaves us a long way short of understanding the basis of variation.

Constancy and variation are indelibly intertwined in evolution through hereditary information (the **genotype**) that primarily determines organismal features (the **phenotype**) that, in multicellular organisms, arise during embryonic development. However, there is no one-to-one correspondence between genotype and phenotype. Developmental and environmental signals influence (and may determine) where and when genes, gene pathways, or gene networks are expressed. This is amply illustrated by sex determination, in which environmental signals such as temperature, presence of absence of signaling chemicals, or even infection with a bacterium can dictate the sex of individual offspring.

Because the molecular basis of genetics long remained unknown, fundamental genetic principles were derived primarily from observations on the transmission of obvious biological features such as size, shape, and color. From such observations under conditions of controlled breeding, Mendel (1866) formulated the basic laws of genetics. The scientific world was unaware of Mendel's work, however, until it was discovered independently in 1900 by three botanists/plant breeders: Hugo De Vries, Carl Erich Correns, and Erich Von Tschermak-Seysenegg.[5] Soon after 1900, discovery of the association

---

[2] Within animals, germ cells are specified either through inheritance of maternal "determinants" (a form of preformation) or by inductive interaction between cells in the developing gonads (epigenesis). Epigenetic determination is the ancestral condition (Extavour and Akam, 2003).

[3] For more in-depth discussions, see West-Eberhard (2003) and the chapters in Hall et al. (2004) and Hallgrímsson and Hall (2005).

[4] For a discussion of breeding experiments conducted before Mendel, see R. J. Wood and V. Orel, 2006. Scientific breeding in Central Europe during the early nineteenth century: background to Mendel's later work. *J. Hist. Biol.*, **38**, 239–272.

[5] Hugo De Vries (1848–1935) was Professor of Botany at the University of Amsterdam, Carl Erich Correns (1864–1933) was a tutor at the University of Tübingen (and, from 1913, first director of the *Kaiser Wihelm Institut für Biologie* in Berlin-Dahlem), and Erich Von Tschermak-Seysenegg (1871–1962) was conducting plant breeding experiments for his doctorate at the Martin Luther University of Halle-Wittenberg University.

TABLE 12.1	Timeline of Major Discoveries in Genetics and the Nature of the Gene from Gregor Mendel to James Watson and Francis Crick

Year(s)	Discovery
1856	Gregor Mendel's first crossbreeding experiments with the garden pea
1866	Publication of Mendel's paper, *Versuche über Pflanzen-hybriden* (*Experiments on Plant Hybridization*)
1871	Isolation by Friedrich Miescher of nuclein from nuclei, the first research with what we now know to be nucleic acids
1900	Mendel's 1866 experiments rediscovered independently by three plant breeders
1902/03	Chromosomal theory of inheritance proposed independently by William Sutton and Theodor Boveri
1909	Demonstration by Archibald Garrod that some human diseases are inborn errors of metabolism inherited as Mendelian recessive characters
1910–11	X-chromosomes, sex linkage, and mutant gene for eye color in *Drosophila* discovered by Thomas Hunt Morgan, who proposes exchange of chromosome segments by crossing over
1913	Chromosomes shown to contain genes in a linear array
1927	Discovery by Hermann Muller that x-rays induce mutations in *Drosophila* genes
1941	Discovery that genes code for enzymes in the mold *Neurospora*; proposal by George Beadle and Edward Tatum of the "one gene–one enzyme" hypothesis
1943–44	Evidence that bacteria contain genes; isolation of DNA and demonstration that DNA is genetic material
1952	Proteins eliminated as basis of genes
1953	Proposal by James Watson and Francis Crick of the three-dimensional double helical molecular structure of DNA, based in part on unpublished x-ray crystallographic data obtained by Rosalind Franklin and Maurice Wilkins

An annotated English translation of MENDEL'S 1866 PAPER *is available at http://www.mendelweb.org/Mendel.html.*

between Mendelian factors and the distribution of chromosomes during meiosis (**BOX 12.1**) formed the basis for a new science: **genetics**.[6] Mendel's exceptional contribution was to demonstrate that organisms have a distinct hereditary system (now known as the genotype), which transmits biological characteristics through discrete units (now known as genes) that remain undiluted in the presence of other units (genes).

MENDEL'S EXPERIMENTS

Gregor Mendel (1822–1884) was born into a German family who resided in the small village of Heinzendorf bei Odrau in Austrian Silesia, then part of the Austrian Empire. From the ages of 18 to 21 Mendel studied in the philosophy faculty of the University of Olomouc, the oldest university in Moravia, established by a Jesuit order in 1573. After graduating, Mendel entered the Augustinian Abbey of St. Thomas in Brno,[7] where, after his training, he spent his life as a monk, teacher, breeder of plants, and from 1867 on, abbot of the monastery (**FIGURE 12.3a**).

Mendel bred pea plants (*Pisum sativum*) for many generations, observing the appearance and counting the numbers of each trait in every individual of every generation in a large number of plants. He experimented with a number of characters (Figure 12.3b), each of which had two alternative appearances (traits): smooth or wrinkled seeds, yellow

---

[6] For the development of Mendelian genetics see Olby (1966), Hartl and Jones (1998), Griffiths (2002), and the Mendel website (http://www.mendelweb.org/).

[7] Heinzendorf bei Odrau and Brno both are parts of the modern Czech Republic. Born Johann Mendel, he adopted the name Gregor when he became a monk.

BOX 12.1   Mitosis and Meiosis

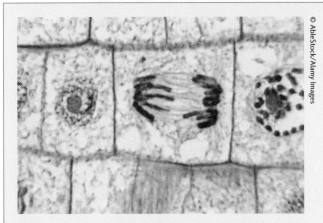

FIGURE B1.1 Separation of chromosomes across the mitotic spindle (dark purple in middle image) during the mitotic division of cells at the tip of the root in an onion

Genetic traits, changed or unchanged, are transmitted from one generation to the next through cell division. Transmission of hereditary information is coordinated with cell division so that parental and daughter cells carry copies of the same information. Because hereditary information is coded as long-chain nucleic acid molecules, the chromosomes on which these molecules are localized must replicate and divide (**FIGURE B1.1**).

Prokaryotic cells, which usually have only a single chromosome, divide by **binary fission** (**FIGURE B1.2**), the two products of chromosomal replication attaching to different locations on the cell membrane. As the cell elongates to form two daughter cells, the two chromosome products separate, each becoming enclosed in a separate daughter cell (Figure B1.2).

Eukaryotic cells have at least two chromosomes, with most eukaryotes having many more, as many as 1260. With

The highest known CHROMOSOME NUMBER *in diploid organisms, 1260, is found in the adder-tongue fern Ophioglossum reticulatum. The lowest chromosome number known is the theoretical minimal number of 1, in this case in a species of bull ant, the jack jumper ant Myrmecia pilosula from Australia. The haploid males of this species have a single chromosome (Crosland and Crozier, 1986). The yellow fever mosquito Aedes aegypti has six chromosomes, the fruit fly Drosophila melanogaster has eight.*

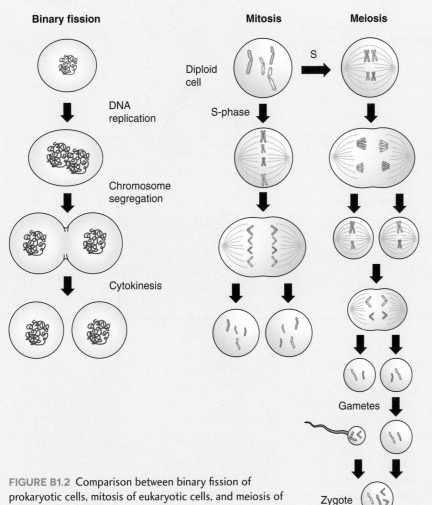

FIGURE B1.2 Comparison between binary fission of prokaryotic cells, mitosis of eukaryotic cells, and meiosis of eukaryotic gametes.

BOX 12.1   **Mitosis and Meiosis (Cont...)**

increased numbers of chromosomes comes more complex cell and chromosomal division:

- **Mitosis**: the division of general body (somatic) cells, including the supporting and nutritive cells of the sex organs that do not form sex cells (gametes).
- **Meiosis**: the division of the cells that produce gametes—eggs and sperm—and which involves both mitotic and reduction (meiotic) divisions (**FIGURES B1.3** and **B1.4**).[a]

Dividing cells contain one or two sets of chromosomes, that is, are haploid or diploid. Animals are diploid but their gametes—eggs and sperm—are haploid; fertilization restores the diploid number of chromosomes. Plants alternate a haploid with a diploid generation, a process known as **alternation of generations** and outlined in **END BOX 12.1**.

MITOSIS

Mitotic cell division provides all the body cells of an organism with the same chromosome constitution, or **karyotype;** cells have precisely the same numbers and kinds of chromosomes before and after each cell division.

As illustrated in Figure B1.3, early in mitosis the chromosomes replicate, each chromosome consisting of two **chromatids** connected at their centromeres. Chromatids arrange across the equator in a phase of mitosis known as metaphase, in which the two members of a pair of chromatids connect to spindle fibers that go to opposite poles of the nucleus. Chromatids of a replicated chromosome then separate (anaphase; Figure B1.2) and move to the two poles in telophase, which is followed by cell division to produce two daughter body cells, each with a complete set of chromosomes identical to the parent cell.

---

[a] **Cleavage**, a third form of division found in animal embryos after fertilization, is characterized by lack of growth between each division, resulting in cell size being halved with each division.

FIGURE B1.3 Diagrammatic presentation of various stages of mitosis in a somatic cell. During early, medium, and late prophase stages chromosomes thicken and condense. Each chromosome with its attached replicate lines up on a "metaphase" plate. During anaphase, the replicates (daughter chromosomes) separate, going to opposite poles. During telophase a nuclear membrane reforms around each polar group of daughter chromosomes, and these chromosomes revert to the more extended interphase state. Division of the cytoplasm (cytokinesis) is also completed during this final mitotic stage.

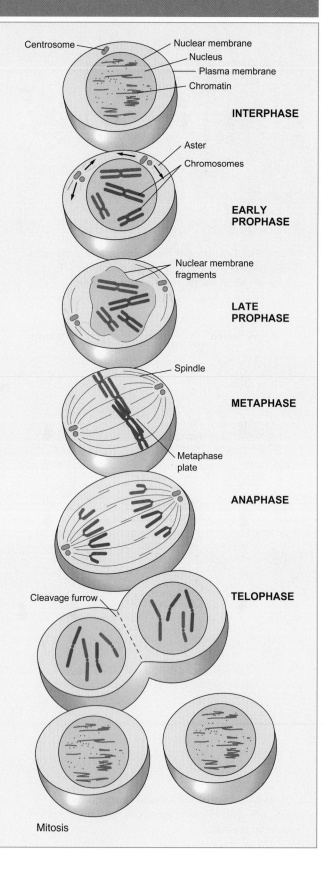

Centrosome — Nuclear membrane — Nucleus — Plasma membrane — Chromatin

INTERPHASE

Aster — Chromosomes

EARLY PROPHASE

Nuclear membrane fragments

LATE PROPHASE

Spindle

METAPHASE

Metaphase plate

ANAPHASE

Cleavage furrow

TELOPHASE

Mitosis

BOX 12.1   Mitosis and Meiosis (Cont...)

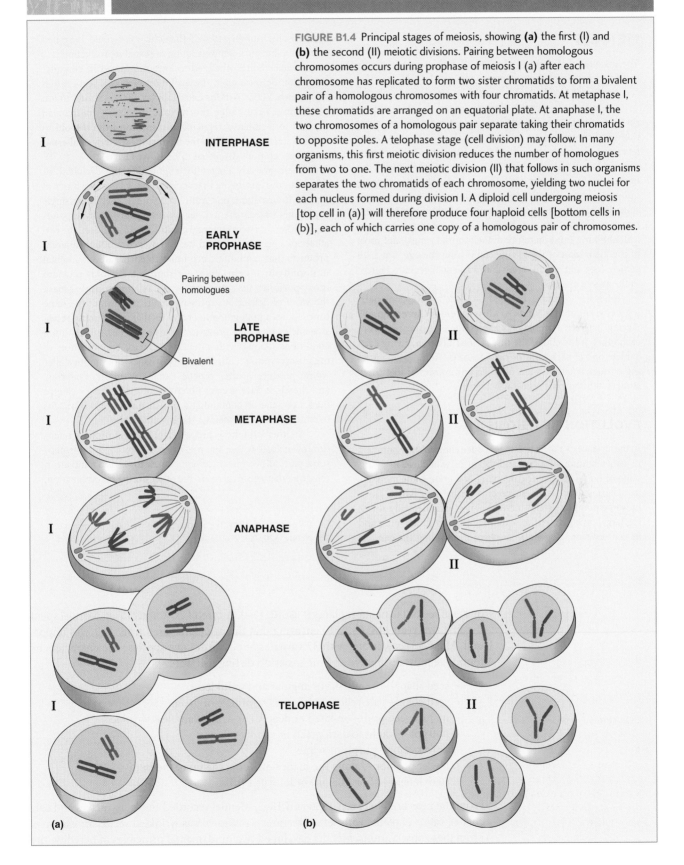

FIGURE B1.4 Principal stages of meiosis, showing **(a)** the first (I) and **(b)** the second (II) meiotic divisions. Pairing between homologous chromosomes occurs during prophase of meiosis I (a) after each chromosome has replicated to form two sister chromatids to form a bivalent pair of a homologous chromosomes with four chromatids. At metaphase I, these chromatids are arranged on an equatorial plate. At anaphase I, the two chromosomes of a homologous pair separate taking their chromatids to opposite poles. A telophase stage (cell division) may follow. In many organisms, this first meiotic division reduces the number of homologues from two to one. The next meiotic division (II) that follows in such organisms separates the two chromatids of each chromosome, yielding two nuclei for each nucleus formed during division I. A diploid cell undergoing meiosis [top cell in (a)] will therefore produce four haploid cells [bottom cells in (b)], each of which carries one copy of a homologous pair of chromosomes.

INTERPHASE

EARLY PROPHASE

Pairing between homologues

LATE PROPHASE

Bivalent

METAPHASE

ANAPHASE

TELOPHASE

(a)

(b)

BOX 12.1 | **Mitosis and Meiosis (Cont...)**

MEIOSIS

The gametes that form at the end of meiosis contain only one of each pair of chromosomes (Figure B1.4). How is this accomplished?

Meiosis is characterized by *one round* of DNA replication—resulting in every chromosome having two chromatids—and *two* rounds of cell divisions (Figure B1.4). A major feature of meiosis is the orderly separation of one of each pair of homologous chromosomes to each daughter cell during the **first cycle of meiotic division**, thus reducing the chromosome number by half. This first cycle is followed by a second division. The two chromatids of a replicated chromosome separate to yield two cells that develop as male or female gametes. Mitochondria (and therefore mitochondrial genes) and most of the cytoplasm of the zygote come from the egg, but each gamete (egg and sperm) contributes an equal genetic element to the zygote nucleus.

Meiosis, therefore, is based on close homologous pairing between similar kinds of chromosomes (**homologues**). Pairing between homologues allows them to separate from each other at the end of the first meiotic division, giving each gamete one of each kind of chromosome. Homologous pairing ensures equal division between chromosomes that carry similar, although not necessarily identical, genes.

EVOLUTION OF MEIOSIS

In the absence of a mechanism to reduce chromosome numbers in gametes, union of sperm and egg nuclei would lead to a doubling of nuclei and, hence, doubling of chromosome numbers in offspring. With each succeeding sexual generation, chromosome numbers would increase almost exponentially, forming large, unwieldy nuclei with difficulties in functioning and in coordination. A meiotic mechanism reducing the gametic chromosome number to half (haploid = n) would have had selective value during the evolution of sexual reproduction.

Meiosis is hypothesized to have originated in conjunction with, or soon after, the evolution of sexual fertilization (Bogdanov, 2003; Wilkins and Holliday, 2009). Sexual union could only occur after the meiotic reduction mechanism evolved to facilitate a regularized transition from diploid to haploid stages, as is seen in the life history strategy of alternation of generations discussed in End Box 12.1.[b]

There are advantages to lengthening the diploid (2n) stage, not the least being that such cells may have two kinds of genetic information, one from each parent, enabling a single organism to use different developmental pathways in responding to different environmental conditions. In addition, the two alleles of a gene in a diploid stage may each produce different products that can buffer each other to ensure developmental uniformity in any particular environment. Diploidy provides the opportunity for dominant genetic relationships that mask the effect of deleterious recessive alleles and yet, at the same time, allows a population to evolve further by retaining recessive alleles that may prove to be advantageous under future conditions. Along with other advantages, a population of organisms whose diploid meiotic tissues are multicellular would produce greater genetic variability among offspring, and therefore have greater potential for evolutionary change than a population containing a similar number of organisms in which the diploid meiotic stage is unicellular.

Another hypothesis proposed for the evolutionary retention of meiosis is that the presence of homologous chromosome pairs allows one member of a pair to act as a template in repairing damage incurred in the other during recombination.

---

[b] A. S. Kondrashov, 1994. The asexual ploidy cycle and the origin of sex. *Nature*, **370**, 213–216.

---

or green seeds, tall or short plants (**FIGURE 12.4**). A careful experimentalist, Mendel spent the first two years (1856–1858) ensuring that the pea plants bred true. This was important; Mendel was working with 37 varieties of peas from one species. The characters he found the most distinctive and the easiest to distinguish were:

- seed shape—round(ish) or angular and wrinkled
- seed color—pale yellow/bright yellow/orange or green
- mature seed pods—inflated or deeply constricted and less wrinkled
- pod color—light to dark green or vividly yellow
- flower color—purple or white
- flower position—along the main stem, or at the tip of the stem
- stem length—short (0.3 m) or long (2.1 m)

From the peas bred between 1856 and 1863, Mendel recorded these characters and calculated ratios of plants with sets of features—round versus wrinkled seeds, for example. From these observations, Mendel developed two fundamental principles of heredity: the principle of segregation, and the principle of independent assortment.

(a)   (b)

FIGURE 12.3 **(a)** Gregor Mendel. **(b)** The pea plant, *Pisum sativum*, used by Mendel in his experiments.

PRINCIPLE OF SEGREGATION

One of the characters used by Mendel was the appearance and "feel" of the seed's coat (its outer covering). Pea seed coats are either smooth or wrinkled (Figure 12.4a). When two plants with different expressions of a character are crossed, the first filial (F_1) generation carries the genes for each of these traits.

We now know that each pea plant has a pair of genes for seed coat texture as well as pairs of genes for other characters, such as seed color and plant size. Like most multicellular organisms, pea plants are diploid, containing one copy of the gene from the female parent and one from the male parent. In current terminology, the **two members of a**

(a)   (b)   (c)

FIGURE 12.4 Features (traits) used by Mendel in his experiments on pea plants **(a)** included the color of the seeds **(b)**, and other traits such as height of the plant, round or wrinkled seeds and white or purple flowers **(c)**.

[(a) Courtesy of Kurt Steuber, www.biolib.de; (b) Courtesy of Søren Holt, In the Toad's Garden.]

particular gene pair (the gene for smooth or wrinkled seeds, for example) are **alternative alleles of the gene**. Each gene has at least two alleles but may have many more, greatly increasing the source of variation over that provided by a pair of alleles.

Geneticists use symbols to distinguish alleles, a system introduced by Mendel himself; the notation *S* for the dominant character, *s* for the recessive, and *Ss* for the hybrid is used throughout Mendel's 1866 paper. Individuals with two different alleles for any particular character (*Ss*) are **heterozygotes**. Those with two identical alleles (*SS*, *ss*) are **homozygotes**. Like Mendel, we will label the smooth and wrinkled alleles *S* and *s*, respectively. These alleles segregate in the gametes of the F_1 generation, which, if crossed, unite to form a second filial generation (F_2) in predictable proportions of 1 *SS* : 2 *Ss* : 1 *ss*. In this experiment Mendel obtained an average of 2.96 smooth seeds for every one wrinkled seed. In a second experiment involving flower color—purple or white—he obtained an average of 2.89 purple flowers for every one white flower. Experiments involving these two characters, along with five other characters, gave an **average ratio of 2.98:1**, or rounding off, 3:1. Why a 3:1 ratio?

As shown in **FIGURE 12.5**, these F_2 proportions arise because the allele for one trait, *S*, is **dominant** over the allele *s* (known as the **recessive** allele) when both are present in *Ss* (heterozygote) individuals. That is, *SS* and *Ss* seeds both are smooth. The only wrinkled seeds are those with the *ss* genotype. Hence, the 3:1 ratio. In general, the alleles found in nature in the majority of individuals (the **wild type alleles**) are dominant, presumably because they produce useful products in the presence of other alleles whose products may not be as useful or may even be deleterious.

Mendel's experiments showed that the factors (genes) responsible for heredity are neither changed nor blended in the heterozygote, but segregate from each other to be transmitted as discrete and constant particles between generations. Within a few years of the rediscovery of his studies in 1900, many other studies in a variety of plants and animals provided support for Mendelian ratios and their variations, for the principle of segregation and for Mendel's second principle of independent assortment.[8]

PRINCIPLE OF INDEPENDENT ASSORTMENT

When Mendel cross-pollinated plants with two different characters, the results could be understood and then predicted if one character had no effect on the segregation of the other. The characters behaved as expected of particulate, non-blending hereditary factors. As discussed earlier in this chapter, we now say that genes for different characters segregate independently of one another. The cellular explanation for such independent assortment is the localization of the genes for each of the two characters on different pairs of chromosomes, shown in **FIGURE 12.6**. During meiosis (Box 12.1) the two halves of a pair of homologous chromosomes move toward opposite poles of the nucleus, independently of any other pair of chromosome. As a result, the segregation of genes on one chromosome is independent of the segregation of genes on other chromosomes (Figure 12.6).

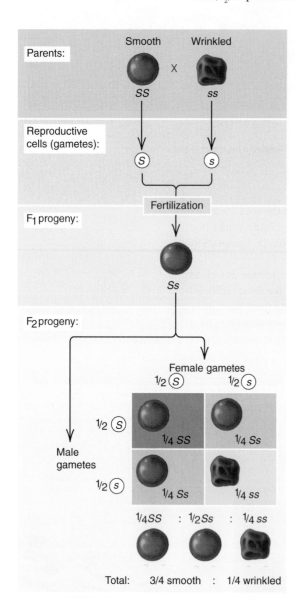

FIGURE 12.5 Explanation of Mendel's results for the inheritance of seed shape—smooth (*S*) or wrinkled (*s*)—in pea plants (*Pisum sativum*) beginning with a parental cross of homozygous smooth (*SS*) and homozygous wrinkled (*ss*) plants, where *S* alleles are dominant and *s* alleles recessive. Self-fertilization of *Ss* gametes from the first filial (F_1) generation results in a 3:1 ratio of smooth to wrinkled seeds in the second filial (F_2) generation.

[8] For the history of Mendelian genetics, see Olby (1966), Dunwell (2007), and the essay on Mendel, Mendelism and Genetics by Robert C. Olby at MendelWeb (http://www.mendelweb.org/MWolby.html).

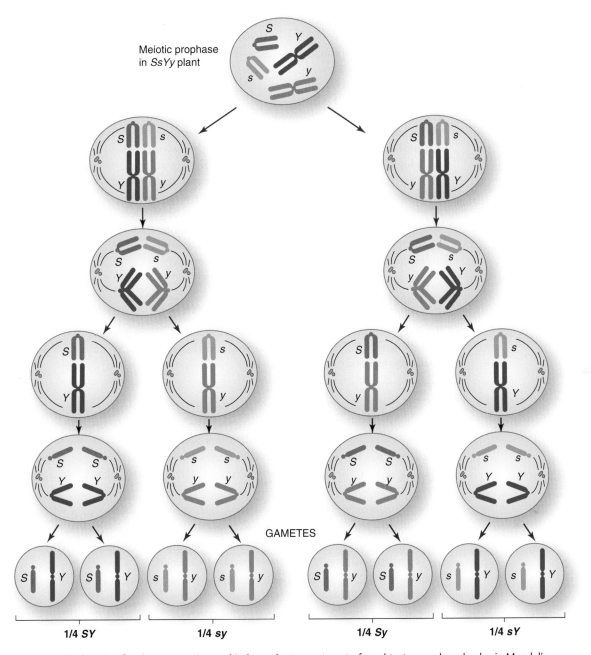

Meiotic prophase in *SsYy* plant

GAMETES

1/4 *SY*   1/4 *sy*   1/4 *Sy*   1/4 *sY*

FIGURE 12.6 Explanation for the segregation and independent assortment of seed texture and seed color in Mendel's experiments in terms of genes (*S*, *s*, and *Y*, *y*) on different chromosomes. Because of independent assortment all four combinations of chromosomes in the gametes correspond to all four possible combinations of the genes *S* and *s* (smooth and wrinkled seeds) and *Y* and *y* (yellow or green seeds).

[Adapted from *Genetics, Third Edition,* by Monroe W. Strickberger. © 1985 by Monroe W. Strickberger. Reprinted by permission of Prentice Hall, Inc., Upper Saddle River, NJ.]

Inheritance proved to be much more complex than the interpretation of Mendel's experiments indicated. For example, many alleles are neither completely dominant nor completely recessive, many genes have more than two alleles, alleles interact with one another, and many genes may be located on the same chromosome, all of which complicate both selection and the independence of traits.

Furthermore, replication of biological information is not always constant or exact; changes (mutations) in the DNA of the genes in gamete-producing cells are passed on to the next generation. The significance of genetic variation lies in its potential to fuel evolution.

■ KEY TERMS

alternation of generations	karyotype
binary fission	mate choice
blending inheritance	meiosis
cleavage	mitosis
chromatid	pangenesis
dominance allele	phenotype
eukaryotic cells	prokaryotic cells
gametophyte	recessive allele
genotype	reproductive isolation
germ plasm theory	segregation
heterozygote	sporophyte
homologous chromosomes	saltation
homozygote	variation
independent assortment	wild type

■ EVOLUTION ON THE WEB

Explore evolution on the Internet! Visit the accompanying website for *Strickberger's Evolution, Fifth Edition,* at **go.jblearning.com/Evolution5eCW** for exercises and links relating to topics covered in this chapter.

■ REFERENCES

Bogdanov, Y. F., 2003. Variation and evolution of meiosis. *Russ. J. Genet.*, **39**, 363–381.

Churchill, F. B., 1968. August Weismann and a break from tradition. *J. Hist. Biol.*, **1**, 91–112.

Crosland, M. W. J., and R. H. Crozier, 1986. *Myrmecia pilosula*, an ant with only one pair of chromosomes. *Science*, **231**, 1278.

Darwin, C., 1868. *The Variation of Animals and Plants under Domestication.* John Murray, London.

Dunwell, J. M., 2007. 100 years in: a century of genetics. *Nat. Rev. Genet.*, **8**, 231–235.

Extavour, C. G., and M. Akam, 2003. Mechanisms of germ cell specification across the metazoans: epigenesis and preformation. *Development*, **130**, 5869–5884.

Griffiths, A. J. F., 2002. *Modern Genetic Analysis: Integrating Genes and Genomes. Second ed.* W. H. Freeman and Co, San Francisco.

Hallgrímsson, B., and B. K. Hall (eds.), 2005. *Variation: A Central Concept in Biology.* Elsevier/Academic Press, New York.

Hall, B. K., Pearson, R. D., and G. B. Müller (eds.), 2004. *Environment, Development, and Evolution; Toward a Synthesis.* MIT Press, Cambridge, MA.

Hartl, D. L., and E. W. Jones, 1998. *Genetics: Principles and Analysis.* Jones and Bartlett, Burlington, MA.

Howard, J. C., 2009. Why didn't Darwin discover Mendel's laws? *J. Biol.*, **8**, 15.1–15.8 (doi:10.1186/jbiol123).

Liu, Y., 2008. A new perspective on Darwin's pangenesis. *Biol. Rev.*, **83**, 141–149.

Mendel, G., 1866. Versuch über Pflanzenhybriden. *Verh. Natur. Brünn*, **4**, 3–47.

Olby, R. C., 1966. *Origins of Mendelism.* Constable, London.

West-Eberhard, M. J., 2003. *Developmental Plasticity and Evolution.* Oxford University Press, Oxford, UK.

Wilkins, A. S., and R. Holliday, 2009. The evolution of meiosis from mitosis. *Genetics*, **181**, 3–12.

This is Mendel's CLASSIC PAPER, *originally published in the* Proceedings of the Brünn Natural History Society. *It was translated into English by William Bateson in 1901 under the title* Experiments in Plant Hybridization.

END BOX 12.1

Alternation of Haploid and Diploid Generations in Plants

© Photos.com

SYNOPSIS: Those organisms in which a diploid phase or stage alternates with a haploid phase or stage are discussed. Animals alternate haploid gametes with diploid embryos and adults. Many plants show alternation of diploid and haploid generations. Hypotheses for the evolutionary advantages of *alternation of generations* are discussed.

As presented in this chapter, animals are diploid but their gametes—eggs and sperm—are diploid. Fertilization of egg by sperm restores the diploid number of chromosomes to the zygote. Plants alternate haploid with diploid generations, not as gametes but as mature phases of the life cycle in an **alternation of generations**. Evolutionary biologists have asked why the haploid stage has persisted over such long periods of evolutionary time.

PERSISTENCE OF THE HAPLOID LIFE HISTORY STAGE

Meiosis is hypothesized to have originated in conjunction with, or soon after, the evolution of sexual fertilization; sexual union could only occur after the meiotic reduction mechanism evolved to facilitate a regularized transition from diploid to haploid stages. This transition is seen in gametes and adults in animals, and in alternation of generations in plants.

Persistence of haploidy as a major life cycle stage may be related to the speed at which haploids eliminate deleterious alleles, which are protected from elimination in the diploid stage. Such advantages favor haploids "if (i) sex is rare, (ii) recombination is rare, (iii) selfing is common, or (iv) assortative mating is common" (Mable and Otto, 1998). However:

> Once a certain ploidy level has become dominant within a taxonomic group, it may be difficult to expand the alternate ploidy phase, either because the necessary mutations simply do not arise or because individuals with atypical ploidy levels are unable to develop normally . . . [A]n organism may evolve developmental pathways that depend on having the appropriate ploidy level (Mable and Otto, 1998).

In animals, the lengthened diploid stage became the dominant feature of the life cycle; the haploid stage is mostly restricted to the gametes. In land plants and some green algae, the lengthened diploid stage, or sporophyte, also produces meiotic products as in animals, though these are spores rather than gametes. Meiotically produced spores develop into haploid gametophytes, which only later produce gametes by mitosis.

ALTERNATION OF GENERATIONS

All plants share a lifestyle characterized by alternation of generations between a free-living haploid **gametophyte generation** and a diploid and parasitic **sporophyte generation**. "All plants" includes liverworts such as the genus *Marchantia*, ferns such as the common polypody *Polypodium vulgare*, and all vascular plants (**FIGURE EB1.1a–c**). Distinctive features of gametophyte and sporophyte life history stages in each of these major lineages are illustrated in **FIGURES EB1.2 to 1.4** and outlined in **TABLE EB1.1**.

Fossil liverworts, which are known from as early as the Devonian Era 420 Mya, preceded vascular land plants by 10–20 My. Liverworts are the sister taxon to vascular plants (Donoghue, 2008) and were important in the origin of leaves, in the transition from water to land, and in the shift from a haploid gametophyte-dominated life cycle to a diploid sporophyte-dominated life cycle (Niklas, 1997; Gensel and Edwards, 2001). Despite this knowledge, the sporophyte–gametophyte alternation of generations in plants (Figure EB1.3) remains puzzling.

An obvious advantage of the sporophyte is that, as the diploid stage of the life cycle, its retention allows genetic recombination to influence future generations. The sporophyte also produces dispersible spores that can resist desiccation. These meiotically-produced spores develop into haploid gametophytes, which later produce gametes (Table EB1.1). Vulnerability of gametes to terrestrial conditions may account for the persistence of the gametophyte stage in plants; sporophytes resist desiccation and so facilitate life on land.

END BOX 12.1

Alternation of Haploid and Diploid Generations in Plants (Cont...)

© Photos.com

FIGURE EB1.1 A typical liverwort, *Marchantia* **(a)**, fern, the common polypod *Polypodium vulgare* **(b)**, and vascular plant, the Pohutukawa or New Zealand Christmas tree, *Metrosideros excelsa* **(c)**.

In some lineages the sporophyte has become independent of a gametophyte. This independence has been regarded as a trade-off between the advantages of diploidy and the needs of spore production. For example, many plants have lost the ability to reproduce sexually. Sexual reproduction through spores has been replaced with asexual reproduction such as vegetative propagation from roots or stems, or reproduction through unfertilized eggs. This process of *parthenogenesis* is seen also in some animals. Although some asexual plants are found over wide geographical ranges, their success is often restricted to specific environments or to conditions that severely limit cross-fertilization. Because they lose genetic diversity, asexual groups rarely survive over long evolutionary periods.

ORIGIN OF SPOROPHYTES

Two theories for the origin of the sporophyte have been investigated for some time (Blackwell, 2003). According to the **antithetic** (interpolation) **theory**, all early plants were gametophytes that produced diploid zygotes. Some of these diploid zygotes underwent a period of delayed meiosis, yet

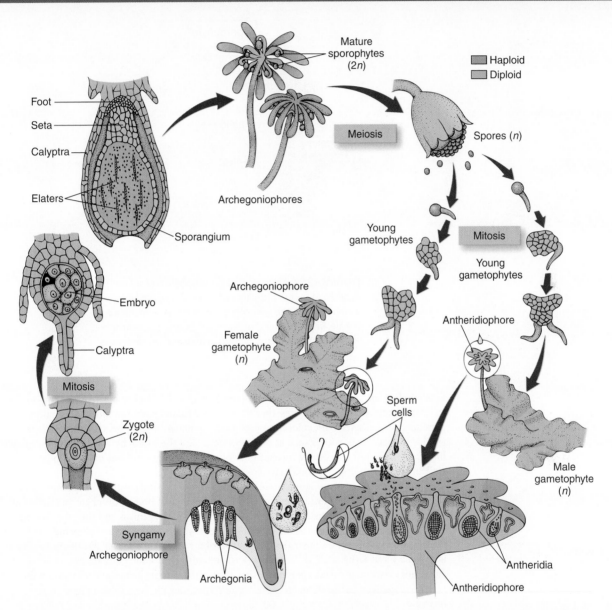

FIGURE EB1.2 Life cycle of a liverwort in the genus *Marchantia*. The gametophyte is the haploid (*n*) stage shown at the bottom of the cycle. Fertilization of the "egg" by a "sperm" produces the diploid (2*n*) zygote, embryo, and sporophyte (left-hand side). The sporophyte produces sporocytes, each of which divides by meiosis to form haploid (*n*) spores. Germination of the spore (lower right) leads to the gametophyte, and the cycle is reinitiated. Also see Figure EB1.3 for alternation between gametophyte and sporophyte.

END BOX 12.1

Alternation of Haploid and Diploid Generations in Plants (Cont...)

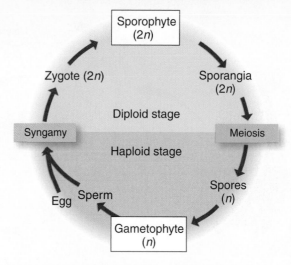

FIGURE EB1.3 Alternation of gametophyte and sporophyte generations in the plant life cycle (see also Figure EB1.2). In some plants, the gametophyte is unisexual, either male or female. In others, the gametophyte is bisexual or hermaphroditic (as shown here), producing male and female tissues.

kept dividing mitotically, thus perpetuating the diploid state. These parasitic diploid tissues were the initial rudimentary sporophytes; thus they differed functionally and morphologically from the parental gametophyte. With colonization of land, an increasing proportion of sporophyte tissues assumed vegetative purposes such as photosynthesis, and the sporophyte became an independent life stage. The antithetic theory invokes developmental changes in a set of evolutionary steps to explain how the sporophyte could have arisen.

The contrasting **homologous** (transformation) **theory** postulates that the sporophyte initially showed little difference from the gametophyte. In this theory, both sporophyte and gametophyte share the same genetic constitution derived from the same organism and therefore should have shared the same patterns of growth. The similarity between sporophyte and gametophyte in many algae, as well as structural similarities of the stems of the (extant) terrestrial fernlike genus *Psilotum* and some basal ferns such as *Stromatopteris*, support the homologous theory. Developmental similarities also are invoked: sporophyte tissue can arise in some gametophytes without the intervention of gametic fertilization, and some sporophytes may produce gametophytes without spore formation.

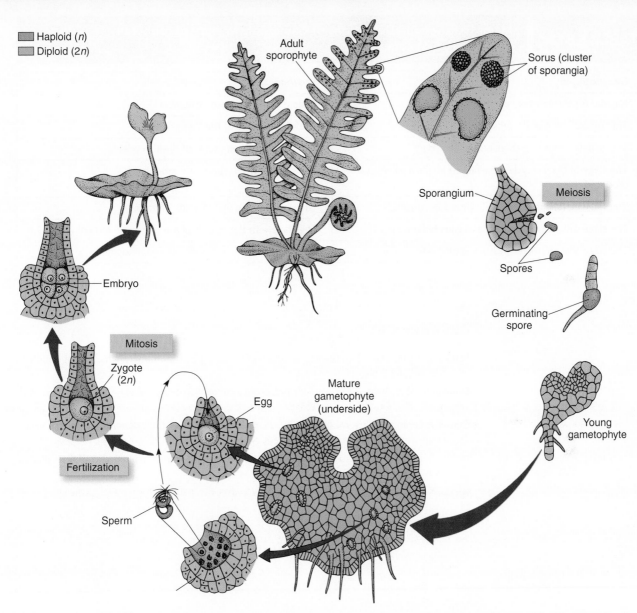

FIGURE EB1.4 Life cycle of a fern, the common polypod, *Polypodium vulgare*, showing alternation between haploid gametophyte and diploid sporophyte. Compare with liverworts (Figure EB1.2) and vascular plants (Figure EB1.3).

END BOX 12.1

Alternation of Haploid and Diploid Generations in Plants (Cont...)

© Photos.com

TABLE EB12.1 — **Major Features and Definitions of Gametophyte and Sporophyte Life History Stages in Plants**

Gametophyte	Sporophyte
Haploid (*n*) chromosome number	Diploid (*2n*) chromosome number
The haploid stage of the life cycle	The diploid stage of the life cycle
The gamete-producing stage of the life cycle	The spore-producing stage of the life cycle
The stage of the life cycle (generation) in which gametes are produced	The stage in which spores are produced
The stage containing the gamete-producing organs	The stage formed by the union of gametes
The stage with the mitotic phase of gamete formation	The stage with the meiotic phase of gamete formation
The stage that alternates with the sporophyte	The stage that alternates with the gametophyte

REFERENCES

Blackwell, W. H., 2003. Two theories of origin of the land-plant sporophyte: which is left standing? *Bot. Rev.*, **69**, 125–148.

Donoghue, M. J., 2008. A phylogenetic perspective on the distribution of plant diversity. *Proc. Natl. Acad. Sci. USA*, **105**, 11549–11555.

Gensel, P. G., and D. Edwards (eds.), 2001. *Plants Invade the Land: Evolutionary and Environmental Perspectives.* Columbia University Press, New York.

Mable, B. K., and S. P. Otto, 1998. The evolution of lifecycles with haploid and diploid phases. *BioEssays*, **20**, 453–462.

Niklas, K. J., 1997. *The Evolutionary Biology of Plants.* University of Chicago Press, Chicago.

CHAPTER

13

Genes, Environment, and Inheritance

CHAPTER SUMMARY

Inheritance has proved to be much more complex than Mendel's experimental results and two principles indicated. For example, dominance of an allele is not always "all or nothing." Many genes have multiple alleles. Genes can, and usually do, interact with one another in what is known as *epistasis*. Furthermore, the first two to three decades of the twentieth century witnessed an explosion of discoveries regarding additional modes of inheritance. We say "modes" because some inheritance is extranuclear and some linked to genes that determine sex. Presence of two sexes in many species and the discovery of varied mechanisms of sex determination demonstrated roles for genes, the environment, and environment by gene interactions in establishing the fundamental features of organisms. Gene by environment interactions emerged as an important mechanism determining major aspects of the phenotype, including whether an individual would become male or female. Surprisingly, in some organisms, sex is not determined by genes on sex chromosomes, but by a diverse array of environmental signals that function under the umbrella name of environmental sex determination. Furthermore, genotypes in many organisms respond differently to diverse environments and environmental signals to elicit aspects of the phenotype other than sex.

DEVIATIONS FROM MENDELIAN GENETICS

Following the rediscovery in 1900 of Mendel's experiments and his two principles of the segregation of alleles and independent assortment, geneticists realized that a single phenotypic difference could be due to differences in more than one gene and that dominance relationships between alleles could range widely; it is not "one gene, one feature." **Incomplete dominance** and three other important sources of variation—**codominance**, **multiple alleles**, and **epistasis**—are discussed below, in each case using the inheritance of flower color as the example.[1]

INCOMPLETE DOMINANCE OF ALLELES

Whether WHITE *is the presence of all colors or the absence of color has been debated. White reflects all the colors of visible light, and so is a color. Black, on the other hand, absorbs all the colors of the visible spectrum—absorbs all light—and so, black is not a color.*

In a dominant-recessive single-gene situation in which red is dominant over white, only red and white flowers are expected to occur when a red-flowered plant is crossed with a white-flowered plant (FIGURE 13.1). Incomplete dominance is the situation in which red (in this example) is only partly or incompletely dominant over white. As a consequence of incomplete dominance, the homozygote *AA* produces red flowers, the homozygote *aa* produces white flowers, but the heterozygote *Aa* produces flowers with an intermediate color, pink.

Mendel was fortunate in working with species—pea plants—possessing external features such as smooth or wrinkled seeds that are readily distinguished from one another (they are said to be discontinuous characters) and controlled by genes that produce features appearing to the naked eye as dominant or recessive traits. With the higher resolution available today, we now know that the recessive allele causing the wrinkled pea seeds studied by Mendel prevents normal expression of an enzyme that makes branched starch molecules.[2] Heterozygotes now can be distinguished from homozygotes because aspects of the phenotype (numbers of starch granules and enzyme concentrations in heterozygotes) are intermediate between homozygote dominant or recessive individuals. Dominance appears complete when assessed by seed texture but is incomplete when assessed by other aspects of the phenotype. This is not an isolated example. It is seen in sickle cell anemia, which is recessive at the disease (phenotypic) level, incompletely dominant at the red blood cell level, and codominant at the level of hemoglobin. Individuals who are diagnosed as recessive for sickle cell anemia show no symptoms of the disease but have abnormal red blood cells.

MULTIPLE ALLELES AND CODOMINANCE OF ALLELES

A gene determining a particular character usually consists of more than two alleles. The genetic basis for α- and β-globins in humans consists of hundreds of alleles; that is, globins are a *multiallelic trait*. Codominance occurs when two or more alleles are expressed *and* influence the character. Different dominance relationships may exist between the alleles. For example, A^1 may be codominant with A^2, or A^1 and A^2 may be dominant over A^3. Codominance is seen therefore in heterozygotes.

Such multiple allelic systems arise because a gene consists of a linear array of hundreds or thousands of nucleotides, and because allelic difference may arise from change in one or only or few nucleotides. For example, individuals heterozygous for the sickle cell trait produce both mutant and normal versions of β-globin in their red blood cells. In ABO blood group antigens in humans, alleles for the A– and B–antigens fail to produce enzymes.

---

[1] See Nijhout (2003) for an elegant discussion of modern genetics, placed in the context of Mendel's experiments.

[2] The genetic basis of the wrinkled pea seeds studied by Mendel is even more complex. It has been shown by Bhattacharyya and colleagues to be based on the insertion of a transposon into the gene (M. K. Bhattacharyya et al., 1990. The wrinkled-seed character of pea described by Mendel is caused by a transposon-like insertion in a gene encoding starch-branching enzyme. *Cell* **60**, 115–122).

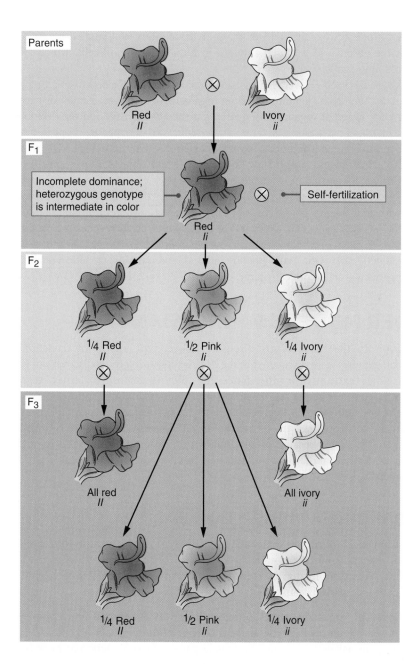

FIGURE 13.1 If red is
incompletely dominant over
ivory, a cross with plant with
ivory colored flowers yield
red, ivory, and pink flowers in
the F_3 generations.

EPISTASIS

Interaction between genes or between the products of different genes is known as epistasis, or **epistatic interaction**. Because of epistatic interactions, one (or more than one) gene can change the qualitative phenotypic effect produced by another (or more than one other) gene. For example, because of epigenetic interactions, each of a number of genes may be able to produce more than one product, as seen in pigment production and flower color (Figure 13.1).

Epistatic interactions include those that **modify the dominance relations of a particular allele**. Expression of allele A^1 changes in the presence of alleles H^1 or H^2 of another gene. In this example, A^1 is dominant over A^2 in certain genotypes (say, H^1) but is recessive when the genotype is different (say, H^2).

At the molecular level, the products of wild-type alleles (which usually are dominant) are often functional proteins such as enzymes. Products of mutant alleles may be nonfunctional or partly functional proteins, or the mutant (null) allele may fail to form

10^{100} *is known as a* GOOGOL. *A googol is larger than the estimated number of protons in the universe, which British astrophysicist Arthur Eddington (1882–1944) estimated to be* 1.575×10^{79}.

the gene product (either the RNA or the protein) at all. A well-studied example of null alleles is the absence of A and B antigens in human red blood cells, which results in type O red blood cells.

Has epistasis itself evolved? That is, do the genomes of some organisms contain more interactions than do the genomes of other organisms? A large-scale analysis carried out across a wide range of organisms suggests that the answer is yes; viruses exhibit fewer epistatic interactions (mutations have smaller effects than expected if epistasis is common) than eukaryotes (mutations have greater effects than expected).[3] The many possible allelic differences—dominance, interaction effects, and epistasis—mean that the amount of genetic variation available as the source of evolutionary change is staggering (Hallgrímsson and Hall, 2003; Duveau and Félix, 2010). To take a simple example, a gene with four alleles (G^1, G^2, G^3, G^4) can produce 10 different diploid combinations (G^1G^1, G^1G^2, G^1G^3, G^1G^4, G^2G^1, G^2G^2, and so on). A hundred such genes can produce 10^{100} combinations. As any cellular organism carries more than 100 genes (and most carry 12,000–30,000, each with many possible alleles), the potential genetic variability in a single organism is astronomically large.

OTHER MODES OF INHERITANCE

In addition to the mechanisms discussed above, two important deviations from Mendelian genetics were discovered in the decades following the rediscovery of Mendel's research. One is **extranuclear inheritance**. The other results from patterns of segregation in alleles linked to other genes on the chromosome, including chromosomes involved in sex determination (**sex-linked genes**). Both deviations have considerable genetic and evolutionary consequences, the former because it counters inheritance confined to the nucleus, and the latter because it compromises the recombination of alleles and raises important questions about how two sexes evolved and are maintained (Lucchesi, 1998; Charlesworth, 1996, 2010).

EXTRANUCLEAR INHERITANCE

The first unusual finding was that some traits do not follow a nuclear pattern of inheritance but are transmitted through the cytoplasm of the egg or through cytoplasmic organelles. This phenomenon is often termed **cytoplasmic inheritance**. Such extranuclear inheritance arises because

1. cytoplasmic organelles such as mitochondria in eukaryotes and chloroplasts in plants have their own genetic material (DNA), and
2. during animal development, the female parent deposits gene products (proteins, mRNAs) into the egg as it is being formed.[4]

The term *extranuclear* usually is applied to the inheritance of mitochondrial and chloroplast DNA, although some of the genes that control organelle activity moved to the nucleus early in organismal evolution and so are no longer extranuclear. Inheritance of maternal gene products is "extranuclear" as far as the zygote is concerned, but not in the context of the female parent; the products are produced under the control of the maternal nuclear genome.

---

[3] R. Sanjuán and S. F. Elena, 2006. Epistasis correlates to genomic complexity. *Proc. Natl. Acad. Sci. USA*, **103**, 14402–14405. See also Peck (1994).

[4] Nurse cells in the ovaries of some animals play important roles in depositing RNA or other gene products into the egg, including the gene products that establish the anterior–posterior body axis. These are also products of the female parent. See **http://www.sdbonline.org/fly/aimorph/oocyte.htm**.

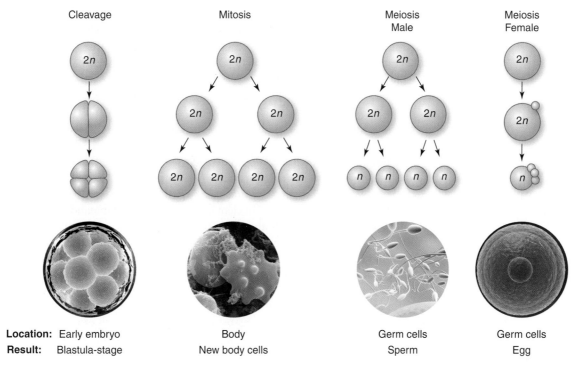

| Cleavage | Mitosis | Meiosis Male | Meiosis Female |

FIGURE 13.2 Essential differences between three types of cell division (cleavage, mitosis, and meiosis) in animals, with emphasis on cell size and chromosome number. Cleavage of the fertilized egg is a modified form of mitosis in which cell size is reduced at each division, most simply by equal division of cytoplasm of cells that remain attached to form the multicellular embryo. All cells are diploid (shown as $2n$); that is, the chromosomal events in cleavage are the same as in mitosis of body cells in which cell size is restored after each division by synthesis of new cytoplasm (growth). Although the mechanism of chromosome reduction during meiosis is the same in germ cells of males and females, the fate of the cells differs. Equal distribution of cytoplasm in the male germ line results, in the example shown, in the production of four sperm-producing cells of equal size. In contrast, unequal distribution of cytoplasm in the female germ line results, in the example shown, in the production of one large and three small cells. Only the large cells go on to become an egg. In both male and female germ lines, the reduction (meiotic) division reduces chromosome number from diploid ($2n$) to haploid (n).

Mitochondria are the energy powerhouses of cells. They contain their own DNA, which is separate and different from nuclear DNA. Because it is not subject to recombination during sexual reproduction, mitochondrial DNA (mtDNA) shows much less variation than does nuclear DNA. mtDNA has proven to be an ideal molecule to use when reconstructing evolutionary history. Furthermore, a number of human disorders and inherited diseases have their origin in defective mtDNA.[5] Many organisms also contain other organisms (endosymbionts), which have their own DNA and can interact with host DNA or the products of host genes The recent and unexpected discovery of an alga acting as an endosymbiont in the embryos, larvae, and adults of the spotted salamander, *Ambystoma maculatum*, opens a window on previously unknown endosymbioses in vertebrates (Kerney et al., 2011).

During animal development, practically all the cytoplasm of the zygote, including organelles such as mitochondria and ribosomes, comes from the egg, not from the sperm (**FIGURE 13.2**). Sperm contribute nuclear genetic material and, at most, a few mitochondria. Mitochondria, the nucleus, nuclear membrane, and the microstructure of the zygote

---

[5] Depending on the populations, mutations in mtDNA are associated with diseases affecting the brain, heart, lungs, gastrointestinal tract, and other organs in 1 in 3,500 to 1 in 6,000 individuals (R. W. Taylor and D. M. Turnbull, 2005. Mitochondrial DNA mutations in human disease. *Nat. Rev. Genet.*, **6**, 389–402).

are all preformed in the cytoplasm of the egg before the start of every generation. Fertilized eggs therefore contain organelles and organelle DNA that are produced under the control of maternal DNA. Such inherited structures provide evidence that we should not abandon entirely the notion of preformation that was popular in the eighteenth century.

SEX-LINKED GENES AND SEXUAL REPRODUCTION

Sexual reproduction provides two important sources of new genetic variation: recombination and gene linkage.

Genes on the same chromosome tend to remain together unless **recombination** moves them to other chromosomes. In recombination, sections of chromosomes exchange material by crossing over, resulting in different linear arrays of nucleotides on *both* chromosomes. As genes are nucleotide sequences, recombination can produce **different combinations of genes along a chromosome**. In humans with 23 pairs of chromosomes, there may be millions of different possible genetic combinations in gametes as a result of recombination alone. Genetic and phenotypic variation produced by sexual reproduction allows for the long-term adaptation/survival of groups facing changing environments.

The second departure from Mendelian genetics—actually an elaboration rather than a departure—is that genes do not assort independently of each other if they are linked together on the same chromosome; genes on the same chromosome tend to remain together during meiosis. **Linkage** of genes occurs on all chromosomes, but is discussed here in the context of sex chromosomes, in which linkage was first demonstrated.

Gene linkage can be appreciated in several contexts, perhaps the most relevant being the **evolution and possession of two sexes** (male and female) in a population, and the **sexual reproduction** that accompanied the evolution of male and female sexes. (Whether separate sexes and genetic recombination evolved together, and why both persist, is discussed in BOX 13.1.) Lack of independent assortment because of linkage was

---

BOX 13.1	Evolution and Persistence of Sexual Reproduction and Genetic Recombination

Despite close to two dozen different proposals, we do not fully understand whether the origination of sex and the evolution of genetic recombination were connected, or why the two conditions persist. We expect natural selection to preserve associations between alleles that benefit the fitness of an individual. However, while sexual reproduction creates new associations, genetic recombination breaks up allele associations. Maynard Smith (1978) is a classic treatment of the topic.[a]

With respect to evolutionary origins, one proposal is that sex (here considered as the origin of meiosis) derives from recombinational **DNA repair mechanisms** that arose as a way of overcoming and repairing DNA damage. Any genetic variation that resulted would have been an accidental byproduct. However, even if recombination arose as a DNA repair mechanism, production of genetic variation is now the primary selective advantage of the persistence of sex. It also has been hypothesized that sex is advantageous to organisms that exist in variable environments, and that parasitic elements, such as transposons and plasmids, initiated or promoted sexual fusion as a mechanism to infect other cells.

Of the many hypotheses to explain the persistence of sexual reproduction (and as a consequence, genetic recombination), **removing deleterious mutations** is perhaps the best supported (Peck, 1994). In the water flea, *Daphnia pulex*—which can modulate its mode of reproduction from sexual to asexual—more mutations accumulate in individuals with asexual reproduction than in individuals with sexual reproduction.[b] Another well-supported proposal is that mating and genetic recombination allows individuals in a population to **incorporate beneficial mutations** more rapidly. Without sexual reproduction the accumulation of beneficial mutations would be much slower.

---

[a] For other insightful studies, see J. Maynard Smith, 1988. The evolution of recombination. In *The Evolution of Sex*, R. E. Michod and B. R. Levin (eds.). Sinauer Associates, Sunderland, MA, pp. 106–125; the papers in Michod and Levin (1988); Ellegren (2011).

[b] S. Paland and M. Lynch, 2006. Transitions to asexuality result in excess amino acid substitutions. *Science*, **311**, 990–992.

discovered when it was found that some genes—sex-linked genes—are localized to the X (female in mammals) or Y (male in mammals) chromosomes, which are not partitioned equally to females and males, females having two Xs, males having one X and one Y. The heterogametic sex (XY)[6] expresses all alleles lying on the X chromosome without regard to dominance relationships.

SEXUAL REPRODUCTION

An important consequence of sexual reproduction is the introduction of new variation that enables a population to produce a wide array of genotypes and so translate natural selection into hereditary change. This variation arises because the two parents each contribute equal numbers of chromosomes to each offspring. Chromosomes reshuffle in the offspring's germ line to produce chromosomal and genetic combinations that differ from those in either parent. Gametes containing these new combinations combine to form the zygotes of the next generation, which therefore have new genetic combinations. Chromosomes reshuffle in the following generation, and the process continues. Importantly, the genome is reformed every generation. The organelles that result from the extranuclear inheritance discussed above, however, are constant between generations.

For populations continually interacting with changing environments, sexual reproduction allows for the production of new combinations of alleles, some of which may be more advantageous and permit the lineage to thrive (even survive) under new or changed environmental conditions. Lineages without the variability introduced by such reproduction can more easily become extinct, should circumstances change. Species that reproduce asexually usually have higher extinction rates than do sexually reproducing species.[7]

In a long-standing population that continually experiences the *same* environment—which is an extremely unusual situation, especially with modification of environments by human activity—genotypes have evolved for phenotypes adapted to (and often able to respond to) that environment. Most, if not all, new genetic combinations will have lower relative fitness than do those of the parental generation. Under such stable circumstances, the advantages of having two sexes are not apparent, especially when asexual reproduction can produce more offspring for a given population size. Many plants can switch from sexual reproduction to reproduction by the spread of vegetative somatic tissues or *parthenogenesis* (reproduction through unfertilized ovules—seen also in unfertilized eggs in some animals). Although some asexual plants are found over wide geographical ranges, their success is often restricted to specific environments or to conditions that severely limit cross-fertilization; only small, inbred populations survive. However, environmental conditions are *usually not stable* and eukaryotes generally use meiotic forms of sexual reproduction for at least part of their life cycle. **Asexuality** only rarely appears as a prevailing character in large taxonomic groups of animals, the rest of which reproduce sexually. Asexual reproduction is more common among smaller groups, although ferns, mosses, and fungi do not fit this pattern. The scarcity of asexual families or higher taxonomic groups indicates that asexuality rarely survives long enough to become the predominant character of a lineage.

SEX DETERMINATION

As discussed above, sexual reproduction involves differentiation into two different sexual forms in which reproduction occurs only as a result of union between gametes of the different sexes to form a zygote (Figure 13.2). This simple statement, however,

---

[6] The heterogametic sex differs in various lineages; in mammals, males are heterogametic (XY), in birds females are heterogametic.

[7] See Wiley (1978) for a discussion in the context of evolutionary species, and see Kokko et al. (2008) for a recent study.

fails to reveal the amazing range of mechanisms for sex determination, a few of which are introduced below.[8] Surprisingly, some mechanisms are not based on sex-determining genes but on specific environmental cues. Broadly, we identify two modes of sex determination: **Sex-chromosome-based** and **environmentally induced** sex determination.

GENETIC SEX DETERMINATION

Early in the twentieth century, geneticists discovered that particular genes were located on chromosome(s) associated with sex determination, which thus became known as **sex chromosomes**. Human and mouse X and Y chromosomes are among the most well known sex chromosomes (Ellegren, 2011).

For many organisms, especially mammals, sex determination is associated with **chromosomal differences between the two sexes**, typically XX females and XY males. The Y chromosome is often smaller and mostly inactive, except for male-determining and male-fertility genes. In some cases, only a single copy of a sex-determining gene is required to establish the sex of an individual. In mammals, the *sex-determining region Y* (*Sry*) gene on the Y chromosome has this function for male development, encoding for a protein known as the testis-determining factor, which is an upstream signaling gene in male sex determination (**FIGURE 13.3**; Kashimada and Koopman, 2010). An independently evolved XX:XY system of sex chromosomes exists in *Drosophila* (below and see **END BOX 13.1**).

SRY AND MAMMALIAN SEX DETERMINATION

Sry is the master switch in mammalian sex determination. The master role played by testis-determining factor was demonstrated dramatically in 1991 when XX mice were converted to males following transgenic introduction of *Sry*. A recent study shows that such sex reversal is preceded by the formation of transient gonads containing testicular and ovarian regions—an ovotestis.[9]

The testis-determining factor produced by the *Sry* gene functions by regulating expression of a related gene, *Sox-9*. The protein product of *Sox-9*, a transcription factor, initiates the differentiation of Sertoli cells[10] that will become the supporting cells of the testes and which are required for testes to form. Mouse *Sox-9* has a gonad-specific enhancer *Tesco* (*testis-specific enhancer of Sox-9 core*) to which testis-determining factor binds.

Y CHROMOSOMES IN MAMMALS

The human Y chromosome has been studied especially intensively. While human X chromosomes are some 165 Mb in size with around 1,000 genes, the Y chromosome is only 60 Mb with 50 functional genes. Genes on the Y chromosome have

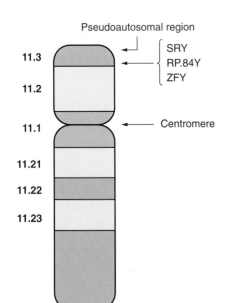

FIGURE 13.3 Basic features of human Y-chromosome including location of the *SRY* gene.

---

[8] The classic book by Bull (1983) provides many examples of sex determination that arise from diverse genetic and chromosomal changes.

[9] For conversion of XX mice to males, see P. Koopman et al., 1991. Male development of chromosomally female mice transgenic for *Sry*. *Nature*, **352**, 117–121. For the ovotestis phase, see E. P. Gregoire et al., 2011. Transient development of ovotestes in XX *Sox9* transgenic mice. *Dev. Biol.*, **349**, 65–77. For ovotestes across the animal kingdom, see Davison (2006).

[10] Named after Enrico Sertoli, (1842–1910), an Italian physiologist and histologist who discovered them in 1865, Sertoli cells secrete at least eight hormones required for spermatogenesis.

Courtesy of Walter Just, Institute of Human Genetics

FIGURE 13.4 Male mole voles, Ellobius lutescens (shown here) and male Japanese spinous country rats *Tokudaia osimensis* have both lost their Y chromosomes. The sex-determining genes have translocated to autosomes.

been and continue to be subject to selection and show higher levels of mutation, deletions, and insertions than other genes in any mammalian genome.[11]

Restricted recombination of Y and X chromosomes, which is favored by selection, reduces the proofreading and repair capabilities of the Y chromosome, explaining, in part, the small size and low numbers of genes on the human Y chromosome. Mutations also accumulate in *Sry* at a greater rate than in other genes, resulting in high nucleotide sequence divergence between mammalian *Sry* genes. This fragility of the *Sry* gene may be why *Sox-9* plays such a dominant role in the initiation of testis differentiation (Kashimada and Koopman, 2010).

Given their importance in sex determination in mammals, it seems paradoxical to find that mammalian Y chromosomes are undergoing systematic degeneration to the point that the human Y chromosome and the *Sry* gene may eventually be lost. Loss of genes from the Y chromosome has proceeded independently in different lines of mammals (Graves, 2006), some of which have gone even further down the road to loss of mammalian Y-chromosomes. Males of two species of mole voles (*Ellobius lutescens* and *E. tancrei*) and males of two species of Japanese spinous country rats (*Tokudaia osimensis* and *Tokudaia osimensis osimensis*) have **lost their Y chromosomes**, both males and females being XO (**FIGURE 13.4**). Sex-determining genes in males have translocated to non-sex chromosomes (autosomes). *Sry*, which is not a sex-determining gene in any of these species, has been lost from all four.[12]

Not all mammals possess the XX:XY system of sex determination (Scherer and Schmid, 2001). Monotremes and placental mammals diverged between 170 and 210 Mya. The platypus, *Ornithorhynchus anatinus*, has five X and five Y chromosomes, a system thought to be partly a modification of the mammalian X chromosome system and partly a carryover from a more ancient reptilian (and avian) sex-determining system.[13]

One of the PLATYPUS X CHROMOSOMES *carries the master sex-determining gene Dmrt1 (*d*ouble*s*ex and* m*ab-3* r*elated* t*ranscription factor* 1*), which is located on the Z chromosome in male birds. Thus, the platypus X chromosome is equivalent to the avian Z chromosome.*

---

[11] Perhaps telling us something quite important about sex-specific mutation, the mutation rate in sperm nuclei in the pollen grains of flowering plants is higher than in ovules (T. M. Knight et al., 2005. Pollen limitation of plant reproduction: pattern and process. *Annu. Rev. Ecol. Evol. Syst.,* **36**, 467–497).

[12] W. Just et al., 1995. Absence of *Sry* in species of the vole *Ellobius. Nat. Genet.,* **11**, 117–118; S. Sutou et al., 2001. Sex determination without the Y chromosome in two Japanese rodents, *Tokudaia osimensis osimensis* and *Tokudaia osimensis* spp. *Mamm. Genome,* **12**, 17–21.

[13] F. Grützner et al., 2004. In the platypus a meiotic chain of ten sex chromosomes shares genes with the bird Z and mammal X chromosomes. *Nature,* **432**, 913–917.

Other animals have different systems of sex chromosomes. In the nematode *Caenorhabditis elegans*, females are XX and males XO. Snakes and birds independently evolved a ZZ:ZW system in which females are ZW and males ZZ.[14] Among the many questions these various findings raise are: how sex determination became tied to sex chromosomes (Box 13.1); how the sex with only a single X (or Z) chromosome (XY, XO, ZW) compensates for the other sex having two copies of the X (or Z) chromosome; and whether there is a selective advantage to the sex ratio in most species being 1:1, a topic taken up in points 5 and 9 in End Box 13.1.

AUTOSOMES AND SEX DETERMINATION

The almost universal presence of sex chromosomes among sexual species does not necessarily mean these are the only chromosomes affecting sexual development. Sex is a complex developmental character affected by numerous genes on non–sex chromosomes (**autosomes**) as well. In *Drosophila*, for example, the sex of an individual is determined by the *ratio of X chromosomes to sets of autosomes (A)*. Females have an X/A ratio of 1 (2X/2A), males an X/A ratio of 0.5 (1X/2A). The function of sex chromosomes (or the X/A ratio) is to act as part of a "switch" mechanism that directs development into male or female. Ratios that differ from 1 or 0.5 are associated with sexual abnormalities; for example, triploids with an X/A ratio of 0.66 (2X/3A) are intersexes (End Box 13.1).[15] An example of autosomal control of sex reversal in fish is discussed below under "Fish that Change Sex as Adults."

ENVIRONMENTALLY INITIATED SEX DETERMINATION

In contrast to genetic mechanisms of sex determination are sex-determining mechanisms based on environment signals. **Environmentally initiated sex determination** occurs in many animals and in most of those plants in which the sexes are separate (Hall et al., 2004).

Environmental sex determination demonstrates the importance of gene by environment interactions.[16] The examples introduced below provide a flavor of the extraordinarily wide range of environmentally dependent sex-determining mechanisms. The discussion includes situations in which sex is determined genetically but environmental cues selected by the female either determine whether a male or female will develop (parasitoid wasps) or suppress development of females (honeybees).[17]

LARVAL BEHAVIOR AND SEX DETERMINATION

Some 15 species of marine tubeworm in the genus *Osedax* consist of populations of females feeding on decaying whale bones at depths of 3,000 m or more (FIGURE 13.5).[18]

---

[14] K. Matsubara et al., 2006. Evidence for different origin of sex chromosomes in snakes, birds, and mammals and stepwise differentiation of snake sex chromosomes. *Proc. Natl. Acad. Sci. USA*, **103**, 18190–18195.

[15] See Ellegren (2011) for further details of the evolution of sex determination.

[16] Interactions between genes and the environment are sometimes written as gene X environment interactions.

[17] See also Hall (1999), West-Eberhard (2003), and Gilbert and Epel (2008) for further examples and theoretical discussions of the integration of development, evolution, and environment.

[18] A recent published study shows that whale-eating worms arose with the origination of whales some 30 Mya (S. Kiel et al., 2010. Fossil traces of the bone-eating worm *Osedax* in early Oligocene whale bone. *Proc. Natl. Acad. Sci. USA*, **107**, 8656–8659).

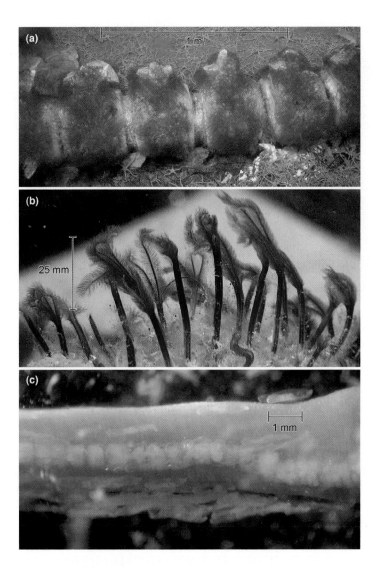

FIGURE 13.5 Tubeworms in the genus *Osedax* consist of large (50 mm long) females that obtain nutrition through digestion of whale bone at depths of 3,000 m or more **(a, b)**. Tiny 1-mm–long dwarf males reside within a gelatinous capsule that surrounds the female **(c)**.

[(a) and (b) Reproduced from Deep Sea Research Part II: Topical Studies in Oceanography, vol. 56, Robert C. Vrijenhoek, Cryptic species, phenotypic plasticity..., pp. 1713–1723. Copyright 2009, with permission from Elsevier. (http://www.sciencedirect .com/science/article/pii/ S0967064509001805). Photo courtesy of Robert C. Vrijenhoek. (c) Courtesy of Greg Rouse.]

The females lack a digestive system but contain endosymbiotic bacteria that produce enzymes for digesting bone. Females are surrounded by a gelatinous capsule, which is colonized by hundreds of dwarf (< 1 mm long) non-feeding males (Figure 13.5). Larvae that settle on bone develop into females. Larvae that settle on females become dwarf males with development arrested at a larval stage (Vrijenhoek et al., 2008; Whiteman, 2008). Unfortunately, our knowledge of the mechanisms operating in these (and other) animals to set sex environmentally is rudimentary.

BONELLIA

In the green spoon worm, *Bonellia viridis*, larvae that are free-swimming and settle on the sea bottom develop into females with a 10- to 20-cm-long body and a meter-long proboscis (FIGURE 13.6). Larvae that land on the proboscis of a female metamorphose into tiny, 1-mm–long males that lack digestive organs, existing as parasites embedded within the genital ducts of the female, producing virtually nothing but sperm.

PENIS FENCING IN FLATWORMS

Even though they are hermaphrodites—both male and females organs are found within each individual—marine flatworms of the species *Pseudobiceros hancockanus* hunt and

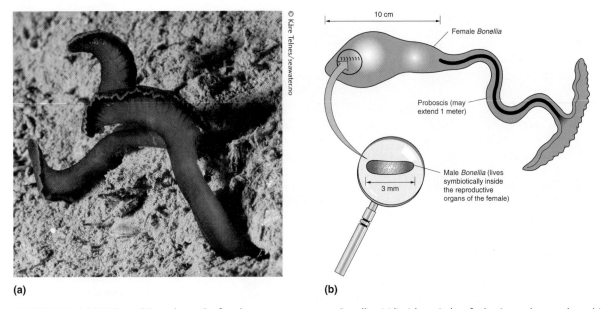

(a)   (b)

FIGURE 13.6 **(a)** At 10- to 20-cm long, the female green spoon worm *Bonellia viridis* (shown) dwarfs the 1-mm-long males, which are shown in **(b)** in the position of the genital ducts of the female.

fight for mates. Having found another individual with which to mate, they begin penis fencing (**FIGURE 13.7**). Indeed, some individuals have two penises, giving them a decided advantage.

Why bother, when you can fertilize yourself? Aside from avoiding the disadvantage of inbreeding, the sex of a penis worm depends on whether it is the first to pierce its opponent (mate) with its penis. The first to do so functions as a male, delivering a package of two-tailed sperm to the other individual, which, perforce, becomes the female, investing her resources in egg production.

TEMPERATURE-DEPENDENT MECHANISMS

The sex of alligators and of many other reptiles is set during embryonic development by the temperature at which the eggs develop. In some species, a higher temperature determines maleness while in other species maleness is set at lower temperatures. An entire clutch therefore may consist of only a single sex (Crews, 2003; Warner and Shine, 2008).

FIGURE 13.7 Penis fencing in flatworms.

[Courtesy of Dr. Nico K. Michiels, Institute for Evolution and Ecology, University Tuebingen.]

FISH THAT CHANGE SEX AS ADULTS

Many chemicals released into the environment by humans lead to the appearance of female characteristics (feminization) in male animals, such as frogs exposed to atrazine. However, sex changes also occur "naturally." The majority of tropical reef fish change sex during their lifetimes, often in a matter of hours, some switching from male to female, others from female to male, and some becoming hermaphrodites. Steroid hormones are involved in these transitions.[19]

Social behavior can influence the sex of some fish. Loss of socially dominant males from a group is followed by conversion of the dominant female into a male. This situation is reversed in some other marine fish, such as clownfish, in which the dominant male becomes the female in a group. In some other coral reef fish, sex changes seem to occur as a result of the size of other individuals in the school; females become males when the surrounding fish in the group are relatively small. Steroid hormones are involved here as well.[20]

In other species such as the Japanese medaka, *Oryzias latipes*, which has an XX/XY sex-determining mechanism, around one percent of wild fish are sex reversed, either XX males or XY females. The gene *Sda-1* (*sex-determining autosomal factor-1*) is located on one of the autosomal chromosomes, suggesting that this autosomal recessive gene regulates sex reversal in these fish.[21]

FERTILIZED VERSUS UNFERTILIZED EGGS AND SEX DETERMINATION

Male honeybees—indeed, all insects in the Order Hymenoptera—develop from unfertilized eggs and so are haploid; females (usually one per colony) develop from fertilized eggs and so are diploid. This mode of sex determination, known as **haplodiploidy**, facilitates the evolution of sociality in which only one diploid female becomes a queen and lays all of the eggs for the colony. All other females—which are haploid, having developed from unfertilized eggs—help raise the queen's eggs and so contribute to her reproductive success and indirectly to their own, a phenomenon known as **kin selection**. Remarkably, the queen is not merely a passive recipient of the contributions of these sterile workers; she constructs this social environment by releasing a hormone that suppresses fertility in the workers.

Parasitoid wasps are a further interesting example. Sex is genetically determined but choice of the environment by a female determines whether she will lay a fertilized egg that will develop into a female or an unfertilized egg that will develop into a male. In *Lariophagus distinguendus* (**FIGURE 13.8**), which parasitizes the larvae of common granary weevils, *Sitophilus granarius*, the female lays differently sexed eggs in response to the size of the wheat grain containing the host larva. Large versus small host larvae seem to be recognized by the wasp as a relative property of the larvae available; as long as there is a size difference, wasps can differentiate between them. If the wheat grain is relatively large, the wasp inserts a single fertilized (female) egg into the weevil larva; if the grain is relatively small, an unfertilized (male) egg is injected. A larval host in a larger grain enables female offspring to produce more offspring; the larger grain supplies the additional resources needed by females. A larval host in a smaller grain supplies enough resources to enable a male offspring to produce a large number of viable sperm.

HYMENOPTERANS, one of the largest orders of insects, comprises over 130,000 species of bees, ants, wasps, and their allies, all of which have two pairs of membranous wings and a specialized egg-laying organ known as an ovipositor.

[19] See M. Nakamura et al., 2005. Sex change in coral reef fish. *Fish Physiol. Biochem.*, **31**, 117–122. Also see Godwin et al. (2003) and Cole (2010) for environmental sex determination in fish.
[20] A. N. Perry and M. S. Grober, 2003. A model for social control of sex change: interactions of behavior, neuropeptides, glucocorticoids, and sex steroids. *Horm. Behav.*, **43**, 31–38.
[21] A. Shinomiya et al., 2010. Inherited XX sex reversal originating from wild medaka populations. *Heredity*, **105**, 443–448.

FIGURE 13.8 A female parasitoid wasp, *Lariophagus distinguendus*, probing a wheat grain to find a larva of the common granary weevils *Sitophilus granarius* in which to deposit an egg.

Courtesy of Simon Haarder.

TRANSFORMATION OF CHROMOSOMAL TO ENVIRONMENTAL SEX DETERMINATION

Interestingly, chromosomal sex determination can transform into environmental sex determination when the sex of one or both of the XY and XX individuals reverses because of sensitivity to agents such as temperature or hormones. Various instances of sex reversal exist and might well lead to an environmental sex-determining system when the environment is "patchy" rather than uniform. For example, a greater food supply in some patches may lead to the production of larger females; other less nutritional patches are associated with the production of smaller males. Such differences can act as important selective factors in establishing sex determination based on environmental sensitivity, rather than on a genetic or chromosomal basis that does not distinguish among environments.

Whether genetically or environmentally determined, if defined as the formation of separate male and female organisms, sex determination does not apply to most plants; some 90 percent of seed plants produce both male and female gametes (i.e., are not male or female themselves). Furthermore, only a minority of the five percent of plant species with separate male and female plants has sex chromosomes. Even in the absence of sex chromosomes, sex determination may be genetic or environmental. For the vast majority of plant species, we just do not know.

Interestingly, one of the most primitive of living plants and the oldest living tree species, GINKGO BILOBA, *has sex chromosomes.*

ENVIRONMENT DICTATES PHENOTYPE

It may be surprising that something so fundamental as sex can be set environmentally. It is even more astounding to find that adults can change their sex and that the mechanism responsible for sex determination can switch from a genetic basis to environmental determination. These transformations tell us that the genomes of organisms are not independent of the environment in which the organisms live. Furthermore, that environment can be dramatically different at different stages of the life cycle. Just think of tadpoles and frogs, in which a single genotype produces two vastly different phenotypes as life-history stages adapted to aquatic and terrestrial environments, respectively.

The biphasic lifestyle of frogs is constitutive, by which we mean that environmental triggers are not required to elicit the transition from tadpole to frog. Metamorphosis is regulated by internal factors, largely changes in production of the hormone thyroxine. Low levels of thyroxine initiate premetamorphosis, in which resorption of the tail begins and the long intestine of the herbivorous tadpole begins to transform into the short intestine of the carnivorous frog. After several weeks of slow change, a rapid rise in thyroxine concentration accelerates tail loss and initiates development of the limbs. These and many other changes are regulated internally and do not require environmental regulation or control.

Diverse changes in other organisms, aside from sex determination, reflect direct response to the environment. In these cases, the environment sets how the genotypic potential of the individual will be expressed. Flower color is controlled genetically but is influenced by the pH of the soil in which the plant grows. The color of mammalian fur is genetically controlled, but the genes that produce the enzymes involved are sensitive to temperature—fur at the tips of the limbs and tail or ears is often darker, a consequence of the lower temperature experienced by those portions of the body. Human height is under genetic control but the actual height reached by each individual is a function of diet. For example, Japanese individuals from the generation after World War II were taller and heavier than Japanese individuals born in the generation before, reflecting major changes in diet, nutrition, and lifestyle.[22] In these examples, environment influences the amount of genetic potential that is expressed.

A final important way in which environment influences phenotype is the ability of an organism with a single genotype to produce more than one life history stage as a direct and facultative response to environmental cues; one genotype → more than one phenotype.[23] Such plasticity is usually referred to as **phenotypic plasticity** and the different phenotypes as **morphs** or sometimes as **ecomorphs** (Losos, 2009). Importantly, even if the environmental signal is not encountered during the life of a single generation, the ability to respond to the signal is retained and transmitted to future generations. Phenotypic plasticity has emerged as a potentially important first step in speciation initiated within a population, which is known as sympatric speciation (West-Eberhard, 2003).

An important lesson therefore is that genes are not independent of the environment. It is not one genotype → one phenotype. Sex can be set environmentally. Alternate phenotypes can arise as an organisms' response to (an) environmental signal(s). All these genotype by environment interactions persist because of the operation of natural selection upon countless individuals, generation after generation.

■ KEY TERMS

asexual reproduction	heterochromatin
autosomes	incomplete dominance
codominance	kin selection
cytoplasmic inheritance	linkage (genetic)
DNA repair mechanisms	morph
dosage compensation (chromosomal)	parasitism
ecomorph	phenotypic plasticity
epistasis	recombination (genetics)
euchromatin	sex chromosomes
extranuclear inheritance	sex determination
Haldane's rule	sex linkage
haplodiploidy	sexual reproduction

---

[22] Both body mass and height of Japanese school children decreased during World War II (Sugiura et al., 2010).

[23] For discussion and examples of such plasticity, see Hall (1999), Hall et al. (2004), and Gilbert and Epel (2008).

■ EVOLUTION ON THE WEB

Explore evolution on the Internet! Visit the accompanying website for *Strickberger's Evolution, Fifth Edition,* at **go.jblearning.com/Evolution5eCW** for exercises and links relating to topics covered in this chapter.

■ REFERENCES

Bull, J. J., 1983. *Evolution of Sex Determining Mechanisms.* Benjamin/Cummings, Menlo Park, CA.

Charlesworth, B., 1996. The evolution of chromosomal sex determination and dosage compensation. *Curr. Biol.,* **6**, 149–162.

Charlesworth, B., 2010. Don't forget the ancestral polymorphisms. *Heredity,* **105**, 509–510.

Cole, K. S., 2010. *Reproduction and Sexuality in Marine Fishes: Patterns and Processes.* University of California Press, Berkeley, CA.

Crews, D., 2003. Sex determination: where environment and genetics meet. *Evol. Dev.,* **5**, 50–55.

Davison, A., 2006. The ovotestis: an undeveloped organ of evolution. *BioEssays,* **28**, 642–650.

Duveau, F., and M.-A. Félix, 2010. Hidden variation mapped. *Heredity,* **105**, 423–425.

Ellegren, H., 2011. Sex-chromosome evolution: recent progress and the influence of male and female heterogamety. *Nat. Rev. Genet.,* **12**, 157–166.

Gilbert, S. F., and D. Epel, 2008. *Ecological Developmental Biology: Integrating Epigenetics, Medicine, and Evolution.* Sinauer Associates, Sunderland, MA.

Godwin, J., J. A. Luckenbach, and R. J. Borski, 2003. Ecology meets endocrinology: environmental sex determination in fishes. *Evol. Dev.,* **5**, 40–49.

Graves, J. A. M., 2006. Sex chromosome specialization and degeneration in mammals. *Cell,* **124**, 901–914.

Hall, B. K., 1999. *Evolutionary Developmental Biology,* 2nd ed. Kluwer Academic Publishers, Dordrecht, The Netherlands.

Hall, B. K., Pearson, R. D., and G. B. Müller (eds.), 2004. *Environment, Development, and Evolution; Toward a Synthesis.* MIT Press, Cambridge, MA.

Hallgrímsson, B., and B. K. Hall (eds.), 2003. *Variation: A Central Concept in Biology.* Elsevier/Academic Press, Burlington, MA.

Kashimada, K., and P. Koopman, 2010. *Sry*: the master switch in mammalian sex determination. *Development,* **137**, 3921–3930.

Kerney, R., E. Kim, R. P. Hangarter, A. A. Heiss, C. Bishop, and B. K. Hall, 2011. Intracellular invasion of green algae. *Proc. Natl. Acad. Sci. USA,* **108**, 6497–6502.

Kokko, H., K. U. Heubel, and D. J. Rankin. 2008. How populations persist when asexuality requires sex: the spatial dynamics of coping with sperm parasites. *Proc. Biol. Sci.* **275**, 817–825.

Losos, J. B., 2009. *Lizards in an Evolutionary Tree. Ecology and Adaptive Radiation of Anoles.* University of California Press, Berkeley, CA.

Lucchesi, J. C., 1998. Dosage compensation in flies and worms: the ups and downs of X-chromosome regulation. *Curr. Opin. Genet. Dev.,* **8**, 179–184.

Maynard Smith, J., 1978. *The Evolution of Sex.* Cambridge University Press, Cambridge, UK.

Michod, R. E., and B. R. Levin (eds.), 1988. *The Evolution of Sex.* Sinauer Associates, Sunderland, MA.

Nijhout, H. F., 2003. The importance of context in genetics. *Am. Sci.,* **91**, 416–423.

Peck, J. R., 1994. A ruby in the rubbish: beneficial mutations, deleterious mutations and the evolution of sex. *Genetics,* **137**, 597–606.

Scherer, G., and M. Schmid (eds.), 2001. *Genes and Mechanisms in Vertebrate Sex Determination.* Birkhäuser Verlag, Basel.

Sugiura, Y., S. Ju, J. Yasuoka, and M. Jimba, 2020. Rapid increase in Japanese life expectancy after World War II. *Biosci. Trends,* **4**, 9–16.

Vrijenhoek, R. C., S. B. Johnson, and G. W. Rouse, 2008. Bone-eating *Osedax* females and the 'harems' of dwarf males are recruited from a common larval pool. *Mol. Ecol.*, **17**, 4535–4544.

Warner, D. A., and R. Shine, 2008. The adaptive significance of temperature-dependent sex determination in a reptile. *Nature*, **451**, 566–568.

West-Eberhard, M. J., 2003. *Developmental Plasticity and Evolution.* Oxford University Press, Oxford, UK.

Whiteman, N. K., 2008. Between a whalebone and the deep blue sea: the provenance of dwarf males in whale-eating tubeworms. *Mol. Ecol.*, **17**, 4395–4397.

Wiley, E. O., 1978. The evolutionary species concept reconsidered. *Syst. Zool.*, **27**, 17–26.

END BOX 13.1

Evolution of Sex-Determining Mechanisms

© Photos.com

SYNOPSIS: Here the discussion of genetic and environmental mechanisms of sex determination in animals is expanded, providing an introduction to the evolution of sex-determining mechanisms and of two sexes. Nine chromosomal and genetic changes associated with the evolution of chromosomal sex determination are outlined. We discuss how a particular sex ratio is maintained in a population, and introduce *Haldane's rule* for situations when the sex ratio is unbalanced.

For **chromosomal sex determination** to occur, organisms that are hermaphroditic—male and female reproductive systems in each individual—must first evolve two sexes.[a] The likely selective pressure is increased genetic diversity from cross-fertilization (essentially, outbreeding) compared to low genetic diversity associated with self-fertilization (inbreeding).

As shown in the top sequence in **FIGURE EB1.1**, the transition to independent females may be as simple as a mutation in a hermaphrodite that causes male sterility. The presence of females, in turn, offers advantages to individuals that specialize in producing male gametes. (Mutations in hermaphrodites may also generate independent males by causing female sterility; Figure EB1.1, bottom sequence.) The result, in this example, is a population with individuals bearing two kinds of chromosomes devoted to sex determination, the X distinguished by recessive male sterility, and the Y by dominant female sterility.

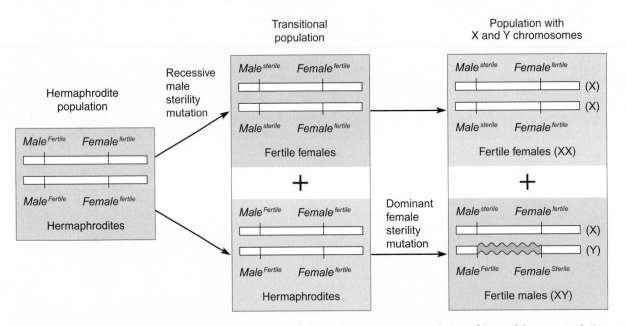

FIGURE EB1.1 Sequence of mutations leading to a transition from a hermaphrodite population of *Drosophila* to a population with separate sexes. Each horizontal bar represents a sex chromosome. In this example, the Y chromosome determines male sex development because of dominant *Male^fertile* and *Female^Sterile* alleles, whereas the X chromosome carries recessives at these loci (*Male^sterile*, *Female^fertile*). Absence of the Y chromosome produces females. The wavy section between male and female genes on the Y chromosome indicates a recombination deficient that interferes with crossovers that might otherwise lead to sterile individuals carrying *Male^sterile* and *Female^Sterile* alleles, or lead back to the hermaphroditic condition, *Male^fertile Female^fertile*.

[Based on B. Charlesworth, 1996. The evolution of chromosomal sex determination and dosage compensation, *Curr. Biol.*, **6**, 149–162 with modifications.]

---

[a] For the evolution of sex determination, see Maynard Smith (1978), Bull (1983), Michod and Levin (1988), Charlesworth (1996), and Ellegren (2011).

A **cascade of consequences** (nine of which are outlined below) follows from the findings that individuals that are both male- and female-sterile cannot reproduce; and that such individuals can arise from crossing over between sex-determining genes (Figure EB1.1):

1. **Selection to prevent crossing over** promotes genes or chromosome arrangements that interfere with recombination between the X and Y and help preserve the tight linkage between genes on the Y chromosome and formation of the heterogametic (for example, male) sex. Thus, sexually antagonistic genes such as those used exclusively for male mating success can be selected to localize on the Y chromosome, and are passed only to Y-bearing offspring. To counter effects that can harm females, it has been proposed that selection of X-linked and autosomal genes would enhance female fitness, resulting in an evolutionary seesaw of male-female "antagonistic coevolution."[b]

2. Once X–Y recombination diminishes, its absence allows **harmful mutations on the Y chromosome to accumulate**, as demonstrated experimentally in association with lack of recombination.[c]

3. To eliminate such deleterious mutant effects, **inactivation of the Y chromosome**—with the exception of its sex-determining genes—**becomes selectively advantageous**. Chromosome inactivation is evidenced when entire chromosomes or sections of chromosomes are made up of **heterochromatin**, which stains differently than normally active, unmodified chromosome sections (**euchromatin**) when viewed under the microscope. Y chromosomes are often heterochromatic and X chromosomes euchromatic.

4. As larger sections of the Y chromosome become inactive because of reduced recombination with the X, the **Y chromosome can decrease in size**. Selection no longer preserves the Y chromosome, a feature common to many XX–XY species.

5. At some stage in this progression, the **Y chromosome has few if any functional genes** other than the sex-determining genes. If these are translocated elsewhere or if their functions are assumed by genes in another chromosome, the result can be an XX female/XO male species, in which sex determination depends on the *Drosophila*-like X/A ratio (discussed in this chapter).

   A key regulatory gene involved in sex determination in *Drosophila* is *Sex lethal (Sxl)*, which results in the development of a female when turned on and of a male when turned off. Turning *Sxl* on and off derives from the relationship among genes on the X chromosome and on autosomes. An X chromosome/autosome ratio of 1 turns *Sxl* on; a ratio of 1/2 turns *Sxl* off. Conversely, translocation between the Y chromosome and an autosome produces a "neo-Y" chromosome and a translocated autosome (X_2). The neo-Y chromosome pairs in meiosis with X_2 and the former X (X_1), leading to an X_1X_2Y male/$X_1X_1X_2X_2$ female sex chromosome constitution (**FIGURE EB1.2;** see Lucchesi, 1994).

6. In some cases, **only a single copy of a sex-determining gene is required** to embark on a pathway of sexually distinctive development. In mammals, the *Sry* gene on the Y chromosome serves such a purpose for male development. Fish, amphibians, and reptiles do not share *Sry*'s mammalian sex-chromosome locus; its role as the sex-determining gene arose with mammals.

7. The difference in X (or Z) **chromosome gene dosage** between the heterogametic and homogametic sexes **can be compensated** for in various ways. Two major systems of **dosage compensation** are known to function when X chromosomes are large, with many active genes. In *Drosophila*, the single X chromosome in the male is about twice as active as each of the two X chromosomes in the female, while in placental mammals, only one X chromosome of an XX female is functional in each cell. The result of both systems is that an XX female has essentially the same level of X chromosome gene activity as an XY male.[d]

As discussed in this chapter, Y CHROMOSOMES *are undergoing degeneration in lineages of mammals.*

[b] W. R. Rice, 1998. Male fitness increases when females are eliminated from gene pool: implications for the Y chromosome. *Proc. Natl. Acad. Sci. USA*, **95**, 6217–6221.
[c] W. R. Rice, 1994. Degeneration of a nonrecombining chromosome. *Science*, **263**, 230–232.
[d] See Lucchesi (1994) and Parkhurst and Meneely (1994).

END BOX 13.1

Evolution of Sex-Determining Mechanisms (Cont...)

© Photos.com

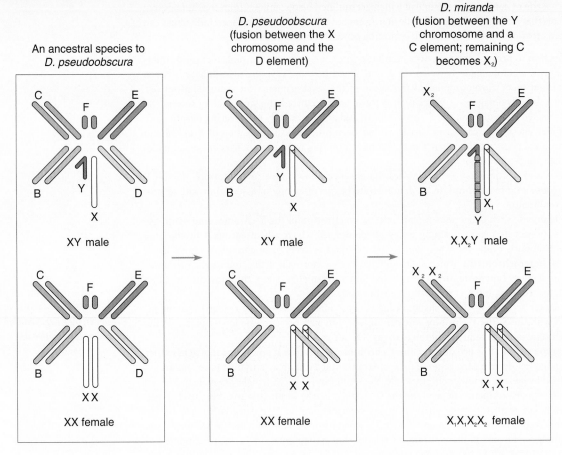

FIGURE EB1.2 Diagrammatic view of sex chromosome changes in a lineage leading to *Drosophila miranda*. The original chromosomes are designated by the six letters A to F, with the A element serving as the sex chromosome. The species on the left represents a hypothetical ancestor in which the common XY chromosome situation has been established. This is followed by fusion of the X with the D chromosome in *D. pseudoobscura* and fusion of the Y with the C chromosome in *D. miranda*. The sex chromosome constitution of *D. miranda* is therefore X_1X_2Y males and $X_1X_1X_2X_2$ females, each sex carrying three pairs of autosomes (elements B, E, and F).

[Adapted from J. C. Lucchesi, 1994. The evolution of heteromorphic sex chromosomes. *BioEssays*, **16**, 81–83.]

8. **Haldane's rule**, which is often used in describing defective hybridization between animal species, states: "When in the F_1 offspring of two different animal races one sex is absent, rare, or sterile, that sex is the heterozygous [heterogametic] sex."[e] Geneticists most often ascribe this rule to sex-linked deleterious recessive alleles; interspecific hybrids carrying autosomal alleles from one species that are incompatible with sex-linked recessive alleles from the other species show deleterious effects more readily in the heterogametic sex. X-linked deleterious recessive alleles in XY males, for example, are not masked by dominant alleles on the other X chromosome as they are in homogametic XX females.

[e] J. B. S. Haldane, 1922. Sex-ratio and unisexual sterility in hybrid animals. *J. Genet.*, **12**, 101–109. For an evaluation of Haldane's rule in terms of modern genetics, see M. Schilthuizen et al., 2011. Haldane's rule in the 21st century. *Heredity*, **107**, 95–103.

Subsequent research has added two concepts to this explanation: (1) that the sexual traits of males are produced by specific sets of genes that evolve rapidly, making some or many such males more sensitive to hybrid sterility; and (2) that sex-biased hybrid inferiority can act as an isolating mechanism capable of impeding gene flow.[f]

9. Although the heterogametic sex can theoretically produce equal numbers of male- and female-determining gametes, **equal sex ratios need not necessarily follow**. Some nuclear genes may affect the viability of male or female zygotes. Other genes may alter the ability of the X or Y chromosomes to segregate normally during meiosis ("segregation distorters"). Either type of abnormality, whether in viability or gametic segregation, can affect male or female frequencies. Y-chromosome inactivation has the advantage of inactivating segregation distortion genes resulting in more Y- than X-carrying gametes, and therefore more males than females.

One way to account for the prevalence of an equal male:female sex ratio in many species is to consider what happens when a particular sex is rare (Basolo, 1994). Should a population have a scarcity of females (for example, a male:female sex ratio of 5:1), females, being more frequently mated, would on average produce more offspring than males, many of which cannot find mates, providing a reproductive advantage to genotypes that produce more females than males, thus tending to correct the distorted sex ratio.

On the other hand, assuming that males are now relatively rare, the advantage would shift to genotypes that produce more males than females, again tending to correct the distorted sex ratio. At some point, the population will approach the stable sex ratio of 1:1, a value that no longer provides a benefit for a genotype to produce more of one sex than the other. Experimentally, it has been shown that populations of the platyfish *Xiphophorus maculatus* and the fruit fly *Drosophila* that are composed of different sex ratio genotypes, evolve in this direction.

Other factors influence sex ratio. An example of **environmental impact on sex ratio** has been described in Seychelles warblers, *Acrocephalus sechellensi*, birds that commonly use their daughters as "helpers" in raising additional offspring. When food is plentiful, helper daughters increase their parents' reproductive success, producing broods with a female:male ratio of about 6:1. When food is scarce, such daughters hinder their parents' reproductive success by competing for the limited supply, and the female:male offspring ratio drops to about 1:3.[g]

REFERENCES

Basolo, A. L., 1994. The dynamics of Fisherian sex-ratio evolution: theoretical and experimental investigations. *Am. Nat.*, **144**, 473–490.

Bull, J. J., 1983. *Evolution of Sex Determining Mechanisms*. Benjamin/Cummings, Menlo Park, CA.

Charlesworth, B., 1996. The evolution of chromosomal sex determination and dosage compensation. *Curr. Biol.*, **6**, 149–162.

Cole, K. S., 2010. *Reproduction and Sexuality in Marine Fishes: Patterns and Processes*. University of California Press, Berkeley, CA.

Ellegren, H., 2011. Sex-chromosome evolution: recent progress and the influence of male and female heterogamety. *Nat. Rev. Genet*, **12**, 157–166.

Lucchesi, J. C., 1994. The evolution of heteromorphic sex chromosomes. *BioEssays*, **16**, 81–83.

Maynard Smith, J., 1978. *The Evolution of Sex*. Cambridge University Press, Cambridge, UK.

Michod, R. E., and B. R. Levin (eds.), 1988. *The Evolution of Sex*. Sinauer Associates, Sunderland, MA.

Parkhurst, S., and P. M. Meneely, 1994. Sex determination and dosage compensation: lessons from flies and worms. *Science*, **264**, 924–932.

---

[f] For hybrid sterility, see M. Turelli, 1998. The causes of Haldane's rule. *Science*, **282**, 889–891. For impeded gene flow, see R.-X. Wang, 2003. Differential strength of sex-biased hybrid inferiority in impeding gene flow may be a cause of Haldane's rule. *J. Evol. Biol.*, **16**, 353–361.

[g] J. Komdeur et al., 1997. Extreme adaptive modification in sex ratio of the Seychelles warbler's eggs. *Nature*, **385**, 522–525.

PART 5

Natural Selection in Action

© Photos.com

CHAPTER

14

Types of Natural Selection

CHAPTER SUMMARY

Over the 150 years after Charles Darwin and Alfred Wallace proposed their theories of evolution by natural selection, researches have studied extensively the responses of organisms to selection reflecting changes in their environments, both in nature and under experimental conditions. Changes in the genotype can track selection on the phenotype, leading to heritable changes in genotype and phenotype.

Natural selection acts on all organisms whether unicellular or multicellular. Natural selection modifies structural, physiological, and/or behavioral features of the phenotype, can act at any stage during the life history of multicellular organisms, and leads to differential survival and reproductive output. Because it results in differential reproduction, selection has the greatest influence when it operates before organisms reproduce. At least five types of selection have been discovered. Three types of natural selection are recognized: stabilizing, directional, and disruptive (diversifying) selection. A further but different type of selection—sexual selection—has been known since it was recognized by Darwin. More recently and more controversially, selection on groups or populations—group and kin selection—has been proposed.

INTRODUCTION

Selection acts on the fitness differences among different phenotypic traits of individuals within populations. Although selection changes the phenotype, the transmission of phenotypic changes between generations is genetic and epigenetic.

After systematic and persistent selection on the phenotype from generation to generation, allele frequencies can change; that is, **genotypes** can change. The genotype need not be only that of the individual under selection. Genes of the parent (maternal gene effects) and mitochondrial DNA transmit change between generations, and the ability to learn is transmitted both genetically and culturally (environmentally).

Genetic variation in populations facilitates the genotypic response that can follow rounds of selection.[1] Indeed, selection also acts on variation itself. Selection can preserve genetic variation if the heterozygous genotype is more fit than either homozygote, as in sickle cell anemia, or when the frequency of an allele affects fitness. Allele frequency may vary in populations of the same species in different environments, as in populations of the British peppered moth, *Biston betularia*, in England or of the rock pocket mouse, *Chaetodipus intermedius*, in the American southwest, two examples that are discussed elsewhere in this text.[2] Advantageous genotypes are said to occupy *adaptive peaks* of varying value according to their degree of fitness, using the analogy of a landscape with mountains of different height. Finally, selection does not only act on adult phenotypes. In multicellular organisms, selection can elicit changes in gametes, embryos, and/or larval stages, depending on the life history stage at which selection acts. In this way, selection can influence survival, fertility, or mating success (**FIGURE 14.1**).

In most animals and plants SELECTION *takes place primarily in the diploid zygote and on post-zygotic embryonic, larval, and adult stages.*

SURVIVAL OF THE FITTEST

Evolution has been described as the **survival of the fittest** because of the operation of natural selection. The term *survival of the fittest*, coined by Herbert Spencer, was adopted by Alfred Wallace from the outset but by Charles Darwin only in later editions of *The Origin of Species*. Darwin thought *struggle for reproduction* was a more apt term than survival of the fittest for evolution by natural selection.

Darwin's concern over the term survival of the fittest has been reiterated by many others. The argument is that, because it defines those that survive as the fittest and the fittest as those that survive, "survival of the fittest" is a circular, tautological, or unprovable statement, and so should be abandoned (Gould, 1976). The term is not tautological but it is incomplete. Why? Because it does not embrace the requirement for heritable change for evolution to occur, and equates fitness with survival rather than with differential reproduction; selection changes the range of phenotypes that survive to reproduce in a single generation. Also, the term fails to capture that selection on a phenotypic character can be followed by change in gene frequency but is *not the same as* a change in gene frequency.

The relative fitness of a genotype is certainly not the only factor contributing to its survival. For example, if there is a heterozygote advantage, the frequency of a harmful recessive allele may remain stable despite selection against homozygotes for that gene. Evolutionary theory certainly includes more than survival of the fittest. R.A. Fisher

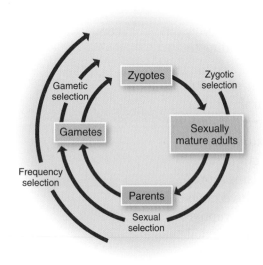

FIGURE 14.1 Simplified diagram of the types of selection acting on the life stages of a multicellular organism during a single generation, shown as from zygote to zygote.

[Adapted from The definition and measurement of fitness. In *Evolutionary Ecology*, B. Shorrocks (ed.), Blackwell, 1984.]

[1] For analyses of the importance of variation, see Fisher (1930), Williams (1996), and Hallgrímsson and Hall (2005).

[2] B.S. Grant, 2004. Allelic melanism in American and British peppered moths. *J. Hered.*, **95**, 97–102; C.R. Linnen et al., 2009. On the origin and spread of an adaptive allele in deer mice. *Science*, **325**, 1095–1098.

began his classic 1930 treatise, *The Genetical Theory of Natural Selection*, with the statement "Natural Selection is not Evolution." Natural selection is the umbrella term for a range of evolutionary processes, the outcome of which is differential survival and reproduction. So, what are the evolutionary processes and, therefore, what modes or types of natural selection are recognized?

TYPES OF SELECTION

It is commonly assumed that selection is always for the optimum, preserving the strong and eliminating the weak. As discussed above, selection is measured by survival of the fittest and reproductive success. Three types of natural selection that affect the phenotypes of individuals are recognized:

- **Stabilizing selection** maintains phenotypic traits close to a mean value.
- **Directional selection** is seen when an extreme phenotype has adaptive value or in the rapid elimination of deleterious mutations.
- **Disruptive (diversifying) selection** occurs when particular phenotypes are better adjusted to some aspects of the environment than are other phenotypes that may be better adapted to other aspects of the environment.

As environmental conditions are often variable, even on short time scales, these different types of selection do not exist alone but combine in different ways. As a consequence of environmental variability or changing biological conditions such as the arrival of a new predator, one type of selection may cease and be followed by or overlap another.[3]

Sexual selection, which differs from the three modes of natural selection, occurs when selection operates differently on males and females of the same species to influence the likelihood of mating.

A further, previously controversial but now accepted form of selection, **group selection**, was proposed for selection on traits that benefit a group (usually a population) at the expense of individuals within the group. Situations in which an individual "sacrifices" itself for the population (**altruism**) are among the clearest example of the operation of a particular type of group selection known as **kin selection**. These six modes of selection are outlined and discussed in the remainder of the chapter.

STABILIZING SELECTION

When selection has operated on particular characters over many generations, most populations achieve phenotypes that are well, if not optimally adapted to their surroundings; many features of the phenotype cluster around some value at which fitness is highest. Individuals who depart from these optimum phenotypes are expected to show lower fitness than those closer to the optimal values.

A classic study documenting stabilizing selection in the wild came about as the result of a major storm, followed by a very cold first of February morning in 1898 in Providence, Rhode Island. Hermon Bumpus (1862–1943), a comparative anatomist at Brown University, found 136 dead or stunned house sparrows (*Passer domesticus*) on the ground. Seventy-two sparrows survived but the remaining 64 died. Bumpus had the foresight to weigh the birds, measure their wingspans, and prepare their skeletons to measure the lengths of some of the major bones (Bumpus, 1899).

Measurements of the sparrows that *survived* the storm *clustered around the mean*, shown by the dashed line in FIGURE 14.2a. Measurements of the same features on the sparrows that *failed to survive* the storm *clustered around the extremes* of the variation

---

[3] For analyses of types of selection see Endler (1986), Grant and Grant (2008), and McKinnon and Pierotti (2010).

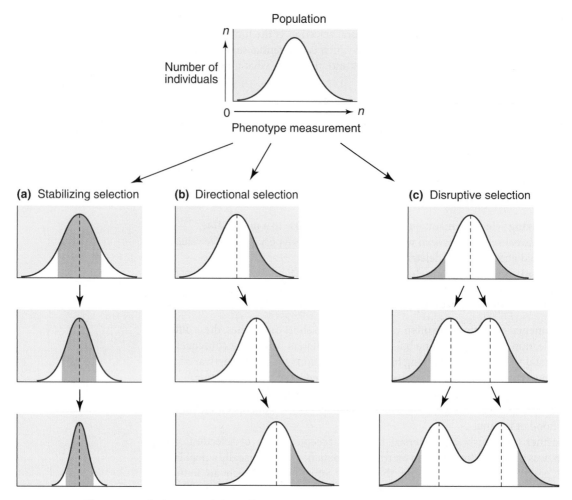

FIGURE 14.2 Three types of selection and their effects on the mean (dashed vertical lines) and variation (brown) of a normally distributed quantitative character. The horizontal axis of each bell-shaped curve represents measurements of the character. The vertical axis represents the number of individuals found at each measurement. Shaded areas represent individuals selected as parents in experiments to demonstrate each type of selection.

A recent example of STABILIZING SELECTION *is a study of hatchery-raised Atlantic salmon (Salmon salar) in which the development rate of fry over the first 30 days of life is subjected to strong stabilizing selection (Bailey et al., 2010).*

for each feature, shown by the shaded portions of the curve in **FIGURE 14.2c**. As Bumpus concluded from these data, "it is quite as dangerous to be conspicuously above a certain standard of organic excellence as it is to be conspicuously below the standard." Bumpus kept all the measurements, which have been reanalyzed in several independent studies, separating males from females and using more powerful statistical analyses than were available in the 1890s.[4] We now realize that stabilizing selection operated primarily upon the females, which were, on average, smaller than the males and had greater variation in body size; stabilizing selection favored the intermediate body size of the females, maintaining body size closer to the norm (Figure 14.2a).

DIRECTIONAL SELECTION

Not all character selection is stabilizing. A second form of selection, directional selection, operated on the male house sparrows discussed above; males are larger than the females and show less variation in body size (Figure 14.2b). Sensitivity of stabilizing

[4] See, for example, B. H. Pugesek and A Tomer, 1996. The Bumpus house sparrow data: a reanalysis using structural equation models. *Evol. Ecol.*, **10**, 387–401, and earlier studies cited therein.

FIGURE 14.3 Male (*upper*) and female (*lower*) house sparrows (*Passer domesticus*), which appear to be the same size to our eyes, have different average body weights with males showing less variation (± 3%) around mean body weight than females (± 6%), differences that reflect stabilizing selection for male body weight and directional selection for female body weight.

and directional modes of selection is reinforced when the size differences between males and females are considered. Males weigh 29.41 ± 0.86 g, females 28.49 ± 1.72 g, differences that are indistinguishable to our eyes (**FIGURE 14.3**) but discriminated by selection.

As organisms move in one or the other direction of a phenotypic distribution, selection can result in an extreme phenotype (**FIGURE 14.2b**). Animal and plant breeders, who select for extremes of yield, productivity, or resistance to disease, commonly practice such directional selection (**FIGURE 14.4**).[5] Directional selection is especially evident when

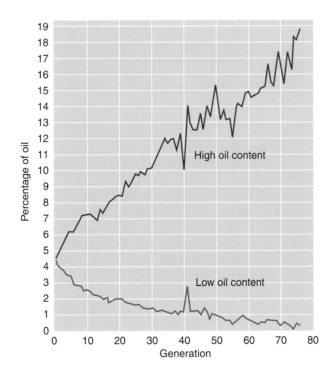

FIGURE 14.4 Results of selection for high and low oil content in corn kernels in an experiment begun in 1896 at the University of Illinois and continuing for 80 years. Selection for high oil content still continues to yield increases, whereas the effect of selection for low oil content tapered off on reaching the 0 percent lower limit.

[Adapted from J. W. Dudley, 1977. Seventy-six generations of selection for oil and protein percentages in maize. In *Proceedings of the International Conference on Quantitative Genetics*, E. Pollak, O. Kempthorne, and T. B. Bailey, Jr. (eds.), Iowa State University Press, Ames, IA, pp. 459–473.]

[5] J. W. Dudley, 1977. Seventy-six generations of selection for oil and protein percentages in maize. In *Proceedings of the International Conference on Quantitative Genetics*, E. Pollak, O. Kempthorne, and T. B. Bailey, Jr. (eds.), Iowa State University Press, Ames, IA, pp. 459–473.

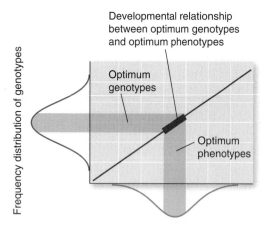

(a) Before canalizing selection

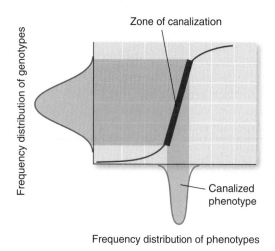

(c) After canalizing selection is completed

(b) Selection for canalization

FIGURE 14.5 Sequence of selection for a canalized phenotype. The diagonal, running from lower left to upper right, represents the developmental relationship between genotype (*vertical axis*) and phenotype (*horizontal axis*); the steeper this developmental curve, the greater the number of different genotypes that can produce the same phenotype. **(a)** Before canalizing selection only a small section of the genotype produces the optimum phenotype. **(b)** and **(c)** As selection for canalization proceeds, the developmental curve assumes more of an "S"-shape as larger portions of genotypic variation produce the optimum phenotype, shown as a zone of canalization in **(c)**.

[Adapted from Strickberger, M.W., *1985. Genetics, Third Edition,* New York: Macmillan.]

the environment of a population is changing and when one or only few phenotypes are adapted to the new conditions.

Directional *and* stabilizing selection have been demonstrated after the separation of one lineage into two during evolution. A classic example is the range of bird species from the Galápagos Islands known as Darwin's finches. Directional and stabilizing selection can operate sequentially, as demonstrated in situations where different genotypes produce a single phenotype. **FIGURE 14.5** diagrams such a scheme devised by English-born (but Australian-based) geneticist James M. Rendel (1915–2001) and described in his informative book, *Canalization and Gene Control* (Rendel, 1967). In a *Drosophila* population selected for a change in bristle number following a mutation in the *scute* gene, directional selection led to alteration in bristle number. Stabilizing selection (Figure 14.2) then maintained the new bristle number. Selection affected both variation of the trait and the pattern of variability of the trait.

Many transcription factors found in humans, chimps, orangutans, and rhesus macaques show similar levels of expression over some 70 My, a pattern interpreted by Gilad et al. (2006) as resulting from stabilizing selection. Other transcription factors show similar patterns in non-human primates but altered expression in the human lineage, taken as the result of directional selection in the human lineage.

DISRUPTIVE SELECTION

Selection, whether stabilizing or directional, may act in a constant fashion if the selective environment is uniform. However, when conditions are changing, a population may be subjected to divergent or cyclically changing (oscillating) environments to which different phenotypes among its members are most suited. Such selection—known as disruptive or diversifying—favors the extremes of the distribution of a phenotype and so establishes multiple optima for the phenotype within a population, as illustrated in Figure 14.2c.[6] Reflecting changing environmental or biological conditions, disruptive selection may be followed by directional selection, which may yield to stabilizing selection, as has been documented for beak size in relation to seed size in Darwin's finches on the Galápagos Islands. Consequently, disruptive selection has been shown to lead to sexual dimorphism and/or to speciation.

SEXUAL SELECTION

Mate choice is one of the most important influences on animal species behavior. When selection operates preferentially and differentially on males or females, we follow Darwin in referring to such selection as sexual selection (Møller, 1994; Eberhard, 1996).

Although both sexes benefit if their offspring survive, male and female animals often differ in the cost of reproduction; females invest more resources in producing large eggs or ovules than do males in producing small sperm or pollen. Accordingly, a female's genes benefit if she discriminates in her choice of mates (female choice) or male gametes (sperm competition) to maximize the chances of success of her relatively expensive production of eggs. For this same reason, there is an advantage for females to seek increased male parental investment in their offspring, a strategy especially notable among bony fish but also seen in mammals and birds, in which reproductive success can depend on a relatively long-term commitment of one parent to their progeny. For example, male Emperor Penguins (*Aptenodytes forsteri*) incubate a single egg for an average of 115 days (**FIGURE 14.6**).

© SuperStock/Age Fotostock

FIGURE 14.6 Male Emperor Penguins incubate the single egg for an average of 115 days.

---

[6] For studies on disruptive selection see J. M. Thoday and J. B. Gibson, 1962. Isolation by disruptive selection. *Nature*, **193**, 1164–1166; T. B. Smith, 1993. Disruptive selection and the genetic basis of bill size polymorphism in the African finch *Pyrenestes. Nature*, **363**, 618–620; I. S. Magalhaes et al., 2009. Divergent selection and phenotypic plasticity during incipient speciation in Lake Victoria cichlid fish. *J. Evol. Biol.*, **22**, 260–274. For disruptive selection associated with color polymorphism see McKinnon and Pierotti (2010).

FIGURE 14.7 Two examples of sexual dimorphism. **(a)** Mature males (right) of the extinct Irish elk, *Megaloceros giganteus* compared with the female (left). Males were 2 m tall at the shoulders, with antlers more than 3.3 m wide and weight about 45 kg. **(b)** Male (right) South American hummingbirds, *Spathura underwoodi,* have long symmetrical tail feathers not found in the female (left).

[(a) and (b) Adapted From Darwin, C. 1871. *The Descent of Man and Selection in Relation to Sex.* John Murray, London.]

(a)   (b)

Males may be more extravagant in disposing of their relatively inexpensive and more plentiful sperm, although competition between sperm for access to eggs is an effective means of increasing their reproductive success (Parker, 1970; Snook, 2005). Genes carried by a male are spread more widely through the population if he fertilizes as many females as possible. This conflict of interest between the sexes leads to a variety of mating patterns depending on a variety of factors, including the degree of parental care necessary for egg or infant survival, and which sex (or whether both sexes) provides such care (Trivers, 1972). In groups where females are primarily responsible for parental care—including many vertebrate species—males are likely to compete with each other for success in mating. As a result, selection can occur for traits that increase the combative abilities of males and/or that increase their attraction to females. As Darwin put it,

> Sexual selection depends on the success of certain individuals over others of the same sex, in relation to the propagation of the species; whilst natural selection depends on the success of both sexes, at all ages, in relation to the general conditions of life. The sexual struggle is of two kinds; in the one it is between the individuals of the same sex, generally the males, in order to drive away or kill their rivals, the females remaining passive; whilst in the other, the struggle is likewise between the individuals of the same sex, in order to excite or charm those of the opposite sex, generally the females, which no longer remain passive, but select the more agreeable partners.[7]

Much evidence exists for competition between males in those species where one male mates with many females (polygony). In order to gain access to females, males may have special competitive armaments such as horns and antlers that can lead to considerable sexual dimorphism (FIGURE 14.7a). Longer horns in male dung beetles are a well-studied example (FIGURE 14.8).[8] Males are generally larger than females in such groups; male elephant seals (*Mirounga leonina*) are eight times larger than females, and male northern fur seals (*Callorhinus ursinus*) are, on average, seven times heavier than females (FIGURE 14.9), both species with extreme polygony.

---

[7] Darwin (1871), p. 558.
[8] See Emlen (2000), Emlen et al. (2005), and Snell-Rood et al. (2010) for approaches to the study of beetle horns.

FIGURE 14.8 Examples of sexual dimorphism in the presence or absence of horns in dung beetle as seen in males of three species **(a)** through **(c)** and **(d)** in a hornless female (left) and a horned male (right).

Because of female choice, male ornaments can become quite conspicuous, as in the dramatic plumage of peacocks, birds of paradise, and hummingbirds (**FIGURE 14.7b**). Such exaggerated traits, like the peacock's tail, are adaptive and selected for because they signal to others that the individual with the phenotype is in sufficiently good condition to expend the extra energy to produce and maintain the trait (**FIGURE 14.10**), a concept known as the **Handicap Principle**.[9]

FIGURE 14.9 Northern fur seals (*Callorhinus ursinus*) showing sexual dimorphism between the large, darkly colored male (up to 340 kg) and the 30 to 50 kg females in his harem.

[Seals photographed on St. Paul Island in the Bering Sea by Dr. Sara Iverson, who kindly supplied the image, taken under MMPA Permit No. 782-1708-05]

FIGURE 14.10 Indian Peafowl (*Pavo cristatus*) showing sexual dimorphism between the female with young **(a)** and the male **(b)**.

(a)

(b)

MONOGAMOUS *refers to mating for life with a single partner.*

Selection for such features can track environmental change rapidly. In barn swallows (*Hirundo rustica*), which are monogamous, the outermost tail feathers are 20 percent longer in males than in females. Over 20 years (1984–2003), in response to changing climatic conditions at sites on the spring migration route, Møller (1994) reported that the lengths of these feathers in males increased by 11 mm but in females only by 3 mm, indicating that characters present in both sexes can respond differentially to the same selective pressures. As concluded by Ronald Fisher in 1930, such situations exist because females continually mate with individuals whose attractiveness in such populations is passed on to their sons, and because the daughters of these females inherit their mother's preference for such phenotypically exaggerated males.

At a behavioral extreme of sexual selection are male Australian red back spiders, which place themselves within reach of the females' jaws to be cannibalized by the female while they are mating or have just mated. This suicidal trait has adaptive features, allowing such males to copulate longer and fertilize more eggs than their non-cannibalized male competitors.[10]

GROUP SELECTION

Selection for the differential reproduction of groups, now known as group selection, was proposed by Darwin (1871) for situations in which social adaptations in a group of organisms appeared to place individual organisms in the population at a disadvantage.

---

[9] The handicap principle is a comprehensive theory of animal display and signaling based on the proposition that the effectiveness of a signal depends on the cost, or handicap, borne by the individual generating the signal (Zahavi, 1975; see also Zahavi and Zahavi, 1997).

[10] M. C. B. Andrade, 1996. Sexual selection for male sacrifice in the Australian red back spider. *Science,* **271,** 70–72.

For many decades, evolutionary biologists accepted that group selection was theo-retically possible but debated the extent to which it occurs. Group selection normally counters the effects of selection on individuals; selection *within* groups is usually found to be stronger than selection *among* groups, and group adaptations can and often do arise from selection on individuals.[11] Selection on individuals and group selection can operate in the same direction, as demonstrated in animals that forage cooperatively. This was demonstrated in a classic experimental test of evolution and of group selec-tion in laboratory populations of the flour beetle *Tribolium castaneum* (Wade, 1976, 1980). Within three or four generations of group selection in which all populations were exposed to the same selection regime, populations differed from one another in size by as much as 200 percent. Group selections occurred despite the fact that individual selec-tion on control populations was reducing population size by as much as 75 percent over the same time.

Mathematical models show that selection within groups necessitates high extinc-tion rates, practically no gene flow, and large population sizes (i.e., is frequency depen-dent). Group selection is receiving increased attention, in part because it has been broadened to consider selection at other levels (multilevel selection), including selection between species (**species selection**) or accidental factors such as catastrophes resulting in survival differences among species (**species sorting**). Other mass interactions, such as susceptibility or resistance to parasitic infection, also can involve group survival or extinction.

The two major examples of group selection are:

- **Altruism** and **reciprocal altruism** occur in situations in which individuals coop-erate only with those who also cooperate with them; the benefits of the altruism lie in the future.
- **Kin selection** has been proposed to explain social behavior in some lineages of ants, bees, and wasps (hymenopteran insects).

ALTRUISM

Although the importance of group selection has been much debated, in a population with sexual reproduction and altruistic social behaviors selection seems to occur for the benefit of the group, even though it may harm an individual's own genetic future (R. A. Wilson, 2005).

Information provided by an individual affecting the survival of other members of the group is dramatically revealed when a monkey encounters a leopard and reacts with a loud scream, signaling other monkeys in the group to take refuge. Although this warning signal/alarm call gives us the impression that it would call the predator's attention to "the screamer" and diminish its chance for survival, alarm signals are usually given by an individual in a safe location or at a distance from the group. Regardless, the signal helps preserve other members of the group to which it is related. Such altruistic relation-ships, which are discussed further in BOX 14.1, remind us that fitness can be affected by behavior in various ways; courtship display and mating behaviors, feeding and care of offspring by individuals other than the parents (as occurs in birds), altruism, behavioral mimicry, schooling or swarming, and aggression are some examples. Genetic and social factors interact to influence fitness in many organisms, especially mammals, as elegantly demonstrated in a recent study of female calving success in wild bottlenose dolphins (*Tursiops* sp.) in Shark Bay, Western Australia.[12]

---

[11] See D. S. Wilson (1992), Williams (1996), and Eldakar and Wilson (2011) for overviews of group selection.

[12] C. H. Frère et al., 2010. Social and genetic interactions drive fitness variation in a free-living dolphin population. *Proc. Natl. Acad. Sci. USA*, **107**, 19949–19954.

BOX 14.1	Altruism and Reciprocal Altruism

ALTRUISM

Social interactions can produce results that are positive or negative for individuals involved in the interactions. Consequences can range from:

- a mutual benefit, to
- a positive advantage to the performer but not to the recipient, to
- negative effects of altruistic sacrifice to the performer but advantages to the other individual(s) (Axelrod and Hamilton, 1981; R. A. Wilson, 2005).

Fitness effects of altruism on the altruist may be negative compared to selfishness, but are positive for others in the population.

On the genetic level, altruism benefits any genes the altruist shares with relatives, even if the altruist is killed and so is unable to pass along its own genes. Thus, individuals in a socially interacting group containing many altruists are better off (achieve higher fitness); the effect on others is more positive than in a group with fewer altruists. So, the success of a group depends on a *group property*—the frequency of individuals expressing certain behaviors—rather than on characters confined to only single individuals. As Darwin (1871) put it for humans,

> It must not be forgotten that although a high standard of morality gives but a slight or no advantage to each individual man and his children over the other men of the same tribe, yet that an increase in the number of well-endowed men and an advancement in the standard of morality will certainly give an immense advantage to one tribe over another.[a]

Since the early 1930s, population geneticists have proposed that there are genetic advantages in altruistic behavior, in which individuals endanger their own survival for those who carry closely related genotypes. In the flour beetle, *Tribolium confusum*, egg-eating cannibalism by larvae declines in groups in which larvae only feed on genetically related eggs. This altruistic behavior of refraining from cannibalism is selected because it enhances the survival and reproduction of related individuals that would be considered prey in the absence of altruism.

RECIPROCAL ALTRUISM

The situation in which individuals cooperate only with those who also cooperate with them is known as reciprocal altruism, a concept introduced by Robert Trivers (Trivers, 1971).[b]

Trivers hypothesized that altruism can become established in a group where the frequency of interaction among individuals is high and their life span sufficiently long to enable recipients of altruistic acts to return favors to the altruists. Although altruism can be dangerous for the altruist, it may have significant benefits to the altruist when it is reciprocated; thus the benefits to individuals who partake in reciprocal altruism can far outweigh the costs. Cooperative defense roles taken on by members of the group, whether lions, primates, cattle, birds, and so forth, involve shared genetic survival. Fish aggregate in schools, the schools offering increased defense against predators when compared to isolated individuals. Group and individual selection reinforce one another in such situations.

---

[a]Darwin (1871), p. 159.
[b]Trivers originally called his principle "delayed return altruism;" see, R. Trivers, 1985. *Social Evolution.* Benjamin/Cummings, Menlo Park, CA. Also see R. Riolo et al., 2001. Evolution of cooperation without reciprocity. *Nature*, **414**, 441–443.

KIN SELECTION

In populations with sexual reproduction and altruistic social behavior, a form of group selection known as kin selection has been proposed to occur for the benefit of the group. Such a form of selection would operate on the population level.

Kin selection has been proposed to occur in species that display altruistic behavior (Box 14.1). Advantages of kin selection are seen in the alarm calls of vervet monkeys (*Chlorocebus pygerythrus*) that provide the individual with indirect benefits by helping its genetic relatives. A similar type of selection has been proposed to occur in distasteful prey species in which individuals with warning (**aposematic**) colors make a poisonous or distasteful animal conspicuous and recognizable to a predator. In some species, all individuals are aposematic, such as yellow and black striped wasps, or brightly colored poisonous frogs and snakes (**FIGURE 14.11**). The warning pattern makes them more susceptible to predation, but through their death they protect related genotypes that carry the same pattern, and the patterns are recognized as toxic by potential (experienced) predators.

(a) (b)

© Milan Vasicek/ShutterStock, Inc.

© Mcds1030/Dreamstime.com

FIGURE 14.11 Warning (aposematic) coloration as seen in yellow and black striped wasps **(a)**, and yellow-banded poison dart frog, *Dendrobates leucomelas* **(b)**.

The life history strategy in which one fertilized female lives cooperatively with many non-breeding individuals and one or more breeding males is known as **eusociality**. In 1964, British evolutionary biologist William Hamilton (1936–2000) provided mathematical formulae by which some of the benefits of such cooperative-social behavior could be evaluated. It is to the genetic advantage of females to invest their energy in raising sisters rather than producing daughters, who are more distantly related (Hamilton, 1964, and see below).

In evolutionary terms, this means that the altruistic behavior of sterile female workers helping raise more sterile worker progeny for their mother (the queen) would be expected to arise repeatedly in hymenoptera. Independent evolution of altruism at least once in ants, eight times in bees, and twice in wasps, bears this out. The hymenopteran system of sex determination and altruism, coupled with parental care, led John Maynard Smith (1920–2004) to coin the term "kin selection." As Maynard Smith pointed out, "the main reason for thinking that kin selection has been an important mechanism in the evolution of cooperation is that most animal societies are in fact composed of relatives."[13]

All ants, bees, and wasps have haploid males and diploid females, a situation known as **haplodiploidy**. Relatedness is greatest in species in which females mate with a single male, and lower in honeybees and ants in which females mate with multiple males. Males develop from unfertilized (haploid) eggs and females from fertilized (diploid) eggs. All the females (sisters) from a single pair of parents are more closely related than are mothers to their own daughters, the diploid female offspring of a queen sharing more genes with their sisters (75%) than with their daughters (50%). Genotypes of the sisters as a group (*kin*) benefit from female workers who, instead of reproducing, rely on their mother's reproductive ability by helping raise sisters rather than producing their own daughters. This hymenopteran system of sex determination has been shown to be especially prone to the evolution of eusociality, the benefits of which include communal nesting, joint protection against predators and parasites, and sharing foraging for food (Khila and Abouheif, 2010).

Genetic relationship leads to a conflict of interest for a colony's sex ratio. Being more related to sisters, it is to the workers' genetic benefit that the colony supports more females and fewer distantly related males. Queens, by contrast, being related equally to sons and daughters, benefit genetically when males and females are produced in equal numbers. When the queen mates more than once, genetic relationships between female workers decline, but a worker is still more closely related to her mother's female offspring than to her sister's female offspring; workers (which only produce male offspring) destroy eggs laid by other workers but not eggs laid by the queen.

---

[13] J. Maynard Smith, 1983. Game theory and the evolution of cooperation. In *Evolution from Molecules to Man*, D. S. Bendall (ed.). Cambridge University Press, Cambridge, UK, pp. 445–456. The quotation is from p. 447. For an enthralling account of George R. Price (1922–1975) who was one of the major contributors to altruism and kin selection, see O. Harman, 2010. *The Price of Altruism: George Price and the Search for the Origins of Kindness*. W. W. Norton & Co, New York.

■ KEY TERMS

altruism	haplodiploidy
aposematic coloration	kin selection
directional selection	reciprocal altruism
disruptive (diversifying) selection	sexual selection
eusociality	species sorting
genotype	stabilizing selection
group selection	survival of the fittest

■ EVOLUTION ON THE WEB

Explore evolution on the Internet! Visit the accompanying website for *Strickberger's Evolution, Fifth Edition,* at **go.jblearning.com/Evolution5eCW** for exercises and links relating to topics covered in this chapter.

■ REFERENCES

Axelrod, R., and W. D. Hamilton, 1981. The evolution of cooperation. *Science*, **211**, 1390–1396.

Bailey, M. M., K. A. Lachapelle, and M. T. Kinnison, 2010. Ontogenetic selection on hatchery salmon in the wild: natural selection on artificial phenotypes. *Evol. Applic.*, **3**, 340–351.

Bumpus, H. C., 1899. The elimination of the unfit as illustrated by the introduced sparrow. *Biol. Lect. Woods Hole*, pp. 209–226.

Darwin, C., 1871. *The Descent of Man and Selection in Relation to Sex.* John Murray, London.

Eberhard, W. C., 1996. *Female Control: Sexual Selection by Cryptic Female Choice.* Princeton University Press, Princeton, NJ.

Eldakar, O. T., and D. S. Wilson, 2011. Eight criticisms not to make about group selection. *Evolution,* **65**, 1523–1526.

Emlen, D. J., 2000. Integrating development with evolution: a case study with beetle horns. *BioScience*, **50**, 403–418.

Emlen, D. J., J. Hunt, and L. W. Simmons, 2005. Evolution of male- and sexual-dimorphism in the expression of beetle horns: phylogenetic evidence for modularity, evolutionary lability, and constraint. *Am. Nat.*, **166**, S42–S68.

Endler, J. A., 1986. *Natural Selection in the Wild.* Princeton University Press, Princeton, NJ.

Fisher, R. A., 1930. *The Genetical Theory of Natural Selection.* Clarendon Press, Oxford, UK. (2nd ed., 1958, Dover, New York.)

Gilad, Y., A. Oshlack, G. K. Smyth, T. P. Speed, and K. P. White., 2006. Expression profiling in primates reveals a rapid evolution of human transcription factors. *Nature*, **440**, 242–245.

Gould, S. J., 1976. Darwin's untimely burial. In *Philosophy of Biology*, M. Ruse (ed.), pp. 93–98. Prometheus Books, New York.

Grant, P. R., and Grant B. R., 2008. *How and Why Species Multiply: The Radiation of Darwin's Finches.* Princeton University Press, Princeton, NJ.

Hallgrímsson, B., and B. K. Hall (eds.), 2005. *Variation: A Central Concept in Biology.* Elsevier/Academic Press, Burlington, MA.

Hamilton, W. D., 1964. The genetical evolution of social behaviour. I. *J. Theor. Biol.* **7**, 1–16, 17–52.

Khila, A., and E. Abouheif, 2010. Evaluating the role of reproductive constraints in ant social evolution. *Phil. Trans. R. Soc. B.*, **365**, 617–630.

McKinnon, J. S., and M. E. R. Pierotti, 2010. Color polymorphism and correlated characters; genetic mechanisms and evolution. *Mol. Ecol.*, **19**, 5101–5125.

Møller, A. P., 1994. *Sexual Selection and the Barn Swallow.* Oxford University Press, Oxford, UK.

Parker, G. A., 1970. Sperm competition and its evolutionary consequences in the insects. *Biol. Rev.*, **45**, 525–567.

Rendel, J. M., 1967. *Canalization and Gene Control.* Logos Press, London.

Snell-Rood, E. C., A. Cash, M. V. Han., 2010. Developmental decoupling of alternate phenotypes; insights from the transcriptomes of horn-polyphenic beetles. *Evolution*, **65**, 231–245.

Snook, R, R., 2005. Sperm in competition: not playing by the numbers. *Trends Ecol. Evol.*, **20**, 46–53.

Trivers, R. L., 1971. The evolution of reciprocal altruism. *Quart. Rev. Biol.*, **46**, 35–57.

Trivers, R. L., 1972. Parental investment and sexual selection. In *Sexual Selection and the Descent of Man*, B. Campbell (ed.), Aldine-Atherton, Chicago, pp. 136–179.

Wade, M. J. 1976. Group selection among laboratory populations of *Tribolium*. *Proc. Natl. Acad. Sci. USA*, **73**, 4604–4607.

Wade, M. J., 1980. Group selection, population growth rate, and competitive ability in the flour beetles, *Tribolium* spp. *Ecology*, **61**, 1056–1064.

Williams, G. C., 1996. *Adaptation and Natural Selection.* Princeton University Press, Princeton, NJ.

Wilson, D. S., 1992. Group selection. In *Keywords in Evolutionary Biology*, E. Fox Keller and E. A. Lloyd (eds.), Harvard University Press, Cambridge, MA, pp. 145–148.

Wilson, R. A., 2005. *Genes and the Agents of Life. The Individual in the Fragile Sciences*, Cambridge University Press, Cambridge, UK.

Zahavi, A. 1975. Mate selection: a selection for a handicap. *J. Theor. Biol.*, **53**, 205–214.

Zahavi, A., and A. Zahavi. 1997. *The Handicap Principle: A Missing Piece of Darwin's Puzzle.* Translated by N. Zahavi-Ely. Oxford University Press, New York.

Natural Selection, Phenotypes, and Genotypes

CHAPTER SUMMARY

In this chapter we examine some of the major consequences of natural selection for the species under selection and for other species with which they interact; selection on one species can influence the evolution of other species through evolutionary processes such as mimicry and co-evolution. Selection acts on the fitness differences among different phenotypic traits of individuals within populations. Selective change is not unbounded, however. Selection is facilitated by the large amount of allele variation (polymorphism) in populations, variation itself being subject to selection. Response to selection is influenced by constraints, canalization (the ability to produce the same phenotype in response to environmental or genotypic variation), and plasticity arising from genetic and developmental processes. How available genetic polymorphisms, responses to changing environments, and phenotypic plasticity mold selection and influence fitness are discussed.

INTRODUCTION

Selection acts on the fitness differences among different phenotypic traits of individuals within populations. Species are bound together as potential competitors, as predators and prey, through mutualistic associations such as mimicry, and as communities through ecological interactions. Consequently, selection on one species influences other species. However, selection is not limitless. Dynamics and constraints of genetic and embryonic developmental pathways work with selection both to provide plasticity and to narrow (canalize) evolutionary change. These interactions can be observed in nature and mimicked in laboratory experiments in which an organism's response to environmental change is measured as a **reaction norm.** The greater the number of reaction norms, the more responsive the organism is to the environment.

SELECTION ON FEATURES OF ONE SPECIES CAN AFFECT OTHER SPECIES

Species compete with each other for resources and through competition in ways that affect fitness. Species also co-evolve; an advantage or increase in fitness for one species may bring about deterioration in the environment for other species and a potential decline in their fitness, or it may enhance the fitness of the other species (Martin, 1996).

In 1976, American evolutionary biologist Leigh van Valen (1935–2010) drew the analogy between species survival and a remark made by the Red Queen whom Alice meets in Lewis Carroll's *Through the Looking Glass.* "Here, you see, it takes all the running you can do to keep in the same place." According to the **Red Queen hypothesis**, selection on one of two competing or co-evolving species gives that species a competitive advantage, reducing the fitness of the second species and eliciting selection in that species.[1] Co-evolving species in a changing biological environment therefore face an "arms race," a race most obviously seen in the co-evolution of predator and prey. Even though the second species would seem to be at a disadvantage, the long-term consequence of the Red Queen's reign is to increase the competitive fitness of each interacting species. As Darwin stated in *Origin of Species*, "If some of these many species become modified and improved, others will have to be improved in a corresponding degree or they will be exterminated."

SELECTION AND CONSTRAINT

For populations to evolve—that is, to change their gene frequencies—mutations must introduce nucleotide differences in the alleles of individuals within the population, adding to the existing genetic variation. The mere *appearance* of new alleles, however, is no guarantee that they will *persist* or *prevail* over others. Explanations for the persistence of many mutations and their increase in frequency must be sought in processes other than the original mutational event.

The existence of developmental systems with their regulated pathways raises the question of whether development constrains or enhances (through plasticity) structural and functional variation, thereby influencing natural selection and future evolution (Hallgrímsson and Hall, 2005). For example, one can ask why insects have six legs and land vertebrates four, when almost any even number of legs, or even none at all, can suffice for locomotion. What role, if any, do the dynamics of development play in maintaining such phylogenetic constraints as the number of legs in insects and in vertebrates, and how do such mechanisms relate to natural selection? The answer is not entirely simple, although development, on the one hand, limits the range of forms that natural selection can bring forth, and on the other, provides plasticity for changes in the phenotype.

Although many developmental changes are possible, certain of these changes, especially those of large degree, are mostly lethal; their effects cannot be successfully integrated with other aspects of embryonic development. The need for integration generation after generation results in **developmental and phylogenetic constraints**.

---

[1] L. Van Valen, 1976. Ecological species, multispecies, and oaks. *Taxon*, **25**, 233–239. The Red Queen hypothesis also has been applied to cycles of evolutionary change between individual classes of molecules such as glycans, which are sugars that attach to large numbers of proteins and lipids (Varki, 2006).

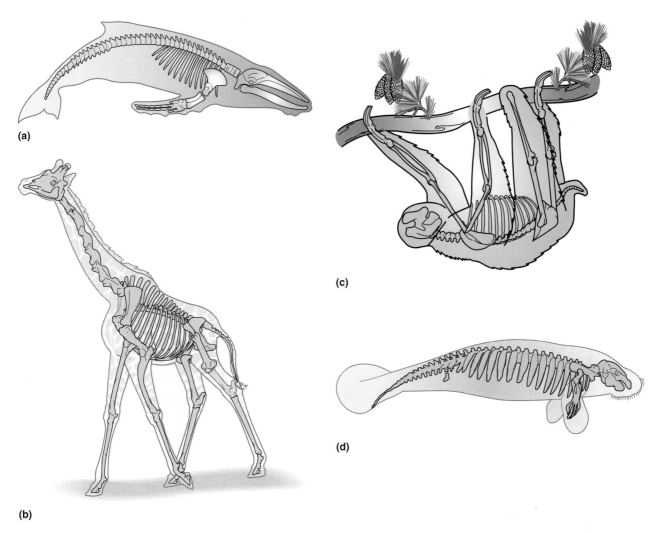

(a)

(b)

(c)

(d)

FIGURE 15.1 The number of neck (cervical) vertebrae in mammals is conserved at seven in most mammals, shown here in two species with very different lifestyles, a whale **(a)** and giraffe **(b)**. Variation can occur, however, as shown in the three-toed sloth **(c)** which has eight or 9, and the manatee **(d)**, which has between five and seven neck vertebrae.

A **constraint** is any influence that acts to canalize or channel the effects of selection. An example of a constrained feature is the presence of seven neck vertebrae in practically all mammals, whether short-necked (whales) or long-necked (giraffes; **FIGURE 15.1a, b**). In order to understand evolutionary conservation or change of phenotypes, the processes that underlie constraint must be integrated with the role of natural selection. Constraints and selection interact to canalize evolutionary change.

A comparison between evolutionary change in frogs and placental mammals illustrates differences in the **evolvability** of phenotypes reflecting as yet unknown differences in constraints on evolution between lineages. Frogs appear in Triassic deposits about 200 Mya. Placental mammals do not appear until the Cretaceous, about 90 Mya. Despite their much longer fossil history and their large number of species, frogs have undergone fewer phenotypic changes in evolution compared to the enormous adaptive radiations of placental mammals. The more than 3,000 species of frogs are so much alike that herpetologists place them in a single clade, the order Anura. The 4,300 species of placental mammals (of which about 2,000 are rodents) diverged widely and usually are classified into 18 lineages (orders), ranging from bats to primates to whales (**FIGURE 15.2**).

FIGURE 15.2 Phylogenetic tree for placental mammals.

The Molecular Tree of Mammalian Relationships

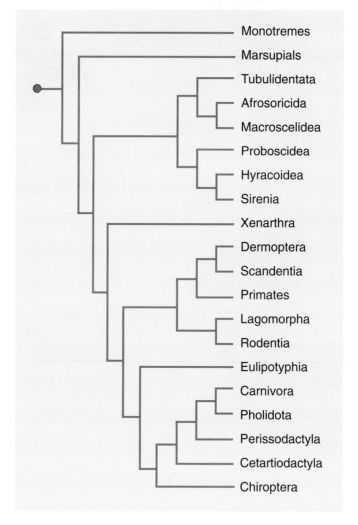

Previous selection history and networks of developmental genetic interactions canalize subsequent evolution. Compatibility of new phenotypic features with established genetic networks channels selection. Furthermore, selection itself can act as a constraint; the potential to respond to subsequent selection regimes may be compromises by past selection. Selection toward becoming a "master of one trade" prevents an organism from becoming a "jack of all trades," and *vice versa*. As one of many examples, selection for increased efficiency of swimming performance impacts traits that enhance running on land, as seen in crocodiles, for example. Maintaining phenotypes that can successfully face all possible future eventualities is unattainable in a single lineage; selection "sees" what is current, not what might be optimal for an unknown future.

Nevertheless, constraints are not impassible barriers. Some mammals escaped the constraint of seven cervical vertebrae; three-toed sloths have eight or nine, and manatees have six (**FIGURE 15.1c, d**). Given sufficient time and new opportunities, vertebrate lineages that evolved from swimming fish into crawling terrestrial forms later evolved into different marine swimming forms (ichthyosaurs, whales) and into entirely novel winged aerial forms (birds, bats, and pterosaurs). Therefore, developmental constraint **narrows variation but does not prevent evolutionary change**, illustrating the importance of understanding exactly how the limits to variation are set.

SELECTION, CHANGING ENVIRONMENTS, AND PLASTICITY

When exposed to altered environmental conditions some species respond by producing more than one phenotype (West-Eberhard, 2003; Hall et al., 2004). This ability goes under various names, including **phenotypic plasticity**. According to terminology introduced early in the development of Mendelian genetics, the range of phenotypic expression that can be produced by a single genotype in response to environmental change can be expressed as a reaction norm.

Reaction norms apply to phenotypes encountering different environments (Schlichting and Pigliucci, 1998; Pigliucci, 2001). They can be compared for the same species under different environmental conditions or for different species under the same conditions, as shown in FIGURE 15.3. Reaction norms are detected experimentally by raising individuals in different environments—different temperatures, predator densities, and so forth—measuring the trait of interest under each set of conditions, and plotting the phenotype against the environmental variable. A reaction norm, therefore, is an index of organismal-environmental interaction on the one hand, and flexibility/plasticity–canalization/constraint on the other. A single line—a single reaction norm—indicates a tightly canalized response and limited responsiveness to environmental changes. Multiple slopes indicate multiple reaction norms and a plastic response of organisms to their environment.

In 1966, Waddington and Robertson reported the results of experiments in which developmental canalization was selected for (**canalizing selection**) and resulted in enhanced phenotypic variation:

> … canalizing selection limits the expression of a trait, presumably by eliminating genotypes that would broaden phenotypic responsiveness to the environment and so broaden expression of the character, which could occur by eliminating genotypes that would permit an individual to respond to any genetic, epigenetic or environmental influences that would lead to greater variability in the phenotypic expression of the trait.

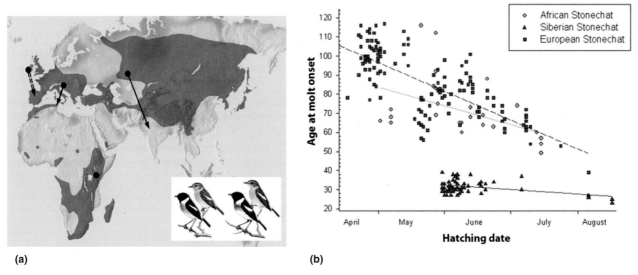

(a) (b)

FIGURE 15.3 The distribution and migration patterns of four congeneric species of Stonechats in Europe and Africa differ **(a)**. The insert shows male and female European stonechats (*Saxicoloa rubicola*) on the left and African stonechats (*S. torquatus*) on the right. Reaction norms of age at molt against time of year for three species analyzed also differ **(b)**. The Siberian Stonechat (*S. maurus*) shows much less plasticity in this feature than do the other two species.

[Courtesy of Max Planck Institute for Ornithology, Department of Biological Rhythms and Behaviour.]

Among Waddington's experiments on canalizing selection were demonstrations in *Drosophila* that phenotypically uniform expression of the normal *Ultrabithorax* gene in the face of environmental stress depends on other "background" genes. The genetic basis for such constant expression became apparent when specific genetic loci that support *Ultrabithorax* transcriptional stability were identified.[2]

New features such as an additional pair of wings seen in *Drosophila* as a result of mutations in genes such as *Ultrabithorax* can be elicited by environmental stimuli and then selected for in the absence of the stimulus. The traits become genetically incorporated and appear in subsequent generations that have not been exposed to the environmental stimulus, an evolutionary process known as **genetic assimilation**. This is not a Lamarckian process of direct instruction by the environment. It occurs because of the presence of unexpressed genetic variation and selection for genotypes that are capable of responding to the environmental change.

The close fit between organismal flexibility and environmental change is a product of underlying genetic components (Wilkins, 2003). This was well demonstrated by Braendle and Flatt (2006) using color polymorphism in the tomato hornworm *(Manduca quinquemaculata)*. Presence of more than one color morph (polymorphism) is regulated by juvenile hormone and is sensitive to environmental signals that influence the action of juvenile hormone. A polymorphic population of hornworms exposed to a heat shock and then selected for color morphs over multiple generations produced the environmentally-sensitive color morph in the *absence of further heat shock*.

SELECTION AND FITNESS

Although hereditary transmission is through the genotype, selection acts upon the phenotype. To the extent that phenotypic differences depend upon genetic variation, phenotypes differ in viability and fertility and thus influence the frequencies of their genotypes.

The consequences of selection on a particular trait in a population are measured in terms of reproductive success or **fitness**. In their simplest form, fitness and the intensity of selection are measured by the number of descendants produced by one phenotype relative to those produced by another. However, in the case of plasticity and reaction norms, there can be differences in fitness among phenotypes produced by the same genotype. Sustained selection on the phenotype can result in change in gene frequencies from generation to generation, especially if genetic assimilation is operating. Effects of selection are recorded as **selection coefficients** (s), which give the greatest effect on fitness a value of 1, with adaptive value being defined as 1-s. The lowest selection coefficient would be 0, indicating no effect on fitness, as seen in mutations on third position codons.

HETEROSIS *and its synonym hybrid vigor are used for hybrids that are heterozygous at many loci.*

We refer to the situation where the heterozygote has superior reproductive fitness over both homozygotes as **heterozygous advantage**. Because heterozygous advantage is often expressed in characters that affect fitness—longevity, number of offspring produced, resistance to disease—it has been studied in some detail as **heterosis**. An oft-cited example of heterosis is the dramatic increase in the agricultural yield of hybrid corn achieved by crossing selected inbred lines. Beginning with an average yield of about 25 bushels per acre in the 1920s, selective breeding of hybrid corn enabled increases in yield to as much as 140 bushels per acre.[3]

[2] G. Gibson and D. S. Hogness, 1996. Effect of polymorphism in the *Drosophila* regulatory gene *Ultrabithorax* on homeotic stability. *Science*, **271**, 200–203; T. Sanchez-Elsner et al., 2006. Noncoding RNAs of Trithorax response elements recruit *Drosophila* Ash1 to *Ultrabithorax*. *Science*, **311**, 1118–1123.
[3] C. J. Kucharik and N. Ramankutty, 2005. Trends and variability in U.S. corn yields over the twentieth century. *Earth Interact.*, **9**, 1–29. Because hybrid corn is sterile, continuous selection is required to maintain these yields.

An example of the interplay of factors affecting fitness is the observation that fitness of the phenotype can be **frequency dependent**, especially in situations where a variant allele is either common or rare. Frequency dependent fitness allows single evolutionary events—the origin and spread of a single mutation—to be modeled and compared. Understanding frequency dependent fitness has the potential to inform us about branching points in evolution. So too does the approach discussed that considers selection and fitness in terms of populations occupying different adaptive peaks rather than valleys in an **adaptive landscape**. In this approach, gene frequencies change as the result of what is termed **genetic drift**.

SELECTION AND BALANCED POLYMORPHISM

Persistence of different genotypes through heterozygote advantage is an example of **balanced polymorphism**, a term coined by Oxford ecological geneticist E. B. Ford (1901–1988) to describe the preservation of genetic variation through selection. A gene locus is defined operationally as being polymorphic if at least two alleles are present and if the second allele—or the least frequent allele when more than two alleles are present—has a frequency of at least one percent. Polymorphisms are ubiquitous in practically all populations examined.

One well-known example of polymorphism is the *sickle cell* gene in humans, in which the sickle cell homozygote is selected against because it cannot carry oxygen efficiently. Because individuals who are heterozygous at the sickle-cell locus are more likely to survive the malarial parasite than are those who are homozygous for either of the two alleles, homozygotes remain in the population. This gene—and others also shown in FIGURE 15.4 that offer protection against malaria—persists in notable frequencies in geographical areas where malaria is endemic. Of course, while retaining the ability to induce sickle cell anemia, the malarial parasite has evolved multiple genes to evade the immune response mounted by the host. Individual *Plasmodium falciparum* parasites contain some 60 genes for one protein involved in transmission. Different individuals contain different variants of the genes, an impressive arsenal mounted by the parasite in the parasite-host arms race.

A second example at the phenotypic level is selection of cryptic moth prey by blue jays (*Cyanocitta cristata*). Jays capture and consume more of the abundant prey types than of the rare types, switching capture strategy if the rare prey type becomes more abundant.[4]

As a third example, selection and balanced polymorphism can be demonstrated in laboratory experiments. When two types of *E. coli* are co-cultured in glucose-limited minimal media, the relative abundance of each is frequency dependent, each type having an advantage when rare, resulting in maintenance of a balanced polymorphism in the population over time.

An influential alternative theory to selection and balanced polymorphism was proposed by Motoo Kimura (1924–1994) in 1979 and expanded into a book in 1983. The **neutralist hypothesis**, which is discussed briefly in BOX 15.1, prompted much controversy that led to research and development of evolutionary theory.

SELECTION AND MIMICRY

Other conditions responsible for polymorphism include a change in selection coefficients, reflecting that genes detrimental at one time are advantageous at another. Selection against a gene may depend on its frequency and may change or even be reversed when the gene is at low frequency and before it can be eliminated.

---

[4] For the experimental approach to such studies, see A. B. Bond and A. C. Kamil, 2002. Visual predators select for crypticity and polymorphism in virtual prey. *Nature*, **415**, 609–613.

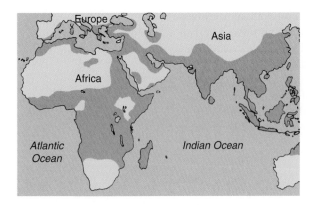
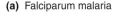

(a) Falciparum malaria

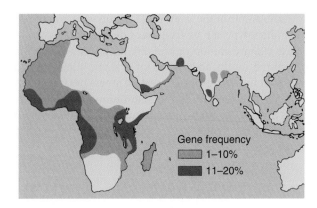

(b) Sickle cell anemia

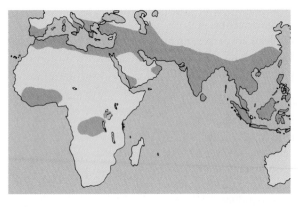

(c) β-thalassemia

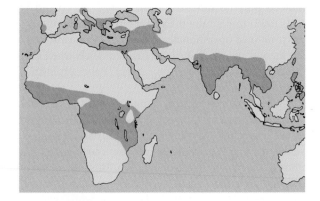

(d) G-6-pd deficiency

FIGURE 15.4 Relationship between the geographic distributions of the most dangerous form of malaria, falciparum malaria initiated by the protozoan parasite *Plasmodium falciparum,* and genes that confer resistance against malaria. **(a)** Distribution of falciparum malaria in Eurasia and Africa before 1930. **(b)** Distribution of the gene for sickle cell anemia (*Hb*^{s}). **(c)** Distribution of the gene for β-thalassemia. **(d)** Distribution of the sex-linked gene for glucose-6 phosphate dehydrogenase (G-6-pd) deficiency in males in frequencies above 2 percent.

[Adapted from Strickberger, M. *Genetics, Third Edition,* New York: Macmillan, 1985.]

An example of such selection is **Batesian mimicry**, a mechanism proposed by explorer Henry Bates in which palatable species that mimic distasteful models are protected against predators (see Brower, 1988). In general, the more frequent the mimic and the less frequent its model, the greater the chance that predators will attack the mimic. Conversely, the less frequent the mimic compared to the model, the greater the chance that the mimic will be protected.

Mimicry in insects evolved early, and, as expected if mimicry is effective, has been remarkably stable. An Eocene leaf-mimicking stick insect, *Eophyllium messelensis,* from 47 Mya closely resembles extant insect mimics (**FIGURE 15.5**) and so is interpreted as a mimic rather than as an example of camouflage not involving mimicry. Furthermore, this ancient insect mimicked the leaves from several types of plants including myrtle and laurel trees and alfalfa, indicating that sophisticated forms of mimicry were already present in the Eocene.[5]

[5] S. Wedmann, S. Bradler, and J. Rust, 2007. The first fossil leaf insect: 47 million years of specialized cryptic morphology and behavior. *Proc. Natl. Acad. Sci. USA,* **104**, 565–569.

BOX 15.1	The Neutralist Hypothesis

Kimura (1979, 1983) argued that the rate and amount of evolution at the molecular level is far more rapid and more pervasive than had been appreciated. Consequently, he proposed that not all polymorphisms result from selection, asserting that the presumed high cost of selection necessitated that most amino acid changes be neutral—that is, that they have no effect on the phenotype. Kimura argued that as selection does not act on neutral mutations, the fixation of such alleles incurs no genetic cost and depends only on their mutation rates and on random changes in gene frequency—**genetic drift**.

This proposition became known as the **neutralist hypothesis**, although it was called non-Darwinian evolution by some because of its dependence on mutation and not on selection (Hey, 1999; Nei, 2005). The extensive enzyme and protein polymorphisms in natural populations, demonstrating that multiple alleles exist at many thousands of loci in many species, seemed to support the neutral hypothesis. High frequencies of selectively neutral mutations in rapidly mutating strains of *E. coli* provide one line of support for the neutralist position.[a]

However, and in contrast to the neutralist hypothesis, recent extensive studies have demonstrated functional consequences of polymorphism. From a survey of insects, mollusks, vertebrates, and plants from Israel it was concluded that the amount of allozyme polymorphism[b] in a species correlates to factors such as life habit and climate (Nevo, 1978). Even more striking is the finding that clones of the bacterium *E. coli* isolated from the intestinal tracts of animals as diverse as lizards and humans, and from localities as widespread as New Guinea and Iowa, share common allozyme frequencies. For each of five different enzymes, one particular allele is present in almost all samples (Milkman, 1973). As other allozymes of these proteins exist in *E. coli*, such a narrow distribution of allozymes is difficult to explain on any basis other than selection on a favorable allozyme.

We know now that mutations in coding genes are strongly selected against, and that many mutations occur in the large amount of non-coding DNA, much of which is involved in gene regulation. Nucleotide sequence analyses show that polymorphisms are significantly more frequent in those DNA sequences that do not determine amino-acid sequences than in those that do. For a given region of the genome, more polymorphisms occur in introns (intervening sequences) that do not code for amino acids, than in exons (expressed sequences), which do. The alcohol dehydrogenase gene in *Drosophila melanogaster* has six percent polymorphism in introns, seven percent polymorphism in third-position codons, and almost no polymorphism in exons. Because random events can hardly explain such pronounced differences in polymorphism, such findings strongly suggest that selection must be the discriminating agent that determines which nucleotide base substitutions will become established in functionally different DNA sequences.

---

[a] J. E. LeClerc et al., 1996. High mutation frequency among *Escherichia coli* and *Salmonella* pathogens. *Science*, **274**, 1208–1211.
[b] Allozymes are different forms of the same enzyme coded for by different alleles of the same gene.

As shown in **FIGURE 15.6**, *mimicry between different species*—**Müllerian mimicry**—benefits both species by enabling predators to learn a single warning pattern that applies to all these potential but distasteful prey. When rare, conspicuous warning patterns on unpalatable individuals offer little protection; predators have few chances to learn their distastefulness. Distinctive patterns, however, offer greater protection to unpalatable individuals when they are at higher densities, as shown by a Texan grasshopper (*Schistocerca emarginata*) that develops colors that blend in with the vegetation when individuals develop alone. When the grasshoppers develop in groups, however, they develop yellow and black warning colors if they consume a plant that is toxic to predators, providing a further example of phenotypic plasticity.[6]

A particularly detailed analysis of the evolution of mimicry concerns the black Calla lily, *Arum palaestinum*, which attracts fruit flies (*Drosophila* spp.) for pollination by producing chemicals that mimic odors produced by yeast. Fruit flies have evolved a set of odor receptors that detect volatile chemicals released by yeast, a major part of the fruit fly diet. Of eight *Drosophila* species examined, seven share the same olfactory response to the volatile chemicals, despite 40 million years of independent evolution separating species pairs (Stökl et al., 2010).

Strictly, selection does not act on mutations but on phenotypes. In this sense, KIMURA'S HYPOTHESIS can be said not to depend on selection.

As a consequence of their composition, particular DNA SEQUENCES—for example, guanine-cytosine rich sequences that replicate early during DNA synthesis—may mutate at different rates and in different directions from other sequences.

---

[6] G. A. Sword, 1999. Density dependent warning coloration. *Nature*, **397**, 217. Also see S. Tanaka, 2004. Environmental control of body-color polyphenism in the American grasshopper, *Schistocerca americana*. *Ann. Entomol. Soc. Am.*, **97**, 293–301.

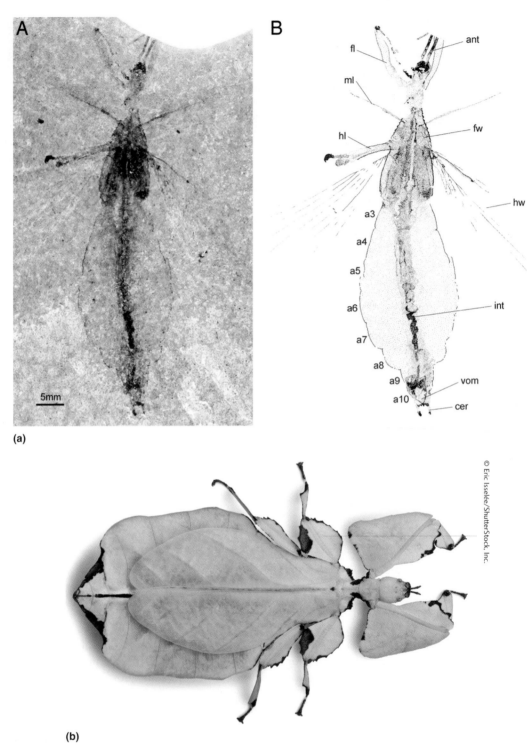

FIGURE 15.5 This Eocene stick insect, *Eophyllium messelensis*, a leaf mimic **(a)** bears a strong resemblance to extant leaf-mimicking stick insects **(b)**.

[(a) Reproduced from Wedmann, Sonja, et al., 2007, The first fossil leaf insect: 47 million years of specialized cryptic morphology and behavior. Proc. Natl Acad. Sci. USA, 104, 565–569.]

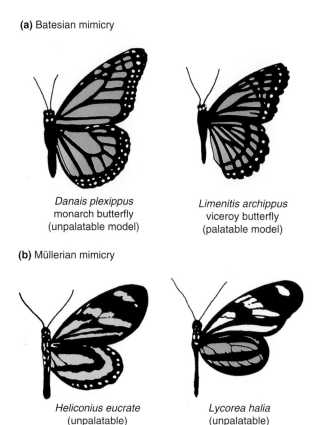

(a) Batesian mimicry

Danais plexippus
monarch butterfly
(unpalatable model)

Limenitis archippus
viceroy butterfly
(palatable model)

(b) Müllerian mimicry

Heliconius eucrate
(unpalatable)

Lycorea halia
(unpalatable)

FIGURE 15.6 Mimicry in different species of butterflies. **(a)** Batesian mimicry by a North American species, in which the more palatable viceroy butterfly (*right*) mimics the more unpalatable monarch (*left*). Resemblance between the two South American unpalatable species in **(b)** provides a common warning pattern to predators and helps protect both prey species (Müllerian mimicry). Mimicry is one of the most obvious examples of convergent evolution.

SELECTION IN CHANGING ENVIRONMENTS: MELANISM

A population sufficiently widespread to occupy many environments may maintain a variety of phenotypes, each of which is superior in a particular habitat. Spatial and temporal organization of the environment may significantly affect the extent to which a population will rely on genetic polymorphism as an adaptive strategy. Coarse-grained environments, in which different individuals in a population are exposed to different experiences, promote greater genetic polymorphism than do fine-grained environments, in which all individuals experience relatively similar conditions. A prominent example is polymorphism associated with **industrial melanism** (Majerus, 1998).

In industrial areas of England in the nineteenth century, vegetation was darkened by deposition of soot from the burning of coal and by exposure of the underlying bark when the soot killed off surface lichens. First observed in Manchester and later spreading to other areas, populations of certain moths and butterflies were found to have increased frequencies of a dark-colored (melanic) morph (**FIGURE 15.7**).

In the early 1970s in the industrial city of Birmingham, Bernard Kettlewell (1907–1979) and others released known numbers of light and melanic morphs of the British peppered moth, *Biston betularia* (Kettlewell, 1973). A significantly greater proportion of melanic than light forms were recaptured at a later date. Kettlewell's data suggested that areas covered in soot offer greater protection to melanic than to light-colored morphs; the light forms are at a selective disadvantage. Thus, more melanic than light morphs survive to be recaptured.[7] The relative fitness of the melanic morphs may lie, at

---

[7] T. D. Sargent, C. D. Millar, and D. M. Lambert, 1998. The "classical" explanation of industrial melanism. *Evol. Biol.*, **30**, 299–322. For an engaging read of the individuals involved and the history of these experiments, see Hooper (2002).

(a) (b)

FIGURE 15.7 Dark-colored and light-colored tree trunks, one with a melanic **(a)** and one with a non-melanic **(b)** peppered moth (*Biston betularia*) placed to show the contrast between them. The light-colored trunks derive their appearance from lichens. Although trees are commonly shown as resting sites for these moths, their actual resting habits are unknown.

least partly, in their ability to remain concealed from bird predators on darkened twigs or tree trunks (Figure 15.7**a**). In non-industrial areas, in contrast, trees covered with normal gray lichens offer decided advantages to the light-colored moths (Figure 15.7**b**).

Whether via environmental camouflage or other unidentified factors, English populations of the peppered moth show various degrees of polymorphism, ranging from high frequencies of the melanic form in industrial areas to almost zero melanic forms in many rural areas. Researchers were able to determine how long it took the frequency of non-melanic moths to decrease in industrial areas using the dates when specimens were collected by amateurs and records in museum collections. Assuming one generation per year, they estimate it took about 40 generations for the frequency of non-melanic phenotypes to decrease from 98 to 5 percent. The selection coefficient (0.2 or even higher) is indicative of intense selection against a non-melanic gene(s) in these industrial areas.

Passage of clean air legislation in Britain in 1956 reduced industrial smoke in many formerly polluted areas. This reduction in pollution altered selection and the frequency of melanic forms of the peppered moth and of other insects declined dramatically (Brakefield, 1998). The difference between the light-colored and the melanic forms of the British peppered moth is now known to be due to allelic differences at the single locus *carbonaria*, recently identified by genetic mapping (van't Hoff et al., 2011). Cross breeding with the melanic form of the moth found in the United States demonstrated that the same locus is involved in these widely separated populations (B. S. Grant, 2004).

A parallel situation is seen in light and dark forms of the rock pocket mouse *Chaetodipus intermedius* found in the American southwest. Mice with light, sandy colored fur lacking melanin granules occur on sandy substrates; mice with dark (melanin-containing) fur are found on black lava flows (**FIGURE 15.8**). These differences in coloration reduce predation from birds and mammals that are visual hunters. Ability to maintain the two fur colors has been traced to the gene *Mc1R* that codes for the melanocortin-1-receptor. Mc1R plays a key role in whether melanin granules will be deposited into the fur in rock pocket mice and in many other animals (Linnen et al., 2009).

(a)

(b)

FIGURE 15.8 Rock pocket mice (*Chaetodipus intermedius*) that forage on sandy substrates have sandy colored fur **(a)** while those that forage on black lava have black fur **(b)**.

[(a) and (b) © R B Forbes and the Mammal Images Library of the American Society of Mammalogists.]

Mutations in *Mc1R* also underlie the melanic form of the beach mouse *Peromyscus polionatus* found in the southeastern United States. Four amino acid differences separate the *Mc1R* gene in light and dark rock pocket mice. A single amino acid substitution in the beach mice from Florida is associated with adaptation to its changing environment (Hoekstra et al., 2006).[8] In contrast, *63* differences separate wild house mice (*Mus musculus*) indigenous to Europe from rock pocket mice. We can conclude that small genetic changes can lead to significant phenotypic change, illustrating selection and evolution in action.

Furthermore, the pathways regulating the deposition of melanin granules lie at the basis of color changes in other species, and not only rodents; mutations in *Mc1R* are associated with change to pale coloration in several species of wild cats, lizards, the snow goose, Arctic skua, and Kermode black bear, all of which also have melanic forms. Pale forms of two species of lizards that live on white gypsum dunes of the White Sands formation in New Mexico result from mutations in *Mc1R*; however, the origins of their coloration are different. In one species, it is traced to a mutation in receptor signaling, in the other to integration of the receptor into the melanocyte membrane; same gene, different mechanism for loss of melanin.[9] Similar phenotypes can arise by evolutionary modification in: (1) different aspects of the function of a single gene, or (2) different aspects of the functions of different genes.

■ KEY TERMS

adaptation	genetic drift
adaptive landscape	heterosis
balanced polymorphism	heterozygous advantage
Batesian mimicry	industrial melanism
canalizing selection	Müllerian mimicry
constraint	neutral theory of molecular evolution
developmental phylogenetic constraint	phenotypic plasticity
evolvability	phylogenetic constraint
fitness	reaction norm
frequency dependent selection	Red Queen hypothesis
genetic assimilation	selection coefficient (*s*)

■ EVOLUTION ON THE WEB

Explore evolution on the Internet! Visit the accompanying website for *Strickberger's Evolution, Fifth Edition,* at **go.jblearning.com/Evolution5eCW** for exercises and links relating to topics covered in this chapter.

■ REFERENCES

Braendle C., and Flatt T., 2006. A role for genetic accommodation in evolution? *BioEssays,* **28**, 868–873.

Brakefield, P. M., 1998. Receding black moths. *Trends Ecol. Evol.,* **13**, 376.

Brower, L. P. (ed.), 1988. *Mimicry and the Evolutionary Process.* University of Chicago Press, Chicago.

---

[8] For a recent study on the role of the *Agouti* allele in the development and evolution of color patterns in both mainland and beach subspecies of *Peromyscus,* see Manceau et al. (2011). For a review of the types of selection acting on color polymorphisms see McKinnon and Pierotti (2010). Also see Endler (1986).

[9] E. B. Rosenblum et al., 2009. Molecular and functional basis of phenotypic convergence in white lizards at White Sands. *Proc. Natl. Acad. Sci. USA,* **107**, 2113–2117.

Endler, J. A., 1986. *Natural Selection in the Wild.* Princeton University Press, Princeton, NJ.

Grant, B. S., 2004. Allelic melanism in American and British peppered moths. *J. Hered.,* **95**, 97–102.

Hall, B. K., R. D. Pearson, and G. B. Müller (eds.), 2004. *Environment, Development, and Evolution; Toward a Synthesis.* MIT Press, Cambridge, MA.

Hallgrímsson, B., and B. K. Hall (eds.), 2005. *Variation: A Central Concept in Biology.* Elsevier/Academic Press, Burlington, MA.

Hey, J., 1999. The neutralist, the fly, and the selectionist. *Trends Ecol. Evol.,* **14**, 35–38.

Hoekstra, H. E., R. J. Hirschmann, R. A. Bundey, et al., 2006. A single amino acid mutation contributes to adaptive color pattern in beach mice. *Science,* **313**, 101–104.

Hooper, J., 2002. *Of Moths and Men. Intrigue, Tragedy and the Peppered Moth.* Fourth Estate, London.

Kettlewell, H. B. D., 1973. *The Evolution of Melanism.* Clarendon Press, Oxford, UK.

Kimura, M., 1979. The neutral theory of molecular evolution. *Sci. Am.,* **241**, 94–104.

Kimura, M., 1983. *The Neutral Theory of Molecular Evolution.* Cambridge University Press, Cambridge, UK.

Linnen, C. R., E. P. Kingsley, J. D. Jensen, and H. E. Hoekstra, 2009. On the origin and spread of an adaptive allele in deer mice. *Science,* **325**, 1095–1098.

Majerus, M. E. N., 1998. *Melanism: Evolution in Action.* Oxford University Press, Oxford, UK.

Manceau, M., V. S. Domingues, R. Mallarino, and H. E. Hoekstra, 2011. The developmental role of Agouti in color pattern evolution. *Science,* **331**, 1062–1065.

Martin, T. E., 1996. Fitness costs of resource overlap among coexisting bird species. *Nature,* **380**, 338–340.

McKinnon, J. S., and M. E. R. Pierotti, 2010. Color polymorphism and correlated characters; genetic mechanisms and evolution. *Mol. Ecol.,* **19**, 5101–5125.

Milkman, R. D., 1973. Electrophoretic variation in *Escherichia coli* from natural sources. *Science,* **182**, 1024–1026.

Nei, M., 2005. Selectionism and neutralism in molecular evolution. *Mol. Biol. Evol.,* **22**, 2318–2342.

Nevo, E., 1978. Genetic variation in natural populations: patterns and theory. *Theor. Popul. Biol.,* **13**, 121–177.

Pigliucci, M., 2001. *Phenotypic Plasticity: Beyond Nature and Nurture.* The Johns Hopkins University Press, Baltimore and London.

Schlichting, C. D., and M. Pigliucci, 1998. *Phenotypic Evolution: A Reaction Norm Perspective.* Sinauer Associates, Sunderland, MA.

Stökl, J., A. Strutz, A. Dafni, et al., 2010. A deceptive pollination system targeting drosophilids through olfactory mimicry of yeast. *Curr. Biol.,* **20**, 1846–1852.

van't Hoff, A. E., N. Edmonds, M. Dalikpvá, et al., 2011. Industrial melanism in British peppered moths has a singular and recent mutational origin. *Science,* **332**, 958–960.

Varki, A., 2006. Nothing in glycobiology makes sense, except in the light of evolution. *Cell,* **126**, 841–845.

Waddington, C. H., and E. Robertson, 1966. Selection for developmental canalization. *Genet. Res. Camb.,* **7**, 303–312.

West-Eberhard, M. J., 2003. *Developmental Plasticity and Evolution.* Oxford University Press, Oxford, UK.

Wilkins, A. S., 2003. Canalization and genetic assimilation. In *Keywords and Concepts in Evolutionary Developmental Biology.* B. K. Hall and W. M. Olson (eds.). Harvard University Press, Cambridge, MA, pp. 23–30.

PART 6

Sources of Variation in Individuals and in Populations

© Photos.com

CHAPTER

16

Chromosomes and Genomes as Sources of Individual Variation

CHAPTER SUMMARY

Evolutionary potential and genetic variation are two sides of every evolutionary coin. Without variation there would be no evolutionary change. Without evolution, variation would be unnecessary. How variation at one level (the genotype) relates to variation at other levels (e.g., structures/behavior—the phenotype) is a major unresolved question in evolutionary biology. Within individuals, variation at the genetic level arises from multiple alleles, change in genome, chromosome or gene numbers, and mutations and modifications of gene regulation (END BOX 16.1). Nucleotide sequences (transposons) that promote their own movement (transposition) between genomes are also important sources of variation and regulation of gene activity discussed in this chapter and in END BOX 16.2.

Prokaryotic cells acquire new genetic information from (i) gene/chromosome duplications, (ii) mutation, (iii) changes in gene regulation, (iv) transposons, and (v) horizontal gene transfer. Add the origin of eukaryote organelles (and their genes) by endosymbiosis to (i) to (v) and we have six major means by which eukaryotic cells acquire new genetic information. This chapter outlines four critical questions concerning the origin and maintenance of variation, and then elaborates one source of variation: genome duplication and variations based in changes affecting entire chromosomes.

An important issue raised by gene duplications, especially whole genome duplications, is whether such events facilitated the origin of novel organs, organisms, and/or adaptive radiation. Duplication of the entire genome facilitated the origination of some lineages of organisms but not others. New species of plants arose by duplication of chromosome number; various species of wheat cultivated for at least 30,000 years are a well-researched example of speciation by chromosome duplication. So too are land plants and bony fish.

INTRODUCTION

Study of the mechanisms producing variation is important because variation is the raw material upon which natural selection acts.

Natural selection acts on existing phenotypic variation. One of the leading twentieth century students of evolution, Ernst Mayr, saw this early in his career and pursued it over 60 years.

"It is amazing to what extent variation in natural populations has been neglected in the study of evolution. Amazing, because natural selection would be meaningless without variation. This conclusion gave me the idea to consider the production of variation as a step in the process of natural selection."[1]

How variation arises in genes and populations is central to evolution but is one of the least understood sets of evolutionary mechanisms. We say "sets" because variation can arise in a number of different ways, which are outlined as four questions below.

1. What is the relationship between genetic variation of the **genotype**, which is transmitted from generation to generation, and variation of the **phenotype**, which is the subject to selection? This question is a major preoccupation of evolutionary developmental biology (evo-devo).[2]
2. What are the mechanisms by which **mutations and modifications of gene regulation** serve as sources of variation? What are their respective roles in maintaining the phenotype, modifying the phenotype, and/or in the origin of new phenotypes? This challenging area seeks to understand why some structures/organisms/species change more than others over the course of their evolution.
3. What **other sources of variation are available to populations?** Sources include gene flow from other populations and random drift of genes within a population.
4. What are the **ecological** and **developmental** determinants of phenotypic variation among species? This topic includes such questions as how body size, geographic range, home range size, niche width, lifespan, environmental stress, population size, and density affect the tendency of populations to exhibit phenotypic variation.

As many as three quarters of the gene loci in most populations consist of more than one allele. With this great *reservoir of variation*, populations can respond to evolutionary pressures without new variants having to arise by mutation. Additional sources of genetic variation come from (1) alterations in chromosome number (especially in plants), either via changes in entire sets of chromosomes or in individual chromosomes (which can include duplications of an entire genome); (2) deletions, duplications, inversions, or translocations of nucleotide sequences that affect one or more genes; or (3) from gene duplications and from modifications in the regulation of gene function, at DNA, RNA, or protein levels. Let's begin with variation at the level of entire chromosomes.

[1] Cited from the Foreword to Hallgrímsson and Hall, 2005, p. xvii. This edited volume, *Variation: A Central Concept in Biology,* is a recent study of the nature and importance of variation.
[2] For an excellent account of how classes of variation were brought under the two headings of genotype and phenotype, see N. Roll-Hansen (2009). Sources of Wilhelm Johannsen's genotype theory. *J. Hist. Biol.,* **42,** 457–493.

VARIATION IN CHROMOSOME NUMBER, INCLUDING WHOLE GENOME DUPLICATION

Variations in chromosome number are of two major kinds

- changes in the **number of entire sets** of chromosomes (**genome duplication**), and
- changes in the **numbers of single chromosomes** within a set.

Repetitive doubling can take **chromosome number** well beyond the diploid condition, an outcome known as **polyploidy**, meaning many times the normal ploidy; a triploid is 3*n*, a tetraploid 4*n*, and so on. Polyploidy provides evolutionary advantages with respect to cell and organism size, stability of the genome, and tissue/organ-specific gene expression. As discussed below, changes in chromosome number—under very different circumstances—led to the evolution of wheat and of tobacco. Repeated doubling of chromosome number has lead to lineages of vascular plants with greater than 80-fold ploidy.[3] Indeed, somewhere between 40 and 70 percent of all plant species are polyploid.

A larger number of chromosomes per cell may allow more genetic recombination and therefore more genetic variation, which would be advantageous in changing environments (Wagner, 2011). Smaller chromosome numbers may allow genetic combinations to persist, and so be associated with long periods of occupation of a specialized environment. As measured either by number of bases or amount of DNA, the largest genome belongs to a flowering plant, *Paris japonica*, with c. 150,000 Mbp (i.e., 150 billion base pairs) and 152.34 picograms of DNA. The smallest is that of a mammalian gut parasitic fungus, *Encephalitozoon intestinalis*, which causes microsporidiosis in humans, with 2.25 Mbp and 0.0023 pg of DNA.[4]

Chromosome number has been **reduced** in some lineages. For example, in some populations of European wild mice, *Mus musculus*, the normal number of 20 pairs of chromosomes has been reduced to as few as 12 pairs. In contrast, conservation of chromosome number has been a persistent feature in other mammalian lineages; extant species of the 30-My-old lineage that includes Asian (Bactrian) and African (dromedary) camels as well as South American guanaco, vicuna, llama, and alpaca, all have the same number (74) of morphologically similar chromosomes.[5]

Variation can also be introduced **when a single chromosome is added or deleted**. In animals, such changes often lead to abnormalities. In plants, however, such changes—especially additions—are common, more tolerated than chromosomal variation in animals, and often associated with the origin of new strains or even new species, as discussed below. Variation in chromosome number between similar animal species sometimes may be associated with the origin of adaptive features, although this has proven hard to demonstrate. Stressful ecological conditions such as periodic aridity and other unpredictable hardships *correlate* with chromosome numbers in Israeli and Turkish populations of the mole rat, *Spalax*. It is hypothesized, but not proven, that increase in chromosome number provided the potential for increased genetic diversity and enhanced potential to respond to ecological variation.[6]

An important issue raised by the existence of gene duplications, especially whole genome duplications, is whether such events facilitated the origin of novel organs, organisms, and/or adaptive radiation.[7] Three examples of genome duplication, all from the plant kingdom are considered below.

---

[3] For chromosome doubling (polyploidy) in plants, see Adams and Wendel (2005), Otto (2007), and Wood et al. (2009).

[4] *P. Japonica* arose from two rounds of hybridization events involving four species (J. Pellicer, M. F. Fay, and I. J. Leitch, 2010. The largest eukaryotic genome of them all? *Bot. J. Linn. Soc.*, **164**, 10–15).

[5] M. B. Qumsiyeh, 1994. Evolution of number and morphology of mammalian chromosomes. *Heredity*, **85**, 455–465.

[6] E. Nevo et al., 1994. Chromosomal speciation and adaptive radiation of mole rats in Asia Minor correlated with increased ecological stress. *Proc. Natl. Acad. Sci. USA*, **91**, 8160–8164.

[7] For duplication, especially of homeobox (*Hox*) genes and the origin of major groups, see Bharathan et al. (1997), Erwin and Davidson (2002), Hughes and Kaufman (2002), and Carroll et al. (2005).

THE EVOLUTION OF WHEAT

The generation of polyploidy as a consequence of hybridization is a common mode of evolution in many plants, including mosses, fruit trees, tomato plants, and corn. Polyploids can arise when different species interbreed naturally or are crossed under field or laboratory conditions (Soltis and Soltis, 1995). A classic example of both natural and forced interbreeding is seen in the evolution of **wheat** (*Triticum aestivum*). Evolution of wheat following chromosome duplication is linked to the origin and expansion of human agriculture. Consequently, the evolution of wheat has attracted many researchers from fields as diverse as genetics, plant breeding, agriculture, archaeology, and anthropology.

As the story goes, at least 30,000 years ago in the Fertile Crescent of southwest Asia, a natural hybrid formed between two grasses, *Triticum monococcum* (wild einkorn) and a species of *Aegilops* (goat grass). Each had 14 pairs of chromosomes. Hunter-gatherers harvested the seeds of this new 28-chromosome plant for millennia. We know it as *Triticum dicoccoides* (**wild emmer**). Around 10,000 years ago, by which time the Ice Age had ended, humans began cultivating wild emmer. By 9,500 years ago, harvesting of the best plants year after year selected desirable qualities and led to a new form of emmer, **cultivated emmer**, now regarded as a new species, *T. dicoccum*.

Cultivated emmer was an important crop for 7,000 years, spreading throughout the Near East and into Egypt. When it reached an area southwest of the Caspian Sea around 9,000 years ago, cultivated emmer came into contact and interbred with another species of goat grass (*Aegilops squarrosa*) with 14 pairs of chromosome. The new hybrid, known as **spelt** (*Triticum spelta*), had 42 chromosomes (28 + 14; **FIGURE 16.1**). Neither emmer nor spelt was easy to harvest, but around 8,500 years ago, a fortuitous mutation changed the nature of the spike or ear, producing a shell that would allow threshing. An unfortunate side effect was that the plant, a new species, *Triticum aestivum* (**bread wheat**), could no longer readily be propagated naturally.

The FERTILE CRESCENT *is a half-moon-shaped region of some 500,000 square miles adjacent to the Mediterranean Sea. Extending from the Nile River Valley to the valleys of the Tigris and Euphrates rivers, it is the site of the origin of agriculture.*

Many grocery stores carry the grain TRITICALE. *This, too, is a new species (Triticale hexaploide) produced by crossing wheat (Triticum) and rye (Secale) to produce a sterile polyploid hybrid with double the chromosome numbers of wheat and rye.*

© Rob/Dreamstime.com

FIGURE 16.1 Spelt (hybrid wheat) to show the habit and nature of the seed heads.

About 7,000 years ago, cultivated emmer evolved in a second direction when new mutations resulted in free-threshing grain, facilitating the spread of agriculture. Having evolved directly from spelt, *Triticum durum* (**macaroni** or **durum wheat**) also has 28 chromosomes. Since then, wheat has undergone considerable change. For millennia, farmers sowed seed they had collected the previous year, creating in the process a multitude of "land races" adapted to local conditions. More recently, agronomists have bred improved varieties of wheat, enormously raising yields.

BREEDING THE EVIL WEED *NICOTIANA TABACUM* IN THE LABORATORY

Another interesting study in the evolution of new species is found in the first (perhaps the only) species produced in the laboratory/greenhouse. It is the product of a cross between two species of tobacco plants: *Nicotiana tabacum* with 48 chromosomes and *N. glutinosa* with 24 chromosomes. The hybrid created by the two plants (**FIGURE 16.2**) was sterile;

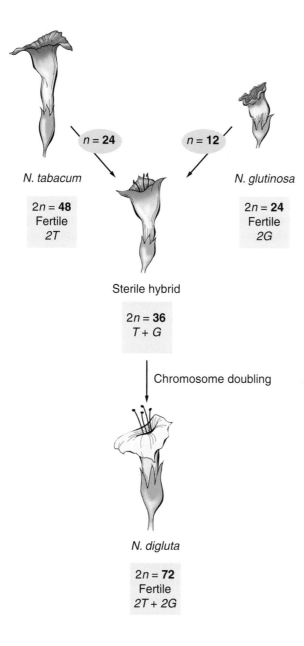

FIGURE 16.2 Flowers and diploid gene numbers of the tobacco plants *Nicotiana tabacum*, *N. glutinosa*, their sterile diploid hybrid, and the new species, *N. digluta*, formed in association with chromosome doubling in the sterile hybrid.

N. tabacum

$n = 24$

2n = **48**
Fertile
2T

N. glutinosa

$n = 12$

2n = **24**
Fertile
2G

Sterile hybrid

2n = **36**
T + G

Chromosome doubling

N. digluta

2n = **72**
Fertile
2T + 2G

the unequal number of chromosomes contributed by each species did not allow normal chromosome pairing during meiosis. However, the hybrid tobacco plants could be propagated by vegetative cuttings.

The next event could neither have been anticipated nor planned. Eventually, and by chance, a chromosome-doubling event produced a fertile plant with 72 chromosomes and normal meiosis. The resulting plant, a new species of tobacco, *Nicotiana digluta*, was self-crossed and found to be fertile.

DOMESTICATED APPLES

Publication in October 2010 of the draft genome of "Golden Delicious" apples, a common variety of the domesticated apple (*Malus domestica*), has provided evidence for the changes associated with a genome-wide duplication that occurred some 50 Mya, transforming a nine-chromosome ancestor to a 17-chromosome descendant ($9 \times 2 = 18$ with fusion of two chromosome = 17; Velasco et al., 2010). Phylogenetic analysis revealed the ancestor of the cultivated apple to be the wild apple *Malus sieversii*, which is native to the mountains of Central Asia but endangered in its native habitat.

Furthermore, flower and fruit development in plants is under the control of MADS-box genes, a family of transcription factors widespread in the three major lineages of multicellular organisms.[8] Genome duplication in apples resulted in the evolution of a 15-member subfamily of MADS-box genes that regulates development of—and presumably facilitated the evolution of—the pome, which is the specialized fruit found in the Pyreae tribe to which apples belong.

OTHER CHANGES IN THE PHENOTYPE OF CHROMOSOMES

A variety of other changes in chromosome structure are shown in FIGURE 16.3. Because these are changes in the appearance of the chromosome, we can speak of them as the **phenotype of the chromosome**, just as we speak or the phenotype of the individual when comparing characters.

Deletions and **deficiencies** remove chromosomal material (Figure 16.3a). If functional genes are removed, deletions can be harmful in diploids and haploids, but not necessarily in polyploids, where such genes may be present on more than one chromosome. Gene deletions are involved in several forms of muscular dystrophy in humans.

Duplications are segments of extra chromosome material originating from duplicated sequences within the genome. Often they result from unequal crossing-over during chromosome pairing (Figure 16.3b and FIGURE 16.4). Duplications have been common during evolution, resulting in the formation of numerous **gene families** of similar or identical genes in many species of all multicellular organisms, as in the MADS-box genes discussed above. Families of myoglobin and globin genes are discussed in this context later in the chapter. As discussed above, entire genomes have been duplicated in various lineages and at various times during evolution (Van de Peer et al., 2009).

Inversions are reversals in chromosomal gene order (Figure 16.3c, d). Genes included within an inversion tend to remain together as a nonrecombinant block, called

---

[8] The acronym MADS is taken from the first letters of the first four genes in the family to be synthesized, which were MCM1 (yeast, *Saccharomyces cerevisiae*), AGAMOUS (watercress, *Arabidopsis thaliana*), DEFICIENS (snapdragon *Antirrhinum majus*), and SRF (humans, *Homo sapiens*).

Chromosome structural change

Meiotic pairing between changed and unchanged chromosomes in heterozygote

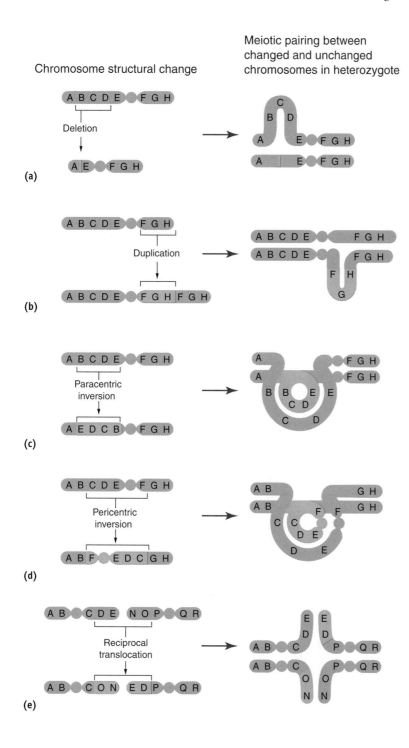

FIGURE 16.3 Major kinds of structural chromosomal changes (red) and their effects on chromosome pairing in heterozygotes carrying both changed and unchanged homologues. (See text for details.)

by some a **supergene**. Because they are linked genetically the loci in a supergene are inherited together, although they may be regulated independently.

Translocation moves genes from one chromosome to another (Figure 16.3e). Translocations may change the number and structure of chromosomes and thereby introduce variation. Among Asiatic muntjac deer, the Indian muntjac, *Muntiacus muntjac*, has only three pairs of large chromosomes, the black muntjac, *Muntiacus crinifrons*, four pairs, and the Chinese muntjac, *Muntiacus reevesi*, 23 pairs (**FIGURE 16.5**). Relative amounts of DNA in the three species, however, are about the same, the large muntjac chromosomes arising from successive translocations combining the smaller muntjac chromosomes. Translocation between human chromosomes is discussed in the following section.

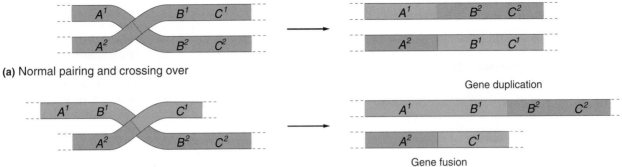

(a) Normal pairing and crossing over

(b) Unequal pairing and crossing over

FIGURE 16.4 The results of equal and unequal crossing-over for three gene segments on a chromosome. **(a)** When pairing between homologous sections on two chromosomes is equal, the crossover products have the same amounts of chromosomal material. **(b)** When pairing between the two chromosomes is unequal, one of the crossover products carries a gene duplication (the *B* gene segment in this illustration), and the other product shows a fusion between gene segments (*A*–*C*) that were formerly separated by the intervening *B* segment.

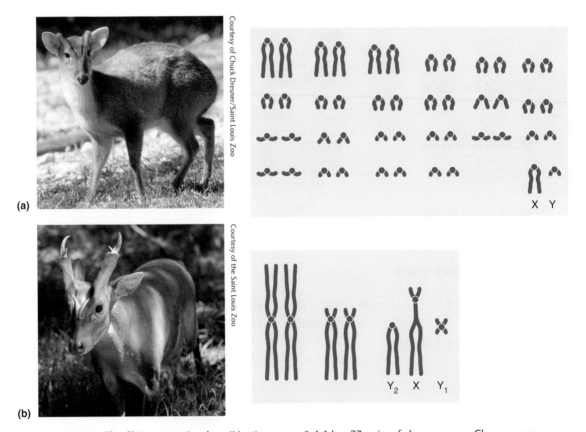

FIGURE 16.5 The Chinese muntjac deer (*Muntiacus reevesi*) **(a)** has 23 pairs of chromosomes. Chromosome number has been reduced enormously in the congeneric species *Muntiacus muntjac*, the Indian muntjac deer **(b)**, in which females have two pairs of autosomes and one pair of sex chromosomes (the lowest known chromosome number for any mammal) and males have three pairs of autosomes and a Y chromosome. (See Ellegren (2011) for the evolution of sex chromosomes.)

[Chromosome spreads are adapted from Austin, C. R., and R. V. Short, 1976. *The Evolution of Reproduction*. Cambridge University Press, Cambridge, UK.]

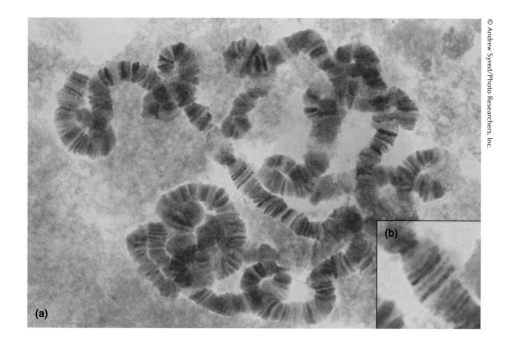

FIGURE 16.6 Phase contrast image of *Drosophila melanogaster* polytene chromosomes **(a)**. The end of the X-chromosome is marked with an arrow. Bands and inter-band regions are shown in **(b)**.

CHROMOSOMAL EVOLUTION IN *DROSOPHILA* AND PRIMATES

In those instances in which linear sections of chromosomes can be identified by their distinctive bandings, chromosomal evolution can be charted in considerable detail. Interest exists at two levels: the evolution of chromosomes themselves as characters of the organism, and how chromosomal evolution influences the evolution of other characters (Dobzhansky, 1944; Qumsiyeh, 1994).

In species of *Drosophila* and other similar insects, the chromosomes of salivary gland cells and other tissues have replicated many times, but the replicates remain attached. The resulting giant chromosomes, called **polytene chromosomes**, are enlarged enormously and show highly detailed banding configurations that enable even minor chromosomal changes to be identified (**FIGURE 16.6**). Geneticists have documented practically all the chromosomal changes in the evolution of hundreds of these fruit fly species.[9]

Although polytene chromosomes are absent in many organisms, chromosomal staining techniques enable detailed comparisons, even between relatively small mammalian chromosomes. The G-banding technique illustrated in **FIGURE 16.7** was used to compare chromosomes from modern humans, chimpanzee, gorilla, and orangutan. These bandings show that some chromosomes—numbers 6, 13, 19, 21, 22, and X—are almost identical in all four species. Difference in chromosome number between humans ($n = 23$) and apes ($n = 24$) derives from a fusion event that combined homologues of two chromosomes found in chimpanzees into the single chromosome number 2 in humans (Figure 16.7). This fusion must have occurred after the human line separated from a human–chimpanzee common ancestor. These banding arrangements provide a source of evidence that modern humans have a closer evolutionary relationship with chimpanzees than with gorillas and a more distant one with orangutans.

[9] P. Lawrence, 1992. *The Making of a Fly: The Genetics of Animal Design*. Blackwell Scientific, Oxford, UK; G. Maroni, 1993, *An Atlas Of* Drosophila *Genes: Sequences And Molecular Features*. Oxford University Press, Oxford, UK.

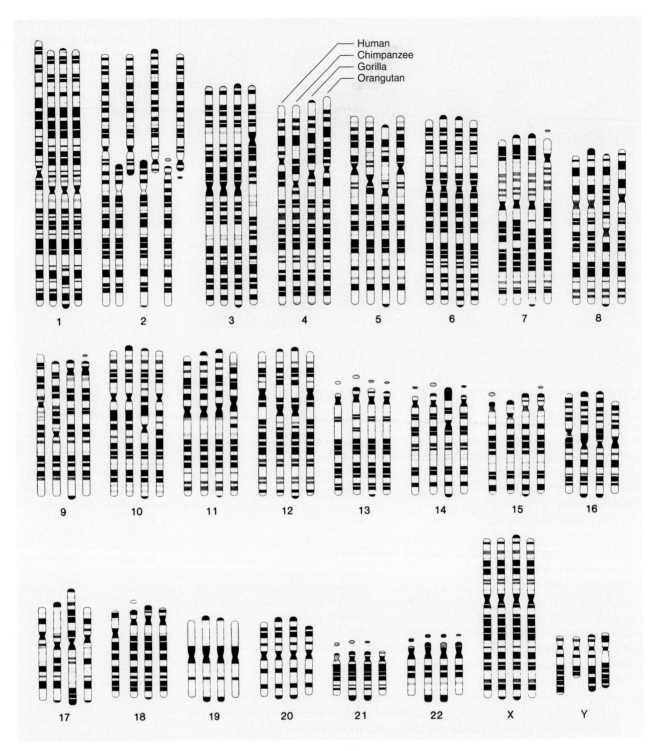

FIGURE 16.7 Banding arrangements of the chromosomes of humans, chimpanzees, gorillas, and orangutans, in respective order from left to right for each chromosome. These banding arrangements indicate that humans have a closer evolutionary relationship with chimpanzees than with gorillas and a more distant one with orangutans.

[Reproduced from J. J. Junis and O. Prakash, *Science*, 1982. **215**, 1525–1530. Reprinted with permission from AAAS.]

GENE DUPLICATION AND DIVERGENCE

Observations of the sequence similarities of different hemoglobin chains strongly suggest that, rather than arising from different genes that converged in sequence and function, hemoglobin and the other **globin genes**, such as the gene for myoglobin, arose as duplications of an original globin-type gene, followed by divergence. Terminology has been adopted to reflect two different types of gene duplication:

- **paralogous genes (paralogues)** for duplications within a species (for example, α, β, and γ hemoglobins in humans), and
- **orthologous genes (orthologues)** for genes shared between species resulting from shared species ancestry (for example, α hemoglobins genes in horses and humans).

The time during the evolution of globin genes when the duplications occurred can be calculated from amino acid differences between the molecules; the greater the amino acid differences between any two chains, the further back in time the duplication occurred. FIGURE 16.8 portrays the genetic phylogeny of the five globins in terms of the numbers of nucleotides necessary to account for the amino acid differences. Just as we can calculate how long ago mutations occurred, we can infer when each of the duplications occurred (Figure 16.8).

Three data sets enable us to order the stages in the evolution of these five globin genes: (1) The myoglobin chain with distinctive amino acids at more than 100 sites differs most from all other globin genes. (2) The α- and β-chains of hemoglobin differ from each other at 77 amino acid sites. (3) The β-chain of hemoglobin differs from the γ-chain at 39 sites but differs from the δ-chain at only 10 sites. As the most divergent, the gene for myoglobin arose from an early duplication. A later duplication produced two α hemoglobin genes, one of which evolved into the β hemoglobin gene. Because they differ least, the separation between β and γ hemoglobin chains arose from the most recent duplication (FIGURE 16.9). Subsequent duplications of a different type, known as **in-tandem** (side-by-side) **duplications,** resulted in two copies of the α gene (α1 and α2). Several tandem duplications of the β gene produced seven β genes in what is known as a **gene or multigene family** (Figure 16.9).

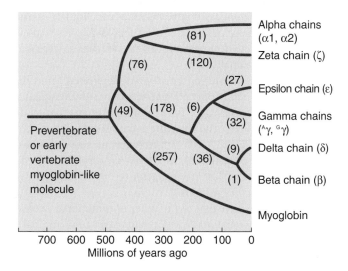

FIGURE 16.8 Phylogenetic relationships between globin-type proteins in humans to show the estimated times when the proteins diverged from each other. Estimated number of nucleotide replacements necessary to produce the observed amino acid changed in each branch of the lineage is given in parentheses.

[Adapted from Strickberger, M. W. *Genetics, Third Edition.* Macmillian, New York, 1985.]

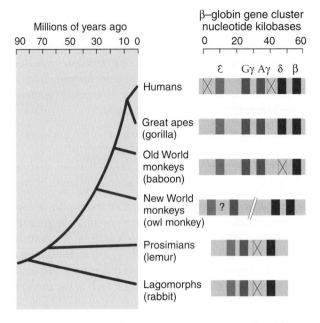

FIGURE 16.9 Clustered organization of the β-globin-type genes in five different primates and in a rabbit, along with an evolutionary tree of the genes. Genes in each of the clusters are linked together on the same chromosome. Each gene, denoted by a small rectangle, is transcribed from left to right. Genes responsible for embryonic and fetal development are on the left (*lighter shading*), genes that produce adult β-globins are on the right (*darker shading*). Genes marked with crosses are nonfunctional pseudogenes.

[Adapted from Strickberger, M. W. *Genetics, Third Edition*. Macmillian, New York, 1985.]

FUNCTIONAL DIVERGENCE WITHIN GENE FAMILIES

Sometimes, all the members of a gene family remain active and retain similar or identical functions. Sometimes the duplicated gene becomes functionless but is retained as a **pseudogene** (Figure 16.9); deleterious mutations in gene duplicates responsible for causing pseudogenes were as likely to occur among early vertebrates as were mutations allowing duplicates to diverge functionally.[10]

Most significantly from an evolutionary view are duplicated genes that evolve new functions, a process often termed **neofunctionalization** or subfunctionalization. A classic example is found in genes coding for the crystallin proteins that lend transparency to vertebrate eye lenses. Although gene structure is conserved, each of the 10 crystallins known in animals is associated with a different enzyme and so a different function. Two independent instances of domestication of carp (*Cyprinus carpio*) involved similar mutations in the two copies of the fibroblast growth factor receptor gene *Fgfr-1*, each of which has a separate function (Rohner et al., 2009).[11]

Duplication of a gene can have one of a number of evolutionary consequences. The second gene (sister copy) may remain as a redundant copy of the original gene. Because it is redundant, the sister copy may escape selection and deteriorate (or not, as seen in the example below). Slight changes in the sister copy may lead it to perform the same

---

[10] A mechanism for the reactivation of pseudogenes (codon reassignment) by a nonsense suppressor mutant transfer RNA has been proposed by L. J. Johnson, 2010. Pseudogene rescue: an adaptive mechanism of codon reassignment. *J. Evol. Biol.*, **23**, 1623–1630.

[11] Fibroblast growth factor is one of a large number of growth factors known, each of which binds to a specific receptor on the cell surface.

function as the original gene but in a different region of the body. Finally, the sister copy may acquire a new function through neofunctionalization.

An elegant example of redundancy and deterioration of the original gene has been demonstrated for the evolution of the *feminizer* (*fem*) gene in species of the honey bee genus *Apis* (Hasselmann et al., 2010). The original function of *fem* is in sex determination. Tandem (side-by-side) duplication of *fem* resulted in the formation of a new sister gene, *sex determiner* (*csd*). The sister gene provided an independent and competing mechanism of sex determination; indeed *csd* interferes with the functioning of *fem*. Purifying selection on *csd* constrained the evolution of *fem*, which has accumulated a higher than usual level of nucleotide variation. Although this appears deleterious for the sex-determining function of *fem* (and it is) it effectively increases the generic variation in the species, which is of longer evolutionary advantage. You can have your cake and eat it too.

How long duplicates take to diverge in function depends, of course, on how many amino acid substitutions are necessary, but also on changes occurring in other genes and in the adaptation/evolution of the organism itself. A striking example of the speed of evolutionary change is the single amino acid mutation that converts the metabolic enzyme lactate dehydrogenase (LDH) to another enzyme, malate dehydrogenase (MDH). Here the functional change would be synchronous with the amino acid change. Most functional changes are much slower because changes in many amino acids are required for the structure and so the function of the gene product to change.

Now we move to a radically different source of genetic variation, the transfer of genetic information between organisms.

TRANSPOSONS

Another source of genetic variation and evolutionary change in prokaryotes and eukaryotes is **transposons**, nucleotide sequences that promote their own transposition within a genome. The history of Barbara McClintock's discovery of transposons says much about the difficulty of gaining acceptance for ideas ahead of their time (End Box 16.1).

A transposon produces special transposase enzymes that allow it to insert copies of itself into various target sites in an organism's nuclear genome. For example, the 768-base pair *E. coli* insertion sequence 1 (IS1) makes staggered cuts at each side of a nine-nucleotide base pair sequence in the maize genome. A copy of the IS1 sequence is then inserted into the gap these cuts produce (FIGURE 16.10).

Transposons can move from one prokaryotic species to another and from prokaryotic to eukaryotic cells. They can also move between eukaryotic organisms; the *P* transposon, originally a transposable element common to *Drosophila willistoni*, spread to *Drosophila melanogaster* about 50 years ago and now is found in all wild populations of this species. The *P* transposon was acquired horizontally; the two species of *Drosophila* differ significantly in their nuclear genes—indicating a separation of about 20 My between the two species—yet their *P* transposons differ by only a single nucleotide. This evidence shows that the transposons are of much more recent origin than are the nuclear genes. More recent evidence shows that *P* elements have repeatedly crossed species barriers in *Drosophila* and spread to related genera, providing new sources of genetic variation.[12] Evidence now supports transposon-mediated horizontal gene transfer between animals, between the mitochondrial genes of plants, and, most recently, between the nuclear genes of millet and rice.

An important advantage transposons gain by horizontal transfer is circumventing the barriers of reproductive isolation among species, thereby escaping inevitable extinction when species die out. As a consequence, we might expect to find large numbers of individual

---

[12] S. B. Daniels et al., 1990. Evidence for horizontal transmission of the P transposable element between *Drosophila* species. *Genetics*, **124**, 339–355; M. G. Kidwell, 1994. The evolutionary history of the P family of transposable elements. *J. Hered.*, **85**, 339–346; J. C. Silva and M. G. Kidwell, 2000. Horizontal transfer and selection in the evolution of P elements. *Mol. Biol. Evol.*, **17**, 1542–1557; X. Diao et al., 2006. Horizontal transfer of a plant transposon. *Public Lib. Sci. Biol.*, **4**, 119–128.

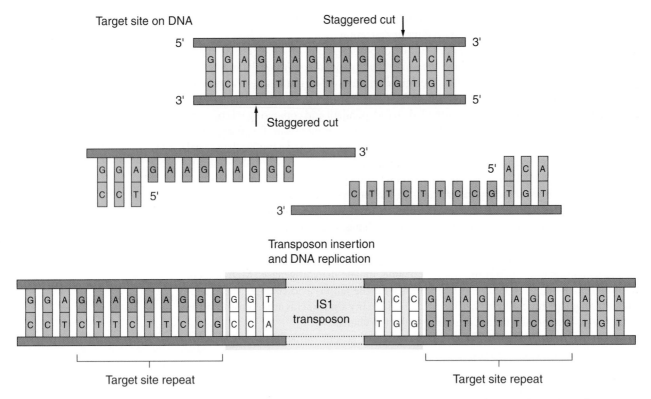

FIGURE 16.10 Mode of insertion of the *E. coli*-derived IS1 transposon into the maize genome. The transposon recognizes the nine-base pair DNA sequence, shown in brown, and cleaves it at the positions marked by the arrows, creating a gap. The IS1 transposon (yellow) inserts into the gap, and DNA sequences are synthesized complementary to the former single-strand sections of the target site. This process produces identical but inverted nine-base pair repeats at each end of the transposon.

[Adapted from Strickberger, M. W. *Genetics, Third Edition*. Macmillian, New York, 1985.]

RESTRICTION ENZYMES *cleave DNA into fragments at specific sites, allowing comparisons of small pieces of DNAs from different species. Different restriction enzymes recognize specific sequences and so produce distinct DNA cleavage products.*

transposons within a genome, and in some cases, we do. In primates, for example, more than one million copies of a 300 base-pair sequence with transposon-like features—named an *Alu* sequence because it is recognized by the Alu restriction enzyme—are present in each diploid human cell. The human genome contains many hundreds of transposable elements in the Alu families that affect gene expression, genetic recombination, and unequal crossing over of chromosomes, some of which entered the genomes of human ancestors 4 Mya.[13]

Smaller repetitive sequences are widely prevalent in various eukaryotes. But this is not always so in all organisms. *Drosophila* genomes carry only about 30 to 50 copies of the *P* element and a similar number of another family of transposons responsible for mutation of a gene controlling eye color (Silva and Kidwell, 2000; Vu and Nuzhdin, 2011). Regulatory agents within transposons and miRNAs (End Box 16.1) control transposon number and so limit their mutagenic effects, a feature that may have been selected to ensure survival of their hosts, and therefore their own survival.

A surprising recent discovery is that a heat-shock protein, Hsp90, in *Drosophila* interacts with piRNAs (which are RNAs that regulate the insertion of transposable elements into genomes; End Box 16.1) to activate transposons and increase genetic variation.[14] Heat shock proteins play a role in modulating the effects of environmental influences to reveal previously hidden and unexpressed genetic variation.

[13] M. A. Batzer et al., 1994. African origin of human-specific polymorphic Alu insertions. *Proc. Natl. Acad. Sci. USA*, **91**, 12288–12292; R. J. Britten, 2010. Transposable elements insertions have strongly affected human evolution. *Proc. Natl. Acad. Sci. USA*, **107**, 18845–19948.
[14] V. Specchia et al., 2010. Hsp90 prevents phenotypic variation by suppressing the mutagenic activity of transposons. *Nature*, **463**, 662–665.

KEY TERMS

alternate splicing	neofunctionalization
chromosomes	orthologous genes (orthologues)
cis-regulation	paralogous genes (paralogues)
deletion (genetics)	phenotype
duplication (genetics)	Piwi-interacting RNA (piRNA)
ecology	polyploidy
exon	polytene chromosomes
exon shuffling	pseudogene
gene family	RNA editing
gene regulation	RNA interference (RNAi)
genome duplication	small interference RNA (siRNA)
genotype	translocation
globin genes	transposons
intron	*trans*-regulation
intron removal	*trans*-regulatory elements
inversion (genetics)	wheat (*Triticum aestivum*)
microRNA (miRNA)	wild emmer (*Triticum dicoccoides*)
modularity	

EVOLUTION ON THE WEB

Explore evolution on the Internet! Visit the accompanying website for *Strickberger's Evolution, Fifth Edition,* at **go.jblearning.com/Evolution5eCW** for exercises and links relating to topics covered in this chapter.

REFERENCES

Adams, K. L., and J. F. Wendel, 2005. Polyploidy and genome evolution in plants. *Curr. Opin. Plant Biol.,* **8**, 135–141.

Bharathan, G., B.-J. Janssen, E. A. Kellogg, and N. Sinha, 1997. Did homeodomain proteins duplicate before the origin of angiosperms, fungi, and metazoa? *Proc. Natl. Acad. Sci. USA,* **94**, 13749–13753.

Carroll, S. B., J. K. Grenier, and S. D. Weatherbee, 2005. *From DNA to Diversity. Molecular Genetics and the Evolution of Animal Design,* 2nd ed. Blackwell Publishing, Malden, MA.

Dobzhansky, Th., 1944. Chromosomal races in *Drosophila pseudoobscura* and *D. persimilis.* Carnegie Inst. Wash. Publ. No. 554, Washington, DC, pp. 47–144.

Ellegren, H. 2011. Sex-chromosome evolution: recent progress and the influence of male and female heterogamy. *Nat. Rev. Genet.,* **12**, 157–166.

Erwin, D. H., and E. H. Davidson, 2002. The last common bilaterian ancestor. *Development,* **129**, 3021–3032.

Hallgrímsson, B., and B. K. Hall (eds.), 2003. *Variation: A Central Concept in Biology.* Elsevier/Academic Press, Burlington, MA.

Hasselmann, M., S. Lechner, C. Schulte, and M. Beye, 2010. Origin of a function by tandem gene duplication limits the evolutionary capability of its sister copy. *Proc. Natl. Acad. Sci. USA,* **107**, 13378–13383.

Hughes, C. L., and T. C. Kaufman, 2002. *Hox* genes and the evolution of the arthropod body plan. *Evol. Devel.* **4**, 459–499.

Mayr, E., 2001. *What Evolution Is.* Basic Books, New York.

Otto, S. P., 2007. The evolutionary consequences of polyploidy. *Cell,* **131**, 452–462.

Qumsiyeh, M. B., 1994. Evolution of number and morphology of mammalian chromosomes. *Heredity,* **85**, 455–465.

Rohner, N., M. Bercsényi, L. Orbán, et al., 2009. Duplication of fgfr1 permits Fgf signaling to serve as a target for selection during domestication. *Curr. Biol.,* **19**, 1642–1647.

Silva, J. C., and M. G. Kidwell, 2000. Horizontal transfer and selection in the evolution of *P* elements. *Mol. Biol. Evol.,* **17**, 1542–1557.

Soltis, D. E., and P. S. Soltis, 1995. The dynamic nature of polyploid genomes. *Proc. Natl. Acad. Sci. USA,* **92**, 8089–8091.

Van de Peer, Y., S. Maere, and A. Meyer, 2009. The evolutionary significance of ancient genome duplications. *Nat. Rev. Genet.,* **10**, 725–732.

Velasco, R., A. Zharkikh, J. Affourtit, et al., 2010. The genome of the domesticated apple (*Malus* × *domestica* Borkh.). *Nat. Genet.,* **42**, 833–839.

Vu, W. and S. Nuzhdin, 2011. Genetic variation of *copia* suppression in *Drosophila melanogaster. Heredity,* **106**, 207–217.

Wagner, A., 2011. *The Origins of Evolutionary Innovations. A Theory of Transformative Changes in Living Systems.* Oxford University Press, Oxford, UK.

Wood, T. E., N. Takebayashi, M. S. Barker, et al., 2009. The frequency of polyploid speciation in vascular plants. *Proc. Natl. Acad. Sci. USA,* **106**, 13875–13879.

END BOX 16.1

Major Mechanisms of Gene Transcription Regulation in Eukaryotic Cells

© Photos.com

SYNOPSIS: This discussion provides further details on the mechanisms by which genes are regulated at the level of DNA transcription into RNA(s). Two major means by which *transcription* is regulated—cis- and trans-*regulation*—are named on the basis of whether the regulatory sequences are adjacent to (*cis*) or distant from (*trans*) the promoter of the genes they regulate. Transcription is also regulated by small interference RNAs (*siRNAs*) acting in RNA interference pathways. *Posttranscription* regulation occurs by differential splicing and by differential degradation of mRNA (including degradation by microRNA [*miRNAs*]), all of which result in the production of different mRNAs from the same sequence of DNA. A final, recently discovered mechanism is a class of small RNA molecules known as Piwi-interacting RNAs (*piRNAs*) that regulate the insertion of transposable elements into genomes.[a]

REGULATION OF TRANSCRIPTION

Three major mechanisms control the transcription of DNA → mRNA in eukaryotic cells. The three are *cis*-, *trans*-, and RNAi-regulation (Carroll et al., 2005; Davidson, 2006; Rebeiz et al., 2011).

CIS- AND *TRANS*-REGULATION

Virtually every gene is regulated by one or more transcription factors. Regulation of transcription occurs via DNA sequences either adjacent to (*cis*) or apart from (*trans*) the gene they regulate. A method developed to tag RNA with a fluorescent probe enabled Larson et al. (2011) to follow transcription in yeast cells (*Saccharomyces cerevisiae*) in real time, opening a new window on the dynamics of the initiation of gene expression and the role played by factors regulating transcription.

cis-**regulatory elements** are short regions of DNA adjacent to the promoter of the gene they regulate (**FIGURE EB1.1a**). Gene regulation is by interaction between promoter and *cis*-regulatory element(s). Modification of *cis*-regulation is an important genetic mechanism leading to morphological change in evolution.[b]

Transcription factors also may be transcribed on different chromosomes from the genes they regulate and are therefore referred to as ***trans*-regulatory elements**. In this mechanism, transcription factors bind to special sites on DNA sequences, called CAAT and TATA boxes (**FIGURE EB1.1b**).

RNA INTERFERENCE

A much more recently discovered mechanism of gene transcription regulation is **RNA interference**, which is based on <u>**small interference RNA**</u> (**siRNA** or more often **RNAi**). It was first reported in the nematode *C. elegans* by Fire et al. (1998). RNAi has been documented experimentally by inserting *extra* copies of a gene that produces pigment into a plant. These extra copies *block* the expression of pigment formation, a totally unexpected result. **FIGURE EB1.2** shows the mechanism by which RNAi interferes with gene expression. In this example, the RNAi came from a virus that entered a mammalian cell.

The 2006 NOBEL PRIZE IN PHYSIOLOGY AND MEDICINE *went to U.S. researchers Andrew Fire and Craig Mello for the discovery of RNAi.*[c]

---

[a] Expression of piRNAs has been used to propose ancient relationships between the two major cell types in one of the major lineages of sponges, the demosponges (N. Funayama et al., 2010. *Piwi* expression in archaeocytes and choanocytes in demosponges: insights into the stem cell system in demosponges. *Evol. Dev.,* **12**, 275–287).

[b] See Prud'homme et al. (2007) for an overview of *cis*-regulation, Wray (2007) for the significance of *cis*-regulatory mutations for evolution, Rebeiz et al. (2011) for how changes in *cis*-regulation co-opted latent potential of regulatory sequences to produce novel gene expression in *Drosophlia*, and Peter and Davidson (2011) for *cis*-regulation of gene regulatory networks and the evolution of animal body plans.

[c] J. Pak, and A. Fire, 2007. Distinct populations of primary and secondary effectors during RNAi in *C. elegans. Science,* **315**, 241–244; P. D. Zamone, 2006. TNA interference: big applause for silencing in Stockholm. *Cell,* **127**, 1083–1086.

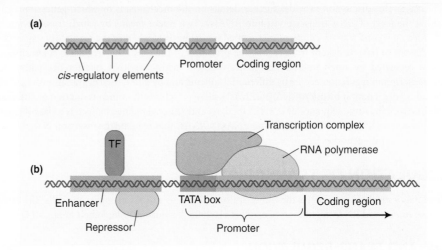

FIGURE EB1.1 *cis-* and *trans*-regulatory elements and gene transcription in animals. **(a)** Location of three *cis*-regulatory elements upstream of the promoter and coding region. **(b)** Transcription factors (TF) in combination with coactivators (not shown) bind to *cis*-regulatory elements (as do repressors in combination with corepressors). The transcription factor(s) with RNA polymerase II (RNA) forms a transcription complex on the TATA box of the promoter, initiating transcription (arrow).

[Adapted from Carroll et al., 2005. *From DNA To Diversity, Second Edition*, Blackwell Publishing, Malden, MA.]

RNAis are now known to be abundant and highly conserved; several hundred RNAi classes have been identified. *Drosophila* genomes contain more than 100 RNAis and human genomes may encode more than 800.[d] Importantly, a single siRNA can regulate many transcripts through **RNA interference pathways**. Because RNAis regulate two of the most important cellular processes—cell division and differentiation—their regulatory role is enormous. Mutations affecting the activity of RNAis and their rapid evolution in response to selection are emerging as important sources of evolutionarily significant variation (Nakayashiki et al., 2006; Obbard et al., 2006).

POST-TRANSCRIPTIONAL REGULATION

A number of different genetic mechanisms, including alternate splicing, RNA editing, and degradation of mRNA by miRNAs, regulate mRNA once it has been transcribed. These mechanisms increase greatly the number of gene products that can form.

DIFFERENTIAL SPLICING OR DEGRADATION OF MRNA

Posttranscription modification of messenger RNA occurs in two ways:
 (1) through **alternate splicing**, which is a mechanism that results in the production of *different* mature mRNAs from the *same* precursor molecule, or
 (2) by modification of mRNA nucleotides (**RNA editing**) by transitions, deletions, or insertions of individual nucleotides.

Modified intron splicing during the evolution of domesticated rice (*Oryza sativa*) produced a single base change in the *Waxy* gene, leading to less waxy protein.[e] Single genes can produce

---

[d] M. Boutros et al., 2004. Genome-wide RNAi analysis of growth and viability in *Drosophila* cells. *Science*, **303**, 832–835.
[e] H.-Y. Hirano et al., 1998. A single base change altered the regulation of the Waxy gene at the posttranscriptional level during the domestication of rice. *Mol. Biol. Evol.*, **15**, 978–987.

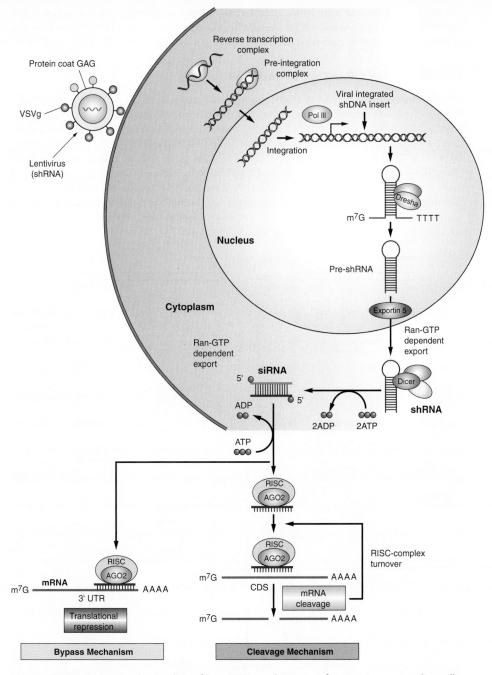

FIGURE EB1.2 Current understanding of how RNAi regulates gene function in a mammalian cell.

different functional proteins by arranging their exons in several different ways through alternate splicing. **Exon arrangement or intron removal** allow the exons coding for these protein subunits to act as **modules** or domains, combining in various ways to form new genes.

Translation of messenger RNA into protein can be regulated by altering the rate at which mRNA is degraded, or by binding proteins or complementary RNA sequences to the mRNA molecule to prevent translation. Differential translation of the ribosomal protein L38 plays an important

END BOX 16.1

Major Mechanisms of Gene Transcription Regulation in Eukaryotic Cells (Cont...)   © Photos.com

and tissue-specific role in mammalian embryos (Kondrashov et al., 2011). A novel mechanism of post-translational regulation of mRNA degradation discovered in the early 1990s is now known to be a major mechanism of gene regulation.

DEGRADATION OF mRNA BY mIRNAS

As with transcriptional regulation, post-transcriptional regulation is controlled by a special class of RNAs known as **microRNAs** (**miRNAs**) that degrade mRNA. miRNAs do not produce proteins, but are 20 to 22 nucleotide sequences of non-coding RNA present in the three eukaryote lineages—plants, animals, and fungi. Paradoxically, and in contravention of the central dogma of DNA → RNA → protein, miRNAs are encoded by RNA genes that are transcribed from DNA (**FIGURE EB1.3**). miRNAs bind to matching target mRNAs, leading to degradation of the mRNA itself. You can think of miRNA as killer RNAs. Over 600 miRNAs have been identified in the neural tubes of mouse embryos (Mukhopadhyay et al., 2011) and over 6,000 in animals and plants. Because each miRNA regulates a number of genes (sometimes a large number) and because mul-

FIGURE EB1.3 Current understanding of the mode of formation of miRNA.

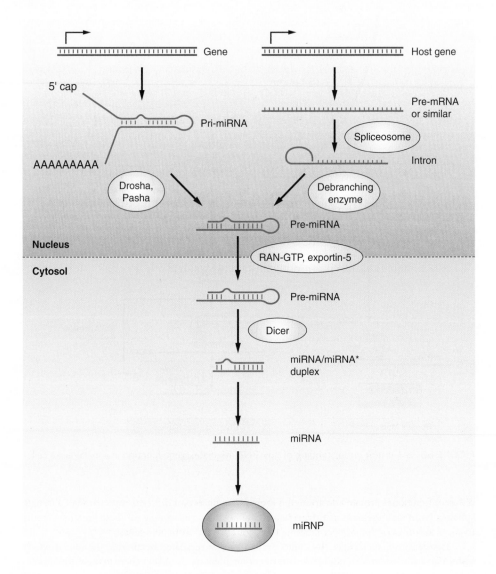

tiple copies of each MiRNA often are present—for example, 90 copies of miR-430 in the zebrafish genome—the role of miRNAs in gene regulation is enormous; 1,000 miRNAs are involved in the regulation of one-third or more of human genes (Kosik, 2009).

Research into miRNA is at an active stage, with descriptions of 447 new miRNA genes in chimpanzee and human brains, the discovery of a core set of miRNAs in all animals except sponges and jellyfish, and evidence that human miRNAs respond to selection.[f] Heimberg et al. (2008) provided evidence for the presence of 41 miRNA families at the base of the vertebrates and for involvement of miRNAs in the diversification of vertebrates. Landgraf et al. (2007) produced an atlas of 250 mammalian miRNA libraries, and began the process of cataloging and classifying miRNAs for 26 different organ systems and cell types.

PROTECTION FROM TRANSPOSON INSERTION BY PIRNAS

To miRNAs we can add recently discovered <u>**piRNAs**</u> (<u>**Pi**</u>wi-interacting <u>**RNAs**</u>), which are 25 to 30 nucleotides long. piRNAs protect the genome of mammalian and *Drosophila* germ cells from transposon insertion. Mechanisms of piRNA action are under vigorous investigation. As discussed at the end of the chapter, a heat-shock protein (Hsp90) in *Drosophila* interacts with piRNAs to activate transposons and increase genetic variation; the roles of piRNAs in evolution may be considerable and varied.[g]

REFERENCES

Carroll, S. B., J. K. Grenier, and S. D. Weatherbee, 2005. *From DNA to Diversity. Molecular Genetics and the Evolution of Animal Design*, 2nd ed. Blackwell Publishing, Malden, MA.

Davidson, E. H., 2006. *The Regulatory Genome. Gene Regulatory Networks in Development and Evolution*. Elsevier/Academic Press, Burlington, MA.

Fire A., S. Xu, M. Montgomery, et al., 1998. Potent and specific genetic interference by double-stranded RNA in *Caenorhabditis elegans*. *Nature*, **391**, 806–811.

Heimberg, A. M., L. S. Sempere, V. N. Moy, P. C. J. Donoghue, and K. J. Peterson, 2008. MicroRNAs and the advent of vertebrate morphological complexity. *Proc. Natl. Acad. Sci. USA*, **105**, 2946–2950.

Kondrashov, N., A. Pusic, C. R. Stumpf, et al., 2011. Ribosome-mediated specificity in Hox mRNA translation and vertebrate tissue patterning. *Cell*, **145**, 383–397.

Koski, K. S., 2009. MicroRNAs tell an evo–devo story. *Nat. Rev. Neurosci.*, **10**, 754–759.

Landgraf, P., M. Rusu, R. Sheridan, et al., 2007. A mammalian microRNA expression atlas based on small RNA library sequencing. *Cell*, **129**, 1401–1414.

Laron, D. R., D. Zenklusen, B. Wu, J. A. Chao, and R. H. Singer, 2011. Real-time observation of transcription initiation and elongation on an endogenous yeast gene. *Science*, **332**, 475–478.

Mukhopadhyay, P., G. Brock, S. Appana, et al. 2011. MicroRNA gene expression signatures in the developing neural tube. *Birth Def. Res. (Part A)*, **91**, 744–762.

Nakayashiki, H., N. Kadotani, and S. Mayama, 2006. Evolution and diversification of RNA silencing proteins in fungi. *J. Mol. Evol.* **63**, 127–135.

---

[f] E. Berezikov et al., 2006. Diversity of microRNAs in human and chimpanzee brain. *Nat. Genet.*, **38**, 1375–1377; L. F. Sempere et al., 2006. The phylogenetic distribution of Metazoan microRNAs; insight into evolutionary complexity and constraint. *J. Exp. Zool. (Mol. Dev. Evol.)*, **306B**, 575–588; K. Chen and N. Rajewsky, 2006. Natural selection on human microRNA binding sites inferred from SNP data. *Nat. Genet.*, **38**, 1452–1456.

[g] V. Specchia et al., 2010. Hsp90 prevents phenotypic variation by suppressing the mutagenic activity of transposons. *Nature*, **463**, 662–665. Hsp90 and other heat shock proteins are discussed elsewhere in this text.

END BOX 16.1

Major Mechanisms of Gene Transcription Regulation in Eukaryotic Cells (Cont...)   © Photos.com

Obbard, D., F. Jiggins, D. Halligan, and T. Little, 2006. Natural selection drives extremely rapid evolution in antiviral RNAi genes. *Curr. Biol.* **16**, 580–585.

Peter, I. S., and E. H. Davidson, 2011. Evolution of gene regulatory networks controlling body plan development. *Cell*, **144**, 970–985.

Prud'homme, B., N. Gompel, and S. B. Carroll, 2007. Emerging principles of regulatory evolution. *Proc. Natl. Acad. Sci. USA*, **104**, 8605–8612.

Rebeiz, M., N. Jikomes, V. A. Kassner, and S. B. Carroll. 2011. Evolutionary origin of a novel gene expression pattern through co-option of the latent activities of existing regulatory sequences. *Proc Natl. Acad. Sci. USA*, **108**, 10036–10043.

Wray, G. A. 2007. The evolutionary significance of *cis*-regulatory mutations. *Nat. Rev Genet.*, **8**, 206–216.

END BOX 16.2

Barbara McClintock and the Discovery of Transposons

© Photos.com

SYNOPSIS: The career of American geneticist *Barbara McClintock* (1902–1992) is outlined, highlighting her discovery of transposable elements in maize (for which she received the 1983 Nobel Prize in Physiology and Medicine). Also, although much less appreciated, in the 1940s McClintock developed a theory of gene regulation for multicellular organisms that predates Jacob and Monod's 1961 publication on the *lac* operon in *E. coli*. McClintock's life teaches us much about how new ideas/theories are accepted in science.

Transposons were introduced elsewhere in this text in relation to their role as agents of horizontal gene transfer and regarding their function in **gene regulation**. Transposons and gene regulation were discovered over a nine-year period (1944–1953) by Barbara McClintock (1902–1992), an American plant geneticist working on maize (**FIGURE EB2.1**). This statement may surprise you for three reasons.

1. Discovery of transposons often is attributed to research on bacteria in the early 1970s, 30 years after McClintock's research.
2. Knowledge that prokaryotic genes are regulated (a topic discussed in End Box 16.1) is attributed to Jacob and Monod's research on the *lac* operon in *E. coli* published in 1961 (**TABLE EB2.1**).
3. Knowledge that eukaryotic genes are regulated is usually attributed to research from the 1980s onwards, on animals such as the fruit fly *Drosophila* and the nematode *C. elegans*.

McClintock's name should be known to every molecular biologist, and to an extent it is, though only occasionally do you find citations to her pioneering work on gene regulation. In 1983, she was belatedly awarded the Nobel Prize in Physiology and Medicine for her discovery of genetic transposition (Keller, 1993, Comfort, 2011).

McClintock's discoveries concerning the transposition and insertion of genetic material into the genome were made between 1944 and 1953 when she was investigating unstable inheritance of the mosaic color patterns of maize seeds (**FIGURE EB2.2**; McClintock, 1950, 1951, 1953). McClintock identified two dominant and interacting gene loci, *Dissociator* (*Ds*) and *Activator* (*Ac*), discovered that *Ds* affects nearby genes only when *Ac* is present, and, perhaps most unexpectedly, that Ds and Ac can change their positions in the chromosome. Because movement within the genome (transposition) is random, some cells produce pigment while others do not, resulting in the mosaic color patterns characteristic of maize seeds (**FIGURE EB2.3**).

By the late 1940s, McClintock had proposed that these two mobile elements (now known to be transposons) were controlling elements distinct from genes. McClintock extended her results to a **theory of gene regulation** applicable to all multicellular organisms. Biology was pre-DNA and pre-molecular in the 1940s. McClintock's genome was much too dynamic to fit the then-current view that genes provided preprogrammed instructions to cells.

In 1961, when the discovery of the *lac* operon was published, McClintock responded with an informed comparison of gene regulation in bacteria and maize (McClintock, 1961). But only after the (re)discovery of transposons in the late 1960s did McClintock's prior findings on genetic transposition begin to receive the credit they deserved (Keller, 1993). Her pioneering work on gene regulation, however, remains underappreciated to this day.

Barbara McClintock was decades ahead of the field. Indeed, her results anticipated the discipline of

FIGURE EB2.1 Barbara McClintock, June 1968.

© Photo Researchers/Alamy

END BOX 16.2

Barbara McClintock and the Discovery of Transposons (Cont...)

© Photos.com

TABLE EB2.1	Highpoints in Knowledge of Gene Function in the 50 Years Following the Discovery of the Structure of DNA in 1953*
1953	Proposal by James Watson and Francis Crick of the three-dimensional double helical molecular structure of DNA, based in part on unpublished x-ray crystallographic data obtained by Rosalind Franklin and by Maurice Wilkins
1957–58	Proposal by Francis Crick of the Central Dogma of molecular biology: DNA → RNA → protein
1950s	François Jacob and Jacques Monod establish control functions located on chromosomes that turn the expression of genes on or off
1958	Isolation of the first DNA replicating enzyme
1960	Discovery of messenger RNA, in bacterial cells
1961	Determination by Jacob, Crick, Monod, Brenner and others of the triplet nature of the genetic code, transcription of information in DNA into messenger RNA, and translation of mRNA into protein
1962	Use of synthetic RNA to unravel the genetic code (genetic information is carried in three-nucleotide sets of 64 codons, each coding for one of the 20 amino acids). Discovery of repressor and transcriptional control of genes
1964	Crick's Central Dogma shown not to hold for viruses
1966	Complete genetic code translating codons into amino acids established
1975	DNA copied from messenger RNA (cDNA)
1983	Discovery of homeobox as basic element of homeotic genes. Direct detection of the nucleotide sequence of a specific mutant allele (sickle cell)
1989	First human gene sequenced, and a defect in the gene product shown to "cause" cystic fibrosis
1999	Sequencing of a human chromosome and of complete genome of the fruit fly *Drosophila melanogaster*
2001	First draft sequences of human genome completed independently by the Human Genome Project and Celera Genomics
2003	Sequencing of 99 percent of the human genome completed to 99.9 percent accuracy
2006	Sequencing of the human genome completed

*Barbara McClintock's discoveries of transposons and gene regulation do not appear in this "traditional" timeline.

FIGURE EB2.2 Mosaic color patterns of seeds in cobs of maize **(a)** and heirloom Indian corn **(b)** reflecting differential production of pigment in different seeds.

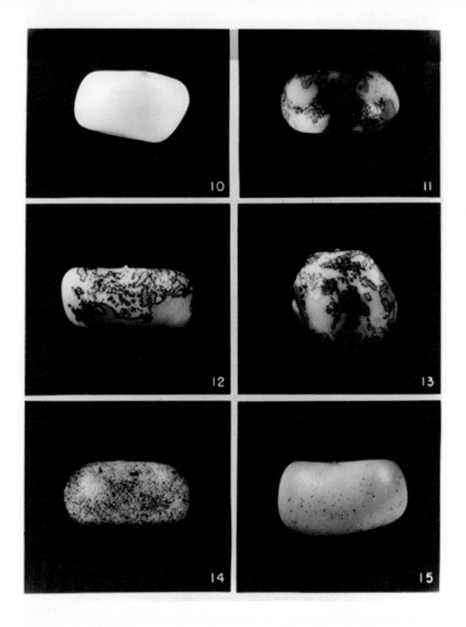

FIGURE EB2.3 Interactions between the dominant gene loci, *Dissociator* (*Ds*) and *Activator* (*Ac*) and their affects on color of maize seeds. Seed 10 is colorless; *Ds* inhibits the synthesis of pigments in the absence of *Ac*. Seeds 11 to 13 have a single copy of *Ac*. In the presence of *Ac*, *Ds* moves within the chromosome resulting in the production of some pigment, and the mosaic color pattern so typical of maize seeds. Seeds 14 and 15 have two and three *Ac* elements respectively, resulting in either more (14) or less (15) pigment production.

[From an image in McClintock, 1951. Reproduced with permission from American Philosophical Society. Library. Barbara McClintock Papers.]

molecular genetics, which did not yet exist. Although her results were published in major journals, she spoke at scientific meetings, and trained and mentored students, her "colleagues" were so skeptical of her findings that in 1953 she stopped publishing on transposition, turning instead to cytogenetics and ethnobotanical studies of maize plant races in South America. Among her many achievements, McClintock discovered genetic recombination by crossing-over during meiosis and produced the first genetic map for maize, a map that allowed physical traits to be linked to specific regions of the chromosomes.

McClintock articulated her frustration in a letter to a geneticist at the University of Leeds who had invited her to participate in a workshop to be held in September 1973, an invitation she declined. Writing of "my attempts during the 1950s to convince geneticists that the action of genes had to be and was controlled," she continued, "It is now equally painful to recognize the fixity of assumptions that many persons hold on the nature of controlling elements in maize and the manners of their operation. One must await the right time for conceptual change."

END BOX 16.2

Barbara McClintock and the Discovery of Transposons (Cont...)

© Photos.com

REFERENCES

Comfort, N., 2011. When your sources talk back: toward a multimodal approach to scientific biography. *J. Hist. Biol.*, **44**, 651–69.

Diao, X., M. Freeling, and D. Lisch, 2006. Horizontal transfer of a plant transposon. *Public Lib. Sci. Biol.*, **4**, 119–128.

Keller, E. F., 1993. *A Feeling for the Organism: The Life and Work of Barbara McClintock,* 10th Anniversary Edition. W. H. Freeman, New York.

McClintock, B., 1950. The origin and behavior of mutable loci in maize. *Proc. Natl. Acad. Sci. USA,* **36**, 344–355.

McClintock, B., 1951. Chromosome organization and genic expression. *Cold Spring Harbor Symp. Quant. Biol.*, **16**, 13–47.

McClintock, B., 1953. Induction of instability at selected loci in maize. *Genetics*, **38**, 579–599.

McClintock, B., 1961. Some parallels between gene control systems in maize and in bacteria. *Am. Nat.*, **95**, 265–77.

CHAPTER

17

Mutations and Gene Regulation as Sources of Individual Variation

CHAPTER SUMMARY

Although it is abundantly clear that there is no one-to-one map between the genotype and the phenotype—genes do not make structures—how variation at the level of the gene relates to variation of the phenotype, or to the origin of new phenotypes, are major unresolved questions in evolutionary biology. In addition to genetic polymorphism and whole genome and chromosomal sources of variation, variation at the genetic level arises from mutations and modification of gene regulation, both of which are discussed in this chapter. Genes are regulated by multiple mechanisms that affect transcription, translation, and posttransitional modification. Indeed, gene regulatory changes of all kinds are among the key agents of organismal evolution. Evidence presented in this chapter is supported elsewhere in this text and documented in END BOX 17.1 through an examination of the origin of animals in the Precambrian and the appearance of almost all animal phyla in the Early Cambrian—the "Cambrian explosion."

INTRODUCTION

Natural selection acts on existing phenotypic variation. Numerous sources of genetic variation enable population genotypes to evolve and phenotypic variation to be translated into heritable evolutionary change. Sources of genetic variation include the vast reservoir of allele polymorphism[1] and new mutations within populations. Variation arises from duplication of entire genomes, chromosomes, or parts of genomes or chromosomes, from gene transfer, and from modifications in the regulation of gene function at DNA, RNA. or protein product levels. Changes in allele frequency resulting from mutation, gene duplication, and exon shuffling are additional sources of variation. Transposable elements provide a further source of genetic variation, as does RNA interference, differential splicing of mRNA, *cis*-regulation, and posttransitional modification.

MUTATIONS AS A SOURCE OF GENETIC VARIATION

Mutations are small, localized changes in nucleotide sequences or regulation that can (but need not) result in changes in a gene or gene product. Without mutations, we would all be mindless blobs of unicellular protoplasm, no more than a few molecules floating in the primordial soup. New mutations are usually detected because we observe a harmful effect(s) on a structural, functional, or behavioral aspect of the phenotype or on the fitness of the organism (Eyre-Walker and Keightley, 2007). Even what appear to be small mutational changes may have considerable phenotypic consequences, as outlined below. But mutations can also be silent with no visible effect.

The enormous range of possible mutations has important consequences for organisms. Too many or too few mutations can interfere with the effect that natural selection has on genetic variation. Too many mutations will allow errors to accumulate in organisms already selected for their environment. Too few mutations will reduce the opportunity for natural selection to initiate changes that could lead to adaptations to changing environments.

Mutations are normally expressed at one of two levels of gene activity,

- changes within a **gene product**, for example, in the amino acid constitution of a particular protein, or
- changes in the **regulation of a gene or its product**.

A **silent mutation** is one that has no effect on the gene or gene product but is retained in the population. Mutations detected because they affect structural aspects of the phenotype include point mutations, insertions, or deletions of sequences. Mutations such as loss- and gain-of-function mutations and dominant-negative mutations that counteract the role of the wild type allele(s) are detected because they affect a function of the organism. Frame-shift, nonsense, or missense mutations affect gene products. Mutations that affect fitness are usually classified as beneficial, harmful, or neutral.

Mutations can affect whether a gene product, such as a protein, is produced. *Thalassemias* are genetic diseases in which, because of a mutation that changes a gene product, α- or β-hemoglobin chains are either not produced or are produced in diminished numbers. Like the **sickle cell allele** discussed below, such mutations often result in death of **homozygotes** but offer protection for **heterozygotes** against something else that has selective advantage. While homozygous sickle cell individuals suffer from various complications stemming from abnormally-shaped red blood cells, individuals heterozygous for the sickle cell allele gain protection against malarial parasites, a protection that explains the frequency of the allele in modern human populations, particularly in tropical and sub-tropical areas. Even without mutations, however, genetic variation is maintained; individual parasites express different combinations of more than 60 variants of the gene for a cell membrane protein used to counteract responses by the host (Chookajorn et al., 2006).

More importantly for evolutionary change, mutations can affect the **amount or rate at which a gene or gene product is produced**. An example is a mutation in the

---

[1] Allele frequency rather than gene frequency is the basic genetic unit in population genetics.

insulin-dependent growth factor-2 (*Igf-2*) gene in pigs. Pigs carrying this mutation have a threefold increase in the production of Igf-2, resulting in a more muscular and lean phenotype.[2] Breeders take advantage of such mutations to produce strains that fit the current tastes of the marketplace, sometimes even creating strains with the hope of influencing the market place.

SINGLE NUCLEOTIDE SUBSTITUTIONS

The simplest type of mutation at the molecular level is a change in a nucleotide that substitutes one base for another. Such changes, known as **base substitutions**, may occur spontaneously or from the action of mutagenic agents such as x-rays. Depending on its position, a single nucleotide mutation can have important consequences for a species. A well-studied example, **sickle cell anemia** in humans, results from a mutated allele of the β-hemoglobin gene that produces the sickle cell mutation *HbS*.

The hemoglobin molecule in adult human blood cells consists of four polypeptide globin chains, two αs and two βs, each about 140 amino acids long and each with its own specific sequence. In individuals homozygous for the sickle cell allele, a mutation resulting in a single amino acid substitution has caused all β-globin chains to differ from normal βs at the number 6 position. The effects of this single genetic mutation are profound, causing a variety of phenotypic changes (**pleiotropy**), often resulting in death of the individual (FIGURE 17.1).[3]

The highest frequency of sickle cell alleles is present in African individuals; in the United States, the mutant allele is almost entirely confined to those with African ancestry. Sickle cell anemia kills, before the age of 20, more than 10 percent of African Americans who are homozygous for the allele. The death rate is even higher in Africa, where the frequency of the disease is correspondingly much higher (0.2% of individuals in Africa are homozygous for the allele vs. 0.04% in African Americans; TABLE 17.1) and medical facilities are more limited.

High frequency of the sickle cell mutation in these populations is related to the selective advantage of sickle cell heterozygotes in regions where malaria is endemic (Ferreira et al., 2011). Disadvantages of the allele when homozygous (anemia) are balanced by advantages of the gene when heterozygous (protection against malaria). Upon infection with the malarial parasite, the malformed red blood cells caused by sickle cell rupture prematurely, preventing the parasite from completing its life cycle and potentially cause the death of the host individual. Because of similar beneficial aspects, mutations for other human diseases, some of which are listed in Table 17.1, also have failed to "die out."

Heterozygote advantage is not true for all disease-causing mutations, however; the high incidences of achromatopsia (several forms of hereditary blindness) among the Pingelapese of the Caroline Islands in the western Pacific Ocean (a gene frequency of 0.22) and the rare Ellis-van Creveld syndrome (which causes abnormal skeletal development) among the Lancaster County Amish, with a gene frequency of 0.07 (Table 17.1) seem to confer no advantage on either their homozygous or heterozygous carriers.[4] Such examples of unique gene frequencies seem best explained by founder or bottleneck effects.

A single gene that results in multiple traits (phenotypes) that are apparently unrelated is called a PLEIOTROPIC GENE. POLYGENIC GENES, *each of small effect, influence a single aspect of the phenotype.*

ACHROMATOPSIA *is a general term for a class of at least five different human diseases, the symptoms of which are the inability to tell colors from one another and to see well in bright light. Ellis-van Creveld syndrome is characterized by short-limb dwarfism, extra fingers or toes, and heart defects (Table 17.1).*

---

[2] A. S. Van Laere et al., 2003. A regulatory mutation in IGF2 causes a major QTL effect on muscle growth in the pig. *Nature*, **425**, 832–836. For discussion of the evolution of mutation rates, see Sniegowski et al. (2000).

[3] For the role of pleiotropy in human diseases see R. L. Perlman, 2005. Why disease persists: an evolutionary nosology. *Med. Health Care Philos.*, **8**, 343–350.

[4] For a recent expansion of the theory underlying heterozygote advantage see D. Sellis et al., 2011. Heterozygote advantage as a natural consequence of adaptation in diploids. *Proc. Natl. Acad. Sci. USA*, **108**, 20666–20671.

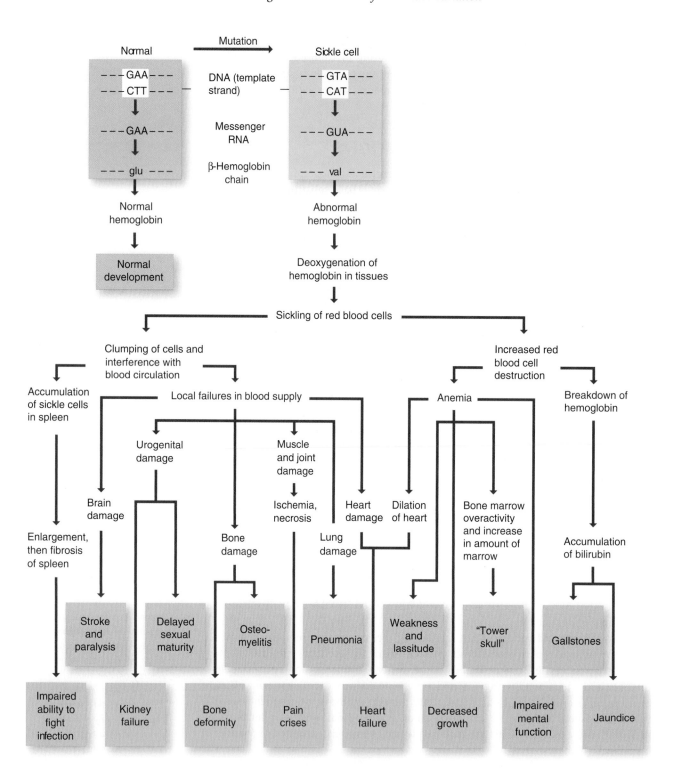

FIGURE 17.1 Varied (pleiotropic) effects of the sickle cell mutation, beginning with the substitution of a thymine by an adenine on the DNA template strand of the β-hemoglobin gene. The resultant GUA triplet coding sequence on messenger RNA translates into a valine (val) amino acid instead of the normal glutamic acid (glu), producing developmental consequences that can seriously affect sickle cell homozygotes.

[Adapted from Strickberger, M. W. *Genetics, Third Edition.* Macmillian, New York, 1985.]

TABLE 17.1	Genotype Frequencies for Some Human Diseases Associated with Recessive Alleles[a]			
Disease	Population	Gene Frequency	Frequency of Homozygotes	Frequency of Carriers
Achromatopsia	Pingelap (Caroline Islands)	0.22	1 in 20	1 in 2.8
Sickle cell anemia	Africa (some areas)	0.2	1 in 25	1 in 3
Albinism	Panama (San Blas Indians)	0.09	1 in 132	1 in 6
Ellis-van Creveld syndrome	Old Order Amish	0.07	1 in 200	1 in 8
Sickle cell anemia	African-Americans	0.04	1 in 625	1 in 13
Cystic fibrosis	European-Americans	0.032	1 in 1,000	1 in 16
Tay-Sachs disease	Ashkenazi Jews	0.018	1 in 3,000	1 in 28
Albinism	Norway	0.010	1 in 10,000	1 in 50
Phenylketonuria	United States	0.0063	1 in 25,000	1 in 80
Cystinuria	England	0.005	1 in 40,000	1 in 100
Galactosemia	United States	0.0032	1 in 100,000	1 in 159
Alkaptonuria	England	0.001	1 in 1,000,000	1 in 500

[a]The diseases are arranged in order of decreasing gene frequency.
[Adapted from Strickberger, M. W. *Genetics, Third Edition.* Reprinted by permission of Pearson Education, Inc., Upper Saddle River, NJ.]

REGULATORY MUTATIONS

The term **regulatory mutation** refers to those changes that affect genes controlling genetic pathways or networks. A classic example, discussed below, is a mutation in the regulatory gene that governs the expression of genes coding for enzymes that digest the sugar lactose in bacteria. Discovery of such mutations in the middle of the twentieth century laid the foundation of our current understanding of gene regulation in prokaryotic and eukaryotic cells.

In multicellular organisms, regulatory mutations affect when, and in which tissue or organ, a gene product is produced during embryonic development. As with any other trait in which genes and their alleles have been selected, a multicellular organism's development is the outcome of a historical evolutionary process (Hall, 1999, 2002, 2003). Whether new mutations incorporate successfully depends on how they interact with existing genetic pathways and developmental processes. Such interactions canalize development, place limits on, and create opportunities for evolutionary change.

From the plant kingdom comes a dramatic example of regulatory evolution associated with hybridization. Genes in the *RAY* locus from a diploid species of a member of the aster family, the Oxford ragwort, *Senecio squalidus*, were incorporated into a new tetraploid species, the common groundsel, *S. vulgaris*. Regulatory genes in the *RAY* locus are major players in establishing the symmetry of flowers. *RAY* genes in *S. vulgaris* promote flower asymmetry and outcrossing of the plants, thus affecting the phenotype, reproductive potential, accumulation of variation, and evolutionary potential of the new species.[5]

GENE REGULATION

A frequent finding from comparing different organisms on the molecular level is that organisms share many of the same genes. Whether prokaryotic or eukaryotic, organisms share similar enzymes involved in basic biochemical processes such as glycolysis, amino acid synthesis, DNA replication, and protein synthesis. This fundamental conservation raises the question of how genetic control varies between different organisms and how such control relates to evolutionary change (Carroll, 2005; Carroll et al., 2005). "The triumph of our DNA [is that] it makes us without determining us." (Lehrer, 2008, p. 46)

When we examine them closely, the distinctive structural features of different organisms within any group are found to depend less on differences among the kinds of genes they have and the proteins they produce than on how shared genes are regulated. Which genes are transcribed into messenger RNA, which transcripts are translated into proteins, and which proteins are activated are the result of interactions that often begin with signals that regulate different genetic regulatory pathways or networks (Carroll et al., 2005; Davidson, 2006). Basic elements of gene regulation in prokaryotic and eukaryotic cells are outlined below and elsewhere in this text.

GENE REGULATION IN THE BACTERIUM *E. COLI*

Although often credited as the first report of gene regulation, McCLINTOCK REPORTED gene regulation in maize before it was discovered in bacteria.

Viruses and prokaryotes have provided fundamental information on mechanisms of gene regulation. A pioneering study leading to the discovery of gene regulation was by François Jacob (1920–) and Jacques Monod (1910–1976) in 1961 on the genes governing the production of enzymes involved in lactose sugar metabolism in the bacterium *E. coli*. For this fundamental discovery described in Jacob and Monod (1961), they and André Lwoff (1902–1994), the director of the Pasteur Institute where the research was conducted, received the 1965 Nobel Prize in Physiology and Medicine.[6]

When the sugar lactose is not in the medium a repressor protein in *E. coli* prevents the genes used in lac enzyme synthesis from being transcribed into mRNA (**FIGURE 17.2a,b**). However, when bacteria encounter lactose in the medium, some lactose molecules are converted to a form called allolactose, which acts as an inducer and binds with the repressor, allowing transcription of the genes necessary to metabolize lactose (**FIGURE 17.2c**). The repressor acts as a regulatory protein; binding of the inducer to the repressor changes the configuration of the DNA binding site, making the repressor inactive. Bacteria such as *E. coli* exploit their ability to adapt to changes in their environment when they acquire resistance to drugs. *lac z* in *E. coli* codes for the enzyme β-lactamase. In the presence of the drug ampicillin, *E. coli* can synthesize β-lactamase, inactivate the drug, and so acquire resistance to the drug.

GENE REGULATION IN EUKARYOTIC CELLS

In comparison to prokaryotic gene regulation, genes in eukaryotic cells are regulated at a surprisingly diverse number of levels and by a wide range of mechanisms (**TABLE 17.2**). Mechanisms known to play a major role in generating protein and functional diversity in eukaryotic cells establish a **regulatory code** that lies between the DNA transcriptional network

---

[5] N. Kim, M.-L. Cui, P. Cubas et al., 2008. Regulatory genes control a key morphological and ecological trait transferred between species. *Science* **322**, 1116–1119.

[6] As summarized by Morange (2007) "One has the feeling that Monod and Jacob were on the verge of attributing to a small family of regulatory genes a major role in the evolution of organisms . . . but just stopped short of developing this hypothesis." (p. 151)

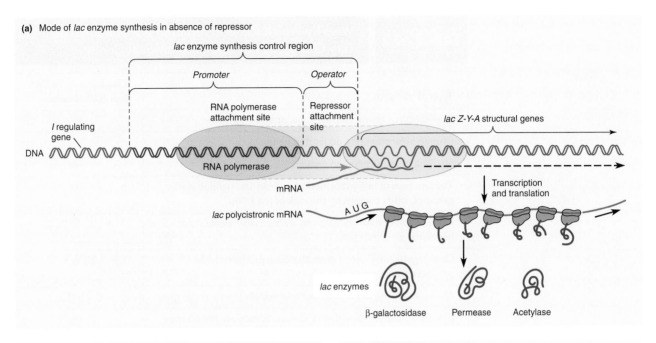

(a) Mode of *lac* enzyme synthesis in absence of repressor

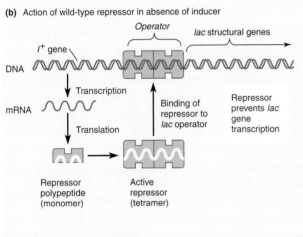

(b) Action of wild-type repressor in absence of inducer

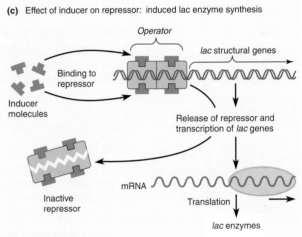

(c) Effect of inducer on repressor: induced lac enzyme synthesis

FIGURE 17.2 General scheme of lac enzyme synthesis in *E. coli* and the effects of repressor function or dysfunction on this inducible system. **(a)** The DNA region involved in controlling transcription of the *lac* structural genes Z, Y, and A consists of two major regulatory sites—an operator and a promoter—each of which binds specific proteins. In the absence of the repressor protein, RNA polymerase begins transcribing at the operator, which is also the site to which the repressor attaches. Transcription of the *lac* genes, followed by translation, leads to synthesis of three *lac* enzymes. **(b)** Transcription and translation of the *I+* gene produces a normal repressor protein that binds to the operator site of the *lac* locus, blocking transcription of *lac* genes. This repressed state appears in normal *E. coli* cells not growing on a lactose medium. **(c)** Transferring bacteria to a medium containing lactose leads to the introduction of allolactose inducer molecules, which results in the repressor dissociating from its DNA binding site on the operator. This allows transcription of the *lac* structural genes, followed by their translation and the synthesis of *lac* enzymes. The repressor acts as a regulatory protein. Binding of the inducer to the repressor changes the configuration of the DNA binding site, making the repressor inactive.

[Adapted from Strickberger, M. W. *Genetics, Third Edition.* Macmillian, New York, 1985.]

and posttranscription and translational regulation of RNA. Major mechanisms of gene regulation are introduced in this section. Three major regulatory mechanisms control **transcription of DNA → mRNA** and are known as *cis-, trans-,* **and RNAi-regulation.** A fourth mechanism, insertion of **transposons** into the genome, is discussed elsewhere in this text.

Work by Sean Carroll and colleagues has shown how evolutionary changes in *Drosophila* wing pigmentation and male abdominal pigmentation patterns arise through regulatory

TABLE 17.2	Ways by Which Genetic Variation Is Maintained or Can Change at the Level of Individual Genes

Type of Variation	Chapter and/or End Box (EB) Number
Duplication of genes, chromosome, or entire genomes	16
Change in function of a duplicated gene	17
Shuffling of exons between genes	6
The presence of many repeated sequences throughout most genomes, often comprising the bulk of the DNA	6
Introduction of novel variation by transposable elements or horizontal gene transfer	6, 16, EB 16.2
Gene regulation whereby more than one product can be produced from a single "gene"	EB 16.1
Alternate splicing to produce multiple mRNAs from a single gene	6, EB 16.2
RNAi and posttransitional modification to increase the number of products a gene can provide	EB 16.1

mutations. Changes in *cis*-regulatory elements determine the spatial distribution of expression of the *yellow* gene on the wing. Gain of a homeobox-protein binding site in a *cis*-regulator of the *yellow* gene operates in abdominal cells. Over time, these changes resulted in five independent losses of wing spots within species of *Drosophila melanogaster*, and two independent gains in species of *D. obscura*.[7] The extent to which mutations in *cis*- and *trans*-acting regulatory elements or posttranscription regulation contribute to the phenotypic differences we see among individuals both within and between species is becoming increasingly apparent. To quote Eric Davidson, one of the major researchers in the field, writing in 2006,

> In my view, *cis*-regulatory information processing, and information processing at the gene regulatory network circuit level, are the real secret of animal development. Probably the appearance of genomic regulatory systems capable of information processing is what made animal evolution possible.

Eukaryotic genes also are regulated after mRNA has been produced by what is known as **post-transcription modification** or translational modifications. Differential splicing or posttranscription editing of mRNAs can result in changes in the protein produced or allow more than one protein to be produced from a single mRNA. The last mechanism, **posttransitional modification** of proteins, can occur through different splicing patterns that remove amino acid sequences, or by chemically modifying amino acid residues.

A class of small RNAs known as **microRNAs (miRNA)** regulate the translation of mRNA into protein in plants, animals, and fungi.[8] MiRNAs have assumed considerable importance in analyses of the evolution of major lineages of organisms: Heimberg et al. (2008) provided evidence for the presence of 41 MiRNA families at the base of the vertebrates, concluding that miRNA origin/diversification facilitated vertebrate evolution. Peterson et al. (2009) documented addition of miRNAs to animal genomes during their evolution and presented evidence that miRNAs increase heritability, miRNAs

---

[7] N. Gompel et al., 2005. Chance caught on the wing: *cis*-regulatory evolution and the origin of pigment patterns in *Drosophila*. *Nature*, **433**, 481–487. Homeobox genes are discussed later in the chapter. For the importance of *cis*-regulation in evolution, see Prud'homme et al., 2007; Wray, 2007; Peter and Davidson, 2011; and Rebeiz et al., 2011

[8] A further class of 25- to 30-nucleotide long **piRNAs (Piwi-interacting RNAs)** that prevent insertion of transposons into eukaryotic genomes are discussed elsewhere in this text.

representing the "'dark matter' of the metazoan genome" (p. 745). Kosik (2010) proposed that miRNAs counteract adverse environmental impacts on cells. Given our advancing knowledge of the roles miRNAs play in gene regulation, "to focus solely on the transcriptional side of the equation [*cis-* and *trans-*regulation] is to miss a significant part of the [evolutionary] process" (Peterson et al., 2009, p. 745).

GENE REGULATION AS A SOURCE OF VARIATION

As outlined above, there is compelling evidence for the hypothesis that changes in regulatory genes are of great importance in evolution. Evolutionary changes in gene regulation are increasingly appreciated as being central to the origin of new features and/or lineages. This is why gene regulation is discussed in various contexts throughout the book.

One of the more fascinating developments in evolutionary biology over the past two decades has been the realization that mutations that affect the DNA sequence of genes and the amino acid composition of their protein products may not be as important as mutations that affect when, where, and how much of a gene product is expressed in the organism. Regulatory pathways and networks, and their specific activity in time and place, play major roles in our explanations for the variety of cells, tissues, and organs in multicellular organisms. Indeed, we are at the stage where maps of transcriptional regulation are being built up for individual species such as humans and compared with other species, such as mice. Ravasi et al. (2010) screened for physical interactions among transcription factors to identify 762 human and 877 mouse interactions, about half of which differ between the two species.

The emerging picture is of all levels of gene activity, from DNA replication onwards, as exquisitely and sensitively controlled and regulated (Table 17.2). One class of evidence is the relatively high degree of conservation between species of structural gene sequences and therefore gene products. This is contrary to what one would expect if structural and functional variation in genes and their protein products accounted for most organismal diversity. Secondly, there are many fewer genes in metazoan genomes than we had thought and many of these genes perform multiple functions.

Thirdly, mutations that affect the structure of a gene product are uncommon because such mutations tend to affect several, often unrelated, aspects of the phenotype, and/or are lethal or selected against. Mutations that influence transcription and regulation can affect the role of a gene in one part of the phenotype without interfering with other functions of the gene in other parts of the organisms (pleiotropy) or at other times during the life cycle (heterotopy). Analysis of pleiotropy across yeast, nematode, and mouse genomes by Wang et al. (2010) reinforces the modular nature of pleiotropy, shows how patterns of pleiotropy across lineages influence the evolution of those lineages, and how a small number of genes can initiate enormous phenotypic variation (Wagner and Zhang, 2011).

GENE REGULATION AND ANIMAL ORIGINS

Much of evolution is about staying the same, although you don't read much about this in textbooks: maintenance of classes of genes or morphologies over hundreds of millions of years of evolution. Aspects of evolution that excite us are those involving the evolution of new features or new types of organisms, be they feathers, birds, turtles, mammals, humans, or whatever. Like a child's Lego set, with which you can produce differently shaped structures by rearranging the modular blocks, regulatory changes can initiate new functions and features. A single regulatory change in a gene that controls other genes can change whether a gene network is turned on or off, with dramatic consequences for the phenotype. An excellent example discussed below, the gene *Ubx* (*Ultrabithorax*), regulates the development of paired structures (halteres) in the abdominal segment in *Drosophila* by regulating signals and pathways involving at least 30 target genes. Target genes can fall into different classes and serve different functions.

FIGURE 17.3 A hierarchical network of four classes of genes involved in the specification of body regions in embryos of the fruit fly, *Drosophila*.

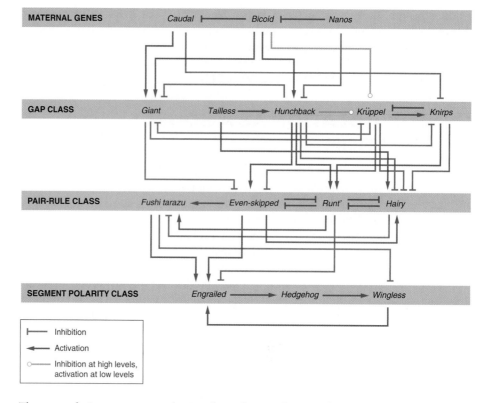

HALTERES *are a pair of balancing organs found on the abdominal segment in flies, a location where hind wings are found in other insects (Figure 17.4). Although different in structure, halteres are homologues of hind wings.*

End Box 17.1 *can be read as a stand alone discussion of the evidence for the* EVOLUTION OF ANIMALS.

The example in **FIGURE 17.3** depicts four classes of genes shown to underlie the organization of the body regions in *Drosophila*. Modifications of such networks of genes are emerging as a fundamental genetic mechanism underlying organismal evolution.[9]

We use the **origin of animals** to illustrate the role played by the regulation of conserved genes in evolutionary change. Because the evidence for the origins of animals is considerable and diverse it has been gathered into End Box 17.1. Evidence discussed in the End Box and illustrated in Figure EB1.1 includes the astonishing preservation of animal fossils from the Early Cambrian, and geological, paleoecological, and paleoclimatological evidence for the conditions associated with the first appearance of animals.

AXES OF SYMMETRY

The most obvious and distinctive morphological features of almost all adult animals are axes of symmetry, paired appendages, and similar tissues and organs.

Many animals are organized around an **anterior–posterior** (**A–P**) axis, usually with a well-developed head at the anterior end (**FIGURE 17.4**). Having an anterior end automatically means these organisms have a **left–right** (**L–R**) axis. When viewed from above, the surface that is uppermost is dorsal, that which is out of view is ventral, the **dorso-ventral** (**D–V**) axis being the third body axis (Figure 17.4). Such organisms, which represent the majority of animal lineages are bilaterally symmetrical, a symmetry that has deep roots in animal evolution (Matus et al., 2006).[10]

Many lineages of animals without a definite head nevertheless have a main body axis running from mouth (oral) to anus (aboral). The **oral–aboral** axis shown in Figure 17.4**c** is equivalent to the A–P axis of the fish shown in Figure 17.4a. Indeed, given that the same

---

[9] For a recent demonstration of the insertion of a transcriptional level of control into an existing network in yeast (intercalary network evolution), see L. N. Booth et al., 2010. Intercalation of a new tier of transcription regulation into an ancient circuit. *Nature*, **468**, 959–963.

[10] Other organisms such as starfish and sea anemones (echinoderms) are radially symmetrical but share genes involved in establishing bilateral symmetry in bilaterians (Martinez et al., 1999; Martindale et al., 2002; Hibino et al., 2006; Saina et al., 2009).

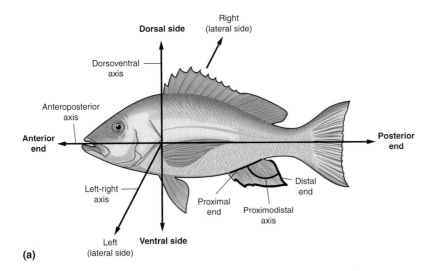

(a)

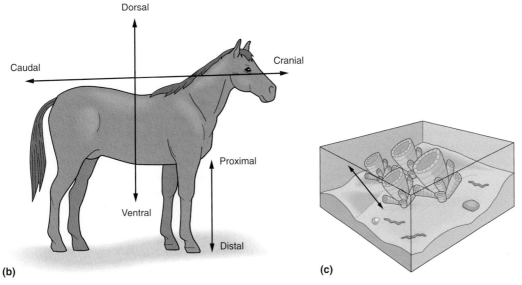

(b)

(c)

FIGURE 17.4 Major body axes (A–P, D–V, L–R) in a bony fish **(a)** and in a horse **(b)** and the oral–aboral axis (red axis) in the Venus flower basket glass sponge *Euplectella aspergillum* **(c)**. In (b) alternate names for the A–P axis, cranial and caudal, are used. These images also show the proximal-distal (P–D) axis of the anal fin (a) and of the fore limbs (b), where proximal is closest to the body and distal at a distance from the body. The P–D axis is important in specification of the parts of the paired limbs.

basic sets of genes specify A–P and oral–aboral axes, these two axes have been identified as homologous (Scholtz, 2010).

Such precise and repeatable features arise from specifically ordered embryonic development based on controlled genetic pathways. One of the most informative approaches to comprehending animal origins lies in understanding the genetic and cellular bases of embryonic development in different animals. From analyses of patterns and mechanisms of gene expression and of natural or induced mutations we now know that genes governing developmental processes in all animals share the same nucleotide sequences, similar linkage orders, and even similar developmental targets.

HOMEOTIC TRANSFORMATIONS

The unequal distribution of maternal gene products deposited into the fruit fly egg by the female parent establishes localized differences at the very outset of development. Many of these molecules are known as **morphogens** because they activate pathways leading to specific morphologies. How morphologies are distributed in three-dimensional morphospace is discussed in End Box 17.1.

---

[11] M. A. Crickmore and R. S. Mann, 2006. Hox control of organ size by regulation of morphogen production and mobility. *Science*, **313**, 63–68.

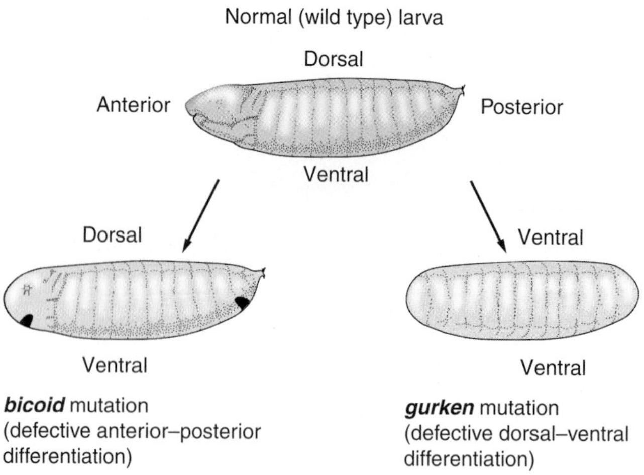

FIGURE 17.5 Effects on formation of the body axis of the fruit fly *Drosophila melanogaster* of mutations in the genes *bicoid* and *gurken*. Bicoid mRNA establishes the posterior end of the egg at the high point of a concentration gradient that decreases along the long axis of the egg. Mutations in the *bicoid* gene interfere with A–P patterning, producing a headless individual; the A–P axis becomes a P–P axis. Gurken mRNA acts in a similar manner to establish the D–V axis. Mutations in *gurken* produce a ventralized larva lacking dorsal tissues as the D–V axis is turned into a V–V axis.

[Adapted from Ephrussi, A., and R. Lehmann, 1992. *Nature*, **358**, 387–392.]

Normal (wild type) larva

Dorsal

Anterior — Posterior

Ventral

Dorsal — Ventral

Ventral

***bicoid* mutation** (defective anterior–posterior differentiation)

Ventral

***gurken* mutation** (defective dorsal–ventral differentiation)

An example of morphogens, using the fruit fly genes *bicoid* and *gurken* that control A–P and D–V patterning, respectively, is shown in FIGURE 17.5.[11] These **homeobox genes**, which are discussed in the following section, provide the primary signals and initiate the pathways that pattern body regions and the formation of particular structures such as wings and legs. As shown in FIGURE 17.6 the bodies of embryonic insects are subdivided into compartments (segments) that arise in sequence along the A–P axis.

FIGURE 17.7 depicts how some cells or tissues obtain information on their position in relation to chemical gradients. Uniformity of the sequence of developmental

Such mutations are called HOMEOTIC *because they change a particular organ in a particular body region to resemble an organ normally found in a different region along the body axis. This type of change is known as* HOMEOSIS.

FIGURE 17.6 Proposed evolutionary steps in insect segmental organization as typified by different insect groups. The three thoracic segments are numbered consecutively from the head boundary as T1, T2, and T3. Body segments, at first, relatively uniform, became more complex as different lineages evolved. In dipteran insects such as *Drosophila* (bottom figure) the segments bear different complex structures; the second thoracic segment (T2) has wings, the third (T3) has the halteres (balancing organs). A1–A10 are eight abdominal segments numbered consecutively from the thoracic boundary.

[Adapted from Strickberger, M. W. *Genetics, Third Edition.* Macmillian, New York, 1985.]

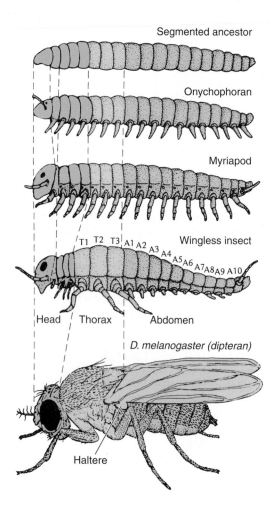

Segmented ancestor

Onychophoran

Myriapod

Wingless insect

T1 T2 T3 A1 A2 A3 A4 A5 A6 A7 A8 A9 A10

Head / Thorax / Abdomen

D. melanogaster (dipteran)

Haltere

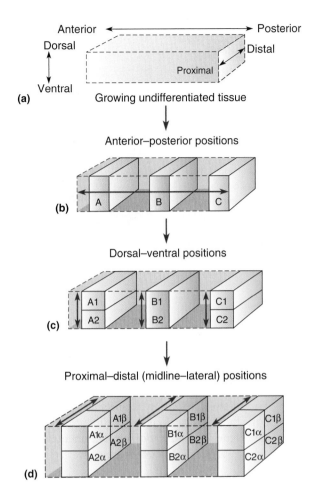

(a) Anterior ↔ Posterior
Dorsal, Distal, Proximal, Ventral
Growing undifferentiated tissue

(b) Anterior–posterior positions
A B C

(c) Dorsal–ventral positions
A1/A2 B1/B2 C1/C2

(d) Proximal–distal (midline–lateral) positions
A1β A1α A2β A2α B1β B1α B2β B2α C1β C1α C2β C2α

FIGURE 17.7 Representation of how cells or tissues along the major body axes respond to gradients during development. A gradient **(a)** provides cells or tissues A, B, and C **(b)** with information as to their relative anterior–posterior positions. Further growth and differentiation confers dorsal–ventral information producing A1/A2, B1/B2, and C1/C2 **(c)**. Subsequent cell division of these tissues provides α and β subclones with information as to their proximal–distal positions **(d)**. As a result of their geographical position and of the activity of selector genes—of which homeotic genes are a major group—cells of tissue A come to possess a specific set of developmental responses and develop differently from tissues B or C, or from other subclones within A.

[Adapted from Strickberger, M. W. *Genetics, Third Edition*. Macmillian, New York, 1985.]

events in different individuals of the same species lies in the fact that the succession of regulatory events is identical from individual to individual. Change a step—for example, by a mutation—and the outcome changes (unless the mutation is of slight effect and/or can be compensated for). The *Drosophila* mutation *Antennapedia* results in a leg developing from the embryonic region (the imaginal disc) that normally forms an antenna (**FIGURE 17.8**). Should expression of the normal *Ubx* gene fail in the anterior abdomen, this region develops as though it was thoracic (more anterior) rather than abdominal (**FIGURE 17.9**). A cluster of six *Antennapedia*-linked genes, called the **antennapedia complex**, possess similar regulatory functions.

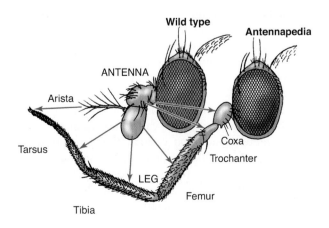

Wild type **Antennapedia**
ANTENNA
Arista
Tarsus
LEG
Tibia
Femur
Coxa
Trochanter

FIGURE 17.8 Positional correspondence between antennae and leg shown in *Drosophila melanogaster* carrying the homeotic *Antennapedia* mutation. Mutational substitution of leg tissues for antenna tissues is shown by the blue arrows; distal cells, for example, interpret their position as tarsal segments, and proximal cells as coxa or trochanter.

(a)

(b)

FIGURE 17.9 A four-winged *Drosophila* can be produced following a mutation at the *bithorax* locus. **(a)** As in all dipteran insects, *Drosophila* has only a single pair of wings, which arise from the second of the three thoracic segments; the second set of paired appendages evolved into balancing organs, *halteres* (blue structure below the wings). **(b)** Certain *bithorax* mutations result in the third thoracic segment transforming into a second thoracic segment complete with wings, a condition similar to ancestral four-winged flies.

Had a mutation leading to a second pair of wings occurred IN NATURE, *it may have led entomologists to erect a special* ORDER *(a higher level of taxon).*

As with the *Drosophila* antennapedia complex, a **bithorax** complex of three genes can initiate homeotic changes. This complex regulates many genes that control the fates of various structures from the posterior part of the second thoracic segment to the tip of the abdomen. Mutations in such gene complexes can alter the development of entire parts of organisms. For example, in *Drosophila melanogaster* the *bithorax* locus specifies a particular segment, the thorax. Mutations of the *bithorax* locus result in the production of an extra set of wings (Figure 17.9); the abdominal segment is converted into a thoracic segment and abdominal appendages (halteres) are converted into thoracic appendages (wings; Figure 17.9b).

As A–P and D–V axes and body segments are present in animals in the Early Cambrian Burgess Shale, the mechanisms for producing these features in animals such as *Drosophila* must have originated in the Precambrian. Insect evolution provides a well-studied example in which we can infer the sequence of changes from the uniform segments in the insect ancestor to the more complex segments found in fruit flies such as *Drosophila* (Figure 17.6). Comparative molecular studies provide insights into the time of origin of homeobox genes associated with axis formation, two anterior genes having arisen at the base of animal evolution (Chourrout et al., 2006).

HOMEOBOX GENES

An implication of the ancient evolutionary origin of the *bithorax* and *antennapedia* genes that specify axial organization of *Drosophila* embryos is that the mechanisms that establish the body axis could have a common genetic basis in all animals. If so, the common genetic system is expressed in different ways in different lineages: wings in some, legs in others; segments in some, lack of segments in others. How could such a genetic system be structured (Scholtz, 2010)?

The discovery in 1983 and publication in 1984 of a family of closely related DNA nucleotide sequences in DNA called the **homeobox** answered this question. Homeobox genes (genes that contain the homeobox) were found in various locations in the *Drosophila* genome, including loci within the *bithorax* and *antennapedia* complexes. Each 180-nucleotide-long homeobox gene codes for a polypeptide sequence of 60 amino acids that encodes for a protein domain known as the **homeodomain**. The homeodomain, in turn, is part of a transcription factor

BOX 17.1 Hox Genes and Vertebrate Origins

First a note about terminology: Homeobox-containing genes in vertebrates are known as **Hox genes**, Ho from homeodomain and x for vertebrate. Thus, *Pax-6* is a vertebrate homologue of the *Drosophila paired rule* (*Pa*) gene, which is combined with *x* to make *Pax*. The 6 indicates that five other genes in the Pax family were known when *Pax-6* was named. Convention also dictates (although not all follow the convention) that gene names are italicized (*Pax-6*), but the protein for which the gene codes is not, and that Hox genes in humans are italicized and capitalized (*PAX-6*). Recognition of the importance of Hox genes marks a fascinating episode in the history of the search for vertebrate origins (Akam, 1998; Carroll et al., 2005).

Hox genes with sequence homology to such *Drosophila* genes as *Ubx* and *Antennapedia* are a series of vertebrate transcription factors organized as homeobox clusters on four chromosomes. As shown for *Drosophila* in Figure 17.10, the order of Hox genes within a cluster is paralleled by an anterior–posterior sequence of expression. The patterning role of Hox genes is demonstrated in mouse embryos by knocking-out or knocking-in a Hox gene to eliminate or enhance its function. This actions results in a transformation of skull, vertebral, or other features into a more anterior element in the sequence—a homeotic transformation. The patterning role also can be observed in comparative studies across the vertebrates, such as analysis of the evolution of vertebral number (Müller et al., 2010 and see text). Regenerating amphibian limbs can be duplicated following manipulation of Hox genes. In some species of frogs, an amputated tail can be made to homeotically duplicate the posterior portion of the body and to regenerate a limb complete with pelvic girdle—not a tail. Finally, conservation of the homeotic roles of vertebrate and *Drosophila* genes has been demonstrated by research showing that, for example, after being inserted into the *Drosophila* genome, the homologous mouse *Hoxb-6* gene elicits leg formation in the place of antennae in *Drosophila*.

The number of Hox clusters varies among vertebrates (Figure 17.11): four clusters of 39 genes in mice, three clusters in lampreys, and up to seven in teleost fish. Duplication of the genome at the origin of the chordates is the most likely explanation for the four clusters; duplication sets up the possibility of future structural and functional divergence and *specialization of function among copies of the gene*. Four possibilities, which are not mutually exclusive, could explain this evolutionary change in gene function.

Two involve change in gene number, either in the number of Hox gene clusters or the number of genes per cluster. Duplication of *Hox* clusters before the teleosts arose—perhaps associated with duplication of large portions of chromosomes or entire chromosomes—would have taken the number from four to eight. This, coupled with subsequent loss of one cluster, would explain the seven clusters found in organisms such as zebra fish.

The *other two possibilities involve* modification of the function of individual Hox genes (*neofunctionalization*), *or increasing the complexity of interaction* between gene networks. Either of these changes could come about by alteration in genes that regulate a Hox gene (are "upstream") or in genes that are regulated by the Hox gene (are "downstream").

---

that binds to DNA, thereby regulating mRNA production (McGinnis et al., 1984). Fifty or more families of homeobox genes have been identified in animal genomes. (Homeobox genes in vertebrates, which are known as the **Hox gene family**, are introduced in **BOX 17.1**).

Although expressed in different structures in various taxa, patterning of the A–P and D–V axes throughout the animal kingdom is controlled by homeobox genes. Conserved homeobox genes also are present in flowering plants and fungi, in both of which homeobox genes produce homeodomain proteins much like those in animals. As discussed in the following section, homeobox genes must have functioned early in eukaryotic history, having arisen before plants, fungi, and animals diverged. It is hypothesized that homeobox genes diversified during the evolution of multicellular organisms by duplication from a common ancestral gene. Early duplications of this gene or gene complex, and subsequent divergent function(s) of each duplicate, provided a mechanism for different regions of the body to specialize.[12]

The HOMEOBOX *is a transcriptional sequence of 180 base pairs shared by all homeobox and Hox genes.*

---

[12] See Bharathan et al. (1997) and Van de Peer et al. (2009) for gene duplication and see Box 17.1 for duplication of Hox genes.

FIGURE 17.10 Homeobox gene relationships in representative animals. Almost all the genes illustrated show a highly conserved relationship between their position along the chromosome (shown in **b**) and developmental function along the anterior–posterior axis (shown in **[a]** and **[c]**). **(a)** Regions of action of homeobox genes in *Drosophila melanogaster* along the anterior–posterior axis of the embryo with *labial* acting most anteriorly and *Abdominal B* group genes acting most posteriorly. **(b)** Clusters of homeobox (Hox) genes from *Drosophila*, amphioxus (a cephalochordate), mammals (Hox-B from mice and humans), a beetle (*Tribolium castaneum*), and a nematode (*Caenorhabditis elegans*). Homologous (orthologous) genes are aligned vertically. **(c)** Expression of Hoxa and Hoxb genes in the anterior region of a mouse embryo. The left hand end of each bar marks the extent of anterior (rostral) expression of each gene, which determine important boundaries such as the position of pharyngeal arches 1 to 7, divisions 1 to 8 within the hindbrain, and boundaries between populations of migrating neural crest cells (*green*).

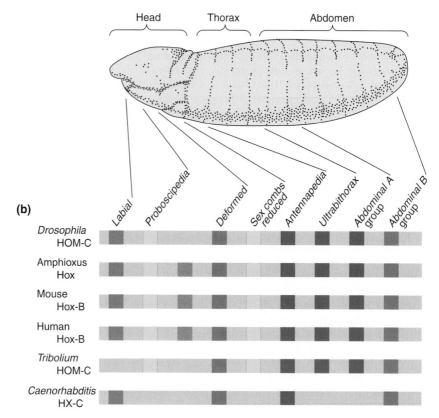

(a) *Drosophila* embryo

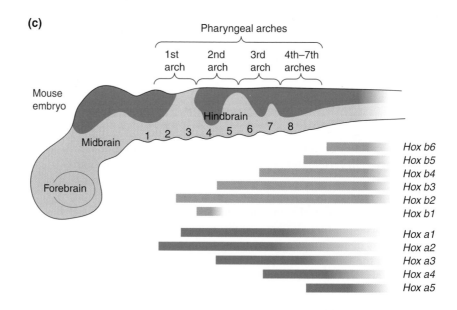

(c)

Surprisingly, the linkage order of these homeobox-containing genes in the *Drosophila antennapedia–bithorax* clusters accords with their expression along the A–P axis of *Drosophila* embryos (**FIGURE 17.10**). Equivalent homeodomains with equivalent linkage orders of homeobox genes occur across the animal kingdom (Figures 17.10 and 17.11)[13],

[13] A comprehensive analysis of the role of homeobox genes in the evolution of the arthropod body plan may be found in C. L. Hughes and T. C. Kaufman, 2002. Hox genes and the evolution of the arthropod body plan. *Evol. Devel.*, **4**, 459–499. For a recent analysis of what they term "preadapted genomes" and the origin of animal body plans, see Marshall and Valentine, 2010.

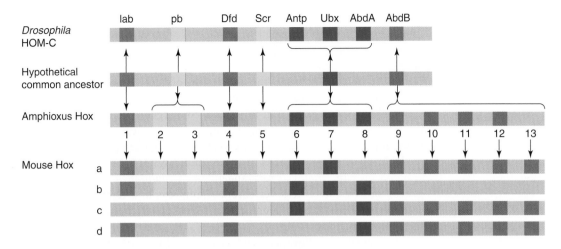

FIGURE 17.11 Relationships between homeotic gene clusters in *Drosophila*, Hox genes in amphioxus and mice, and homeotic genes, and in the hypothetical common ancestor of arthropods and chordates. Expansion of the six gene clusters in the common ancestor gave rise to eight clusters in *Drosophila* and to 13 in mice.

[Adapted from Hall, B. K., 1999. *Evolutionary Developmental Biology, Second Edition.* Kluwer Academic Publishers; modified from Holland, P. W. H., and J. Garcia-Fernàndez, 1996. *Dev. Biol.,* **173**, 382–395.]

indicating that specification of body plans in different lineages has a common and ancient evolutionary origin.

Finally, a particular homeobox protein can perform a similar function in different organisms; for example, much or all of the function of the *Drosophila* homeobox gene *Antennapedia* can be assumed by a protein produced by a homologous *Hoxb* gene in mice. Homologous developmental genes (such as *engrailed*, which specifies compartment distinctions within segments) function throughout the animal kingdom.

Several major conclusions come from these studies of comparative gene structure and function in the animal kingdom:

- These important developmental (regulatory) genes all **share** a common, highly conserved, and evolutionarily **ancient role** as transcription factors.
- A **common genetic evolutionary origin** underlies the conservation of the basic developmental pathways that establish animal body plans.
- What has **changed with evolution is context**; the specific function of these conserved regulators varies from cell lineage to cell lineage, tissue to tissue, organ to organ, as well as from time to time during development.
- Regulation of gene activity must play a major role in evolution.

GENE REGULATION AND NATURAL SELECTION

As outlined above, embryonic development is embedded in the differential regulation in time and space of genetic and cellular activity. Genetic changes can modify each developmental stage by affecting regulatory agents and processes, from signal reception to transcription and translation. François Jacob named this process "**tinkering**" (Jacob, 1977; Goode, 2007). Although Jacob's proposal is sometimes discussed as if he saw all evolutionary change as the result of tinkering, this is not the case. Jacob distinguished evolutionary change by tinkering from "the 'few big steps' in evolution which can only result from the acquisition of new genes and new genetic information. For Jacob, major steps in evolution could not be explained by changes in the regulation of preexisting genes" (Morange, 2007, p. 152).

Evolution by gene regulation helps explain why **evolutionary convergence** is more common than expected by chance. Conservation of gene pathways or networks, and changes in their regulation, explain the repeated occurrence of similar evolutionary changes. Complex developmental pathways can be tinkered with by up- or down-regulating the expression of individual genes or proteins, or by turning on or off cascades of events that determine the fates of individual body parts.[14] [A word of caution at this stage: convergent evolution of a phenotypic feature does not imply or require common underlying genetic pathways. As one example, Cooper and Alder (2006) review a fascinating example of convergent evolution of the antigen-receptor genes in the immune systems of jawed vertebrates (which is based on combinatorial rearrangement of *immunoglobins* or portions of T cell genes) and extant jawless vertebrates (lampreys and hagfish), which is based on combinatorial assembly of leucine-rich-repeat genetic modules that encode variable *lymphocyte receptors*.]

As regulatory pathways extend, change, and interact, organismal integration and complexity increases. This is one of the key reasons proposed for the observations that major changes in embryonic development occurred at the outset of multicellular organismal evolution, changes that would not be compatible with major modifications in the development of extant organisms (Hall, 2002, 2003). Cambrian and even late Precambrian **cleavage-stage animal embryos** are being found and analyzed. Interpretation of these embryos as fossils rather than, for example, giant bacteria, is becoming more firmly based as novel technological innovations such as x-ray tomography are applied to the specimens. While not numerous, these specimens have given us a glimpse of jellyfish development 530 Mya, and of the presence of segments in Cambrian embryos. Absence of any fossilized **larval stages** strongly suggests that early animals developed without a larval stage; that is, they had direct development. These fossils, combined with genetic and paleontological analyses, enable us to discuss with some confidence the nature of the last common animal or bilaterally symmetrical animal ancestor.[15]

DIRECT DEVELOPMENT *refers to a life cycle without a larval stage. Humans are a good example.* **INDIRECT DEVELOPMENT** *indicates a life cycle with a larval stage. Caterpillars/butterflies and tadpoles/frogs are good examples.*

CAVEFISH AS AN EXAMPLE

To be effective as an evolutionary mechanism gene regulation (tinkering) has to be responsive to selection. One class of evidence comes from the finding that similar gene networks can be used by different organs in the same individual yet respond differentially to selection. An example is a gene network regulated by the genes *Sonic hedgehog* (*Shh*), *Pax-6*, and the homeobox gene *Prox1* (*prospero-related homeobox 1*) and used in the development of sensory organs in animals. We discuss its operation in two forms (morphs) of a fish, the Mexican tetra, *Astyanax mexicanus*. One morph found in surface pools (the surface morph) has eyes, can see, and is pigmented. The other morph, found in caves, is blind (the result of reduced and vestigial eyes) and lacks pigmentation (**FIGURE 17.12**).[16]

---

[14] For elaboration of these points, see Wilkins, 2002, 2007; Carroll et al., 2005, and Davidson, 2006.

[15] For entry into the literature, see X.-P. Dong et al., 2004. Fossil embryos from the middle and late Cambrian period of Hunan, South China. *Nature*, **427**, 237–240; P. C. J. Donoghue, 2007. Embryonic identity crisis. *Nature*, **445**, 155–156; J.-Y. Chen et al., 2009. Complex embryos displaying bilaterian characters from Precambrian Doushantuo phosphate deposits, Weng'an, Ghizhou, China. *Proc. Natl. Acad. Sci. USA*, **106**, 19056–19060; and D. H. Erwin and E. H. Davidson, 2002. The last common bilaterian ancestor. *Development*, **129**, 3021–3032.

[16] For a recent and insightful analysis of evolution in caves, including influences of plate tectonics, and modes of colonization and speciation, see Juan et al., (2010).

(a)

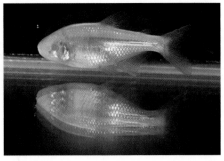
(b)

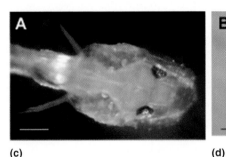
(c)

(d)

FIGURE 17.12 Surface **(a)** and cave **(b)** morphs of the Mexican cave fish, *Astyanax mexicanus* showing the failure of eye development and lack of pigmentation in the cave morph. Compare the reduced eyes in a blind cave morph 9 days post-fertilization with **(c)** the prominent eyes in a sighted surface morph 6 days post-fertilization **(d)**.

[(a) © Nature's Images, Inc.; (b) Courtesy of Dr. Simon Walker, Department of Zoology, University of Oxford; (c) and (d) Images courtesy of Tamara Franz-Odendaal and Megan Dufton, Mount Saint Vincent University.]

Sense organs other than the eyes also differ between cave and surface morphs. Fish use taste buds around the mouth and a sensory system found in a line that runs along the body, known as the *lateral line system*, to detect prey, movement, and vibration. Both taste buds and the lateral line system are enhanced in cave morphs in comparison to surface morphs. Eyes, taste buds, and the lateral line system are developmental **modules**. Modules and the concept of **modularity** are discussed in **END BOX 17.2**.

The gene network initiated by *Shh/Pax-6* operates in the eye and in the taste buds that surround the mouth (**FIGURE 17.13**). As an upstream regulatory gene, *Shh* elicits different responses from the two developing organ systems. Expression of *Shh* suppresses eye development but enhances taste bud development, whereas non-expression of *Shh* results in eyes but fewer taste buds. In laboratory experiments, over-expressing *Shh* in surface fish (in which *Shh* is normally repressed) results in diminished eye development but increases the numbers of taste buds. Knocking out *Shh* enhances eye development and results in fewer taste buds. Selection could operate on the adult cavefish phenotype, on the sensory modules, or on the common gene network (Figure 17.13). Shared but pleiotropic action of *Shh* enables these modules to evolve in a coordinated fashion in response to selection to change the phenotype.[17]

---

[17] See Jeffery et al. (2000), Franz-Odendaal and Hall (2006), and Jeffery (2010). A recent study of a cave-dwelling crustacean, the water slater *Asellus aquaticus* (which also has surface forms), shows that eye loss and eye reduction are regulated by different genetic pathways (Protas et al., 2011).

FIGURE 17.13 **(a).** Three modules—lateral line, taste buds, eyes—in Mexican tetra *Astyanax mexicanus* share an upstream control (*Prox1*) for gene pathways. **(b)** *Prox1* signals through hedgehog (*Shh*) in two of the pathways; the equivalent gene for lateral lines modules is not known. **(c)** The common upstream genetic control shown in (b) operates whether selection is on the adult phenotype, development (sensory modules), or common gene pathway, resulting in expansion of the lateral line and taste bud modules and reduction in the eye module.

[Adapted from Franz-Odendaal, T. A., and B. K. Hall, 2006: *Evol. Dev.*, **8**, 94–100.]

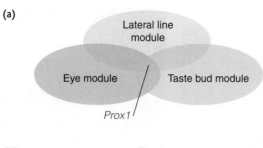

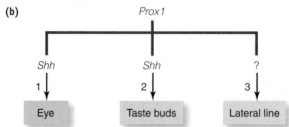

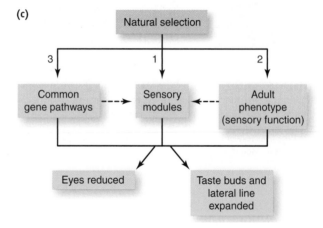

■ KEY TERMS

antennapedia complex
anterior–posterior (A–P) (axis)
arms race
base substitutions
Burgess Shale fossils
Cambrian explosion
cis-, *trans-*, and RNAi-gene regulation
cleavage (animal development)
coevolution
convergence (convergent evolution)
co-option
crown group
dorso–ventral (D–V) (axis)
duplication (genetics)
Ediacaran Biota
Ediacaran Period
fractal
heterozygote

heterozygote advantage
homeobox
homeodomain
homeotic mutations (homeosis)
homozygote
Hox gene
larvae
micro RNA (miRNA)
modularity
molecular clock
morphogenesis
morphospace
Piwi-interacting RNAs (piRNAs)
pleiotropy
predation
rangeomorphs
regulatory gene
sickle cell allele

sickle cell anemia
snowball Earth
stem group

tinkering (bricolage)
transcription
Vendozoa

■ EVOLUTION ON THE WEB

Explore evolution on the Internet! Visit the accompanying website for *Strickberger's Evolution, Fifth Edition,* at **go.jblearning.com/Evolution5eCW** for exercises and links relating to topics covered in this chapter.

■ REFERENCES

Akam, M., 1998. Hox genes: from master genes to micromanagers. *Curr. Biol.*, **8**, R676–R678.

Bharathan, G., B.-J. Janssen, E. A. Kellogg, and N. Sinha, 1997. Did homeodomain proteins duplicate before the origin of angiosperms, fungi, and metazoa? *Proc. Natl. Acad. Sci. USA*, **94**, 13749–13753.

Carroll, S. B., 2005. *Endless Forms Most Beautiful*. W. W. Norton & Co., New York.

Carroll, S. B., J. K. Grenier, and S. D. Weatherbee, 2005. *From DNA to Diversity. Molecular Genetics and the Evolution of Animal Design*, 2nd ed. Blackwell Publishing, Malden, MA.

Chookajorn, T., R. Dzikowski, M. Frank, et al., 2006. Epigenetic memory at malaria virulence genes. *Proc. Natl. Acad. Sci. USA*, **104**, 899–902.

Chourrout, D., F. Delsuc, P. Chourrout, et al., 2006. Minimal protoHox cluster inferred from bilaterian and cnidarian Hox complements. *Nature*, **442**, 684–687.

Cooper, M. D., and M. N. Alder, 2006. The evolution of adaptive immune systems. *Cell*, **124**, 815–822.

Davidson, E. H., 2006. *The Regulatory Genome. Gene Regulatory Networks in Development and Evolution*. Elsevier/Academic Press, Burlington, MA.

Eyre-Walker, A., and P. Keightley, 2007. The distribution of fitness effects of new mutations. *Nat. Rev. Genet.*, **8**, 610–618.

Ferreira, A., I. Marguti, I. Bechmann, et al. 2011. Sickle hemoglobin confers tolerance to *Plasmodium* infection. *Cell*, **145**, 396–409.

Franz-Odendaal, T. A., and B. K. Hall, 2006. Modularity and sense organs in the blind cavefish, *Astyanax mexicanus. Evol. Devel.* **8**, 94–100.

Goode, J. (ed.), 2007. *Tinkering: The Microevolution of Development*. Novartis Foundation Symposium No. 284. John Wiley and Sons, Chichester, UK.

Hall, B. K., 1999. *Evolutionary Developmental Biology*, 2nd ed. Kluwer Academic Publishers, Dordrecht, Netherlands.

Hall, B. K., 2002. Palaeontology and evolutionary developmental biology: a science of the 19th and 21st centuries. *Palaeontology*, **45**, 647–669.

Hall, B. K., 2003. The emergence of form: the shape of things to come. *Dev. Dyn.*, **228**, 292–298.

Hallgrímsson, B., and B. K. Hall (eds.), 2003. *Variation: A Central Concept in Biology*. Elsevier/Academic Press, Burlington, MA.

Heimberg, A. M., L. S. Sempere, V. N. Moy, P. C. J. Donoghue, and K. J. Peterson, 2008. MicroRNAs and the advent of vertebrate morphological complexity. *Proc. Natl. Acad. Sci. USA*, **105**, 2946–2950.

Hibino, T., A. Nishino, and S. Amemiya, 2006. Phylogenetic correspondence of the body axes in bilaterians is revealed by the right-sided expression of *Pitx* genes in echinoderm larvae. *Dev. Growth Differ.*, **48**, 587–595.

Jacob, F., 1977. Evolution as tinkering. *Science*, **196**, 1161–1166.

Jacob, F., and J. Monod, 1961. Genetic regulatory mechanisms in the synthesis of proteins. *J. Mol. Biol.*, **3**, 318–356.

Jeffery, W. R., 2010. Pleiotropy and eye degeneration in cavefish. *Heredity*, **105**, 496–497.

Jeffery, W. R., A. G. Strickler, S. Guiney, et al., 2000. *Prox 1* in eye degeneration and sensory organ compensation during development and evolution of the cavefish *Astyanax. Dev. Genes Evol.*, **210**, 223–230.

Juan, C., M. T. Guzik, D. Jaume, and S. J. B. Cooper, 2010. Evolution in caves: Darwin's 'wrecks of ancient life' in the molecular era. *Mol. Ecol.*, **19**, 3865–3880.

Kosik, K. S., 2010. MicroRNAs and cellular phenotypy. *Cell*, **143**, 21–26.

Lehrer, J., 2008. *Proust Was a Neuroscientist*. Houghton Mifflin, Boston.

Marshall, C. R., and J. W. Valentine, 2010. The importance of preadapted genomes in the origin of the animal body plans and the Cambrian explosion. *Evolution*, **64**, 1189–1201.

Martindale, M. Q., J. R. Finnerty, and J. Q. Henry, 2002. The Radiata and the evolutionary origins of the bilaterian body plan. *Mol. Phylogenet. Evol.*, **24**, 358–365.

Martinez, P., J. P. Rast, C. Arena-Mena, and E. H. Davidson, 1999. Organization of an echinoderm Hox gene cluster. *Proc. Natl. Acad. Sci. USA*, **96**, 1469–1474.

Matus, D. Q., K. Pang, H. Marlow, et al., 2006. Molecular evidence for deep evolutionary roots of bilaterality in animal development. *Proc. Natl. Acad. Sci. USA*, **103**, 11195–11200.

McGinnis W., M. S. Levine, E. Hafen, A, Kuroiwa, and W. J. Gehring, 1984. A conserved DNA sequence in homoeotic genes of the *Drosophila* Antennapedia and bithorax complexes. *Nature*, **308**, 428–433.

Morange, M., 2007. French tradition and the rise of evo-devo. *Theory Biosci.*, **126**, 149–153.

Müller, J., T. M. Scheyer, J. J. Head, et al., 2010. Homeotic effects, somitogenesis and the evolution of vertebral numbers in recent and fossil amniotes. *Proc. Natl. Acad. Sci. USA*, **107**, 2118–2123.

Peter, I. S., and E. H. Davidson, 2011. Evolution of gene regulatory networks controlling body plan development. *Cell*, **144**, 970–985.

Peterson, K. J., M. R. Dietrich, and M. A. McPeek, 2009. MicroRNAs and metazoan macroevolution: insights into canalization, complexity, and the Cambrian explosion. *BioEssays*, **31**, 736–747.

Protas, M. E., P. Trontelj, and N. H. Patel, 2011. Genetic basis of eye and pigment loss in the cave crustacean, *Asellus aquaticus. Proc. Natl. Acad. Sci. USA*, **108**, 5702–5707.

Prud'homme, B., N. Gompel, and S. B. Carroll, 2007. Emerging principles of regulatory evolution. *Proc. Natl. Acad. Sci. USA*, **104**, 8605–8612.

Ravasi, T., H. Suzuki, C. V. Cannistraci, et al., 2010. An atlas of combinatorial transcriptional regulation in mouse and man. *Cell*, **140**, 744–752.

Rebeiz, M., N. Jikomes, V. A. Kassner, and S. B. Carroll, 2011. Evolutionary origin of a novel gene expression pattern through co-option of the latent activities of existing regulatory sequences. *Proc. Natl. Acad. Sci. USA*, **108**, 10036–10043.

Saina, M., G. Genikhovich, E. Renfer, and U. Technau, 2009. Bmps and chordin regulate patterning of the directive axis in a sea anemone. *Proc. Natl. Acad. Sci. USA*, **106**, 18592–18597.

Scholtz, G., 2010. Deconstructing morphology. *Acta Zool. (Stockh.)*, **91**, 44–63.

Sniegowski, P., P. Gerrish, T. Johnson, and A. Shaver 2000. The evolution of mutation rates: separating causes from consequences. *BioEssays*, **22**, 1057–1066.

Van de Peer, Y., S. Maere, and A. Meyer, 2009. The evolutionary significance of ancient genome duplications. *Nat. Rev. Genet.*, **10**, 725–732.

Wagner, G., and J. Zhang, 2011. The pleiotropic structure of the genotype–phenotype map: the evolvability of complex organisms. *Nat. Rev. Genet.* **12**, 204–213.

Wang, Z., B.-Y. Liao, and J. Zhang, 2010. Genomic patterns of pleiotropy and the evolution of complexity. *Proc. Natl. Acad. Sci. USA*, **107**, 18034–18039.

Wilkins, A. S., 2002. *The Evolution of Genetic Pathways*. Sinauer Associates, Sunderland, MA.

Wilkins, A. S., 2007. Between "design." *Proc. Natl. Acad. Sci. USA*, **104**, 8590–8596.

Wray, G. A. 2007. The evolutionary significance of cis-regulatory mutations. *Nat. Rev. Genet.*, **8**, 206–216.

END BOX 17.1

Animals Arise

SYNOPSIS: This extensive discussion on the **origin and evolution of animals** will be especially appropriate for those wanting information on how we recognize different types of animals, how we know when animals first arose, and how we describe and evaluate patterns and processes of organismal evolution. To these ends, this end box:

(1) introduces the animals found as fossils in early Cambrian deposits worldwide, especially those of the Burgess Shale formation in British Columbia;

(2) discusses what looks to have been an alternate Precambrian experiment in organismal evolution, the *Ediacaran Biota*;

(3) assesses how the earliest representatives of a lineage (*stem taxa*) are recognized and distinguished from *crown taxa*, and why related organisms do not occupy all the regions of *morphospace* that theories suggest they could;

(4) evaluates whether *evolutionary molecular clocks* can be calibrated from rates of change in molecules over time and used to determine times of origination, branching, and divergence in animal evolution;

(5) considers evidence that animals must have originated in the Precambrian, even though few Precambrian animal fossils exist; and

(6) discusses proposed causes for the origin and diversification ("explosion") of animals perhaps as few as 10 million years at the base of the Cambrian Period.

Within a relatively short geological time span in the late Precambrian, 700 to 750 Mya, and in what may have been as little as 10 My, an explosive radiation marked the emergence of several dozen distinctive and different animal body plans representing virtually every known animal phylum (**FIGURES EB1.1 AND EB1.2**). We can identify these organisms as animals and place them into phyla and lower taxa because they possess the phylum-specific characters seen in their descendants. However, the major groups of animals originated and were well established before they appear in the fossil record. Indeed, as much as 200 My of evolution may have taken place in the Precambrian before the fossil record of animals appeared. By the beginning of the Cambrian, 545 Mya, many different types of animals were present, recognizable, and today can be identified and classified on the basis of their different body forms (Valentine, 2004; Marshall and Valentine, 2010).

Before discussing the origin of animals, however, we must introduce a group(s) of multicellular organisms that remain enigmatic to this day. They are known as the **Ediacaran Fauna** after the Ediacara Hills in South Australia where they were discovered. In hindsight, "fauna" is not the most appropriate term to describe this and similar associations of organisms. Fauna refers to the animal life found in a specific region or at a particular time. Flora is a similar term used for the plants of a region or period. As the Ediacaran organisms are not animals (with one possible exception) the term fauna is inappropriate. *Biota* is a more appropriate term that applies to all the organisms found in a region or at a geological period. So we will speak of the **Ediacaran Biota**.

A new geological period, the EDIACARAN PERIOD, *635 to 542 Mya, has been named for this geological age.*

EDIACARAN BIOTA

A diverse and inscrutable assemblage of organisms, the Ediacaran Biota existed from early in the Precambrian into the Early Cambrian (**FIGURES EB1.3**). The soft-bodied fossils found in the Precambrian Ediacaran strata of South Australia—and those discovered more recently in deposits at other locations around the globe—are strange indeed. The fossils at Mistaken Point, Newfoundland, are preserved beneath a thick layer of volcanic ash that has been dated at 565 ± 3 My. Amongst the oldest Ediacaran organisms, these fossil communities provide amazing insights into a *Precambrian ecosystem*.[a]

[a] M. E. Clapham et al., 2003. Paleoecology of the oldest-known animal communities: Ediacaran assemblages at Mistaken Point, Newfoundland. *Paleobiology*, **29**, 527–544; M. Brasier and J. Antcliffe, 2004. Decoding the Ediacaran enigma. *Science*, **305**, 1115–1117. Even older forms from 577–635 Mya and with no apparent relationships to other Ediacarans have now been described from China by Yuan et al. (2011).

END BOX 17.1

Animals Arise (Cont...)

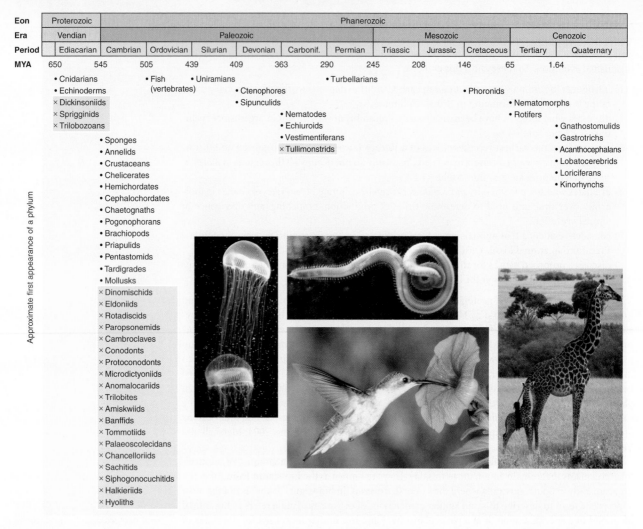

Eon	Proterozoic		Phanerozoic										
Era	Vendian		Paleozoic							Mesozoic			Cenozoic
Period	Ediacarian	Cambrian	Ordovician	Silurian	Devonian	Carbonif.	Permian	Triassic	Jurassic	Cretaceous	Tertiary	Quaternary
MYA	650 545	505	439	409	363	290	245	208	146	65	1.64

Approximate first appearance of a phylum

- Cnidarians
- Echinoderms
- × Dickinsoniids
- × Sprigginids
- × Trilobozoans
- Sponges
- Annelids
- Crustaceans
- Chelicerates
- Hemichordates
- Cephalochordates
- Chaetognaths
- Pogonophorans
- Brachiopods
- Priapulids
- Pentastomids
- Tardigrades
- Mollusks
- × Dinomischids
- × Eldoniids
- × Rotadiscids
- × Paropsonemids
- × Cambroclaves
- × Conodonts
- × Protoconodonts
- × Microdictyoniids
- × Anomalocariids
- × Trilobites
- × Amiskwiids
- × Banffids
- × Tommotiids
- × Palaeoscolecidans
- × Chancelloriids
- × Sachitids
- × Siphogonocuchitids
- × Halkieriids
- × Hyoliths

- Fish (vertebrates)
- Uniramians
- Ctenophores
- Sipunculids
- Nematodes
- Echiuroids
- Vestimentiferans
- × Tullimonstrids

- Turbellarians
- Phoronids
- Nematomorphs
- Rotifers
- Gnathostomulids
- Gastrotrichs
- Acanthocephalans
- Lobatocerebrids
- Loriciferans
- Kinorhynchs

FIGURE EB1.1 Approximate times at which various major metazoan lineages first appear in the fossil record. Unshaded groups marked with filled-in circles have extant descendants, although the original species representing these groups are long extinct. Shaded groups marked with xs represent extinct clades (orders or phyla) that have no known surviving descendants.

[Adapted from Conway Morris, 1993. *Nature*, **361**, 219–225 and from Conway Morris, 1998. *The Crucible of Creation: The Burgess Shale and the Rise of Animals.* Oxford University Press, Oxford, UK; INSETS: © Dwight Smith/ShutterStock, Inc.; Courtesy of NOAA, National Estuarine Research Reserve; © AbleStock.]

Any relationship between Ediacaran and other multicellular organisms, such as the animals discussed in the next section, has been hard to uncover. This may be because the Ediacaran Biota was an independent evolutionary experiment(s) in multicellularity. We say "experiment(s)" rather than lineage(s) because these assemblages contain multiple lineages of organisms whose relationships to one another are as obscure as are their relationships to other multicellular organisms. We cannot even be sure whether they are eukaryotic in their cellular organization; they may be as much an independent experiment in cellular evolution as in organismal design.

As depicted in Figure EB1.3, some Ediacaran fossils superficially resemble jellyfish. Others bear likenesses to segmented worms and, more distantly, to mollusks and echinoderms (see Figure EB1.1). Whether these resemblances mean that Ediacaran organisms (or some Ediacaran lineages) were animals is another matter entirely. Paleobiologists have not detected any basic

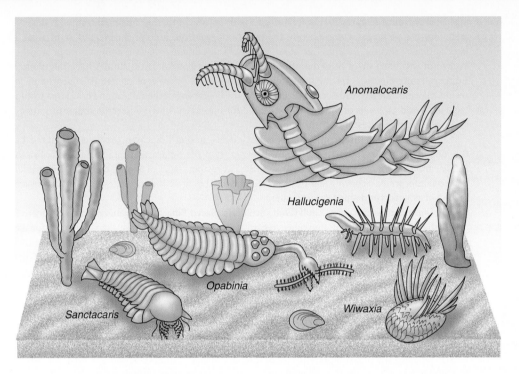

FIGURE EB1.2 Part of the diversity of Cambrian animals from the Burgess Shale of British Columbia, Canada. Many are arthropods, including some trilobite-like animals (*Sanctacaris*), the giant *Anomalocaris*, and *Opabinia* with its anterior feeding appendage. *Aysheaia*, seen feeding on a sponge, is considered in the ancestral arthropod lineage related to onychophora (velvet worms).

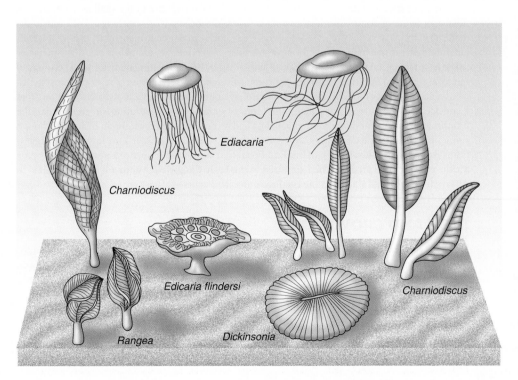

FIGURE EB1.3 A panorama of soft-bodied animals found in Ediacaran tidal flat deposits of South Australia and 10 or so other places throughout the globe, most occurring in the Precambrian, 545 to 650 Mya. *Ediacara* resembles extant jellyfish. Others, like *Rangea* (see also Figure EB1.4) and *Charniodiscus*, are frond-like and bear no resemblance to extant animals or plants.

END BOX 17.1

Animals Arise (Cont...)

© Photos.com

animal anatomical features in any Ediacaran fossil: no eyes, mouths, anuses, intestinal tracts, or locomotory appendages. Consequently, the relationship of Ediacaran organisms to Cambrian and later animals remains unclear. That said, we should not assume that we know all there is to be known of this biota or of the Ediacaran Period. Recent paleontological studies have interpreted Ediacaran microfossils as resting stages of animal embryos (Cohen et al., 2009). Paleogeochemical results may indicate a rise in ocean circulations at the beginning of the Ediacaran Period that would have increased O_2 levels of the deep sea-ocean floor boundary where Ediacaran organisms arose, and where animal ancestors may be found.[b]

The existence of divergent lineages of organisms among Ediacaran fossils is illustrated by the identification and reconstruction of *Kimberella* from the Ediacaran Biota of the White Sea in Russia, and by what Cohen et al., (2009) identified as resting stages of animal embryos. *Kimberella* has been interpreted as having features consistent with it being a common ancestor (stem taxon; see below) of mollusks and allied invertebrates.

A German paleontologist, Adolf (Dolf) Seilacher, placed Ediacaran organisms into a distinctive group, the **Vendozoa**, which he and others regard as unrelated to any animals; a separate evolutionary experiment.[c] The absence of features usually associated with prey capture, and the absence of digestive tracts led to proposals that many Ediacaran organisms may have depended on photosynthetic symbionts, or on other types of symbiosis, as an alga and a fungus do today, cooperating to form a lichen.

The Canadian paleontologist, Guy Narbonne, invoked repeated structures (**modularity**; see End Box 17.2) based on fractal repeats of a frond structure in his analyses of early Ediacaran organisms to identify a major group which he named **rangeomorphs** (FIGURE EB1.4). Rangeomorphs existed in complex ecological communities with more than one trophic level (Narbonne, 2004). One species, *Charnia wardi*, found in the Drook Formation in Newfoundland, grew to heights of two meters. It is the oldest representative of this assemblage, dating to 570 to 575 Mya. Narbonne is cautious when it comes to saying whether rangeomorphs were animals or an alternative (and earlier) evolutionary experiment in multicellularity.

Another species, *Funisia dorothea*, is the most abundant species in the shallow water marine Ediacaran fauna of South Australia. *Funisia* consists of tubes up to 30 cm long and 12 mm wide, made up of modular serially repeated elements (segments?) anchored to a microbial mat by holdfasts (as you might see in a living kelp). The closest similarity of *Funisia* to animals is as a stem sponge or jellyfish, but that is no more than conjecture at this stage.[d] The contrasting body form and ecology of rangeomorphs and *Funisia* attest to the enormous diversity of Precambrian species and ecosystems.

Whatever their affinities, the morphological diversity seen in the Ediacaran biotas shows that a considerable diversity of body plans and types of organisms existed up to, or even across, the Cambrian boundary, and that these organisms had a considerable Precambrian history. Although some survived into the Cambrian, they soon became extinct. Why? We don't know. One suggestion, discussed in the last section of this End Box is that lack of armor made these animals easy prey for the many mobile predators that appeared in the Early Cambrian. Another is that they failed to survive the "**snowball Earth**" in the late Precambrian, discussed elsewhere in text.

STEM AND CROWN TAXA

Numerous interpretations of the Ediacaran Biota have been proposed. Any interpretation must distinguish the morphological features that define phyla and distinguish **stem** from **crown taxa**. A stem taxon is the earliest representative of a lineage. There can only be one stem taxon for a

A FRACTAL *is a shape that can be split into parts, each of which has the same shape as the original. A fern leaf is an example of a fractal.*

FIGURE EB1.4
A rangeomorph frondlet to show the organization based on repetitive branching from a central stalk. Compare with the frond-like forms in Figure EB1.3.

[Reproduced from Narbonne, Guy, *Science* 305 (2004): 1141–1144. Reprinted with permission from AAAS.]

[b] For evaluations of these lines of evidence, see Shen et al. (2008), Cohen et al. (2009), and Li et al. (2010).

[c] A. Seilacher, 1989. Vendozoa: organismic construction in the Proterozoic biosphere. *Lethaia*, **22**, 229–239.

[d] M. L. Droser and J. C. Gehling, 2008. Synchronous aggregate growth in an abundant new Ediacaran tubular organism. *Science*, **319**, 1660–1662.

lineage. Crown taxa are the terminal branches of a lineage arising from a stem taxon. A single lineage may have more than one crown taxon. Stem taxa do not show all the features of the crown groups into which they evolve. Crown taxa are at the top of the tree, stem taxa at the bottom.

Some crown taxa went extinct before the present day. Trilobites, which are derived arthropods (a crown taxon), went extinct 248 Mya at the end of the Permian. Because trilobites did not give rise to another lineage, Cambrian trilobites are crown taxa. Other arthropod crown taxa, of different lineages than trilobites, exist today. Cambrian crustaceans tell a different story. Some are crown taxa, while others show some, but not all of the features of a crown taxon crustacean (Adamowicz et al., 2008).

MORPHOSPACE

The only evidence we possess to recognize and classify Early Cambrian organisms is morphological. How do we separate individuals into crown taxa, such as phyla, and decide whether the range of morphologies in Cambrian animals is similar to the range of morphologies in living members of the same groups?

One approach measures features and plots them in three dimensions. Imagine plotting the height, weight, and waist measurements of all modern humans in three dimensions, using the height, weight, and diameter of all mammals as the three axes. If some humans were as big as elephants or as small as mice, the entire three-dimensional (3-D) space would be filled. However, as humans represent only a small portion of animal body sizes, they occupy only a small region of the 3-D space. Elephants would cluster at one extreme, mice at another extreme.

In 1966, American paleontologist and paleobiologist David Raup developed such an approach to plot the morphologies of organisms in three dimensions in what he termed **morphospace**, represented as a cube with different features of the organisms along each axis (FIGURE EB1.5). Morphospace represents the universe of all possible forms. If all morphologies were possible, the cube (morphospace) would be filled. Occupancy of only a portion of morphospace by known extinct and extant forms suggests that only a subset of morphologies is possible. Raup's analysis showed that the known morphologies of shelled invertebrates (bivalves, brachiopods, cephalopods, and gastropods) cluster in one region of morphospace (Figure EB1.5), a clustering that demarks the range of body plans in these organisms. Unoccupied morphospace

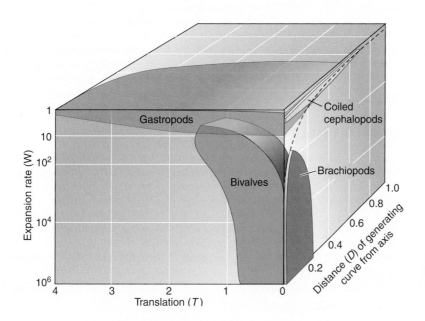

FIGURE EB1.5 Only a portion of morphospace (represented as three-dimensional space) is occupied by known shell morphologies of bivalves, gastropods, brachiopods, and coiled cephalopods. The axes represent three measures of shell morphology/growth.

[From Hall, B. K., 1999. *Evolutionary Developmental Biology, Second Edition* Kluwer Academic Publishers, Dordrecht, The Netherlands; modified from Raup, D. M., *J. Paleontol.*, **40**, 1966. 1178–1190.]

could represent impossible morphologies, nonadaptive morphologies, and/or constraints on morphology.[e]

BURGESS SHALE FAUNA

The most well known formation in which Cambrian organisms are preserved is a 160 m deep and more than 20 km long limestone reef in the **Burgess Shale** in Yoho National Park, British Columbia.

Animals of the Burgess Shale were neither an isolated evolutionary experiment nor unrepresentative of the general situation in the Early to Middle Cambrian. At least 12 other sites contain animals with an equivalent range of body plans, including faunas in Pennsylvania, north Greenland, Spain, Poland, and China, including the Chengjiang fauna, which is some 10 My older than the Burgess Shale fauna.[f] Recent discoveries in southeastern Morocco of some 1,500 specimens representing 50 or more taxa reveal that Burgess Shale forms, including stem groups, persisted into the Lower Ordovician (Van Roy et al., 2010).

Although discovered in 1909, not until an expedition in the late 1960s did the Burgess Shale reveal its true story, one with a cast of 124 genera and 140 species. A little over a third of the genera are arthropods (especially trilobites), but sponges, brachiopods, polychaete worms, echinoderms, cnidarians, and mollusks all are represented. Discoveries continue to be made; a soft-bodied, shell-less cephalopod *Nectocaris pteryx* from the Middle Cambrian and described in 2010, extends the fossil record for cephalopods back 30 My.[g] So well known is the Burgess Shale fauna that the scientific names of some species are known outside paleontology—*Marrella splendens* and *Canadapsis perfecta* (arthropods) and the velvet worm *Hallucigenia sparsa* may be familiar to you.

Our ability to assign organisms from the Burgess Shale to clades, such as phyla, is *prima facie* evidence for three important and interrelated conclusions.

1. These organisms already had evolved the characters that define phyla of living animals.
2. Some of these lineages were crown taxa. Others had some, but not all, of the features of the phylum to which they are assigned.
3. Origination of phyla from stem taxa has to be sought in earlier forms in older deposits.

The diversity of Cambrian organisms could be interpreted as adaptive radiation in which many new ecological opportunities were made available for organisms with the capacity to evolve in diverse ways and to occupy and exploit changing environments (Figure EB1.2). If resources were limited, increasing competition for those resources among so many groups would have led to selection among them. Additionally, entire populations of organisms, or entire ecosystems, could have been lost through geological events. As discussed under "Causes of the Cambrian Radiations" below, just as at many other times in the history of life, both selective and accidental factors led to extinction or to adaptation, survival, and possible diversification. One way to find the *time of origin* of these (or indeed of any) forms is to use nucleotide sequence data as an **evolutionary molecular clock**.

EVOLUTIONARY MOLECULAR CLOCKS

Phylogenetic trees based on molecular characters provide a means of establishing relationships independently of morphology. Reconstructing ancestral states using combined morphological *and*

---

[e] For evaluations of the evidence for how morphological variation is constrained/bounded, see Hall (2003, 2007), Alfaro and Santini (2010), and Scholtz (2010).
[f] See Gould (1989), Briggs et al. (1994), Conway Morris (1998), Hou et al. (2004), Erwin et al. (2011), and Liu et al., (2011).
[g] M. R. Smith and J.-B. Caron, 2010. Primitive soft-bodied cephalopods from the Cambrian. *Nature*, **465**, 469–472.

molecular characters provides further hypotheses of phylogenetic relationships that can be tested. Thirdly, fossils and the ages of geological strata can be used to calibrate molecular clocks.[h]

Inherent in the reconstruction of any phylogenetic tree based on molecular characters is:

(a) that evolutionary differences between organisms are related to molecular differences, and
(b) that the greater the number of molecular differences between organisms, the greater the evolutionary distance between the organisms.

In many past phylogenetic reconstructions, mutational differences were used to provide evolutionary time scales. The underlying premise was, and still is, that mutations are incorporated (fixed) at fairly regular rates over time, and so can be used to calculate the molecular clock.

A basic assumption is that molecular clocks can be calibrated for any protein (or gene) for which nucleotide data are available. This means that the proportional rate of fixation of one protein relative to the rates of fixation in other proteins should stay the same throughout the history of any lineage. But molecular clocks for different proteins do not run at the same speed (Dietrich and Skipper Jr., 2007; Ho, 2007). Averaging rates among different genes to obtain a single rate has been done, but caution has to be exercised; no single molecular clock applies to every nucleotide sequence. The molecular clock does not tick at the same rate in all taxonomic groups, nor for all genes or all proteins.

PRECAMBRIAN ORIGIN OF ANIMALS

Although ancestral connections are unclear, many, if not all, of the forms present in the Cambrian are interpreted as new adaptive radiations of Precambrian stem taxa; each form is too specialized when first seen in the fossil record to have originated in the Cambrian. Soft-bodied stem taxa are not preserved in the fossil record, with the possible exception of *Kimberella* in the Ediacaran Biota.

Based on an enormous amount of morphological and molecular data, we can conclude that divergences among major Cambrian lineages began in the Precambrian between 700 Mya and 750 Mya. Modifications during the Cambrian added hard parts and mineralized skeletons that provided leverage for evolving muscles, support for body organs, enclosures for gills and filtering systems, and protective shells and spines (Figure EB1.1).[i]

How did so many body plans arise and why have so few, if any, arisen since the Early Cambrian? Evidence supporting theories based on physical factors—geological, environmental, climatic, and/or atmospheric change—and ecological factors such as predator–prey associations are outlined in the last section of this End Box. Evidence that the origin of animals reflects the origin of major regulatory genes and gene networks is summarized in the text.

CAUSES OF THE CAMBRIAN RADIATIONS

As outlined above, essentially all animal phyla were present at the base of the Cambrian, 545 Mya. At least five hypotheses have been produced and supporting evidence amassed to explain the evolutionary burst of multicellular animals in what is known as the **Cambrian radiation** or **Cambrian explosion** (although from the discussion above, you realize that this was a Precambrian radiation).

1. Geological, environmental, and climatic conditions
2. Changing O_2 levels

---

[h] For molecular clocks in various contexts, see Thorpe (1982), Feng et al. (1997), Donoghue and Smith (2004), Müller and Reisz (2005), Dietrich and Skipper Jr. (2007), and Parfrey et al. (2011). For whether molecular clocks "run at all," see Schwartz and Maresca (2006).
[i] For when animal groups diverged, see S. Conway Morris, 1993. The fossil record and the early evolution of the metazoa. *Nature*, **361**, 219–225; D. H. Erwin and E. H. Davidson, 2002. The last common bilaterian ancestor. *Development*, **129**, 3021–3032; Ayala et al. (1998), Bromham et al. (1998), Hall (1999), Donoghue and Smith (2004), Valentine (2004), and Erwin et al. (2011).

3. Predator–prey relationships
4. Evolution of embryonic development and body plan specification
5. New sources of genetic variation and changes in gene number and regulation

Evidence for numbers 1 to 4 is discussed below and by Conway Morris (1998) and Knoll (2003). Sources of genetic variation (number 5) are discussed elsewhere in this chapter.

GEOLOGICAL CONDITIONS

Change in **oxygen concentration** is one of several factors hypothesized to have facilitated the Cambrian Explosion (Knoll, 1992; Dahl et al., 2010).

Animal fossils may be absent from Precambrian rocks because geological conditions prevented fossilization or destroyed any fossils present; the heat and pressure involved in Precambrian mountain building has been proposed as an important factor limiting fossilization. However, prokaryotic and eukaryotic organisms (and the Ediacaran Biota) were preserved in the Precambrian. Therefore, it is unlikely that the Cambrian discontinuity in animals is unreal and merely the consequence of geological metamorphism or imperfect fossilization.

Another physical mechanism offered to explain the Cambrian Explosion is plate tectonics (discussed elsewhere in this text) and the resulting changes in the shape, extent, and latitude of shorelines, climate, and environment. Sea level changes that accompany glaciation would have played a role by opening up new environments also.

RISING OXYGEN LEVELS

As discussed elsewhere in this text, coincident with the rise and spread of cyanobacteria, O_2 began to accumulate in the atmosphere a billion years before the Cambrian. Evidence also exists for major increases in oxygen during the Precambrian.

Oxygen forms a protective blanket of ozone that would have facilitated the expansion and radiation of multicellular animals in shallow waters, tide pools, and nearby rocky surfaces. Most recent evidence is based on analyses of molybdenum isotopes in sedimentary rocks (black shale) that detect global rises in oxygenation 550–560 Mya (when animals diversified) and around 400 Mya (when land plants and large predatory fish diversified; Dahl et al., 2010).

Extrapolating from our knowledge of the descendants of Cambrian organisms we can conclude that aerobic metabolism, which is dependent on oxygen, facilitated the use of new sources of energy, permitting increase in body size. Aerobic metabolism would have required an oxygen atmosphere that had reached some one percent of current atmospheric oxygen. The Early Cambrian atmosphere is hypothesized to have reached such levels. It is further hypothesized that animals capable of exploiting Early Cambrian oxygen would have possessed a battery of common genes, including those for hemoglobin, a conveyor of molecular oxygen.

PREDATOR–PREY RELATIONSHIPS

Another biological factor that could have driven increased organismal diversity is changing modes of feeding facilitated by rising oxygen levels.

Predators feed on the most abundant **prey** species, reduce the numbers of prey, and so allow other species to use resources formerly monopolized by the dominant prey. We see this in modern-day plant communities where removal of a dominant plant species by a predator provides an opportunity for other species previously present in low numbers or excluded from the habitat to expand or migrate into it. Diversification of prey species leads in turn to diversification of predator species. Prey–predator interactions escalate into what has been characterized as a persistent **co-evolving arms race**: successive rounds of selection for predator responses to their prey's protective devices, followed in turn by adaptations by the prey to their predators' devices, promoting diversity.

This theoretical discussion of predator–prey interactions has been bolstered by a groundbreaking analysis by Dunne et al. (2008) of food webs in the Burgess Shale and Chengjiang Cambrian faunas and their comparison with eight extant biotas. When compared using the numbers of organisms at each trophic level, all food webs show a similar structure. Differences between the two extinct faunas indicate a higher level of stability and hierarchy in the younger Burgess Shale fauna, which Dunne and colleagues interpret as evidence for more prolonged evolution of the Burgess Shale food web, which is 15 My younger than the Chengjiang fauna.

REFERENCES

Adamowicz, S. J., A. Purvis, and M. A. Wills, 2008. Increasing morphological complexity in multiple parallel lineages of the Crustacea. *Proc. Natl. Acad. Sci. USA*, **105**, 4786–4791.

Alfaro, M., and F. Santini, 2010. A flourishing of fish forms. *Nature*, **464**, 840–842.

Ayala, F. J., A. Rzhetsky, and F. J. Ayala, 1998. Origin of the metazoan phyla: molecular clocks confirm paleontological estimates. *Proc. Natl. Acad. Sci. USA*, **95**, 606–611.

Briggs, D. E. G., D. H. Erwin, and F. J. Collier, with photographs by Chip Clark, 1994. *The Fossils of the Burgess Shale*. Smithsonian Institution Press, Washington and London.

Bromham, L., A. Rambaut, R. Fortey, A. Cooper, and D. Penny, 1998. Testing the Cambrian explosion hypothesis by using a molecular dating technique. *Proc. Natl. Acad. Sci. USA*, **95**, 12386–12389.

Cohen, P. A., A. H. Knoll, and R. B. Kodner, 2009. Large spinose microfossils in Ediacaran rocks as resting stagers of early animals. *Proc. Natl. Acad. Sci. USA*, **106**, 6519–6524.

Conway Morris, S., 1998. *The Crucible of Creation: The Burgess Shale and the Rise of Animals*. Oxford University Press, Oxford, UK.

Dahl, T. W., E. U. Hammarlund, A. D. Anbar, et al., 2010. Devonian rise in atmospheric oxygen correlated to the radiations of terrestrial plants and large predatory fish. *Proc. Natl. Acad. Sci. USA*, **107**, 17911–17915.

Dietrich, M. R., and R. A. Skipper, Jr., 2007. Manipulating underdetermination in scientific controversy: the case of the molecular clock. *Perspect. Sci.*, 15, 295–326.

Donoghue, P. C. J., and M. P. Smith (eds.), 2004. *Telling the Evolutionary Time. Molecular Clocks and the Fossil Record*. CRC Press, Boca Raton, FL.

Dunne, J. A., R. J. Williams, N. D. Martinez, R. A. Wood and D. H. Erwin, 2008. Compilation and Network Analyses of Cambrian Food Webs. *PLoS Biol* **6**(4), e102. doi:10.1371/journal.pbio.0060102.

Erwin D. H., M. Laflamme, S. M. Tweedt, et al., 2011. The Cambrian conundrum: early divergence and later ecological success in the early history of animals. *Science*, **334**, 1091–1097.

Feng, D.-F., G. Cho, and R. F. Doolittle, 1997. Determining divergence times with a protein clock: update and reevaluation. *Proc. Natl. Acad. Sci. USA*, **94**, 13028–13033.

Gould, S. J., 1989. *Wonderful Life. The Burgess Shale and the Nature of History*. W. W. Norton & Co., New York.

Hall, B. K., 1999. *Evolutionary Developmental Biology*, 2nd ed. Kluwer Academic Publishers, Dordrecht, Netherlands.

Hall, B. K., 2003. The emergence of form: the shape of things to come. *Devel. Dyn.*, **228**, 292–298.

Hall, B. K., 2007. From Marshalling yards to landscapes to triangles to morphospace. *Evol. Biol.* **35**, 97–99.

Ho, S. Y. H., 2007. Calibrating molecular estimates of substitution rates and divergence times in birds. *J. Avian Biol.*, **38**, 409–414.

Hou, X.-G., R. J. Aldridge, J. Bergström, D. J. Siveter, and X.-H. Feng, 2004. *The Cambrian Fossils of Chengjiang, China: The Flowering of Early Animal Life*. Blackwell Science, Oxford, UK.

END BOX 17.1

Animals Arise (Cont...)   © Photos.com

Knoll, A. H., 1992. The early evolution of eukaryotes: a geological perspective. *Science*, **256**, 622–627.

Knoll, A. H., 2003. *Life on a Young Planet: The First Three Billion Years of Evolution on Earth.* Princeton University Press, Princeton, NJ.

Li, C., G. D. Love, T. W. Lyons, et al., 2010. A stratified redox model for the Ediacaran ocean. *Science*, **328**, 80–83.

Liu, J., M. Steiner, J. A. Dunlop, et al., 2011. An armoured Cambrian lobopodian from China with arthropod-like appendages. *Nature*, **470**, 526–530.

Marshall, C. R., and J. W. Valentine, 2010. The importance of preadapted genomes in the origin of the animal body plans and the Cambrian explosion. *Evolution*, **64**, 1189–1201.

Müller, J., and R. R. Reisz, 2005. Four well-constrained calibration points from the vertebrate fossil record for molecular clock estimates. *BioEssays*, **27**, 1069–1075.

Narbonne, G. M., 2004. Modular construction of early Ediacaran complex life forms. *Science*, **305**, 1141–1144.

Parfrey, L. W., D. J. G. Lahr, A. H. Knoll, and L. A. Katz. 2011. Estimating the timing of early eukaryotic diversification with multigene molecular clocks. *Proc. Natl. Acad. Sci. USA*, **108**, 13624–13629.

Scholtz, G., 2010. Deconstructing morphology. *Acta Zool. (Stockh.)*, **91**, 44–63.

Schwartz, J. H., and B. Maresca, 2006. Do molecular clocks run at all? A critique of molecular systematics. *Biol. Theory*, **1**, 357–371.

Shen, Y., T. Zhang, and P. F. Hoffman, 2008. On the coevolution of Ediacaran oceans and animals. *Proc. Natl. Acad. Sci. USA*, **105**, 7376–7381.

Thorpe, J. P., 1982. The molecular clock hypothesis: biochemical evolution, genetic differentiation and systematics. *Ann. Rev. Ecol. Syst.*, **13**, 139–168.

Valentine, J. W., 2004. *On the Origin on Phyla.* The University of Chicago Press, Chicago.

Van Roy, P., P. J. Orr, J. P. Botting, et al., 2010. Ordovician faunas of Burgess Shale type. *Nature*, **465**, 215–218.

Yuan, X., Z. Chen, S. Xiao, C. Zhou, and H. Hua, 2011. An early Ediacaran assemblage of macroscopic and morphologically differentiated eukaryotes. *Nature*, **470**, 390–393.

END BOX 17.2

Modularity

SYNOPSIS: **Modularity** is the concept that biological units from genes to organisms are composed of subunits (modules) that develop and evolve independently. Modules are subject to selection and so allow parts of a feature to change without changes in other parts of the same feature—one form of mosaic evolution; the number of fingers can decrease or increase without changes to the bones of the lower arm, as one example.

As a consequence of their multi-component subunit structure, hierarchies are more stable and more easily constructed than a nonhierarchical system in which all parts must be simultaneously assembled and in which the absence of any single component would prevent assembly at all (Gass and Bolker, 2003). A cell composed of multiple subsystems (**modules**) is less vulnerable to accident and more easily synthesized than a similar structure that has no subsystems but can function only when all of its many chemical components match perfectly and aggregate simultaneously.

Modularity has emerged as an organizing principle at all levels from gene networks, through cell populations to organ primordia. For example, Vickaryous and Hall (2006) identified the numbers and types of human cells and used phylogenetic tree-building approaches in order to understand relationships among these cell types. Galvão et al. (2010) then analyzed these data sets to create modular maps of human cell types. Modularity is being applied in developmental genetics, developmental biology, evolution, and evolutionary developmental biology. Because modules are subunits of what would otherwise be large biological units, and because modules are subject to selection, the concept of modularity is fundamental to understanding how the phenotype arises; modules link the genotype with the phenotype.[a]

Modularity allows changes in timing of the development of correlated features (Hall, 1999, 2003); degeneration of the eyes in larval ostracods is a recent example of changes in different species resulting from changes in the timing of development of the units (modules) from which the eye is constructed (Kaji and Tsukagoshi, 2010). Modularity allows building blocks of organisms to vary independently in response to selection but in a coordinated way. Two of the many examples are the coordinated changes in eyes, taste buds, and the lateral line system in blind Mexican cavefish, and the modular organization of insect body parts, both discussed in this chapter. The metabolic pathways that form the basis of energy transfer in organisms are modular in their organization.[b] Finally, modular organization is ancient; identification of rangeomorphs from the Ediacaran biota, which flourished during what is now known as the Ediacaran Period between 635 and 542 Mya, is based on modular repeated units (Narbonne, 2004).

The existence of modules provides a basis for the linkage between genes and gene networks that underlie morphology (Atchley and Hall, 1991; Hall, 2003, Cheverud, 2004). The involvement of gene networks in the modularity of sensory module development in surface and blind Mexican cavefish is discussed in this chapter (see Figures 17.12 and 17.13; Franz-Odendaal and Hall, 2006). As the cavefish example illustrates, a gene network may operate in several tissue organs or even regions of an organ, resulting in change in one but not in the others.

Gene- and cell-based modules are revealed in population-level analyses in which multiple genes influencing a region of the phenotype are quantified (Beltrao et al., 2010). Therefore, modularity provides a means for independent yet integrated development of embryos and for independent (mosaic) evolution of modules in response to selection. Indeed, modules evolve in much the same ways as genes, by:

- **duplication**, with one of the duplicates acquiring a new function,
- **dissociation**, in which a module separates either in space or in time and acquires a new function, and/or by
- **co-option**, the incorporation of one module into another (Raff and Raff, 2000; Allen, 2008).

Modules have a distinctive fate, internal integration, and the ability to interact with other modules (Atchley and Hall, 1991; Allen, 2008). A limb bud is a developmental module in a chicken

[a] For an early discussion of the concept of modules (condensations) and modularity, see Atchley and Hall (1991). For modularity across all levels of biological organization, see the papers in Schlosser and Wagner (2004) and in Callebaut and Rasskin-Gutman (2005).
[b] E. Ravasz et al., 2002. Hierarchical organization of modularity in metabolic networks. *Science*, **297**, 1551–1555.

END BOX 17.2

Modularity (Cont...)

FIGURE EB2.1 The single dentary bone of the rodent lower jaw is comprised of six modules that form a unified single, but complex element. Molar- and incisor-alveolar units are derived from cell populations that forms the molar and incisor teeth, respectively. The ramal population, a separate cell population not associated with tooth formation, forms the body of the dentary. The coronoid, condylar, and angular processes each derive in part from the ramal population and in part from a cartilage-forming population of cells at the tip of each process.

[Courtesy of Brian Hall.]

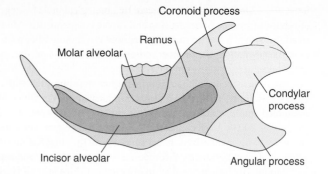

embryo and forms either a wing or a leg, both of which are modular components of the adult. Anterior and posterior limb buds share some modular gene networks that, for example, result in the differentiation of cartilage or muscle, but differ in other networks, such as those that specify whether a limb will be a wing or a leg.

The existence and role of modules can be tested for. If a region of an early chicken embryo from where a wing is known to develop is grafted to another region of the embryonic body, the transplanted piece forms a wing in the ectopic position. The gene network for "wing" that is activated in the ectopic position region, and which initiates wing formation, both functions as a module and confers modular properties onto cells in the new location.

Another example of modularity at cellular and genetic levels is illustrated in **Figures EB2.1** and **EB2.2**. What appears to be a single bone (the dentary) making up the mammalian lower jaw is comprised of six modules, each of which arises from a separate population of cells (Figure EB2.2). Each module is subject to independent genetic control, as shown by the effects of knocking out single genes, in which case one module is prevented from developing or is diminished in size, while other modules develop normally (Figure EB2.2). An example of the result of selection on such modules is shown in the extremely reduced dentary of a marsupial, the Western Australian honey possum *Tarsipes rostratus*. Honey possums feed by licking the nectar and pollen from *Banksia*

FIGURE EB2.2 The modules in the single dentary bone of the rodent lower jaw are subject to independent genetic control as illustrated by the effect of knocking out individual genes. The wild type shows six modules. Modules that fail to develop or are underdeveloped in knockout mice are colored dark brown. Lack of the gene *Msx-1* [shown as *Msx-1* (–)] prevents the molar and incisor alveolar units from developing; other modules develop normally. Lack of *Goosecoid* results in smaller coronoid and angular units but does not affect the growth or shape of the other modules. Lack of *TGF* β*-2* results in underdevelopment of all three posterior processes.

[Courtesy of Brian Hall.]

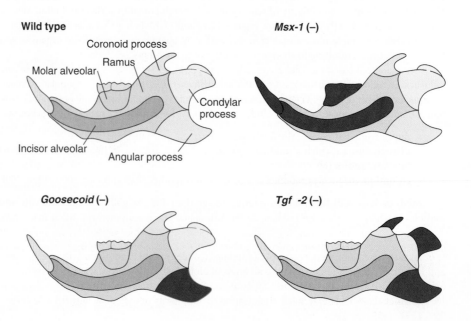

flowers (**FIGURE EB2.3**). Mice, other rodents, and voles chew and gnaw their food. Comparison of the lower jaws of honey possums with those of rodents (Figure EB2.2) illustrates the extreme reduction of the honey possum dentary resulting from reduction of specific modules. A fascinating example of adaptive response to functional demands acting on similar modules, which attests to the evolutionary importance of modules, is found in the lower jaw of the toothless blue whale, *Balaenoptera musculus*, the largest animal that ever existed. Its lower jaw is remarkably similar to that of the honey possum (**FIGURE EB2.4**), although adult blue whales retain no evidence of the teeth they had as embryos.

© ANT Photo Library/Photo Researchers, Inc.

(a)

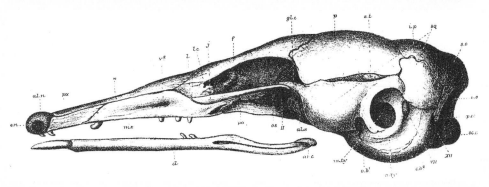

(b)

FIGURE EB2.3 The Western Australian honey possum, *Tarsipes rostratus*, which feeds on *Banksia* nectar and pollen (**a**) has the most reduced lower jaw and teeth of almost any mammal (**b**). The teeth consist of peg-like molars situated midway along the dentary bone (n) and a forward-projecting incisor in the slender and much reduced dentary bone. Compare with the rodent shown in Figure EB2.1.

[(b) Reproduced from Parker, W. K. On the skull of Tarspies rostratus. Stud. Mus. Zool. Univ. Coll. Dundee, 1890:79–83.]

END BOX 17.2

Modularity (Cont...)

© Photos.com

FIGURE EB2.4 The lower jaw of the largest of all animals, the blue whale *Balaenoptera musculus*, which feeds by swimming slowly with mouth open to capture krill, shows remarkable similarity to the lower jaw of one of the smallest mammals, the honey possum shown in Figure EB2.3. Both represent consequence of selection on similar developmental/evolutionary modules.

© Danita Delimont/Alamy

REFERENCES

Allen, C. E., 2008. The "eyespot module" and eyespots as modules: Development, evolution, and integration of a complex phenotype. *J. Exp. Zool. (Mol. Dev. Evol.)*, **310B**, 179–190.

Atchley, W. R., and Hall, B. K., 1991. A model for development and evolution of complex morphological structures. *Biol. Rev. Camb. Philos. Soc.*, 66: 101–157.

Beltrao, P., G. Cagney, and N. J. Krogan, 2010. Quantitative genetic interactions reveal biological modularity. *Cell*, **141**, 739–745.

Callebaut, W., and D. Rasskin-Gutman (eds.), 2005. *Modularity: Understanding the development and Evolution of Natural Complex Systems*. The MIT Press, Cambridge. MA.

Cheverud, J. M., 2004. Modular pleiotropic effects of quantitative trait loci on morphological traits. In *Modularity in Development and Evolution*. G. Schlosser and G. P. Wagner (eds.). The University of Chicago Press, Chicago, pp. 132–153

Franz-Odendaal, T. A., and B. K. Hall, 2006. Modularity and sense organs in the blind cavefish, *Astyanax mexicanus*. *Evol. Dev.*, **8**, 94–100.

Galväo, V., J. G. V. Miranda, R. F. S. Andrade, et al., 2010. Modularity map of the network of human cell differentiation. *Proc. Natl. Acad. Sci. USA*, **107**, 5750–5755.

Gass, G. L., and J. A. Bolker, 2003. Modularity. In *Keywords and Concepts in Evolutionary Developmental Biology*, B. K. Hall and W. M. Olson (eds.). Harvard University Press, Cambridge, MA, pp. 260–267.

Hall, B. K., 1999. *Evolutionary Developmental Biology*, 2nd ed. Kluwer Academic Publishers, Dordrecht, Netherlands.

Hall, B. K., 2003. Unlocking the black box between genotype and phenotype: Cell condensations as morphogenetic (modular) units. *Biol. Philos.*, **18**, 219–247.

Kaji, T., and A. Tsukagoshi, 2010. Heterochrony and modularity in the degeneration of maxillopodian nauplius eyes. *Biol. J. Linn. Soc.*, **99**, 521–529.

Narbonne, G. M., 2004. Modular construction of early Ediacaran complex life forms. *Science*, **305**, 1141–1144.

Raff, E. C., and R. A. Raff, 2000. Dissociability, modularity, evolvability. *Evol. Dev.*, **2**, 235–237.

Schlosser, G., and G. P. Wagner (eds.), 2004. *Modularity in Development and Evolution*. University of Chicago Press, Chicago.

Vickaryous, M. and B. K. Hall, 2006. Human cell type diversity, evolution development classification with special reference to cells derived from the neuronal crest. *Biol. Rev. Camb. Philos. Soc.*, **81**, 425–477.

CHAPTER

18

Genetic Variation in Populations

CHAPTER SUMMARY

Genetic variation provides the raw material enabling evolutionary change in response to natural selection. Without variation there would be no evolutionary change. Natural selections acts on individuals but individuals do not evolve. Populations evolve.

Mutations, which provide one source of genetic variation, are not random; some parts of the genome are more susceptible to mutation than are others, while some kinds of mutations are more common than others (Colegrave and Collins, 2008). Large numbers of alleles at a single gene locus (polymorphism) provide a much greater source of genetic variation than do the relatively few new mutations that arise each generation. Frequencies of individual alleles (END BOX 18.1) and the gene pool (all the alleles of all individuals in the population) are two major attributes of a population (a group of potentially interbreeding organisms). The gene pool represents all the variation available in the current generation, and, setting mutation aside, all the variation than can contribute to the next generation.

The genetics of quantitative traits are significant in evolution, and quantitative trait loci (QTLs) are an important tool for investigating available variation. QTLs consist of regions of the genome that influence specific phenotypic characters. Once identified, such regions can be probed for precise genes that influence quantitative traits. Additional sources of genetic variation in populations are the loss through death or emigration and/or the gain through immigration of individuals. The population-level processes responsible are gene flow and genetic drift.

INTRODUCTION

When coupled with the multiple levels of gene regulation within individuals, we see that genetic variation can be introduced into, maintained in, or lost from populations in multiple ways: mutations; maintenance of genetic variation through multiple alleles (genetic polymorphism); continuous variation through genes, each of which has a small effect and which behaves as quantitative trait locus; random drift in gene frequencies (genetic drift) within a population; and gene flow between populations all contribute to enhanced variation at the population level. The

- randomness of mutation,
- variation in mutation rates among genes and among species,
- frequencies of alleles in populations, and
- ability of DNA to repair itself (and so to compensate for otherwise deleterious effects of mutation)

ALLELE FREQUENCY *rather than* GENE FREQUENCY *is the basic genetic unit in population genetics*

all contribute to maintaining genetic diversity within populations.

Phenotypic variation is the basis upon which natural selection operates on individuals. Genetic variation within populations provides the raw material enabling hereditary evolutionary change. Consequently, it is import to be aware of the different roles played by individuals and populations in evolution.

INDIVIDUALS, POPULATIONS, AND EVOLUTION

Natural selection reflects the differential survival and reproduction of individual organisms with particular features. Differential survival and reproduction are the mechanism of evolution; natural selection is the outcome. It is important to remember that natural selection takes place *within a generation and affects individuals of that generation. Heritable response to natural selection is by populations and occurs between generations*, and only for those features that are heritable.

Although selection reflects the fate of *individuals*, response to selection lies in the information content of the genomes of all individuals in the *population*, information that can change as a result of mutation or random processes such as genetic drift. Natural selection acts on individuals, populations evolve. Consequently, "questions about differences between groups require a different kind of analysis than do questions about differences between individuals,"[1] as was evident to R. A. Fisher when he published the first population level analyses of Mendelian inheritance (Fisher, 1918). It is important, therefore, to have a clear view of individuals and populations with respect to evolution (**TABLE 18.1**), not merely because use of the word evolution has shifted from individuals to populations, but because individuals respond to selection but populations evolve. As this is important, the essential differences between individuals and populations as far as evolution is concerned are outlined as 12 points below.[2] Each is presented as a statement

---

[1] The quotation is from Keller, 2010, p. 53 in the context of an analysis of the nature-nurture "debate."
[2] For a recent perspective on individuals in the context of levels of selection, see H. J. Folse III, and J. Roughgarden, 2010. What is an individual organism? A multilevel selection perspective. *Q. Rev. Biol.*, **85**, 447–472.

TABLE 18.1	**Comparison of Characteristics of Individuals and Populations**	

Characteristic	Individual	Population
Life span	One generation	Many generations
Spatial continuity	Limited	Extensive
Genetic characteristics	Genotype	Gene frequencies
Genetic variation	Expressed in one lifespan	Expressed in evolutionary change
Evolutionary characteristics	No changes, because an individual has only one genotype and is limited to a single generation	Can evolve (change in gene frequency), because evolution occurs between generations
Selection	Operates on phenotype in one generation	With mutation, leads to change in genotype from generation to generation
Mutation	Somatic mutation transferred through reproduction	Gametic inheritance/gametic
Variation	Phenotypic not inherited Transferred via reproduction	Genotype inherited/ genotypic

of fact. You can regard them as conclusions, the evidence for which is provided in the remainder of the book. You also can read them as a summary of the evolutionary process.

1. Organisms exist as **individuals**. Individual multicellular organisms (animals, plants, fungi) develop, grow, mature, reproduce, senesce (in most cases), and die. Individual unicellular organisms reproduce and die.
2. Natural selection acts on individuals but individuals do not evolve. Individuals pass on their genes, along with mutations in those genes, to individuals of the next generation.
3. In most species of uni- and multicellular organisms, individuals exist in **populations** that inhabit discrete ecological niches. Populations of multicellular organisms have a structure that usually includes different age classes of a single generation and often includes overlapping generations (grandparents, parents, grandchildren), especially in species with short reproductive cycles and long lives.
4. Populations of sexually reproducing organisms consist of individuals that are **not identical** to one another—they are not clones. In a population of a species in which individuals reproduce asexually, for example by budding or fission, all individuals in a population may be identical.
5. Resources are often limited, with the consequence that not all individuals in a population will survive to reproduce and contribute offspring to the next generation.
6. Populations do not reproduce, individuals reproduce. Populations pass to the next generation(s) a gene pool from those individuals that reproduce.
7. Variation is an essential prerequisite for evolution to act: natural selection allows some variants to survive and others not.
8. Resources are limited so that natural selection results in survival to the next generation through reproduction of the individuals (variants) that are best suited to the conditions of their existence.
9. Because the genetic background of individual sexually reproducing organisms differs, those that are selected on the basis of their fitness in the environment will preferentially pass their genes to the next generation. Those individuals that are less well-fitted to the environment will tend not to pass their genes to the next generation.

10. Because of differential reproduction, the genetic composition of a population will change gradually from generation to generation. Genetic changes also accumulate through random drift of genes from generation to generation and/or from spontaneous mutations that change the genetic composition of populations.

11. Populations may subdivide into smaller groups. Differences that emerge in the subgroups can provide the basis for speciation.

12. Populations or subsets or populations may "crash" or become extinct in response to environmental catastrophes.

FEATURES OF MUTATIONS IN POPULATIONS

Because they accumulate in individuals, mutations supply an important source of variation in populations. A population that has long been established in a particular environment will have many genes adapted for prevailing conditions. New mutations that arise, if not neutral in effect, will rarely be better, and likely will be worse than the genes already present—a consequence not much different from the damage we would expect if a random change was introduced into any intricately organized and integrated system, such as a computer or a car engine.

Because multicellular organisms are constrained by their evolutionary history, advantageous mutations are generally confined to few of the many intricate developmental processes and functions. For example, although plants and animals are separated by more than a billion years of evolution, there are only two differences between them in the 100 amino acids of histone 4, which is the protein that binds and folds DNA. For such phylogenetically crucial genes, conserved molecular sequences are commonplace, and viable changes occur only rarely. For other genes and functions, a change in environmental conditions can elicit a genetic response based on the available genetic variation (Barrett and Schluter, 2007). Alleles formerly in low frequency may now have high relative fitness. We see this in

In an extensive analysis of the incorporation of MUTATIONS into the genome, Nei (2007) concluded that mutations provide a driving force for evolutionary change that exceeds that of natural selection.

- rapid genetic changes in many insect populations exposed to pesticides such as dieldrin or DDT (dichloro-diphenyl-trichloroethane; **FIGURE 18.1**), where resistant alleles appear on all major chromosomes (**FIGURE 18.2**);
- large increases in the frequencies of black alleles in populations of the peppered moth *Biston betularia* in industrialized regions and of a receptor involved in pigment cells in light and dark strains of the rock pocket mouse, *Chaetodipus intermedius* in southwestern USA;

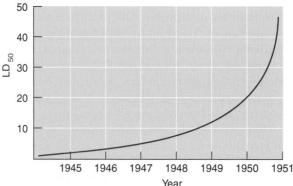

FIGURE 18.1 Resistance to DDT in common houseflies (*Musca domestica*) collected from farms in Illinois between 1945 and 1951 when use of DDT as an insecticide was increasing. The measure is the lethal dose necessary to kill 50 percent of the flies (known as the LD_{50}).

[Adapted from Strickberger, M. W. *Genetics, Third Edition.* Macmillian, New York, 1985. (Based on data from Decker, G. E., and W. N. Bruce, 1952. House fly resistance to chemicals. *Am. J. Trop. Med. Hyg.*, **1**, 395–403 © Frank B. Yuwono/ShutterStock, Inc.) INSET: © Frank B. Yuwono/ShutterStock, Inc.]

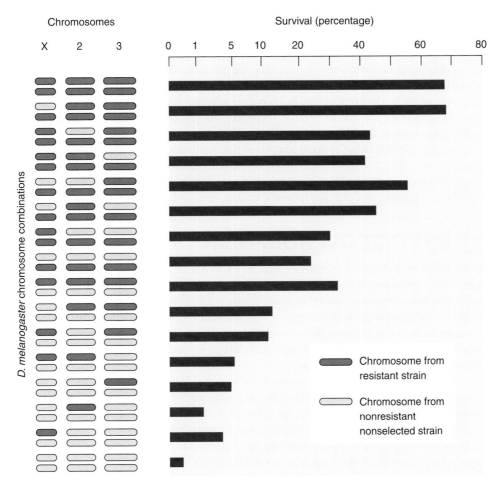

FIGURE 18.2 Percent survival of fruit flies (*Drosophila melanogaster*) with 16 chromosomal patterns of resistance/nonresistance to the insecticide DDT after exposure to a uniform dose of DDT. Each population of flies carries a set of chromosomes X, 2, or 3 derived from DDT-resistant and DDT-nonresistant strains. DDT resistance (shown as % survival) increases with the increased number of chromosomes in resistant strains.

[Adapted from Crow, J. F., 1957. Genetics of insect resistance to chemicals. *Ann. Rev. Entomol.*, **2**, 227–246.]

- increased frequencies of resistant genes in some plant populations exposed to herbicides and metallic toxins; and
- genes that modify red blood cell physiology offering protection against malaria in the human populations, such as the sickle cell allele.

In most populations, many individuals die, leaving those (and it may be those few) individuals with the alleles for resistance, melanism, or protection to pass their alleles to the next generation. In essence, these individuals become the *founders of the next generation.*

MUTATION RATES

Although you might have the idea that mutations are "bad," optimal mutation rates in a population are advantageous (Orzack and Sober, 2001). **Mutation rates** and **mutational equilibrium** both can be calculated using the approach outlined in End Box 18.1.

Mutation rates are generally low, on the order of one per 100,000 copies of a gene, although there is much variation. In humans, the mutation rate for achondroplasia, a form of dwarfism, is 0.6 to 1.3 mutations/100,000 gametes. Mutation rate for neurofibromatosis is 5 to 10 mutations/100,000 gametes.[3] In modern humans, carrying an estimated 25,000 genes per haploid genome, each sperm and egg may well carry less than one new mutation, or an average of <0.4 new mutations in a diploid fertilized zygote.

Mutation rates are not only low, they are also not constant. As with other essential traits, mutation rates seem mostly selected for optimum values, balancing on the delicate adaptive line between retaining prevailing adaptive features yet facilitating the origin of new features. Specific nucleotide sequences may be the site of higher than average mutation rates or they may increase mutation rates among adjacent nucleotides. In such sequences, known as **hot spots of mutation**, mutations may be to 100 times the "normal" rate. Oftentimes, hot spots are sites where the nucleotide change is less readily repaired or compensated for than is a sequence change at another site; specific patterns of DNA coiling may result in DNA polymerase enzymes producing replication errors.

GENETIC POLYMORPHISM

Evolutionary potential and genetic variation are two sides of the same evolutionary coin.

New mutations with an immediate beneficial effect on an organism generally seem to be rare. Some mutations are either *neutral* in their effect or harmful *only* when they occur in relatively rare homozygotes. Neutral mutations or deleterious but recessive mutations accumulate in a population and provide a reservoir of potential genetic variation. Such genetic variation, expressed in a population as two or more genetically distinct forms, is known as **genetic polymorphism**.

In the fruit fly *Drosophila pseudoobscura*, populations in different localities in the western United States are polymorphic for a wide variety of gene arrangements on the third chromosome (FIGURE 18.3a). Frequencies of such arrangements may change seasonally (FIGURE 18.3b), indicating that genetic polymorphism is generally adaptive in this species and that certain polymorphic variations are preferentially adaptive in specific seasons.[4]

About two-thirds to three-quarters of all protein loci in many species are polymorphic. Of the three billion nucleotides in the haploid human genome, one human may differ from another at an average of about five million sites. Therefore, genetic polymorphism provides a much greater source of genetic variation than do the relatively few new mutations that arise each generation. For example, exposure of insect populations to the pesticide DDT results in a widespread increase in the presence of various DDT-resistant mechanisms, including changes in the permeability of the insect to absorption of DDT, and increased frequency of those enzymes that break down DDT into relatively less toxic products (Roush and McKenzie, 1987; Denholm et al; 2002). It is therefore not surprising that insecticide resistance is associated with numerous genes, and that many genetic changes have arisen in response to insecticide exposure (Figure 18.2). Hundreds of examples have been described.

MASKING VARIATION

A remarkable way in which some organisms accumulate mutations and maintain variation without experiencing their immediate effects is to bind their gene products with **heat shock proteins**, such as heat shock protein-90 (Hsp-90), which mask genetic variation (Hallgrìmsson and Hall, 2005).

Heat shock proteins are molecular chaperones that help (chaperone) other proteins to maintain their normal 3-D conformation and prevent them from degrading. Because of these protective roles, heat shock proteins mask some, perhaps many, of the effects

---

[3] This range of mutation rates may partly reflect the inclusion under one name of several syndromes with distinct genetic bases but similar phenotypes.
[4] Dobzhansky, Th. 1947. A directional change in the genetic constitution of a natural population of *Drosophila pseudoobscura*. *Heredity*, **1**, 53–64.

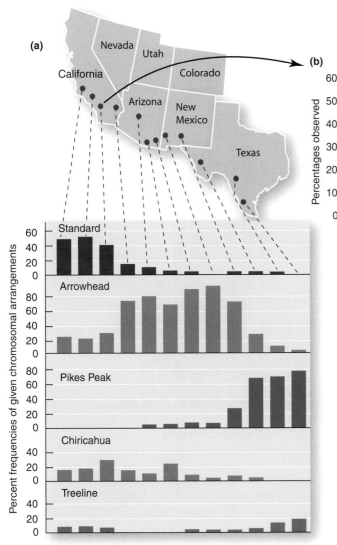

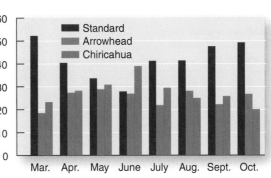

FIGURE 18.3 (a) Frequencies of five different third chromosome gene inversions in *Drosophila pseudoobscura* in 12 localities on an east–west transect along the United States–Mexican border. **(b)** Percentages of different third chromosomal inversions found at different months during the year in one locality, Mount San Jacinto, California. Each of these chromosome inversions maintains a specific gene combination that enables adaptive interactions between component alleles (epistasis).

[(a) Adapted from Dobzhansky, Th., 1944. Chromosomal races in *Drosophila pseudoobscura* and *D. persimilis*. *Carnegie Inst. Wash. Publ.* No. 554, Washington, DC, pp. 47–144; (b) Adapted from Dobzhansky, Th., 1947. A directional change in the genetic constitution of a natural population of *Drosophila pseudoobscura. Heredity,* **1**, 53–64.)]

that mutations in protein-coding genes otherwise would have on the phenotype. After organisms are exposed to an *environmental* shock such as a sharp change in temperature or salinity, many alleles are unmasked as protective heat shock proteins are disabled. New patterns and combinations of proteins expressed can result in significant changes in development, some of which will be adaptive and open up new evolutionary opportunities.[5]

CONTINUOUS VARIATION

From Charles Darwin onwards, many evolutionary biologists have suggested that rather small heritable changes provide most of the variation on which natural selection acts. In Darwin's words from *The Origin of Species*

> Extremely slight modifications in the structure and habits of one species would often give it an advantage over others; and still further modifications of the same kind would often

---

[5]For mechanisms of the action of Hsp-90 see Rutherford and Lindquist (1998), Suzuki and Nijhout (2006), and Connolly and Hall (2008). As In *Drosophila*, Hsp-90 interacts with piRNAs to activate transposons and increase genetic variation.

FIGURE 18.4 Distribution of the heights of 1,000 Harvard College students aged 18 to 25.

[Adapted from Castle, W. E., 1932. *Genetics and Eugenics, Fourth Edition.* Harvard University Press, Cambridge, MA.]

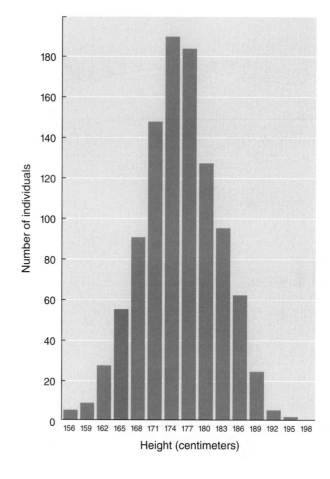

still further increase the advantage . . . Under nature, the slightest differences of structure or constitution may well turn the nicely balanced scale in the struggle for life, and so be preserved.

For many measurable traits such as crop size and yield, animal height/length or body weight, researchers usually focus on small changes or **continuous variation**. Evidence of this variation is found in characters for which plots of the distribution of variation form a bell-shaped curve or a normal distribution. An example is seen in the case of human height (**FIGURE 18.4**). In terms of evolution, it is clear that many genetic differences of small phenotypic effect can accumulate through selection to give large quantitative differences. Selecting for the presence or absence of white spotting in Dutch rabbits can lead to completely colored or completely white strains (**FIGURE 18.5**), results that illustrate the presence of previously hidden genetic variation.

SELECTING CONTINUOUS VARIATION

Artificial selection in agriculture for commercially important traits such as body size or milk yield has provided important insights into how selection produces evolutionary change in quantitative traits. Darwin relied extensively on examples from artificial selection and they remain relevant to evolutionary biology today. **Milk yield in cattle**, for example, is a trait that has changed enormously during the past century. Even though cattle have undergone millennia of selection for milk production since their domestication, in Europe and North America, today the average dairy cow produces about 2.5 times as much milk as cows did 60 years ago. Selection on dairy cows has many phenotypic effects, including increased udder size and decreased muscle mass, the opposite of the changes seen in beef cattle selected for increased meat production (**FIGURE 18.6**). Economic

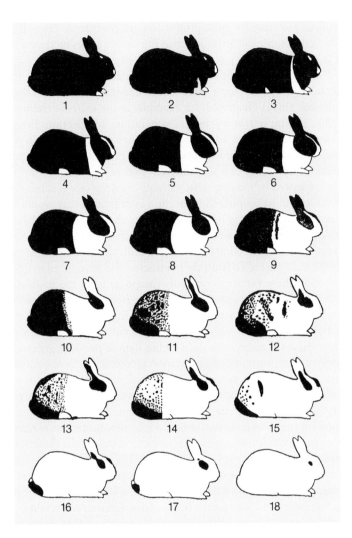

FIGURE 18.5 Selection for white spotting in Dutch rabbits, beginning with white only on the tips of the forepaws (grade 1) results in all ranges of spotting to animals that are totally white (grade 18).

[Reprinted by permission of the publishers from *Genetics and Eugenics: A Text-book for students of biology and a reference book for animal and plant breeders, Fourth Revised Edition*, by W. E. Castle, p. 281, Cambridge, Mass.: Harvard University Press, Copyright @ 1916, 1920, 1924, 1930 by the President and Fellows of Harvard College. Copyright © renewed 1958 by William Earnest Castle.]

(a)

(b)

FIGURE 18.6 A Holstein dairy cow **(a)** selected for milk production compared with a cow selected for beef **(b)** to show the opposite trends in udder size and muscle mass.

impacts of artificial selection for milk production are enormous. According to the U.S. Department of Agriculture, milk production in the United States increased between 1959 and 1990 but the number of dairy cows declined by 40 percent.[6]

There is a tremendous amount of variation in natural populations for most such traits and selection on this variation can produce changes well beyond that seen in the original population. This is illustrated, for example, in the extreme change in body size produced in dogs during only a few tens of generations of selection. Even without the addition of new genetic variation, selection can produce dramatic shifts in phenotypes by retaining individuals with genetic variants that were present in the original population but in combinations that produce a very different phenotype.

If a trait is influenced by many genetic variants, then the rare combinations that produce extreme values can be produced by selection even though they would have virtually no chance of appearing randomly in the original population. Milk yield illustrates some important principles about how selection produces changes in continuous phenotypic traits. In each generation of selection for increased milk yield, variants that produce lower than average milk yield are culled from the population and so reduced in frequency. With continuing selection these variants eventually disappear from the population. Variants associated with higher than average milk yield rise in frequency and become increasingly present in combinations that are associated with ever higher milk yield. With continuing selection, the population increasingly exhibits combinations of such high yield variants that act jointly to push the mean milk yield higher than in previous generations.

This process of selecting on existing variation accounts for the majority of the change in the trait from generation to generation. Eventually, however, if no new variants are introduced, selection will run out of genetic variation. Although they are a very small contributor to the variation in each generation, new mutations are critical to variation over long periods of time. Variation accounted for by new mutations is referred to as the **mutational variance**. While mutations that influence a trait in the direction of selection are fairly rare, they do occur.

To return to the milk yield example, new mutations that increase milk yield have arisen in selected dairy cows. One such mutation with a fairly large positive effect on milk yield occurs in the gene that codes for the bovine growth hormone receptor. This example illustrates another important lesson from artificial selection: that desired changes in a phenotypic trait usually come at a cost. Genes rarely influence only the trait under selection. Almost always, they have multiple effects. Changes in the selected character often produce changes in other characters that may not be adaptive or, in the case of artificial selection, desired. Selection for milk yield illustrates this beautifully. These same cows that produce more than two and a half times as much milk as their ancestors did in the 1950s are significantly less fertile and have tremendous difficulty giving birth without human assistance. Such undesired "side-effects" occur because so many genetic variants that influence milk yield have multiple phenotypic effects (pleiotropy; Wagner and Zhang, 2011). Selection also acts on the correlations or pleiotropic relationships themselves, favoring correlations among functionally related traits above those that are not.

THE GENETICS OF CONTINUOUS VARIATION

The genetic basis for continuous phenotypic variation such as that shown in Figure 18.5 is not well understood. It is not known, for example, to what extent continuous variation in complex traits is determined:

- by a moderate number of common genetic variants each of small effect;
- by many rare variants each with a small effect;

[6]U.S. Department of Agriculture, 2011. NAHMS Dairy Studies. Accessed July 2, 2012 from http://www.aphis.usda.gov/animal_health/nahms/dairy/index.shtml.

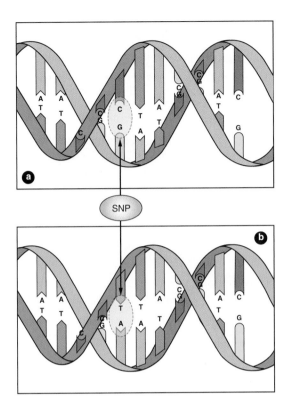

FIGURE 18.7 Example of a single-nucleotide substitution (SNP) DNA in which the region of DNA in **(a)** differs from that in **(b)** molecule at a single base-pair location.

- by rare variants that have moderate to large effects;
- or by a combination of all three.

It is also not known to what extent interactions among the effects of different variants contributes to the total genetic variance of most traits. These are important questions that impinge directly on our understanding of how selection acts on phenotypic variation to produce inherited changes in the underlying genome.

Recent years have seen dramatic advances in the technology and methods enabling us to analyze the genetics of continuous phenotypic traits. Human stature is an example of a trait with a complex genetic basis. Using a combination of **genomics** and **bioinformatics**, recent studies have yielded important insights into the genetic basis for variation in human stature. One genomic technique is to use gene chips that allow molecular geneticists to genotype individuals for multiple sequence variants at the same time. Such chips typically have 500,000 or even 1 million sequence variants. Studies using this approach have revealed that variation in human stature is determined by many variants of small effect. One study identified 20 variants that collectively explain three percent of the variation in human height; the researchers estimated that 93,000 single nucleotide polymorphisms (SNPs; FIGURE 18.7) would be required to explain eighty percent of the genetic variation in human height (Weedon and Frayling 2008; Weedon et al., 2008). Genotyping sequence variants in this way, however, misses copy number polymorphisms or variation in the extent to which genes or genomic regions are present in multiple copies. A recent study suggests that this is a significant source of genetic variation in humans (Sebat et al., 2004).

A single-nucleotide polymorphism (SNP, pronounced snip) is a one base difference between the DNA strands of a pair of chromosomes between two individuals or species (Figure 18.7). SNPs may occur within the coding regions of a gene or between coding regions.

QUANTITATIVE TRAIT LOCI

Quantitative trait loci (QTLs) are an important tool for investigating the genetic basis of continuous traits. A QTL is a region of a chromosome that affects a *particular* quantitative aspect of the phenotype, such as size or shape, or processes such as growth and morphogenesis that influence size and shape (Mackay, 2004; Mackay and Lyman, 2005; Wagner and Zhang, 2011). Individual QTLs may affect the entire individual (as in

FIGURE 18.8 Individual threespine sticklebacks, *Gasterosteus aculeatus*, cleared and stained to reveal the skeleton. From **a** to **c** note the reduction in the pelvic girdle to a nubbin in **c**, and the reduction of the dermal plates to six and five in **b** and **c**, respectively, compared with the complete set of plates in **a**.

[Reproduced from Foster, S.A., and Baker, J.A. 2004. Evolution in parallel: New insights from a classic system. *Trends in Ecology & Evolution,* **19**, 456–459.) © 2004, with permission from Elsevier. Photographs courtesy of Dr. W. A. Cresko, University of Oregon.]

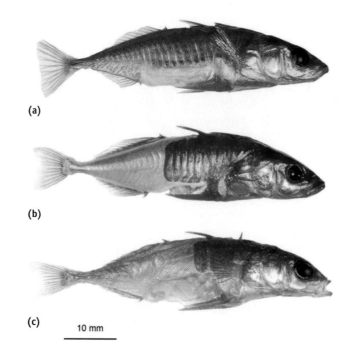

(a)

(b)

(c)

10 mm

growth), different regions of the organisms (tail growth versus leg growth), or different characters (liver versus kidney).[7]

QTL analysis enables us to isolate suites of genes acting on particular parts—modules—of the phenotype at particular stages in ontogeny and to determine their relative effects (Beltrao et al., 2010). For instance, in recent studies with selected inbred lines of rats, four QTLs were shown to influence tail growth and body weight. One QTL has a substantial effect on both tail growth and body weight throughout development. A second has a large effect late in development but only a small effect early. A third shows the reverse pattern, while the fourth has minor effects at one or two discrete developmental stages only (Cheverud, 2004). Analysis of these QTLs has identified genomic regions whose influence varies by feature and over developmental time.

A second example comes from natural populations of a lake fish found in British Columbia, the threespine stickleback, *Gasterosteus aculeatus*. Sticklebacks have complex sets of skeletal elements including spines, bony plates in the skin (a dermal armor), a three-part pelvic girdle, and bony supports (rakers) in the gills (Ostlund-Nilsson et al., 2007). Illustrated in FIGURE 18.8 is the variation expressed under different conditions in the length of the dorsal spines, the number of bony plates, and the number of gill rakers, each of which is influenced by QTLs. One QTL explains much of the variation in plate number, two QTLs explain 66 percent of the variation in gill raker number (Peichel et al., 2001; Bell et al., 2004).

In the past, it was assumed that the cumulative actions of many genes of small effect were responsible for the entire **QTL effect**. More recent analyses in *Drosophila*, mice, and fish have revealed that a small number of genes within the QTL contribute disproportionately to the QTL effect; the QTL shows a region of the chromosome with many genes of small effect, but may also alert us to a gene or genes of large effect located in the same region of the chromosome. Such a locus has been discovered in fresh water sticklebacks that arose from marine populations. An ancestral allele (*complete*) of the gene *Ectodysplasin* in marine sticklebacks is associated with maintaining the bony plates shown in Figure 18.8. The derived allele *low* found in freshwater fish is associated with plate loss. *low* has been fixed independently in different freshwater populations, and

---

[7]For discussions, see the book-length treatments by Falconer and Mackay, 1996; Lynch and Walsh, 1998; Roff, 1997, Hedrick, 2000, and Charlesworth and Charlesworth, 2010.

may well be a major gene in the QTL responsive to selection resulting in reduced plate number.[8] You can see that QTL analysis enables us to match environmental conditions to evolution quite precisely.

QTL analysis can lead to unexpected conclusions, however, reflecting the differential evolution of QTLs in different lineages. Thus, a recent study of thousands of QTLs in a cross between two strains of the nematode *Caenorhabditis elegans* led to the conclusion that the pattern of distribution of many of the QTLs was independent of the feature in which they were found. This pattern of *trait-independent variation* seemed best explained by the number of genes (or at least the size of the genome) with which a QTL was associated.[9] Sorting out the various influences of environmental conditions, the genetic response to different environmental conditions, and gene X environment interactions is a major preoccupation of many evolutionary biologists and quantitative geneticists.

POPULATION GENETICS AND GENE (ALLELE) FREQUENCIES IN POPULATIONS

Darwin proposed that natural selection operates on small, continuous, hereditary variations. His cousin, Francis Galton, and others accepted evolution, but maintained that variations are sharp and discontinuous. The controversy was resolved when it was shown that several genes, each with small effect (**polygenes**), can have a large effect when they influence the expression of a single phenotypic trait. By the 1930s, it became accepted through genetics that

1. Evolution is a population phenomenon.
2. It can be represented as a change in gene (now allele) frequencies resulting from the action of various natural forces such as mutation, selection, and genetic drift.
3. These changes can lead to differences among populations, species, and higher clades.[10]

This population genetics view of evolution became known as the **neo-Darwinian theory**, a theory of evolution that emphasizes the frequency of genes in populations as the basis of evolutionary change.

Population genetics deals with genes as alleles and gene frequencies as allele frequencies. Allele frequencies—the frequencies of individual alleles (End Box 18.1)—and the gene pool—all the alleles of all individuals in the population—are two major attributes of a population, the latter being defined here as a group of potentially interbreeding organisms (below). The gene pool represents all the variation available in the current generation, and, excepting mutation, all the variation than can contribute to the next generation.

The union of population genetics and evolution as the neo-Darwinian theory of evolution (NEO-DARWINISM) is not same as the MODERN SYNTHESIS of evolution in which systematics and paleontology were added to population genetics as necessary components of a theory of evolution.

THE HARDY-WEINBERG PRINCIPLE (EQUILIBRIUM)

Population genetics is sophisticated statistically, mathematically, and conceptually. Below you will find an introduction to population thinking, the concept of population frequencies, the gene pool, and the genetic attributes of populations. The concepts are expanded further in END BOX 18.2.

According to a principle devised independently by Geoffrey Hardy (1877–1947) in England and Wilhelm Weinberg (1863–1937) in Germany and known as the **Hardy–Weinberg principle**, in a population of randomly mating individuals (a deme),

---

[8] D. Schluter et al., 2010. Natural selection and the genetics of adaptation in threespine stickleback. *Phil. Trans. R. Soc. B*, **365**, 2479–2486.

[9] M. V. Rockman, S. S. Skrovanek, and L. Kruglyak, 2010. Selection at linked sites shapes heritable phenotypic variation in *C. elegans*. *Science*, **330**, 372–376.

[10] Crow (1988) provides an excellent summary of the beginnings of population genetics. The three editions of *Genetics and the Origin of Species* (Dobzhansky, 1937, 1941, 1951) contain the classic foundations of population genetics and the origin of species. Gillespie (2004) is an excellent recent text.

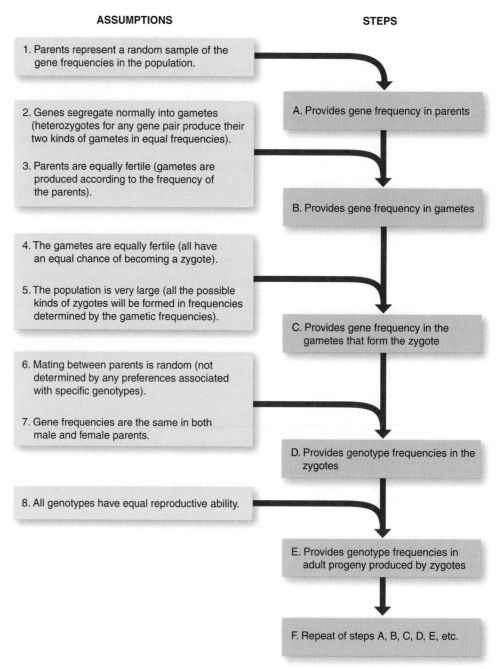

FIGURE 18.9 Assumptions and steps in the Hardy–Weinberg equilibrium.

[Based on Falconer, D. S., and T. F. C. Mackay, 1996. *Introduction to Quantitative Genetic*, 4th ed. Longman, Harlow, Essex, UK.]

allele frequencies are conserved and in equilibrium unless external forces act on them (**FIGURE 18.9**). As mutation rates are usually observed to be of the order of 5×10^{-5} or less, the shift toward equilibrium is slow, so slow that mutational equilibrium is rarely if ever reached, especially because, as discussed above, mutation rates are not constant.

The Hardy–Weinberg principle is the founding theorem of population genetics. Assumptions underlying this principle as it operates in populations are outlined in Figure 18.9 and in End Box 18.2. Such principles include: random mating in large populations, equilibrium allele frequencies, absence of gene flow into the population, and stepwise changes in gene frequencies.

In natural populations, provided that the genes are far apart on the chromosome and the number of alleles is limited, ideally to two, we can determine genotype frequencies and Hardy–Weinberg equilibria for single genes. The more gene pairs, the longer it takes to achieve equilibrium.

MEASURING ALLELE FREQUENCIES

Allele frequencies as a measure of evolution were discovered soon after the rediscovery of Mendelian genetics in 1900. This measure arose directly from Mendel's 3:1 ratio of dominant to recessive phenotypes. Dominant alleles, so the argument went, would reach a stable equilibrium frequency of three dominant individuals to one recessive.

Hardy and Weinberg independently demonstrated that allele frequencies do not depend upon dominance or recessiveness (that is, on genotype frequencies) but remain essentially unchanged from one generation to the next, *provided that* mating is random and all genotypes are equally viable. Therefore, by confining our attention to alleles rather than genotypes, we can predict allele and genotype frequencies in future generations. After the first generation, the genotype frequencies will remain at Hardy–Weinberg equilibrium, all else being equal. That said, the ideal conditions under which the Hardy–Weinberg principle was established—the genes have to be on separate chromosomes, and there have to be at least two alleles, a large effective population size and no inbreeding—turn out rarely to be seen in nature. (See End Box 18.2 for inbreeding.)

The general relationship between allele and genotype frequencies can be described in algebraic terms using the Hardy–Weinberg principle: If p is the frequency of an allele T at a given locus in a panmictic population, and q the frequency of another allele at that locus (t), so that $p + q = 1$ (that is, there are no other alleles), then the equilibrium frequencies of the genotypes are given by the terms $p^2(TT)$, $2pq(Tt)$, and $q^2(tt)$. If the allele frequencies of T and t are $p = 0.6$ and $q = 0.4$, respectively, the equilibrium frequencies will be

$$(0.6)^2(TT) + 2(0.6)(0.4)(Tt) + (0.4)^2(tt) =$$

$$0.36\ TT + 0.48\ Tt + 0.16\ tt.$$

This relationship can be visualized by drawing a checkerboard in which the genotype frequencies stem from random union between alleles that are in the frequencies of p and q (**FIGURE 18.10**). The same results derive from the **binomial expansion**

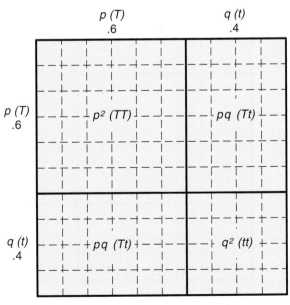

Total: p^2 *(TT)* + 2 *pq (Tt)* + q^2 *(tt)*

FIGURE 18.10 Genotypic frequencies generated under conditions of random mating for two alleles, *T* and *t*, at a locus when their respective frequencies are p = .6 and q = .4. Equilibrium genotypic frequencies are therefore .36 *TT*, .48 *Tt*, and .16 *tt*.

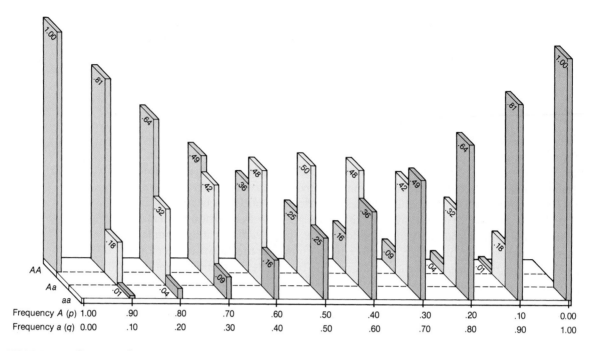

FIGURE 18.11 Genotypic frequencies at Hardy–Weinberg equilibrium for a variety of allele frequencies of *A* (*p*) and *a* (*q*).

[Adapted from Wallace, B., 1970. *Genetic Load: It's Biological and Conceptual Aspects.* Prentice Hall, Englewood Cliffs, NJ.]

$(p + q)^2 = p^2 + 2pq + q^2$. Therefore, with any given p and q and random mating between genotypes, one generation of a population in which generations do not overlap is enough to establish equilibrium for the frequencies of alleles and genotypes. Once established, the equilibrium will persist until the allele frequencies change.

FIGURE 18.11 shows the genotypic frequencies at Hardy–Weinberg equilibrium for a two-allele locus, where the frequency of each allele ranges from 0 to 1. Note that the frequency of heterozygotes never exceeds 0.50 but is significantly higher than the frequency of homozygotes for a rare allele (for example, when a is 0.1, *aa* is 0.01 but *Aa* is 0.18).

FIGURE 18.12 Genotypic frequencies generated under conditions of random mating when there are three alleles, A_1, A_2, and A_3, present at a locus. For illustration, the respective frequencies of these alleles are given as $p = .2$, $q = .5$, and $r = .3$. Equilibrium genotypic frequencies are therefore .04 A_1A_1, .20 A_1A_2, .12 A_1A_3, .25 A_2A_2, .30 A_2A_3, and .09 A_3A_3.

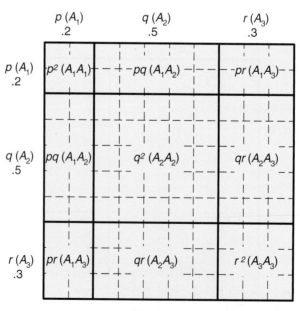

Total:
$p^2 A_1A_1 + 2pq A_1A_2 + 2pr A_1A_3 + q^2 A_2A_2 + 2qr A_2A_3 + r^2 A_3A_3$

When more than two alleles are present at a locus (a common situation), the frequency of each allele must be considered as an element in a multinomial expansion. For example, with three possible alleles at a locus, *A1*, *A2*, and *A3*, with respective frequencies *p*, *q*, and *r*, so that $p + q + r = 1$, the trinomial expansion $(p + q + r)^2$ determines the genotypic equilibrium frequencies. The six genotypic values are

$$p^2\, A_1A_1 + 2pq\, A_1A_2 + 2pr\, A_1A_3 + q^2\, A_2A_2 + 2qr\, A_2A_3 + r^2\, A_3A_3.$$

Because each haploid gamete contains only a single allele for any gene locus, zygotic combinations will depend only on the frequency of each allele (**FIGURE 18.12**) and, as when only two alleles are present, equilibrium is established in a single generation of random mating. Non-random mating, which is often a consequence of self-fertilization, and which complicates attainment of Hardy–Weinberg equilibrium, is discussed in End Box 18.2.

■ KEY TERMS

bioinformatics
continuous variation
genomics
genetic polymorphism
genetic variation
Hardy–Weinberg equilibrium (principle)
heat shock proteins
heterozygote
inbreeding
inbreeding coefficient (*F*)
inbreeding depression
independent assortment
individual
mutation
neo-Darwinism
polygene
population
population genetics
QTL effect
quantitative trait loci (QTLs)

■ EVOLUTION ON THE WEB

Explore evolution on the Internet! Visit the accompanying website for *Strickberger's Evolution, Fifth Edition,* at **go.jblearning.com/Evolution5eCW** for exercises and links relating to topics covered in this chapter.

■ REFERENCES

Barrett, R. D. H., and D. Schluter, 2007. Adaptation from standing genetic variation. *Trends Ecol. Evol.*, **23**, 38–44.

Bell, M. A., W. E. Aguirre, and N. J. Buck, 2004. Twelve years of contemporary armor evolution in a threespine stickleback population. *Evolution*, **58**, 814–824.

Beltrao, P., G. Cagney, and N. J. Krogan, 2010. Quantitative genetic interactions reveal biological modularity. *Cell*, **141**, 739–745.

Charlesworth, B., and D. Charlesworth, 2010. *Elements of Evolutionary Genetics*. Roberts and Company Publishers, Greenwood Village, CO.

Cheverud, J. M., 2004. Modular pleiotropic effects of quantitative trait loci on morphological traits. In *Modularity in Development and Evolution*. G. Schlosser and G. P. Wagner (eds.). The University of Chicago Press, Chicago, pp. 132–153.

Colegrave, N., and S. Collins, 2008. Experimental evolution: experimental evolution and evolvability. *Heredity*, **100**, 464–470.

Connolly, M. H., and B. K. Hall, Embryonic heat shock reveals latent *hsp90* translation in zebrafish (*Danio rerio*). *Int. J. Dev. Biol.*, **52**: 71–79.

Crow, J. F., 1988. Eighty years ago: the beginnings of population genetics. *Genetics*, **119**, 473–476.

Denholm, I., G. J. Devine, and M. S. Williamson, 2002. Evolutionary genetics. Insecticide resistance on the move. *Science*, **297**, 2222–2223

Dobzhansky, Th., 1937, 1941, 1951. *Genetics and the Origin of Species* (three eds.). Columbia University Press, New York.

Falconer, D. S., and T. F. C. Mackay, 1996. *Introduction to Quantitative Genetics*, 4th ed. Longman, Harlow, Essex, UK.

Fisher, R. A., 1918. The correlation between relatives on the supposition of Mendelian inheritance. *Phil. Trans R. Soc. Edinb.*, **52**, 399–433.

Gillespie, J. H., 2004. *Population Genetics: A Concise Guide,* 2nd ed. Johns Hopkins University Press, Baltimore, MD.

Hallgrímsson, B., and B. K. Hall (eds.), 2005. *Variation: A Central Concept in Biology*. Elsevier/Academic Press, Burlington, MA.

Hedrick, P. W., 2000. *Genetics of Populations*, 2nd ed. Jones and Bartlett Publishers, Burlington, MA.

Keller, E. F., 2010. *The Mirage of a Space Between Nature and Nurture*. Duke University Press, Durham, NC.

Lynch, M., and B. Walsh, 1998. *Genetics and Analysis of Quantitative Traits*. Sinauer Associates, Sunderland, MA.

Mackay, T. F. C., 2004. The genetic architecture of quantitative traits: lessons from *Drosophila. Curr. Opin. Genet. Dev.*, **14**, 253–257.

Mackay, T. F. C., and R. F. Lyman, 2005. *Drosophila* bristles and the nature of quantitative genetic variation. *Proc. R. Soc. Lond. B*, **360**, 1513–1527.

Nei, M., 2007. The new mutation theory of phenotypic evolution. *Proc. Natl. Acad. Sci. USA*, **104**, 12235–12242.

Orzack, S. H., and E. Sober (eds.), 2001. *Adaptation and Optimality*. Cambridge University Press, Cambridge, UK.

Ostlund-Nilsson, S., I. Mayer, and F. A. Huntingford, 2007. *Biology of the Three-Spined Stickleback*. CRC Press, Boca Raton, FL.

Pagel, M. (ed. in chief), 2002. *Encyclopedia of Evolution*, 2 Volumes. Oxford University Press, New York.

Peichel, C. L., K. S. Nereng, K. A. Ohgi, et al., 2001. The genetic architecture of divergence between threespine stickleback species. *Nature*, **414**, 901–905.

Roff, D. A., 1997. *Evolutionary Quantitative Genetics*. Chapman and Hall, New York.

Roush, R., and D. R. McKenzie, 1987. Ecological genetics of insecticide and acaricide resistance. *Ann. Rev. Entomol.*, **32**, 361–380.

Rutherford, S. L., and S. Lindquist, 1998. Hsp90 as a capacitor for morphological evolution. *Nature*, **396**, 336–342.

Suzuki, Y., and H. D. Nijhout, 2006. Evolution of a polyphenism by genetic accommodation. *Science*, **311**, 650–652.

Sebat, J., B. Lakshmi, J. Troge, et al., 2004. Large-scale copy number polymorphism in the human genome. *Science*, **305**, 525–528.

Wagner, G. P., and J. Zhang, 2011. The pleiotropic structure of the genotype-phenotype map: the evolvability of complex organisms. *Nat. Rev. Genet.*, **12**, 204–213.

Weedon, M. N., and T. M. Frayling, 2008. Reaching new heights: insights into the genetics of human stature. *Trends Genet.*, **24**, 595–603.

Weedon, M. N., H. Lango, C. M. Lindgren, et al., 2008. Genome-wide association analysis identifies 20 loci that influence adult height. *Nat. Genet.*, **40**, 575–583.

END BOX 18.1

Calculation of Mutation Rate and Mutational Equilibrium

SYNOPSIS: *Mutation rates* and *mutational equilibrium* can be calculated, provided that we assume few alleles for each gene and simple dominance–recessive relationships between alleles. This end box defines mutation rate and mutational equilibrium and outlines how they are calculated.

Most genetic changes do not involve a simple mutation of one allele to another, but the accumulation of many new mutations in a population. In a population whose structure is stable, an equilibrium level of heterozygosity will be reached, reflecting nucleotide diversity of the original and the new alleles.[a] Sequence lineages can be recognized when gene sequences are analyzed. One example involves the control regions of mitochondrial genes from 191 extant domestic horses, and from ancient horse DNA found in archaeological sites and Late Pleistocene deposits. These data sets reveal widespread integration of matrilines (mother–daughter lineages) over long periods of horse evolution.[b]

Allele frequency is influenced by the frequency of mutation. Although most real life situations involve infinite alleles and/or unique mutations, it is customary to simplify the discussion to the classic Mendelian case: a gene with a dominant allele *A* and a recessive allele *a*, with mutation affecting *a* so that its frequency increases. This simplified allele situation illustrates the principles that operate, though not the known complexity of, allele interactions. In reality, codominance and linkage of complex and interacting loci are much more common than the situation of a single dominant and single recessive allele acting alone. The simplified situation is used here to illustrate how mutation rates and mutational equilibria are calculated.

If allele *A* continually mutates to *a*, the chances improve that *a* will increase in frequency with each generation. Given a long enough time and a persistent mutation rate in a population of constant size, *a* can eventually replace *A*. Of course, the mutation rate does not always occur in only one direction, especially at loci with multiple alleles. For example, if *u* is the mutation rate of *A* to *a*, the allele *a* may mutate back to *A* with frequency *v*. We can estimate these effects quantitatively by calling the initial frequencies of alleles *A* and *a*, p_0 and q_0, respectively, and noting that a single generation of mutation will produce a frequency of *A* equal to $p_0 + vq_0$ and a frequency of a equal to $q_0 + up_0$.

If we confine our attention to only one of the alleles, *a*, clearly it has gained the fraction up_0 (new *a* alleles) but lost the fraction vq_0 (new *A* alleles). In other words, the change in the frequency of *a*, which we call delta *q* (Δq), can be expressed as $\Delta q = up_0 - vq_0$. Thus, if *p* was relatively large and *q* small, Δq would be large and *q* would increase rapidly; when *q* became larger and *p* became smaller, Δq would diminish. The point at which Δq is zero—that is, the point where there is no further change and *p* and *q* are balanced in relation to their mutation frequencies—we call the mutational equilibrium (frequency $a = \hat{q}$): $\Delta q = 0 = up - vq$, or $up = vq$ at $\hat{q}$. However, because there are only two alleles, *A* and *a*, $p = 1 - q$, which leads to:

$$up = vq$$
$$u(1-q) = vq$$
$$u = uq + vq = q(u+v)$$
$$\hat{q} = \frac{u}{u+v}.$$

The same procedure applied to the frequency of *A* gives $\hat{p} = v/(u+v)$, so that $\hat{p}/\hat{q} = [v/(u+v)]/[u/(u+v)] = v/u$. Thus, when the mutation rates are equal ($u = v$) the **equilibrium gene frequencies** $\hat{p}$ and $\hat{q}$ will be equal. If the mutation rates differ, so will the equilibrium frequencies. For example, if $u = 0.00005$ and $v = 0.00003$, the equilibrium frequency $\hat{q}$ equals $5/8 = 0.625$ and $\hat{p} = 3/8 = 0.375$. The rate at which mutation reaches this equilibrium frequency, however, is usually quite slow. We can derive it by calculus from Δq as,

$$(u+v)n = \ln\left[(q_0 - \hat{q})/(q_n - \hat{q})\right],$$

---

[a] See Hedrick (2000) and Gillespie (2004) for more extended treatments and discussions of allele frequencies in populations.

[b] C. Vila et al., 2001. Widespread origins of domestic horse lineages. *Science*, **291**, 474–477.

END BOX 18.1

Calculation of Mutation Rate and Mutational Equilibrium (Cont...)

© Photos.com

where n is the number of generations required to reach a frequency q_n when starting with a frequency q_0. In this example, the number of generations necessary for q to increase from a frequency of one-eighth to three-eighths is,

$$(.00008)n = ln \frac{.125 - .625}{.375 - .625}$$

$$= ln\ 2.00 = .69315$$

$$n = \frac{0.69315}{.00008} = 8,664 \text{ generations.}$$

In terms of mutation rates calculated for organisms in the wild, a rate of .00008 is low, akin to that found in mutations in the amino acid leucine in *E. coli*.[13] Mutation rates are much higher in eukaryotes: 10/100,000 gametes in corn (and for the neurofibromatosis gene[s] in humans); 12 and 6/100,000 gametes for mutation of y^+ to *yellow* and of ey^+ to *eyeless* in *Drosophila*, respectively; 3/100,000 gametes for mutation of *agouti* to *nonagouti* in the mouse, *Mus musculus*. Thus, the shift toward equilibrium is slow, so slow that mutational equilibrium is rarely if ever reached.

REFERENCES

Gillespie, J. H., 2004. *Population Genetics: A Concise Guide*, 2nd ed. Johns Hopkins University Press, Baltimore, MD.

Hedrick, P. W., 2000. *Genetics of Populations*, 2nd ed. Jones and Bartlett Publishers, Burlington, MA.

---

[13] For *E. coli* and other unicells, mutation rate is expressed as mutations/100,000 cells.

END BOX 18.2

Inbreeding and the Hardy–Weinberg Equilibrium

© Photos.com

SYNOPSIS: *Inbreeding* and the *Hardy–Weinberg equilibrium* of gene frequencies can be calculated under assumptions of random mating between individuals in large populations, absence of gene flow between populations, and lack of inbreeding within populations. This end box outlines how the inbreeding coefficient (F), inbreeding depression, and the Hardy–Weinberg equilibrium of gene frequencies are calculated.

As discussed briefly in this chapter, the Hardy–Weinberg principle is the founding theorem of population genetics. Simply put, this principle states that in a population of randomly mating individuals allele frequencies are conserved and in equilibrium unless external forces act on them (Figure 18.9). The assumptions underlying attainment of equilibrium between alleles—random mating, large populations, equilibrium allele frequencies, and absence of gene flow—are attained only rarely in natural populations. These and related issues are explored in this end box.

Nonrandom mating interferes with the Hardy–Weinberg equilibrium. Nonrandom mating occurs when related individuals of similar genotype mate preferentially with each other in **inbreeding**.[a] Although inbreeding does not change the overall allele frequency, it leads to an excess of homozygous genotypes. Thus, inbreeding results in a rare recessive allele appearing in greater homozygous frequency than under random mating, offering increased opportunity for selection on rare recessives.

INBREEDING COEFFICIENT

Inbreeding is quantified by an **inbreeding coefficient**, *F*, which measures the probability that the two alleles of a gene in a diploid zygote are identical; that is, descended from a single ancestral allele (Wright, 1921). For example, beginning with a heterozygous diploid, A^1A^2, normal Mendelian segregation will confer a 1/2 probability that each F_1 offspring will receive the same A^1 allele. The allele transmitted by an F_1 individual to its offspring, in turn, has a 1/2 chance of being the same as the ancestral allele. This means that if two such F_1 offspring mate, the chances that the two alleles in one of their offspring (an F_2) are identical by descent (A^1A^1), is $1/2 \times 1/2 \times 1/2$, or the inbreeding coefficient is 1/8.

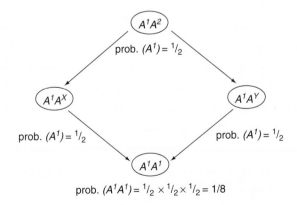

Having identical alleles means homozygosity, and *F* can range from one (complete homozygosity) to zero (complete **heterozygosity**). Of the inbred proportion measured by *F*, some will be *AA* and some *aa*, the frequencies of each depending on their respective population allele frequencies *p* and *q*. Thus, inbreeding will produce *pF AA* and *qF aa* genotypes. In addition to these, however, the remaining individuals in this population (1–F) will bear genotypes with

---

[a] An extreme form of inbreeding, *self-fertilization*, when two gametes of a single individual unite to form a fertile zygote, is discussed both here and elsewhere in this text.

END BOX 18.2

Inbreeding and the Hardy-Weinberg Equilibrium (Cont...)

© Photos.com

frequencies determined according to the Hardy–Weinberg equilibrium of p^2 *AA*, *2pq Aa*, and q^2 *aa*. The three genotypes will have the following frequencies:

$$AA = -p^2(1-F) + pF = p^2 - p^2F \pm pF =$$
$$p^2 + pF(1-p) = p^2 + pqF$$
$$Aa = 2pq(1-F) = 2pq - 2pqF$$
$$aa = -q^2(1-F) + qF = q^2 - q^2F + qF =$$
$$q^2 + qF(1-q) = q^2 + pqF.$$

The increase in the frequency of each homozygote type by a factor of pqF flows from an equivalent decrease in the heterozygote frequency ($-2pqF$). Note also that this reduction in **heterozygotes** affects the allele frequencies p and q equally, so that only the genotypic frequencies change. *When inbreeding is absent*, $F = 0$, and the preceding equations reduce to the Hardy–Weinberg frequencies p^2 *AA*, *2pq Aa*, and q^2 *aa*. *When inbreeding is complete*, $F = 1$, $2pq - 2pqF = 0$, and the only remaining genotypes are pAA and qaa.

SELF-FERTILIZATION

Inbreeding is greatest under self-fertilization where F is equal to 0.5 in the first generation and approaches 1 within four or five generations. As shown in **FIGURE EB2.1**, any other mating scheme slows the rate of inbreeding. Similarly, as seen in **FIGURE EB2.2**, population size can affect inbreeding; the smaller the size, the greater the opportunity for related individuals to mate.

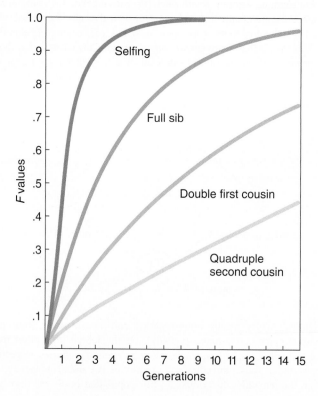

FIGURE EB2.1 Inbreeding coefficients at generations 1 to 15 for four different systems of inbreeding (pedigrees given in Strickberger M. W., 1985. *Genetics*, 3rd ed. Macmillan, New York.). Formulae for calculating F in other inbreeding systems can be obtained from Wright's basic work in this field (Wright, 1921).

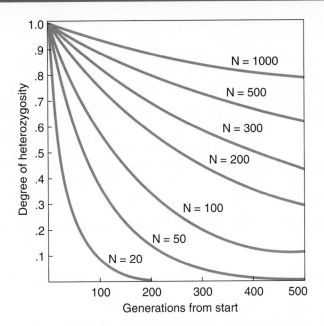

FIGURE EB2.2 Degrees of heterozygosity (heterozygote frequency) remaining in populations of different sizes after given generations of random union between gametes. Calculations are based on 1.00 as the initial degree of heterozygosity (or when *F* = 0).

[Adapted from Strickberger, M. W. *Genetics, Third Edition*. Macmillian, New York, 1985.]

Genetic assortative mating is the name given to systems such as brother–sister mating and first cousin mating, in which individuals mate on the basis of their genetic relationship.

Phenotypic similarity, however, also may lead to preferential mating; in many human societies mates are chosen who share characteristics such as height, color, facial form (especially symmetry according to recent studies), muscular build, and intelligence. Homozygosity can increase in such **phenotypic assortative mating**, but only for those loci involved in the trait(s) on which the preferred matings are based. This is in contrast to the genetic assortative mating of inbreeding, which tends to increase homozygosity at all loci.

Disassortative mating occurs when individuals of unlike genotype or phenotype mate, thereby preventing inbreeding and helping to maintain heterozygote frequency (heterozygosity). Examples of such systems include alleles in plants that result in sterility of male gametes when they interact with ova of the same genotype (*self-sterility alleles*).

INBREEDING DEPRESSION

Although some degree of inbreeding occurs in most outbreeding populations, significant amounts of inbreeding can result in **inbreeding depression**, in which rare deleterious recessives may appear with increased homozygous frequency. If a recessive disease occurs with frequency q^2 in a random outbred population, its frequency will increase by pqF in an inbred population. Therefore, the ratio of inbred to outbred frequency for the homozygous recessive will be,

$$\frac{q^2 + pqF}{q^2} = \frac{q(q + pF)}{q^2} = \frac{q + pF}{q}.$$

Obviously, if *q* is large and *F* is small, the inbreeding increment *pF* will be relatively small, and the increased frequency of homozygous recessives will hardly be noticeable. However, if *q* is rare and *p* is large, then *pF* provides a notable increase in recessives even when *F* is fairly small.

END BOX 18.2

Inbreeding and the Hardy-Weinberg Equilibrium (Cont...)

For example, if q is 0.5, first cousin mating ($F = 0.0625$) will produce an inbred-to-out bred ratio of homozygotes of,

$$\frac{q + pF}{q} = \frac{0.5 + (0.5)(0.0625)}{0.5} = \frac{0.53125}{0.5} = 1.06.$$

However, if q is 0.005, this ratio increases to 0.067/0.005 = 13.4. When $q = 0.0005$, the increase of homozygotes resulting from first cousin mating is 0.0630/0.0005, or 126 times that of randomly bred populations.

If homozygous recessives have a quantitative effect on one or more traits, inbreeding results in the measured values of these traits moving towards recessive values. Thus, in various outbred populations, inbreeding depression can reduce traits like organisms height, crop yield, and other characteristics. Inbreeding depression, however, is not a universal phenomenon in all species, certainly not in many species that are normally self-fertilized and have eliminated most or all of their deleterious recessives. On the whole, ample evidence shows that most normally cross-fertilizing species deteriorate with consistent inbreeding, leading even to extinction,[b] although some strains may escape extinction because they carry relatively few deleterious recessive alleles.

REFERENCES

Gillespie, J. H., 2004. *Population Genetics: A Concise Guide*, 2nd ed. Johns Hopkins University Press, Baltimore, MD.

Hedrick, P. W., 2000. *Genetics of Populations*, 2nd ed. Jones and Bartlett Publishers, Burlington, MA.

Wright, S., 1921. Systems of mating. *Genetics*, **6**, 111–178.

[b] See I. Saccheri et al., 1998. Inbreeding and extinction in a butterfly metapopulation. *Nature*, **392**, 491–494 for one study.

CHAPTER

19

Demes, Gene Flow, and Genetic Drift

CHAPTER SUMMARY

Without heritable variation, there would be no evolution; genetic variation provides the raw material enabling evolutionary change following natural selection. Mutations are a source of genetic variation that occur in individuals and accumulate within a population as allelic (genetic) variants. Additional sources of genetic variation arise from genetic drift within a population and from gene flow between populations. In these two processes, hereditary variation arises not from mutation but from loss of individuals (and their genes) from a population or from immigration of individuals (and their genes) into a population. These individuals constitute the local interbreeding group (deme) among which random mating occurs. Animal and plant populations share common features in their response to inbreeding, selection, mutation, migration, and genetic drift. Even with concepts such as the adaptive landscape (END BOX 19.1), in the overwhelming majority of populations, however, it is a challenge to discover the genetic information associated with phenotypic variation and how changes in phenotypes correlate with the evolutionary mechanisms of mutation, selection, and genetic drift. Large-scale geographical patterns of species distribution (phylogeography) can be determined using the genetic history of populations.

387

INTRODUCTION

Through theoretical advances and the development of new methods of computation such as bioinformatics, population ecology has combined with population genetics to enhance our understanding of the bases of population structures, and of the interactions that occur within and between populations. Population geneticists measure the amounts and kinds of genetic variation in populations and unite observations on natural populations with models of population genetic structure.[1] Such studies received special attention from the 1920s and 1930s onward as the mathematical bases of population genetics were established.

Geneticists define a **population** as a group of sexually interbreeding or potentially interbreeding individuals. Size of the interbreeding population may vary, but is usually taken to be a local interbreeding group (a **deme**), within which random mating occurs. Structural features of the environment—a river that individuals cannot cross, a landslide—reduce effective deme size. As a consequence, variation in "the population" is effectively variation in each deme. Furthermore, selection may take demes in different directions, potentially leading to isolation of one or more groups, a condition that can lead to speciation.

DEMES AND THE GENE POOL

All members of a species share a common **gene pool**, although populations may vary genetically from one another. We expect widely separated populations to have less opportunity to share gene pools than those closer together. Many species consist of genetically diverse local populations and demes. Oftentimes, species comprise many demes that do not exchange alleles. Demes have the potential to contribute to the gene pool of the entire species, but are sufficiently separated geographically from each other to maintain specific gene frequencies.

Because the forces acting on demes may change among localities, differences among populations arise and are maintained. A transect across central California shows that populations of the yarrow plant, *Achillea*, differ significantly in such traits as height and growing season (**FIGURE 19.1**). The adaptive nature of such traits is evidenced by the differential responses of individuals from populations at one locality when moved to a different locality. Coastal plants are weak when grown at higher altitudes; high-altitude plants grow poorly at lower altitudes (**FIGURE 19.2**).[2] Adaptive features of many such plant populations reflect the accumulated genetic response of a population to a particular ecological habitat.

The considerable genetic variability in human groups shows that populations of modern humans are demes. On the basis of blood types (A, B, O, Rh$^+$, Rh$^-$) five groups of humans can be distinguished: African, Caucasian, Greater Asian, Amerindian, and Australoid. Members of each group are not genetically homogeneous in the sense of sharing a uniform genetic identity. Differences among these groups have not reached the point where each population is fixed for a different allele. Rather, an allele fixed in one population is usually polymorphic in other populations. Genetic uniformity does not even apply to members of the same family or related individuals. About 84 percent of the genetic variation among humans comes from differences among individuals and populations of the same group. Only 16 percent comes from differences between groups.

RH *is shorthand for Rhesus-factor. Rh-positive blood type is the blood group (approximately 85% of people) whose red cells have the Rh antigen.*

---

[1] For a recent treatment of population genetics, see Roff (1997), Hedrick (2000), and Charlesworth and Charlesworth (2010).

[2] J. C. Clausen, D. D. Keck, and W. M. Hiesey, 1948. Experimental studies on the nature of species. III. Environmental responses of climatic races of *Achillea. Carnegie Inst. Wash. Publ.* No. **581**, 1–219 (supp. 391–392).

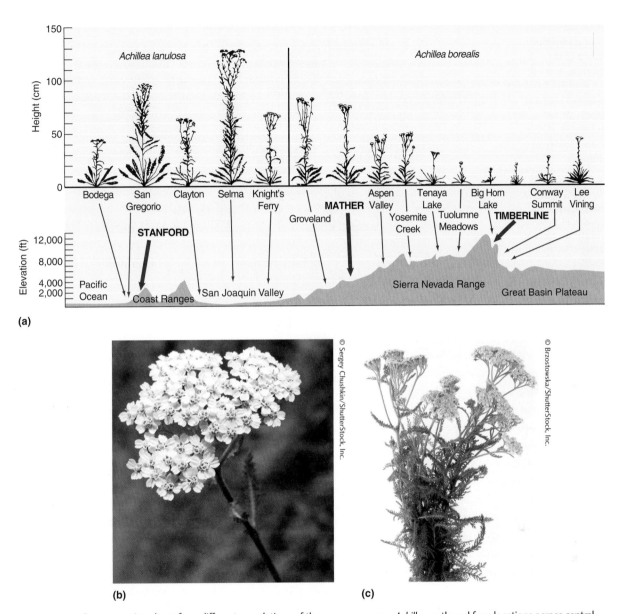

FIGURE 19.1 Representative plants from different populations of the common yarrow, *Achillea*, gathered from locations across central California and grown under uniform conditions in Stanford, California. The differences in plant size, leaf shape and other characteristics, and the transition between *A. lanulosa* and *A. borealis* shown in **(a)** relate to the locations of the populations and indicate that genetic differences have evolved among them. Examples of *Achillea borealis* and *A. lanulosa* are shown in **(b)** and **(c)**, respectively.

[(a) Adapted from Clausen, J. D., D. Keck, and W. M. Hiesey, 1948. Experimental studies on the nature of species. III. Experimental responses of climatic races of *Achillea*. *Carnegie Inst. Wash. Publ.*, **No 581**, 1–219.]

STANDING GENETIC VARIATION

Large amounts of genetic (allelic) variation exist in practically all natural populations, providing the **standing genetic variation** that allows genetic evolutionary changes to proceed (see Barrett and Schluter, 2007 for an excellent overview). However, many genes maintained in natural populations may compromise the survivability and fecundity of their carriers. Selection can account for a significant loss of non-optimal individuals even in long-standing populations (Orzack and Sober, 2001). The extent to which a population departs from a perfect genetic constitution is called its **genetic load**, and is marked by the loss of some individuals through **genetic death**. Genetic death is not necessarily an actual death before reproductive age. Genetic death can be sterility, inability to find a mate, or any means that reduces reproductive ability. We express estimates of these values as

FIGURE 19.2 Differential responses of clones from representative of the common yarrow, *Achillea*, originating from five localities in California and grown at three different altitudes: sea level (Stanford), 1,400 m (Mather), and 3,050 m (Timberline).

[Adapted from Clausen, J. D., D. Keck, and W. M. Hiesey, 1948. Experimental studies on the nature of species. III. Experimental responses of climatic races of *Achillea. Carnegie Inst. Wash. Publ.,* **No 581**, 1–219.]

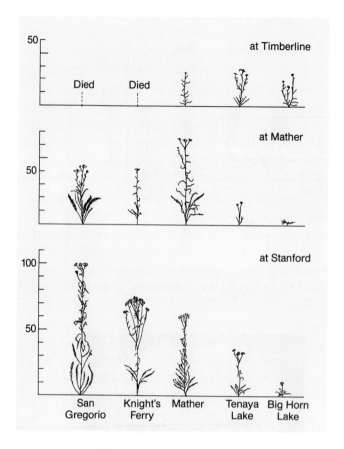

the proportion of individuals eliminated by selection, high and low selection coefficients reflecting rapid or slow elimination of a gene, respectively.

Most, if not all, populations carry genetic loads of one kind or another. Environments change with time and the advantages of different genotypes change accordingly. A population with a relatively large genetic load may encounter a new environment in which formerly deleterious alleles endow it with adaptations that benefit survival. Populations also obtain new alleles from other populations through a process known as **gene flow**.

GENE FLOW

Mutation is not the only mechanism by which new alleles enter a population. A population may receive alleles from a nearby population in a process known as gene flow, or sometimes **gene migration**, as individuals, gametes, parasites, viruses, or bacteria move from population to population, taking their genotypes with them.[3] Gene flow may be between populations with similar allele frequencies or between populations with different allele frequencies. In either case, gene flow results in some mixing of the gene pools and slows further differentiation between the populations.

At least three factors impact the recipient population,

- the **difference in gene frequencies** between the two populations;
- the **proportion of migrating genes** incorporated into each generation; and
- the **pattern of gene flow**, whether occurring once or more than once.

If we designate q_0 as the initial gene frequency in the recipient, or hybrid, population, Q as the frequency of the same allele in the migrant population, and m as the proportion of newly introduced genes each generation, the allele frequency in the hybrid population will experience a loss of q_0 equal to mq_0 and a gain of Q equal to mQ. Over n generations

[3]Gene flow also occurs between species under conditions of hybridization or lateral gene transfer.

of migration, when the gene frequency of the hybrid population becomes q_n, one can calculate that the relationship between these factors will reach:

$$q_n - Q = (1 - m)^n(q_0 - Q)$$

$$\text{or } (1 - m)^n = \frac{q_n - Q}{q_0 - Q}.$$

For populations where this equation can be applied, four of these factors must be known in order to calculate the fifth.

A classic example of gene flow, which preceded our extensive knowledge of DNA sequences, is **blood group gene frequencies** in African Americans and European Americans, two populations between which gene exchange has occurred over many generations. In one study, Parra et al. (1998) genetic exchange was estimated by analyzing the genetic contribution of Europeans to 10 populations of African descent in the United States, using nine autosomal DNA markers.[4] Contributions from European ancestry range from seven percent to 23 percent in the 10 populations of African descent. Accordingly, the proportion of African ancestry in the 10 populations ranges from 77 to 93 percent.

The consequences of gene flow are complex: gene flow can hinder local evolutionary changes by infusing genes from populations that are not adapted to local conditions. In contrast, gene flow can have beneficial effects, or can produce no phenotypic effects at all. For example, populations of the checkerspot butterfly, *Euphydryas editha*, often show no phenotypic changes after they are the recipients of gene flow from phenotypically different populations.

GENETIC DRIFT

Mutation and gene flow share one important quality; if they continue from generation to generation they can result in what seem to be progressive changes in gene frequencies that give the appearance of directionality. Other types of genetic change have no predictable constancy or direction from generation to generation. **Genetic drift**, one of the most important of such random or non-directional forces, is a consequence of random fluctuations in gene frequencies that arise in small populations. In larger populations, genetic drift is countered by selection. In one study, Hallatschek et al., 2008 demonstrated that an established and expanding colony of two strains of *E. coli* developed well-defined regions driven by genetic drift at the boundaries between the two strains. Such random fluctuations can dramatically alter the gene pools in the boundary regions, resulting in genetic differentiation within the colony.

WEAK SELECTION *only slowly leads to change in allele frequencies, often no more effectively than* GENETIC DRIFT.

CALCULATING GENETIC DRIFT

For practical reasons, mutation, selection, migration, and drift cannot always be analyzed in the same population/study. A reminder that these four processes act in concert comes from a set of simulations mimicking interaction between genetic drift and gene flow in a fragmented population. The traditional assumption that any connections between subpopulations in the species range *inhibit* local adaptation was not borne out; low migration rates improved fitness in marginal populations, reflecting a complex interplay between migration rate, mutation rate, subpopulation size, and species range (Alleaume-Benharira et al., 2006).

Genetic drift arises from variable sampling of the gene pool each generation; in the absence of directional forces to change allele frequencies, there is always a strong likelihood of obtaining an accurate sample of the previous generation's alleles as long as the number of parents in a population is consistently large. However, as real populations are limited in size, genetic drift will change gene frequency due to sampling variation. For example, if only a few parents begin a new generation, the small sample of alleles may deviate widely from the allele frequency in the previous generation.

[4] Parra, E., et al., 1998. Estimating African-American admixture proportions by use of population-specific alleles. *Am. J. Hum. Genet.*, **63**, 1839–1851.

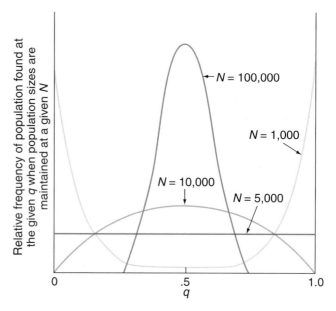

FIGURE 19.3 Distribution of equilibrium gene frequencies for populations of different sizes when selection is zero and a small amount of migration occurs into each population ($m = .0001$) from a population with a gene frequency of $q = .5$. Despite this migration, populations of sizes $N = 1,000$ and $N = 5,000$ show considerable genetic drift, many reaching elimination ($q = 0$) or fixation ($q = 1$). Only populations of relatively large sizes ($N = 10,000$, $N = 100,000$) maintain the initial gene frequency $q = .5$ in appreciable proportions.

[Adapted from Wright, S, 1951. The genetic structure of populations. *Ann. Eugen.*, **15**, 323–354.]

The extent of the deviation for all population sizes can be measured mathematically by the standard deviation of a proportion $\sigma = \sqrt{pq/N}$, where p is the frequency of one allele, q of the other, and N the number of genes sampled. For diploid parents, each carrying two alleles, $\sigma = \sqrt{pq/2N}$, where N is the number of actual parents. For example, if we begin with a large diploid population, where $p = q = .5$, and continue this population each generation by using 5,000 parents, then $\sigma = \sqrt{(.5)(.5)/10,000} = \sqrt{.000025} = .005$. The values of such populations will fluctuate mostly around $.5 \pm .005$, or between .495 and .505. A choice of only two parents as founders will produce a standard deviation of $\sqrt{(.5)(.5)/4} = \sqrt{.625} = .25$, or values of $.5 \pm .25$ (from .25 to .75).

In other words, **sampling bias** because of smaller population size can easily yield allele frequencies that depart considerably from the initial .5 values in a single generation. If the population remained small and the next generation began with either of these extremes—that is, a frequency of .25 or .75 for a particular allele—in the following generation the frequency of that allele may fall to almost zero ($.25 \pm \sqrt{(.25)(.75)/4} = .256 \pm .22$; a range of .03 to .47) or increase almost to 1 ($.75 \pm \sqrt{(.75)(.25)/4} = .75 \pm .22$, a range of .53 to .97). Should such small populations continue each generation, the likelihood increases that one or more alleles will eventually reach fixation. The proportion of such populations that attain fixation (that is, the *rate of fixation*) will eventually reach $1/2N$. Obviously, if N is large, fixation proceeds slowly, but even large populations can show some degree of drift, as diagrammed in **FIGURE 19.3**.

WHAT EXACTLY IS GENETIC DRIFT?

Despite this simple summary, genetic drift means different things to different people. Seven definitions of genetic drift (with comments on them) are listed in **TABLE 19.1**. None specifies genetic drift as a process that can produce evolutionary change. Several see drift as fluctuating changes in gene frequencies in groups, populations, or species (definitions 1, 2, and 4 to 7), without specifying whether that chance is a factor in evolutionary change.

| TABLE 19.1 | Definitions of Genetic Drift and Comments on Them |

Definition	Comment
1. Random change in gene frequencies in populations	Far too broad and nonspecific
2. Variation in the genetic makeup of a species over time, often as a consequence of environmental change or isolation	How does environmental change affect gene frequency? Isolation of the species, a population, a subset of individuals?
3. Stochastic fluctuations produced by sampling in finite populations	Assumes that genetic drift is an anomaly of sampling
4. Changes in gene frequencies in a small population resulting from chance processes and not from natural selection	Sets genetic drift apart from selection, which is good, but confines the process to small populations
5. Gene frequency changes in populations resulting from random events in the gene pool of small populations rather than natural selection, especially the effects of sampling error	Confines drift to random events but then sees drift as a sampling problem not a natural process
6. Random fluctuations in gene frequencies, most evident in small populations	Does not confine the process to small populations
7. Change in frequency of a gene in a population through mutation, regardless of the adaptive value of the mutation	Does not distinguish genetic drift from mutation

Two definitions (3 and 5) regard genetic drift as an artifact of sampling a small number of individuals from small populations.

Despite the differences in these approaches, this inclusion of population size in all definitions of genetic drift emphasizes the importance of what is called the **effective population size** (N_e), which is calculated as $4N_f N_m/(N_f + N_m)$, where N_f is the number of female parents and N_m the number of male parents. Effective population size is the number of parents who *contribute* offspring to the next generation. If, out of a total population of 1,000, three males mated to 300 females produce the next generation, the effective population size is more than six (three males mating with three females) but less than 303 (three males mating with 300 females). In this example, N_e is 11 [4(300)(3)/303]. Sewall Wright (1889–1988) proposed that genetic drift is quite important in changing allele frequencies among populations when the effective size of the population is small. His proposal is known as the **shifting balance theory**.

Sewall Wright and the English evolutionary geneticists R. A. Fisher (1890–1962) and J. B. S. Haldane (1892–1964) are the acknowledged founders of the NEO–DARWINIAN THEORY OF EVOLUTION *based on population genetics (Provine, 1986).*

THE SHIFTING BALANCE THEORY

The following five statements provide an outline of the operation of the shifting balance theory of population, proposed by Sewall Wright.[5]

1. Genetic drift, acting on genetic variation at various loci, allows a number of demes to change their allele frequencies. As fitness changes, such demes are modeled as moving across nonadaptive valleys to different parts of an adaptive landscape, a model discussed in End Box 19.1.
2. Selection pushes some of these demes up the nearest available adaptive peak by changing allele frequencies even further.
3. Variation at other loci provides further opportunity for genetic drift to move a population to a higher adaptive peak.
4. A deme that has a high fitness coefficient (has attained a high adaptive peak) tends to displace other demes of lower fitness by expanding in size or dispersing outward and changing the genetic structure of other demes through migration.

Much of SEWALL WRIGHT'S *long career—he died in 1988 and published for 75 of his 99 years—was spent providing evidence for, explaining, and defending the shifting balance theory.*

---

[5] See Wright (1951, 1963, 1978, 1988) for elaborations of the theory, Coyne et al. (1997) and Ruse (2004) for critiques of the theory from the perspective of evolutionary biologists and a philosopher of science, and Provine (1986) for a biography of Sewell Wright.

5. Environmental change, such as a new stream or seismic earthquakes, can act on populations by changing the environment (the adaptive landscape; End Box 19.1) to which populations must adjust or perish. Channeling selection in new directions encourages populations to continually shift their genetic structures. As concluded from a long-term study of selective changes in Darwin's finches on the Galapagos Islands: "The population tracks a moving peak (in an adaptive landscape) caused by environmental fluctuations, and there is more than one individual fitness optimum within the range of phenotypes in the population" (Grant and Grant, 1989, p. 300).

An important consequence of this theory is its emphasis on differences in survival or extinction among populations rather than only among individuals. **Selection among individuals** in a population is a conservative force that pushes the population up a single adaptive peak (End Box 19.1 and Figure EB1.1). **Selection among populations** (accompanied by genetic drift) leads to the occupation of higher adaptive peaks and the replacement, extinction, or colonization of populations at lower adaptive peaks.

An important study supporting the shifting balance theory began with 107 separate lines of *D. melanogaster*, each carrying two alleles at the *brown* locus (*bw* and *bw*[75]) at initially equal allele frequencies of 50 percent.[6] The lines were continued for 19 generations by randomly selecting 8 males and 8 females as parents from each preceding generation ($N = 16 = 32$ *brown* alleles) and scoring the frequency of the two different *brown* alleles. As shown in FIGURE 19.4, by the first generation, a number of populations already had departed from the original 50 percent *bw*[75] frequency. Genetic drift continued to increase successively so that by generation 19 more than half the 107 populations reached fixation for either the *bw* or *bw*[75] alleles. Note that despite these genetic differences, the *overall average frequency of the brown alleles in all populations combined is 0.5, which was the initial gene frequency.* We can conclude that genetic drift increases variation between populations, but on the average, not in any particular direction. These results also indicate that, because of genetic drift, selection in small populations, unless intense, may have little or no effect on the frequency of a deleterious allele.

VARIATION AND SELECTION FOR INCREASED FITNESS

Selection on the phenotype will only result in heritable change if genetic variation is present (Barrett and Schluter, 2007). Change in gene frequency in response to selection is a constant feature of most or all populations facing changing environments. Populations tend to change genetically in directions that enhance fitness for their environment, as has been repeatedly demonstrated in natural populations. R. A. Fisher, a founding father of population genetics, formulated this principle mathematically in 1930 as a fundamental theorem: the greater the genetic variation upon which selection for fitness may act, the greater the expected improvement in fitness.[7]

As a consequence, we expect populations long subjected to selection to have little remaining variation for genes affecting fitness; selection would have diminished such variation. Traits such as fecundity, for example, often have very low heritabilities because they have little genetic variation. Continued existence of selection therefore implies that **variation itself is subject to selection**, and so the propensity to vary (variability) is an important attribute of organisms.[8] Continuous changes in the environment (resources,

---

[6] P. Buri (1956). Gene frequency in small populations of mutant *Drosophila. Evolution*, **10**, 367–402.
[7] For experimental validation of Fisher's principle see A. B. Carvalho et al. (1998). An experimental demonstration of Fisher's principle: evolution of sexual proportion by natural selection. *Genetics,* **148**, 719–731. For expansion of Fisher's principle to accommodate heterozygote advantage and adaptive change, see D. Sellis et al. (2011). Heterozygote advantage as a natural consequence of adaptation in diploids. *Proc. Natl. Acad. Sci. USA*, **108**, 20666–20671.
[8] See the papers in Hallgrímsson and Hall (2005) for analyses of the many factors affecting phenotypic variation and variability, and for a proposal to further our understanding of these factors and how they interact.

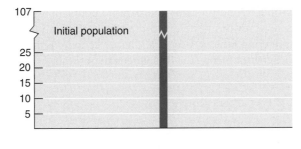
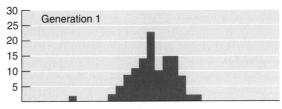
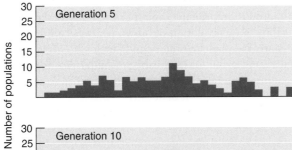
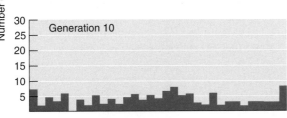
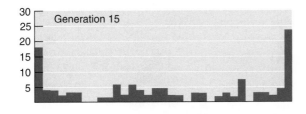
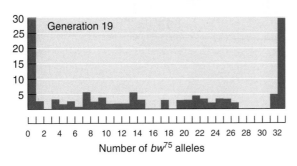

FIGURE 19.4 Distributions of the numbers of *brown* (*bw*^75) alleles in 107 populations of *D. melanogaster*, each with an initial frequency of 0.5 *bw*^75 (top panel). The populations were continued for 19 generations (bottom panel), using 16 parents to start each generation. By generation 19, the *bw*^75 allele had been eliminated (0 alleles) from 30 populations and had been fixed (32 alleles) in 28 populations.

[Data from Buri, P., 1956. Gene frequency in small populations of *Drosophila. Evolution*, **10**, 1956: 367–402.]

supplies, waste products, and predator and parasite populations, etc.) can bring about changes in genes not previously subject to the effects of selection and so provide new opportunities for altering fitness. Also recall that selection on one species affects other species in the same environment—the **Red Queen hypothesis**.

PHYLOGEOGRAPHY

Large-scale geographical patterns of species distribution can be determined using the genetic history of populations. A recent approach to understanding the evolutionary processes regulating the geographic distributions of groups is reconstruction of the genealogies of individual genes, groups of genes, or populations. Differences among populations of a species are detected by differences in gene frequencies. The field is known as **phylogeography**: the reconstruction of the geographical history of (a) lineage(s) using genetics.[9]

Phylogeography reveals group, population, and species histories. Rather than sampling a single population, as is often the case in standard population genetics analyses, different populations within a species are examined. Consequently, information about past patterns of migration can be used to explain the current distribution and subdivision of species into groups. In principle, any species group can be studied this way. Past studies include a study of the evolution of brown bears (*Ursus arctos*) in Alaska, which revealed that bears are paraphyletic (Talbot and Shields, 1996), and a study of the geographical, ecological, and evolutionary history of butterflies in the genus *Lycaeides* in North America (Gompertz et al., 2010). Determination of the patterns responsible for the current geographical distribution of strains of wheat is an excellent example of the application of phylogeography.

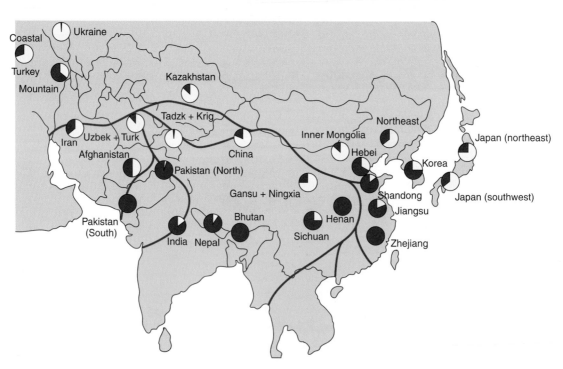

FIGURE 19.5 Phylogeographic analysis of Asian common wheat, *Triticum aestivum*, based on isozyme analysis. Collection sites of the populations used in the study discussed in the text range across all of Asia. The major trade route (the "Silk Road" is shown in brown).

[Adapted from Ghimire, S. K., Y. Akashi, C. Maitani, M. Nakanishi, and K. Kato, 2005. Genetic diversity and geographical differentiation in Asian common wheat (*Triticum aestivum* L.), revealed by the analysis of peroxidase and esterase isozymes. *Journal of Plant Breeding and Crop Science*, **55**, 75–185.]

[9] See Knowles (2004) and Holsinger (2010) for assessment of the field of phylogeography.

SPREAD OF WHEAT

Strains of the Asian common wheat, *Triticum aestivum*, reflect adaptations of a single species to a large number of environments, growing conditions, and artificial selection pressures across Asia, from the Ukraine, and Turkey in the west to Japan in the east. Using the argument that the Tibetan plateau and large areas of desert in west China form effective geographical barriers to migration, three routes for the spread of wheat along trade routes to East Asia have been proposed (FIGURE 19.5).

- From Turkey to Sichuan China along an ancient Myanmar route,
- along the Silk Road, and
- from the coastal area of China and Korea.

This interpretation of the distribution of the strains of wheat, which is based on geographical barriers and trade routes developed and maintained by humans, is supported by two studies based on the distribution of genes for two enzymes (Ghimire et al., 2005). Six hundred and forty-eight races of wheat revealed 33 populations that could be grouped into six clusters that originated from three lineages (FIGURE 19.6).

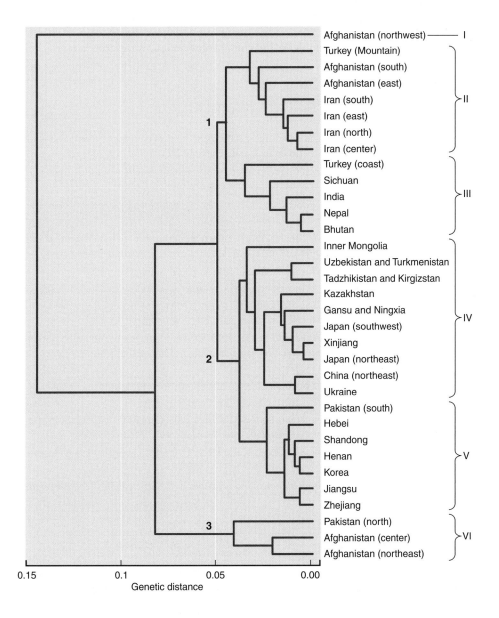

FIGURE 19.6 The 33 populations of Asian common wheat shown in Figure 19.5 were resolved into six clusters (I–VI) derived from three lineages 1–3.

This study revealed genetic and geographical differentiation of Asian wheat and allowed the possible sources of origin and routes of migration of populations to be verified. This is phylogeography at its best.

■ KEY TERMS

adaptive landscape
altruism
blood group gene frequencies
deme
fitness
gene flow
gene migration
gene pool
genetic drift

genetic variation
phylogeography
population
population size
red queen hypothesis
selection
shifting balance theory
spread of wheat
variation

■ EVOLUTION ON THE WEB

Explore evolution on the Internet! Visit the accompanying website for *Strickberger's Evolution, Fifth Edition,* at **go.jblearning.com/Evolution5eCW** for exercises and links relating to topics covered in this chapter.

■ REFERENCES

Alleaume-Benharia, M., I. R. Pen, and O. Ronce, 2006. Geographical patterns of adaptation within a species' range: interactions between drift and gene flow. *J. Evol. Biol.,* **19**, 203–215.

Barrett, R. D. H., and D. Schluter, 2007. Adaptation from standing genetic variation. *Trends Ecol. Evol.,* **23**, 38–44.

Charlesworth, B., and D. Charlesworth, 2010. *Elements of Evolutionary Genetics.* Roberts and Company Publishers, Greenwood Village, CO.

Coyne, J. A., N. H. Barton, and M. Turelli, 1997. Perspective: a critique of Sewall Wright's shifting balance theory of evolution. *Evolution,* **51**, 643–671.

Fisher, R. A., 1930. *The Genetical Theory of Natural Selection.* Clarendon, Oxford, UK.

Ghimire, S. K., Y. Akashi, C. Maitani, M. Nakanishi, and K. Kato, 2005. Genetic diversity and geographical differentiation in Asian common wheat (*Triticum aestivum* L.), revealed by the analysis of peroxidase and esterase isozymes. *Plant Breeding and Crop Sci,* **55**, 75–185.

Gompertz, Z., M. L. Forister, J. A. Fordyce, et al., 2010. Bayesian analysis of molecular variance in pyrosequences quantifies population genetic structure across the genome of *Lycaeides* butterflies. *Mol. Ecol.,* **19**, 2455–2473.

Grant, B. R., and P. R. Grant, 1989. Natural selection in a population of Darwin's finches. *Am. Nat.* 133, 377–393.

Hallatschek, O., P. Hersen, S. Ramanathan, and D. R. Nelson, 2008. Genetic drift at expanding frontiers promotes gene segregation. *Proc. Natl Acad. Sci. USA.,* **105**, 19926–19930.

Hallgrímsson, B. and B. K. Hall (eds.), 2005. *Variation: A Central Concept in Biology.* Elsevier/ Academic Press, Burlington, MA.

Hedrick, P. W., 2000. *Genetics of Populations,* 2nd ed. Jones and Bartlett Publishers, Burlington, MA.

Holsinger, K. E., 2010. Next generation population genetics and phylogeography. *Mol. Ecol.,* **19**, 2361–2363.

Knowles, L. L., 2004. The burgeoning field of statistical phylogeography. *J. Evol. Biol.,* **17**, 1–10.

Orzack, S. H., and E. Sober (eds.), 2001. *Adaptation and Optimality.* Cambridge University Press, Cambridge, UK.

Pagel, M. (ed. in chief), 2002. *Encyclopedia of Evolution*, 2 Volumes. Oxford University Press, New York.

Provine, W. B., 1986. *Sewall Wright and Evolutionary Biology*. The University of Chicago Press, Chicago.

Roff, D. A., 1997. *Evolutionary Quantitative Genetics*. Chapman and Hall, New York.

Ruse, M., 2004. Adaptive landscapes and dynamic equilibrium. The Spencerian contribution to twentieth-century American evolutionary biology. In *Darwinian Heresies*, A. Lustig, R. J. Richards, and M. Ruse (eds.). Cambridge University Press, Cambridge, UK, pp. 131–150.

Talbot, S. L., and G. F. Shields, 1996. Phylogeography of brown bears (*Ursus arctos*) of Alaska and paraphyly within the Ursidae. *Mol. Phylogenet. Evol.*, **5**, 477–494.

Wright, S., 1951. The genetic structure of populations. *Ann. Eugen.*, **15**, 323–354.

Wright, S., 1963. Genic interaction. In *Methodology in Mammalian Genetics*, W. J. Burdette (ed.). Holden-Day, San Francisco, pp. 159–192.

Wright, S., 1978. *Evolution and the Genetics of Populations: Vol. 4. Variability Within and Among Natural Populations*. University of Chicago Press, Chicago.

Wright, S., 1988. Surfaces of selective value revisited. *Am. Nat.*, **131**, 115–123.

END BOX 19.1

Adaptive Landscapes and the Shifting Balance Theory

© Photos.com

SYNOPSIS: Sewall Wright's *shifting balance theory* for how alleles are maintained in a population, which in introduced in this chapter, is explored more fully here. Particular features discussed are **demes**, as the breeding units within populations, fluctuations in gene frequencies because of **genetic drift**, how genotypes move to *adaptive peaks*, and altruism.

Sewall Wright's shifting balance theory for maintaining alleles in a population is outlined in this chapter. According to this theory each genotype at a locus occupies an adaptive peak in an adaptive landscape, an adaptive peak being a position of high fitness associated with a specific environment (**FIGURE EB 1.1**). As long as no other factors change the fitness of these genotypes, each of the six adaptive peaks shown in Figure EB 1.1 will be of equal height. A population consisting entirely of any one of these genotypes would achieve maximum fitness for this phenotype.*

Movement of a population from peak to peak depends on various factors, especially the size of the population. Small populations are more subject to random fluctuations in gene frequencies (genetic drift), so their frequencies vary more readily than in large populations. As the environment changes, the relative fitness of genotypes changes. As new alleles are introduced, interaction with other alleles (epistasis) changes their relative fitness. What is a "valley" (a state of low fitness) at one time need not be a valley at another, and a population's position on the landscape can fluctuate accordingly. Once such an adaptive landscape has evolved, further evolution will depend on the origin of a new selective environment and new adaptive peaks. However, if conditions are not changing rapidly, the same set of adaptive peaks may remain for long periods of time.

How can a population located on a relatively low adaptive peak evolve so that it occupies the highest or near-highest peaks on the adaptive landscape? To answer this problem, Sewall Wright

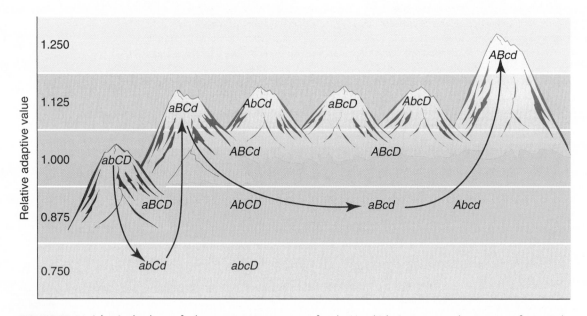

FIGURE EB 1.1 Adaptive landscape for homozygous genotypes at four loci in which six genotypes homozygous for capital letter alleles at two loci (*AbcD, AbCd, aBcD,* and so on) attain relatively high relative fitness (adaptive) values (peaks). Further differences among them result from the fitnesses of different alleles at the *Aa* and *Bb* loci. Genotypes bearing more capital letter alleles at these loci have higher relative fitness values than those that do not (for example, *aBCd, abCD*).

[Adapted from Wright, S., 1963. Genic interaction. In *Methodology in Mammalian Genetics*. W. J. Burdette (ed.). Holden-Day, San Francisco.]

*For Wright's development of the shifting balance theory, see Wright, 1951, 1963, 1978, 1988. As judged by scientific papers, Sewall Wright's first paper was published when he was 15 and his last when he was 90 (Provine, 1986).

proposed that many populations break into small groups of subpopulations (demes). These demes are small enough to differ genetically because of random fluctuations in gene frequencies (genetic drift) but are not so widely separate as to completely prevent gene exchange and the introduction of new genetic variation. Adaptive landscapes therefore are occupied by a network of demes, some at higher peaks than others. Selection occurs between genotypes competing within demes, and between demes competing within a general environment. By contrast, Ronald Fisher suggested that most populations are large and fairly homogeneous, and that selection tests each new allele independently in competition with all other alleles in the population (Fisher, 1930) This large population primarily increases its fitness by small, incremental ("additive") selective steps.

The number of adaptive peaks increases astronomically with the number of loci and alleles; one locus with four alleles gives 10 gene combinations; 100 loci with four alleles each gives 10,100 gene combinations. Thus, even if only a small portion of gene combinations is adaptive, there are many more possible adaptive peaks than a species can occupy at any one time. To move from peak to peak until the highest peak is reached, a population must travel through genotypes that occupy lower adaptive valleys of the adaptive landscape. Arrows in Figure EB 1.1 indicate such reductions in fitness, showing the general route a population at *abCD* might take to reach the highest peak at *ABcd*.

A potentially high adaptive peak for a population need not coincide with a high selective peak for a genotype(s) within the population. This discrepancy arises because the selective values of genotypes are based on competition with other genotypes but may not indicate their effect on the population. One important consequence of Wright's shifting balance theory was to emphasize differences in survival or extinction among populations rather than only among individuals, an emphasis that seems to require selection on the group (group selection) rather than on the individual.

Altruists that sacrifice themselves for the benefit of shared genotypes may have low selective value as individuals, although a population bearing such altruistic genotypes may have higher reproductive values than one without them. Sterile castes among insects that forgo reproduction represent an example of **altruism**. Situations in which the altruist is not disadvantaged have been proposed also. When alarmed, many species of antelope "bounce" up and down with the joints of the legs locked and the legs in an extended posture. This behavior, called stotting or pronking, warns other individuals of danger, but, it has been argued, also alerts the predator that the individual "pronker" is aware of the presence of the predator, and so may not be an easy target for predation.

Investigation into the theories proposed by Fisher and Wright continues, with evidence interpreted as supporting or contradicting the theory. The genetic drift, selection, and gene flow required for Wright's theory to hold are not found in many populations. Structures of demes and evidence for intergroup selection have been used to support the theory. Until further evidence appears, both Fisherian and Wrightian populations should be assumed to exist, large and homogeneous at one time and subdivided at another. As Wright stated, different aspects of populations may demand different theoretical approaches.

REFERENCES

Coyne, J. A., N. H. Barton, and M. Turelli, 1997. Perspective: a critique of Sewall Wright's shifting balance theory of evolution. *Evolution*, **51**, 643–671.

Fisher, R. A., 1930. *The Genetical Theory of Natural Selection*. Clarendon, Oxford, UK.

Provine, W. B., 1986. *Sewall Wright and Evolutionary Biology*. The University of Chicago Press, Chicago.

Wright, S., 1951. The genetic structure of populations. *Ann. Eugen.*, **15**, 323–354.

Wright, S., 1963. Genic interaction. In *Methodology in Mammalian Genetics*, W. J. Burdette (ed.). Holden-Day, San Francisco, pp. 159–192.

Wright, S., 1978. *Evolution and the Genetics of Populations: Vol. 4. Variability Within and Among Natural Populations*. University of Chicago Press, Chicago.

Wright, S., 1988. Surfaces of selective value revisited. *Am. Nat.*, **131**, 115–123

PART 7

Populations, Speciation, and Extinctions

20

Competition, Predation, and Population Biology

CHAPTER SUMMARY

A population is a group of organisms that interbreed. Species are composed of populations that, at least potentially, interbreed. Evolution occurs through changes in populations. The effective unit within a population is the interbreeding group known as the deme. Individuals interact within populations and with their environment to occupy an ecological niche. Different reproductive strategies and modes of exploiting the environment result in different species producing many or few offspring. Growth of populations is an important parameter used in ecological and evolutionary studies. Population size is often measured as the rate of increase (r) from one time to another. Populations modify their physical and biological environments, and so influence the use of the environment by other species. Analysis of the many interactions between populations and their environment allows us to calculate the carrying capacity (K) of the environment. In this chapter, we discuss three of the most important biological interactions between populations: competition between species, predation of one species on another, and symbiosis.

INTRODUCTION

In nature, the structure and relationships of **populations** depart from many of the ideal conditions that would make their evolutionary behavior simple to understand. Populations are neither of constant size nor uniformly distributed in space. Populations are subject to variable degrees of mutation, migration, selection, genetic drift, and environmental influences such as floods or droughts. Neither the physical nor the biological environment—temperature, amount of light, prey, predators, and so forth—remain constant over time. Furthermore, and importantly, populations are not passive recipients of environmental or ecological information. A population modifies its physical and biological environment in ways that can diminish or enhance its own resources and those of other populations. As a consequence of all these interactions, it has been said that populations "must continue running in order to keep in the same place" (the Red Queen hypothesis). In general, the biological and ecological approaches briefly reviewed in this chapter emphasize how populations respond to their environment and to other species as assessed by the numbers of individuals in the population.[1]

POPULATION GROWTH

The structures of populations are complex. One characteristic that can be measured is **population growth**. As Malthus, Darwin, and Wallace were well aware, unlimited population growth becomes exponential. Rate of growth of a population is typically measured as the **rate of increase** (r). However, because the environment imposes restrictions on the rate at which the potential growth of a population can be realized, populations stabilize at a size (number of individuals), called the **carrying capacity** (K), a size that can be reset if the environment or interactions with other species changes. These two aspects of population growth are discussed further in BOX 20.1.

BIOLOGICAL INTERACTIONS WITHIN POPULATIONS

FECUNDITY *is the term used for the number of offspring produced by an individual, usually a female.*

In terms of survival, growth, or **fecundity**, a slight increase in the numbers of individuals in one species may result in an increase, a decrease, or have no discernible effect on the numbers of individuals in another species. The analysis of interactions among individuals and between individuals and their environment has been central to the study of evolution for almost a century. These interactions occur in populations. Interactions between two species have been classified according to the terminology set out in **TABLE 20.1**, with categories ranging from those where both species benefits (mutualism) to interactions with maximal effects on both species (competition).

Three broad and overlapping types of interactions affecting populations can be categorized as *biological*, and can be an initial step in the reproductive isolation that leads to speciation. These interactions—**competition**, **predation**, and **symbiosis**—influence the maintenance and the selection of particular traits. For practical reasons of analysis, each interaction must be disentangled from others and explored separately. Each interaction may be seen only in one combination of species, reflecting the past evolutionary history of the populations, spatial limitations, climatic conditions, soil nutrients, and the effects of other species in the community. In their various forms and through interactions between them, these processes determine the dynamics of **demes**, **populations**, and **species**.

Competition results in **niche** and character distinctions between demes, in some cases eliminating one group entirely. Predation has complex effects on the size and structure of the populations of both prey and predator. Below we discuss the essential features of competition and predation.

---

[1] See Rockwood (2006), Rose and Mueller (2006), and Morris and Lundberg (2011) for excellent treatment of evolutionary ecology.

BOX 20.1   Population Growth

Although reproductive modes differ among organisms, the evolutionary potential of a population's **reproductive power** can be determined.

Many annual plants and insects breed only once during their lifetimes. Many perennial plants and vertebrates breed repeatedly, although not necessarily annually. The number of offspring a female produces at any one time varies significantly between species, ranging from a single offspring in many larger mammals to thousands of eggs produced by a salmon. Furthermore, as an individual may die before reaching reproductive age, the sooner reproduction begins, the greater the chances of producing offspring. As a consequence, in some organisms, such as many insects, reproduction begins soon after hatching or birth.

The consequence of birth rate minus death rate—the primary factor determining population size—is normally presented as the rate of increase (r). In most species, natural selection favors genotypes that confer survival to reproductive age, with little or no selection for individuals to survive after reproducing. As Richard Dawkins pointed out, "we inherit whatever it takes to be young, but not necessarily whatever it takes to be old." How selection influences lifespan but not post-reproductive lifespan has been demonstrated in experiments with guppies (*Poecilia reticulata*), which are among the most popular fish maintained in aquaria. Populations with high mortality rates, because they co-exist with predators, mature at an earlier age, and have longer life spans because the reproductive phase is lengthened, not because of lengthened post-reproductive lifespan (Reznick et al., 2006).

Under ideal conditions the early stages of population growth in an environment well supplied with resources can occur geometrically in each generation ($2 \rightarrow 4 \rightarrow 8 \rightarrow 16$, and so on), closely following the growth curve shown in **FIGURE B1.1a**. Such a pattern of population growth would be limitless if reproduction rate was unchecked and if space and resources were limitless. Environmental resources and space are not limitless, however. Nor is population size, which may stabilize at some near-constant value or may "crash" in response to an environmental event, expansion of a predator, or other circumstances. Malthus's popularization of the idea that war, famine, disease, and other influences check the exponential growth of human populations led Darwin and Wallace independently to the concepts of the struggle for existence and natural selection.

To evaluate such impacts on a population we need to determine how many individuals the environment can support. This value, K, is the carrying capacity. Using the population of the yeast *Saccharomyces cerevisiae* shown in **FIGURE B1.1b**, we see that the population levels off at a plateau of about 665 individuals, the rate of increase displaying an

SACCHAROMYCES CEREVISIAE, sometimes known as Brewer's yeast, has been used for thousands of years to make bread and brew beer.

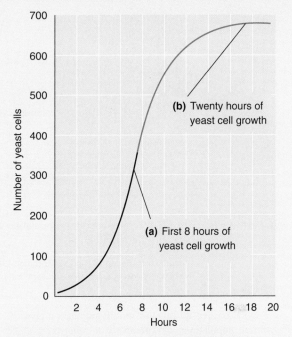

FIGURE B1.1 Numbers of cells of the budding yeast, *Saccharomyces cerevisiae*, in a defined volume of culture medium for two growth periods, beginning with approximately 10 cells per volume. **(a)** Geometric growth during the first 8 hours. **(b)** S-shaped growth curve for the 20-hour growth period.

[Data from Carlson, T., 1913. *Biochemische Zeitschrift* **57**, 313–335.]

S-shaped growth curve. In reality, populations rarely follow such smooth growth curves.

The size of a population can change dramatically over short period of time, as shown in **FIGURE B1.2** for the great tit, *Parus major*, in Holland. Many parameters affect such fluctuations in numbers within a population. *Density independent* agents such as climate influence population size but are not affected by population size or crowding. *Density dependent* agents reflect effects of a combination of population size and space available.

Organisms such as some bacteria and plants display rapid population growth in the face of fluctuating environments and have a rapid rate of increase (r) at low population densities. As stated by Darwin, "A large number of eggs is of some importance to those species, which depend on a rapidly fluctuating amount of food, for it allows them rapidly to increase in numbers." Other organisms, such as large vertebrates, face more uniform or predictable environments with population sizes that are close to carrying capacity (K). Selective advantages of this pattern of reproduction are increased efficiency in resource use and a greater likelihood that offspring can be raised to the stage when they themselves can reproduce.

| BOX 20.1 | Population Growth (Cont...) |

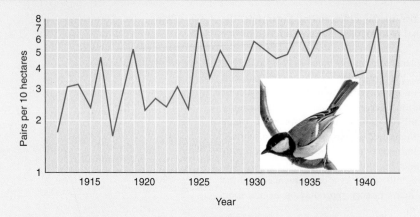

FIGURE B1.2 Fluctuations in population size of the great tit, *Parus major*, in Holland between 1912 and 1943.

[Adapted from Begon, M., and M. Mortimer, 1986. *Population Ecology, Second Edition*, Blackwell; INSET: © Sebastian Knight/ShutterStock, Inc.]

Population density can also modify population growth. In natural populations of *Drosophila melanogaster*, allele differences in a foraging gene can affect feeding behavior. The produces a protein kinase used in signal transduction pathways that dictate the foraging strategy of the organisms. In this case, "rovers" move longer distances than "sitters." These differences in foraging activity are selected in response to population density: rovers are adaptive at high density where larvae must travel longer distances to obtain food; sitters are adaptive at lower densities where food is more available.*

* K. A. Osborne, A. Robichon, and E. Burgess, 1997. Natural behavior polymorphism due to a cGMP-dependent protein kinase of *Drosophila. Science*, **277**, 834–836; [INSET] © Sebastian Knight/ShutterStock, Inc.

COMPETITION

Competition arises when two groups depend on the same limited environmental resource(s) so that changes in one group result in changes in the other's numbers, often with important ecological or behavioral consequences. Three consequences of competition are outlined below: resource partitioning, character displacement, and competitive exclusion.

| TABLE 20.1 | Potential Interactions that Can Occur Between Populations of Two Species |

Type	Nature of Interaction
Mutualism	Both species benefit.
Predator–prey-induced polymorphism	Both species benefit, as seen in phenotypic plasticity.
Commensalism	One species is unaffected while the second (commensal) species benefits.
Predation	The predator benefits at the expense of the prey.
Parasitism	The parasite benefits at the expense of the host.
Competition	All species involved may be affected, positively or negatively.

RESOURCE PARTITIONING

Competition often leads to ecological diversity. It can be to the advantage of competing groups to minimize the harmful effects of direct competition by using different aspects of their common environmental resources. Among the many examples of such **resource partitioning** is the one illustrated in FIGURE 20.1 for five species of warblers, each using different parts of their spruce tree habitat. Should such resource partitioning be disrupted

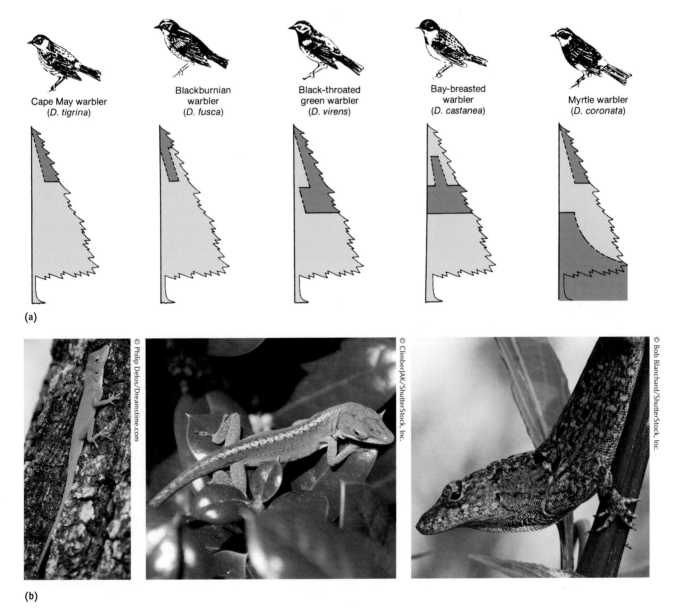

(a)

(b)

FIGURE 20.1 **(a)** Resource partitioning as demonstrated by the most common feeding zones (dark green) in spruce trees for five species of northeastern U.S. warblers of the genus *Dendroica*. The Cape May warbler (*Dendroica tigrina*) is quite rare unless there is a large outbreak of insects. The Myrtle Warbler—which is the eastern sub-species of the Yellow-Rumped warbler, *Setophaga coronata*—is both the most abundant and the least specialized warbler in most coniferous forests. More common warblers—Blackburnian, Bay-Breasted and Black-Throated Green—are different enough in feeding zone preferences to explain their coexistence. **(b)** Species of *Anolis* lizards that inhabit different parts of the habitat (tree, bush, grass) have distinctive morphologies.

[(a) Adapted from Krebs, C. J., 2001. *Ecology: The Experimental Analysis of Distribution and Abundance,* 5th ed. Benjamin Cummings]

and habitats overlap, the fitness of the individuals will decline because of competition associated with sharing the resources. As an example, nest predation increases when nesting sites overlap between different species.[2] Depending on their interactions, changes in fitness may not affect all species equally: raccoons (*Procyon lotor*) and Virginia opossums (*Didelphis virginiana*) share habitats and food preferences. Removal of raccoons increased the fitness of the possums, which had been restricted by sharing resources with the raccoons (Kasparian et al., 2002).

Different species may occupy different parts of the habitat, as illustrated by tree-dwelling anoles (species of lizards in the genus *Anolis*) from the Caribbean Islands. Each of the species shown in Figure 20.1b occupies a different microhabitat. Each displays features associated with the particular microhabitat: the trunk species *Anolis distichus* has a snout-vent length of 40–58 mm, a short tail and is gray in color; the crown giant, *Anolis ricordii* has a snout-vent length of 130–190 mm, a long tail and is green (Figure 20.1a).[3]

CHARACTER DISPLACEMENT

An important evolutionary response to competition is **character displacement**, in which changes in morphology accompany resource partitioning among coexisting groups. When facing competition with a species utilizing the same food supply, for example, changes in structures associated with feeding permit two species to coexist by "dividing-up" the food supply and so avoiding further competitive interactions. *Anolis* lizards discussed above provide an excellent example (Losos, 2000) as do Darwin's finches.

In their recent book HOW AND WHY SPECIES MULTIPLY: THE RADIATION OF DARWIN'S FINCHES, the Grants provide an outstanding overview of speciation (Grant and Grant, 2008).

The 40-year-long (and ongoing) studies on Darwin's finches (*Geospiza* spp.) in the Galápagos Islands by Peter and Rosemary Grant of Princeton University (BOX 20.2) demonstrate how our understanding of evolutionary processes can be advanced through close and continuous observations of natural populations. Among numerous important discoveries (Grant, 1986, Grant and Grant, 2006, 2008), the Grants documented a series of processes, beginning with competitive interaction and character displacement, and culminating in evolutionary change in one of the participating species. Where two species of finches coexist on the same island, both species show character displacement in measurable differences in bill sizes (8 and 12 mm), enabling each to feed on differently sized seeds. In contrast, species isolated on different islands possess intermediate bill sizes (10 mm) enabling them to feed without partitioning seed resources (FIGURE 20.2; Grant and Grant, 2006).

COMPETITIVE EXCLUSION

When competition is not checked by partitioning or fluctuation of resources, and when two competing species use exactly the same resources in the same environment, they occupy the SAME NICHE.

The **niche** or **ecological niche** outlined in BOX 20.3 represents all the environmental resources used by individual organisms as well as the ways in which they use those resources and interact with other individuals of their own or of other species.

In laboratory experiments in which two species occupy the same niche, one species commonly dies out (FIGURE 20.3), supporting the principle of **competitive exclusion**: two species cannot continue to coexist in the same environment if they use it in the same way. This principle was foreshadowed by Darwin in *On the Origin of Species.*

> Owing to the high geometrical rate of increase of all organic beings, each area is already fully stocked with inhabitants; and it follows from this, that as the favored forms increase in number, so generally will the less favored decrease and become rare.

---

[2] T. E. Martin, 1996. Fitness costs of resource overlap among coexisting bird species. *Nature*, **380**, 338–340.

[3] For *Anolis* lizards and colonization, biogeography and speciation on islands, see Losos (2000, 2009) and Losos and Ricklefs (2009a,b). For current analyses of the interplay between ecology and evolution—eco-evolutionary dynamics, the eco-evolutionary process—as illustrated by various species, including anole lizards, see Morris and Lundberg (2011) and Schoener (2011).

BOX 20.2 Darwin's Finches and the Jaws of Fish

Darwin's finches are an important cluster of species used by Charles Darwin as evidence for species change (see Figure 20.2). The importance of this cluster of 13 species to our understanding of evolution has been greatly enhanced by the work of Peter and Rosemary Grant, who have studied Darwin's finches continuously since 1973.[a]

The Grants' work has documented evolution in action by relating environmental changes leading to evolutionary changes in beak size and body size over the span of several finch generations. An early study by Peter Grant and colleagues published in 1976 confirmed a prediction made by Leigh Van Valen that the degree of variation in important environmental factors should be reflected in the amount of variation in the morphological traits relevant to those factors.[b] Thus, species encountering a food that varies greatly in size and hardness should show higher variation in beak dimensions. Within each species, individual birds choose food that is of the appropriate size and hardness for their beaks.

As their work progressed and data from marked individuals and their offspring built up, the Grants showed that natural selection can produce evolutionary change very rapidly. They also showed, surprisingly, that the direction of selection in particular populations can change frequently and unpredictably as environmental conditions fluctuate (see Figure 20.2). By documenting evolution in action in natural populations, this work has made profound contributions to our understanding of the evolutionary process.

A question raised by the results of the Grants' study of evolution in Darwin's finches is how, in genetic and developmental terms, evolution can proceed so rapidly. If changes to beak size and shape required selection to alter the expression or function of many genes in some coordinated fashion, it is unlikely that evolutionary changes could happen as quickly as they clearly have over the past 30 years in these finches. An insight into this question has come from a study by Arhhat Abzhanov and his colleagues on the developmental–genetic basis for variation in beak size and shape in Darwin's finches (Abzhanov et al., 2004). Remarkably, the amount and area of expression in embryonic beaks of a single gene, *Bmp-4*, is correlated with variation in beak size and shape across species of Darwin's finches (**FIGURE B2.1**).[c] The same has since been show in a comparison of house finches (*Carpodacus mexicanus*) in Sonoran desert and urban habitats; levels of Bmp protein correlate with the size and hardness of the seeds on which these populations feed (Badyaev et al., 2008).

In a fascinating parallel to both studies, levels of *Bmp-4* are higher in the jaws of a species of cichlid fish from Lake Malawi that feeds by biting and crushing than in another species from the same lake that takes in plankton from the

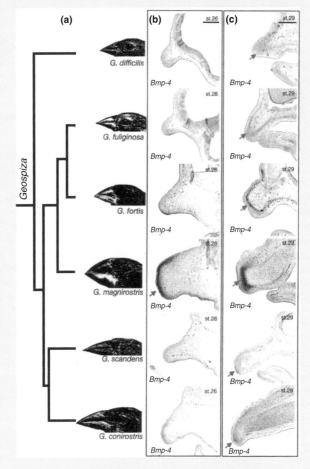

FIGURE B2.1 Relationship between **(a)** evolutionary relationship and beak morphology and **(b, c)** variation in *Bmp-4* expression in six species of Darwin's finches. The histological sections are from the developing beak at two embryonic stages, 26 (b) and 29 (c). Darkly stained areas indicate high levels of *Bmp-4* expression.

[Reproduced from Abzhanov, A., Protas, M., Grant, B. R., Grant, P. R., and C. J. Tabin, 2004. *Bmp4* and morphological variation of beaks in Darwin's finches. *Science*, **305**, 1462–1465. Reprinted with permission from AAAS.]

water column using suction feeding. Furthermore, although expression of *Bmp-4* is conserved in tooth-forming regions in three species of fish (zebrafish, Japanese medaka, and Mexican tetra), *Bmp* genes are not expressed in toothless regions of the jaws, implicating *Bmp* in evolutionary tooth loss.[d]

These results suggest that the evolution of the beak in Darwin's finches (and perhaps in many other lineages of birds), and jaw and tooth morphology associated with feeding in fish, is facilitated by a relatively simple developmental-genetic mechanism responsible for variation in beak, jaw and/or tooth sizes and shapes.

[a] For a perspective on being a naturalist while also making major contributions to evolutionary theory, read Peter Grant's paper on being a naturalist in the second half of the twentieth century (P. R. Grant, 2000).
[b] P. R. Grant et al., 1976. Darwin's finches: population variation and natural selection. *Proc. Natl. Acad. Sci. USA*, **73**, 257–261.
[c] Abzhanov and his colleagues have since expanded this study to other genes that regulate the development of the upper beak, providing evidence for distinctive yet integrated upper- and lower-beak modules involved in regulating beak shape in Darwin's finches (Mallarino et al., 2011).

[d] Albertson, R. C., J. T., Streelman, and T. D. Kocher, 2000. The beak of the fish: genetic basis of adaptive shape differences among Lake Malawi cichlid fishes. Assessing morphological differences in an adaptive trait: a landmark-based morphometric approach. *J. Exp. Zool.*, **289**, 385–403.

FIGURE 20.2 Character displacement illustrated by beak depth (mm) in two species of Darwin's finches in the Galápagos Islands. The scale on the X axis of **(a)** applies to all three graphs. **(a)** *Geospiza fuliginosa* (tan) and *G. fortis* (blue) coexisting on four islands show large differences in beak sizes (8 versus 12.5 mm), enabling each to feed on seeds of different sizes. However, when either species exists alone on different islands **(b)** and **(c)**, beak size is intermediate (about 10 mm, although note the bimodal distribution in c), enabling feeding without partitioning seed resources.

[Adapted from Givnish, T. J. and K. J. Sytsma (eds). 1977. *Molecular Evolution and Adaptive Radiation.* Cambridge University Press, Cambridge, UK; based on Grant, P. R., 1986. *Ecology and Evolution of Darwin's Finches.* Princeton University Press, Princeton, NJ.]

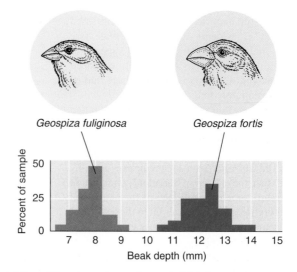

(a) Two different species on the same island (Abingdon, Bindloe, James, Jervis Isls)

(b) A single species on an island (Daphne)

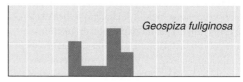

(c) A single species on an island (Crossman)

The example shown in Figure 20.3 is taken from a classic study by Russian microbiologist and pioneer of the concept of competitive exclusion, Georgii F. Gauss (1910–1986). In this study, one paramecium strain, *Paramecium caudatum*, is more sensitive to metabolic byproducts than is the other species, *P. aurelia*, resulting in competitive exclusion (Gause, 1934).

To cite only one recent study from many that have been undertaken, bacterial samples were taken from the cecum of chickens that had been exposed to the bacterium *Salmonella typhimurium*. These resulting mixed cultures of 29 bacterial isolates were established in a continuous flow culture system. Less than one percent of the salmonella produced colonies, demonstrating almost complete competitive exclusion of *S. typhimurium* by gut bacteria. This cell line is now sold under the trade name PREEMPT™ and used to reduce salmonella infections of broiler chickens.[4]

Despite such studies, evolutionary biologists have questioned whether competitive exclusion alone accounts for the differences observed among coexisting species; some invoke character displacement as an additional mechanism. Furthermore, closely related and potentially competitive species often occupy different habitats. Morphological differences among species that coexist today may have evolved in the past in places where these species did not compete. In this case, related species occupying different habitats

---

[4] D. Nisbet, 2002. Defined competitive exclusion cultures in the prevention of enteropathogen colonisation in poultry and swine. *Antonie van Leeuwenhoek*, **81**, 481–486.

BOX 20.3 **The Niche Concept**

Organisms do not live in isolation. They interact and often compete with other organisms, inhabit particular environments (which they can modify) and coevolve with those other organisms and with changing environments. These relationships are captured by the **niche concept**.

The word niche (from the French *nicher*, to nest) was introduced into biology in 1917 by Joseph Grinnell (1877–1939), a pioneering ecologist and the first of a distinguished line of directors of The University of California (Berkeley) Museum of Vertebrate Zoology.[a] Grinnell's paper, "The niche-relationships of the California Thrasher" (Grinnell, 1917) demonstrated that the distribution of this bird species (*Toxostoma redivivum*), the largest in the thrasher family, coincides with the chaparral shrub land in California and Baja California, a habitat with which it specifically interacts in a niche relationship, akin to the relationship of a bird with its nest. The habitat—chaparral shrub land—has adapted to a Mediterranean climate and to frequent fires; the shrubs tolerate drought and fire to form dense thickets. Feeding on berries and hiding in the dense brush characteristic of the habitat, individual birds are rarely seen (**FIGURE B3.1**).

Grinnell's analysis of these and other interactions provided us with the concept of the niche or ecological niche—the place where an organism lives, the relation of the organism to that place, to the resources present, and to other organisms of its own or different species in that place. An organism may limit access of other organisms to resources in the niche, slow its growth rate when resources are limited, occupy a different part of the niche when a predator is present, provide food for the predator, and so forth. The niche therefore represents all the environmental resources used by individual organisms, as well as the ways in which they use those resources and the ways in which they interact with other individuals of their own or of other species (Rockwood, 2006).

In an important paper published in 1957, American ecologist G. Evelyn Hutchinson (1903–1991) outlined a niche concept that placed the emphasis on the organism rather than the habitat, as proposed by Grinnell (Hutchinson, 1957).[b] Following the direction set by Hutchinson, American evolutionary biologist Richard Lewontin captured the essence of the niche and of the relationships of an organism to its environment in his book, *The Triple Helix: Gene, Organism, and Environment*. "Niches do not preexist organisms but come into existence as a consequence of the nature of the organisms themselves . . ." (Lewontin, 2000, p. 51).

The definition of a niche as "a unique ecological role of an organism in a community," comes closest to what is now known as **niche construction**, an ecological and evolutionary concept pioneered by F. John Odling-Smee in which modification of its environment by a species influences not only its life history but also its evolution (Odling-Smee et al., 2003). Dams built by beavers are classic examples of niche construction and of the accompanying concept of **ecological inheritance**, which is akin to social evolution.

(a)

(b)

FIGURE B3.1 Chaparral shrub land **(a)** with its specialized vegetation evolved in response to the Mediterranean climate and frequent fires of California and Baja California. Chaparral shrub is the habitat of the Chaparral thrasher, *Toxostoma redivivum* **(b)**.

[a] Grinnell, who served as director from 1908 when the museum was founded until his death in 1939, is credited as a co-founder of the principle of competitive exclusion discussed in the text.

[b] For a recent excellent analysis of how Hutchinson changed our view of the niche, see R. D. Holt (2009) Bringing the Hutchinsonian niche into the 21st century: ecological and evolutionary perspectives. *Proc. Natl Acad. Sci. USA*, **106** (Suppl 2), 19659–19665.

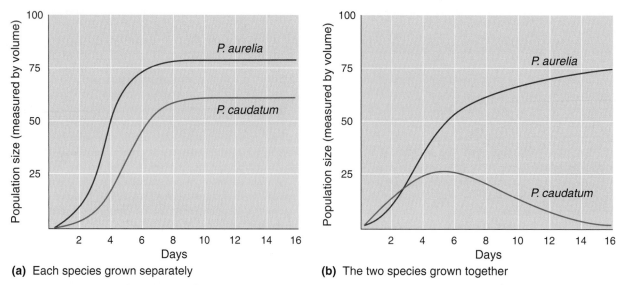

FIGURE 20.3 Growth of two species of *Paramecium* **(a)** in separate cultures, and **(b)** in mixed cultures. Although *P. aurelia* replaces *P. caudatum* **(b)**, in some mixed culture conditions *P. caudatum* multiplies faster than *P. aurelia*.

[Adapted from Gause, G. F., 1934. *The Struggle for Existence*. Williams & Wilkins.]

may not have diverged because of competition but because of different food preferences, nesting sites and so on.[5]

NICHE SEGREGATION

Partitioning of the resources in an environment was discussed above as one of the consequences of and responses to competition. An alternative model to competition goes under the name **niche segregation**. In this situation, potentially overlapping species effectively subdivide available niches either by (i) random selection of niches that do not overlap those of other species, or (ii) by reinforcing reproductive barriers to prevent interbreeding or hybridization. The niches may be large, such as regions occupied by giant rhinoceros beetles (*Chalcosoma* spp.), which occupy all of Southeast Asia except Java, or as small as the gills of fish occupied by parasites.[6] How or whether niche segregation and competition are related has been an active topic of discussion in ecology since Connell (1980) reviewed the topic and concluded that: "the notion of coevolutionary shaping of competitors' niches has little support at present" (p. 131). Different conceptions of whether most niches are full or empty, whether organisms select niches randomly, and/or whether selection narrows or widens niches all feed into the continuing discussion.

PREDATION

Predation is an important factor diminishing competitive exclusion among competing, coexisting species; predators can reduce the possibility of a dominant species reaching its carrying capacity, thereby making room for other competitors. A predator entirely or partially consumes its prey, affecting the numbers of those organisms it feeds on.

---

[5] See Lu et al 2011 for an example of the types of analyses required to differentiate between various mechanisms of niche segregation.

[6] K. Kawano, 2002. Character displacement in giant rhinoceros beetles. *Am. Nat.*, **159**, 255–271; A. Simkova et al., 2000. Co-existence of nine gill ectoparasites (*Dactylogyus*: Monogenea) parasitising the roach *Rutilus rutilus* (L.): history and present ecology. *Int. J. Parasit.*, **30**, 1077–1088.

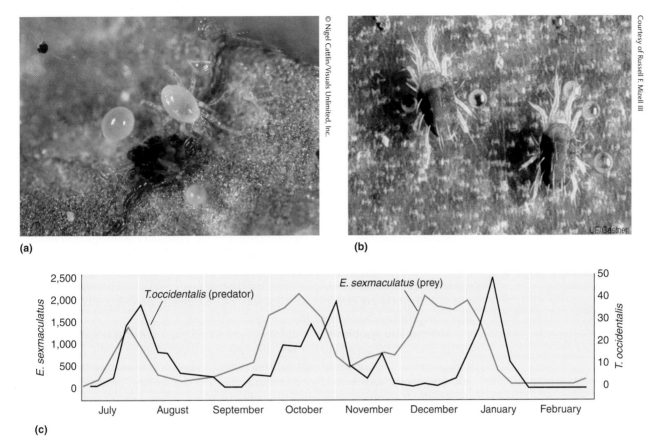

(a)

(b)

(c)

FIGURE 20.4 Three cycles of oscillating population numbers for two species of mites in a controlled environment. One species, the western predatory mite *Typhlodromus occidentalis* **(a)** is the predator. All stages in the life cycle of a herbivorous species the six-spotted mite, *Eotetranychus sexmaculatus* are the prey. **(b)**. For each cycle **(c)**, as numbers of prey increase (brown) predator numbers follow (black) causing a "crash" in the prey population, followed by a crash in the predator population.

[(c) Adapted from C. B. Huffaker, *Hilgardia*, **27**, 1958. 343–383.]

Intimate dependence of predators on their prey often leads to a coupling of their relative abundances; increase in numbers of prey is followed by an increase in predators, which, in turn, can reduce prey, causing a reduction in predator numbers—an arms race. As shown in **FIGURE 20.4**, cyclic oscillations in population numbers can be correlated, especially for predators that prey on a single species.

Such simplicities are not the rule, however. In some interactions, presence of the predator does not lead to a substantial drop in the abundance of their prey. As seen in the predation of wolves on caribou herds, wolves prey mostly on young caribou or those weakened by age or disease (which have less reproductive potential than do healthier individuals). Moreover, predator numbers may be buffered when more than one species of prey is being exploited. Buffering reduces large oscillations in predator population size and spreads the effects of predation so that the size of the prey population also remains fairly stable. When wolves (*Canis lupus*) or mountain lions (*Puma concolor*) were introduced into populations of mule deer, *Odocoileus hemionus*, the mountain lions influenced the mortality rate of the deer but the wolves did not.[7]

[7] W. B. Ballard et al., 1997. Ecology of wolves in relation to a migratory caribou herd in northwest Alaska, *Wildlife Monog.* **135**, 3–47; B. M. Pierce et al., 2000. Selection of mule deer by mountain lions and coyotes: effects of hunting style, body size, and reproductive status. *J Mammal.*, **81**, 462–472.

Different combinations of factors produce differences greater than predicted from the actions of each factor alone in populations of Canadian snowshoe hares, *Lepus americanus*. Reducing predation *or* increasing the food supply *each* result in a two- to threefold increase in snowshoe hare numbers—*a cumulative four- to six fold* increase. In combination, reduced predation and increased food supply resulted in *an 11-fold* increase in numbers. Obviously, different interactions occur when predation is low and food is plentiful than when food is plentiful and predators are prevalent, or when predation is low and food is meager.

SYMBIOSIS

The concept of interactions between different organisms can be discussed in the context of the evolutionary origin of eukaryotic cellular organelles. The term used for such interactions is **endosymbiosis**, meaning the incorporation of an entire prokaryotic organism by another prokaryote and the transformation of the engulfed organisms into cellular organelles. Those interactions that occurred early in the history of life to produce the first mitochondrion-containing eukaryotic cells are referred to as examples of *primary endosymbiosis*. There are also examples of *secondary endosymbiosis* in which some eukaryotic algae acquired chloroplasts from a eukaryotic cell. Secondary here means later in evolution than primary endosymbiosis.

Here we use **symbiosis** in a different (and its original) sense to describe an association and interactions between two different organisms that are to the advantage of both partners. The organisms may be from different kingdoms, as seen in the symbiotic relationships between individual protozoan ciliates of the species *Paramecium bursaria* and hundreds of intracellular aerobic and photosynthetic green algae (*Chlorella vulgaris*). This association represents one of the two major types of symbiosis—**mutualism**—in which both species benefit through positive effects on survival, growth, reproduction, and/or fitness (Table 20.1). The presence of hundreds of millions of bacteria in the human gut is becoming a well-researched example of mutualism and the evolution of mutualistic relationships (Yang et al., 2011; Smillie et al., 2011).

A second type of mutualism, pertaining to phenotypic plasticity, refers to interactions between a predator and one or more prey species in which a chemical released by the predator induces some of the eggs of the prey species to development into forms (morphs) that prevent them from being eaten by the predator. The prey benefits through survival of some individuals in the presence of the predator, while the predator benefits by not eliminating all its prey. Such **predator–prey-induced polymorphism** represents a form of phenotypic plasticity that may be an initial step in speciation. The other form of symbiosis listed in Table 20.1, **commensalism** ("at table together"), refers to an association in which one species is unaffected while the second (commensal) species benefits. Many examples of commensalism are known. Familiar examples are the barnacles that adhere to the surfaces of whales, birds that pick parasites from the skin of cattle, and tropical fish that shelter among the tentacles of sea anemones.

Such close associations, whether through competition/predation, or because of mutualism/commensalism, inevitably take us into a discussion of coevolution.

■ KEY TERMS

carrying capacity (*K*)	endosymbiosis
character displacement	fecundity
commensalism	mutualism
competition	niche
competitive exclusion	niche concept
deme	niche construction
ecological inheritance	polymorphism
ecological niche	population

population growth
predation
polymorphism
rate of increase (*r*)

reproduction
resource partitioning
symbiosis

■ EVOLUTION ON THE WEB

Explore evolution on the Internet! Visit the accompanying website for *Strickberger's Evolution, Fifth Edition,* at **go.jblearning.com/Evolution5eCW** for exercises and links relating to topics covered in this chapter.

■ REFERENCES

Abzhanov, A., M. Protas, B. R. Grant, et al., 2004. *Bmp4* and morphological variation of beaks in Darwin's finches. *Science*, **305**, 1462–1465.

Badyaev, A., R. L. Young, K. P. Oh, and C. Addison, 2008. Evolution on a local scale: developmental, functional, and genetic bases of divergence in bill form and associated changes in song structure between adjacent habitats. *Evolution*, **62**, 1951–1964.

Connell, J. H. 1980. Diversity and the coevolution of competitors, or the ghost of competition past. *Oikos*, **35**, 131–138.

Gause, G. F., 1934. *The Struggle for Existence.* Williams & Wilkins, Baltimore.

Grant, P. R., 1986. *Ecology and Evolution of Darwin's Finches.* Princeton University Press, Princeton, NJ. [Reissued in 1999 with a Foreword by J. Weiner.]

Grant, P. R., 2000. What does it mean to be a naturalist at the end of the twentieth century? *Am. Nat.*, **155**, 1–12.

Grant, P. R., and Grant, B. R., 2006. Evolution of character displacement in Darwin's finches. *Science*, **313**, 224–226.

Grant, P. R., and B. R. Grant, 2008. *How and Why Species Multiply: The Radiation of Darwin's Finches.* Princeton University Press, Princeton, NJ.

Grinnell, J., 1917. The niche-relationships of the California thrasher. *Auk*, **34**, 427–433.

Hutchinson, G. E. 1957. Concluding remarks. *Cold Spring Harbor Symp.*, **22**, 415–427.

Kasparian, M. A., Hellgren, E. C., and S. M. Ginger, 2002. Food habits of the Virginia opossum during raccoon removal in the Cross Timbers ecoregion, Oklahoma. *Proc. Okla. Acad. Sci.*, **82**, 73–78.

Lewontin, R., 2000. *The Triple Helix: Gene, Organism, and Environment.* Harvard University Press, Cambridge, MA.

Losos, J. B., 2000. Ecological character displacement and the study of adaptation. *Proc. Natl. Acad. Sci. USA*, **97**, 5693–5695.

Losos, J. B., 2009. *Lizards in an Evolutionary Tree. Ecology and Adaptive Radiation of Anoles.* University of California Press, Berkeley, CA.

Losos, J. B., and R. E. Ricklefs, 2009a. Adaptation and diversification on islands. *Nature*, **457**, 830–836.

Losos, J. B., and R. E. Ricklefs (eds.), 2009b. *The Theory of Island Biogeography.* Princeton University Press, Princeton, NJ.

Lu, X., G. Gong, and X. Ma, 2011. Niche segregation between two Alpine Rose finches: to coexist in extreme environments. *Evol. Biol.*, 38, 79–87.

Mallarino, R., P. R. Grant, B. R. Grant, et al., 2011. Two developmental modules establish 3D beak-shape variation in Darwin's finches. *Proc. Natl. Acad. Sci. USA*, **108**, 4057–4062.

Morris, D. W., and P. Lundberg. 2011. *Pillars of Evolution. Fundamental Principles of the Eco-Evolutionary Process.* Oxford University Press, Oxford, UK.

Odling-Smee, F. J., K. N. Laland, and M. W. Feldman, 2003. *Niche Construction. The Neglected Process in Evolution.* Princeton University Press, Princeton, NJ.

Reznick, D., M. Bryant, and D. Holmes, 2006. The evolution of senescence and post-reproductive lifespan in guppies (*Poecilia reticulata*). *PloS Biol.* **4**(1), e7 (DOI: 10.1371/journal.pbio.0040007).

Rockwood, L. L., 2006. *Introduction to Population Ecology.* Blackwell, Malden, MA.

Rose, M. R., and L. D. Mueller, 2006. *Evolution and Ecology of the Organism.* Prentice-Hall, Upper Saddle River, NJ.

Schoener, T. W., 2011. The newest synthesis: understanding the interplay of evolutionary and ecological dynamics. *Science*, **331**, 426–429.

Smillie, C. S., M. B. Smith, J. Freidman, et al. 2011. Ecology drives a global network of gene exchange connecting the human microbiome. *Nature*, **480**, 214–244.

Yang, L., L. Jelsbak, R. L. Marvig, S. Damkiær, et al., 2011. Evolutionary dynamics of bacteria in a human host environment. *Proc. Natl Acad. Sci. USA*, **108**, 7481–7486.

Coevolution

CHAPTER SUMMARY

Competition between species, and predation of one species on another, binds species into coevolving units. Often as a consequence of competition and predation we find that coevolution dominates evolution. Indeed, at some level of change, much of evolution is coevolution. Coevolution between parasites and their hosts, organisms and viral pathogens, and between pollinating insects and their host plants is extensive in nature. Communities of species can coevolve, especially as evident in terrestrial ecosystems, plankton, and the bacteria in the human gut. Coevolution of species from different lineages, including different kingdoms, is illustrated using parasitism, fungi, and insect–host plant interactions as case studies. Although the many instances of specificity between species of insects, moths, or bees and their host plant are interpreted as examples of coevolution, in some cases, extensive evolution of one lineage preceded coevolution with the other lineage.

INTRODUCTION

Interactions between species have various consequences, one of which, **coevolution**, is enormously common. Coevolution of features occurs when two species evolve in parallel, when changes in a species and its environment co-occur, or when male and female of a species evolve in parallel. Intimate ecological relationships among many species derive from coevolutionary events in which adaptive changes in one species force adaptive changes in one or more other species. What often begin as competitive interactions between different species develop into obligatory coevolving associations often maintained by the competition that initiated the interactions.[1]

Coevolution applies to features of species and to the features of interacting groups of organisms above the level of species; flowering plants and insects and lineages of parasites and their hosts are two well-studied examples.

In this chapter, our discussion concentrates on coevolution at the three levels of *organisms, species, and lineages.* We could have approached coevolution at other levels in the biological hierarchy, however. Transcription factors and the nucleotide sequences to which they bind is one level. *Cis*-regulation of genes is another. Coevolution of protein–protein interactions is an example from the level of gene products. Evolution of the immune system is another.[2] Coevolution of signaling molecules such as hormones or growth factors and their cell surface or nuclear membrane cell receptors are examples from the cell and cell-to-cell signaling levels of organization.[3]

EXTENT OF COEVOLUTION

Coevolution is most commonly driven by competition. The Red Queen hypothesis was developed for situations of coevolution in relation to sexual selection, and expanded to coevolution in response to competition for resources (energy) in changing ecosystems. Evidence from fossil deposits in which populations, rather than individuals, have been preserved provide evidence for competition between herbivorous and carnivorous mammals and among and between dinosaurs and small mammals. Over the time periods shown in the fossils, size, speed, and protective devices increase sequentially in competing lineages (Weishampel et al., 2004). In such well-preserved sites, the comparatively small numbers of bones of predators and the much larger numbers of bones of other species can be used to determine predator–prey ratios. Such patterns of coevolution have been consistent trends in many prey–predator interactions, as well as in parasitism.

Although competition, predation, and parasitism reflect the popular concept of "nature red in tooth and claw," instances of cooperation and mutualism can modify the impact of these processes.[4] Examples include cooperative relationships in group selection and symbiotic relationships, including those between cellular organelles and their eukaryotic hosts, algae, and fungi, and cellulose-digesting protozoans in termite intestinal tracts.

Human intestines are populated by as many as 100 trillion unicellular organisms that have evolved in parallel in response to selection on the human hosts and on the unicells, either as individual species or communities (Ley et al., 2006; Yang et al., 2011). Long periods of coevolution made many such associations obligatory, but others such as between some pollinators and plants may be facultative, neither species relying on the other for survival. Others have been forced upon species by human methods of cultivation; the dwarf palm, *Chamaerops humilis*, has been forced into a mutualistic relationship (in this case, forced pollination) by being cultivated in botanical gardens with the host-specific palm flower weevil, *Derelomus chamaeropsis* (Dufaÿ and Anstett, 2004).

MUTUALISM *is a reciprocally beneficial symbiotic relationship between two species.*

[1] For discussions and examples of coevolution, see Hall (1999), Fox et al. (2001), Thompson (2005), and Futuyma (2010).

[2] See Hershberg and Margalit (2006) for coevolution of transcription factors, Lovell and Robertson (2010) for protein-protein interactions, and Danilova (2006) for immune systems.

[3] Coevolution of steroidal hormones such as estrogen involved in vertebrate reproduction and their receptors, one example among many, are discussed by M. E. Baker, 2004. Coevolution of steroidogenic and steroid-inactivating enzymes and adrenal and sex steroid receptors. *Mol. Cell Endocrinol.*, **215**, 55–62.

[4] For discussions of mutualism from various perspectives see Ley et al. (2006), Clutton-Brock (2009), and Leigh Jr. (2010). Costs associated with mutualism are discussed by Bronstein (2001).

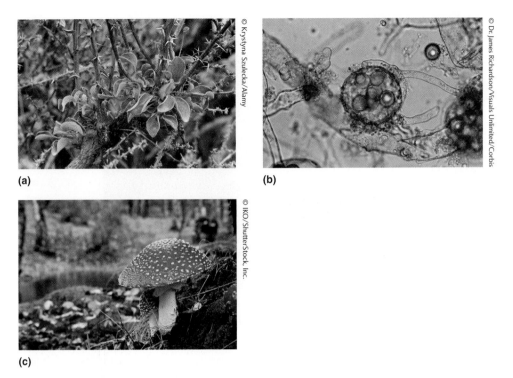

(a)

(b)

(c)

FIGURE 21.1 Part of the diversity of fungi. **(a)** The rust fungus, *Aecidium magellanicum* produces rust and growth defect known as "witches broom" in plants. **(b)** Hyphae of the water mold, *Saprolegnia sp*. **(c)** A toadstool.

PARASITISM AND VIRAL PATHOGENS

Notable examples of coevolution have been documented between parasites or pathogens and their hosts. Twenty-seven genes in the flax plant, *Linum usitatissium*, confer resistance against a fungal rust pathogen (**FIGURE 21.1a**, and below). The pathogen, in turn, has a similar number of genes allowing it to overcome the resistance these host genes confer. In such cases, as in many others such as resistance of plants to pathogens, increased frequency of a resistant mutation in the host is followed by selection for increased frequency of one or more mutant genes in the parasite or pathogen to overcome that resistance.[5]

Studies that link development, life history, and evolution of parasites are beginning to appear. A recent example is an analysis of the platyhelminth *Polystoma gallieni*, an internal parasite of the Mediterranean tree frog *Hyla meridionalis*. The parasite has two alternate development pathways. One, the branchial morph, is found in or on the gills of the tadpole; the other, the bladder morph, is found in the bladder. Branchial morphs reproduce continuously until the tadpole metamorphoses. Bladder morphs only reproduce when adult frogs are ready to mate. The pathway taken by the parasite, which is a response to the physiological state of the tadpoles when the parasite larva attaches, is regulated by the homeobox gene *Pg-Lox4*, a gene whose orthologue *Lox4* is involved in segment specification and regeneration in other animals (Badets et al., 2010).

[5] See Thompson (2005) and Poulin (2006) for ecological aspects of coevolution, and see Aranzana et al. (2005) for genome-wide analysis in *Arabidopsis*.

FIGURE 21.2 Three-dimensional reconstructions of a myxoma virus **(a)** based on transmission electron microscopic analysis **(b)**.

[Reproduced from ICTVdB Management (2006). 00.058.1.05.001. Myoma Virus. In: ICTVdB - The Universal Virus Database, version 4. Büchen-Osmond, C. (Ed), Columbia University, New York.]

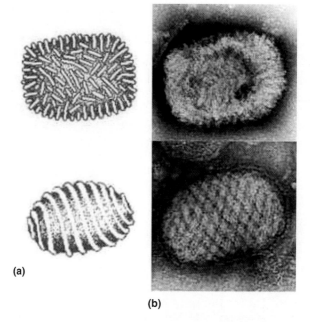

(a)

(b)

MYXOMA VIRUSES, *which are closely related to the smallpox virus, affect mammals, including humans.*

Some SMALL DNA VIRUSES *are genetically stable compared to rapidly evolving RNA viruses, and can persist and coevolve with their hosts without causing disease.*

Although viruses are non-living, they too coevolve with the organisms they infect. A prominent example is a **myxoma virus** (FIGURE 21.2), imported from South America to Australia to control a phenomenal population growth of the European rabbit, *Oryctolagus cuniculus* (Fenner and Fantini, 1999). Although the virus resulted only in a mild disease in cottontail rabbits, *Sylvilagus brasilensis*, in their native South American habitat, it acted as a highly lethal pathogen among the Australian rabbits, being transmitted primarily by mosquitoes. Viral-induced death in 1950–1951 was as high as 99 percent among infected rabbits.

It seemed as though the Australian rabbit population either would be eradicated by the virus or persist only at low numbers. Some Australian rabbits were resistant to the virus and so survived. Unexpectedly, however, the virus itself became less virulent. Because mosquitoes feed only on live rabbits, the rate at which they transmit the virus falls if the virus kills the mosquitoes (its immediate host) too quickly. Highly virulent, rapidly replicating strains of virus were therefore selected against. Strains with reduced virulence were selected for; they allow infected rabbits to live long enough for the virus to spread more readily. This host–pathogen relationship is not static; as rabbit resistance to the virus increases, increased viral virulence becomes a selected trait and is selected for. Thus, the relationship between virus and rabbit in Australia may eventually emulate the relationship in South America, in which a virulent virus has only a partially harmful effect on its native host.[6] Such evolutionary outcomes are not unusual.

Reduced virulence, however, is not always a successful option for the virus. Success among competing strains of virus depends on their ability to replicate rapidly; rapid replication is a major factor in causing virulence. Viruses that incorporate into host chromosomes and are transmitted between generations by *vertical transmission* have been selected for *reduced virulence* because they depend on host reproduction for survival.[6] On the other hand, viruses that enter organisms exclusively through infection from other hosts (*horizontal transmission*) are selected to *increase infectivity* by replicating rapidly, and thus almost always destroy the host. The sexually transmitted human immunodeficiency virus (HIV) increases in virulence in populations in which individuals have multiple sexual partners (Ewald, 1994; Mindell, 2006; Poulin, 2006).

---

[6]F. F. Shadan and L. P. Villarreal, 1993. Coevolution of persistently infecting small DNA viruses and their hosts linked to host-interactive regulatory domains. *Proc. Natl Acad. Sci. USA*, **90**, 4117–4121.

We turn now to kingdom of multicellular organisms—fungi—for which parasitism is the way of life.

FUNGI AND ADOPTING A PARASITIC LIFE STYLE

Fungi are no longer classified as simple plants without chlorophyll but are the third major lineage of multicellular organisms. A sister group to animals, fungi and animals share a choanoflagellate ancestor.[7] Fungi are usually recognized as having initially appeared with the first flowering land plants in the Silurian, 440 Mya, although according to some interpretations of the fossil record, fungi can be traced back to the Precambrian. By the Paleozoic (554 Mya), fungi had evolved into most of the lineages represented by extant forms (Figure 21.1).

Lack of chlorophyll is reflected in fungal life styles; fungi obtain nutrition either from parasitizing live hosts (a chytrid fungus, *Batrachochytrium dendrobatidis*, is a major contributor to the worldwide decline in amphibian populations) or by digesting dead organic matter. Adopting a parasitic existence on live hosts offered the opportunity for early aquatic fungi—which may be related to extant forms, the water molds shown in Figure 21.1b—to resist desiccation in the host tissues of land plants and evolve subsequently into forms that use aerially dispersed spores.

A comprehensive analysis of six gene regions in 200 species of fungi showed that the spore flagellum used in swimming may have been lost four times independently during fungal evolution. Each of these losses was accompanied by the evolution of new mechanisms of spore dispersal (Bruns, 2006; James et al., 2006). Parasitic fungi are still actively evolving on the gene level, in what has been called an "**arms race**" between host and parasite. Each new genetic variant of a host that confers resistance against a fungal parasite is often overcome by selection for increased frequency of a fungal genetic variant that increases host susceptibility. This study by Wade (2007) is a fine example of the emerging field of "coevolutionary genetics" in which molecular genetics, comparative genomics, and evolutionary theory provide the basis to detect and analyze coadaptation, coevolution, and cospeciation.

An even more recent series of studies on fungal pathogens of plants highlights the role that comparative genomics can play in revealing parallel evolutionary trends. Several different lineages of fungi and allied organisms all experienced rapid evolution of effector proteins and rapid speciation associated with shifts in the range of their host plants.[8] One study examined six genomes from four species of water molds from the genus *Phytophthora*, which produce potato blight. These genomes contain *hundreds of mutations* in disease-effector genes in dynamic genomic compartments with differential rates of genomic evolution in species with 99.9 percent identified in other portions of the genome (Raffaele et al., 2010). Intriguingly, most of the genes in this and other studies are found in gene-poor but transposon-rich regions of the genomes, strongly suggesting that transposons facilitated the changes.

INSECTS AND HOST PLANTS

Changes in insects and the flowers on the plants they pollinate provide one of the most well studied examples of coevolution, discussed here in the context of the origin of insects and flowering plants.[9]

Not all FUNGI *are multicellular. Some of the 120,000 fungal species, such as yeasts, are unicellular. Other fungi have vegetative stages of branched multicellular or multinuclear filaments called* HYPHAE *that aggregate into a mass called a* MYCELIUM *(Figure 21.1b).*

The term ARMS RACE *originated for the buildup of military weapons by the Soviet Union and the United States after the Soviet Union tested an atomic bomb on August 29, 1949. The term has expanded (evolved) to any competition in which the goal is to stay ahead of the competition. Winning may never be an option, as illustrated so convincingly by the 1983 movie* WAR GAMES, *starring Matthew Broderick and Ally Sheedy.*

---

[7]S. L. Baldauf and J. D. Palmer, 1993. Animals and fungi are each other's closest relative: congruent evidence from multiple proteins. *Proc. Natl Acad. Sci. USA*, **90**, 11558–11562; C. Borchiellini et al., 1998. Phylogenetic analysis of the Hsp70 sequences reveals the monophyly of metazoa and specific phylogenetic relationships between animals and fungi. *Mol. Biol. Evol.*, **15**, 647–655.
[8]See Dodds (2010) for an overview of four studies published in the journal *Science*.
[9]For insect–plant coevolution, see Feinsinger (1983) and Waser and Ollerton (2006).

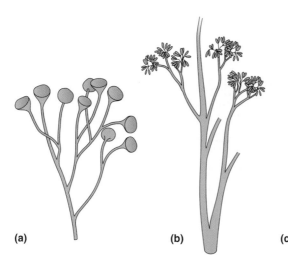

FIGURE 21.3 Two early fossil plants and an extant member of the whisk ferns. **(a)** This diminutive Upper Silurian plant, *Cooksonia caledonica,* had naked, dichotomously branched axes with terminal sporangia. **(b)** *Psilophyton princeps*, a spiny, leafless fossil plant that first appeared less than 10 My after *Cooksonia,* had a main stem with lateral branches terminating in sporangia. **(c)** The extant epiphytic whisk fern, *Psilotum nudum*, has simple stems, nondiscernible leaves, and absence of a root system that resemble the fossil forms.

PLANT EVOLUTION

The earliest Silurian plant fossils include a number of simple vascular plants with leafless stems classified in the genus *Cooksonia*.[10] A diminutive plant about 25 mm or so high, *Cooksonia caledonica,* had no distinctive leaves or roots (**FIGURE 21.3a**). The first leafless land plants also appeared in the Devonian (**FIGURE 21.3b**). The fossil record shows rapid evolutionary radiation of vascular plants after their appearance more than 400 Mya. The oldest known insect fossil also is from the Devonian.

By the Mid- to Late Devonian the evolution of the **cambium** permitted plants to increase in size by an order of magnitude over those present earlier in the Devonian.[11] By the end of the Devonian forests containing a great variety of woody trees were established. By 350 Mya (the beginning of the Carboniferous), lush and extensive forests had developed in vast tropical swamps along the east coast of North America and similar coastal regions of Europe and North Africa (Kenrick and Davis, 2004); death, submergence, and decay of these forests produced enormous Carboniferous coal seams. These successful land plants were vascular with xylem to transport water and phloem to distribute nutrients. Both xylem and phloem arise from the cambium, which was a key innovation to land plant diversification.[12] By the Late Carboniferous–Early Permian 290 Mya, insects as diverse as dragonflies, cockroaches, mayflies, and grasshoppers were abundant. Giant dragonfly-like forms such as *Meganeura* and *Meganeuropsis* had wingspans in excess of 60 cm (23.6 in), some as wide as 70 cm (27.6 in; **FIGURE 21.4**).

GROWTH *of stem and root tips in vascular plants derives from terminal apical meristems of dividing cells. Growth of side branches is from lateral meristems.*

PLANT-INSECT COEVOLUTION

The many instances of specificity between species of insects, such as moths or bees, and their host plant are interpreted as examples of coevolution; evolution and diversification of insects

[10] Fossil pollen is known from 125–127 Mya. The oldest fossil plant, the recently described *Leefructus mirus*, is 123–127 My old (Sun et al., 2011).

[11] The cambium is the tissue from which new cells are produced to increase the diameter of plant stems (Cronk, 2009).

[12] For the evolution of plants and plant tissues/organs, see Niklas (1997), Gensel and Edwards (2001), Crepet (2000), Frohlich and Chase, 2007, and Cronk (2009).

© Stocktrek Images, Inc./Alamy

© Cathy Melloan/Alamy

(a) **(b)**

FIGURE 21.4 Giant dragon-fly like insects from the Late Carboniferous included *Meganeura* sp. **(a)** and *Meganeuropsis* sp **(b)**.

paralleled vascular plant evolution, while coevolution with insects contributed to the rapid evolution of flowering plants.[13] From an analysis of basal extant and early fossil vascular plants, Hu et al. (2008) concluded that flowering plant–pollinator coevolution was present at or near the base of vascular plant evolution (Early Cretaceous). Coevolution of insect and plant implies, and may even require, that the plant not be able to pollinate itself, a topic discussed in BOX 21.1.

BOX 21.1   Self-Incompatibility

Coevolution of insect and plant implies, and may even require, that the plant avoids **self-fertilization**—that is, it is not able to pollinate itself.

A novel mechanism of self-fertilization came to light in 2006 from observations of the wild pink-flowered orchid, *Holcoglossum amesianum*. The orchid fertilizes itself by rotating the anther of its bisexual flower through a full circle, against gravity, to place pollen grains on its own stigma. *H. amesianum* is found at elevations of 1,200 to 2,000 m (3,937 to 6562 ft) in China's Yunnan province. It grows on tree trunks where wind and potential insect-pollinators are scarce during the dry months of February to April, when the orchid flowers.[a]

However, to avoid close inbreeding and to protect the genetic variation produced by sexual reproduction, the evolution of **self-incompatibility** (or self-sterility) **alleles** was an extremely important means to prevent self-fertilization. In such cases, for example pollen bearing the same combination of alleles found in the ovule, cannot grow on the female style. What follows is an abbreviated summary of self-incompatibility: eight or more different mechanisms operate in flowering plants to prevent inbreeding, and to encourage outbreeding, both of which facilitate coevolution with insects.[b]

In the typical mechanism of self-incompatibility seen in flowering plants, development of the next generation is stopped at one of several stages of germination, fertilization, or embryo development. A haploid pollen grain carrying the self-sterility allele S^1 will not germinate, initiate pollen tube growth, or fertilize the ovule (or if fertilization does occur, will not initiate embryo development) on a female plant carrying the same allele. The simplest S-locus codes for two basic proteins that function as female and male determinants. The male determinant is expressed in the anther and/or pollen, the female determinate in the pistil, but the alleles are inherited as a linked genetic unit. Interaction between the two proteins arrests pollen germination or growth of the pollen tube into the style.

Once an allele such as S^1 spreads throughout a population, its frequency is reduced by the many sterile mating combinations to which it is exposed. Rare alleles, in contrast, will successfully fertilize almost every female plant they meet until they too become common. Such frequency-dependent selection can lead to self-sterility systems comprised of many alleles, for example, 200 or more in red clover (*Trifolium pratense*).

[a] K. Liu et al., 2006. Pollination: self-fertilization strategy in an orchid. *Nature*, **441**, 945–946. For self-fertilization in a lineage of fish, a mechanism that may have originated hundreds of thousands of years ago, see A. Tatarenkov et al., 2009. Long-term retention of self-fertilization in a fish clade. *Proc. Natl Acad. Sci. USA*, **106**, 14456–14459.

[b] S. J. Hiscock and D. A. Tabah, 2003. The different mechanisms of sporophytic self-incompatibility. *Philos. Trans. R. Soc. Lond. B. Biol. Sci.*, **358**, 1037–1045; V. E. Franklin-Tong, and F. C. H. Franklin, 2003. The different mechanisms of gametophytic self-incompatibility. *Philos. Trans. R. Soc. Lond. B. Biol. Sci.*, **358**, 1025–1032; D. Charlesworth et al., 2005. Plant self-incompatibility systems: a molecular evolutionary perspective. *New Phytologist*, **168**, 61–69; K.-i. Kubo et al., 2010. Collaborative non-self recognition system in S-RNAse-based self-incompatibility. *Science*, **330**, 796–799.

[13] For perspectives on insect–plant coevolution, see Darwin (1862), Grimaldi and Engel (2005), and Waser and Ollerton (2006).

FIGURE 21.5 Diagram of a generalized flower after fertilization. The female gametophyte is within the ovule; the male gametophyte is the pollen tube. The male gametophyte produces two sperm nuclei. One fertilizes the egg nucleus to produce the zygote (2*n*). The other fertilizes the two polar nuclei to produce the endosperm (3*n*). Petals are usually the more colorful part of the flower, and are organized in a whorl called the *corolla*.

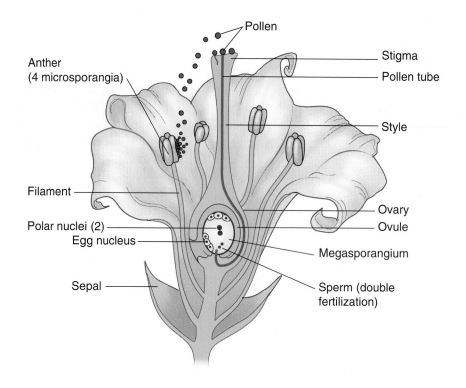

Features contributing to the rapid evolution of flowering plants included (a) evolution and elaboration of **flowers** as reproductive organs (**FIGURE 21.5**), (b) evolution of flower structures that enable insects or birds to **pollinate** them and to **disperse** their seeds (**FIGURE 21.6**), and (c) a deep origin of genes involved in the formation and evolution of reproductive cones in gymnosperms and flowers in angiosperms (flowering plants) (Chanderbali et al., 2010). With the evolution of leaves into the petals and sepals of flowers, competition for pollinators arose. Large-scale analyses of insect pollinator and plant evolution support origination and/or radiation of insect pollinators soon after the origin of the plant group that is pollinated by insects. Pierid butterflies, which are pale-colored members of a family that includes the cabbage white butterfly *Pieris rapae*, lay their eggs on plants such as mustard and cabbages. The plants evolved insecticidal chemicals that are released when the leaves are damaged. The butterflies, in turn, evolved a chemical pathway that renders the plant chemical nontoxic, a coevolution that arose 10 Mya.[14]

Insect pollinators coevolved mechanisms that facilitated feeding on specific flowers (Figure 21.6).[15] A recent combined molecular and morphological analysis of bees, of which there are more than 16,000 extant species, is consistent with the earliest Mid-Cretaceous lineages having included host-plant specialists. These findings support **host-plant specificity** as an ancient feature in plant evolution.[16]

We do have to be careful, however, when interpreting associations as having arisen by coevolution. Some of the *associations* between insects and plants that have been thus interpreted are *not examples of coevolution,* a conclusion reached after detailed comparisons of the phylogenies of insects and plants. In one study, nuclear DNA was used to construct a molecular phylogeny of 100 species of leaf-mining moths to obtain estimates of the moth

[14] S. P. Courtney, 1982. Coevolution of pierid butterflies and their cruciferous food plants. *Oecologia*, **51**, 91–96; C. W. Wheat et al., 2007. The genetic basis of a plant-insect coevolutionary key innovation. *Proc. Natl Acad. Sci. USA*, **104**, 20427–20431. For the evolution of scent in flowers, see N. Dudareva and E. Pichersky (eds), 2006. *Biology of Floral Scent*. CRC Press, Atlanta, GA.
[15] For overviews of plant-insect coevolution, see Grant and Grant (1965), Alcock (2006), and Waser and Ollerton (2006).
[16] B. N. Danforth et al., 2006. The history of early bee diversification based on five genes plus morphology. *Proc. Natl Acad. Sci. USA*, **103**, 15118–15123.

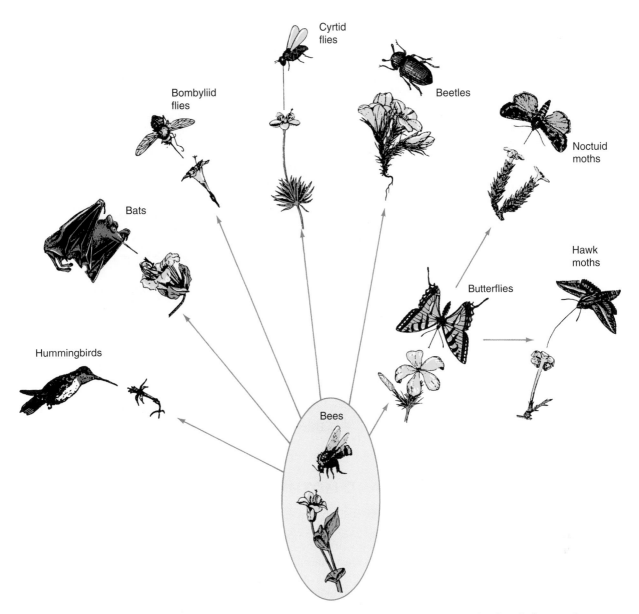

FIGURE 21.6 The range of flower types and their animal pollinators in phlox plants (Polemoniaceae). Only relatively few mutations are required for the transition from a species pollinated by bees to a new species pollinated by hummingbirds. In more exotic cases, different plant species deposit pollen on different parts of the pollinator's body, allowing the same animal species to pollinate different plant species.

[Adapted from Grant, V., and K. A. Grant 1965. *Flower Pollination In The Phlox Family*. Columbia University Press.]

species' origin. Comparison with the distributional ages in the fossil record of the moth's host plants led to the conclusion that the main radiation of the moth genus (*Phyllonorycter*) took place 27.3 to 50.8 Mya, which was well after the radiation of their host plants 84 to 90 Mya. Provided that the ages obtained from the molecular and fossil data are both accurate, the moth radiation would not be an example of coevolution (in the strict sense of that word) but of delayed colonization of host plants. Similarly, a phylogenetic analysis on a data matrix of three genes and 1,900 species of beetles revealed that the beetles diversified into more than a hundred lineages in the pre-Cretaceous, before the rise of angiosperms.[17]

[17]C. Lopez-Vaamonde et al., 2006. Fossil-calibrated molecular phylogenies reveal that leaf-mining moths radiated millions of years after their host plants. *J. Evol. Biol.*, **19**, 1314–1326; T. Hunt et al., 2007. A comprehensive phylogeny of beetles reveals the evolutionary origins of a superradiation. *Science*, **318**, 1913–1916.

FIGURE 21.7 A diversity of orchid flowers as drawn by Ernst Haeckel.

[Reproduced from E. Haeckel, 1974. *Art Forms in Nature.* Dover Publications, Inc., New York.]

ORCHIDS AND MOTHS

Charles Darwin extensively discussed coevolved adaptations of plants and insects that facilitated pollination of flowering plants, especially in relation to orchids and moths. Aware of the diversity of exotic flowers amongst orchids (FIGURE 21.7), Darwin postulated that even the most unusual orchid flower known to him would have a matching pollinator, although no pollinator was then known (Darwin, 1862; Alcock, 2006).

The orchid Darwin had in mind was the Madagascar (Malagasy) star (Christmas) orchid, *Angraecum sesquipedale,* the nectary of which is at the base of a 25-cm (nearly 10 in) long corolla tube (FIGURE 21.8). Darwin's prediction that a pollinator with a

FIGURE 21.8 The Madagascar (Malagasy) star (Christmas) orchid, *Angraecum sesquipedale,* has a 25-cm-long corolla tube.

© Sami Sarkis Studio/Alamy Images

FIGURE 21.9 The Morgan's Sphinx moth (*Xanthopan morgani praedicta*) has a 30-cm-long tongue.

similarly long proboscis must exist was borne out a century later with the discovery of a giant hawk moth, Morgan's Sphinx moth (*Xanthopan morgani praedicta*). This moth has an appropriately long (30 cm; 11.8 in) proboscis, which is longer than the rest of its body (FIGURE 21.9). Other species of hawk moths have an even longer proboscis, allowing them to pollinate star orchids that have been selected for even longer corolla tubes.[18]

An astonishing parallel to the coevolution of the hawk moth and the Malagasy star orchid was described in 2006. It involves the coevolution of the tongue of the tube-lipped nectar bat, *Anoura fistulata,* and the length of the corolla of a plant, *Centropogon nigricans,* found in the cloud forests of the Ecuadorian Andes Mountains. Close relatives of the nectar bat—species in the same genus, in fact—have tongues that can extend to 37 to 39 mm (1.45 to 1.53 in). The tongue of the nectar bat can extend 85 mm (3.35 in), 1.5 times its body length (FIGURE 21.10; a feat second only to the chameleon, which can extend its tongue approximately twice its body length). When the tongue is not in use, the nectar bat stores it inside the thorax in a special tube.[19]

The PROBOSCIS *of a hawk moth can attain such a prodigious length because it can be coiled up beneath the head when not in use.*

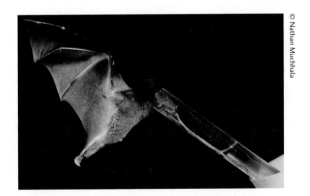

FIGURE 21.10 The tongue of the tube-lipped nectar bat *Anoura fistulata* can extend to 85 mm, which is 1.5 times its body length.

[18] A detailed analysis of this and related examples of coevolution may be found in Arditti et al. (2012).
[19] N. Muchhala, 2006. Nectar bat stows huge tongue in its rib cage. *Nature,* **444,** 701–702.

PLANT–HERBIVORE COEVOLUTION

A well-documented example of plant–herbivore coevolution on the North American continent is the coevolution of Great Plains grasses with American bison. Before European settlement, millions of American bison (*Bison bison*) dominated the continent from New York to California and from the desert grasslands of northern Mexico to the meadows of Alaska, feeding on grasses and sedges. Generations of coevolution of bison and plants shaped and sustained the landscape. With European settlement, however, came horses and cattle. In browsing grass to the ground these introduced domesticated animals had a major impact on the habitat and on bison numbers (Isenberg, 2000), as did the killing of bison for meat and fur by European settlers.

A pattern of coevolution that is as ancient as the plants and organisms (usually insects) that eat them (fossils from 420 Mya show evidence of the activity of herbivores[20]), is the evolution by plants of mechanisms to prevent, limit, or recover from feeding by herbivores. As herbivores evolved mechanisms to use plants as food, the plants evolved defenses against being eaten. The herbivores, in turn, evolved protective traits against the plants' defenses, the herbivores sometimes even using defensive chemicals for their own benefit. A pioneering study on the coevolution of butterflies and plants, published in 1964 by Ehrlich and Raven, was the first study to show that herbivores can accumulate the defensive chemicals produced by the plant and use them to protect themselves against predation.

Secondary metabolites are defined as byproducts of metabolism that are not required by the plants producing them. Individual plant species generate products effective against herbivores that are specialized to feed on that plant species, as well as chemicals that protect against non-specialized feeders (generalist herbivores). Many secondary metabolites are now used as drugs or in agriculture. Opium, aspirin, quinine, and atropine evolved in plants to provide protection against damage from herbivores.

Other examples of coevolution abound, including mimics that evolve in step with the evolution of their models, ants that cospeciate with fungi they "farm" for food,[21] competing species that evolve changes between them to reduce competition (for example, character displacement), and many others.

■ KEY TERMS

arms race
cambium
coevolution
flower
host-plant specificity

pollen
pollination
seed disposal
self-fertilization
self-incompatibility

■ EVOLUTION ON THE WEB

Explore evolution on the Internet! Visit the accompanying website for *Strickberger's Evolution, Fifth Edition,* at **go.jblearning.com/Evolution5eCW** for exercises and links relating to topics covered in this chapter.

---

[20] C. Labandeira, 2007. The origin of herbivory on land: initial patterns of plant tissue consumption by arthropods. *Insect Sci.*, **14**, 259–275.

[21] R. D. Reed et al., 2011. *optix* drives the repeated convergent evolution of butterfly wing pattern mimicry. *Science*, **333**, 1137–1141; U. G. Mueller et al., 2005. The evolution of agriculture in insects. *Ann. Rev. Ecol. Evol. Syst.*, **36**, 563–595; T. R. Schultz and S. G. Brady, 2008. Major evolutionary transitions in ant agriculture. *Proc. Natl Acad. Sci. USA*, **105**, 5435–5440.

■ REFERENCES

Alcock, J., 2006. *An Enthusiasm for Orchids: Sex and Deception in Plant Evolution*. Oxford University Press, Oxford, UK.

Aranzana, M. J., S. Kim, K. Zhao, et al., 2005. Genome wide association mapping in *Arabidopsis* identifies previously known flowering time and pathogen resistance genes. *PloS Genet.*, **1**, 531–539.

Arditti, J., J. Elliott, I. J. Kitching, and L. T. Wasserthal. 2012. 'Good heavens what insect can suck it', Darwin, *Angraecum sesquipedale* and *Xanthopan morgani praedicta*. *Bot. J. Linn. Soc.* **169**, 403–432.

Badets, M., G. Mitta, R. Galiner, and O. Verneau, 2010. Expression patterns of *Abd-A/Lox4* in a monogenean parasite with alternate developmental paths. *Mol. Biochem. Parasitol.*, **173**, 154–157.

Bronstein, J. L., 2001. The costs of mutualism. *Am. Zool.*, **41**, 825–839.

Bruns, T., 2006. A kingdom revised. *Nature*, 443, 758–761.

Chanderbali, A. S., M.-J. Yoo, L. M. Zahn, et al., 2010. Conservation and canalisation of gene expression during angiosperm diversification accompany the origin and evolution of the flower. *Proc. Natl Acad. Sci. USA*, **107**, 22570–22575.

Clutton-Brock, T. 2009. Cooperation between non-kin in animal societies. *Nature*, **462**, 51–57.

Crepet, W. L., 2000. Progress in understanding angiosperm history, success, and relationships: Darwin's abominably "perplexing phenomenon." *Proc. Natl Acad. Sci. USA*, **97**, 12939–12941.

Cronk, Q. C. B., 2009. *The Molecular Organography of Plants*. Oxford University Press, Oxford, UK.

Danilova, N., 2006. The evolution of immune mechanisms. *J. Exp. Biol.* (*Mol. Dev. Evol.*), **306B**, 496–520.

Darwin, C., 1862. *On the Various Contrivances by Which British and Foreign Orchids Are Fertilized by Insects, and on the Good Effects of Intercrossing*. Murray, London.

Dodds, P. N., 2010. Genome evolution in plant pathogens. *Science*, **330**, 1486–1487.

Dufaÿ, M., and M.-C. Anstett, 2004. Cheating is not always punished: killer female plants and pollination by deceit in the dwarf palm, *Chamaerops humilis*. *J. Evol. Biol.*, **17**, 862–868.

Ehrlich, P. R., and P. H. Raven, 1964. Butterflies and plants: a study of coevolution. *Evolution*, **18**, 586–608.

Ewald, P. W., 1994. *Evolution of Infectious Disease*. Oxford University Press, Oxford, UK.

Feinsinger, P., 1983. Coevolution and pollination. In *Coevolution*, D. J. Futuyma and M. Slatkin (eds.). Sinauer Associates, Sunderland, MA, pp. 282–310.

Fenner, F., and B. Fantini, 1999. *Biological Control of Vertebrate Pests: The History of Myxomatosis—An Experiment in Evolution*. CABI Publishing, Oxford, UK.

Fox, C. W., D. A. Roff, and D. J. Fairbairn (eds.), 2001. *Evolutionary Ecology. Concepts and Case Studies*. Oxford University Press, New York.

Frohlich, M. W., and M. W. Chase, 2007. After a dozen years of progress the origin of angiosperms is still a great mystery. *Nature*, **450**, 1184–1189.

Futuyma, D. J., 2010. How species affect each other's evolution. *Evo. Edu. Outreach*, **3**, 3–5.

Gensel, P. G., and D. Edwards (eds.), 2001. *Plants Invade the Land: Evolutionary and Environmental Perspectives*. Columbia University Press, New York.

Grant, V., and K. A. Grant, 1965. *Flower Pollination in the Phlox Family*. Columbia University Press, New York.

Grimaldi, D., and M. S. Engel, 2005. *Evolution of the Insects*. Cambridge University Press, New York.

Hall, B. K., 1999. *Evolutionary Developmental Biology*, 2nd ed. Kluwer Academic Publishers, Netherlands.

Hershberg, R., and H. Margalit, 2006. Coevolution of transcription factors and their targets depends on mode of regulation. *Genome Biol.*, **7**, R62. DOI: 10.1186/gb-2006-7-7-r62.

Hu, S., D. L. Dilcher, D. M. Jarzen, and D. W. Taylor, 2008. Early steps of angiosperm-pollinator coevolution. *Proc. Natl Acad. Sci. USA*, **105**, 240–245.

Isenberg, A. C. 2000. *The Destruction Of The Bison: An Environmental History, 1750–1920*. Cambridge University Press, New York.

James, T. Y., F. Kauff, C. L. Schoch, et al., 2006. Reconstructing the early evolution of fungi using a six-gene phylogeny. *Nature*, **443**, 818–822.

Kenrick, P., and P. Davis, 2004. *Fossil Plants*. Natural History Museum, London.

Leigh, E. G. Jr., 2010. The evolution of mutualism. *J. Evol. Biol.*, **23**, 2507–2528.

Ley, R. E., D. A. Peterson, and J. I. Gordon, 2006. Ecological and evolutionary forces shaping microbial diversity in the human intestine. *Cell*, **124**, 837–848.

Lovell, S. C., and D. L. Robertson, 2010. An integrated view of molecular coevolution in protein–protein interactions. *Mol. Biol. Evol.*, **27**, 2567–2575.

Mindell, D. P., 2006. *The Evolving World: Evolution in Everyday Life*. Harvard University Press, Cambridge, MA.

Niklas, K. J., 1997. *The Evolutionary Biology of Plants*. University of Chicago Press, Chicago.

Poulin, R., 2006. *Evolutionary Ecology of Parasites*. 2nd ed. Princeton University Press, Princeton, NJ.

Raffaele, S., R. A. Farrer, L. M. Cano, et al., 2010. Genome evolution following host jumps in the Irish potato famine pathogen lineage. *Science*, **330**, 1540–1543.

Sun, G., G. L. Dilcher, H. Wang, and Z. Chen, 2011. A eudicot from the Early Cretaceous of China. *Nature*, **471**, 625–628.

Thompson, J. N., 2005. *The Geographic Mosaic of Coevolution*. The University of Chicago Press, Chicago.

Wade, M. J., 2007. The co-evolutionary genetics of ecological communities. *Nat. Rev. Genet.*, **8**, 185–195.

Waser, N. M., and J. Ollerton, 2006. *Plant-Pollinator Interactions: From Specialization to Generalization*. The University of Chicago Press, Chicago.

Weishampel, D. B., P. Dodson, and H. Osmólska (eds.), 2004. *The Dinosauria*. 2nd ed. University of California Press, Berkeley, CA.

Yang, L., L. Jelsbak, R. L. Marvig, et al., 2011. Evolutionary dynamics of bacteria in a human host environment. *Proc. Natl Acad. Sci. USA*, **108**, 7481–7486.

Phenotypic Plasticity and Sympatric Speciation

CHAPTER SUMMARY

An important consequence of competition, predation, and specialized feeding (the latter especially of insects on plants) is the association of two or more species as coevolving units. Additionally, many species display different morphologies as a response to signals received from prey, predators, or the environment. This property, known as phenotypic plasticity, links communities of organisms together, integrates the evolution and ecology of populations with genetic and developmental changes in individuals, and may provide a basis for species diversification leading to speciation. The origin of new species of sexually reproducing organisms is grounded in the concept of species as reproductively isolated units. Two major modes of speciation in sexually reproducing organisms are recognized. One, sympatric speciation, is initiated *within a population*. The other, initiated *between populations* following geographical isolation and known as allopatric speciation, is discussed elsewhere in this text. Speciation in organs that reproduce asexually is discussed in BOX 22.1.

INTRODUCTION

Species are bound together as predators and prey, as well as through mimicry, ecological interactions, and coevolution. Consequently, selection on one species can, and almost inevitably does, influence other species. Selection is not limitless, however. Dynamics

BOX 22.1	Speciation in Organisms with Asexual Reproduction

Our discussion of speciation deals with mechanisms in **sexually reproducing organisms**. **Asexual reproduction** is found in many organisms, including fungi, green algae, many bacteria, excavates, and those parthenogenetic animals and plants that reproduce from unfertilized eggs/ovules. How do species that reproduce asexually undergo speciation? Do such species show less variation and therefore lower rates of speciation than sexually reproducing organisms?

Asexual reproduction need not mean lack of genetic isolation or lack of genetic variation. By using population and evolutionary genetics approaches (linkage disequilibrium, gaps in the presence of genes, phylogenetic analyses) some asexual organisms have been shown to group into species that are reproductively isolated. Such populations are subject to selection and genetic drift and so their speciation can be understood using the same principles known to apply in sexually reproducing populations/species. Some organisms, such as rotifers in the genus *Brachionus*, discussed in the text under the heading of "Plasticity in Aquatic Organisms," switch from asexual to sexual reproduction in response to both an unknown factor within the eggs and the level of crowding of the population (Gilbert, 2003).

Birky and Barraclough (2009) gathered and analyzed evidence for both allopatric and sympatric speciation in asexually reproducing species. A well-documented example is reproduction among the 400 species of fresh-water dwelling bdelloid rotifers, a group that has survived and diversified for 80 My. All individuals are females, and all reproduce by producing clones of daughters from unfertilized eggs. Genetic variation is not obtained through fertilization but through the evolution of different functions for each copy of a pair of genes. Variation also accumulates from the high number of transposons transmitted into bdelloid rotifers by horizontal gene transfer from plants, bacteria, and fungi (Gladyshev et al., 2008).

Birky et al. (2005) undertook a phylogenetic analysis on bdelloid rotifers using the mitochondrial gene *Cox-1*, which codes for subunit 1 of cytochrome c oxidase, a transmembrane protein essential to oxygen transport across the mitochondrial membrane. Their analysis resolved 21 terminal clades that fitted the criteria of species. They concluded that species in the genus *Rotaria* had been subjected to divergent selection and had undergone adaptive radiation.[*] When followed by geographical isolation, allopatric speciation can take place in such situations (Birky and Barraclough, 2009).

Despite ongoing horizontal gene transfer, species-specific regions of the genomes of such single-celled organisms as *E. coli* and *Salmonella enterica* have diverged from one another, albeit slowly (on the order of tens of millions of years). Such isolation does, in time, lead to complete genetic isolation of the genomes, that is, to speciation (Fraser et al., 2007). Furthermore, *E. coli* maintained in the laboratory under conditions of selection, diversify to the stage where they no longer recognize their parent populations.

---

[*] D. E. A. Fontaneto et al. (2007). Independently evolving species in asexual bdelloid rotifers. *PLoS Biol.*, **5**, 1–8.

A REACTION NORM, *a way to measure the response of a phenotype to different environments, is an index of organismal-environmental interaction on the one hand and of flexibility/plasticity-canalization/constraint on the other.*

and constraints of genetic and developmental pathways work with selection to canalize evolutionary change. These interactions can be observed in nature and mimicked in laboratory experiments in which an organism's response to environmental change is measured as a **reaction norm**; the greater the number of reaction norms, the more responsive the organism is to the environment.

For populations to evolve—that is, to change their gene frequencies—(i) mutations must introduce nucleotide differences in the alleles of genes of individuals within the population, and/or (ii) gene regulatory networks must change. Mere appearance of new alleles, or of a gene duplication (to take only two examples) is no guarantee that these changes will persist or prevail over others. Explanations for the persistence of such changes and their increase in frequency in the population must be sought elsewhere than in the original mutational event.

Although evolution is marked by phenotypic changes, the transmission of such changes between generations is genotypic. The genotype need not only be that of the individual under selection; genotypic inheritance is both direct and indirect. Genes of the parent (maternal gene effects) and influences from the gene products of other organisms or from the environment can transmit change between generations. As one example, the ability to learn is transmitted both genetically and culturally. Finally, selection is not only on the adult phenotype. In multicellular organisms, selection can elicit changes in gametes, embryos, and/or larval stages, depending upon the life history stage at which selection influences survival or fertility. Developmental variation itself is a feature of the phenotype that is subject to selection (Vogt et al., 2008; Hallgrímsson and Hall, 2011; Hallsson and Björklund, 2012).

All this being said, and before moving into discussions of modes of speciation, species can persist for long periods of time with apparently little change, or can change without the changes leading to speciation. Maintenance of species is as important an evolutionary question, and a much more common event, as origination of species.

SPECIES CAN CHANGE WITHOUT SPECIATION BEING INITIATED

Much of evolution is about populations maintaining the integrated features associated with past selection. Even when the phenotype changes over time, speciation may not. Why? Because speciation involves more than phenotypic change, although phenotypic differentiation is often the first sign that a species is changing. One major mode of speciation requires **reproductive isolation** of previously interbreeding populations.

Most species consist of more than one population. Indeed, many species are composed of many populations. Environments inhabited by the different populations of a single species may differ. As a consequence local adaptations may occur between the populations of a species. Such local changes are recognizable by differences in morphology, physiology, and/or behavior. If these local adaptations do not affect the reproductive biology of the populations, and should the populations combine, individuals from different populations will still be able to interbreed; the populations remain as a single species, although **subspecies** or **varieties** may be recognizable.

It should be clear that the discussion of speciation in this chapter employs the BIOLOGICAL SPECIES CONCEPT, *defining species on the basis of their reproductive isolation from other species.*

PHENOTYPIC PLASTICITY

Many species have different body forms at different stages in their life history. Caterpillar/butterfly and tadpole/frog are two familiar examples. An important lesson learned from such species is that the single genome of some species can give rise to more than one phenotype (larval and adult in these two examples). Production of these life history stages is independent of population or environmental cues. They are constitutive, not facultative, with individual stages found only in particular environments; tadpoles are aquatic and not terrestrial.[1]

A second type of plasticity involves the production of alternate, usually adult, body forms (**morphs**) within a single species in response to external cues received from a predator, prey, population density, hormonal, or environmental signal such as temperature, seasonal changes, or calcium levels in the water. Importantly, the external signal may not be present in every generation, with the consequence that one of the phenotypes (morphs) may not be present in every generation. Production of winged and wingless ants in a single population is an example under active investigation (Abouheif, 2004). If individuals are not exposed to the environmental signal in a given generation, the environmentally induced phenotype will not appear, but, importantly, the ability to respond to the signal is passed on to the next generation. We refer to this facultative ability of a single genotype to produce more than one phenotype as **phenotypic plasticity**. Such plasticity integrates the evolution and ecology of populations with genetic and developmental changes in individuals and may lay the variational basis for speciation.[2]

It is important to remember that natural selection takes place within a generation and affects INDIVIDUALS *of that generation. Response to natural selection takes place in* POPULATIONS *between generations, and only for those features or aspects of features that are heritable.*

---

[1] For developmental and evolutionary aspects of life history stages see Hall (2004), Hall et al. (2004), and Gilbert and Epel (2008). For constitutive larval stages see Hall and Wake (1999) and Flatt et al. (2005).

[2] For examples, underlying theory and discussions of phenotypic plasticity see Stearns (1989), Greene (1999), Wilkins (2002), West-Eberhard (2003), Hall (2004), Hall et al. (2004), Sultan and Stearns (2005), and Gilbert and Epel (2008).

Many examples of phenotypic plasticity are known, a few of which are discussed here. The ubiquity and utility of phenotypic plasticity has been the subject of many studies in many fields, especially ecology and life history theory. Indeed, life history theory, which is normally studied by ecologists, has become of major interest to evolutionary, developmental, and molecular biologists as they seek to understand how phenotypes arise and are maintained.[3] Integrating the evolution and ecology of populations with phenotypic plasticity (Hall, 2004; Hall et al., 2004) is proving to be a challenge, although progress is being made:

- Chevin and Lande (2011) demonstrated that plasticity is higher in individuals at the edges of a given geographical range and in environments where developmental variation is maintained (development is not canalized).
- Vogt et al. (2008) and Beldade et al. (2011) evaluated the role of the environment in producing variation during development through regulation of gene expression in examples as varied as seasonal morphs in the butterfly *Bicyclus anynana*, castes in social ants, bees and wasps, and a parthenogenetic strain of the marbled crayfish, *Procambarus alleni*.
- Using a modeling approach, Fierst (2011) showed that populations with a history of phenotypic plasticity showed a higher rate of adaptation to new environments than did populations without such a history, and, perhaps surprisingly, that the rate of adaptation depended on the strength of the signal in the original environment.

PLASTICITY IN AQUATIC ORGANISMS

Some species in aquatic communities are held in balance by chemical interactions between individuals of predator and prey species. This fascinating phenomenon has been studied thoroughly in two situations: (a) interactions between species of predator and prey rotifers (**FIGURE 22.1a**), and (b) interactions between the water flea *Daphnia pulex* (a microscopic crustacean; **FIGURE 22.1b**) and its predators (which include larvae of midges and tadpole shrimps), or reactions to seasonal changes in water temperature or turbulence.[4]

In rotifers (especially well studied is the genus *Brachionus*), the predator releases a chemical that *acts on some but not all* of the eggs of the prey species. Rotifers developing from eggs that respond to the chemical signal produce an *extra* set of spines, preventing them from being eaten by the predator (Figure 22.1a). Importantly, if all rotifer eggs in a population responded to the predator's signal, the predator species would be wiped out. If no rotifer eggs responded to the signal the prey species would be wiped out (Gilbert, 1966). Phenotypic plasticity in some but not all individuals provides a balance that benefits both predator and prey.

Water fleas that develop in the presence of a chemical signal produced by the phantom midge *Chaoborus flavicans* develop enlarged heads and a prominent helmet (Figure 22.1b) that prevent them from being eaten by the predator.[5] Water fleas also show similar change seasonally as a response to water temperature or turbulence; the enlarged helmets enhance buoyancy.

When this plasticity is expressed seasonally in the plankton, it is often referred to in the literature as CYCLOMORPHOSIS.

---

[3] For approaches to phenotypic plasticity from these various perspectives, see Stearns (1989), Hall (1999, 2004), Debat and David (2001), Pigliucci (2001), Hall and Olson (2003), Hall et al. (2004), Gilbert and Epel (2008), and Fusco and Minelli (2010).

[4] Tollrian, R., 1993. Neckteeth formation in *Daphnia pulex* as an example of continuous phenotypic plasticity: morphological effects of *Chaoborus* kairomone concentration and their quantification. *J. Plankton Res.*, **15**, 1309–1318; Repka, S., and K. Pihlajamaa, 1996. Predator-induced phenotypic plasticity in *Daphnia pulex*: uncoupling morphological defenses and life history shifts. *Hydrobiologia*, **339**, 67–71.

[5] A. Petrusek et al., 2009. A "crown of thorns" is an inducible defense that protects *Daphnia* against an ancient predator. *Proc. Natl. Acad. Sci. USA*, **106**, 2248–2252. In a recent study Colbourne et al. (2011) showed that some 43 percent of the genes in *D. pulex* arose by gene duplication. Many of these genes function in interactions with the environment.

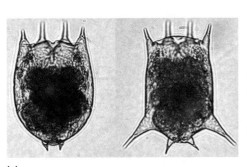
(a)

(b)

FIGURE 22.1 Two examples of phenotypic plasticity in response to chemical signals released by predators (see text for details). **(a)** A rotifer (*Brachionus calyciflorus*) shows the additional spines (right individual) formed in the presence of the predator. **(b)** A water flea (*Daphnia* shows the helmet (left individual) that forms in the presence of the predator.

[(a) Photos courtesy of Susan Bragg, Thomas Massie and Gregor Fussmann; (b) © Christian Laforsch/Photo Researchers, Inc.]

Four other examples of phenotypic plasticity in aquatic organisms or aquatic life history stages are introduced briefly below.[6]

1. In the presence of predatory dragonfly larvae, tadpoles of the Australian brown striped marsh frog *Limnodynastes peroni* develop increasing tail height and muscle mass, features that enhance their chance of escaping predation.

2. When food supply is restricted or population numbers are high, tadpoles of spadefoot toads in the genus *Spea* display even more modified development; they become cannibalistic with much larger heads and more massive jaws than their non-cannibalistic brothers and sisters (which they eat; FIGURE 22.2; Ledón-Rettig and Pfennig, 2011).[7]

3. In a bony fish, the threespine stickleback, *Gasterosteus aculeatus,* one, two, or three bones of the pelvic girdle are reduced in response to a combination of a *biotic factor*—absence of predatory fish—and an *abiotic factor*—the level of dissolved calcium in the ambient water, shown in FIGURE 22.3 in fish from two locations.[8]

4. Plasticity in body forms in the pumpkinseed sunfish, *Lepomis gibbosus*, is reflected in an experimental study of inshore and deep-water morphs that diverged from one another in response to predation by walleye *Sander vitreus*. Januszkiewicz and Robinson (2007) concluded that this diversification was driven by selection related to predation rather than diverse use of available resources by the two morphs; selection for predator-induced polymorphism drove the divergence.[9]

---

[6] For these and many other examples see Hall (1999), the papers in issue number 1 of volume 5 (2003) of the journal *Evolution & Development*, and Gilbert and Epel, 2008.

[7] A recent reassessment of spadefoot toads showed that a small percentage of cannibalistic morphs are present in the absence of dietary or population stress, indicating that this is a more complex form of plasticity than previously thought (Storz et al., 2011).

[8] C. Tickle and N. J. Cole, 2004. Morphological diversity: taking the spine out of threespine stickleback. *Curr. Biol.*, **14**, R422–R424.

[9] A. J. Januszkiewicz and B. W. Robinson, 2007. Divergent walleye (*Sander vitreus*)-mediated inducible defenses in the centrarchid pumpkinseed sunfish (*Lepomis gibbobus*). *Biol. J. Linn. Soc.*, **90**, 25–36. Resource competition between species is a basis for sympatric speciation in other species (Seger, 1985).

(a)

© Ronald Altig

(b)

© Gary Nafis

FIGURE 22.2 Cannibalistic tadpoles of Couch's spadefoot toad, *Scaphiopus couchii*, viewed from the left side to show the body form **(a)** and viewed from the front to show the massive head and tooth rows **(b)**.

PLASTICITY IN TERRESTRIAL ORGANISMS

Phenotypic plasticity is not confined to aquatic communities, as will be illustrated below using three examples, one a species of moth, the others, species of birds.

THE NORTH AMERICAN EMERALD MOTH: SEASONAL POLYMORPHISM

In the life cycle of the North American emerald moth, *Nemoria arizonaria*, caterpillars have different morphologies, ecologies, and reproductive potential depending on whether they hatch in spring or summer. In both seasons, caterpillars hatch on several species of

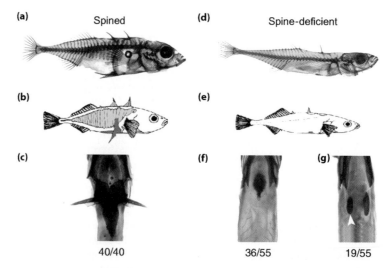

FIGURE 22.3 Images of threespine sticklebacks with pelvic spines and pelvic girdle **(a–c)** and with spine and pelvic girdle reduction **(d–f)**. Panels (a) and (d) are stained for cartilage (blue) and bone (red). In (b) the pelvic girdle is in red (also e), the lateral body armor is yellow and the dorsal spines is blue (also e). Panels (c), (f), and (g) are ventral views. Asymmetrical pelvic girdle reduction is shown by the arrow in (g).

[Reproduced form C. Tickle and N. J. Cole, 2004. Morphological diversity: Taking the spine out of three-spined stickleback. *Current Biol.* **14**, R422–R424.]

 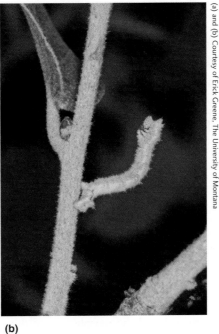

(a) (b)

FIGURE 22.4 Catkin **(a)** and twig **(b)** morphs of the North American emerald moth *Nemoria arizonaria.*

(a) and (b) Courtesy of Erick Greene, The University of Montana

oak trees. Those that hatch in the spring when catkins are present feed on catkins and have a morphology that mimics catkins. They are **catkin morphs**. Those that hatch in the summer, after the catkins have fallen, feed on oak leaves and have a morphology that mimics twigs. They are **twig morphs** (**FIGURE 22.4**). This seasonal polymorphism is triggered by the level of tannin in the food eaten by the caterpillars. Catkins are low in tannin, oak leaves have high tannin content.[10] Tannins were demonstrated experimentally to be the environmental cue that triggers morph development; 94% of larvae fed catkins develop as catkin morphs (Greene, 1989, 1999).

Caterpillars are similar when they emerge, whether in spring or in summer, but later develop into different morphs known as **seasonal polymorphs**. Feeding on a seasonal food supply such as catkins could be disadvantageous if the time when catkins were present was shorter than the time caterpillars require to develop. However, any such disadvantage is balanced by ecological and reproductive advantages that are subject to selection. Relative to twig morphs, catkins morphs pupate more rapidly (30 vs. 40 days), have higher survival to pupation and are larger at pupation (680 vs. 280 mg for females, 840 vs. 640 mg for males). Females that develop from catkin morphs produce more offspring (have higher fecundity) that do females that develop from twig morphs (130 vs. 80 offspring; Greene, 1989). Such diet-induced plasticity could play a major role in the evolution of host specificity—emerald moths hatch on several species of oak trees[11]—or varieties/subspecies.

THE MEDITERRANEAN BLUE TIT AND CHANGING ENVIRONMENTS

Evidence for the role that phenotypic plasticity plays in allowing birds to adjust to changing environments (including food abundance) comes from a 30-year study on phenotypic *and* genetic variation in a small Mediterranean passerine bird, the blue tit, *Cyanistes caeruleus.* Comparisons were made between blue tit populations in deciduous and evergreen forest environments, both on the mainland and on islands.

---

[10] Tannins are polyphenols (compounds made of carbon, hydrogen, and oxygen) produced by plants as the brown pigment seen after leaves lose their color in autumn. Tannins are used to soften animal skins to make leather.

[11] In his survey in southeast Arizona, Greene (1989) found larvae on the Arizona white oak, *Quercus arizonica*, the Emory oak, *Q. emoryi*, the scrub oak, *Q. undulata*, and the gray oak, *Q. grisea*.

TABLE 22.1	Characteristics of Blue Tit (*Cyanistes caeruleus*) Populations from One Mainland and Two Island Habitats in the Mediterranean		

Trait	Mainland	Island 1	Island 2
Blue tit population density (pairs/hectare)	1.0	1.28	0.35
Prey (caterpillar) abundance (mg/m²/day)	23.0	493.0	–
Blue tit body weight (g) Males Females	 11.2 10.7	 9.9 9.7	 9.4 9.4
Clutch size	9.8	8.5	7.2
Number of fledglings	7.5	7.3	5.0
Fledgling weight (g)	10.7	10.4	9.3

[*Source:* Blondel et al. (2006). A thirty-year study of phenotype and genetic variation of blue tits in Mediterranean habitat mosaics. *Bioscience* **56**, 661–673.]

Extensive data are available on population structure, food abundance, breeding performance, and the morphology of this species (Blondel et al., 2006).[12] Phenotypic plasticity emerged as a function of the distances over which birds dispersed; **TABLE 22.1** summarizes some of the habitat-specific differences recorded on the mainland and on two of the deciduous island sites. Short-distance dispersal results in local specializations. Longer-distance dispersal results in phenotypic plasticity, which, when local selection regimes oppose gene flow, leads to local adaptations out of proportion to the geographical separation between the populations. Jacques Blondel and his colleagues distinguished deciduous oak and evergreen oak types but found maladapted populations in which, for example, breeding time did not coincide with peak food (caterpillar) supply (Blondel et al., 2006). Wherever habitat patchiness was a mixture of deciduous and evergreen patches, local specializations reflecting phenotypic plasticity resulted.

MIGRATING EUROPEAN BIRDS AND CLIMATE CHANGE

Migrating species of birds display behavioral plasticity in response to climate change. One of the most well-known studies, conducted by Rubolini et al. (2007), concerns the timing of the arrival in Europe of spring migrant birds between 1960 and 2006. Data of first arrival date were available for 184 species, data for mean arrival date for 113 species. Spring arrival dates were shown to be species-specific and to have advanced significantly—0.37 days/year for first arrival date; 0.16 days/year for mean arrival date—in response to climate change, demonstrating plasticity in response to changing climatic features.

ISOLATING MECHANISMS AND SIBLING SPECIES

Even when a species consists of many populations, gene flow between populations may slow or inhibit local specializations and so promote continuance and stability of the species; the greater the gene flow, the fewer the differences that develop over time and the more stable the species. On the other hand, mechanisms that reduce gene exchange between populations accelerate the formation of distinctive groups (varieties/subspecies). If such groups can no longer interbreed, they are often referred to as **sibling species**. For example, gene flow is greater between populations of many non-migratory species of birds than it is between populations of migratory species; migratory species have, on average, less than half the number of varieties found in non-migratory species.

---

[12] See Losos and Ricklefs (2009) for an analysis of adaptation and speciation on islands.

Male parental care, another example, is seen more often in birds with higher numbers of subspecies than in birds with fewer subspecies. The implication is that behavioral and reproductive patterns, such as migration or the sex providing parental care, influence gene flow to the extent of influencing differentiation *within* a species.

Even where morphological differences between two species are minimal, behavioral differences may prevent interbreeding. *Drosophila melanogaster* and *D. simulans*, designated as sibling species on the basis of their morphological similarity, normally do not mate with each other even when kept together in a single population cage.[13] Normally, however, the barriers separating species do not result from a single isolating mechanism. The sibling species *Drosophila pseudoobscura* and *D. persimilis* are isolated from each other by

- habitat—*D. persimilis* usually lives in cooler regions and at higher elevations;
- courtship period—*D. persimilis* is usually more active in the morning, *D. pseudoobscura* in the evening; and
- mating behavior—females preferentially mate with males of their own species.[14]

Although the distribution ranges of these two species overlap throughout large areas of the western United States, these three isolating mechanisms are enough to maintain them as separate species; to date, only a few individuals resulting from interbreeding have been found in nature among many thousands of flies examined.

MODES OF SPECIATION

Given that new species evolve from existing species, the question of the origin of species is the question of how species change. Darwin devoted himself largely to explaining how, beginning with a single species, a lineage of organisms could change through time in response to natural selection to produce a succession of species over time. This mode of evolutionary change, known as **adaptive radiation**, reflects the origin and establishment of differences (differentiation; Darwin's term) among populations. Also known as **species diversification**, adaptive radiation occurs if splits and divisions within an ancestral line result in the emergence of more than one species or a cluster of species (a *clade*)—that is, a single branch on the tree of life.[15]

Groups of organisms that are evolutionarily related are often, but not always, connected geographically. Large geographical barriers such as oceans and mountain ranges isolate populations from one another and can establish conditions that facilitate the development of considerable differences among the separated groups. Reflecting these geographical factors, two major modes of speciation are recognized (FIGURE 22.5):

1. One, **allopatric speciation**, is initiated *between populations* following **geographical isolation**.
2. The other, **sympatric speciation**, is initiated *within a population*.

In an important synthesis of speciation, Coyne and Orr (2004) outlined four conditions to be met before speciation could be identified as sympatric or allopatric:

1. Speciation must have gone to completion. This criterion may seem obvious, but it is often difficult to determine whether two lineages have become completely separated as two species, in part, because of the various definitions of a species.

---

[13] A. W. Davis and C-I. Wu, 1996. The broom of the sorcerer's apprentice: the fine structure of a chromosomal region causing reproductive isolation between two sibling species of *Drosophila*. *Genetics*, **143**, 1287–1298; A. Civetta and R. S. Singh, 1998. Sex and speciation: genetic architecture and evolutionary potential of sexual versus nonsexual traits in the sibling species of the *Drosophila melanogaster* complex. *Evolution*, **52**, 1080–1092.

[14] Th. Dobzhansky, 1944. Chromosomal races in *Drosophila pseudoobscura* and *D. persimilis*. *Carnegie Inst. Wash. Publ.* No. 554, Washington, DC, pp. 47–144; Th. Dobzhansky, 1947. A directional change in the genetic constitution of a natural population of *Drosophila pseudoobscura*. *Heredity*, **1**, 53–64. For an overview of modes and patterns of speciation in *Drosophila*, see Coyne and Orr, 1997.

[15] For current thinking on speciation, see Schluter (2000), Coyne and Orr (2004), Grant and Grant (2008), Hendry (2009), and Yoder et al. (2010).

FIGURE 22.5 A simplified diagram of two modes of speciation, distinguished from one another by whether geographical isolation occurs. In **allopatric speciation**, a population of species A **(a)** is divided between one or more geographical localities **(b)** allowing genetic differentiation to occur **(c)**. Subsequently, mixture of the groups **(d)** can result in selection for increased reproductive isolation mechanisms among them. Speciation is complete **(e)** when gene flow between the groups can no longer occur, even when the two species (A and B) occupy the same locality. In **sympatric speciation**, a population of species C (a) is divided into one or more groups that occupy different habitats or use different food sources (b). Increased genetic differentiation between the groups (c) permits selection for reproductive isolation mechanisms (d) that eventually lead to complete speciation (e), resulting in species C and D.

[Adapted from Strickberger, M. W, 1985. *Genetics, Third Edition*, Macmillan, New York.]

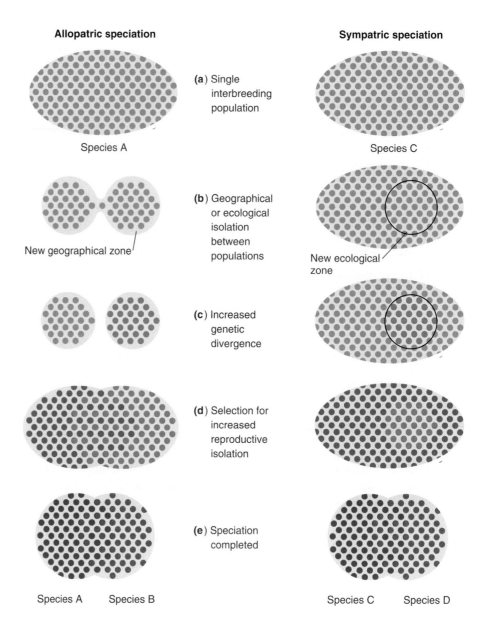

Allopatric speciation

Sympatric speciation

(a) Single interbreeding population

Species A

Species C

(b) Geographical or ecological isolation between populations

New geographical zone

New ecological zone

(c) Increased genetic divergence

(d) Selection for increased reproductive isolation

(e) Speciation completed

Species A   Species B

Species C   Species D

2. Sympatric speciation only occurs between sister species.
3. Sympatric speciation should result in species pairs with overlapping ranges.
4. Knowledge of the past history of the species, especially its biogeographical range, must not allow for past allopatric speciation.

SPECIATION WITHOUT GEOGRAPHICAL ISOLATION: SYMPATRIC SPECIATION

A population in a single locality may be adapted to different habitats within that locality. Such a population would accumulate increased genetic variability through disruptive selection (Scheiner, 2002). Increased genetic variability amplifies the chances that selection could facilitate the evolution of different forms (*polymorphism*) within the species. Examples include (a) polymorphism in the British peppered moth *Biston betularia* associated with industrial pollution, (b) mimicry in the Viceroy butterfly (FIGURE 22.6a), and (c) the evolution of distinct populations or varieties via mimicry in the butterfly *Heliconius erato* (Hines et al., 2011).

FIGURE 22.6 Mimicry among butterflies **(a)** and a species of fly undergoing sympatric speciation **(b, c)**. The photograph in **(a)** shows an unpalatable Monarch butterfly *(Danaus plexippus)* on top and a palatable mimic, the Viceroy butterfly *(Limentitis archippus)* below. The North American fly *Rhagoletis pomonella* **(b)** is undergoing sympatric speciation to produce a new race in which hawthorn-feeding larva have evolved into the apple-feeding larvae known as apple maggots, shown in **(c)** along with the damage done to the apple.

Two examples of sympatric speciation discussed below (apple maggots and cichlid fishes) demonstrate that adaptation and differentiation *within a population* can lead to speciation.[16] Although we saw from the discussion of phenotypic plasticity that the formation of morphs within a species is not speciation, you can see that the presence of morphs establishes conditions under which reproductive isolation could arise and speciation be initiated.

Phenotypic plasticity, which underlies polymorphism, has emerged as a mode of sympatric speciation, especially in situations in which morphological changes are accompanied by changes in behavior. Both of these changes can serve as mechanisms isolating individuals or groups within a population (West-Eberhard, 2003). Such conditions are most likely to be met in animal or plant species whose development is plastic (allowing phenotypic variation to arise), where the environment is subdivided into microhabitats, and, for animals, where behaviors such as aggression, territoriality, or defending breeding sites provide effective reproductive isolation. All these conditions are fulfilled in populations of apple maggots and of cichlid fishes.

SYMPATRIC SPECIATION IN INSECTS

Sympatric speciation among insects can be initiated by adaptation to different kinds of host plants as sources of food (Whitman and Ananthakrishnan, 2009). A well-studied example is the adaptation of apple maggots to feed on apples in North America.

Apple maggot is the name given to the larvae of the North American fly, *Rhagoletis pomonella* (**FIGURE 22.6b**). The natural diet of apple maggots is fruit of native hawthorn bushes (*Crataegus* spp.). Around 1800, cultivated apple trees (*Malus pumila*) from England were introduced to North America. Within less than 50 years a subset of apple maggot fly, usually referred to as the "apple-feeding race," whose larvae fed on apples had arisen (**FIGURE 22.6c**). Differentiation of this new race was sufficient that the hawthorn-feeding

---

[16] For overviews of the evidence and theoretical basis for sympatric speciation, see Otte and Endler (1989), Bush (1975, 1994), Coyne and Orr (2004), and Bolnick and Fitzpatrick (2007).

form would not feed on apples and the apple-feeding form would not feed on hawthorn fruits. Furthermore, the two insect morphs emerged as adults at different times, correlated with the timing of fruit maturation in apples and hawthorns. This differentiation generated reproductive barriers that prevented gene exchange between the two forms. Genetic differentiation between co-occurring populations of the two forms has been taken as evidence of sympatric speciation (Feder, 1998). Wasps that are parasitic on the North American apple maggot also differentiated into new forms sympatrically, one sympatric event establishing conditions for a second sympatric event (Forbes et al., 2009).

If this is sympatric speciation, why is the new form referred to as the apple-feeding race and not as a new species? Because reproductive isolation between the two forms is not complete. The ecological separation is sufficient to prevent them interbreeding in nature and the genetic differentiation is sufficient to show that separation is occurring (even in co-occurring populations). This is sympatric speciation in progress. Similarly, although experimental evidence shows that ecological heterogeneity can reinforce genetic diversity and contribute to reproductive isolation in populations of *Drosophila melanogaster* selected for radically different experimental habitats (Bush, 1975, 1994), such evidence stops short of demonstrating sympatric speciation.

Strong evidence for differentiation in *Rhagoletis pomonella* has now been provided from a study in which Michel et al. (2010) compared genomic divergence in apple- and hawthorn-feeding forms. Genetic differences based on host plant specificity, variation along a north–south axis of distribution, and differential responses to selection in an experimental situation (over-wintering) were related to selection on independent regions of the genome in each form. The apple maggot is an exemplar of the mechanisms underlying separation of a species in a single location.

SYMPATRIC SPECIATION IN CICHLID FISHES

Having originated no later than 3 to 4 Mya, cichlids fishes in the small lakes in the African Rift Valley represent a prime vertebrate example of explosive radiation and sympatric speciation (FIGURE 22.7). Cichlids are diverse ecologically and behaviorally. The combination of developmental and morphological plasticity, the ability to adjust phenotype in rapid response to environmental changes, short generation time, an abundance of potential diet items, territoriality, aggressive defense of nests, and the availability of many microhabitats in the lakes make cichlids a "textbook" example of rapid evolutionary change and the role of phenotypic plasticity in speciation.

One of the most dramatic of such speciation events occurred in Lake Victoria in the time since the last Ice Age, 12,500 to 14,000 years ago. Geological evidence shows that Lake Victoria dried up completely during this Ice Age. Nevertheless, 300 new cichlid species have arisen since that time, an astonishingly high rate of speciation (Goldschmidt, 1996).

How do we know that these cichlids speciated sympatrically? They meet the criteria outlined by Coyne and Orr (2004) and summarized earlier in this chapter. Speciation is complete—the species no longer interbreed. Species ranges overlap. Speciation was between sister species; mtDNA sequence analysis indicates that the many morphologically diverse species of cichlid fish in East Africa are monophyletic (below). No evidence exists for past periods of allopatric speciation. Phenotypic and behavioral similarities in fish in different lakes arose by parallelism or convergence through sympatric speciation (FIGURES 22.5, 22.7, 22.8, and see Kocher et al., 1993).

An important specialization in these fishes, the evolution of a second set of jaws, allowed the fishes to adapt to specialized diets. These **pharyngeal jaws** develop by modification of one or more of the cartilages supporting the gills (BOX 22.2). With two sets of jaws, the pharyngeal jaws take over the function of processing prey, freeing the mandibular jaws to specialize for prey capture (Liem, 1974; Huysseune et al., 1994). Consequently, ability to capture specialized diets has evolved to a spectacular

If recent findings for the South American cichlid genus Apistogramma *hold true for African cichlids,* THESE SPECIES NUMBERS MAY BE CONSIDERABLY UNDERESTIMATED. *Individuals from the same populations of* Apistogramma caetei *that differ only in color and so have been regarded as the same species, fail to interbreed because of female mate choice; that is, they are good biologicalspecies, (Ready et al., 2006)*

FIGURE 22.7 A sample of the diversity of cichlid fish body form, coloration, and patterning.

degree. One clade of seven cichlid species in Lake Tanganyika has asymmetrical jaws, allowing for extreme specialization in feeding behavior in which some species feed by scraping scales from the left sides of prey fishes, other species by scraping scales from the right.

Similarly rapid speciation occurred in cichlids in Nicaraguan lakes and in the Amazon. In a study of cichlid species in Lake Apoyo, a small (5 km [3.1 mi] diameter), shallow (200 m [650 ft] deep), and geologically recent (<23,000 years old) volcanic crater lake in Nicaragua, a combined molecular, morphometric, and ecological approach demonstrated that:

- the lake has been populated only once by the Midas cichlid, *Amphilophus citrinellus*, a benthic, robust, deep-bodied species, now the most common species in the area (**FIGURE 22.9a**);
- within less than 10,000 years, a second species, the arrow cichlid (*A. zaliosus*) evolved from the ancestral species, *A. citrinellus*. *A. zaliosus* is elongate rather than high-bodied (**FIGURE 22.9b,d**) and ecologically and reproductively isolated from *A. citrinellus* (Barluenga et al., 2006).

Analysis of mtDNA sequences clearly shows the evolutionary separation between the two species (**FIGURE 22.9c**), and that both species are separated from populations of the Midas cichlid in other lakes in the region. Both sympatric and allopatric speciation drove speciation of this species complex. The combined actions of these two modes of speciation also have been documented in what are called **ring species** or **ring species complexes.**

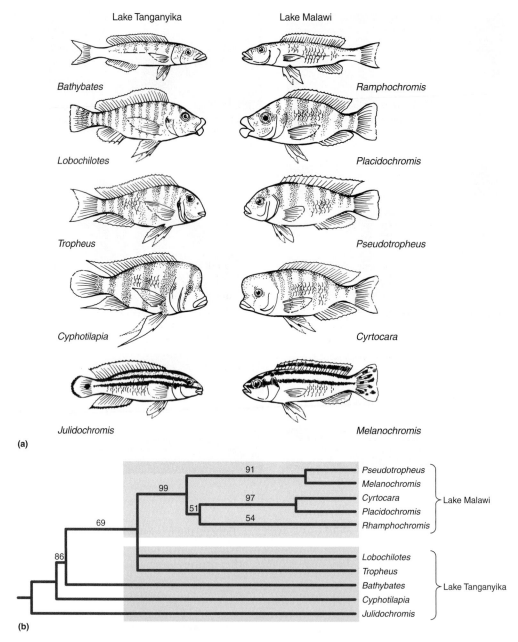

FIGURE 22.8 (a) Phenotypic comparisons among some of the cichlid species from 12 genera found in Lake Tanganyika and Lake Malawi. (b) Phylogenetic tree showing the separate origins of these species. Numbers represent percent bootstrap (confidence) values that are over 50 percent. These species are only a small sample of the thousand or so different cichlid species found in various East African lakes and rivers.

[Adapted from T. D. Kocher et al., 1992. *Mol. Phylog. Evol.,* **2**: 158–165.]

BOX 22.2 Pharyngeal Jaws and Speciation of Cichlid Fish

Evolutionary plasticity has enabled East African lake cichlids to speciate rapidly and to exploit many microhabitats and diets within the lakes. Lakes Victoria, Malawi, and Tanganyika contain 300, 200, and 125 endemic species of cichlids, respectively, all of which have arisen rapidly by sympatric speciation (Goldschmidt, 1996; Magalhaes et al., 2009). Morphological specialization in cichlids is especially evident in the jaws, reflecting ecological specialization in adaptation to a diversity of diets—feeding on plankton, detritus, insects, insect larvae, other fish, and fish embryos; grazing on algae; crushing snails, and more.

An especially fine example of the capacity to exploit a wide variety of environments is the independent evolution in several fish lineages of a **second set of jaws**, known as **pharyngeal jaws**. Especially diversified in cichlids, particularly in those that crush snails, pharyngeal jaws (**FIGURE B2.1**) arise from a pharyngeal arch that lies behind the mandibular (the jaws proper). Both mandibular and pharyngeal jaws support teeth. Pharyngeal jaws and pharyngeal teeth, which are used to process food, enable the "true" jaws to be adapted for food capture. As **FIGURE B2.2** shows, pharyngeal jaws consist of a massive crushing plate, with morphological specializations that relate to whether the food eaten is soft or hard. Such a sub-specialization, when combined with the other attributes outlined in the text, facilitated cichlid diversification and speciation.

Sadly, the decline in biodiversity in Lake Victoria over the past four decades has been the most rapid and devastating of any lake on Earth. The deadly combination of over-fishing, the introduction of exotic species—especially the Nile tilapia, *Oreochromis niloticus*, and the Nile perch, *Lates niloticus*—pollution, and poor management of the surrounding land, all contributed to depleting oxygen levels and the consequent "choking" of the lake with algae to the extent that more than half of the indigenous cichlid species are now extinct. Cichlids are not the only species at risk; 30 million people depend on Lake Victoria to make a living, and ultimately to survive. A crater of speciation has become a crater of doom.

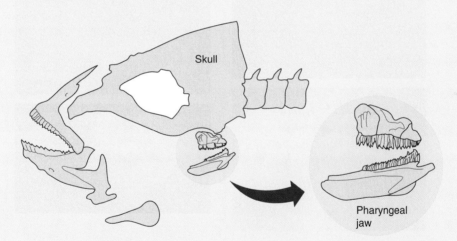

FIGURE B2.1 Head of a cichlid fish showing the location of the upper and lower pharyngeal jaws between the lower jaw and the ventral surface of the skull. The image to the right shows the pharyngeal jaws and their teeth at higher magnification.

BOX 22.2 **Pharyngeal Jaws and Speciation of Cichlid Fish (Cont...)**

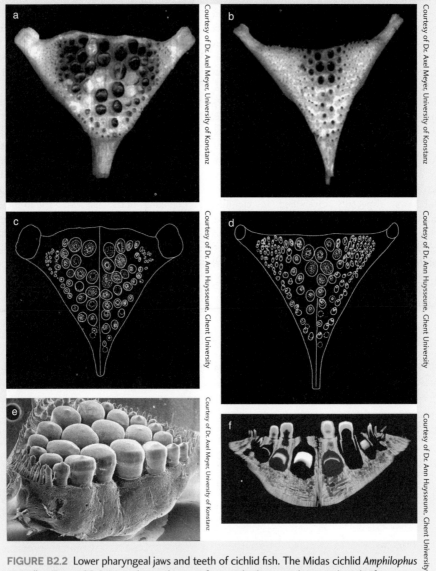

FIGURE B2.2 Lower pharyngeal jaws and teeth of cichlid fish. The Midas cichlid *Amphilophus citrinellus* (Figure 22.9a) is polymorphic, one form with pharyngeal jaws with molariform teeth **(a)**, the other with papilliform teeth **(b)**, reflecting the different diets of the two morphs. The East African cichlid, *Astatoreochromis alluaudi*, also displays two morphs, one molariform **(c)** associated with hard food and a second papilliform **(d)** associated with soft diet, both of which can be modulated by changing the diet (a–d are shown as viewed from inside the mouth). **(e)** Molariform teeth as seen with scanning electron microscopy. **(f)** Molariform teeth as seen in an x-ray through a section of the pharyngeal jaws.

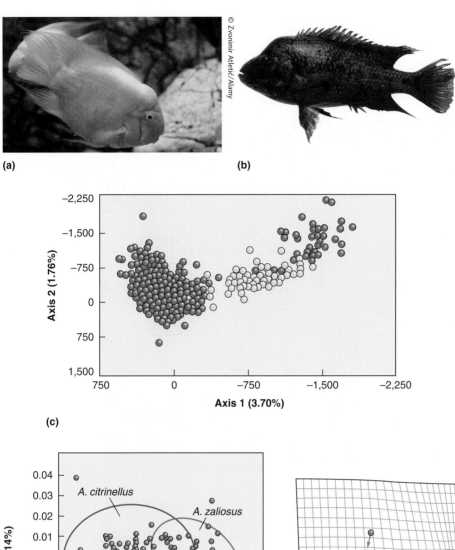

(a)

(b)

(c)

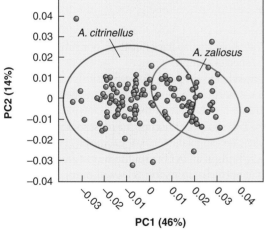

(d)

FIGURE 22.9 The Midas benthic cichlid *Amphilophus citrinellus* **(a)** was the original colonizer of Lake Apoyo in Nicaragua. The limnetic arrow cichlid *A. zaliosus* evolved from *A. citrinellus* **(b)** around 13,000 years ago. **(c)** Haplotypes derived from 637 mtDNA sequence separate *A. zaliosus* (red) from *A. citrinellus* (yellow), species *A. citrinellus* from other lakes (blue). **(d)** Principal component analysis based on nine landmarks shows the different but overlapping body shapes (left) and the directions of change in body shape (right) of the midas and arrow cichlids from Lake Apoyo.

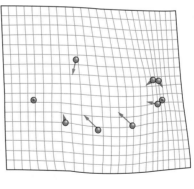

RING SPECIES

A fascinating link between speciation with and without geographical isolation is provided by ring species, which are species in the same genus (congeneric species), or subspecies, whose geographical distribution forms a ring with overlap in one part of the ring (FIGURE 22.10). Ring species have been most fully studied in birds (the Greenish Warbler, *Phylloscopus trochiloides,* in the forests of northern and central Asia) and in salamanders (the redwood salamander, *Ensatina eschscholtzii,* in the mountains of the Central Valley of

FIGURE 22.10 The ring species complex of seven subspecies of the salamander species *Ensatina eschscholtzii* in the mountains of the Central Valley of California.

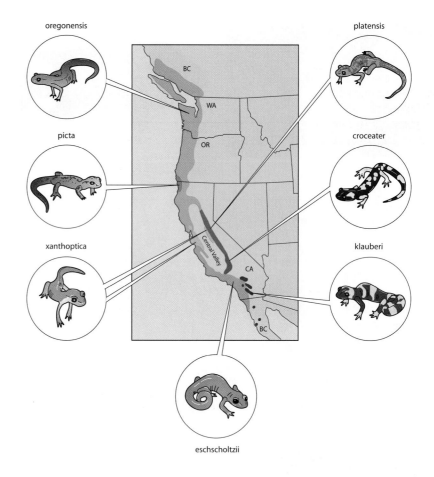

California).[17] We use *Ensatina* to illustrate how ring species illustrate sympatric and allopatric speciation in different parts of the species complex's range.

Divergence in *E. eschscholtzii* has proceeded to the point that seven subspecies are identified (Figure 22.10). *E. e. oregonensis* in the northern part of the species range has zones of overlap with three other subspecies and interbreeds with all three. Phenotypic and genetic divergence increases from north to south. Two subspecies occupy the most southerly portion of the range, a coastal form, *E. e. eschscholtzii,* and an inland form, *E. e. klauberi* (Figure 22.10). These two subspecies interbreed to form hybrids in all but the southernmost area of contact, but they hybridize less frequently than the northernmost subspecies and in much narrower zones of hybridization. This part of the ring species complex's range tells us that hybrid formation and zones of hybridization can play a role in speciation.

What about the most southerly area of contact between these two subspecies? Hybridization has never been reported between E. e. *eschscholtzii* and *E. e. klauberi* in this region of contact; in other words, sympatric speciation is complete. Progressive divergence from north to south has left us with a record of the speciation process as intermediate forms of the two terminal subspecies in the complex—in effect, a map of how speciation without complete geographical isolation can occur. As Pereira and Wake (2009) concluded: "Diversification and consequent genetic interactions in *Ensatina* reveal a continuum between populations, ecological races, and species, where polytypic traits and high genetic differentiation are maintained without reproductive isolation" (p. 2288).

---

[17] See Irwin et al. (2005) for the Greenish Warbler, and Wake (1997) and Pereira and Wake (2009) for *Ensatina*.

■ KEY TERMS

adaptive radiation	reproductive isolation
allopatric speciation	ring species
asexual reproduction	sexual reproduction
cyclomorphosis	sibling species
evolutionary plasticity	species
geographical isolation	species diversification
morph	subspecies
pharyngeal jaws	sympatric speciation
phenotypic plasticity	variation
plasticity	varieties
reaction norm	

■ EVOLUTION ON THE WEB

Explore evolution on the Internet! Visit the accompanying website for *Strickberger's Evolution, Fifth Edition,* at **go.jblearning.com/Evolution5eCW** for exercises and links relating to topics covered in this chapter.

■ REFERENCES

Abouheif, E., 2004. A framework for studying the evolution of gene networks underlying polyphenism: insights from winged and wingless ant castes. In *Environment, Development, and Evolution.* B. K. Hall, R. D. Pearson, and G. B. Müller (eds.). MIT Press, Cambridge, MA, pp. 125–137.

Barluenga, M., K. N. Stölting, W. Salzburger, et al., 2006. Sympatric speciation in Nicaraguan crater lake cichlid fish. *Nature*, **439**, 719–723.

Beldada, P., A. R. A. Mateus, and R. A. Keller. 2011. Evolution and molecular mechanisms of adaptive developmental plasticity. *Mol. Ecol.*, **20**, 1347–1363.

Birky, C. W. Jr., and T. G. Barraclough, 2009. *Lost Sex: The Evolutionary Biology of Parthenogenesis.* Springer, New York.

Birky, C. W. Jr., C. Wolf, H. Maughan, L. Herbertson, and E. Henry, 2005. Speciation and selection without sex. *Hydrobiologia*, **546**, 29–45.

Blondel, J., K. N. Stölting, W. Salzburger, et al., 2006. A thirty-year study of phenotypic and genetic variation of blue tits in Mediterranean habitat mosaics. *BioScience*, **56**, 661–673.

Bolnick, D. I., and B. M. Fitzpatrick, 2007. Sympatric speciation: models and empirical evidence. *Annu. Rev. Ecol. Evol. Syst.*, **38**, 459–487.

Bush, G. L., 1975. Modes of animal speciation. *Annu. Rev. Ecol. Syst.*, **6**, 339–364.

Bush, G. L., 1994. Sympatric speciation in animals: new wine in old bottles. *Trends Ecol. Evol.*, **9**, 285–288.

Chevin, L.-M., and R. Lande. 2011. Adaptation to marginal habitats by evolution of increased phenotypic plasticity. *J. Evol. Biol.*, **24**, 1462–1476.

Colbourne, J. K., M. E. Pfrender, D. Gilbert, et al., 2011. The ecoresponsive genome of *Daphnia pulex. Science*, **331**, 555–561.

Coyne, J. A., and H. A. Orr, 1997. "Patterns of speciation in *Drosophila*" revisited. *Evolution*, **51**, 295–303.

Coyne, J. A., and H. A. Orr, 2004. *Speciation.* Sinauer and Associates, Sunderland, MA.

Debat, V, and P. David., 2001. Mapping phenotypes: canalization, plasticity and developmental stability. *TREE*, **16**, 555–561.

Feder, J. L., 1998. The apple maggot fly, *Rhagoletis pomonella*: flies in the face of conventional wisdom about speciation? In Howard, D. J. and S. H. Berlocher (eds.), *Endless Forms: Species and Speciation.* Oxford University Press, Oxford, UK.

Fierst, J. L. 2011. A history of phenotypic plasticity accelerates adaptation to a new environment. *J. Evol. Biol.*, **24**, 1992–2001.

Flatt, T., M.-P. Tu, and M. Tatar, 2005. Hormonal pleiotropy and the juvenile hormone regulation of *Drosophila* development and life history. *BioEssays*, **27**, 999–1010.

Forbes, A. A., L. L. Stelinski, T. H. Q. Powell, et al., 2009. Sequential sympatric speciation across trophic levels. *Science*, **323**, 776–779.

Fraser, C., W. P. Hanage, and B. G. Spratt, 2007. Recombination and the nature of bacterial speciation. *Science*, **315**, 476–480.

Fusco, G., and A. Minelli, 2010. Phenotypic plasticity in development and evolution: facts and concepts. *Phil. Trans. R. Soc. Lond. (B)*, **365**, 547–556.

Gilbert, J. J. 1966. Rotifer ecology and embryological induction. *Science*, **151**, 1234–1237.

Gilbert, J. J., 2003. Environmental and endogenous control of sexuality in a rotifer life cycle: developmental and population biology. *Evol. Dev.*, **5**, 19–24.

Gilbert, S. F., and D. Epel, 2008. *Ecological Developmental Biology: Integrating Epigenetics, Medicine, and Evolution.* Sinauer and Associates, Sunderland, MA.

Gladyshev, E. A., M. Meselson, and I. R. Arkhipova, 2008. Massive horizontal gene transfer in bdelloid rotifers. *Science*, **320**, 1210–1213.

Goldschmidt, T., 1996. *Darwin's Dreampond. Drama in Lake Victoria.* MIT Press, Cambridge, MA.

Grant, P. R., and B. R. Grant, 2008. *How and Why Species Multiply: The Radiation of Darwin's Finches.* Princeton University Press, Princeton, NJ.

Greene, E., 1989. A diet-induced developmental polymorphism in a caterpillar. *Science*, **243**, 643–646.

Greene, E., 1999. Phenotypic variation in larval development and evolution: polymorphism, polyphenism, and developmental reaction norms. In *The Origin and Evolution of Larval Form*, B. K. Hall and M. H. Wake (eds.). Academic Press, San Diego, pp. 379–410.

Hall, B. K., 1999. *Evolutionary Developmental Biology*, 2nd ed. Kluwer Academic Publishers, Netherlands.

Hall, B. K., 2004. Evolution as the control of development by ecology. In *Environment, Development, and Evolution; Toward a Synthesis*, B. K. Hall, R. D. Pearson, and G. B. Müller (eds.). MIT Press, Cambridge, MA, pp. ix–xxiii.

Hall, B. K., and W. M. Olson (eds.), 2003. *Keywords and Concepts in Evolutionary Developmental Biology.* Harvard University Press, Cambridge, MA.

Hall, B. K., R. D. Pearson, and G. B. Müller (eds.), 2004. *Environment, Development, and Evolution; Toward a Synthesis.* MIT Press, Cambridge, MA.

Hall, B. K., and M. H. Wake (eds.) 1999. *The Origin and Evolution of Larval Forms.* San Diego, CA: Academic Press.

Hallgrímsson, B., and Hall, B. K. (eds.), 2011. *Epigenetics: Linking Genotype and Phenotype in Development and Evolution.* University of California Press, Berkeley, CA.

Hallsson, L. R., and M. Björklund, 2012. Selection in a fluctuating environment leads to decreased genetic variation and facilitates the evolution of phenotypic plasticity. *J. Evol. Biol.*, **25**, 1275–1290.

Hendry, A. P., 2009. Speciation. *Nature*, **458**, 162–164.

Hines, H. M., B. A. Counterman, R. Papa, et al., 2011. Wing patterning gene redefines the mimetic history of Heliconius butterflies. *Proc. Natl. Acad. Sci. USA*, **108**, 19666–19671.

Huysseune, A., J.-Y. Sire, and F. J. Meunier, 1994. Comparative study of lower pharyngeal jaw structure in two phenotypes of *Astatoreochromis alluaudi* (Teleostei: Cichlidae). *J. Morphol.*, **221**, 25–43.

Irwin, D. E., S. Bensch, J. H. Irwin, and T. D. Price, 2005. Speciation by distance in a ring species. *Science*, **307**, 414–416.

Kocher, T. D., J. A. Conroy, K. R. McKaye, and J. R. Stauffer, 1993. Similar morphologies of cichlid fish in Lake Tanganyika and Lake Malawi are due to convergence. *Mol. Phylog. Evol.*, **2**, 158–165.

Ledón-Rettig, C. C., and D. W. Pfennig, 2011. Emerging model systems in eco-evo-devo: the environmentally responsive spadefoot toad. *Evol. Dev.*, **13**, 391–400.

Liem, K. F., 1974. Evolutionary strategies and morphological innovations: cichlid pharyngeal jaws. *Syst. Zool.*, **22**, 425–441.

Losos, J. B., and R. E. Ricklefs, 2009. Adaptation and diversification on islands. *Nature*, **457**, 830–836.

Magalhaes, I. S., S. Mwaiko, M. V. Schneider, and O. Seehausen, 2009. Divergent selection and phenotypic plasticity during incipient speciation in Lake Victoria cichlid fish. *J. Evol. Biol.*, **22**, 260–274

Michel, A. P., S. Sim, T. H. Q. Powell, et al., 2010. Widespread genomic divergence during sympatric speciation. *Proc. Nat. Acad. Sci. USA*, **107**, 9724–9729.

Otte, D., and J. A. Endler (eds.), 1989. *Speciation and Its Consequences*. Sinauer Associates, Sunderland, MA.

Pereira, R. J., and D. B. Wake, 2009. Genetic leakage after adaptive and nonadaptive divergence in the *Ensatina eschscholtzii* ring species. *Evolution*, **68**, 2288–2301.

Pigliucci, M., 2001. *Phenotypic Plasticity: Beyond Nature and Nurture.* The Johns Hopkins University Press, Baltimore and London.

Ready, J. S., I. Sampaio, H. Schneider, et al., 2006. Color forms of Amazonian cichlid fish represent reproductively isolated species. *J. Evol. Biol.*, **19**, 1139–1148.

Rubolini, D., A. P. Møller, K. Raino, and E. Lehikoinen, 2007. Intraspecific consistency and geographical variability in temporal trends of spring migration phenology among European bird species. *Climate Res.*, **35**, 135–146.

Scheiner, S. M., 2002. Selection experiments and the study of phenotypic plasticity. *J. Evol. Biol.*, **15**, 889–898.

Schlichting, C. D., and M. Pigliucci, 1998. *Phenotypic Evolution: A Reaction Norm Perspective.* Sinauer Associates, Sunderland, MA.

Schluter, D., 2000. *The Ecology of Adaptive Radiation.* Oxford University Press, Oxford, UK.

Seger, J., 1985. Intraspecific resource competition as a cause of sympatric speciation. In *Evolution: Essays in Honour of John Maynard Smith,* P. J. Greenwood, P. H. Harvey, and M. Slatkin (eds.). Cambridge University Press, Cambridge, UK, pp. 43–53.

Stearns, S. C., 1989. The evolutionary significance of phenotypic plasticity. *BioScience*, **37**, 436–445.

Storz, B. L., J. Heinrichs, A. Yazdani, et al., 2011. Reassessment of the environmental model of developmental polyphenism in spadefoot toad tadpoles. *Oecologia*, **165**, 55–65.

Sultan, S. E., and S. C. Stearns, 2005. Environmentally contingent variation: phenotypic plasticity and norms of reaction. In *Variation. A Central Concept in Biology*, B. Hallgrímsson and B. K. Hall (eds.). Elsevier/Academic Press, New York, pp. 303–332.

Vogt, G., M. Huber, M. Thiemann, et al., 2008. Production of different phenotypes from the same genotype in the same environment by developmental variation. *J. Exp. Biol.*, **211**, 510–523.

Wake, D. B., 19797. Incipient species formation in salamanders of the *Ensatina* complex. *Proc. Natl Acad. Sci. USA*, **94**, 7761–7767.

West-Eberhard, M. J., 2003. *Developmental Plasticity and Evolution*. Oxford University Press, Oxford, UK.

Whitman, D., and T. N. Ananthakrishnan (eds.), 2009. *Phenotypic Plasticity of Insects: Mechanisms and Consequences.* Science Publishers, Enfield, NH.

Wilkins, A. S., 2002. *The Evolution of Developmental Pathways.* Sinauer and Associates, Sunderland, MA.

Yoder, J. B., E. Clancey, S. des Roches, et al., 2010. Ecological opportunity and the origin of adaptive radiations. *J. Evol. Biol.*, **23**, 1581–1598.

Image © Johan Swanepoel/ ShutterStock, Inc.

CHAPTER

23

Allopatric Speciation and Hybridization

CHAPTER SUMMARY

Although many species persist for long periods of time, changes occur over that time in both genotypes and phenotypes. Speciation is grounded in the biological species concept and requires more than phenotypic change; speciation requires reproductive isolation, which can occur within a population—sympatric speciation—or can follow geographical isolation—allopatric speciation. Preconditions for allopatric speciation can be established when a population subdivides or when a few individuals found a new population in a new location (the latter known as the founder effect). Natural and sexual selection will lead to adaptation to the new environment as the isolated population differentiates from the original population. If there exists a barrier to interbreeding with the original population, the isolated population can become reproductively isolated as a new species. Behavioral, seasonal, or habitat differences can facilitate reproductive isolation, as can hybridization. Although the role of hybrid formation in speciation has been controversial, the work of Rosemary and Peter Grant on Darwin's finches has produced compelling evidence for the role of hybridization.

INTRODUCTION

In our discussions of speciation, the BIOLOGICAL SPECIES CONCEPT is employed; species are defined on the basis of their reproductive isolation from other species.

Given that new species evolve from existing species, the question of the origin of species is the question of how species change.[1] Allopatric speciation requires **geographical isolation** as the initiating process, although isolation is only the first of many steps required for species to diverge allopatrically. Geographical isolation, local adaptation in response to natural selection (most commonly divergent selection), and reproductive isolation preventing interbreeding with the parent population or with other species, all are essential elements of allopatric speciation. Populations accumulate genetic differences when they are sufficiently separated spatially or temporally to prevent gene exchange that could (and probably would) eradicate those differences. Allopatric speciation is usually followed by sympatric speciation within the isolated populations, as we see in the example of ring species.

Colonizers capable of crossing geographical barriers and moving into new environments can be the ancestors of entirely new groups, as demonstrated most graphically in the wide evolutionary radiation of species that descended from finches reaching the Galápagos Islands. Beginning with an ordinary mainland finch, new species of finches evolved on the 13 major and six smaller islands (Grant, 1986, and Huber et al., 2007) of the Galápagos. Such adaptive radiation signifies the rapid evolution of one or a few forms into many different species occupying a variety of habitats and creating new niches within a new geographical area. Radiation of marsupial mammals in Australia shows how protection from competition by the isolation of a continent—in this case, the absence of placental mammals from Australia—can lead to an array of species with widely divergent features and functions, ranging from herbivores to carnivores, and paralleling placental mammals on other continents.

SPECIATION INITIATED BY GEOGRAPHICAL ISOLATION: ALLOPATRIC SPECIATION

The initiating event in allopatric speciation is almost always the geographical splitting of a population or the movement of one or a few individuals to a new location that may be a great distance away, as with the ancestors of Darwin's finches (above) or of the Hawaiian fruit flies (below).

The **founder effect**, described by Ernst Mayr in 1952[2], is a concept of geographical isolation developed in the nineteenth century by Reverend John Gulick (1832–1923) to explain speciation of land snails on Hawaii (BOX 23.1). If the environment occupied by isolated populations differs from that experienced by the original population, the isolates are likely to adapt through natural selection to the new environmental conditions. Over time, the isolate[3] may differentiate phenotypically from the original species. Concurrently or subsequently, **reproductive-isolating mechanisms** may arise and provide a barrier to gene exchange with the original population and/or with other populations in the new environment. Isolating mechanisms that prevent interbreeding may be seasonal, behavioral, or based on habitat preference (below). Changes in reproductive biology may prevent interbreeding, fertilization, or embryonic development, all of which are effective reproductive-isolating mechanisms (BOX 23.2).

---

[1] For recent discussions of speciation, see Coyne and Orr (2004), Grant and Grant (2008), and Sobel et al. (2009).

[2] See the papers in Hey et al. (2005) for analyses of Mayr's role in our understanding of speciation; see Barton and Charlesworth (1984) for the founder effect and speciation; and see Sobel et al. (2009) for an analysis of the neglected role of geographical isolation in speciation.

[3] The term *isolate* is used as shorthand for a subpopulation that has separated from the main population. The isolate may range from a few individuals to a large fraction of the original population. In general, the larger the fraction, the greater the amount of variation in the isolate.

BOX 23.1   The Father of Geographical Isolation

Although almost unknown today, during the late nineteenth and early twentieth centuries Reverend John Gulick (1832–1923) was one of the world's most well-known and influential evolutionary biologists (FIGURE B1.1a; A. Gulick, 1932). An extensive correspondence with George Romanes (1848–1894), and a 20-year-long confrontation over the adaptive nature of evolution, which pitted Romanes and Gulick against Alfred Wallace, kept Gulick in the spotlight.[a]

In the early 1850s, while still in his early twenties, Gulick collected over 200 species of land snails of the genus *Achatinella* from Oahu, Hawaii (FIGURE B1.1b), where his father was a missionary. Gulick, too, spent his life as a missionary. His analysis of the restricted geographical distributions of each species—often to a single valley or region within a valley—and publication of the results beginning in 1872, provided what many considered the missing mechanism in Darwin's theory of evolution, namely *geographical isolation* (Hall, 2006a,b). One or more individuals become isolated from the rest of the population, and their descendants subsequently diverge from that population to the extent that they become separate species. As these snails are hermaphroditic, a single individual could potentially establish a new population (Gulick, 1872, 1873, 1883).

In reviewing a paper on geographical isolation, read by Gulick to the Linnean Society on December 15, 1887, Romanes was enthusiastic in his acceptance of the paper for publication in the society's journal: "It [Gulick, 1888] is not only without comparison the best essay that has been written on the subject of isolation but in my opinion one of the most important to the whole philosophy of evolution."[b] Almost a century later (1952) the evolutionary biologist Ernst Mayr (1904–2005) called this process the founder effect, acknowledging Gulick's origination of the concept and recognizing Gulick as, "the first author to develop a theory of evolution based on random variation" (Mayr, 1988, p. 139).

(a)

(b)

FIGURE B1.1 **(a)** Rev. John Gulick. **(b)** Tree snails (*Achatinella mustelina*) on a leaf of a *Nestegis sandwicensis* (a native Hawaiian plant in the olive family) in Makaha Valley, island of Oahu.

[(a) Frontispiece of the book "Darwin and After Darwin" volume 3, 1897, The Open Court Publishing Company; (b) Photo courtesy of Amy Tsuneyoshi.]

[a]Canadian born, Romanes spent his career at University College London. An influential evolutionary biologist and physiologist, his research into cognition laid the basis for comparative psychology. Romanes coined the term "neo-Darwinism" in 1895 for the theory of the segregation of germ plasm proposed by August Weismann.
[b]Letter of January 1888 from Romanes to Gulick, cited from A. Gulick, 1932, p. 405.

REPRODUCTIVE INCOMPATIBILITY

The inability of species to interbreed may be due to mechanical incompatibility between the sexual organs or because species breed at different seasons, in different habitats, or require different reproductive behaviors to be able to mate. Such ways of preventing interbreeding are known as **prezygotic mechanisms** of reproductive incompatibility. In other species, mating occurs but **postzygotic mechanisms** operate to prevent what would otherwise be the formation of a hybrid offspring. Common postzygotic mechanisms (in rough order of decreasing effectiveness) are failure of fertilization; death of gametes, zygotes, or early embryos; formation of a hybrid with low viability and/or production of a hybrid with low sterility.

Pollen grains in plants, for example, may be unable to grow pollen tubes in the styles of other species, with the consequence that development cannot be initiated. Crosses among 12 frog species of the genus *Rana* revealed a wide range of inviability. In some crosses, no egg cleavage occurs. In others, the cleavage and blastula stages are normal but gastrulation fails. In still others, early development is normal but later stages do not develop.[c]

In some *Drosophila* crosses, an insemination reaction in the vagina of the female produces swelling and prevents successful fertilization of the egg. As discussed in the text, *Drosophila pseudoobscura* and *D. persimilis* are virtually completely isolated reproductively. Even when interbreeding occurs, gene exchange is impeded by reproductive incompatibility; the F1 hybrid male is sterile and the progenies of fertile F1 females backcrossed to males of either species show markedly lower viabilities than the parental stocks (a process known as hybrid breakdown).

---

[c] J. A. Moore, 1949. Patterns of evolution in the genus *Rana.* In *Genetics, Paleontology, and Evolution,* G. L. Jepsen, E. Mayr, and G. G. Simpson (eds.). Princeton University Press, Princeton, NJ, pp. 315–355. For recent confirmation of the validity of *Rana* species, see M. Sumida et al., 2003. Reproductive isolating mechanisms and molecular phylogenetic relationships among Palearctic and Oriental brown frogs. *Zool. Sci.,* **20,** 567–580.

Although geographical isolation is the first step in allopatric speciation, speciation is by no means an automatic consequence of geographical isolation of a subset(s) of the species. Conditions encountered by the isolated population may be identical to those associated with the original population, in which case the isolate would remain as a population of the original species. Even if the environments differ, populations can adapt to changing conditions without becoming reproductively isolated from other populations of the same species. For speciation to be initiated, a geographically isolated population has to adapt phenotypically to the new environment and accumulate genetic change. Without reproductive isolation, however, the isolated populations will remain members of the same species, albeit phenotypically distinct. Only with reproductive isolation can speciation be said to be complete.

FORMS OF GEOGRAPHICAL ISOLATION

Potential for reduced gene exchange and reproductive isolation occurs when populations subdivide to occupy different areas or habitats. Once isolated geographically, a population can become further isolated by processes involving seasonal, behavioral, or habitat changes that may lead to restricted gene flow and reproductive isolation.

Subdivision may: (a) split a population in two, as when a new river cuts through a region, (b) isolate a small portion of the original population, or (c) be initiated if a few individuals leave or become isolated from the original population, the latter being how many new species have arisen on islands. We recognize three modes of speciation that differ with respect to the degree of geographical isolation and therefore the potential for continued gene exchange (**FIGURE 23.1**):

- **allopatric speciation**, when a *population is divided* by a natural physical barrier or because intervening geographical populations become extinct;
- **peripatric speciation**, when a *population is divided* by the budding off of a small, completely isolated founder colony from a larger population; and

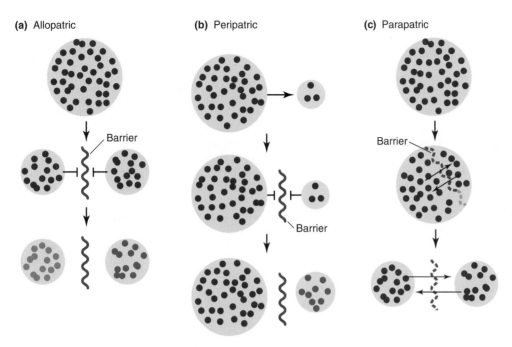

FIGURE 23.1 Comparison between three patterns of speciation: **(a)** Allopatric, where a barrier divides a population into two, preventing gene flow (—I); **(b)** Peripatric, where a small population is isolated and forms a founder population, again with no gene flow (—I) between the populations; **(c)** Parapatric, where a barrier isolates a group but where gene flow between the two groups still occurs (→ ←).

- **parapatric speciation**, when a population at the periphery of a species adapts to different environments but *remains contiguous* with its parent so that gene flow is possible between them.[4]

The isolated population could be small. A single female carrying fertilized eggs/ovules or a single pair of individuals in a species with separate sexes could establish a new population that could form the basis of a new species. Theoretically, one individual is sufficient to found a new population, if that individual is hermaphrodite (such as land snails; Box 23.1). A classic example of the founder effect, speciation of fruit flies in the genus *Drosophila* on the Hawaiian Islands, is discussed in BOX 23.3. Radiation and speciation of Darwin's finches on the Galápagos Islands, also initiated by the founder effect, is discussed below.

MECHANISMS FACILITATING REPRODUCTIVE ISOLATION AND SPECIATION

Reproductive isolation can be facilitated in several ways, of which seasonal isolation and/or changes in behavioral characters are the most common. Developmental mechanisms resulting in reproductive incompatibility in sexually reproducing organisms are outlined in Box 23.2.

In **seasonal isolation**, potential mates do not meet because they mature and/or reproduce in different seasons. For example, some plant species, such as the spiderworts

---

[4]Parapatric speciation resembles the process of sympatric speciation, which can be initiated if gene flow is interrupted within a population *without* geographical isolation. The species or subspecies pairs in a ring species complex show both parapatric and sympatric speciation.

BOX 23.3 Speciation of Fruit Flies in Hawaii

Tectonic events formed, and continue to form, the Hawaiian Islands. A localized "hot spot" in Earth's mantle, under what is now Hawaii, pierced the lithosphere and produced volcanic eruptions that formed a succession of islands as the Pacific plate moved northwestward (**FIGURE B3.1**). The movement of the plate is evident in the relative ages of the islands: Kauai is over five million years old, Hawaii (the youngest island) less than half a million years old. Hawaii, which is larger than all the other Hawaiian Islands combined, is currently being built by five volcanoes, generated by the "hot spot" (Figure B3.1). The Kilauea volcano erupted for 36 days early in 1960. An eruption in 1955 buried 1,600 hectares of land under lava. In time various older islands eroded, first becoming atolls, then seamounts (submerged volcanoes) to form the existing series of islands that extend from Hawaii to Midway to a point near the far-western Aleutians.

The Hawaiian Islands are home to more than 800 native species of *Drosophila*, the most ancient of which arose over 30 Mya (Edwards et al., 2007). Among these are 100 species in the "picture-winged" species group, so named because of

their large decorated wings (**FIGURE B3.2**). Careful analysis of salivary chromosome banding patterns has shown that these 100 species were derived from founder events in which each Hawaiian island was settled by relatively few individuals whose descendants evolved into different species (Figure B3.1). Each successful founder is presumed to have been a fertilized female. An analysis of the five major Hawaiian *Drosophila* species groups, using data from eight gene regions, generated a most-parsimonious phylogenetic tree showing that the *adiastola* picture-winged subgroup (Figure B3.2**a**) are sister taxon to the other picture-winged subgroups (Figure B3.2**b,c**).*

The oldest Hawaiian island, Kauai, is 5.6 My old. Ten of the 12 species of picture-winged *Drosophila* on Kauai are the most ancient species on any Hawaiian island. Looking at other islands in order of descending age, 41 species endemic to the Maui island complex (Figure B3.1) derive from only 12 founders, ten from Oahu and two from Kauai. The youngest island, Hawaii, was colonized entirely by founders from the older islands, 26 species evolving in what may have been less than 500,000 years (Carson, 1986, 1992).

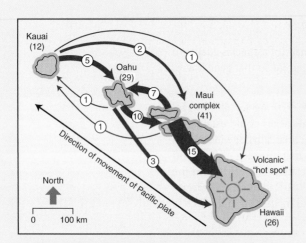

FIGURE B3.1 Colonization pattern showing the founder events associated with speciation of the "picture-winged" group of *Drosophila* in the Hawaiian Islands. The width of each arrow is proportional to the number of founders (circled). The number of *Drosophila* species present on each island in 1992 is given in parentheses.

[Adapted from Carson, H. L., 1992. Inversions in Hawaiian *Drosophila*. In *Drosophila Inversion Polymorpism*, C. B. Krimbas and J. R. Powell [eds.] CRC Press, Boca Raton, FL, pp. 407–439.]

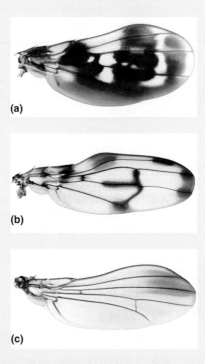

FIGURE B3.2 Wing patterns of representative picture-winged *Drosophila* species groups from Hawaii. Wings of species from the *adiastola* (**a**), *planitibia* (**b**), and *antopocerus* subgroups (**c**).

[Photos modified from Edwards K.A, Doescher L.T., Kaneshiro K.Y., Yamamoto D., (2007) A Database of Wing Diversity in the Hawaiian Drosophila. PLoS ONE 2(5): e487. doi:10.1371/journal. pone.0000487. Courtesy of Kevin Edwards, Illinois State University.]

*R. H. Baker and R. DeSalle, 1997. Multiple sources of character information and the phylogeny of Hawaiian drosophilids. *Syst. Biol.* **46**, 654–673.

FIGURE 23.2 Species-specific courtship displays of **(a)** penguins, **(b)** albatross, and **(c)** grebes.

Tradescantia canaliculata and *T. subaspera,* have the same geographical distribution and yet remain isolated because their flowers bloom in different seasons. Further, one species grows in sunlight and the other in deep shade. The two species have created different niches, although this current situation does not tell us whether the species arose allopatrically or sympatrically.[5]

With **behavioral isolation**, the sexes of two species of animals may occur together in the same locality, but have different patterns of courtship that prevent mating. Distinctive songs of many birds, the special mating calls of some species of frogs and, indeed, the sexual displays of most animals (**FIGURE 23.2**) are attractive only to potential mates of the same species. Many plants have floral displays that discriminate between insect and bird pollinators, or that attract only certain insect pollinators (Coyne and Orr, 2004).

Behavioral isolation can arise allopatrically or sympatrically. Several species clusters arose by sympatric speciation involving behavioral isolation. Among the best examples are sibling species of *Drosophila,* and the evolution of Darwin's finches discussed below.

DARWIN'S FINCHES

Studies by Rosemary and Peter Grant on two of Darwin's finches—*Geospiza fortis,* the medium ground finch, and *Geospiza scandens,* the cactus finch—on Daphne Major Island in the Galápagos, demonstrate that *species-specific song,* a learned trait that is culturally transmitted, can function as a reproductive barrier between species. Only the males sing.

Song in Darwin's finches is learned by both sexes from the male parent during a short, sensitive period between 10 and 40 days after hatching. This mode of learning has a high fidelity in birds; only rarely is a song misimprinted. This species-specific

---

[5] E. Anderson and K. Sax, 1936. A cytological monograph of the American species of *Tradescantia. Bot. Gaz.* **97**, 433–476.

FIGURE 23.3 **(a)** Teosinte (*Zea mays parviglumis*), the wild ancestor of cultivated maize, showing the mature plant and a kernel-bearing ear. **(b)** A mature plant and ear of its descendant, corn (*Zea mays mays*). Although strikingly different in plant and ear architecture, these two forms differ in relatively few genes.

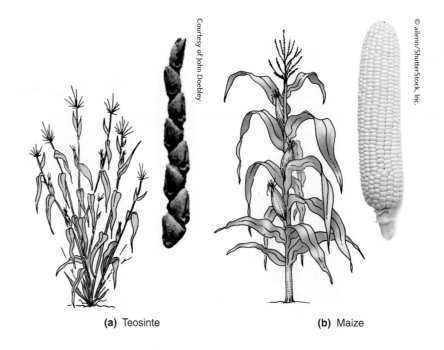

(a) Teosinte   (b) Maize

learning does not provide complete reproductive isolation, however. Misimprinting and interbreeding can lead to hybridization—see below under ***Speciation and Hybridization Zones***—although hybridization is rare, occurring in less than one percent of breeding pairs (Grant and Grant, 1994, 2002).

In a recent report of their ongoing studies, the Grants demonstrated that the song of the medium ground and cactus finches diverged from the song of the more aggressive large ground finch, *G. magnirostris*, after the latter colonized Daphne Major Island in 1983 (Grant and Grant, 2010). Large ground finches, sing in the same frequency range (2–4 kHz) as do the two smaller and less aggressive species. Divergence was traced to imprinting (that is, behavioral/cultural evolution/transmission) of songs in male offspring in both the medium ground and cactus finches. This divergence has produced a new lineage of finch on Daphne Major Island, the *G. fortiscandens* lineage; an incipient species that arose through reproductive isolation (hybridization) resulting from the changed song.[6]

DO SPECIATION GENES EXIST?

Speciation events can occur in various ways and at various rates. In some groups, such as the picture-wing Hawaiian Drosophilidae, speciation has been dramatically rapid (Box 23.3) and may well involve fewer genes with greater phenotypic effects than in the slower speciation events in many other species. Relatively few mutations seem to account for the rapid transition from teosinte to maize (FIGURE 23.3) and from bee-pollinated to hummingbird-pollinated species of monkey flower, *Mimulus spp* (FIGURE 23.4; Bradshaw et al., 2005). Such examples have prompted the question "Are changes in specific classes of genes associated with speciation?" Such genes could either promote speciation or limit hybridization and so maintain species.

---

[6] Song *can* diverge as a result of changes in beak length; beak size influences both the rate and frequency range of song production. Although not shown for Darwin's finches, divergence in song follows changes in beak size in the house finch (*Carpodacus mexicanus*). Habitat also can influence bird song; lower frequency songs transmit better through dense vegetation than through more open habitats (Grant and Grant, 2010).

(a)

(b)

(c)

(d)

FIGURE 23.4 Adaptation in flower structure **(a, b)** associated with the transition from bee-pollination **(c)** to hummingbird-pollination **(d)** of monkey flowers, *Mimulus spp.*

[Photos (a) through (d) courtesy Toby Bradshaw (University of Washington) and Douglas Schemske (Michigan State University).]

Promoting Speciation. The search for the first type of "speciation genes" focuses on sexual traits. Changes in sexual traits correlate with early speciation events and may be instrumental in speciation; sexual traits often are involved in erecting the barriers that isolate species. Variation between sexual and nonsexual traits for a group of *Drosophila* species shows that sexual traits exhibit greater variation between species and less variation within species, consistent with selection acting differently on sexual traits and at different times (Civetta and Singh, 1998a,b).

Maintaining Species. In *Drosophila*, sexual traits undergo greater selection for differences between populations during speciation, and greater selection for uniformity within species after attaining speciation. Studies pointing to similar genetic roles for sexual traits in various organisms include investigations of pheromonal differences between *Drosophila* species, mating preferences in *Heliconius* passion-vine butterflies, evolution of mating type genes in the unicellular green alga, *Chlamydomonas*, and sperm–egg fertilization interaction in animals. Presgraves (2010) identifies 14 different genes associated with sterility or death ("incompatibility genes") in pairs of species, including yeast, flowering plants, fruit flies, and mice. As these genes prevent hybridization they are candidate genes for maintaining species.

Correa et al. (2012) used a novel transgenomics approach in which over 1,100 20-kb-pair genomic clones from the Alabama glade cress (*Leavenworthia alabamica*) were inserted into the genome of the watercress (*Arabidopsis thaliana*) and the seeds produced screened for genes associated with maintenance of the species. Presence of one genomic clone (11-11B) resulted in the death of more than one-third of the seeds. The remaining seeds were smaller than normal. Their conclusion was that "the causal gene within 11-11B has the potential to contribute to reproductive isolation between incipient species . . . [and] is a candidate for being a *potential* speciation gene that has the potential to undergo mutations that could contribute to hybrid infertility between the incipient species" (pp. 495, 502).

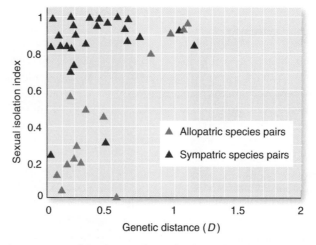

FIGURE 23.5 Measurements of the degree of sexual isolation for pairs of allopatric and sympatric *Drosophila* species plotted against genetic distance. A sexual isolation index of 1 = complete isolation, 0 = no isolation. As the figure shows, when species in a pair are closely related—genetic distance between them is small, for example, 0.5—they are more isolated from each other when they are sympatric than when they are allopatric. Allopatric populations require a much longer genetic distance to achieve reproductive isolation than do sympatric populations; D = 0.54 and 0.04, respectively. As an approximation, species of *Drosophila* take 200,000 years to speciate sympatrically, but 2.7 My to speciate allopatrically.

[Adapted from Coyne, J. A., N. H. Barton, and M. Turelli, 1999. *Evolution,* **51**, 643–671.]

HYBRIDS AND HYBRIDIZATION

Hybrid sterility, long thought to be important in geographical isolation and speciation, has received much attention. The most familiar hybrid may be the mule, which is the progeny of a male donkey (*Equus asinus*) and a female horse (*Equus caballus*). The opposite cross produces a sterile hinny. Sterility comes from chromosomal incompatibility. Horses have 64 pairs of chromosomes, donkeys 62, and mules 63.

In his treatise on Darwinism published in 1889, Alfred Wallace proposed that natural selection might favor the establishment of mating barriers among populations if the hybrids were adaptively inferior (**hybrid inferiority**). According to this hypothesis, subsequently supported with experimental evidence, selection for sexual isolation arises because most groups and species are strongly adapted to specific environments. Maintenance of speciation enables populations to preserve their adaptive advantages from gene flow from non-adapted groups. Hybrids between two highly adapted populations represent a genetic dilution of their parental gene complexes that can be of great disadvantage in the original environments. In such cases, hybrid inferiority would reinforce species separation. The likely sequence of evolutionary changes would be:

- evolution of genetic differentiation between allopatric populations of a single species;
- overlap of these differentiated populations in a sympatric area;
- intensification of sexual isolating mechanisms through selection; and
- selection against the hybrid (**FIGURE 23.5**, left column).

However, and as discussed later under ***Speciation and Hybridization Zones***, hybrids can represent new gene combinations that may be better adapted to changed conditions than those of either parent species, in which case hybridization would reinforce speciation.[7] How then is sexual isolation normally maintained, and is it maintained differently in sympatric and allopatric populations?

Because the MULE IS STERILE, *it is not given a species name. Neither are most modern domestic species. This convention is assumed because almost all domestic "species" have been shown to have polytypic origins.*

[7]For overviews of various aspects of hybridization and speciation, see Levin (1979), Rieseberg et al. (1996), Arnold (1997), Rieseberg (1997), and Ungerer et al. (1998).

SEXUAL ISOLATION IN SYMPATRIC AND ALLOPATRIC POPULATIONS

Because they are close enough together to produce deleterious hybrids, sexual isolation should be strongest among **sympatric populations** of related species. Because they are too distantly separated to produce such hybrids, sexual isolation should be weakest among **allopatric populations**. Evolutionary biologists have sought evidence for these hypotheses by comparing the degree of sexual isolation among different sympatric and allopatric populations, either in the wild or under experimental conditions. Some of the best information has come from studies on fruit flies.

One experiment tested the degree of sexual isolation between two sibling species, *Drosophila arizonensis* and *D. mojavensis*, by attempting crosses using individuals from connected (sympatric) or separated (allopatric) populations. When the individuals came from separated populations, the interspecific cross *arizonensis* × *mojavensis* occurred only in 4% of matings.[8] This contrasted with 25% of matings between individuals from connected populations. The same has been shown experimentally in plants. The most difficult to cross of nine species in the annual herb genus *Gilia* are the allopatric species. Sympatric species, by contrast, show no barriers against intercrossing, even though all F1 hybrids are sterile (V. Grant, 1985). To these observations can be added data from hundreds of moth species, which show that scent-emitting organs used to attract females are significantly more common among males of species associated with the same host plant than among species associated with different host plants; species-specific scents inhibit hybridization.[9]

Selection against hybrids in *Drosophila* is selection for sexual isolation. Although sexual isolation exists between sibling species *Drosophila pseudoobscura* and *D. persimilis*, both in nature and at normal temperatures in the laboratory, cold temperatures can significantly increase interspecific mating.[10] By identifying individuals of each of the two species on the basis of different homozygous recessive alleles, hybrids formed under low-temperature conditions could be recognized and removed from interspecific population cages. When this operation was performed each generation, fewer and fewer hybrids appeared; after five generations, the frequency of hybrids in the mixed populations had fallen from as high as 50% to just 5% percent. This is striking evidence that **selection against the hybrid** results in rapid selection for sexual isolation and reduction in hybrid formation. A somewhat similar experiment involved planting a mixture of yellow sweet and white flint strains of corn. By eliminating plants that produced the greatest proportion of heterozygotes, intercrossing was reduced from about 40% to less than percent in five generations.[11]

An extensive survey of populations in nature also shows that reproductive isolation between pairs of *Drosophila* species is greater for sympatric species than for allopatric species pairs (Figure 23.5). The basis for the analysis is **genetic distance**, which is a measure of the evolutionary divergence between two species, usually determined on the basis of allele frequencies from zero to 1 (Coyne et al., 1997). The lower the genetic distance, the more closely related the species. As shown in Figure 23.5, when species in a pair are closely related—the genetic distance between them is small, for example, 0.5—they are more isolated from each other when the species are sympatric than when they are allopatric.

SWEET CORN, *the corn on the cob sold in supermarkets, is yellow because of the high content of vitamin A in the kernels.*

---

[8] T. A. Markow, 1981. Courtship behavior and control of reproductive isolation between *Drosophila mojavensis* and *Drosophila arizonensis. Evolution*, **35**, 1022–1026.

[9] For moth–plant interactions and speciation, see Feinsinger (1983), Dudareva and Pichersky (2006), and Waser and Ollerton (2006).

[10] Mating tests between *D. pseudoobscura* females and *D. persimilis* males taken from natural populations show that sexual isolation is increased in areas where populations overlap compared to areas where *D. persimilis* is absent.

[11] R. Butlin, 1989. Reinforcement of premating isolation. In *Speciation and Its Consequences*, D. Otte and J. A. Endler (eds.). Sinauer Associates, Sunderland, MA, pp. 158–179; E. Paterniani, 1969. Selection for reproductive isolation between two populations of maize, *Zea mays* L. *Evolution*, **23**, 534–547; M. A. Noor, 1995. Speciation driven by natural selection in *Drosophila. Nature*, **375**, 674–675.

Furthermore, allopatric populations require much greater genetic isolation (greater genetic distance) to achieve reproductive isolation than do sympatric populations. For the species in Figure 23.5, genetic distance is 0.54 for allopatric and 0.04 for sympatric populations.

Genetic distance may not be a sufficiently sensitive measure of genetic separation between species, however. Using nuclear and mitochondrial genes and microsatellites from *Drosophila simulans* and *D. mauritiana*, species that diverged allopatrically 250,000 years ago, Nunes et al. (2010) discovered mtDNA haplotypes that diverged 127,000 years ago. This evidence supports gene exchange and independent hybridization events between these two species.

SPECIATION AND HYBRIDIZATION ZONES

Where species barriers break down to produce viable and fertile hybrids, as often occurs in plants, **zones of hybridization** or **hybrid swarms** can develop in which genotypes and phenotypes differ from both parental species. In some cases, fertile hybrids can act as intermediaries, introducing genes from one species into the other, thereby enhancing a species' ecological range and evolutionary flexibility, as seen in Darwin's finches (below). If an environment exists to which the hybrids are better adapted than are the parents, the new population may eventually become isolated, a process demonstrated in plants and in animals.

PLANTS

A contribution of hybridization to speciation—in this case sympatric speciation—is supported by detailed demonstrations of changes in chromosome number (polyploidy), especially in plants (V. Grant, 1985), but also in animals (Bullini, 1994). Between 40 and 70 percent of all plant species are polyploid and so could have arisen by hybridization, although chromosome number can change within a single species by chromosome doubling without hybridization, and hybridization can occur without polyploidy. Polyploidy through chromosome doubling would be classed as sympatric speciation, polyploidy through hybridization as allopatric speciation.

One well-investigated example of hybridization between plant species occurs in sunflowers of the genus *Helianthus*.[12] Three western United States species show rapid evolution of a hybrid (*Helianthus anomalus*) initially formed from a cross between *H. annuus* and *H. petiolaris* less than 60 generations ago (FIGURE 23.6). Interestingly, "synthetic hybrids" can be made experimentally by crossing the two parental sunflower species and then successively crossing and backcrossing them. Synthetic hybrids generate genomes similar to the natural *H. anomalus* hybrid, incorporating similar parental genes and excluding others. Once an initial hybrid is formed selection becomes an important factor, allowing genes to accumulate that enhance the genetic architecture of the hybrid (Rieseberg et al., 1996).

ANIMALS

Although less common, hybridization has been associated with speciation in animals.[13] As discussed earlier in the chapter, mating in two of Darwin's finches—the medium ground finch and the cactus finch on Daphne Major Island—is based on recognition of song type. Because this cross-species learning does not provide complete reproductive isolation, interbreeding can lead to hybridization, although hybridization occurs in less than one percent of breeding pairs.

Hybrids have beak sizes intermediate between their parental species but only survive when seeds of the appropriate size are present (Grant et al., 2004). Hybrid reproduction

---

[12] M. C. Ungerer et al., 1998. Rapid hybrid speciation in wild sunflowers. *Proc. Natl. Acad. Sci. USA*, **95**, 11757–11762.

[13] For hybridization in animals, see Bullini (1994), Arnold (1997), and Dowling and Secor (1997).

FIGURE 23.6 Three western United States species illustrate rapid evolution of a hybrid *Helianthus anomalus* **(a)** initially formed from a cross between *H. annuus* **(b)** and *H. petiolaris* **(c)** less than 60 generations ago.

and survival is under environmental control. After a year of exceptionally heavy rains, the ecological condition of Daphne Major Island changed and an abundance of small soft seeds were produced. Under these new conditions, hybrids with intermediate beak size enabling them to eat these seeds survived and had the highest breeding success. However, hybrids did not mate with each other (perhaps because there were so few of them). Rather, hybrids backcrossed to one or other of their parental species **according to their imprinted song type**, allowing genes to flow from one species into the other. High survival of several backcross generations resulted in over 30 percent of individual cactus finch with some genes from the medium ground finch.

FREQUENCY AND IMPACT OF HYBRIDIZATION IN NATURE

As Darwin's finches show, the evolutionary consequence of episodic hybridization is to increase the genetic variation on which selection can act. If the environment subsequently changes, rapid evolutionary change may ensue. Episodic genetic exchange of this type between closely related species is widespread in nature; examples are found in insects, cichlid fish, sticklebacks, many species of plants, and birds and mammals with male parental care. It could be an important contributor to rapid adaptive radiation.

It has been difficult to document how often new hybrid species occur. In part this is because of natural variation in the **frequency of hybridization**, which, in vascular plants, varies between families. Some plant families show hardly any hybrid species, while in others 50 percent or more of the species arose by hybridization. With these caveats in mind, it has been estimated that some 25% of plant and 10% of animal species can hybridize with one or more other species.[14] Ability to hybridize is most commonly seen in groups undergoing rapid radiation where species are more likely to be closely related than in lineages that diverged longer ago.

The **impact of hybrids** on evolution, however, may be more significant than their frequency suggests. Some plant hybrids may be the ancestors of entire lineages comprising many species and occupying many different environments. Of course, as discussed earlier, hybrid sterility normally is a barrier to further evolution. Even then, and especially in plants,

---

[14]N. C. Ellstrand et al., 1996. Distribution of spontaneous plant hybrids. *Proc. Natl. Acad. Sci. USA*, **93**, 5090–5093; L. H. Rieseberg, 1997. Hybrid origins of plant species. *Ann. Rev. Ecol. Syst.*, **28**, 359–389.

polyploidy may arise in a vegetatively propagating or self-fertilizing hybrid, enabling it to produce fertile gametes. Because these gametes are diploid relative to the haploid gametes of the two parental species, a new species is born at one stroke, fertile with itself or other such polyploid hybrids, but sterile in crosses with either parental species. A laboratory polyploid hybrid strain of silk worm, produced by crossing the domesticated silk worm (*Bombyx mori*) with the wild silk worm (*B. mandarina*), is viable because triploid gametes were used to backcross with haploid gametes to produce tetraploids that could interbreed.[15] Relationships between hybridization and speciation remain an important topic for future analysis.

■ KEY TERMS

allopatric

allopatric speciation

founder effect

genetic distance

geographical isolation

hybrid

hybrid inferiority

hybridization

peripatric speciation

postzygotic mating barrier

prezygotic mating barrier

reproductive isolation

seasonal isolation

sympatric speciation

■ EVOLUTION ON THE WEB

Explore evolution on the Internet! Visit the accompanying website for *Strickberger's Evolution, Fifth Edition,* at **go.jblearning.com/Evolution5eCW** for exercises and links relating to topics covered in this chapter.

■ REFERENCES

Arnold, M. L., 1997. *Natural Hybridization and Evolution.* Oxford University Press, Oxford, UK.

Barton, N. H., and B. Charlesworth, 1984. Genetic revolutions, founder effects, and speciation. *Ann. Rev. Ecol. Syst.,* **15**, 133–164.

Bradshaw, H. D., Jr., S. M. Wilbert, K. G. Otto, and D. W. Shemske, 1995. Genetic mapping of floral traits associated with reproductive isolation in monkeyflowers (*Mimulus*). *Nature,* **376**, 762–765.

Bullini, L., 1994. Origin and evolution of animal hybrid species. *Trends Ecol. Evol.,* **9**, 422–426.

Carson, H. L., 1986. Sexual selection and speciation. In *Evolutionary Processes and Theory,* S. Karlin and E. Nevo (eds.). Academic Press, Orlando, FL, pp. 391–409.

Carson, H. L., 1992. Inversions in Hawaiian *Drosophila.* In Drosophila *Inversion Polymorphism*, C. B. Krimbas and J. R. Powell (eds.). CRC Press, Boca Raton, FL, pp. 407–439.

Civetta, A., and R. S. Singh, 1998a. Sex-related genes, directional sexual selection, and speciation. *Mol. Biol. Evol.,* **15**, 901–909.

Civetta, A., and R. S. Singh, 1998b. Sex and speciation: genetic architecture and evolutionary potential of sexual versus nonsexual traits in the sibling species of the *Drosophila melanogaster* complex. *Evolution,* **52**, 1080–1092.

Correa, R., J. Stanga, B. Larget, et al., 2012. An assessment of transgenomics as a tool for identifying genes involved in the evolutionary differentiation of closely related plant species. *New Phytol.,* **193**, 494–503.

Coyne, J. A., N. H. Barton, and M. Turelli, 1997. Perspective: a critique of Sewall Wright's shifting balance theory of evolution. *Evolution,* **51**, 643–671.

Coyne, J. A., and H. A. Orr, 2004. *Speciation.* Sinauer and Associates, Sunderland, MA.

Dowling, T. E., and C. L. Secor, 1997. The role of hybridization and introgression in the diversification of animals. *Ann. Rev. Ecol. Syst.,* **28**, 593–619.

[15] B. L. Astaurov, 1969. Experimental polyploidy in animals. *Ann. Rev. Genet.,* **3**, 99–126.

Dudareva, N., and E. Pichersky (eds.), 2006. *Biology of Floral Scent.* CRC Press, Atlanta, GA.

Edwards, K. A., L. T. Doescher, K. Y. Kaneshiro, and D. Yamamoto 2007. A database of wing diversity in the Hawaiian *Drosophila. Plos One* **2**(5), e487. doi:10.1371/journal.pone.0000487.

Feinsinger, P., 1983. Coevolution and pollination. In *Coevolution*, D. J. Futuyma and M. Slatkin (eds.). Sinauer Associates, Sunderland, MA, pp. 282–310.

Grant, B. R., and P. R. Grant, 2010. Songs of Darwin's finches diverge when a new species enters the community. *Proc. Natl. Acad. Sci. USA*, **107**, 20156–20163.

Grant, P. R., 1986. *Ecology and Evolution of Darwin's Finches.* Princeton University Press, Princeton, NJ. [Reissued in 1999 with a Foreword by J. Weiner]

Grant, P. R., and B. R. Grant, 1994. Phenotypic and genetic effects of hybridization in Darwin's finches. *Evolution*, **48**, 297–316.

Grant, P. R., and B. R. Grant, 2002. Unpredictable evolution in a 30-year study of Darwin's finches. *Science*, **296**, 707–711.

Grant, P. R., and B. R. Grant, 2008. *How and Why Species Multiply: The Radiation of Darwin's Finches.* Princeton University Press, Princeton, NJ.

Grant, P. R., B. R. Grant, L. F. Keller, et al., 2004. Convergent evolution of Darwin's finches caused by introgressive hybridization and selection. *Evolution, 58*, 1588–1599.

Grant, V., 1985. *The Evolutionary Process.* Columbia University Press, New York.

Gulick, A., 1932. *Evolutionist and Missionary. John Thomas Gulick, Portrayed through Documents and Discussions.* The University of Chicago Press, Chicago.

Gulick, J. T., 1872. On the variation of species as related to their geographical distributions, illustrated by the *Achatinellinae. Nature*, **6**, 222–224.

Gulick, J. T., 1873. On diversity of evolution under one set of external conditions. *J. Linn. Soc. Lond.*, **11**, 496–505.

Gulick, J. T., 1883. Darwin's theory of evolution applied to Sandwich Island Mollusks. *Chrysanthemum*, **3**, 6–11.

Gulick, J. T., 1888. Divergent evolution through cumulative segregation. *Zool. J. Linn. Soc.*, **20**, 189–224.

Hall, B. K., 2006a. Evolutionist and Missionary, The Reverend John Thomas Gulick (1832–1923). Part I. Cumulative Segregation—Geographical Isolation. *J. Exp. Zool. (Mol. Dev. Evol)*, **306B**, 407–418.

Hall, B. K., 2006b. Evolutionist and Missionary, The Reverend John Thomas Gulick (1832–1923). Part II. Coincident or Ontogenetic Selection—The Baldwin Effect. *J. Exp. Zool. (Mol. Dev. Evol)*, **306B**, 489–495.

Hey, J., W. M. Fitch, and F. J. Ayala (eds.), 2005. *Systematics and the Origin of Species on Ernst Mayr's 100th Anniversary.* The National Academies Press, Washington, DC.

Huber, S. K., L. F. De León, A. P. Hendry, E., Bermingham, and J. Podos, 2007. Reproductive isolation of sympatric morphs in a population of Darwin's finches. *Proc R. Soc. Lond. (B)*, **27**, 1709–1714.

Levin, D. A. (ed.), 1979. *Hybridization: An Evolutionary Perspective.* Dowden, Hutchinson and Ross, Stroudsburg, PA.

Mayr, E, 1988. *Toward a New Philosophy of Biology. Observations of an Evolutionist.* The Belknap Press of Harvard University Press, Cambridge, MA.

Nunes, M. D. S., P. O.-T. Wengel, M. Kreissl, and C. Schlötterer, 2010. Multiple hybridization events between *Drosophila simulans* and *Drosophila mauritiana* are supported by mtDNA introgression. *Mol. Ecol.*, **19**, 4695–4707.

Presgraves, D. C., 2010. The molecular evolutionary basis of species formation. *Nat. Rev. Genet.*, **11**, 175–180.

Rieseberg, L. H., 1997. Hybrid origins of plant species. *Ann. Rev. Ecol. Syst.*, **28**, 359–389.

Rieseberg, L. H., B. Sinervo, C. R. Linder, M. C. Ungerer, and D. M. Arias, 1996. Role of gene interactions in hybrid speciation: evidence from ancient and experimental hybrids. *Science*, **272**, 741–745.

Sobel, J. M., G. F. Chen, L. R. Watt, and D. W. Schemske, 2009. The biology of speciation. *Evolution*, **64**, 295–315.

Ungerer, M. C., S. J. Baird, J. Pan, and L. H. Rieseberg, 1998. Rapid hybrid speciation in wild sunflowers. *Proc. Natl. Acad. Sci. USA*, **95**, 11757–11762.

Wallace, A. R., 1889. *Darwinism: An Exposition of the Theory of Natural Selection with Some of Its Applications.* Macmillan, London.

Waser, N. M., and J. Ollerton, 2006. *Plant-Pollinator Interactions: From Specialization to Generalization.* The University of Chicago Press, Chicago.

CHAPTER

24

Mass Extinctions, Opportunities, and Adaptive Radiations

CHAPTER SUMMARY

Extinctions may wipe out a single species, or they may be massive (mass extinctions) and wipe out an entire biota. In the mass extinction discussed in this chapter most dinosaurs and many other terrestrial and marine organisms died out at the end of the Cretaceous, 65 Mya. This is the only mass extinction that can conclusively be associated with the impact of an asteroid, although other causes also were involved.

Extinctions are devastating but also provide opportunities. Ecological factors—opening up of environments vacated by the dinosaurs in the Late Cretaceous, origination of new ecosystems as a consequence of continental drift—facilitated mammalian evolution and adaptive radiation from stem mammals that survived the Late Cretaceous mass extinction (END BOX 24.1). Mammalian evolution teaches us that dominance at one time is no guarantee of continued dominance; long-term evolutionary success is unpredictable.

Although most dinosaurs became extinct at the end of the Cretaceous Period, the lineage that gave rise to birds persisted. Pterosaurs and birds independently evolved the ability to fly. Feathers evolved only in birds and in avian ancestors, the feathered dinosaurs. Birds radiated as rapidly as insects. Both lineages teach us much about the basis for adaptive radiation, the occupation of new habitats, and the creation of new niches.

INTRODUCTION

The universe is a violent place. It had an abrupt birth. Its stars and galaxies were born in the midst of powerful interactions. Its elements originated from the debris of many explosive episodes, and violent impacts still occur—even in our small solar system.

At the end of the Cretaceous 65 Mya, most dinosaurs, along with other large marine reptiles, fish, and various invertebrates, died out. The best-supported hypothesis to explain these extinctions involves changes initiated by the impact of an asteroid. The rare element iridium, often found in meteorites, is found worldwide in a stratum at the Cretaceous-Tertiary boundary, as are particular types of quartzes that only form as a result of high impact. Also, we have the likely site of impact, the 185–200 km (115–124 mile) diameter Chicxulub crater off the Yucatán Peninsula in Mexico.

Stem mammals survived the mass extinction that removed the nonavian dinosaurs. The habitats vacated by the dinosaurs and the evolution of land plants facilitated mammalian evolution and adaptive radiation. At least 25 lineages of mammals coexisted with, and survived, the dinosaurs (Luo, 2007). As a consequence, mammals have much to teach us about survival, radiation, diversification, adaptations to new environments, and parallel evolution.

Opportunities also were created for reptiles that survived the mass extinction; adaptations for sustained flight appeared independently in two lineages of reptiles: pterosaurs/pterodactyls[1] and birds. Pterosaurs developed hollow bones and flight membranes between the body and the wings, and in some cases lost the tail and teeth, leaving the jaws as a beak. Birds arose from bipedal, ground- or tree-dwelling small carnivorous dinosaurs.

EXTINCTION

Extinction is the flip side of the origin of species; species arise and species disappear.

Extinction has appeared in several contexts through the text. As discussed in a historical context elsewhere in this text, once the realities of fossils and extinction were accepted, it was possible to conceive of a "law of succession"—evolution—in which one form replaced another. The "poster-child" for evolution from the fossil record, the evolution of horse lineages, was characterized by many episodes of speciation *and* of extinction. Extinction was common, if not inevitable. Why? because species could not always adapt to large or rapid environmental changes. The Ediacaran Biota disappeared in the Early Cambrian, 541 Mya. A major group of Cambrian arthropods, the trilobites, went extinct at the end of the Permian, 251 Mya (discussed elsewhere in this text; see Erwin, 2006 and Erwin et al., 2011).

Death of individuals may be local and specific, and not result in extinction of the species, as might occur if a flood or an avalanche wiped out a population of a species but spared other populations. Genetic variation in a species might be reduced by such local events but the species would survive.

Extinction can conveniently be considered at three levels of impact.

- **Extinction** that eliminates an entire species, as might occur if the species consisted of one or few local populations, all of which were wiped out by an environmental event, such as a flood, or by a biological event, such as elimination of a specific food item or a species-wide disease.
- **Mass extinction** that eliminates all the species in a region or ecosystem, as might occur following a volcanic eruption.[2]
- **Global extinction**, an event of such a large scale that it eliminates most of the species on a continent or on Earth. **Snowball Earth**, discussed in **END BOX 24.2**, is an example. Extinction following elevation in CO_2 levels is another.[3]

---

[1] Pterodactyls are a derived lineages of pterosaurs.

[2] In one convention, a mass extinction is an event that eliminates half or more of the species in a region.

[3] For excellent reviews of extinction, see Benton (2003), Erwin (2006), and Hallam (2005).

While they eliminate species or larger taxonomic groups, extinction events open up habitats and so can facilitate the radiation of organisms that survived the mass extinction.

EXTRATERRESTRIAL IMPACTS

As a result of data gathered during the *Apollo* space program, we know that a series of heavy extraterrestrial impacts battered the Moon about 4 Bya. We infer that similar events occurred on Earth's surface (**FIGURE 24.1**). Although their frequency greatly diminished as the solar system evolved, impacts have persisted. Geologists have identified more than 100 craters on Earth; a recent study describes the 45-m (147 ft)–wide Kamil Crater in southern Egypt.[4] Ten meteors, each 1 km in diameter, are estimated to have each produced 20-km–wide craters at a frequency of one every 400,000 years. A 50-km–wide crater is produced every 12.5 My, a 150-km–wide crater every 100 My.

Depending on their size, objects causing such impacts may have enormous and/or long-term environmental effects. The impact crater throws large numbers of particles into the atmosphere, producing dust clouds that interfere with photosynthesis, causing food chains to collapse. Depending on whether the impact occurred on land or at sea, it could have resulted in large atmospheric temperature changes, which may run through a cycle of an immediate "winter" from the dust thrown up, followed by a high-temperature "greenhouse effect" from the water vapor released. Heat generated by the object entering the atmosphere and the heated material it ejects on impact spark raging forest fires and generate nitrous oxides that seed acid rains, destroying vegetation and marine organisms.

© Glock/ShutterStock, Inc.

FIGURE 24.1 Artist's visualization of an impact of an asteroid with Earth.

[4]L Folco et al., 2010. The Kamil Crater in Egypt. *Science*, **329**, 804.

MASS EXTINCTIONS

Mass extinctions are not isolated or singular events. As concluded from fossil data, there have been five major mass extinctions since the Cambrian (**TABLE 24.1**). According to Raup and Sepkoski (1986), each was marked by the relatively abrupt disappearance of at least 75 percent of marine animal species. Extinctions on the order of the type in Table 24.1 are estimated by Fox (1987) to have occurred on average about every 100 My (**FIGURE 24.2**). Extinction events that eliminate five percent of species occur on average once every million years.[5] We may be in the midst of a sixth mass extinction, the elimination of the large animals (megafauna) from many parts of the world by human action.

We have accumulated considerable evidence, allowing us to conclude that a mass extinction resulted from a collision between an asteroid and Earth at the end of the Cretaceous, 65 Mya. The impact and its aftermath resulted in the extinction of many animals (including most dinosaurs) and plants. Their removal opened up habitats for an extensive radiation of mammals. Extinction of one group creates an opportunity for others. Organisms that move into new habitats may be subject to different selection pressures than in their previous environments, facilitating adaptation and radiation.

This mass extinction is often referred to as the K–T EXTINCTION, *from the German words for the Cretaceous and Tertiary Periods (*Kreidezeit, Tertiär*).*

DINOSAURS AND LATE CRETACEOUS MASS EXTINCTIONS

No large dinosaurs survived the end of the Cretaceous Period (Dingus and Rowe, 1998). Indeed, no land vertebrate larger than about 23 kg (50 lbs) survived into the Tertiary. To give some idea of scale, a Labrador dog weighs around 25 kg (55 lbs), a horse about 500 kg (1100 lbs).

Other terrestrial and marine organisms became extinct at the same time; paleontologists estimate that more than half of all species of reptiles became extinct during a relatively short geological period at the end of the Cretaceous. Some groups, such as fish, diversified as they occupied habitats left empty by forms that went extinct (Alfaro and Santini, 2010). Other groups survived but failed to diversify; their species/generic diversity was no higher five to ten millions of years after extinction than it had been before.

Many mechanisms have been proposed to account for such a wide spectrum of extinction and/or slowed speciation. A once popular hypothesis, now abandoned,

TABLE 24.1	Details of the Five Major Mass Extinctions Since the Cambrian		
		Estimated % of Marine Animal Extinctions	
Extinction Period	**Date (Mya)**	**Genera**	**Species**
End of the Ordovician	440	61	85
End of the Devonian	365	55	82
End of the Permian	245	84	96
End of the Triassic	208	50	76
End of the Cretaceous	65	50	76

[Modified from Raup and Sepkoski, Jr., 1986.]

[5] D. Jablonski, 2000. Micro- and macroevolution: scale and hierarchy in evolutionary biology and paleobiology. In *Deep Time: Paleobiology's Perspective*, D. H. Erwin, and S. L. Wing (eds.). *Paleobiology*, 26 (Supplement to No. 4), The Paleontological Society, Lawrence, KS, pp. 15–22; D. Jablonski, 2002. Survival without recovery after mass extinctions. *Proc. Natl Acad. Sci. USA*, **99**, 8139–8144.

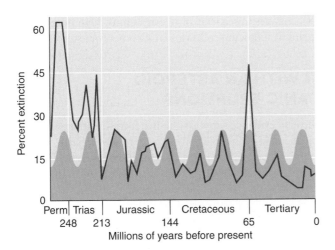

FIGURE 24.2 Repeated extinction events are indicated by the solid line that shows the percentage of marine animal extinctions during the 260 My interval between the Permian and the present. Areas shaded in green are based on a periodicity of an extinction event every 26 My.

[Adapted from Fox, W. T., 1987. *Paleobiology* **13**, 257–271.]

sought to explain the extinction of dinosaurs by internal rather than external factors. This hypothesis suggested that, just as individuals are born, grow old and die, species follow a similar life history driven by internal factors. The hypothesized process, **orthogenesis**, would result in evolution proceeding in a direction unrelated to selection and adaptation. Evidence used to support orthogenesis included the appearance of bizarre, and what appeared to be nonadaptive, characters. Some long considered the large, seemingly clumsy 3.3-m-wide antlers of the Irish elk *Megaloceros* (**FIGURE 24.3**) to have been a factor in its senescence and extinction. However, no evidence supports any biological mechanisms other than the inability to cope with changing environmental or competitive challenges to explain the extinction and replacement of organisms.[6]

There is no shortage of more plausible hypotheses than orthogenesis to explain mass extinctions. Hypotheses include intense volcanic activity, epidemics, changes in plant composition, shifting continental profiles, elevated CO_2 levels (and the subsequent

© Florilegius/Alamy

FIGURE 24.3 Extinct male Irish elks (*Megaloceros giganteus*) stood 2 m (6.5 ft) tall at the shoulders and weighed 45 kg (99 lbs), but the antlers were more than 3.3 m (10.8 ft) across.

[6]See S. J. Gould, 1974. The origin and function of "bizarre" structures: antler size and skull size in the "Irish Elk," *Megaloceros giganteus. Evolution*, **28**, 191–220.

greenhouse effect), changes in sea level or ocean salinity, high doses of ultraviolet radiation, and/or dust clouds following collisions with comets or asteroids.

COLLISION WITH AN ASTEROID AND VOLCANIC ERUPTIONS

A large body of diverse evidence supports the hypothesis that late Cretaceous extinctions were initiated or accelerated by the collision of an asteroid with Earth. The **Collision Theory** gathered considerable support in the 1980s after the discovery of **iridium** deposits in strata marking the Cretaceous–Tertiary (K–T) boundary (Alvarez, 1983; Alvarez and Azaro, 1992).

Discovered in 1803 and named after the Greek Goddess Iris, the metal iridium is one of the rarest elements in Earth's crust—0.001 ppm; for comparison, gold is found at 0.005 ppm. However, iridium is found at much higher levels (0.5 ppm) in asteroids and meteorites. The worldwide presence of iridium at these higher concentrations in Cretaceous–Tertiary boundary strata, along with high-impact particles—glasslike spherules and shocked/fractured quartz—provides strong evidence for collision with an extraterrestrial body at this time (Alvarez, 1983). This anomalous iridium-rich layer at the K–T boundary has now been found at more than 100 different localities (Benton, 2003).

For an impact to produce global effects significant enough to cause the Late Cretaceous mass extinction, its resulting crater would have to be quite large. Among the best candidates for such a crater is the **Chicxulub crater**. Discovered in 1993 off the coast of the Yucatán Peninsula in Mexico, Chicxulub has an outer diameter of between 185 and 200 km (115 and 124 miles; FIGURE 24.4). According to its discovers, Chicxulub "may be one of the largest impact structures produced in the inner solar system since the period of early bombardment ended nearly four billion years ago" (Sharpton et al., 1993, p. 1564, and see Schulte et al., 2010).

Does this impact account for all the Late Cretaceous extinctions? Not necessarily. Volcanic activity may have played a role also. Even when non-explosive, volcanic eruptions can affect world climate; the eruption of a chain of volcanoes in Iceland in 1873 produced a cloud of ash that stretched across Europe into parts of Asia and Africa. Chemicals in the ash, especially sulfur dioxide and hydrogen chloride, produced acid rain that destroyed crops and killed farm animals, causing famines as far away as the Nile

FIGURE 24.4 The Chicxulub crater as visualized using data measuring the field intensity of gravity reveals an oblique impact crater.

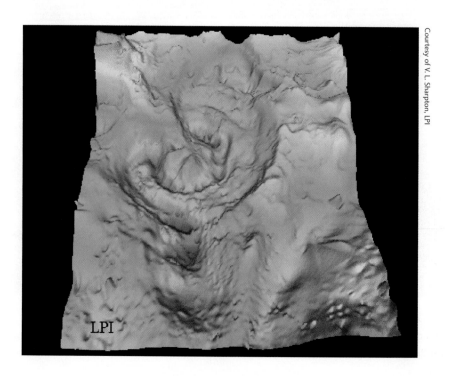

Courtesy of V. L. Sharpton, LPI

Valley in Egypt, where a sixth of the population died. Paleogeologists have proposed that deposition of atmospheric dust by violent volcanic eruptions and ash also could have influenced the Late Cretaceous extinctions.

Secondly, evidence shows that the meteor fell between eruptions of plumes of deep mantle material. An enormous outpouring of nearly 4.1 million km³ (2.5 million mi³) of volcanic lava—the "**Deccan Traps**"—which covered one-third of India during the Late Cretaceous extinctions[7], provides support for the role of volcanic activity in the mass extinction (**FIGURE 24.5**). Furthermore, for 20,000 years before, and for as much as 100,000 years after the asteroid impacted, ocean and land temperatures were 3–4°C (37–39°F) and 7–8°C (44–46°F) higher, respectively, than had prevailed in the preceding

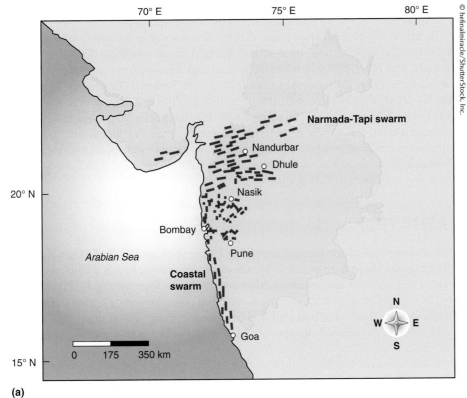

(a)

FIGURE 24.5 **(a)** A map of the location of the Deccan trap basalts. **(b)** Portion of the more than 2,000 meters of basalt lava flows that cover an area of 500,000 km².

© hefinalmiracle/ShutterStock, Inc.

(b)

© hefinalmiracle/ShutterStock, Inc.

---

[7]For comparison, the north and south polar ice caps are estimated to contain between 2 and 3 million km³ (1.2–1.9 million mi³) of ice.

time period. Such a warming period would have been fatal for many organisms, including dinosaurs.

Paleontological evidence tells us that dinosaurs and other animals had already declined in numbers, or may have already disappeared, *before* the asteroid impact. Contrary to the immediate effects of an extraterrestrial impact, dinosaur extinction may have taken a million years or more. One hypothesis is that extinction of the largest dinosaurs followed a combination of stressful environments as detailed above, as well as an extraterrestrial impact. Some species of other groups, however, including small mammals, survived and expanded (Kemp, 2005, 2007; Luo, 2007). Small size and efficient control of body temperature may have been crucial for the survival of terrestrial vertebrates in the Late Cretaceous. These data imply that traits beneficial during extinction were not the traits most advantageous before extinction; they became advantageous when the surviving organisms moved into habitats vacated by newly extinct species.

No firm evidence supports hypotheses that extinctions other than the Late Cretaceous mass extinction were initiated by meteor or asteroid impacts. Iridium deposits in strata associated with other extinctions are not great enough to assume extraterrestrial impact; strata deposited after the greatest of all extinctions, the **Permian–Triassic (P–T) extinction**, 251 Mya (Table 24.1), contain almost no iridium. The P–T extinction is referred to as the **Great Dying**; as much as 96 percent of all marine species, 70 percent of terrestrial vertebrate species, and 85 percent of terrestrial insects became extinct in what may have been three pulses of extinction over 165,000 years. Proposed mechanisms include release of methane from the sea floor, droughts of long duration, and decreased atmospheric oxygen resulting from climate change.[8]

RADIATIONS OF MAMMALS

Some species survived the asteroid impact, including small mammals. Extinction of the large dinosaurs allowed mammals to invade environments previously occupied by herbivorous and carnivorous dinosaurs. Mammals not only survived the K–T boundary, they thrived, radiating into 25 or more forms/lineages and diversifying into terrestrial, aquatic, and aerial habitats (bats are the second largest group of mammals), making use of new adaptations such as limbs specialized for running, swimming, and flying (Kielan-Jaworowska et al., 2004; Kemp, 2005, 2007; Luo, 2007).[9]

Conditions favoring mammalian radiation did not begin with the K–T mass extinction, however. An earlier significant stimulus for mammalian radiation was the breakup of the large Pangaea landmass that began in the Triassic Period, 225 Mya and is discussed in BOX 24.1.[10] Both the breakup of Pangaea and the movements of tectonic plates after the K–T extinctions established new continents (Condie, 1997; Condie and Sloan, 1998). Evolving connections and separations of landmasses facilitated dispersal and/or isolation of major mammalian groups.

To these land movements—with their marked effects on climate and environment, and their production of new geographical and ecological regions—we can add the uplifting of mountain systems that took place from the Cretaceous onward. These mountain-building events lead to chains such as the Rockies, Andes, Alps, and Himalaya,

[8] S. A. Bowring et al., 1998. U/Pb zircon geochronology and tempo of the End-Permian mass extinction. *Science*, **280**, 1039–1045; S. M. Stanley, 2008. Evidence from ammonoids and conodonts for multiple Early Triassic mass extinctions. *Proc. Natl. Acad. Sci. USA*, **106**, 15256–15259.
[9] Phylogenetic relationships of early mammals are a subject of active research, especially whether mammals radiated well before, or primarily contemporaneously with, the K–T mass extinction event, and how mode of preservation influences estimates of divergence times; see Foote et al. (1999), Luo (2007), and Wible et al. (2007) for detailed recent analyses and for access to prior studies.
[10] Early dinosaurs were established in southwestern Pangaea 230 Mya (R. N. Martinez et al., 2011. A basal dinosaur from the dawn of the dinosaur era in southwestern Pangaea. *Science*, **331**, 206–210).

Break-up of the initial giant landmasses on Earth isolated large regions in which speciation and radiation proceeded independently.[a]

Geography in the Devonian, 375 Mya, shows two major landmasses: the **Gondwana** group of continents and a North American–Eurasian group called **Laurasia**. One hundred My later, by the end of the Paleozoic, these two major continental groups had united to form the giant landmass **Pangaea**. In turn, Pangaea began to break up about 225 Mya, during the Triassic.

Fragmentation of Pangaea started with the separation of Western Gondwana (South America and Africa) from Eastern Gondwana (India, Antarctica, and Australia). By the Late Jurassic, 145 Mya, sea-floor spreading had begun to separate North America from Africa. By the Cretaceous, 142 Mya, North America had separated from Greenland, and South America had moved away from Africa.

In the Western Hemisphere, the rapid drift of South America away from Africa, which began about 100 Mya, led eventually to a reunion with North America some 4 to 5 Mya. In the Southern Hemisphere, New Zealand had drifted away from the Australian–Antarctican–South American landmass before the end of the Cretaceous.[b] The other Southern continents were joined until the beginning of the Tertiary about 65 Mya. However, by the Eocene, 20 My later, Australia had begun the northward journey that will eventually unite it with the Asia plate.

[a]Movements of landmasses and continents result from the action of plate tectonics; see Condie (1997) and Condie and Sloan (1998).
[b]T. H. Worthy et al., 2006. Miocene mammal reveals a Mesozoic ghost lineage on insular New Zealand, southwest Pacific. *Proc. Natl. Acad. Sci. USA*, **103**, 19419–19423.

---

submersions and regressions of shallow seas, and delineation of new shorelines. Changes in vegetation, especially the diversification and radiation of ferns, flowering plants, and plant–insect associations, took place during these periods, generating new landscapes of grasslands, savannas, and forests (**FIGURE 24.6**). This origination, modification, and shaping of new and different habitats had a major impact on mammalian adaptation, variation, and distribution, resulting in the evolution of some 4,000 genera of fossil mammals and 5,400 species of extant mammals.

RADIATION OF MAMMALS INTO SOUTH AMERICA

Extinctions also played a role in later stages of mammalian evolution, illustrating again how extinction of one group creates opportunities for others.

A colony of North American marsupials reached Europe during the Early Eocene, 50 Mya, which is some 15 My after the K–T mass extinction. Other short-lived mammalian lineages made their way to Asia and Africa through a North Atlantic–Greenland–Europe connection. During the Miocene, marsupial species of both northern continents became extinct. South and North America reunited in the Pliocene, 5.3 Mya, re-establishing a North American–South American land bridge. An extensive interchange between the mammals of these two continents followed, with marsupials invading North America from the south and placental mammals invading South America from the north (discussed elsewhere in this text, and see Flynn and Wyss, 1998).

Many South American marsupials became extinct as invading North American placental mammals diversified rapidly and took their place. Extinction resulted from competition for the same habitats. As elsewhere, extinction and radiation in South America went hand in hand, testifying again to the basic opportunism of evolutionary change. An important lesson is that the survival or extinction of any group or lineage may be closely connected to the survival or extinction of other groups or lineages: **coevolution and coextinction**.

Mammalian radiation continued through the Pleistocene (1.8 Mya to the present), an epoch that marked the appearance of many mammals in their current forms. Climatically, the Pleistocene also marks a period of at least **seven glaciations** (the Ice Ages), which at times covered one-third of Earth's surface. Woolly mammoths and woolly

You will appreciate that "NORTH AMERICA" (or any other modern continent) did not exist in their current form in the Cretaceous Era. Names such as North America or Australia are used as a convenient shorthand for the location on the globe being referred to.

NEW ZEALAND separated from the other Gondwanan continents during the Cretaceous. Only native marsupial mammals and bats had been found there until recently, when a mouse-size mammal was discovered in Miocene deposits, the age of which is consistent with this mammal having been present before the divergence of marsupial from placental mammals.

(a) Early Eocene—50 millions years ago

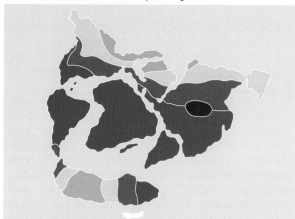

(b) Early Oligocene—32 millions years ago

(c) Late Miocene—10 millions years ago

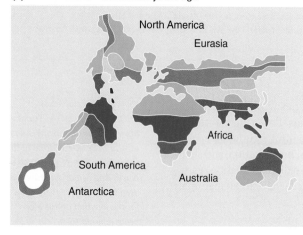

- ■ Tropical forest
- ■ Paratropical forest (with dry season)
- Subtropical woodland (broadleafed evergreen)
- Polar broadleafed deciduous forest
- ■ Woody savanna
- ■ Temperate woodland (broadleafed deciduous)
- Temperate woodland (mixed coniferous and deciduous)
- Grassland/open savanna
- Mediterranean-type woodland/thorn scrub/chaparral
- ■ Tundra
- □ Ice

FIGURE 24.6 Positions of continental landmasses at three stages over 40 My during the Cenozoic Era, showing the distribution of different kinds of vegetation during the **(a)** Early Eocene, **(b)** Early Oligocene, and **(c)** Late Miocene.

[Adapted from Janis, C. M., 1993. *Ann. Rev. Ecol. Syst.*, **24**, 467–500.]

rhinoceroses appeared in the northern continents during this interval, along with giant deer, giant cattle, and large cave bears.

Interestingly, these large mammals, in addition to horses, camels, ground sloths, and various other lineages, all became extinct in North America about 11,000 years ago. Among approximately 79 mammalian species weighing more than 45 kg (99 lbs), 57 became extinct at that time. Limited extinction occurred in Europe, however. One likely explanation for these Late Pleistocene extinctions is rapid deterioration of the climatic advantages for large animals as the ice sheets retreated. Another is the predatory role of humans; stone-age hunters entered formerly glaciated areas of North America and Europe at this time. One hypothesis is that humans initiated this, the first "**man-made extinctions**," by killing these mammals for food. A similar story is found in Australia, where a species of giant kangaroo, *Protemnodon anak*, 2 m (6.5 ft) tall and weighing 100 to 150 kg (220–330 lbs) disappeared from Tasmania within a few thousand years of the arrivals of humans (41,000–43,000 years ago). Its extinction was almost certainly due to hunting, although climate and/or vegetation change contributed to the disappearance of elements of the megafauna on mainland Australia (Turney et al., 2008). The invasion of Australia by humans, both during the Pleistocene and more recently, was accompanied by other placental mammals, including dogs, rabbits, sheep, and rodents, all of which impacted the indigenous fauna.

MEGAFAUNA *refers to mammals with an average body weight of 40 kg (100 lb). Extinct mammoths, mastodons, and giant bison are examples.*

INVADING THE AIR: FLYING REPTILES

As discussed above, escape from extinction is an important hallmark of biological survival, and may at times depend on the ability to move rapidly from threatened environments.

Flight provides a number of advantages: rapid escape from terrestrial predators and menacing conditions, access to feeding and breeding grounds that would otherwise be difficult or impossible to reach, and relatively swift transit between localities. Although forms capable of gliding for short distances arose in various vertebrate lineages (including "flying fish"; FIGURE 24.7), known adaptations for sustained powered flight have appeared only three times in the evolution of terrestrial vertebrates: in pterosaurs, birds, and bats. The two forms that arose within the reptiles—pterosaurs and birds—differ in respect to mechanisms of flight and accompanying adaptations. In pterosaurs, a flight membrane (patagium) of skin stretched between the trunk and wing. An elongated fourth finger was present on each hand. In birds, the flying surface consists of many stiff wing feathers that project posteriorly from a wing with only three digits. Pterosaurs died out as a result of the K–T extinction. Birds, which had originated before the mass extinction, flourished.

FIGURE 24.7 A "flying fish" photographed from above showing adaptation of the fins for gliding through the air.

PTEROSAURS AND PTERODACTYLS

Early pterosaurs appear in fossil record in the Late Triassic, 220 Mya, birds appear in the Late Jurassic, 135 Mya (Buffetaut and Mazin, 2003). The description of a sexually mature Jurassic pterosaur, *Darwinopterus,* discovered together with an egg, provides evidence for sexual dimorphism and the mode of pterosaur reproduction as well (Lü et al., 2011).

A well-described Late Jurassic pterosaur, *Rhamphorhynchus* (FIGURE 24.8a), was about 0.6 m (2 ft) long. The tail was long, with a small, rudder-like flap of skin at the end. The bones of *Rhamphorhynchus* and all other pterosaurs were light and hollow, and its elongated jaws were armed with strong, pointed teeth. The breast bone and its accessory bones provided sufficient surface for attachment of large flight muscles, as is also seen in "flying birds." In Brazil, a newly discovered species of toothless pterosaur from the Lower Cretaceous, *Lacusovagus magnificens* ("magnificent lake wanderer"), is estimated to have had a wingspan of 5 m (16.4 ft) and to have stood 1 m (3.3 ft) tall at the shoulder. Such an organism dwarfs related specimens known from China with wingspans of 0.8 m (2.6 ft):[11] For reference, the wingspan of an extant Wandering albatross is around 3 m, the wingspan of a crow around 0.6 m.

By the late Jurassic, 145 Mya, a new group of pterosaurs—**pterodactyls**—had arisen (FIGURE 24.8b). Pterodactyls survived until almost the end of the Cretaceous, 65 Mya. When compared with basal pterosaurs, the pterodactyl tail is much reduced, the fifth toe is lost and the teeth tend to be reduced in size and number (leading eventually to a long, toothless beak). In *Pteranodon* (FIGURE 24.9) and other forms, a long bony crest extends behind the skull. Bones of the wrist became greatly elongated; while standing on their hind limbs, pterodactyls could reach the ground with their forelimbs. This may explain why at least some of them left quadrupedal track ways.

Several lineages of pterodactyls became quite large, reflecting abundant resources and adaptive success. The largest pterodactyl, the Late Cretaceous *Quetzalcoatlus,* found in Texas, had a wingspan that may have reached 12 m (over 39 ft). Flying animals of that size have only a limited capability of active flight ("flapping"); the energetic costs of flight are too high. Instead, like modern vultures and condors, large pterodactyls must have mostly soared, using sea or land thermals for lift. Pterosaurs survived for 150 million years, but their flight ability did not protect them from extinction and, like

*We say "*FLYING BIRDS*" because some birds have lost the ability to fly. Flightless birds are discussed in End Box 24.3.*

[11] M. P. Witton, 2008. A new azhdarchoid pterosaur from the Crato Formation (Lower Cretaceous, Aptian?) of Brazil. *Palaeontology,* **51**, 1289–1300.

FIGURE 24.8 Comparison between rhamphorhynchoids, a lineage of pterosaurs (left in **a** and **b**) and pterodactyloids, a lineage of pterodactyls (right in **a** and **b**). Rhamphorhynchoids were tailed with wing spans of 0.3 to 2.1 m. Pterodactyloids had very short tails, and wing spans from 15 cm to as much as 12 m.

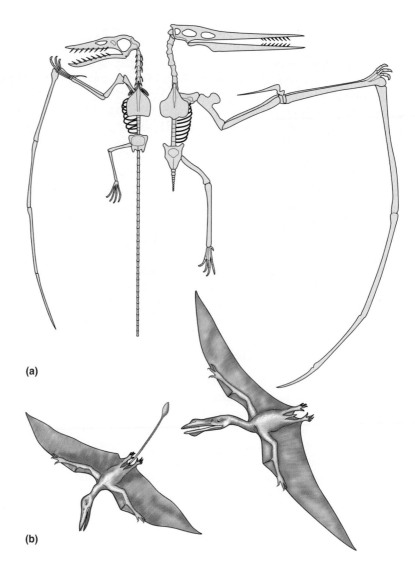

(a)

(b)

FIGURE 24.9 Reconstruction of *Pteranodon*, perhaps the largest Late Cretaceous pterodactyl.

© Paul B. Moore/ShutterStock, Inc.

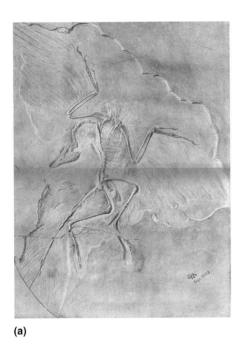

(a)

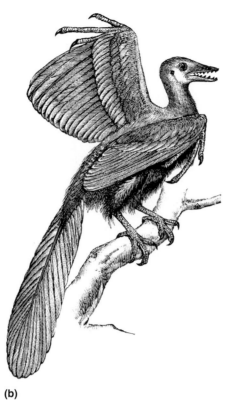

(b)

FIGURE 24.10 **(a)** The Berlin specimen of *Archaeopteryx*. **(b)** Heilmann's reconstruction of what *Archaeopteryx* may have looked like in real life.

[(a) and (b) Reproduced from Heilmann, G, 1927. *The Origin of Birds*. Appleton. (Reprinted Dover Publications, 1972, New York.)]

the non-avian dinosaurs, they did not survive beyond the end of the Cretaceous. Instead, their cousins the birds became the most widely distributed group of Cenozoic flyers.

AVIAN DINOSAURS: BIRDS

Based on cladistic classification of lineages, all birds nest within dinosaurs. Consequently, not all dinosaurs became extinct; birds are living dinosaurs. Several molecular studies have begun to unravel the relationships among the 35 or so avian lineages. Recent discoveries of feathered dinosaurs caused great excitement among paleontologists and others fascinated by the evolution of dinosaurs (**END BOX 24.3**).[12]

ARCHAEOPTERYX

The most famous early bird, and the first bird fossil discovered—though not the first bird, is *Archaeopteryx lithographica*. It was unearthed in Bavaria in 1961, lying in Upper Jurassic limestone 125 to 145 My old. Of the seven known skeletons, four are nearly complete. Some have flight feathers on the wings and tail, preserved in their natural positions (**FIGURE 24.10**). The primary feathers of *Archaeopteryx* are remarkably similar in vane structure to the primary (flight) feathers of today's flying birds, indicating that *Archaeopteryx* could indeed fly. Similarity between *Archaeopteryx* and dinosaurs is so striking that, had it lacked feathers, paleontologists would have classified *Archaeopteryx* as a dinosaur rather than a bird (**FIGURE 24.11**). Most striking are the similarities to small, carnivorous dinosaurs (Figure 24.11), whose teeth, separate clawed fingers, long bony tail, and dozens of other features indicate their status as a link between earlier carnivorous bipedal dinosaurs and extant birds (**FIGURE 24.12**).

ARCHAEOPTERYX *is not a "missing link" between dinosaurs and birds. It is too late in the fossil sequence and too derived to have been the first bird.*

[12] For birds as dinosaurs and for molecular studies on avian relationships, see Dingus and Rowe (1998), Cracraft et al. (2004), Chiappe (2007), and Hackett et al. (2008).

FIGURE 24.11 Skeletons of **(a)** *Compsognathus*, a dinosaur, **(b)** *Archaeopteryx*, a fossil bird, and **(c)** *Gallus*, a chicken. Although *Archaeopteryx* had proportionately longer forelimbs and digits than *Compsognathus*, birds and dinosaurs share some common skeletal features.

[(a) and (b) Adapted from Carroll, Robert, 1997. *Patterns and Processes of Vertebrate Evolution.* Cambridge University Press, Cambridge, UK. (c) Adapted from Dingus, L., and T. Rowe, 1998. *The Mistaken Extinction: Dinosaur Evolution and the Origin of Birds.* W. H. Freeman and Company.]

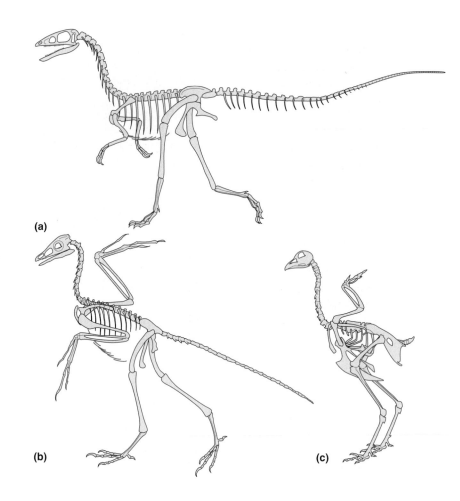

FIGURE 24.12 The phylogenetic relationships between birds and sauropod and ornithischian dinosaurs. Branch **(a)** indicates the presence of simple feathers; Branch **(b)** shows the presence of more complex (essentially modern) feathers.

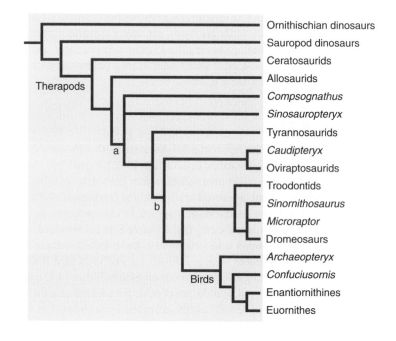

Interpretation of *Archaeopteryx* as a dinosaur and not an early bird has been strengthened over the past decade by discoveries of dinosaurs that share "avian" features found in *Archaeopteryx,* features such as feathers, a wishbone, and a forelimb with only three digits. The 2011 publication of the description of *Xiaotingia zhengi,* a new species of theropod-like dinosaur from China that is closely related to *Archaeopteryx,* coupled with phylogenetic analyses that include this new species, show that both species nest within dinosaurs (Xu et al., 2011, and see the commentary by Witmer, 2011). The transition of dinosaurs to birds and the recognition of living birds as flying dinosaurs is even more firmly grounded by these studies than it was 150 years ago when Thomas Huxley recognized *Archaeopteryx* as an avian cousin of the dinosaurs.

BIRD EVOLUTION

Within about 30 million years, a range of aquatic and shore birds had originated and diversified, indicative of rapid speciation and habitat exploitation. Some of these fossils represent groups such as flamingos, loons, cormorants, and sandpipers. Others still retained the teeth indicative of their reptilian origins. Lineages that can be placed into most of the recognized extant orders of birds appear to have originated about 60 to 90 Mya, that is, just before the Late Cretaceous and into the Tertiary Period. Although bird fossils are not plentiful, they exist in sufficient numbers and kinds of Eocene, Oligocene, and Cretaceous deposits in China to indicate that almost all the major clades of birds (recognized as orders) had evolved by these periods, some lineages having originated in the Late Cretaceous.

Figure 24.12 shows the phylogenetic relationships of birds and dinosaurs. *Archaeopteryx* is shown as a basal bird. *Confuciusornis* is known from 120 My-old (lower Cretaceous) deposits in China. Its beak was toothless, as in modern birds, indicating that teeth were lost early in avian evolution. Large claws on the forelimbs indicate that the several known species of *Confuciusornis* retained the claws seen on *Archaeopteryx.* Enantiornithines (literally, opposite birds), represented by *Nanantius eos,* composed a lineage that had many of the features of extant birds—feet that allowed them to perch, well-developed flight—but they did not survive the mass extinction 65 Mya.

The oldest fossil bird with the greatest resemblance to modern birds is a loon-like shorebird, *Gansus yumenensis,* described on the basis of five specimens found in Cretaceous deposits in China.[13] Analysis of *Gansus* and the oldest (already flightless) penguin, *Waimanu,* from the Paleocene (60 to 62 Mya) on New Zealand, combined with analysis of mitochondrial DNA, have opened new windows on the origin and diversification of birds that could not have been imagined even a few years ago (Slack et al., 2006).

INSECTS CONQUER LAND AND AIR

Many features enable **arthropods** to exploit almost every conceivable ecological habitat on Earth: for example, jointed paired appendages, a hardened exoskeleton with inner projections for muscle attachments, and highly developed sensory structures on the head structures (**FIGURE 24.13a,b**).[14] A hardened exoskeleton that, along with waxy waterproofing, acts as a barrier to desiccation, the ability to burrow into shoreline sand and soil, and protected gills that evolved into lungs are major traits accounting for arthropods' successful invasion of land and subsequent adaptive radiation. Invasion of the land by arthropods resulted in the evolution of terrestrial spiders, some terrestrial crustaceans, insects, millipedes, and centipedes and their relatives. Among arthropods, **insects** underwent what may be the most explosive adaptive radiation of any animals since the

---

[13] H.-I. You et al., 2006. A nearly modern amphibious bird from the early cretaceous of northwestern China. *Science,* **312,** 1640–1643.

[14] Arthropods, nematodes, and other animals with a molting phase in their life cycle were united into a single group, the Ecdysozoa by Aguinaldo et al. (1997).

FIGURE 24.13 Features of arthropods include paired jointed legs as seen in this grasshopper **(a)** and elaborate sensory appendages such as antennae in this bee **(b)**. Dragonflies in the Carboniferous Era had wingspans of half a meter **(c)**, while Ordovician sea scorpions were as much as two meters in length **(d)**.

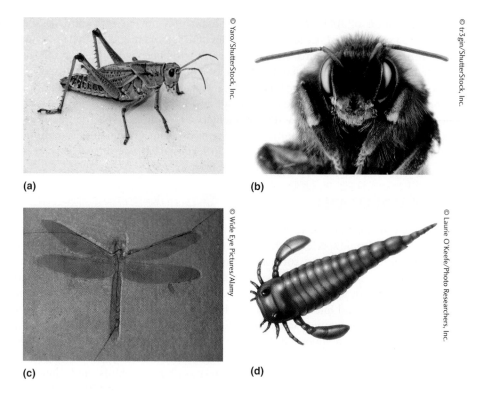

(a)

(b)

(c)

(d)

Cambrian, diversifying into 900,000 known extant species and as many as eight million undescribed species (Grimaldi and Engel, 2005). One group of insects, the beetles (coleopterans), contains 350,000 described species. Alfred Wallace alone collected 80,000 beetles in the Malay Archipelago.

Although some early insects reached relatively large sizes—some Carboniferous dragonflies measured 0.6 m (2 ft) between wing tips (FIGURE 24.13c)—insects have tended to remain small; there are limits to the volume of tissue in which insect respiratory tubules can effectively exchange gases. An even more limiting factor may be that the insect exoskeleton would have had to become much heavier and more unwieldy as insects became larger. Sea scorpions (eurypterids; Figure 24.13d) could get away with it in the ocean because of the buoyancy of water, but arthropods could not do the same on land.

The enormous diversification of insects has been attributed to the **modular organization** of the insect body, in which antennae can evolve independently of wings, mouthparts independently of legs, and so forth—a process known as **mosaic evolution**. Homeotic mutations, so common in insects, attest to the comparative ease with which individual segments can be altered in a body plan that consists of serially repetitive elements. Such partial independence of body parts, or the embryonic units from which they arise, is known as **modularity**. The speed with which gene regulation and development can be modified is referred to as **tinkering**.

Coevolution of flowering plants and insect pollinators was a further major factor in the diversification of insects. So too was **life history evolution**; advantages accrued to those lineages with specialized larval stages that lived and fed differently than adult forms. An aquatic larva with a terrestrial adult or a wingless larva and winged adult can exploit a wide range of aquatic and terrestrial environments and diets.

SOCIAL ORGANIZATION

Insects diversified into both **social** and **nonsocial forms**. Key to the success of insect evolution has been the evolution of **social organization**, which entails a **division of labor** and **phenotypic differentiation** among different members of a group. Social organization in insects goes

(a)

(b)

FIGURE 24.14 A swarm of caterpillars in and on the temporary "tent" they have constructed **(a)** and a termite mound in the Kakadu National Park in northern Australia **(b)**. The interior of a termite mound **(c)** is highly organized into specialized territories, reflecting cooperation between the insects in the mound.

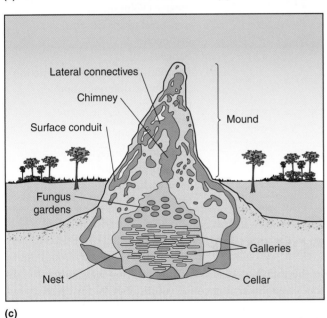
(c)

far beyond the kinds of simple aggregation exemplified by swarms of migrating locusts or the cooperative "tents" that some caterpillars construct (**FIGURE 24.14a**). Social insect societies include termites (**FIGURE 24.14b,c**) and hymenopterans such as various ants, bees, and wasps.

In social insects, different morphological types (castes) or age groups assume different functions. Each colony, for example, usually has only one fertile queen engaged in egg production, often many fertile males for egg fertilization, and one or more classes of sterile workers exclusively engaged in food gathering, cooperative brood care, nest maintenance, and defense of the colony. In worker honeybees, age-related division of labor ties younger bees to caretaking and maintenance, and older bees to food foraging and defense. Ability to develop different sexes or life history stages from fertilized and unfertilized eggs, and the evolution of **kin selection**, also characterize some forms of social insects in which each colony has a single mated queen. The complete sequencing of the genome of the honeybee, *Apis mellifera,* has already revealed differences in patterns of gene expression between different castes and associated with foraging.

Only one example of caste formation is known among vertebrates. It is in the NAKED MOLE RAT, HETEROCEPHALUS GLABER, *from East Africa, in which one female produces all the young and in which there is complete sexual dimorphism between reproductive and helper females.*

Hormones provide a key mechanism linking and integrating environmental changes to gene action and to changes in the phenotype, especially in insects in which behavioral differences between castes correlate with differences in juvenile hormone. Chemical influences on development and behavior also are responsible for determining who will be a queen and who will be the sterile workers. The queen produces a diffusible hormone (a pheromone) that suppresses fertility in workers, a suppression that can be overcome by various environmental influences including special foods, such as royal jelly in honeybees.

The social structure and interactions among and between the various castes in social insects (queen, workers, soldiers), each of which makes a specific contribution to the origin and maintenance of the colony and which could not survive without the presence of other castes, led to the development of the concept of social insects as **superorganisms**.[15]

■ KEY TERMS

arthropods	man-made extinctions
Chicxulub crater	mass extinction
coevolution	modular organization
coextinction	modularity
Collision Theory	mosaic evolution
Deccan Traps	nonsocial forms (insects)
division of labor	orthogenesis
extinction	Pangaea
global extinction	Permian–Triassic (P–T) extinction
Gondwana	phenotypic differentiation
Great Dying	pterodactyls
insects	seven glaciations
iridium	snowball Earth
kin selection	social forms (insects)
K–T extinction	social organization
Laurasia	superorganisms
life history evolution	tinkering

■ EVOLUTION ON THE WEB

Explore evolution on the Internet! Visit the accompanying website for *Strickberger's Evolution, Fifth Edition,* at **go.jblearning.com/Evolution5eCW** for exercises and links relating to topics covered in this chapter.

■ REFERENCES

Aguinaldo, A. M. A., J. M. Turbeville, L. S. Linford, et al., 1997. Evidence for a clade of nematodes, arthropods and other moulting animals. *Nature*, **387**, 489–493.

Alfaro, M., and F. Santini, 2010. A flourishing of fish forms. *Nature*, **464**, 840–842.

Alvarez, L. W., 1983. Experimental evidence that an asteroid impact led to the extinction of many species 65 million years ago. *Proc. Natl Acad. Sci. USA*, **80**, 627–642.

Alvarez, W., and F. Asaro, 1992. The extinction of the dinosaurs. In *Understanding Catastrophe.* J. Bourriau (ed.). Cambridge University Press, Cambridge, UK, pp. 28–56.

Benton, M. J., 2003. *When Life Nearly Died: The Greatest Mass Extinction of All Time.* Thames and Hudson, London.

---

[15] For social insects and social insects as superorganisms, see Grimaldi and Engel (2005), Reeve and Hölldobler (2007), Hölldobler and Wilson (2009), and Johnson and Linksvayer (2010).

Buffetaut, E., and J.-M. Mazin (eds.), 2003. *Evolution and Paleobiology of Pterosaurs*. Geological Society of London Special Publication 217. Geological Society, London.

Chiappe, L. M., 2007. *Glorified Dinosaurs: The Origin and Early Evolution of Birds*. University of New South Wales Press/John Wiley, Coogie, NSW, Australia.

Condie, K. C., 1997. *Plate Tectonics and Crustal Evolution*, 4th ed. Pergamon, New York.

Condie, K. C., and R. E. Sloan, 1998. *Origin and Evolution of Earth: Principles of Historical Geology*. Prentice Hall, Upper Saddle River, NJ.

Cracraft, J., F. K. Barker, M. J. Braun, et al., 2004. Phylogenetic relationships among modern birds (Neornithes): toward an avian Tree of Life. In *Assembling the Tree of Life*. J. Cracraft and M. J. Donoghue (eds.). Oxford University Press, New York, pp. 468–489.

Dingus, L., and T. Rowe, 1998. *The Mistaken Extinction: Dinosaur Evolution and the Origin of Birds*. W. H. Freeman and Company, New York.

Erwin, D. H., 2006. *Extinction: How Life on Earth Nearly Ended 250 Million Years Ago*. Princeton University Press, Princeton, NJ.

Erwin, D. H., M. Laflamme, S. M. Tweedt, et al., 2011. The Cambrian conundrum: early divergence and later ecological success in the early history of animals. *Science*, **334**, 1091–1097.

Flynn, J. J., and A. R. Wyss, 1998. Recent advances in South American mammalian paleontology. *Trends Ecol. Evol.*, 13, 449–454.

Foote, M., J. P. Hunter, C. M. Janis, and J. J. Sepkoski, Jr., 1999. Evolutionary and preservational constraints on origins of biologic groups: divergence times of eutherian mammals. *Science*, **283**, 1310–1314.

Fox, W. T., 1987. Harmonic analysis of periodic extinctions. *Paleobiology*, **13**, 257–271.

Grimaldi, D., and M. S. Engel, 2005. *Evolution of the Insects*. Cambridge University Press, Cambridge, UK.

Hackett, S. J., R. T. Kimball, S. Reddy, et al., 2008. A phylogenomic study of birds reveals their evolutionary history. *Science*, **320**, 1763–1768.

Hallam, A., 2005. *Catastrophies and Lesser Calamities*. Oxford University Press, Oxford, UK.

Hölldobler, B., and E. O. Wilson, 2009. *The Superorganism: The Beauty, Elegance, and Strangeness of Insect Societies*. W. W. Norton, New York.

Johnson, B. R., and T. A. Linksvayer, 2010. Deconstructing the superorganism: social physiology, ground plans, and sociogenomics. *Q. Rev. Biol.*, **85**, 57–78.

Kemp, T. S., 2005. *The Origin and Evolution of Mammals*. Oxford University Press, Oxford, UK and New York.

Kemp, T. S., 2007. The origin of higher taxa: macroevolutionary processes, and the case of the mammals. *Acta Zool.* (Stockh.), **88**, 3–22.

Kielan-Jaworowska, Z., R. L. Cifelli, and Z.-X. Luo, 2004. *Mammals from the Age of Dinosaurs. Origins, Evolution, and Structure*. Columbia University Press, New York.

Lü, J., D. M. Unwin, D. C. Deeming, et al., 2011. An egg-adult association, gender, and reproduction in pterosaurs. *Science*, **331**, 321–324.

Luo, Z.-X., 2007. Transformation and diversification in early mammal evolution. *Nature*, **450**, 1011–1019.

Raup, D. M., and J. J. Sepkoski, Jr., 1986. Periodic extinction of families and genera. *Science*, **231**, 833–835.

Reeve, H. K., and B. Hölldobler, 2007. The emergence of a superorganism through intergroup competition. *Proc. Natl Acad. Sci. USA*, **104**, 9736–9740.

Schulte, P., L. Alegret, I. Arenillas, et al., 2010. The Chicxulub asteroid impact and mass extinction at the Cretaceous-Paleogene boundary. *Science*, **327**, 1214–1218.

Sharpton, V. L., K. Burke, A. Camargo-Zanoguera, et al., 1993. Chicxulub multiring impact basin: size and other characteristics derived from gravity analysis. *Science*, **261**, 1564–1567.

Slack, K. E., C. M. Jones, T. Ando, et al., 2006. Early penguin fossils, plus mitochondrial genomes, calibrate avian evolution. *Mol. Biol. Evol.*, **23**, 1144–1155.

Turney, C. S., M. Flannery, T. F. Roberts, et al., 2008. Late-surviving megafauna in Tasmania, Australia, implicate human involvement in their extinction. *Proc. Natl Acad. Sci. USA*, **105**, 12150–12153.

Wible, J. R., G. W. Rougier, M. Novacek, and R. J. Asher., 2007. Cretaceous Eutherians and Laurasian origin for placental mammals near the K/T boundary. *Nature*, **447**, 1003–1006.

Witmer, L. M., 2011. An icon knocked from its perch. *Nature*, **475**, 458–459.

Xu, X., H. You, K. Du, and F. Han, 2011. An *Archaeopteryx*-like theropod from China and the origin of Avialae. *Nature*, **475**, 465–470.

END BOX 24.1

Moving Continents and South American Mammalian Radiations

© Photos.com

SYNOPSIS: The origin and radiation of mammals in what is now South America provides an excellent example of how evolution was facilitated by the breakup of old landmasses and the creation and of new ones as a result of continental drift.

Because of the pattern of continental drift, introduced elsewhere in this text, it is clear that early monotremes and marsupials entered southern parts of Pangaea by the Late Jurassic and Early Cretaceous (**FIGURE EB1.1a**). Subsequent separation of Australia from other land masses isolated its mammalian fauna from later competition. More derived eutherian lineages evolved in western Pangaea during the Late Cretaceous and Early Tertiary (Figure EB1.1b).

mtDNA analysis of extant Australian marsupials adds a further class of data that can be used to determine when Australian marsupials radiated, and is providing new insights into the phylogenetic relationship among monotreme, marsupial, and placental mammals (Janke et al., 1997). In South America, marsupials and early placentals had replaced monotremes by the Early- and Mid-Tertiary. By that time, the South American continent had drifted considerably from Africa and separated from North America (Figure EB1.1c). Analysis of the first marsupial genome to be sequenced, that of the North American short-tailed opossum, *Monodelphis domestica,* led to the conclusion that much of the 20 percent difference in conserved non-coding elements between eutherians and marsupials represents eutherian elements that arose from insertion of transposable elements after marsupials separated from eutherians (Mikkelsen et al., 2007).

Xenarthrans are also called EDENTATES *because of their reduced or suppressed teeth.*

South American placentals, although beginning only with some ungulates and xenarthrans ("strange-jointed"), radiated even more rapidly than did marsupials on that isolated continent (Flynn and Wyss, 1998; Kemp, 2005). By the Early Eocene, within 15 to 20 My of their initial Late Cretaceous colonization, placentals comprised 75 to 100 new genera in some 15 families. Xenarthrans produced a strange bestiary of armadillos, glyptodonts, sloths, and anteaters (**FIGURE EB1.2a**). Also radiating widely were the hoofed ungulates (Figure EB1.2b). Convergent or parallel evolution produced striking similarities on different continents.

FIGURE EB1.1 Effects of continental drift on the dispersion and isolation of major lineages of mammals over the 140 My spanning the Jurassic to the Pliocene.

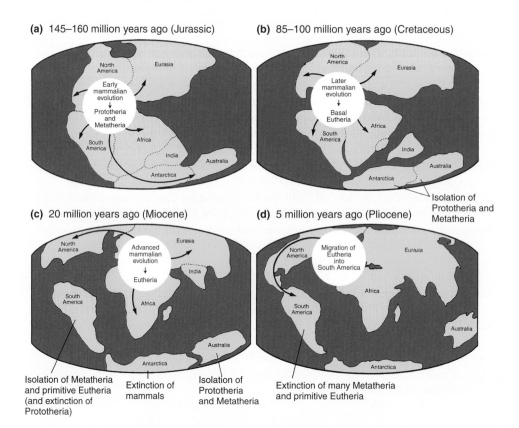

(a) 145–160 million years ago (Jurassic)

(b) 85–100 million years ago (Cretaceous)

(c) 20 million years ago (Miocene)

(d) 5 million years ago (Pliocene)

Isolation of Prototheria and Metatheria

Isolation of Metatheria and primitive Eutheria (and extinction of Prototheria)

Extinction of mammals

Isolation of Prototheria and Metatheria

Extinction of many Metatheria and primitive Eutheria

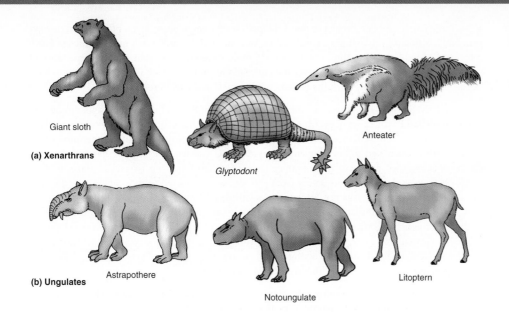

FIGURE EB1.2

Reconstructions of early placental mammals of South America: **(a)** three lineages of xenarthrans (edentates) and **(b)** three lineages of ungulates that arose in the South American placental mammalian radiation.

[Adapted from Steel, R., and A. P. Harvey, 1979. *The Encyclopaedia of Prehistoric Life*. Mitchell-Beazley, 1979.]

In the Oligocene, a similar rapid radiation began among rodents and primates that had reached South America from Africa, most likely by "island hopping" along the island chains on the oceanic ridges cast up in the South Atlantic Ocean. Rodents produced a great diversity of caviomorphs (cavies) distinguished by special jaw muscle attachments. Primates, confined mostly to tropical areas, produced the wide array of New World monkeys.

Evolution of mammals on the isolated continent of South America was largely independent of mammalian evolution elsewhere until South America rejoined North America via the Panama Isthmus during the Pliocene (Figure EB1.1**d**). During the Oligocene, however, island hopping, combined with transport on floating debris ("rafting"), allowed various monkeys and caviomorph rodents to move to South America from Africa or North America.

By the Pliocene, 5 Mya, considerable evolution toward more derived eutherian forms had occurred in Africa and Laurasia (now North America and Eurasia); most of the South American mammalian fauna showed far less change. When the Pleistocene began, massive invasions of northern eutherians across the Central American land bridge into South America resulted in the rapid extinction of many South American mammalian families. Only rarely, as with opossums, did early South American mammals manage to successfully invade North America. Supporting this view of plate tectonics and mammalian evolution is abundant fossil evidence of South American extinctions in the Pliocene and Pleistocene.[*]

REFERENCES

Flynn, J. J., and A. R. Wyss, 1998. Recent advances in South American mammalian paleontology. *Trends Ecol. Evol.*, **13**, 449–454.

Janke, A., X. Xu, and U. Arnason, 1997. The complete mitochondrial genome of the wallaroo (*Macropis robustus*) and the phylogenetic relationship among Monotremata, Marsupialia, and Eutheria. *Proc. Natl Acad. Sci.* USA, **94**, 1276–1281.

Kemp, T. S., 2005. *The Origin and Evolution of Mammals*. Oxford University Press, Oxford, UK and New York.

Mikkelsen, T. S., M. J. Wakefield, B. Aken, et al., 2007. Genome of the marsupial *Monodelphis domestica* reveals innovation in non-coding sequences. *Nature*, **447**, 167–177.

Steel, R., and A. P. Harvey, 1979. *The Encyclopaedia of Prehistoric Life*. Mitchell-Beazley, London, UK.

---

[*]N. Owen-Smith, 1987. Pleistocene extinctions: the pivotal role of megaherbivores. *Paleobiology*, **13**, 351–362.

END BOX 24.2

Snowball Earth

SYNOPSIS: Earth has been subject to many fluctuations in temperature, levels of atmospheric oxygen, bombardment from extraterrestrial bodies, volcanic activity, and much more. *Snowball Earth* refers to 10 My of global cooling that began 635 Mya, resulting in a mean global temperature of −40°C and a 1 km-thick layer of ice over the entire globe. Snowball Earth is hypothesized to have been a major factor holding back the evolution of animals in the late Precambrian.

SNOWBALL EARTH *may have been a more regular feature of Earth's history than previously anticipated. An even earlier snowball Earth, dating from about 2.2 Bya and detected on the basis of glacial deposits in South Africa, would have had profound impacts on life (Kopp et al., 2005).*

A relatively recent hypothesis for the lack of evidence of Precambrian animals goes under the name "**snowball Earth**," a set of events 635 Mya that resulted in the formation of a kilometer-thick layer of ice over the entire Earth. This ice persisted for 10 My, during which the temperature of the entire Earth is calculated to have hovered around −40°C. Subsequent research has revealed evidence for at least two, and as many as four, episodes of snowball Earth between 725 and 635 Mya.[*] Until 1964, when glacial deposits dating from 600 Mya were reported from what are now the baking deserts of Nambia. It was not known that such tropical regions had (an) ice age(s) in their past. The discovery of similar glacial deposits on every present day continent (FIGURE EB2.1) led to the hypothesis of a long period of intense, worldwide cold. Mathematical models developed in the 1970s showed how, under certain conditions, runaway freezing of the whole planet was almost inevitable. How did Earth and its evolving organisms recover from the effects of prolonged universal temperatures as low as −50°C? The hypothesis is as follows:

- The intense heat of emerging volcanoes penetrated the ice, releasing CO_2, which is a greenhouse gas that raises the temperature of the atmosphere.
- CO_2 levels increased because it was too cold for rain to wash the CO_2 from the atmosphere.

FIGURE EB2.1 Carbonaceous rocks in South Australia (indicated by Rudy Raff of Indiana University) deposited following the melting of ice associated with "snowball Earth."

[*]For evidence for and implications of snowball Earth, see Hoffman et al. (1998), Hoffman and Schrag (2000, 2002), Bodiselitsch et al. (2005), Kopp et al. (2005), and Harland (2007).

- Eventually, atmospheric CO_2 reached such high levels—hypothesized to be as much as 10% or 260 times current levels—that the greenhouse effect kicked in, elevating temperatures to levels where the ice began to melt and disappeared over as short a period as a few thousand years.

This is also how geologists explain another puzzle in Namibia: the presence, immediately above the glacial deposits, of thick layers of carbonaceous rock that form only in water and that bear all the marks of having been deposited very quickly (Figure EB2.1). Again, these carbonaceous rocks are found around the globe. With these findings, the hypothesis of recovery from snowball Earth has been extended.

- As temperatures continued to rise from the coldest conditions ever experienced on Earth (−50°C) to the hottest (+50°C), evaporation from the oceans resulted in torrential rains that may have lasted for centuries, washing CO_2 from the air.
- CO_2 dissolved in water forms a weak acid, carbonic acid. Large amounts of carbonic acid would have bathed the globe, interacting with particles of rock eroded by the torrential rains. The resulting carbonaceous deposits formed the cap on the glacial remains seen in the geological record and illustrated in Figure EB2.1. Eventually, conditions would have returned to temperatures and CO_2 levels more typical of Earth over the past 500 to 600 My.
- Although the waters of the deep oceans were anoxic after snowball Earth, recent studies are consistent with atmospheric oxygen and marine sulfate concentrations having been higher after than before snowball Earth. These conditions would have provided a marine environment that could have been colonized by the Precambrian (Ediacaran) biota discussed elsewhere in this text and known to have arisen between 635 and 540 Mya.

As you might expect, alternatives to snowball Earth have been put forth as an explanation for the paucity of fossils in the Precambrian and their "sudden" appearance in the Lower Cambrian. Three physicists at the University of Toronto integrated an analysis of the global carbon cycle with a model for climate change in the Precambrian to conclude that remineralization of dissolved organic carbon in the oceans would have increased atmospheric CO_2, enhanced greenhouse warming, and prevented snowball Earth (Peltier et al., 2007, and see an evaluation of the model by Kaufman, 2007). More recently, an international group of geochemists used data from Neoproterozoic glacial deposits to challenge the proposed high levels of CO_2 required for snowball Earth (Sansjofre et al., 2011).

When you consider that GLOBAL CLIMATE CHANGE *today is being driven by quite small increases in CO_2 levels— from less than 0.03% before the Industrial Revolution to 0.038% today—imagine the impact of 10% CO_2. Runaway global warming would be the only logical outcome.*

REFERENCES

Bodiselitsch, B., C. Koeberl, S. Master, and W. U. Reimold, 2005. Estimating duration and intensity of Neoproterozoic snowball glaciations from Ir anomalies. *Science* **308**, 239.

Harland, W. B., 2007. Origin and assessment of snowball Earth hypotheses. *Geol. Mag.*, **144**, 633–642.

Hoffman, P. F., A. J. Kaufman, G. P. Halverson, and D. P. Schrag, 1998. A Neoproterozoic snowball Earth. *Science*, **281**, 1342–1346.

Hoffman, P. F., and D. P. Schrag, 2000. Snowball Earth. *Sci. Am.*, **282**(1), 68–75.

Hoffman, P. F., and D. P. Schrag, 2002. The snowball Earth hypothesis: testing the limits of global change. *Terra Nova*, **14**, 129–155.

Kaufman, A. J., 2007. Slush find. *Nature*, **450**, 807–818.

Kopp, R. E., J. L. Kirschvink, I. A. Hilburn, and C. Z. Nash, 2005. The Paleoproterozoic snowball Earth: a climate disaster triggered by the evolution of oxygenic photosynthesis *Proc. Natl Acad. Sci. USA*, **102**, 11131–11136.

Peltier, W. R., Y. Liu, and J. W. Crowley, 2007. Snowball Earth prevented by dissolved organic carbon remineralization. *Nature*, **450**, 813–818.

Sansjofre, P., M. Adler, R. I. F. Trindale, et al., 2011. A carbon isotope challenge to the snowball Earth. *Nature*, **478**, 93–96.

END BOX 24.3

Origin of Feathers and Evolution of Flightlessness

© Photos.com

SYNOPSIS: Evidence for the origin of *feathers* in (a) lineage(s) of dinosaurs in the Late Cretaceous, for the origin of birds from a dinosaur lineage and for the subsequent loss of the ability to fly in some lineages of birds is summarized.

ORIGIN OF FEATHERS

Feathers are homologues of and derived from reptilian scales. *Archaeopteryx,* which is feathered, is an early bird but not the first bird. Until recently, the lack of any pre-*Archaeopteryx* fossils with preserved soft tissue kept alive the mystery of when and how feathers first evolved, and left the origin of avian flight unresolved.

Two major hypotheses for the origin of feathers have been presented:
- Feathers were an adaptation for insulating the (presumed) warm-blooded and ground-dwelling reptilian ancestors of birds. Fossils of **feathered dinosaurs** (below) suggest (demonstrate?) an insulation/courtship-display function.
- Ancestral birds were tree-dwelling reptiles that used their developing wings to glide from branch to branch, or were ground-dwelling runners whose feathers formed planing surfaces, enabling them to increase their speed.

ORIGIN OF BIRDS

Considerable evidence supports the conclusion that some SMALL DINOSAURS COULD REGULATE THEIR BODY TEMPERATURES *physiologically, without relying on behavior such as moving into the shade or facing their body profile away from the Sun's rays.*

Overwhelming classes of evidence supports the origin of birds from ground-dwelling dinosaurs.

In the mid 1970s American paleontologist John Ostrom (1928–2005) proposed that feathers evolved primarily as a means of controlling heat loss in warm-blooded dinosaurs. The report in 1998 of a mane of feathers down the back of a basal carnivorous bipedal dinosaur, *Compsognathus* (see Figures 24.11 and 24.12), produced immediate discussion. A mane of feathers on a dinosaur provides strong support for Ostrom's notion that birds originated from a feathered-dinosaur ancestor.[a]

Recent fossil discoveries lead us to ask whether any species of animals with feathers should be called a bird, or whether some derived dinosaurs also had feathers. In the last ten years, species of "feathered dinosaurs" belonging to 14 genera have been discovered. The most basal of these, *Sinosauropteryx* (see Figure 24.12) from the Jurassic–Cretaceous boundary, some 150 to 120 Mya, had hollow, feather-like structures on its entire body. Feathers of *Beipiaosaurus* from 124.6 Mya consisted of single broad filaments, consistent with the first stage of feather evolution (Xu et al., 2009). *Protarchaeopteryx* and *Caudipteryx* from 135 to 121 Mya had feathers that resembled more closely the flight feathers of birds (Currie et al., 2004).

Astonishing as these animals are, to everyone's amazement a dinosaur, *Microraptor,* was discovered with feathers on both fore and hind limbs (**FIGURE EB3.1**), adaptations interpreted to show that this four-winged carnivorous bipedal dinosaur could have glided through the Cretaceous forests. At 40 cm (16 in)-long, including a 25 cm (10 in)-long tail (close to the size of *Archaeopteryx*), *Microraptor* was small enough and sufficiently light to fly (Xu et al., 2003).

Even more astounding is the recent description of *Xiaotinga zhengi,* an *Archaeopteryx*-like theropod dinosaur whose "avian" features led Xu et al. (2011) to reinterpret *Archaeopteryx* as a dinosaur and not an early bird; *Archaeopteryx* was knocked from its perch (Witmer, 2011).

FLIGHTLESSNESS

The extinction of the dinosaurs opened up another dramatic adaptive opportunity—vacant terrestrial habitats where various large, flightless ground birds evolved. Giant forms, such as the 2.1 m (6.9 ft)-tall *Diatryma* (**FIGURE EB3.2**), and others that may have reached a height of 3 m (nearly 10 ft) or

---

[a] See Ostrom (1974, 1991) for warm-blooded dinosaurs, and see Chen et al. (1998) for *Compsognathus.*

FIGURE EB3.1 Reconstruction of a four-winged *Microraptor.*

more, were widely distributed until they became extinct later in the Cenozoic.[b] Few flightless birds now survive. Ostriches of Africa, emus and cassowaries of Australia, rheas of South America, and the smaller flightless species such as kiwis and island rails are mostly confined to diminishing habitats. Marine flightless birds, such as penguins, are more widespread.

We assume that the evolution of flightlessness was associated with changes in selection pressure, either absence of predation or as a response to newly forming marine habitats. In the first case,

3.6 m —

1.8 m —

Diatryma gigantea

Aepyornis maximus
(elephant bird)

Dinornis maximus
(giant moa)

FIGURE EB3.2
Reconstructions of some extinct large flightless birds showing their relative sizes. *Diatryma* was an early Cenozoic bird; the others date from the much later Pleistocene.

[Adapted from Feduccia, A., 1980. *The Age of Birds.* Harvard University Press, Cambridge, MA.]

[b]One group of flightless birds, the "thunder birds" (Dromornithidae) from Australia, includes one of the largest birds ever, *Dromornis stirtoni,* estimated to have weighed 570 kg (over 1520 lbs) (Murray and Vickers-Rich, 2004).

END BOX 24.3

Origin of Feathers and Evolution of Flightlessness (Cont…)

© Photos.com

in protected or island habitats where major carnivorous forms were absent, local birds could evolve to dominate the terrestrial food chain. Some marine birds, such as penguins and steamer ducks, which spend little time flying in air, responded to selection for wing modifications that enhanced underwater propulsion but reduced flight ability. These marine birds "fly underwater." The major difference between flight in air and in water is the density of the medium and not the dynamics or mechanics of this form of locomotion.

REFERENCES

Chen, P-J, Z-M. Dong, and S-N. Zhen, 1998. An exceptionally well-preserved theropod dinosaur from the Yixian Formation of China. *Nature*, **391**, 147–152.

Chiappe, L. M., 2007. *Glorified Dinosaurs: The Origin and Early Evolution of Birds*. University of New South Wales Press/John Wiley, NSW, Australia.

Currie, P. J., Koppelhus, E. B., M. A. Shugar, and J. L. Wright (eds.), 2004. *Feathered Dragons: Studies on the Transition from Dinosaurs to Birds*. Indiana University Press, Bloomington, IN.

Feduccia, A., 1980. *The Age of Birds*. Harvard University Press, Cambridge, MA.

Murray, P. F., and P. Vickers-Rich, 2004. *Magnificent Mihirungs. The Colossal Flightless Birds of the Australian Dreamtime*. Indiana University Press, Bloomington, IN.

Ostrom, J. H., 1974. Archaeopteryx and the origin of flight. *Q. Rev. Biol.*, **49**, 27–47.

Ostrom, J. H., 1991. The question of the origin of birds. In *Origins of the Higher Groups of Tetrapods: Controversy and Consensus*. H.-P. Schultze and L. Trueb (eds.). Cornell University Press, Ithaca, New York, pp. 467–484.

Witmer, L. M. 2011. An icon knocked from its perch. *Nature*, **475**, 458–459.

Xu, X., H. You, K. Du, and F. Han. 2011. An *Archaeopteryx*-like theropod from China and the origin of Avialae. *Nature*, **475**, 465–470.

Xu, X., X. Zheng, and H. You, 2009. A new feather type in a nonavian theropod and the early evolution of feathers. *Proc. Natl. Acad. Sci. USA*, **106**, 832–834.

Xu, X., Z. Zhou, X. Wang, et al., 2003. Four-winged dinosaurs from China. *Nature*, **421**, 335–340.

PART 8

Human Origins, Evolution, and Influence

Image © Taras Vyshnya/ShutterStock, Inc.

CHAPTER

25

Human Origins

CHAPTER SUMMARY

Monkeys and apes are primates, a lineage of mammals that arose 85 Mya. Chimpanzees are the closest relatives (sister group) to humans. Apes are estimated to have diverged from monkeys between 29 and 34 Mya. Hominins (humans of the genus *Homo* and their ancestors) diverged from chimpanzees between 7 and 7.5 Mya. The ancestors of our species, *Homo sapiens*, arose in Africa and from there migrated to colonize the globe. This chapter explores the nature of the evidence that documents human evolution in, and migration out of, Africa. The many fossil species known present the problem of how to identify, separate, and relate the numerous species of bipedal primates. Existence of numerous species in the genus *Homo* and numerous species of the earlier southern apes of Africa (australopithecines) provide the evidence that more than one lineage of hominins (among which humans nest) existed in Africa. Enlarged brains, the ability to walk and run upright on their hind limbs (bipedalism), speech, language, and highly developed social organization characterize the lineage that results in humans; the manufacture and use of tools pre-existed the origin of our species. Analysis of the various lines of available evidence supports the hypothesis that the evolution of bipedalism was the driving force behind other changes in the human lineage, including the evolution of humans as hunter-gatherers.

PRIMATES: A RAPID SURVEY OF 85 MILLION YEARS OF EVOLUTION

This section serves as a summary of primate classification, types of primates, and the lineages of bipedal primates discussed in this chapter.

Primates, a group (order) of mammals comprising lemurs, tarsiers, monkeys, chimpanzees, and other apes, arose some 85 million years ago (**FIGURE 25.1a**). All primates share large brains, a grasping hand, stereoscopic vision, and social organization. The 400 species of living primates range in body size from the smallest lemur (30 g; about 1 oz) to the largest gorillas (227 kg; 500 lbs; Fleagle, 1999).

Primates are **classified** within two monophyletic lineages that diverged about 77 Mya in the Late Cretaceous:

- Wet-nosed primates (strepsirrhines) are small-bodied, nocturnal (active at night) lemurs of Madagascar and lorises of Africa and South East Asia.
- Dry-nosed primates (haplorhines) are classified into three lineages: New World monkeys, Old World monkeys, and apes (**FIGURES 25.2** and **25.3**).

The first **bipedal primates** (and the earliest **hominins**; **BOX 25.1**) are represented by two African species, *Orrorin tugenensis* and *Ardipithecus kadabba*, which are 6.0 and 5.5 My-old, respectively[1] (Figure 25.1**b**). Slightly more recent species in the genera *Ardipithecus* and *Australopithecus* ("southern apes") show long bone changes indicative of predominant bipedal locomotion, and modifications in the teeth and jaws signifying changing diet. As the brain of **australopithecines** was no larger than the brain of the earliest bipedal hominins, we can conclude that increase in brain size *followed* the origin of bipedalism. Although this reads like an obvious conclusion, the issue has been contentious.

The earliest fossils of the genus *Homo*, which are 2.4 to 2.5 My-old, have brains much larger than those of the australopithecines. Four African species are estimated to have existed between 2.4 and 1.6 Mya. *Homo ergaster* is regarded as having given rise to *H. erectus*, the dominant hominin species at the time. *H. erectus* may have persisted in Southeast Asia until 600,000 to 700,000 years ago and in Java until only 50,000 years ago. By 500,000 years ago, sufficient evolutionary change had accumulated that three new species can be identified: *Homo heidelbergensis, H. helmei,* and *H. rhodesiensis.* These may be forerunners of species known as **anatomically modern humans** (*Home sapiens*), which first appear as fossils in Ethiopia 160,000 years ago (Figure 25.1**b**; Wood, 2005, 2010).

Populations of anatomically modern humans expanded rapidly, spreading in multiple migrations from Africa to Europe, Asia, and then to the New World—the "Out of Africa hypothesis" discussed below. It is thought that *Homo heidelbergensis* gave rise to *Homo neanderthalensis* (Neanderthals) in Europe about 250,000 years ago. Neanderthals persisted in Spain and Gibraltar as late as 24,000 years ago, leading to suggestions that they could have coexisted with *Homo sapiens*.

CHIMPANZEES AND HUMANS

African apes, particularly chimpanzees, share many features with humans (Figure 25.3). Chimpanzees have many behavioral traits once thought unique to humans, including tool use and modification, planning, organized aggression, and

Humans and their closest extinct relatives are known as HOMININS *(Tribe* Hominini). *See Figure 25.3 and Box 25.1.*

[1] A third and older (7-My-old) African species and the first hominin, *Sahelanthropus tchadensis,* which may have been bipedal, is discussed in the text.

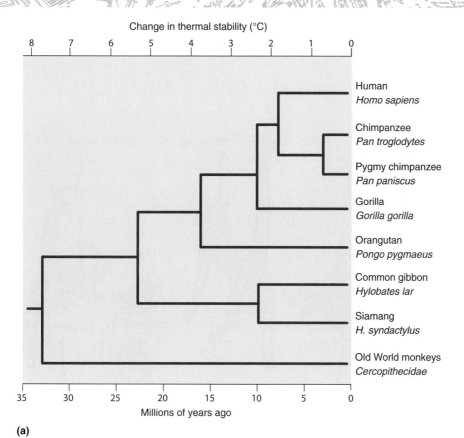

Change in thermal stability (°C)

Millions of years ago

(a)

FIGURE 25.1

(a) A phylogenetic tree and dates of divergence for humans, Great Apes, and Old World monkeys based on DNA–DNA hybridization studies. **(b)** Species from the fossil record identified as hominins, grouped into grades and plotted in relation to their estimated or known ages (Mya). Abbreviations for genera are *Ar, Ardipithecus; Au, Australopithecus; H, Homo; K, Kenyanthropus; O, Orrorin; P, Paranthropus; S, Sahelanthropus. (P. boisei and P. robustus* are discussed as species of *Australopithecus* in the text.)

[(a) Adapted from Sibley, C. G., and J. E. Ahlquist. *J. Mol. Evol.*, 20, 1984: 2–15; (b) Adapted from Wood, B, 2010. Reconstructing human evolution: Achievements, challenges, and opportunities. Proc. Natl Acad. Sci. USA, 107, 8902–8909.]

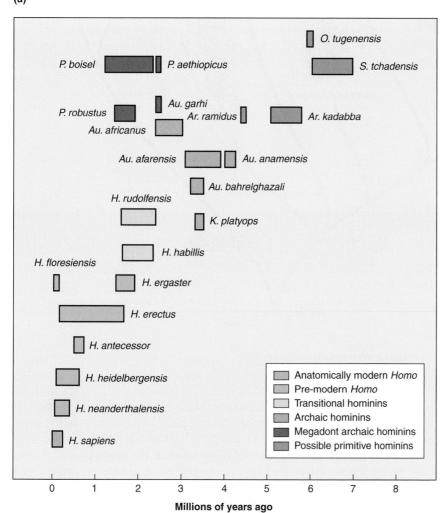

Millions of years ago

(b)

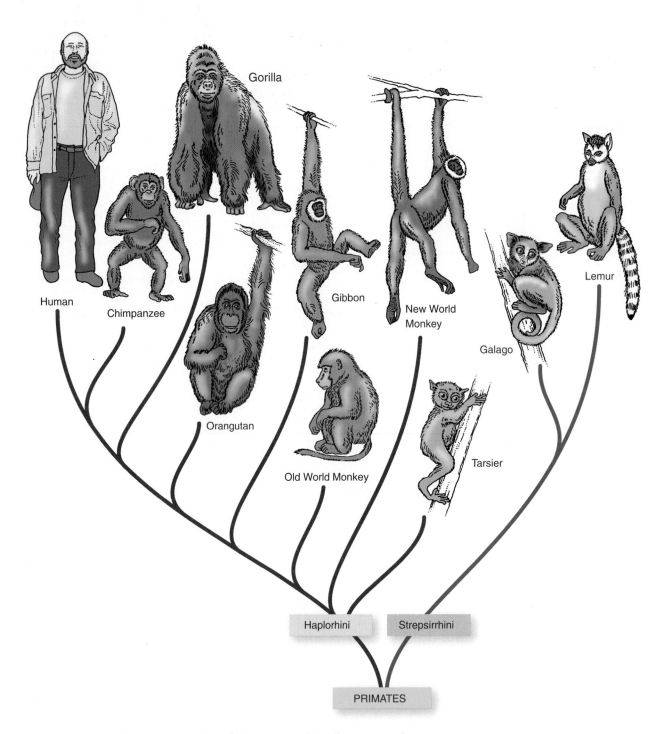

FIGURE 25.2 Various living representatives of primates as traditionally represented.

complex social networks. Phylogenetic analyses based on large data sets—43 nuclear and 15 mitochondrial genes (Fabre et al., 2009)—place chimpanzees as the sister group to modern humans (Figures 25.1a and 25.3). Humans are distinguished from other hominoids by adaptations for bipedal locomotion on the ground (shorter arms in relation to leg length; **FIGURE 25.4**), an enlarged brain, shortened fingers, and opposable thumbs.

    The two extant species of **chimpanzees** found in equatorial Africa are the common chimpanzee, *Pan troglodytes* and the more slender and lightly built bonobo, *Pan paniscus*.

BONOBOS *are often, but incorrectly, called pygmy chimpanzees.*

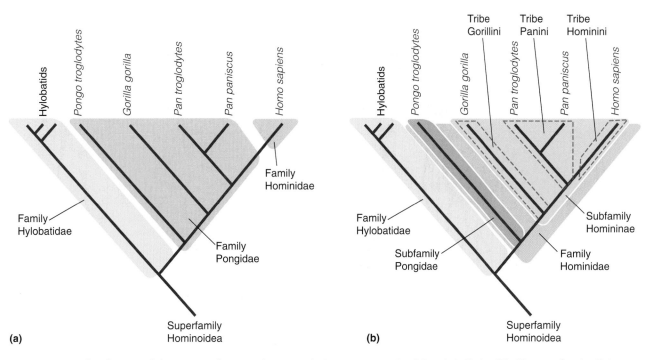

FIGURE 25.3 Classification of the Hominoidea according to evolutionary systematics **(a)** and cladistics **(b)**. (See text for details.)

Individuals average 32 to 45 kg (70 to 100 lbs) in weight, depending on species and sex. Both species have complex social structures and hierarchies, reflected in their large social groups of up to 120 individuals.

We would expect to find substantial genetic differences between humans and chimpanzees. The DNA and protein sequences of chimps and humans differ by around four percent (four base pairs out of every 100). By comparison,

In CLADISTICS, all groups include all descendents and the common ancestor of the group.

BOX 25.1	What's in a Name? Hominids and Hominins

For many decades, apes and humans have been known collectively as **hominoids** (family Hominoidea; Figure 25.3). It has become increasingly clear, however, that humans and African apes (particularly chimpanzees) are more closely related to one another than they are to other apes and that chimps are the sister group to (closest relative of) humans. As the diversity of fossil species demonstrating the relationships between humans and the Great Apes increased,* many paleoanthropologists saw that the traditional classification of humans and our closest relatives was inadequate and inconsistent. For this reason, a cladistic classification has gained favor in recent years.

Unfortunately, in the process of sorting out hominoid relationships, older terms have been given new meanings, which is always confusing.

- Great Apes and humans are now referred to as **hominids** (Family *Hominidae in traditional classification*), a term previously used for extant humans and their closest extinct relatives.
- Humans and their closest extinct relatives are hominins (Family Hominini; Figure 25.1).

When paleoanthropologists use the term "hominin" in the recent literature they mean the same thing as "hominid" in older literature. At stake is more than terminology. This change reflects our shifting view of relationships within this portion of the Tree of Life.

In cladistic classifications, humans and our closest extinct relatives belong to the "tribe" *hominini*, a taxonomic level between subfamily and genus. The two other tribes are Panini for chimpanzees and Gorillini for the gorillas.

---

* The Great Apes, (humans, chimpanzees, gorillas, and orangutans) have been long distinguished in classifications from the lesser apes (gibbons).

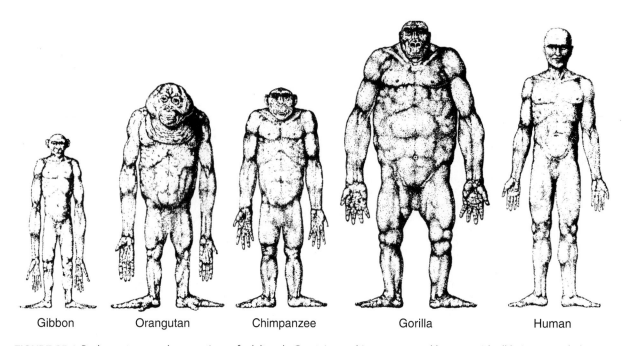

| Gibbon | Orangutan | Chimpanzee | Gorilla | Human |

FIGURE 25.4 Body contours and proportions of adult male Great Apes, chimpanzees, and humans with all hair removed, drawn to the same scale.

[Adapted from Schultz, A. H., 1969 *The Life of Primates*. New York: Universe Books.]

individual humans differ only in one base pair out of 1,000 (0.001%). These differences, along with changes in gene regulation[2] underlie the features that allow us to distinguish humans from other apes. Those features include the comparative lengths of human fore and hind limbs (Figure 25.4), the larger human brain, absence of a protruding face, enlarged sexually dimorphic canine teeth, and sparse body hair.

As chimpanzees are our closest relatives, much effort has gone into comparing the morphology, genetics and behavior of humans and chimpanzees, with the aim of determining when the two lineages diverged. Molecular data shown in Figure 25.1a place the divergence at 7.5 Mya, only a little earlier than the divergence at 7 Mya indicated by the fossil record (Figure 25.1b). A recent analysis involving the alignment of 20 million base pairs of the genomes of five primate species places the divergence at 6 Mya. Y chromosomes have diverged enormously between chimps and humans. While this is not surprising, it is interesting that the divergence occurred through gene loss in chimps, gene gain in humans, and chromosomal rearrangement in both lineages. Finally, 447 microRNAs (discussed elsewhere in this text) found in the human genome are not found in the chimp genome, providing evidence of independent genomic evolution at the level of gene regulation.[3]

---

[2] Y. Gilad et al., 2006. Expression profiling in primates reveals a rapid evolution of human transcription factors. *Nature*, **440**, 242–245; D. Garrigan and M. F. Hammer, 2006. Reconstructing human origins in the genomic era. *Nat. Rev. Genet.*, **7**, 669–680.
[3] N. Patterson et al., 2006. Genetic evidence for complex speciation of humans and chimpanzees. *Nature*, **441**, 1103–1108; J. F. Hughes et al., 2010. Chimpanzee and human Y chromosomes are remarkably divergent in structure and gene content. *Nature*, **463**, 536–539; E. Berezikov et al., 2006. Diversity of microRNAs in human and chimpanzee brain. *Nat. Genet.*, **38**, 1375–1377.

THE EARLIEST HOMININS

From fossil evidence dated between 5 and 6.5 Mya we know that a number of forms of hominin were competing for ground-dwelling habitats and developing the features of bipedalism required for a ground-dwelling existence. Two bipedal species of hominins from this time have been discovered (Figure 25.1b):

- *Orrorin tugenensis* has hind limb features indicating that it was bipedal when on the ground, but forelimb features indicating that it also lived in trees. With an estimated occurrence between 6.1 and 5.8 Mya, the chimpanzee-sized *Orronin* is the second oldest "putative" hominin (putative because experts disagree on where to place *Orronin*). (See Wood and Harrison, 2011 for an insightful analysis.)
- Dated to between 5.8 and 5.2 Mya, the second species, *Ardipithecus kadabba*, is younger than *O. tugenensis*; clearly, dating fossils is important when attempting to determine position within a lineage. *A. kadabba* is regarded as having been bipedal but fossil evidence other than from the skull is minimal.[4]

A third species, *Sahelanthropus tchadensis*, known from a cranium and fragments of jaws and teeth, had a brain estimated at 340 to 360 cm^3 compared to 1,120 to 1,260 cm^3 in modern humans. Whether this species was bipedal is unclear; lack of post-cranial material makes comments on mode of locomotion premature.[5] The large opening (foramen magnum) at the base of the skull provides indirect evidence for bipedalism; the head was held upright. Dated to between 6.5 and 7.4 Mya (Figure 25.1b), if bipedal, *S. tchadensis* would be the earliest known putative hominin.

Of the three species *A. kadabba* has more features in common—shares more derived features—with later australopithecines. However, the presence of large canine teeth (which is a more basal feature), and its age of occurrence, makes *Ardipithecus* difficult to position in the hominin lineage.

Here, PUTATIVE *means that the fossil record is sparse, but given the evidence available,* Sahelanthropus tchadensis *has been recognized as the earliest known hominin ancestor.*

AUSTRALOPITHECINES: THE SOUTHERN APES OF AFRICA

Continued exploration of human origins takes us into the group of apes known as australopithecines—the southern apes of Africa. Australopithecines radiated as at least eight lineages between 2.5 and 4.5 Mya. A Pliocene species from East Africa, **Australopithecus boisei**, is the largest of the australopithecines.[6] Its large molar teeth and powerful jaw muscles are interpreted as reflecting a diet of seeds and fruits with hard husks and pods.

Three species in two genera (*Australopithecus* and *Ardipithecus*) are of special interest (Figure 25.1b, and see Wood 2010). The discussion that follows is centered on origins of genera and on three major evolutionary changes; bipedalism, brain size, and tool use. Either approach, or indeed other approaches, are equally valid ways to tell the story of human evolution.

---

[4] For *Orrorin tugensis,* see B. Senut et al., 2001. First hominid from the Miocene. (Lukeino formation, Kenya). *C. R. Acad. Sci. Paris, (Earth Planetary Sci.),* **332,** 1–9. For *Ardipithecus kadabba,* see Haile-Selassie, 2001. Late Miocene hominids from the Middle Swash, Ethiopia. *Nature,* **412,** 178–181. See Wood and Harrison (2011) for positioning of these and other species in the hominin clade.

[5] M. Brunet et al., 2002. A new hominid from the Upper Miocene of Chad, Central Africa. *Nature,* **418,** 145–151.

[6] Reflecting different experts' opinions on hominin species and separation of robust from gracile australopithecines, *Australopithecus robustus* and *A. boisei* are treated as *Paranthropus robustus and P. boisei* in Figure 25.1b. (See p. 8906 in Wood, 2010 for discussion and literature.)

ARDIPITHECUS RAMIDUS

The earliest australopithecine fossils currently known (mostly as teeth) and the most ape-like hominin ancestor known are relics of an Ethiopian species (and genus) named ***Ardipithecus ramidus***, dated to between 4.3 and 4.5 Mya.[7] *A ramidus* shares relationships with *Australopithecus afarensis* (below) and with extant Great Apes, especially chimpanzees. Links between *A. ramidus* and australopithecines were based on teeth alone until a suite of 11 papers and accompanying news reviews reporting reconstruction of a skeleton was published in 2009.[8]

AUSTRALOPITHECUS AFARENSIS

Fossils from East African sites at Laetoli, Tanzania, and the Afar (Hadar) region of Ethiopia, estimated as 3.9 to 3 My-old, are classified as ***Australopithecus afarensis***, a species with heavy brow ridges and a low forehead. Despite such ape-like features, *A. afarensis* displays a large number of important cranial, dental, and skeletal differences from apes. For example, although the canines are larger in *afarensis* males than in females, this sexual dimorphism is much less pronounced than in apes or early Miocene chimp, gorilla, or orangutan-like fossils. We can draw these conclusions because of a substantial record of *A. afarensis*, including the almost complete skeleton of a muscular, 1 to 1.2 m (3.2 to 3.9 ft) tall bipedal female known as "**Lucy**" (**FIGURE 25.5**). Lucy now can be compared with the skeleton of a large fully upright, 1.5 to 1.7 m (4.9–5.6 ft) tall, bipedal male *A. afarensis* discovered in Ethiopia and dated to 3.6 Mya (Haile-Selassie et al., 2010). Fossil footprints preserved under a layer of 3.7-My-old volcanic ash record the bipedal gait of two individuals over a distance of some 21 m (nearly 70 ft),[9] while the recent discovery of a complete

FIGURE 25.5 Lucy, an amazingly complete skeleton of a female *Australopithecus afarensis*, showing the bones that were discovered **(a)** and the skeleton as it has been reconstructed **(b)**.

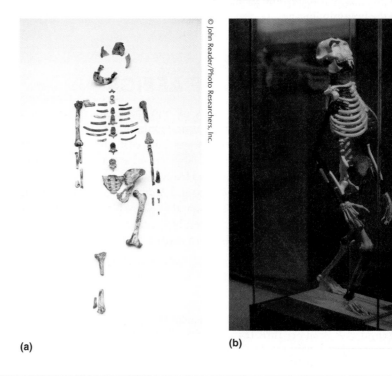

(a) (b)

[7] T. D. White et al., 1995. *Australopithecus ramidus*, a new species of early hominid from Aramis, Ethiopia. *Nature*, **375**, 88.
[8] For these 11 studies from numerous research groups, see *Science*, 2009, **326**, 60–106.
[9] Although rare, fossilized footprints are enormously informative. In addition to those of *A. afarensis* dated to 3.7 Mya are footprints of a 1.7 m (5.6 ft) tall adult *H. erectus* and a 0.9 m (3 ft) tall child dated at 1.5 Mya—the oldest known footprints for any species of *Homo* (Bennett et al., 2009).

fourth metatarsal bone demonstrates that *A. afarensis* had the high arch, and was capable of the striding gait, seen in modern humans (Ward et al., 2011).

AUSTRALOPITHECUS AFRICANUS

From a historical perspective, a skull and lower jaw found in 1912 in England set the stage for what an early hominin should look like. But, that specimen—**Piltdown man**—had not been deposited by the usual modes of fossilization. It was a fake. The story is told in END BOX 25.1.

The story of the discovery of authentic early hominin fossils began in 1925 when the anthropologist Raymond Dart reported an early hominin fossil from a lime quarry at Taung in the Cape Province of South Africa. The fossil consisted of the front part of the skull and most of the lower jaw of a 6-year-old, who would become known as the "**Taung child**" (FIGURE 25.6). Dart named the new species *Australopithecus africanus*. Aspects of the teeth and brain were more like humans than they were like other apes. Adult australopithecine skulls, dated to between 2.5 and 3 Mya, were discovered in the 1940s at *Sterkfontein*, not far from Taung. Further discoveries of postcranial skeletal material reinforced the human-like nature of the vertebral column and pelvic girdle.

FIGURE 25.6 The skull of the "Taung child", *Australopithecus africanus*.

Many more expeditions over the past 70 years have revealed more, and even older, australopithecine species. The latest, published in 2010, represents one of the most complete hominins uncovered. *A. sediba*, dated at 1.95 to 1.78 Mya, has the small brain, long arms, and body form of a chimpanzee, but fingers, thumb, and regions of the brain that are more like humans.[10] Because *Homo* arose just over two million years ago, it is tempting to see *A. sediba* as the last of the australopithecines, or even of the line that gave rise to humans. These fossils tell us that considerable diversification of southern apes occurred over a 2 to 2.5 million-year period between 2.5 and 4.5 Mya. These important finds, and the conclusion that all known early australopithecines were bipedal, allow us to conclude that the human genus, *Homo*, had an immediate ancestor that was bipedal. Other features that characterize *Homo* evolved after bipedal locomotion arose.

Lest these conclusions read as if they are final or as if all agree, interpreting the evolution of bipedal locomotion is complex. Why? Because bipedality may have arisen more than once. Australopithecines were actively arboreal and terrestrial, a duality reflected in their bipedal locomotion, which differed functionally and so could have been derived independently from bipedalism in humans (Kingdon, 2004; Bramble and Lieberman, 2004).

BIPEDALISM AND BRAIN SIZE: HYPOTHESES ABOUT EARLY HUMAN EVOLUTION

Before evaluating the changes that initiated the human lineage assigned to the genus *Homo*, it is helpful to outline the various hypotheses concerning relationships between the origin of bipedalism and increase in brain size. We then will discuss the other major development: the origin of tool making and tool use.

Charles Darwin said virtually nothing about human evolution in *The Origin of Species*. However, his second major book of evolution—*The Descent of Man*, published in 1871—dealt specifically with humans and their evolutionary origins. Ever since, bipedalism, increase in the size of the brain, and the use of tools and language, which in combination are hallmarks of human evolution, have been regarded as a set of related changes, each

Many explanations have been suggested for the evolution of BIPEDALISM, including advantages in foraging, evading predators, carrying provisions, using tools, and manipulating weapons.

[10] L. R. Berger et al., 2010. *Australopithecus sediba*: a new species of *Homo*-like Australopith from South Africa. *Science*, **328**, 195–204.

reinforcing the others. The fossil record, however, informs us that bipedalism arose much earlier than did the other two traits. Therefore, at issue for some time has been whether the evolution of bipedal locomotion was an evolutionary **innovation** that facilitated changes in brain size and tool use (Henke and Tattersall, 2007).

We can reconstruct enough of their **environment** to know that early hominins lived in woodland, dry grassland, and bush where food supplies would have been seasonal. This coupled with the patchy nature of the environment would have selected for hominins who could adapt their diet to a wide range of foods (omnivores) and search for food over long distances. Evidence in the fossil record supports the hypothesis that standing upright and the ability to move rapidly enhanced this lifestyle. The combined advantages of an upright stance and faster locomotion on the ground, while retaining the ability to climb, would have improved avoidance of predators.[11] Bipedalism would have fostered stick wielding and stone throwing, actions enabled by the freeing of the fore limbs from the duties of locomotion. Knowledge that chimpanzee use sticks as tools to obtain food has been used to develop a scenario that tool making in *Homo*, perhaps also in australopithecines, began with using unmodified sticks and stones from which the ability to modify those sticks and stones into more sophisticated tools arose.

ORIGINS OF *HOMO*

The genus *Homo*, which first appears in the fossil record in East Africa around 2.5 Mya, was similar in size to the australopithecines, but had a larger brain (600 to 700 cm^3) and smaller molar teeth.

The prevailing trend in the evolution of *Homo*, as in mammalian evolution as a whole, is of parallel lineages arising contemporaneously or almost contemporaneously. At least six species of *Homo* illustrate adaptive radiation in response to natural selection that took place in Africa over less than a million years, between 1.6 and 2.5 Mya (Figure 25.1b).

HOMO RUDOLFENSIS AND *H. HABILIS*

Even though the numbers of known specimens is small, they reveal that skeletal change occurred rapidly in the *Homo* lineage. Many experts assign early East Africa *Homo* fossils to two species that occupied similar geographical regions, ***Homo rudolfensis*** as early as between 2.4 and 1.8 Mya, and ***H. habilis*** between 2.0 to 1.6 Mya (Figure 25.1b). The anatomical structure of the hands and feet of the early australopithecines and early *Homo* (*H. habilis*) are consistent with the ability to climb trees.

H. rudolfensis and *H. habilis* are the most likely members of the lineage that gave rise to the next two species of *Homo*, which appear in the fossil record 1.9 Mya—that is, no more than 500,000 years, and as little as 100,000 years, after *H. habilis* and *H. rudolfensis*. One of these species lived in Africa, the other in Asia. Both were taller (1.7 m; 5.6 ft) and had larger brains (750 cm^3 or more) than their predecessors. The species are ***H. ergaster*** and ***H. erectus***.

HOMO ERGASTER AND *H. ERECTUS*

The African species, *Homo ergaster*, was taller and leaner with a brain capacity of 750 to 1100 cm^3 and a larger body size than *H. habilis* or *H. rudolfensis*, either of which could have been its ancestor. *H. ergaster* used fire and large hand axes of the Acheulean type (see FIGURE EB3.1, and below). Whether these attributes indicate group living and hunting is hard to determine. As with Lucy, the almost complete skeleton of *A. afarensis*, the

[11] M. H. Day, 1986. Bipedalism: pressures, origins and modes. In *Major Topics in Primate and Human Evolution*, B. Wood, L. Martin, and P. Andrews (eds.). Cambridge University Press, Cambridge, UK, pp. 188–202.

1.6 My-old **Nariokotome skeleton** is the most complete fossil of a juvenile male *H. ergaster* (FIGURE 25.7, and see Walker and Leakey, 1993).

A second species, *Homo erectus*, contemporaneous in time with *H. ergaster*, was first discovered in Java. A thicker skull and heavier brow ridges along with the different geographical distributions allow the two species to be differentiated. Brain volumes of the two species overlapped (750 cm^3 to greater than 1,200 cm^3 in *erectus*) and both species are associated with Acheulean tools (see Figure EB3.1**b**). Later discoveries showed that *H. erectus* was present over a wide geographical area in Africa, China, and Europe, and persisted until as late as 50,000 years ago in Java.

Fossils discovered in 2004 from the island of Flores in Indonesia reveal what has been interpreted either as a relict population of dwarf *H. erectus* or as a new species, ***Homo floresiensis***, that existed as recently as 18,000 years ago (FIGURE EB1.1). Such a recent existence of a human species other than our own is a remarkable suggestion and highly controversial. Some experts claim that the best-preserved individual is more modern, a dwarf human (*H. sapiens*) but with signs of a congenital malformation, a topic taken up in END BOX 25.2.

HOMO SAPIENS AND H. HEIDELBERGENSIS

One of the reasons for *H. erectus* receiving so much attention, and a partial explanation for the controversy over the Flores individuals, (End Box 25.2) is that our species, *Homo sapiens*, has been traced back to a lineage that contained *H. erectus* (Wolpoff and Caspari, 1997). Variation among specimens of *H. erectus*, and the often but subtle differences between *H. erectus* and *H. sapiens*, led some experts to claim that *H. erectus* (or some variants within *H. erectus*) is (are) the first manifestation of *H. sapiens*.

Part of the variation within *H. erectus* reflects individuals in different geographical locations that would have existed as distinct, and potentially non-interbreeding, populations. Some conclude that the variation can in part be resolved by recognizing another species, ***Homo heidelbergensis***, from fossils in Swanscombe, England and Steinheim, Germany, dated to about 200,000 years ago. These specimens have larger brain volumes than, and other features intermediate between, *H. erectus* and *H. sapiens*. To complicate things further, *H. heidelbergensis* may have arisen from a population of *H. ergaster* and not from the more geographically widespread *H. erectus*, which persisted in parts of the world long after *H. heidelbergensis* arose.

The tree of *Homo* evolution, therefore, is much more like a shrub with many branches, than it is like a tree with a single trunk; "Instead of a ladder with humans at the pinnacle, there is a bush with humans as one little twig."[12] As you will see as you read on, the tree is about to become even bushier. Why? For two reasons: (a) another species, ***Homo neanderthalensis***, discovered in 1856 in the Neander valley, Germany, arose from European populations of *H. heidelbergensis* (BOX 25.2); (b) many paleoanthropologists have concluded that *H. sapiens* evolved from *H. heidelbergensis*, separation of the two lines having occurred in Africa around 200,000 years ago.

HOMO SAPIENS (HUMANS)

Many older texts used terms such as "primitive" and "advanced", or "higher" and "lower" when referring to earlier and later forms within a lineage. Such terms are no longer regarded as appropriate, conveying as they do the notions of evolutionary progress on some subjective scale. The term "anatomically modern humans" is now used for extant and fossil individuals of the species *H. sapiens*. The earliest fossil evidence for *H. sapiens* is located in Africa, a finding reflected in the "Out of Africa" hypothesis for the spread of *H. sapiens* around the world, as discussed in the following section.

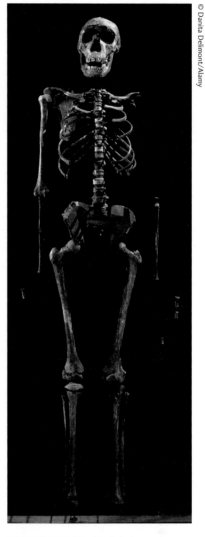

© Danita Delimont/Alamy

FIGURE 25.7 The Nariokotome skeleton, a juvenile male, *Homo ergaster*.

DOZENS OF PAPERS *have been written exploring every possible interpretation of these fossils.*

---

[12]R. A. Foley, 1995. *Humans Before Humanity*. Blackwell Publishers, Oxford, UK.

Neanderthals (*Homo neanderthalensis*) were widespread in Europe and Western Asia from 300,000 to 30,000 years ago. Shorter than modern humans, and with brains that were 10% larger, Neanderthals had distinctive brow ridges, large jaws, small chins, robust skeletons, and other anatomical features not seen in humans (**FIGURE B2.1**).

Like humans, Neanderthals used stone tools to hunt large animals; bears and mammoths are shown in most museum reconstructions and dioramas. From flowers found on the graves of dead individuals it has been concluded that Neanderthals performed rituals (but see below). Seventy-eight bones from six individuals who lived 100,000 to 120,000 years ago provide evidence that Neanderthals practiced cannibalism. The bones, scattered around the hearths of three fires, show cut marks and evidence that the bones had been torn apart, both of which are consistent with dismembering (Defleur et al., 1999).

Because the Mousterian Culture (Figure EB3.1c) is associated with Neanderthals *and* with *H. sapiens*, some have concluded that there may have been cultural overlap between the two species. Parallel tool use does not necessarily mean cultural overlap, however. To add further complication, in 2010 a previously unknown population of archaic humans were described to have inhabited the Denisova Cave in Siberia 30,000 to 50,000 years ago. The cave also contains Neanderthal fossils and Mousterian tools typical of those made and used by Neanderthals, dated to 45,000 years ago. The cave was later occupied by humans also. Genomic sequencing of the "Denisovans," as they are known, confirmed them as archaic humans separate from Neanderthals but also showed that some 5% of the DNA in living Australian aborigines and Melanesians is derived from Denisovans (Reich et al., 2010, 2011).

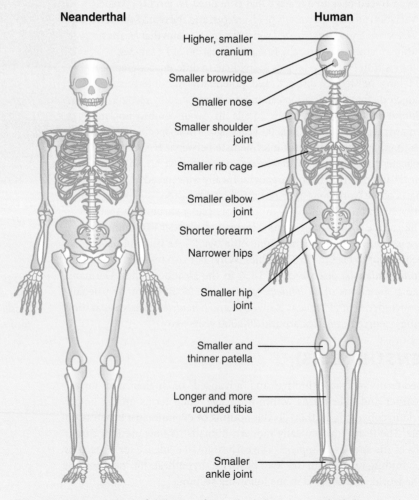

Neanderthal   **Human**

Higher, smaller cranium

Smaller browridge

Smaller nose

Smaller shoulder joint

Smaller rib cage

Smaller elbow joint

Shorter forearm

Narrower hips

Smaller hip joint

Smaller and thinner patella

Longer and more rounded tibia

Smaller ankle joint

FIGURE B2.1 Comparison of skeletons of Neanderthal (*Homo neanderthalensis*) with anatomically modern human (*H. sapiens*).

BOX 25.2 Neanderthals: Another Species of *Homo* (with comment on Denisovans) (Cont...)

SPECIES STATUS

The difficulty of assigning species status was discussed in the text in relation to *H. erectus* and the Flores specimens and is elaborated further in End Box 25.2. Similar difficulties are encountered with Neanderthals. Some physical anthropologists have concluded that skeletal evidence—thicker skulls, differences in the pelvic girdles—and the presumed limited language ability of Neanderthals (see End Box 25.4) support Neanderthals as a subspecies of *H. sapiens*, designated *Homo sapiens neanderthalensis*, although Neanderthal faces and their large, deep-rooted teeth do not align them with any human populations.[a]

REPLACED BY, AND NOT INTEGRATED WITH, *HOMO SAPIENS*

Homo sapiens began to replace Neanderthals in various parts of the world about 40,000 years ago; Neanderthals are estimated to have died out 28,000 to 24,000 years ago (Finlayson et al., 2006; Harvati and Harrison, 2006). How do we know that Neanderthals died out rather than integrating with *H. sapiens*? The two species certainly coexisted between 500,000 years and 24,000 ago, and both overlapped with *H. erectus* between 500,000 and 250,000 years ago. Indeed, mixing of artifacts in the reindeer cave (Grotte du Renne) in France means that assignment of ornament (or rituals) to Neanderthals has to be done with caution (Higham et al., 2010). Such artifacts may have been generated by *H. sapiens*.

Strong molecular evidence *against* integration comes from the recovery and analysis of Neanderthal mtDNA sequences, which show little evidence of any contribution of Neanderthal to human mtDNA (Krings et al., 1997). mtDNA isolated from a 38,000-year-old Neanderthal skeleton is significantly more different from human mtDNA than expected if it were a sample of normal human variation. According to the data from this study, Neanderthal and human mtDNA lineages diverged 660,000 ± 140,000 years ago, and Neanderthals "went extinct without contributing mtDNA to modern humans" (p. 19).

---

[a] The roles of the evolution of speech and language in human evolution are also discussed in End Box 25.4.

NEANDERTHAL GENOMES

In recent years, sufficient Neanderthal DNA has been isolated and sequenced. When analyzed with sophisticated methods to estimate missing DNA, a reasonable first approximation of the Neanderthal genome was constructed (Green et al., 2006; Noonan et al., 2006). One study used one million base pairs of Neanderthal (humans have around 3.2 billion base pairs of DNA), the other 65,000 base pairs. These studies have been described as "perhaps the most significant contributions ... since the discovery of Neanderthals 150 years ago."[b] The one million base pair study produced an estimate of divergence of humans from Neanderthals 516,000 years ago. The 65,000 base pair study produced an estimated divergence time of 370,000 years ago.

DEVELOPING NEANDERTHALS

Information from CT scans (computed tomography) on Neanderthal pre- and postnatal development has been obtained (a neonate from a cave in Russia and two infants from a cave in Syria) and compared with developmental stages of anatomically modern humans. Both species share similarly large brain sizes at birth, with Neanderthal brains growing faster during infancy than human brains and Neanderthal brain size increasing rapidly in the first few years after birth (Ponce de León et al., 2008). An even more recent study using a method that allows very precise aging showed that tooth development was faster in Neanderthals than in humans (Smith et al., 2010). It seems reasonable to conclude that Neanderthal fetuses developed and matured faster than human fetuses, although both species show lengthening of prenatal development when compared with other hominins. Although the weight of evidence supports Neanderthals and humans as separate species, we await even more detailed studies of these and other organ systems.

---

[b] D. M. Lambert and C. D. Millar, 2006. Evolutionary biology: ancient genomics is born, *Nature*, **444**, 275–276. Also see K. M. Weiss and F. H. Smith, 2007. Out of the veil of death rode the one million! Neanderthals and their genes. *BioEssays*, **29**, 105–110.

Three individual specimens of anatomically modern humans discovered in 160,000-year-old deposits at Herto, Ethiopia are the geologically oldest members of *H. sapiens*. More recent fossils from Mount Carmel in Israel date to 90,000 years ago.[13] Mousterian stone tools (see below) are associated with both sites. In even more recent deposits—35,000 years ago during the Upper Paleolithic in Europe—the features of

---

[13] T. D. White et al., 2003. Pleistocene *Homo sapiens* from Middle Awash, Ethiopia. *Nature*, **423**, 742–747; T. Akazawa, K. Aoki, and O. Bar-Yosef (eds.), 1998, *Neandertals and Modern Humans in Western Asia*. Plenum Press, New York.

TABLE 25.1	Tool Use by *HOMO*	
Oldowan tools	*Homo habilis*	2.5 Mya
Acheulean tools	*Homo ergaster, Homo erectus* *Homo heidelbergensis*	1.5 Mya[14] 500,000 years ago
Mousterian tools	*Homo neanderthalensis, Homo sapiens*	200,000 years ago
Upper Paleolithic tools	*Homo sapiens*	90,000 years ago

human fossils and the tools present both had changed. Brow ridges decreased, skull vaults were higher (reflecting changes in brain size and form), faces were smaller and less protruding, and flaked flint tools of the Aurignacian Era (see Figure EB3.1d) were present. Humans were now representing their environment, especially the animals in it, as paintings on cave walls and as sculptures in bone.[15]

TOOL USE BY SPECIES OF *HOMO*

Individuals in the genus *Homo* developed the capacity to make and use stone tools, the oldest of which are dated to 2.5 Mya.

Tool making and use has been classified into a series of four stages, discussed in END BOX 25.3 and depicted in Figure EB3.1. Each stage is associated with a particular species of *Homo*, beginning 2.5 Mya with *H. habilis*, and extending to 200,000 years ago with *H. sapiens*. These tools were used to manipulate plant material; hunt; carve up small reptiles, rodents, pigs and antelopes; and scavenge the remains of larger mammals. **TABLE 25.**1 contains a summary of the stages, species, and oldest appearance of the associated tools, information that is elaborated in the End Box 3.3.

OUT OF AFRICA

There are two main views of the origin and subsequent dispersal of modern humans from Africa (FIGURE 25.8).

- The **single-origin hypothesis**, also called the **Out of Africa** or **Noah's Ark model**, proposes the origin of *H. sapiens* in a single geographical region, Africa, followed by dispersal to other continents (Figure 25.8a).
- The **multiple-origin hypothesis**, also called the **Candelabra model** based on its shape (Figure 25.8**b**), proposes parallel origin of *H. sapiens* in different unconnected localities.[16]

Accurate determination of the date of the last common ancestor of modern humans is a crucial piece of evidence in deciding between the two hypotheses. Existence of the most immediate ancestor one or more Mya would coincide with one of the dispersals of *H. erectus* from Africa. It would indicate that modern humans found in different continents represent evolutionary lineages that each began with *H. erectus* in geographically separated localities, and so support the multiple origin hypothesis. If the most immediate *H. sapiens* ancestor was much more recent—say,

[14]J. F. Hoffecker, 2005. Innovation and technological knowledge in the Upper Paleolithic of Northern Eurasia. *Evol. Anthropol.*, **14**, 185–198.
[15]Lepre et al. (2011) provide evidence for the Acheulean dating back to 1.76 Mya.
[16]For discussions of evidence for these two hypotheses, see Wolpoff (1989), Wolpoff and Caspari (1997), volume 3 of Henke and Tattersall (2007), and Balter (2011).

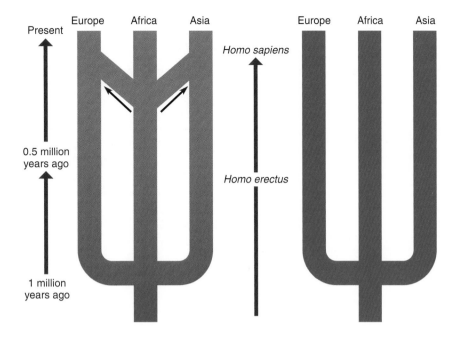

FIGURE 25.8 Diagrammatic representation of two models for the origin of *Homo sapiens*. In **(a)** humans (blue) originated in one locality (Africa) and migrated to other continents where they replaced relict *H. erectus* populations (brown) that had entered these continents 1 My or more years ago. In **(b)** humans (blue) originated in different localities independently of other species.

(a) "Out of Africa" model

Single origin of *Homo sapiens* in Africa, and replacement of *Homo erectus* in Europe, Africa, and Asia

(b) "Candelabra" model

Multiple origins of *Homo sapiens* from *Homo erectus* populations in Europe, Africa, and Asia

100,000 to 500,000 years old—then the dispersal of *H. sapiens* would have occurred in areas occupied by *H. erectus* after a 1- or 2-My-old dispersal of *H. erectus*, and so support the single-origin hypothesis.

MITOCHONDRIAL EVE

An important data set helping to resolve the immediate ancestor of *H. sapiens* is mtDNA from a large number of extant humans, representing many populations. Two major findings are consistent with the single origin hypothesis: (1) all of the non-African mtDNA sequences are variants of the African sequence, and (2) most of the variability in mtDNA sequences occurs among members of African populations. Both these findings are consistent with the oldest mtDNA sequences being in African populations; multiple populations evolving in parallel would be expected to show similar amounts of sequence variability.

In 1987, on the basis of mtDNA nucleotide sequence analysis of geographically diverse human populations, Rebecca Cann and coworkers (Cann et al., 1987) proposed a 200,000-year-old **common mitochondrial DNA ancestor** ("mitochondrial Eve") of modern humans, a finding consistent with the single-origin hypothesis. Subsequent studies, in which calibration of the age of human origination was based on similar mtDNA sequence data, provided further support for the single origin hypothesis (although, as discussed in elsewhere in this text, molecular clocks are difficult to calibrate). One major conclusion from these analyses is that all modern human mtDNA sequences originated with a single ancestral sequence in Africa between 140,000 and 290,000 years ago (FIGURE 25.9a).

Each geographical group outside Africa includes more than one mtDNA branch— that is, each arose from more than one population, each with distinctive mtDNA. Nine major descendant sequences (Figure 25.9b–j) were distributed to other geographical regions through migration. Further sequence changes occurred over time in populations in those regions. Highland peoples of New Guinea have seven maternal origins, meaning that individual geographical regions were colonized more than once. As the rate of

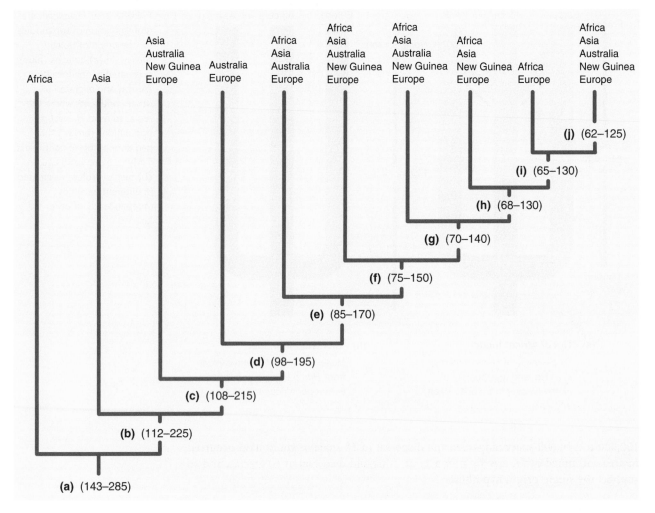

FIGURE 25.9 Phylogeny of mitochondrial DNA from 147 humans from five geographic regions. The ancestral sequence is designated as **(a)**. Each node in the phylogeny **(b–j)** indicates a major descendant sequence. Estimated dates for each node (thousands of years ago) are given in parentheses. Areas colonized by each sequence are indicated at the top of the figure.

(Based on data that Cann et al., 1987, obtained from surveying 370 restriction enzyme sites per individual, covering about 1,500 bases of the mitochondrial genome.)

mtDNA sequence divergence can be estimated (2–4 % nucleotide change/My in vertebrates, though slower in primates), timing of waves of colonization also can be projected (**FIGURE 25.10**). Even though such estimates are influenced by the way in which fossil species are arranged phylogenetically, the hypothesis of a single African origin of *Homo sapiens* remains the best explanation for the origin of anatomically modern humans.

Y-CHROMOSOME ADAM

MtDNA sequences are not complicated by the recombination of genes from the male and female parent; mtDNA sequences track the female genome back in time. Interestingly, perhaps even surprisingly, the Y chromosome, which is inherited through the male line, also contains phylogenetic information, but of a different nature to that seen in mtDNA.[17]

---

[17] Evolutionary history also can be resolved using X chromosomes; see E. E. Harris and J. Hey, 1999. X chromosome evidence for ancient human histories. *Proc. Natl Acad. Sci. USA*, **96**, 3320–3324.

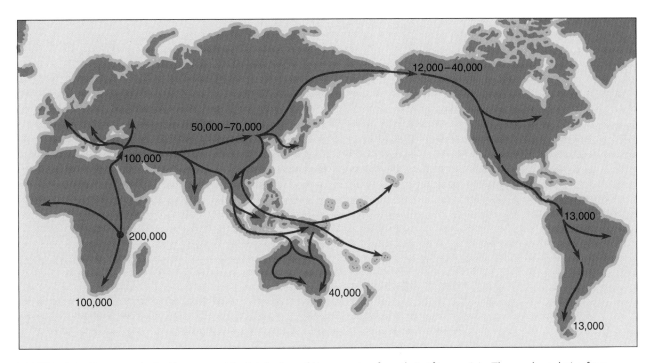

FIGURE 25.10 One scenario for the geographic distribution of *Homo sapiens* from their African origin. The numbers derive from a study of genetic distances between 26 human populations and represent estimated dates (thousands of years ago) when these populations reached their various destinations. These migrations correlate highly with many major patterns of linguistic evolution.

[Adapted from Nei, M., and A. K. Roychoudhury, 1993. *Mol. Biol. Evol.*, **10**, 927–943.]

Ordinarily, sequence changes in nuclear genes would be hard to differentiate into those resulting from recombination (and so potentially recent), and those that are much older. Fortuitously, a large proportion of the human Y chromosome does not participate in recombination. As a consequence, mutations build up in the regions over time. A phylogenetic analysis of more than 1,500 individual humans from all continents traced Y chromosomes to a common African ancestor living about 150,000 years ago. So both "mitochondrial Eve" and "**Y-chromosome Adam**" confirm an African origin of humans (Hammer et al., 1998). Given the margins of errors in such estimates, 150,000 for Y-chromosome Adam and 140,000 years for mitochondrial Eve are remarkably close.

AUTOSOMAL HUMAN

A phylogenetic analysis by Nei and Roychoudhury (1993) of 29 autosomal genes—nuclear genes not located on the sex chromosomes—from 26 populations supports a single African origin of humans around 200,000 years ago. According to these data, dispersal resulted in humans reaching Eastern Europe 50,000 to 70,000 years ago, Australia 40,000 years ago, North America 12,000 to 40,000 years ago, and South America 13,000 years ago (Figure 25.10).

Humans did not undergo such vast migration without adaptation to, and interactions with, their environment, developing social structures and enhancing their ability to communicate using speech and language (END BOX 25.4).

HUMANS AS HUNTER-GATHERERS

Hunting in groups (packs) did not evolve with humans. Jackals and hyenas hunt in packs as their major means of capturing prey. Chimpanzees and baboons also occasionally form groups to hunt smaller animals; groups of two to five male chimpanzees will tree a monkey and cut off its escape by positioning themselves around it.

Jane Goodall observed more than 200 INCIDENTS *where Gombe chimpanzees caught and/or ate colobus monkeys near Lake Tanganyika. The chimpanzee kill rate in Gombe is about 225 to 300 mammals a year.*

Interestingly, the foot of H. FLORESIENSIS *is exceptionally long for a species in the* Homo, *a feature interpreted by Jungers et al. (2009) as basal for hominins. This explanation could call into question the origin of* H. floresiensis *from* H. erectus, *for which see End Box 25.2.*

The earliest members of *H. sapiens* moved from forest to savanna as they roamed over large areas in search of food and prey items that were patchily distributed, spatially and seasonally. Exposure to differing environments during migration, to the food types in those environments, and the discovery of fire and cooking, would have selected for structural and behavioral changes and adaptations.[18] For example, anatomical and inferred physiological evidence has been used by Bramble and Lieberman (2004) to conclude that adaptations for long-distance endurance running separate *H. sapiens* from *H. erectus*. Analysis by Rolian et al. (2010) of covariation and correlations of toe and finger lengths in 89 extant individual common chimpanzee and 202 humans supports coevolution of human, but not chimpanzee, hands and feet. Elongation of human toes facilitated bipedalism and running, elongation of human fingers facilitated changes in the manufacture of tools.[19]

Co-occurrence of human and other animal bones in East African fossil sites are interpreted to mean that scavenging was established 2 Mya. This early date is interpreted as indicating that humans were scavenging kills made by other large predators. Evidence indicates that hunting and gathering arose in humans much more recently than scavenging, on the order of 55,000 years ago.[20] A single kill of a large mammal would have provided food for more than one individual and for more than a day, if kept away from other scavenging mammals. Such skills would have been passed on to the next generation by imitation, and later, as social living became more established, by active teaching of hunting skills and of the locations of seasonal populations of prey. A **hunter-gatherer society** had begun.

■ KEY TERMS

Acheulean tools	*Homo rudolfensis*
anatomically modern humans	human
Ardipithecus ramidus	hunter-gatherer societies
australopithecines	innovation
Australopithecus afarensis	key innovation
Australopithecus africanus	language
Australopithecus boisei	microcephaly
autosomal humans	mitochondrial Eve
bipedalism	Mousterian tools
brain evolution	Oldowan tools
Broca's area	Out of Africa (model)
chimpanzees	Paleolithic Stone Age
environment	Piltdown man
hominin	primate
hominoids	primate evolution
Homo erectus	speech
Homo ergaster	Taung child
Homo floresiensis	tool use
Homo habilis	Upper Paleolithic tools
Homo heidelbergensis	Wernicke's area
Homo neanderthalensis	Y-chromosome Adam

[18] Wrangham (2009) advocates the hypothesis that the discovery of fire and its use to cook food was an important factor in the rapid evolution of brain size in humans.

[19] For an analysis involving additional species—gorilla, gibbon, rhesus macaque, and three species of monkey—see N. M. Young, G. P. Wagner, and B. Hallgrímsson, 2010. Development and the evolvability of human limbs. *Proc. Natl. Acad. Sci. USA,* **107**, 3400–3405.

[20] N. M. Tanner, 1987. The chimpanzee model revisited and the gathering hypothesis. In *The Evolution of Human Behavior: Primate Models,* W. G. Kinzey (ed.). SUNY Press, Albany, NY, pp. 3–27; J. Tooby and I. DeVore, 1987. The reconstruction of hominid behavioral evolution through strategic modeling. In *The Evolution of Human Behavior: Primate Models,* W. G. Kinzey (ed.). SUNY Press, Albany, NY, pp. 183–237.

■ EVOLUTION ON THE WEB

Explore evolution on the Internet! Visit the accompanying website for *Strickberger's Evolution, Fifth Edition,* at **go.jblearning.com/Evolution5eCW** for exercises and links relating to topics covered in this chapter.

■ REFERENCES

Balter, M., 2011. Was North Africa the launch pad for modern human migrations? *Science*, **331**, 20–23.

Bennett, M. R., J. W. Harris, B. G. Richmond, et al., 2009. Early hominin foot morphology based on 1.5-million-year-old footprints from Ileret, Kenya. *Science*, **323**, 1197–1201.

Bramble, D. M., and D. E. Lieberman, 2004. Endurance running and the evolution of *Homo*. *Nature*, **432**, 345–352.

Cann, R. L., M. Stoneking, and A. C. Wilson, 1987. Mitochondrial DNA and human evolution. *Nature*, **325**, 31–36.

Dart, R., 1925. *Australopithecus africanus*: the man-ape of South Africa. *Nature*, **115**, 195–199.

Darwin, C., 1871. *The Descent of Man, and Selection in Relation to Sex*. Murray, London.

Defleur, A., T. White, P. Valens, et al., 1999. Neanderthal cannibalism at Moula-Guercy, Ardèche, France. *Science*, **286**, 128–131.

Fabre, P. H., A. Rodrigues, and E. J. P. Douzery, 2009. Patterns of macroevolution among primates inferred from a supermatrix of mitochondrial and nuclear genes. *Mol. Phylogenet. Evol.*, **53**, 808–825.

Finlayson, C., F. G. Pacheco, J. Rodriguez-Vidal, et al., 2006. Late survival of Neanderthals at the southernmost extreme of Europe. *Nature*, **443**, 850–853.

Fleagle, J. G., 1999. *Primate Adaptation and Evolution*. 2nd Ed. Academic Press, San Diego, CA.

Green, R. E., A. W. Briggs, J. Krause, et al., 2006. The Neanderthal genome and ancient DNA authenticity. *EMBO J.*, **28**, 2494–2502.

Haile-Selassie, Y., B. M. Latimer, M. Alene, et al., 2010. An early *Australopithecus afarensis* postcranium from Woranso-Mille, Ethiopia. *Proc. Natl Acad. Sci. USA*, **107**, 12121–12126.

Hammer, M. F., T. Karafet, A. Rasanayagam, et al., 1998. Out of Africa and back again: nested cladistic analysis of human Y chromosome variation. *Mol. Biol. Evol.*, **15**, 427–441.

Harvati, K., and T. Harrison (eds.), 2006. *Neanderthals Revisited. New Approaches and Perspectives*. Springer, Netherlands.

Henke, W., and I. Tattersall (eds.), 2007. *Handbook of Paleoanthropology*, Vol. 1: *Principles, Methods and Approaches*, Vol 2: *Primate Evolution and Human Origins*, and Vol 3: *Phylogeny of Hominines*. Springer-Verlag, New York.

Higham, T., R. Jacobi, M. Julien, et al., 2010. Chronology of the Grotte du Renne (France) and implications for the context of ornaments and human remains within the Châtelperronian. *Proc. Natl. Acad. Sci. USA*, **107**, 20234–20239.

Jungers, W. L., W. E. H. Harcourt-Smith, R. E. Wunderlich, et al., 2009. The foot of *Homo floresiensis*. *Nature*, **459**, 81–84.

Kingdon, J., 2004. *Lowly Origin: Where, When, and Why Our Ancestors First Stood Up*. Princeton University Press, Princeton, NJ.

Krings, M., A. Stone, R. W. Schmitz, H. Kainitzki, M. Stoneking, and S. Pääbo, 1997. Neanderthal DNA sequences and the origin of modern humans. *Cell*, **90**, 19–30.

Lepre, C. J., H. Roche, D. V. Kent, et al., 2011. An earlier origin for the Acheulian. *Nature*, **472**, 82–85.

Nei, M., and A. K. Roychoudhury, 1993. Evolutionary relationships of human populations on a global scale. *Mol. Biol. Evol.*, **10**, 927–943.

Noonan, J. P., G. Coop, S. Kudaravalli, et al., 2006. Sequencing and analysis of Neanderthal genomic DNA. *Science*, **314**, 1113–1118.

Ponce de León, M. S., L. Golovanova, V. Doronichev, et al., 2008. Neanderthal brain size at birth provides insights into human life history evolution. *Proc. Natl Acad. Sci. USA*, **105**, 13764–13768.

Reich, D., R. E. Green, M. Kitcher, et al., 2010. Genetic history of an archaic hominin group from Denisova Cave in Siberia. *Nature*, **468**, 1053–1060.

Reich, D., N. Patterson, M. Kitcher, et al., 2011. Denisova admixture and the first modern human dispersals into Southeast Asia and Oceania. *Am. J. Hum. Gen.*, **89**, 516–528.

Rolian, C., D. E. Lieberman, and B. Hallgrímsson, 2010. The coevolution of human hands and feet. *Evolution*, **64**, 1558–1568.

Smith, T. M., P. Tafforeau, D. J. Reid, et al., 2010. Dental evidence for ontogenetic differences between modern humans and Neanderthals. *Proc. Natl. Acad. Sci. USA*, **107**, 20923–20928.

Walker, A, and R. E. Leakey (eds.), 1993. *The Nariokotome* Homo erectus *skeleton*. Harvard University Press, Cambridge MA.

Ward, C. V., W. H. Kimbel, and D. C. Johanson, 2011. Complete fourth metatarsal and arches in the foot of *Australopithecus afarensis*. *Science*, **331**, 750–753.

Wolpoff, M. H., 1989. Multiregional evolution: the fossil alternative to Eden. In *The Human Revolution: Behavioural and Biological Perspectives on the Origin of Modern Humans*, P. Mellars and C. Stringer (eds.). Edinburgh University Press, Edinburgh, UK, pp. 62–108.

Wolpoff, M. H., and R. Caspari, 1997. *Race and Human Evolution*. Simon and Schuster, New York.

Wood, B, 2005. *Human Evolution. A Very Short Introduction*. Oxford University Press, Oxford, UK.

Wood, B, 2010. Reconstructing human evolution: achievements, challenges, and opportunities. *Proc. Natl. Acad. Sci. USA*, **107**, 8902–8909.

Wood, B., and T. Harrison, 2011. The evolutionary context of the first hominins. *Nature*, **470**, 347–352.

Wrangham, R., 2009. *Catching Fire: How Cooking Made Us Human*. Basic Books, New York.

END BOX 25. 1

Piltdown Man

SYNOPSIS: This relates the story of how a fossil discovered in 1912 in England influenced our thinking about the origin of humans, until it was revealed that the fossil, known as *Piltdown man*, was a clever fake, a combination of a human skull and the lower jaw of a female orangutan.

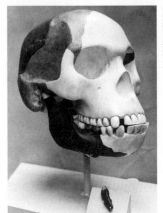

By around 1915 most anthropologists had agreed that early humans had large braincases, ape-like jaws, and large canine teeth. Evidence for this interpretation came from a fossil cranium and lower jaw found in 1912 at Piltdown, England by an amateur British archaeologist, Charles Dawson (1864–1916) and named *Eoanthropus dawsoni* (FIGURES EB1.1 AND EB1.2c).

Many anthropologists accepted the Piltdown fossil as valid, a situation that continued for about 40 years until it was shown that the entire "fossil" was a hoax. The teeth had been artificially ground down, the cranium was of a different age than the jaw, artificial pigmentation had been used to color the bones, and the molar teeth had long roots like those of apes (Figure EB1.1). Moreover, the associated animal fossils at the Piltdown site had a large accumulation of radioactive salts whose origin could be traced to a site in Tunisia. Piltdown man turned out to be a fabrication—a human cranium and the lower jaw of a female orangutan—a hoax perpetrated by someone who knew enough to destroy all

FIGURE EB1.1 Reconstruction of the skull of Piltdown man.

(a) Chimpanzee (*Pan troglodytes*)

(b) Taung child (juvenile *A. africanus*)

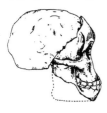

(c) "Piltdown man"

(d) Adult *A. africanus*

(e) *A. robustus*

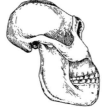

(f) *A. boisei*

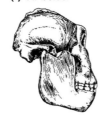

(g) *A. afarensis*

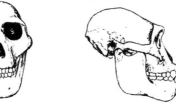

FIGURE EB1.2 Skulls of a chimpanzee **(a)** and fossil hominids **(b-g)** in the genus *Australopithecus* (*A. sp*), including cranial and lower jaw fragments of the Piltdown forgery **(c)**.

[Adapted from Johanson, D. and M. Edey, 1981. *Lucy: The Beginnings of Mankind*. Simon and Schuster, New York.]

END BOX 25.1

Piltdown Man (Cont...)

Because of the female's jawbone, Piltdown man skull should really be named PILTDOWN WOMAN *or Piltdown human (but the jaw is from an orangutan).*

obvious signs of the specimen's true origin by removing the jaw joint and modifying other key features.

In a book on the specimen published in 1990, Frank Spencer argued that the Piltdown forgery was the work of its principal "discoverer" Charles Dawson, in a conspiracy with Arthur Keith (1866–1955), a leading Scottish anatomist and physical anthropologist. Other historians of science suggest other suspects.* Whoever perpetrated this hoax, the result was the preservation of false views with false facts, at least for a time. The events that followed showed that false facts can be challenged in science and false views replaced. Discovery, description, and naming of *Australopithecus africanus* (Dart, 1925) laid the basis for Africa as the geographical location of the origin of humans and the place from which humans radiated to populate the entire globe, the "Out of Africa" hypothesis.

REFERENCES

Dart, R., 1925. *Australopithecus africanus*: the man-ape of South Africa. *Nature*, **115**, 195–199.

Millar, R., 1998. *The Piltdown Mystery: The Story Behind the World's Greatest Archaeological Hoax.* S. B. Publications, Seaford, East Sussex, UK.

Russell, M., 2003. *Piltdown Man: The Secret Life of Charles Dawson and the World's Greatest Archaeological Hoax.* Tempus Publishing, Gloucestershire, UK.

Spencer, F., 1990. *Piltdown: A Scientific Forgery.* Oxford University Press, Oxford, UK.

Weiner, J. S., 1955. *The Piltdown Forgery.* Oxford University Press, Oxford, UK (reprinted 2003).

---

*For three evaluations of Piltdown Man, see Weiner (1955), Millar (1998), and Russell (2003).

END BOX 25. 2

Dwarf Hominins on the Island of Flores

© Photos.com

SYNOPSIS: Further to the discussion of human origins, more information is provided about the discovery of fossils of a population of "pygmy" hominins who lived as recently as 18,000 years ago on the island of Flores in Indonesia. The cause of enormous excitement and controversy, this population has variously been interpreted as a relict population of *Homo erectus*, a new species of humans (*Homo floresiensis*), or a dwarf and congenitally deformed population of *Homo sapiens*. These widely divergent conclusions are based primarily on analysis of the one complete skull (known as LB1), analysis of long bone sizes and shapes, and the presence of tools normally assigned to more than one species of *Homo*.

Discovery of what has been interpreted as a dwarf (one meter—about 3.3 feet—tall) or pygmy population of *Homo erectus* on the island of Flores in Indonesia is so recent (Brown et al., 2004), and the findings so unexpected, that analysis and interpretation is ongoing at a fast pace. Especially controversial is whether these individuals represent a new species of *Homo* (*Homo floresiensis*), a relict population of *H. erectus*, or a dwarf and/or isolated population of our own species, *H. sapiens* (**FIGURE EB2.1**). Assignment to *H. erectus* would mean that a remnant of this species existed 18,000 years ago; specimens from the site date back to 95,000 years ago. The presence of tools more advanced than those *H. erectus* is known to have used is evidence against assignment to *H. erectus*.[a]

Analysis of 140 cranial features led investigators to conclude that these individuals represent an early pygmy population of *Homo sapiens* with signs in the one complete skull (specimen LB1, an adult female unearthed at Luang Bua) of a congenital malformation similar to a malformation called **microcephaly,** known to occur in modern-day humans.[b] Microcephaly is a condition in which individuals have small brains and therefore small skulls, resulting in reduced life expectancy and brain function (**FIGURE EB2.2**). The one complete skull, which enclosed a brain of 400 cm³ (13.1 in³) is comparable in volume to the brain of a chimpanzee and compares in volume and other measures with microcephalic human skulls (Jacob et al., 2006; Martin et al., 2006; Vannucci et al., 2011).

Much of the analysis has been conducted on this one fairly complete skull (LB1). Material from a dozen other individuals remains to be examined, so caution is required when drawing interpretations or conclusions.

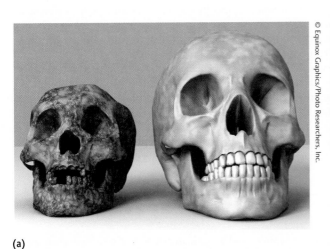

© Equinox Graphics/Photo Researchers, Inc.

© Mark Boulton/Alamy Images

(a)   **(b)**

FIGURE EB2.1 (a) The skull of the most complete hominin from Flores dated at 18,000 before present (left), compared with a human skull on the right. **(b)** This individual is reconstructed as having stood a little over 1 m in height, with long arms in relation to the length of the legs, unusual shoulders, and a chinless mandible.

[a]For some of the many papers on assignment of this species, see Argue et al. (2009), Jungers et al. (2009), and Lieberman (2009).
[b]D. Falk et al., 2005. The brain of LB1, *Homo floresiensis*. *Science*, **308**, 242.

END BOX 25. 2

Dwarf Hominins on the Island of Flores (Cont...)

© Photos.com

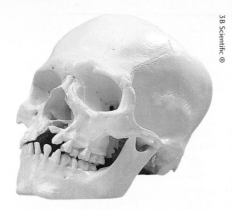

3B Scientific ®

FIGURE EB2.2 Skull of a microcephalic young adult modern human male. Compare with Figure EB2.1.

Enlargement of brain size during primate (including human) evolution has been associated with enhanced function of *MCPH* genes, which regulate the development of nerve cells in the cerebral cortex. To explain the diminutive size of Flores individuals, Richards (2006) invoked mutation in the gene *Microcephalin* (*MCPH1*), one of half a dozen genes in which mutation results in microcephaly.[c]

Another study amassed evidence that a mutation in the gene for the growth hormone receptor made Flores individuals insensitive to growth hormone. Disruption of growth hormone would upset the growth hormone–insulin-like growth factor-1 axis required to regulate growth. Consistent with this hypothesis, individual extant humans with the inherited disorder **Laron syndrome** (Laron-type dwarfism) have a mutation in the growth hormone receptor, making them insensitive to growth hormone, resulting in short stature (Hershkovitz et al., 2007). **Cretinism** also has been invoked to explain the small brain volume and features of the Flores skull. Oxnard et al. (2008) provided evidence for myxoedematous endemic cretinism, resulting from congenital hypothyroidism. Both brain and skeletal size are reduced in individual modern humans with cretinism as a consequence of underactivity of the thyroid gland. Local high incidences of cretinism on Flores have been associated with iodine deficiency, which remains a problem for individuals living on Flores today.

By definition, individuals with microcephaly have small skulls in relation to body size. More recent analyses of bones from seven Flores individuals led to the conclusion that the size of the skull is proportional to the size of the postcranial skeletal elements, a finding that is inconsistent with microcephaly (but not with dwarfism). Extinct hippopotamus from Madagascar, which have 30 percent smaller brains than predicted from hippos on the mainland scaled to the same size, were used by Weston and Lister (2009) to demonstrate that *H. floresiensis* could have evolved its relatively smaller brain in response to isolation. (Additional factors involved in the evolution of the human brain are outlined elsewhere in this text.) The interpretation that the Flores population is "an island adapted population of *Homo sapiens* [hence the small size], perhaps with some individuals expressing congenital abnormalities" was reinforced by the discovery in March 2008 of the remains of a population of small-bodied (1.2 m [3.9 ft] tall) humans who lived in caves in the Western Caroline island of Micronesia 1,400 to 3,000 years ago.[d]

On the other hand, analysis of the feet of Flores individuals lead some investigators to conclude that the feet are those of a basal biped, comparable to those known from 2- to 3-My-old human ancestors. The feet of what Jungers and colleagues classify as *H. floresiensis* are exceptionally long for a species in the genus *Homo*, a feature interpreted by Jungers et al. (2009) as basal for hominins, calling into question the origin of *H. floresiensis* from *H. erectus*. The large, hairy feet of hobbits in J. R. R. Tolkien's epic trilogy *The Lord of the Rings* are invoked in comparison. With such divergent structures (mosaic evolution? pathology?) the jury remains deadlocked on the precise identification of this population.

REFERENCES

Argue, D., M. J. Morwood, T. Sutikna, Jatmiko, and W. E. Saptomo, 2009. *Homo floresiensis*: a cladistic analysis. *J. Human Evol.*, **57**, 623–639.

Brown, P., T. Sutikna, M. K. Morwood, et al., 2004. A new small-bodied hominin from the Late Pleistocene of Flores, Indonesia. *Nature*, **431**, 1055–1061.

Hershkovitz, I., L. Kornreich, and Z. Laron, 2007. Comparative skeletal features between *Homo floresiensis* and patients with primary growth hormone insensitivity (Laron Syndrome). *Am. J. Phys. Anthropol.*, **134**, 198–208.

---

[c]Mutation of *MCPH1* in modern humans results in *autosomal recessive primary microcephaly-1*.

[d]L. R. Berger et al., 2008. Small-bodied humans from Palau, Micronesia. *PLoS ONE*, **3**, e1780.

Jacob, T., E. Indriati, R. P. Soejono, et al., 2006. Pygmoid Australomelanesian *Homo sapiens* skeletal remains from Liang Bua, Flores; population affinities and pathological anomalies. *Proc. Natl Acad. Sci. USA*, **103**, 13421–13426.

Jungers, W. L., W. E. H. Harcourt-Smith, R. E. Wunderlich, et al., 2009. The foot of *Homo floresiensis*. *Nature*, **459**, 81–84.

Lieberman, D. E. 2009. *Homo floresiensis* from head to toe. *Nature*, **459**, 41–42.

Martin, R. D., A. M. MacLarnon, J. L. Phillips, and W. B. Dobyns, 2006. Flores hominid: new species of microcephalic dwarf? *Anat. Rec.* Part A, **288A**, 1123–1145.

Oxnard, C., P. J. Obendorf, and B. J. Kefford, 2008. Post-cranial skeletons of hypothyroid cretins show a similar anatomical mosaic as *Homo floresiensis*. *PLoS ONE*, **5**(9) e13018. doi:10.1371/journal.pone.0013018.

Richards, G. D., 2006. Genetic, physiologic and ecogeographic factors contributing to variation in *Homo sapiens: Homo floresiensis* reconsidered. *J. Evol. Biol.*, **19**, 1744–1767.

Vannucci, R. C., T. F. Barron, and R. L. Holloway, 2011. Craniometric ratios of microcephaly and LB1, *Homo floresiensis*, using MRI and endocasts. *Proc. Natl Acad. Sci. USA*, **108**, 14043–14048.

Weston, E. M., and A. M. Lister, 2009. Insular dwarfism in hippos and a model for brain size reduction in *Homo floresiensis*. *Nature*, **459**, 85–88.

END BOX 25. 3

Stages in the Manufacture and Use of Tools by Species of *Homo*

© Photos.com

SYNOPSIS: As introduced in the accompanying chapter tools used by species in the genus *Homo* dating back to 2.5 Mya are assigned to four stages. Characteristics of tools from these four stages and their assignment to particular species of *Homo* are outlined.

Tool making and tool use is classified into four stages and related to species of *Homo* that made and used each tool type, beginning 2.5 Mya with *Homo* habilis, and extending to *Homo* sapiens (Henke and Tattersall, 2007).

Oldowan tools (FIGURE EB3.1a) consist of *sharp-edged stones* that could be made with a *single blow* of one rock against another. Such tools are sufficient to strip tough fibers from plants, break up animal carcasses, and expose the marrow cavities of long bones. Oldowan tools date back to deposits that are at least 2.5 My-old and are found in deposits over the ensuing million years. Some paleoanthropologists suggest that although Oldowan tools are traditionally ascribed to *H. habilis*, they may have been used by earlier hominins.

Acheulean tools (Figure EB3.1**b**) consist of stones *chipped from both sides* using *multiple strikes* to create the cutting edge. Acheulean tools are known as hand axes used 500,000 years ago and were in use up until 200,000 years ago. Cleavers and hand axes are most abundantly found 1.5 Mya associated with *H. ergaster* and *H. erectus.*[*]

Mousterian tools (Figure EB3.1**c**) require more *complicated working and reworking* (manufacturing?) than the other types of tools. Mousterian tools first appear 200,000 years ago and are found in 40,000 years old deposits. They are associated with *H. neanderthalensis* in Europe and with both Neanderthals and *H. sapiens* elsewhere. The presence of *points* indicates tools made for hunting, with the implication that the points were affixed to long sticks (*spears*). The presence of *scrapers* indicates tools made for scraping animal carcasses, either for food or for skins to be made into clothing.

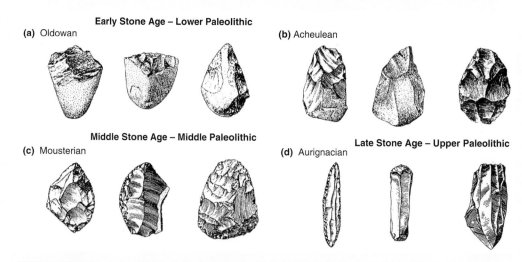

FIGURE EB3.1 Stages in stone tool development in the **Paleolithic Stone Age** begin with choppers and scrapers of the Oldowan stone industry **(a)**. Acheulean tools **(b)** consisting of cleavers (*left*) and hand axes (*right*) were most abundant 1.5 Mya and used by *Homo erectus* and *H. ergaster*. More complex Mousterian tools **(c)** consisting of points, scrapers, awls, hammer stones, and other forms are found between 200,000 and 40,000 years associated with *H. neanderthalensis* in Europe and both Neanderthals and *H. sapiens* elsewhere. Aurignacian **(d)** and Upper Paleolithic tools are associated with *H. sapiens*. Aurignacian tools date to between 34,000 and 29,000 years ago.

---

[*]A recent study by Lepre et al. (2011) places the origin of the Acheulean at 1.76 Mya and provides evidence for overlap of Acheulean and Oldowan tools. This observation is interpreted as evidence for overlap of those hominins using the tools.

Upper Paleolithic tools used by *H. sapiens* include *hooks* used for fishing and *needles* for sewing. These were the dominant tools in Asia and in Africa between 90,000 and 12,000 years ago. Specialists divide the Upper Paleolithic tool period into temporal and species-specific sub-periods (Figure EB3.1**d**), reflecting increasing sophistication and diversified use of tools and other objects. *Ivory beads* and *carved female Venus figurines* are first found between 28,000 and 22,000 years ago, depending on geographical location. Barbed *harpoons*, *spear throwers*, and the first *rock paintings* appear 18,000 to 12,000 years ago.

Following this period are the Mesolithic and Neolithic ages, the latter beginning about 10,000 years ago and marked by polished stone tools, pottery, domesticated animals, cultivated plants, and woven cloth.

REFERENCES

Henke, W., and I. Tattersall (eds.), 2007. *Handbook of Paleoanthropology.* Vol. 1: *Principles, Methods and Approaches,* Vol 2: *Primate Evolution and Human Origins,* and Vol 3: *Phylogeny of Hominines.* Springer-Verlag, New York.

Lepre, C. J., H. Roche, D. V. Kent, et al., 2011. An earlier origin for the Acheulian. *Nature,* **472,** 82–85.

END BOX 25. 4

Speech and Language

© Photos.com

SYNOPSIS: The anatomical and genetic evidence for why humans possess speech and language, why chimpanzees do not, and whether Neanderthals were capable of speech are discussed.

The most symbolic of animal vocalizations is human **speech** and **language**. The sound-producing systems of modern humans and our closest sister group, the chimpanzees, have been investigated to determine why humans have speech and language but chimps do not.

In the **human** sound-producing system, the larynx (**FIGURE EB4.1**) acts like a reed in a woodwind instrument. Sound is turned into the basic units of speech—vowels—in the pharynx. The length and shape of the pharynx determines the frequency pattern of the sound emitted and so modulates vowel sounds (**FIGURE EB4.2**). As the size of the lower jaws was reduced during human evolution, the larynx moved to a lower position and the tongue formed the anterior wall of the pharynx (Figure EB4.1**a**).

The chimpanzee tongue is isolated from the pharynx (Figure EB4.2**a**) so that air is expired through the nostrils rather than through the mouth. When combined with the inability of chimps to

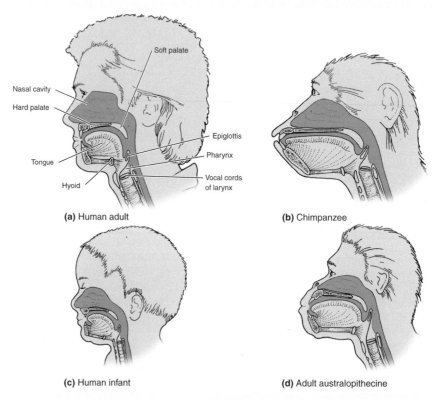

(a) Human adult

(b) Chimpanzee

(c) Human infant

(d) Adult australopithecine

FIGURE EB4.1 Upper respiratory systems of humans, chimpanzee, and australopithecines showing structures associated with vocalization. The pharynx is much longer in human adults **(a)** than in chimpanzees **(b)**, enabling humans to enunciate vowels and syllables by positioning the tongue in both mouth and pharynx. Newborn human infants **(c)** show the same overlap between epiglottis and soft palate as nonhuman primates **(b)** but the pharynx lengthens considerably during infancy and childhood. **(d)** A reconstruction of the presumed upper respiratory system of an adult australopithecine. As in nonhuman primates, the epiglottis overlaps the soft palate, the back of the tongue does not reach the pharynx, and the larynx is relatively high in the vocal tract.

[Adapted from Conroy, G. C., et al., 1998. *Science*, **280**, 1730–1731; and Lieberman, P., 1984. *The Biology and Evolution of Language.* Harvard University Press, Cambridge, MA and Lieberman, P., 1991 *Uniquely Human: The Evolution of Speech, Thought and Selfless Behavior.* Harvard University Press, Cambridge, MA.]

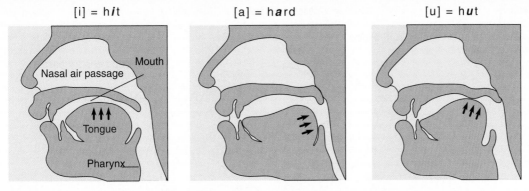

FIGURE EB4.2 Diagrammatic views of how adult humans produce three vowel sounds by positioning the tongue (arrows) in different parts of the oral airway. Note that a sharp bend (formed by the hard palate above the mouth and rear wall of the pharynx) partitions this airway into right-angled mouth and pharyngeal sections that are essential for these vowel sounds.

[Adapted from Aiello, L., and C. Dean, 1990. *An Introduction to Human Evolutionary Anatomy*. Academic Press.]

manipulate the tongue within the pharynx, the anatomical limitations to speech in chimps become evident, although numerous attempts to teach chimps to speak have been made.[a] Chimpanzees do communicate extremely effectively with facial expressions, body language, and sounds, both amongst themselves and with human trainers under controlled conditions (Lieberman, 1991; Bickerton, 1995).

The position of the larynx and the size of the tongue in relation to the length of the pharynx in australopithecines shows relationships similar to those seen in nonhuman primates (Figure EB 4.1**d**), supporting the hypothesis that speech arose only in a later lineage(s). Neanderthals had a chimpanzee-like vocal tract. It has therefore been concluded that Neanderthals were not one of the lineages in which speech arose (Cavalli-Sforza et al., 1992; Mithen, 2006). However, the acquisition of DNA sequences from Neanderthals is consistent with a different interpretation.

The gene *Forkhead Box P-2* (*FOXP-2*) has been linked to speech in humans. *FOXP-2* is found in most mammals, but the human gene contains two mutations not found in other mammals; nucleotide substitutions at positions 911 (threonine replaces aspartic acid) and 977 (arginine replaces serine) of exon 7. Humans with a single copy of the gene have speech impediments, and mice with a single copy have much reduced vocalization. Neanderthal DNA contains the same two mutations found in humans; chimpanzees lack the mutations. Clearly, it is tempting to link these two mutations either to the acquisition of speech/vocalization or to one or more important aspects of speech, and to conclude that Neanderthals had some form of speech, and that chimpanzees lack speech because they lack these mutations.[b]

Various aspects of speech and language have been traced to particular areas of the brain. On the temporal lobe of the left cerebral hemisphere is **Wernicke's area**, concerned with formulating and comprehending intelligent speech (**FIGURE EB 4.3**); individuals who have lesions in this area

---

[a]R. A. Gardner, and B. T. Gardner, 1969. Teaching sign language to a chimpanzee. *Science*, **165**, 664–672; D. Premack, 1971. Language in the chimpanzee? *Science*, **172**, 808–822; D. M. Rumbaugh, 1977. *Language Learning by a Chimpanzee: The Lana Project*. Academic Press, New York.

[b]C. S. L. Lai et al., 2001. A forkhead-domain gene is mutated in a severe speech and language disorder. *Nature*, **413**, 519–523; J. Krause et al., 2007. The derived *FOXP2* variant of modern humans was shared with Neanderthals. *Curr. Biol.*, **17**, 1908–1912; C. Scharff and J. Petri, 2011. Evo-devo, deep homology and FoxP2: implications for the evolution of speech and language. *Phil. Trans. R. Soc. (B)*, **366**, 2124–2140.

FIGURE EB4.3 Wernicke's area in the left temporal lobe and Broca's area in the left frontal lobe of the human brain are both sites associated with speech.

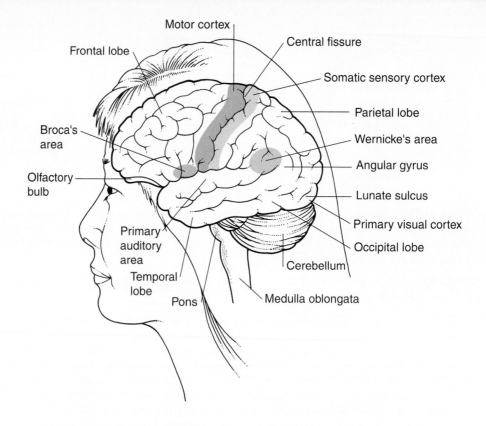

emit wordy babble that conveys no information. Individuals with lesions in **Broca's area** on the left frontal lobe, which coordinates vocal muscular movements, have difficulty speaking (Figure EB4.3). Aspects of language, such as grammatical structure and vocabulary, seem associated with neural circuits in the prefrontal cerebral cortex that lie somewhat forward of Broca's area. Disproportionate enlargement of the prefrontal cortex is an obvious feature of the brain of *Homo sapiens*.

It has been hypothesized that tool use and oral language share a common neurological evolutionary basis. Both language and manual tool manipulation are sequential processes. Appearance of stone tools coincident with the appearance of Broca's area in hominoids is consistent with a functional connection, thought it does not prove such a connection. Linkage between the two functions is consistent with their common localization in the left or dominant hemisphere of right-handed individuals. For those left-handed people whose manual control center lies in the right hemisphere, language control is often found there as well.

REFERENCES

Bickerton, D., 1955. *Language and Human Behaviour.* University of Washington Press, Seattle.

Cavalli-Sforza, L. L., E. Minch, and J. L. Mountain, 1992. Coevolution of genes and languages revisited. *Proc. Natl. Acad. Sci. USA*, **89**, 5620–5624.

Lieberman, P., 1991. *Uniquely Human: The Evolution of Speech, Thought, and Selfless Behavior.* Harvard University Press, Cambridge, MA.

Mithen, S., 2006. *The Singing Neanderthals: The Origins of Music, Language, Mind, and Body.* Harvard University Press, Cambridge, MA.

CHAPTER 26

Cultural and Social Evolution

CHAPTER SUMMARY

Animal behavior is traditionally regarded as instinctive (innate) and hard-wired in the genome, or learned and so subject to experience. Cultural/social evolution has been most thoroughly revealed through the study of instincts, behavior, imitation, learning, and the manufacture and use of tools. In the natural habitat, imitation and the passing on of skills from one generation to the next through learning are seen in chimpanzee tool making and tool use. Individuals, genes, and environment all interact in cultural and social behavior. Modern humans possess two modes of inheritance. One is phenotypic and based on inheritance of DNA, passed down from generation to generation genetically. The second is cultural/social inheritance based on imitation, experience, and learning, passed down from generation to generation through tradition. Cultural evolution has outpaced phenotypic evolution for much of human history. The first recognition of human social evolution came in applying Darwin's evolutionary ideas to society, in what came to be known as Social Darwinism. Cultural evolution is often, though inappropriately, couched as "How much is nature and how much is nurture?" In the mid-1970s, seeking answers to nature–nurture question led to the development of sociobiology, the systematic study of the biological basis of social behavior.

INSTINCTS AND LEARNED BEHAVIOR

Relatively complex innate behaviors, often called **instincts**, may involve many behavioral components; for example, the courtship patterns of most animals and the "dancing" patterns used by honeybees to communicate the direction and distance of food sources. Whether simple or complex, innate behaviors are often uniform within a species and therefore seem to be entirely, or almost entirely, under genetic control. Indeed, genes and their associated neuroanatomical locations associated with courtship behavior have been identified in *Drosophila*, and in these organisms male–female sexual orientation can be reversed genetically.[1]

Genetic components underlying **learned behavior** have been widely investigated. For example, genes in different breeds of dogs influence tameness, playfulness, and aggressiveness. In mice, knocking out the gene *disheveled* has no observable effect other than on social behavior, causing reduced social interaction and deficient nest building. There is much evidence that genes are involved in human nervous system disorders that alter human behavior; schizophrenia and manic depression are two well-studied examples.[2]

The capacities for most behavioral traits, like so many other adaptations, result from evolutionary selective forces. Because in some groups an individual can modify its behavior to suit different environmental or social circumstances, learned behavior can be more responsive to environmental changes than are instinctive or **innate behaviors** (Alcock, 2005; Konner, 2010). Much learned behavior derives from practice, allowing individuals to modify their behavior on the basis of their own or others' past experiences, as seen in chimpanzees and monkeys learning to use and manipulate tools (BOX 26.1).

Tool making by Caledonian crows (*Corvus moneduloides*) on the island of New Caledonia is a fascinating non-mammalian example of learned behavior (FIGURE 26.1). Adults fashion barbed and hooked tools from leaves of *Pandanus* palms. Adult birds in different parts of the island make different tools. Hunt and Gray, 2003 hypothesize that the ability to make tools is passed on to the young by a process involving **social learning**. Certainly, selectable life history benefits accrue from tool use; the number of larvae obtained by crows using tools satisfies 20% of the daily energy need, providing a fitness benefit that would favor social learning and tool use (Rutz et al., 2010). These crows appear to be no less sophisticated tool-makers than chimpanzees, though the latter are capable of further social behaviors, communicating extremely effectively with facial expressions, body language, and sounds, both amongst themselves and with human trainers under controlled conditions.[3]

An equally fascinating aspect of learning in birds is cooperation in food gathering. Indicative of social evolution, cooperation is greatest between those individuals that are more tolerant of other birds. Cooperation has been demonstrated under experimental conditions in which pairs of rooks (*Corvus frugilegus*) have to simultaneously pull on two ends of a rope to raise a platform to obtain food. Birds were classified into those that tolerate other birds and those that do not on the basis of behaviors such as feeding out of the same dish without aggressive encounters. "Tolerant" birds learn to cooperate more quickly and obtain the food much more frequently than intolerant birds, 63% of attempts versus 20% of attempts, respectively.[4]

---

[1] A. Civetta and R. S. Singh, 1998. Sex-related genes, directional sexual selection, and speciation. *Mol. Biol. Evol.*, **15**, 901–909; D. Grimald and M. S. Engel, 2005. *Evolution of the Insects.* Cambridge University Press, New York.

[2] N. Lijam et al., 2007. Social interaction and sensorimotor gating abnormalities in mice lacking *Dvl1. Cell*, **90**, 895–905; J. S. Robert, 2000. Schizophrenia epigenesis? *Theor. Med. Bioeth.*, **21**, 191–215.

[3] Recent studies with chimps have revealed a range of complex emotional states including altruism and lack of spite (Jensen, K., J. Call, and M. Tomasello, 2007. Chimpanzees are vengeful but not spiteful. *Proc. Natl Acad. Sci. USA*, **104**, 13046–13050).

[4] A. M. Seed et al., 2008. Cooperative problem solving in rooks (*Corvus frugilegus*). *Proc. R. Soc. Lond. B* **275**, 1421–1429. For the evolution of cooperation in humans see Ridley (1996).

BOX 26.1 Tool Use by Chimpanzees and Monkeys

Chimpanzee use sticks as tools to obtain food (FIGURE B1.1). The scenario therefore has developed that tool making in *Homo*, perhaps also in australopithecines, began with the use of unmodified sticks and stones, from which the ability to modify sticks and stones arose (Savage-Rumbaugh and Rumbaugh, 1993).

In their natural habitat, imitation and the passing on of skills from one generation to the next through learning are obvious in chimpanzee tool making and use. This is especially seen in the way chimpanzees employ twigs and vines in fishing for termites (Figure B1.1). **Termite fishing** involves:

- being aware of when to fish—October and November—although chimps probably know the season at large rather than the exact months;
- locating the sealed termite tunnels, which often requires importing the necessary supply of tools from as far away as a kilometer;
- shaping some of the tools by removing leaves;
- biting off the ends of the tools to achieve an optimum length;
- inserting the tool with a proper twisting motion to follow the curves of the termite tunnel;
- vibrating the tool gently to "bait" the termite soldiers, and
- retracting it carefully to avoid dislodging the termites.

Learning these tasks takes years; even a male anthropologist, who studied the technique for months, was no better at it than a novice chimpanzee.

Similar demanding techniques used in cracking nuts require a chimpanzee to find a properly sized stone or hardwood club to use as a "hammer," choose a well-shaped tree root as an "anvil," and precisely position each nut on the anvil; nuts from different species must be positioned differently. The hammer must be gripped in its most effective position, swung with the proper force, and aimed so that it hits the nut in exact locations to extract the maximum amount of nutmeat. Although young chimpanzees generally learn this technique from their mothers by imitation, incidents have been recorded in which mothers actively intervene in their offspring's unsuccessful nut-cracking attempts by taking the hammer, positioning the nut, and demonstrating the proper technique.[a] When this occurs in humans we call it "active teaching."

Long thought limited to chimps and humans, monkeys have now been shown to use tools. Tufted capuchin monkeys (*Cebus apella*) that inhabit Brazilian savannah or forests environments use stones to crack open nuts during daily foraging for food. These monkeys also use tools to dig for tubers and to probe holes and crevices. Bearded capuchin monkeys (*Cebus libidinosus*) select the most effective stone to crack open palm nuts, selecting heavier over lighter stones even when the heavier stones are smaller.[b]

(a) (b)

FIGURE B1.1 Tool use in chimpanzees (*Pan troglodytes*). **(a)** A stick fashioned for use in "fishing" termites from their nest. **(b)** Young chimps learn tool use by observation and initiation.

[a]C. Boesch, 1993. Aspects of transmission of tool-use in wild chimpanzees. In *Tools, Language and Cognition in Human Evolution*, K. R. Gibson and T. Ingold (eds.). Cambridge University Press, Cambridge, UK, pp. 171–183.
[b]E. B. Ottoni et al., 2005. Watching the best nutcrackers: what capuchin monkeys (*Cebus apella*) know about others' tool-using skills. *Anim. Cogn.*, **8**, 215–219; I. C. Waga et al., 2005. Spontaneous tool use by wild Tufted capuchin monkeys (*Cebus libidinosus*) in the Cerrado. *Folia Primatol.*, 77, 337–344.

FIGURE 26.1 Caledonian crows (*Corvus moneduloides*) using tools they manufactured from leaves to probe a tree stump for insects.

A fascinating example of how experience influences survival comes from a study of nestling survival in herring gulls (*Larus argentatus*) as a function of the age of the parent birds incubating the young; mortality of chicks, especially females, incubated by young parents is higher than mortality of chicks incubated by older, more experienced parents (Bogdanova et al., 2007).[5]

Surprisingly, behavior can also be inferred from the fossil record (Benton, 2010). Preservation of a fossil in a "living" position or situation, such as a dinosaur on a nest of eggs, speaks volumes about, in this case, dinosaur reproductive behavior. Evolution of sexual dimorphism has been used to infer social behavior in primates.[6] Fossilized footprints illustrate a great deal about social organization as when adult and child prints are found side by side.

LEARNING, SOCIETY, AND CULTURE

OVERVIEW

Unlike most other animals, humans transfer information from generation to generation through genes *and* culture. Speed of human cultural evolution has accelerated so much more rapidly than phenotypic evolution that we each gather new experience at a rate many times that of our ancestors (Stock, 2008). Cultural (cultural and social, sociocultural) evolution has a genetic basis, but has the added component of cultural processes that influence the evolutionary process.

In the nineteenth century, proponents of Social Darwinism maintained that cultural differences evolved primarily by natural selection, as embodied in the concept of the survival of the fittest (Hawkins, 1997). This belief, which "justified" many social inequities, was

---

[5] For the influence of parental care and clutch size in birds, see T. E. Martin et al., 2000. Parental care and clutch sizes in North and South American birds. *Science*, **287**, 1482–1485.

[6] L. M. Chiappe, 2007. *Glorified Dinosaurs: The Origin and Early Evolution of Birds*. University of New South Wales Press/John Wiley; J. M. Plavcan, 2000. Inferring social behavior from sexual dimorphism in the fossil record. *J. Hum. Evol.*, **39**, 327–344.

based on the prevailing philosophy that society, which often incorporates non-biological goals and value systems, is governed by the same laws as govern phenotypic evolution. Phenotypic and cultural evolution interact at several levels; many developments in behavioral and social biology over the past quarter of a century are based on the proposition that there is an inherited basis for human cultural evolution and that genotypes predisposed toward cultural development accumulate through natural selection (Wilson, 1975, 1978).

CULTURAL EVOLUTION

The most distinctive feature of the human species is our mental capacity—intelligence—and our consequent ability to learn from our own experiences or from the experiences of others. However measured, intelligence provides us with flexible adaptive behaviors that are far more complex than those attained by any other species. Humans have the capacity to (a) analyze their environmental experiences, (b) incorporate the lessons learned into their behavior, and (c) create new environments over which considerable control is exercised.

Much human learning follows a pattern of conscious acquisition and transmission of those behavioral responses that meet the needs of specific situations (Konner, 2010). Most other organisms with the capacity to learn rely primarily for survival on automatic responses that appear to be "hard-wired" into their nervous systems. A good example is ants cooperating to "stitch together" the edges of two leaves (FIGURE 26.2). More than any other animal, humans can acquire and transmit practices and behaviors—**culture**—through social exchanges involving language, teaching, and imitation, both among individuals and between generations. Cultural transmission of learned behavior and of the ability to imitate eliminates the hazards encountered when an individual must learn by trial and error to cope with environmental variables. Cultural transmission allows **imitative learning** of adaptive practices to be incorporated more successfully into our social and cultural heritage, often developed over more than a single lifetime.

Some broaden the term *culture* to include any form of socially-transmitted learned behavior, even if the transmission is by imitation alone. This broader definition extends culture to organisms lacking speech; evolution of speech and language is discussed elsewhere in this text.[7] Chimpanzees, for example, can incorporate variations in learned behaviors into their community through imitation, resulting in a variety of social histories ("traditions") that vary from population to population and from group to group (Whiten et al., 1999). Thirty-nine behavior patterns, among them tool usage, grooming, and courtship behaviors, are present in some chimpanzee communities, even though

© Hugh Lansdown/ShutterStock, Inc.

FIGURE 26.2 Ants cooperating in stitching together the edges of two leaves.

---

[7] For analyses of attempts to teach chimps to speak, see R. A. Gardner, and B. T. Gardner, 1969. Teaching sign language to a chimpanzee. *Science*, **165**, 664–672; D. Premack, 1971. Language in the chimpanzee? *Science*, **172**, 808–822; D. M. Rumbaugh, 1977. *Language Learning by a Chimpanzee: The Lana Project*. Academic Press, New York.

genetic differences between the same communities are small. A recent cladistic analysis found that cultural differences between populations of a single subspecies of the common chimpanzee, *Pan troglodytes*, were greater than differences between three subspecies, a finding consistent with a substantial learned component in cultural behavior.[8]

Because of socially mediated transmission, cultural changes—unlike changes in genetic makeup—are not restricted to vertical transmission from one generation to another, but may be developed through interactions between *related and unrelated individuals*; the cultural "parents" of individuals need not be their biological parents, nor need they derive from the same geographical area as their cultural offspring. Consequently, the kinds of isolation barriers that inhibit genetic exchange between biological species do not necessarily prevent transmission of culture between groups of animals. In the case of humans, therefore, we see two hereditary systems:

- a **gene-based system**, which transfers hereditary information from biological parent to offspring, and
- a **culture-based system**, which transfers cultural information from speaker to listener, from writer to reader, from performer to spectator.

Importantly, both systems are **informational and heritable**, the genetic system through the coding properties of DNA, and the cultural system through social interactions coded in language and custom and embodied in records and traditions. Evolution through experience is an extension of the method by which humans learn. It depends on conscious agents—humans with brains—who can modify cultural information in a direction that offers them greater adaptiveness or utility. Transmission occurs from mind to mind rather than through DNA. Information received from ancestors and contemporaries can be purposely changed to provide improved utility for ourselves, our offspring, and for others.

CULTURAL EVOLUTION OUTPACES PHENOTYPIC EVOLUTION

In 1963, D. J. da Solla Price estimated that, of all the scientists who have ever lived, MORE THAN 90 PERCENT WERE ALIVE IN 1963. Peter Gruss (President of the Max Planck Society for the Advancement of Science) made the same estimation in 2005.

One measure of how change continues to affect us is the time it takes to double our collective knowledge, a process that once took many thousands of years but now occurs in a mere handful.[9]

The generation time for cultural evolution is as rapid as communication methods can make it. Humans can now move from place to place faster than the speed of sound, and transfer ideas at the speed of electrons. Stages in human evolution, such as the softer diets that came with the development of agriculture, can be inferred from changes in the skeleton; the lower jaw became shorter and wider with the transition from hunting-gathering to agriculture (von Cramon-Taubadel, 2011). In contrast, phenotypic changes in humans since agriculture arose seem relatively small, if detectable at all. Indeed, the most distinguished possession of *Homo sapiens*, the human brain, shows no change in size over the last 100,000 years (BOX 26.2). Why this difference in speed between cultural and phenotypic evolution? Oversimplifying a little, this contrast can be ascribed to differences between the two distinct types of evolution, both of which are molded by natural selection: inheritance of learned characters through gene-based and cultural evolution; inheritance of the phenotype through gene-based evolution.

The high rate of cultural evolution is in striking contrast to the speed of evolution by natural selection for a number of reasons. Phenotypic (organic) evolution occurs through a process of selection (among other forces) on the phenotype. Because it usually requires

---

[8]S. J. Lycett et al., 2007. Phylogenetic analyses of behavior support existence of culture among wild chimpanzees. *Proc. Natl Acad. Sci. USA*, **104**, 17588–17592; A. Whiten, 2007. *Pan* African culture: memes and genes in wild chimpanzee. *Proc. Natl Acad. Sci. USA*, **104**, 17559–17560.

[9]For four different, but complementary analyses of the acquisition of knowledge, science, and the scientific method see Ziman (1976), Boorstin (1983), Pickstone (2001), and Knight (2009).

fortuitous changes in DNA sequences, their organization into the existing genome and selection of the resulting phenotype, phenotypic evolution can be slow. Each change may take many generations before it is incorporated into the population, although processes ranging from horizontal gene transfer in prokaryotes, phenotypic plasticity in eukaryotes, and evolution of gene regulation in all organisms greatly speed the rate(s) of phenotypic evolution over that resulting from mutation alone.

The incremental process of phenotypic evolution is vastly different from the rapid and conscious selective process humans use to choose among behavioral alternatives, although the hereditary material needed to transmit and use cultural information (memory, perception, language ability) connects them both. By consciously choosing among alternatives because of their consequences, human minds have become agents of a novel selection mechanism. It is "human minds" and not "human mind" because cultural evolution relies on communication among individuals and on group interactions (Konner, 2010). Cultural evolution is vastly more than the sum of its parts. The products of socially coordinated groups of individuals—cities, daily newspapers, factories—are quantitatively and qualitatively more than such individuals could create alone.

It is incorrect to assume that cultural evolution has replaced evolution by natural selection in humans. Studies continue to document ongoing selection on human populations, often isolated populations in which selection is most readily apparent. A recent example is the demonstration of selection favoring an earlier age at first reproduction in an island-based (isolated) French-Canadian population than seen in other populations. Age at first reproduction declined from 26 to 22 years between 1800 and 1939, resulting in increased reproductive success; in this population, more children survive to adulthood now, a well-accepted index of selection (Milot et al., 2011).

The LARGE CRANIAL VOLUME *of human newborns—a cost of brain growth in* utero—*helps explain the difficulties faced by human females in giving birth.*

BOX 26.2   The Human Brain

The average human brain is 340% larger in volume and 290% larger when corrected for body weight than the brain of our closest sister group, chimpanzees. Human brains also differ significantly in size from gorilla brains. The average human brain is 267% larger in volume *and 341% larger when corrected for body weight* than the average gorilla brain. This difference represents evolution of humans *and* gorillas, both having pursued their own line of evolution and both having escaped from the constraints of scaling of brain to body size so common in other mammals (Rilling, 2006).

GROWTH AND METABOLISM

The brains of humans became larger relative to body size because human brains follow a different pattern of development than the brains of other mammals or even other primates. In most mammals (including primates), brain growth is rapid relative to body growth during fetal stages but diminishes after birth. In humans, prenatal brain growth is also quite rapid relative to body growth, but this rate does not significantly diminish until infants are past one year of age. By adding 12 months of extrauterine development—the first year of life—such rapid early postnatal brain growth effectively extends the human gestation period from 9 to 21 months (Martin, 1990; Portmann, 1990).

Big brains relative to body size are metabolically expensive. Although the adult human brain represents only two percent of total body weight, it can consume as much as 20 percent of the daily energy budget.[a] For this reason, strong selection is required to increase brain size relative to body size.

Determining why, or even whether, bigger brains are better has preoccupied students of human evolution for centuries; the hypothesis is that changes in hominin brain size over the last 4 My (**FIGURE B2.1**) must signify some crucial changes in mental capacity. Nevertheless, the literature on increase in brain size relative to body size over evolutionary time (**encephalization**) is difficult to interpret. Encephalization is not universal among mammals. Proportional changes between brain and body size (allometry) has proven insufficient as an explanation. So has metabolic rate. A study by Shultz and Dunbar (2010) of over 500 species of fossil and extant mammals showing a correlation between sociality and larger brain size provides a new hypothesis to explore. So does a study in which endocasts of early fossil mammals were plotted against the encephalization quotient to reveal three pulses of brain evolution within mammals (Rowe et al., 2011).

[a]S. T. Parker, 1990. Why big brains are so rare: energy costs of intelligence and brain size in anthropoid primates. In *"Language" and Intelligence in Monkeys and Apes*, S. T. Parker and K. R. Gibson (eds.). Cambridge University Press, Cambridge, UK, pp. 129–154.

BOX 26.2   **The Human Brain (Cont...)**

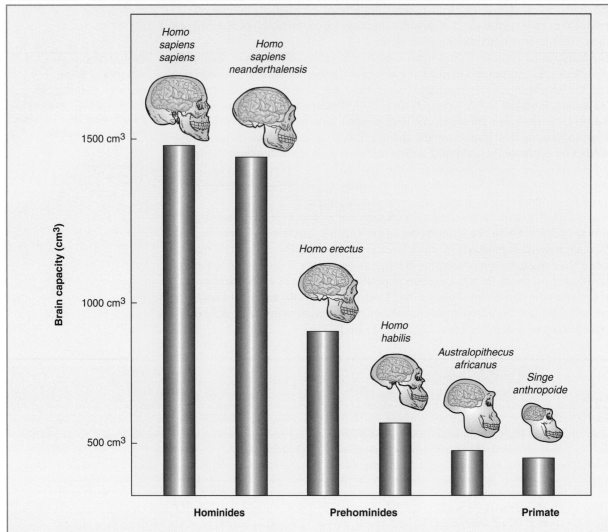

FIGURE B2.1 Evolution of brain capacity (cm³) in the lineage leading to modern humans.

EXPANSION OF SPECIFIC BRAIN REGIONS

The area of the cortex of a rat brain would cover a postage stamp, of a monkey a postcard, of a chimp a page, and of a human, four pages—all subjective measures, but the point is clear. It was assumed in the past that humans were more complex than other organisms and so required more brain cells and bigger brains, and/or that the ability to compartmentalize functions drove human brain evolution as it drove evolution at many other levels, going back to the evolution of cellular organelles.

Comparisons between *regions of the brain* among closely related taxa have been used to correlate increases in specific regions with specific attributes of each taxon. Much of the increase in human brain volume is associated with increased area and thickness of the **cerebral cortex**. Much of the increase in brain volume of great apes is in the cerebellum and frontal lobes, the regions of the brain most involved in perception, coordination, and motor control (**FIGURE B2.2**; and see Deacon, 1990, and Rilling, 2006). Moving beneath the surface to genetic differences, a molecular study identified 49 regions of the genome that regulate neural development (some of which are expressed in the neocortex) that are conserved in many mammals, but that diverged rapidly after humans separated from chimpanzees.[b]

---

[b]K. S. Pollard et al., 2006. An RNA gene expressed during cortical development evolved rapidly in humans. *Nature*, **443**, 167–172.

BOX 26.2	The Human Brain (Cont...)

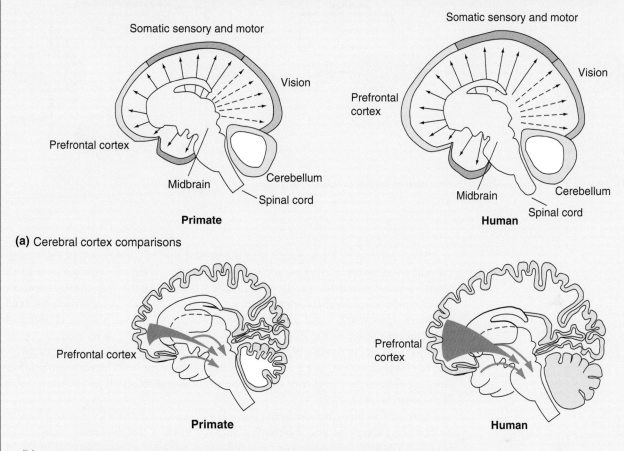

(a) Cerebral cortex comparisons

(b) Proposed differences in the magnitude of prefrontal effects on midbrain structures

FIGURE B2.2 Comparisons between cortical areas in a typical primate (*left side*) and humans with an enlarged cerebrum (*right side*). Most of the disproportionate cortical enlargement in humans is in the prefrontal cortex **(a)**, which affects the midbrain limbic system, which governs many emotional responses **(b)**. According to Deacon (1990), the prefrontal cortex is more than twice as large as one would predict for a comparably sized ape brain.

[Adapted from Deacon, T. W., 1990. Rethinking mammalian brain evolution. *Am. Zool.*, **30**, 629–705.]

SOCIAL DARWINISM

From the discussion so far in the chapter it is clear that the simple dichotomy of nature versus nurture is just that, simple. Human culture has an inherited foundation, and both culture and phenotype arise from informational systems that evolve over time.

Various writers in the late nineteenth and early twentieth centuries suggested that evolutionary concepts can be extended to society, and that nature and culture share similar evolutionary mechanisms, especially natural selection. These ideas were developed into a body of thought now known as **Social Darwinism**,[10] the chief concepts of which are that:

- differences among human individuals and groups arise through natural selection, and
- natural selection is the mechanism that led to social class structures and to national differences with respect to economic, military, and social power.

[10] For seminal studies of Social Darwinism in different contexts see Hofstadter (1955), Hawkins (1997), Bowler (2003).

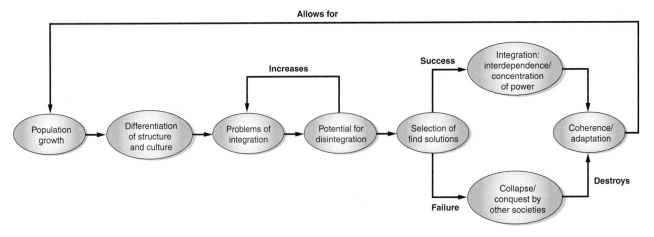

FIGURE 26.3 Herbert Spencer's representation of the forces acting in evolution.

These early writers saw parallels in the pivotal role of competition in the biological and social spheres. Slogans such as "struggle for existence" and "survival of the fittest," when applied to social traits, enabled English Social Darwinists, especially notable proponent Herbert Spencer, to suggest that cultural evolution was proceeding inevitably toward social and moral perfection (and approaching its culmination in Victorian society in England!). Although the views of Malthus on human populations were crucial in developing his theory, Darwin was opposed to injustice in society and to slavery, and was not a Social Darwinist; others applied Darwin's evolutionary writings to society. The theory of evolution by natural selection is a science, Social Darwinism is not.

Spencer's writings for example (1887, 1897), which ranged widely across biology, economics, philosophy, and sociology, continued to exert their influence well into the twentieth century. Interestingly, although Spencer coined the term "survival of the fittest," which implies the action of natural selection, he remained a Lamarckian in respect to phenotypic evolution until relatively late in his life. Whatever his views on the mechanisms, Spencer conceived of evolution as a powerful force that governed all spheres of existence (FIGURE 26.3) and therefore justified social and economic policies that supported those who were most "morally fit." For many Protestant intellectuals, Spencer's belief in the cosmic power of evolution helped reconcile science to their religion and made his writings extremely popular.

In its harsher economic and social forms, the Spencerian approach became popular in various circles in the United States (where more than half a million copies of his books were sold), especially through the teachings of William Graham Sumner (1840–1910), the best-known American Social Darwinist. In 1883 Sumner wrote: "We cannot go outside of this alternative: liberty, inequality, survival of the fittest; not-liberty, equality, survival of the unfittest. The former carries society forward and favors all its best members; the latter carries society downwards and favors all its worst members."[11] Not surprisingly, many wealthy capitalists found such views to their liking (Hofstadter, 1955). John D. Rockefeller, Jr. (1874–1960), whose father forged the gigantic Standard Oil Trust by destroying many smaller

---

[11] W. G. Sumner, 1883. *What Social Classes Owe to Each Other*. Harper & Bros., New York (quotation from p. 25).

(a)

(b)

FIGURE 26.4 John D. Rockefeller and his company Standard Oil **(a)** as the king of capitalism and **(b)** as an octopus devouring all competition.

enterprises (**FIGURE 26.4**), justified such behavior to a Sunday School class with the observation that,

> The growth of a large business is merely a survival of the fittest . . . The American Beauty rose can be produced in the splendor and fragrance which bring cheer to its beholder only by sacrificing the early buds which grow up around it. This is not an evil tendency in business. It is merely the working-out of a law of nature and a law of God.[12]

Similarly, for the United States railway magnate James J. Hill (1838–1916), "the fortunes of railroad companies are determined by the law of the survival of the fittest" (Hofstadter, 1955, pp. 44–45).

Along with objections raised by others, including Thomas Huxley in his book *Evolution and Ethics* (Huxley, 1894), the major problem with Social Darwinism was its assumption that society (economics, politics, government) operates through the same laws as biology and for the same goals.[13] However, the laws governing the inheritance of wealth and power in society are entirely man-made; the laws of biological inheritance are not. Because they can be consciously selected, social goals can be directed toward almost any objective: poverty or wealth, and socialism or capitalism, for example.

[12] Cited on p. 29 in W. Ghent, 1902, *Our Benevolent Feudalism*. Macmillan, New York.

[13] Thomas Huxley's 1894 book contains the text of the Romanes Lectures on evolution and genetics delivered at the University of Oxford in 1893. Fifty years later Huxley's grandson, Julian, delivered the Romanes Lectures on *Evolutionary Ethics*, and published both sets of lectures with an extensive introduction and commentary (T. H. and J. S. Huxley, 1947). Both grandfather and grandson affirm evolution as enhancing our moral progress. See Ayala (2010) for morality and cultural evolution in humans.

Unfortunately, the possibility of implementing social goals attracted individuals and groups who aspired to occupy superior positions over other individuals or groups. Goals for the "good of society" have been used to justify or reinforce racism, genocide, and social and national oppression. An extreme example is the role played during the 1930s and early 1940s by Adolf Hitler in the "racial health" movement—the purposeful destruction of millions of people because they were considered members of "inferior" racial groups. Even in the United States, with its more democratic social heritage, laws were passed during the 1920s restricting immigration from eastern and southern Europe because of their "inferior" or "undesirable" races. Immigration of virtually all Asians into the United States was halted in 1882 by the Chinese Exclusion Acts, which were not repealed until 1943. While economic and political considerations played a large role in the push for immigration restrictions in the United States, racial consideration formed part of the social framework under which the laws were promulgated. The United States was not alone. Australia enshrined a non-white immigration restriction policy that came to be known colloquially as the "White Australia Policy" in the second bill passed after confederation in 1901.

INHERITANCE OF SOCIAL BEHAVIOR

Many examples of social behavior have been discovered in a wide range of animals. A female lion will nurse her own cubs and those of her close genetic relatives. When a new dominant male in a pride of lions kills off his rival's offspring, he is eliminating competition with his own offspring. Both behaviors help to ensure survival of, respectively, her own and his close relatives' genes into the next generation (FIGURE 26.5). Biology exerts influences on social interactions between people as well. The close mother–infant relationship is only one of many behaviors common to all human groups.

All behavior, including learned behavior, has a genetic basis and all behaviors reflect organism- (and therefore gene-) environment interactions. Like morphology, animal behavior is shaped by natural selection: behavioral traits that maximize an individual's reproductive success are more likely to be carried into the next generation than those traits that do not. **Altruism** has been a difficult problem for evolutionary biologists because an animal that benefits another at its own expense may not leave any descendants. But even such social behaviors as altruism have evolutionary significance; selection acts on the fitness of an individual carrier of favorable behavioral genes and on the fitness of the genetic relatives of such individuals.

Drawing on methods used in population genetics, ethology, ecology, and other disciplines, **sociobiology** provides a rationale for assessing how and to what extent biological changes account for human social behaviors. In his book, *Sociobiology: The New Synthesis*, Wilson (1975) defined sociobiology as "the systematic study of the biological basis of all social behavior," and, in a second book on the subject, as "the extension of population biology and evolutionary theory to social organization" (Wilson, 1978, p. x).

Sociobiology deals less with the immediate causes of a particular human behavior than with its "ultimate" underlying evolutionary function, an aspect developed by Wilson (1978). Because of interactions between genes and environment, Wilson said, "there is no reason to regard most forms of human social behavior as qualitatively different from physiological and non-social psychological traits." Human nature is determined as much by heredity as by culture.[14] The challenge posed by sociobiology is to tease out the mix of phenotypic and cultural evolution, an exceedingly difficult but important task.

---

[14]One of the best discussions on genetic determinism remains the essay "Biological Potentiality vs. Biological Determinism" published by Stephen J. Gould in 1976 and reprinted in his first collection of essays (Gould, 1977).

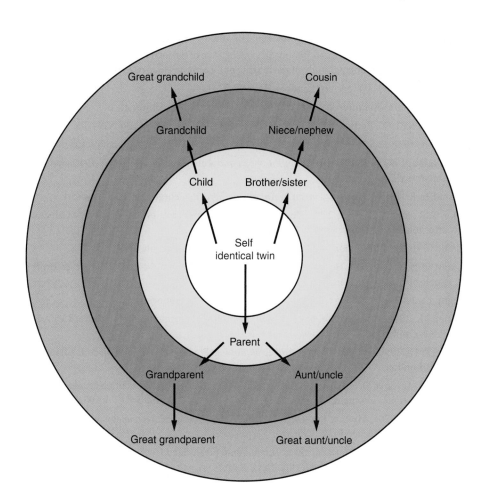

FIGURE 26.5 A schematic to show the percentage of genes shared between individuals of different degrees of relatedness from 100% to self and identical twin (white), through 50% to parent, child, brother/sister (yellow), to 25% (blue) and 12.5% (pink) for more distant relatives.

■ KEY TERMS

altruism
cultural evolution
culture
encephalization quotient (EQ)
imitative learning
inheritance of social behavior
innate behaviors

instinct
learned behavior
phenotypic evolution
social learning
sociobiology
termite fishing

■ EVOLUTION ON THE WEB

Explore evolution on the Internet! Visit the accompanying website for *Strickberger's Evolution, Fifth Edition,* at **go.jblearning.com/Evolution5eCW** for exercises and links relating to topics covered in this chapter.

■ REFERENCES

Alcock, J., 2005. *Animal Behavior: An Evolutionary Approach.* Sinauer Associates, Sunderland, MA.

Ayala, F. J., 2010. The difference of being human: Morality. *Proc. Natl. Acad. Sci. USA,* **107,** 9015–9022.

Benton, M. J., 2010. Studying function and behavior in the fossil record. *PloS. Biol.*, **8**(3), e1000321;

Bogdanova, M. I., R. G. Nager, and P. Monaghan, 2007. Age of the incubating parents affects nestling survival: an experimental study of the herring gull *Larus argentatus. J. Avian Biol.*, **38**, 83–93.

Boorstin, D. J., 1983. *Discoverers*. Random House, NY.

Bowler, P. J., 2003. *Evolution: The History of an Idea. Third edition*. University of California Press.

Deacon, T. W., 1990. Rethinking mammalian brain evolution. *Am. Zool.*, **30**, 629–705.

Gould, S. J., 1977. *Ever Since Darwin: Reflections on Natural History*. W. W. Norton & Co., New York.

Hawkins, M., 1997. *Social Darwinism in European and American Thought: 1860–1945*. Cambridge University Press, Cambridge.

Hofstadter, R., 1955. *Social Darwinism in American Thought*. George Braziller, New York.

Hunt, G. R., and R. D. Gray, 2003. Diversification and cumulative evolution in New Caledonian crow tool manufacture. *Proc. R. Soc. Lond. B.*, **270**, 867–874.

Huxley, T. H., 1894. *Evolution and Ethics, and Other Essays*. Macmillan, London.

Huxley, T. H., and Julian Huxley, 1947. *Touchstone for Ethics 1894–1943*. Harper & Brothers Publishers, New York and London.

Knight, D., 2009. *The Making of Modern Science*. Polity Press, Cambridge, UK.

Konner, M., 2010. *The Evolution of Childhood. Relationships, Emotion, Mind*. Harvard University Press, Cambridge.

Martin, R. D., 1990. *Primate Origins and Evolution: A Phylogenetic Reconstruction*. Chapman & Hall, London.

Milot, E., F. M. Mayer, D. H. Nussey, M. Boisvert, et al., 2011. Evidence for evolution in response to natural selection in a contemporary human population. *Proc. Natl Acad. Sci. USA*, **108**, 17040–17045.

Pickstone, J. V., 2001. *Ways of Knowing. A New History of Science, Technology and Medicine*. The University of Chicago Press, Chicago.

Portmann, A., 1990. *A Biologist Looks at Humankind*. (Translated from an earlier German edition by J. Schaefer.) Columbia University Press, New York.

Ridley, M., 1996. *The Evolution of Virtue: Human Instincts and the Evolution of Cooperation*. Viking, New York.

Rilling, J. K., 2006. Human and nonhuman primate brains: Are they allometrically scaled versions of the same design? *Evol. Anthropol.*, **15**, 65–77.

Rowe, T. B., Macrini, T. E., and Z.-X. Luo, 2011. Fossil evidence on origin of the mammalian brain. *Science*, **332**, 955–957.

Rutz, C., L. A. Bluff, N. Reed, et al., 2010. The ecological significance of tool use in New Caledonian crows. *Science*, **329**, 1523–1526.

Savage-Rumbaugh, E. S., and D. M. Rumbaugh, 1993. The emergence of language. In *Tools, Language and Cognition in Human Evolution*, K. R. Gibson and T. Ingold (eds.). Cambridge University Press, Cambridge, UK, pp. 86–108.

Secord, J. (ed.), 2010. *Charles Darwin. Evolutionary Writings Including the Autobiographies*. Oxford University Press, Oxford, UK.

Shultz, S., and R. Dunbar, 2010. Encephalization in not a universal macroevolutionary phenomenon in mammals but is associated with sociality. *Proc. Natl Acad. Sci. USA*, **107**, 21582–21586.

Spencer, H., 1887. *The Factors of Organic Evolution*. D. Appleton and Co., New York.

Spencer, H., 1897. *First Principles*. D. Appleton and Co., New York.

Stock, J. T., 2008. Are humans still evolving? *EMBO Rep.*, **9**, sp. Issue, S51–S54.

Von Cramon-Taubadel, N. 2011. Global human mandibular variation reflects differences in agricultural and hunter-gatherer subsistence strategies. *Proc. Natl Acad. Sci. USA*, **108**, 19546–19551.

Whiten, A., J. Goodall, M. C. McGrew, et al., 1999. Culture in chimpanzees. *Nature*, **399**, 682–685.

Wilson, E. O., 1975. *Sociobiology: The New Synthesis*. Harvard University Press, Cambridge, MA.

Wilson, E. O., 1978. *On Human Nature*. Harvard University Press, Cambridge, MA.

Ziman, J., 1976. *The Force of knowledge. The Scientific Dimensions of Society*. Cambridge University Press, Cambridge, UK.

Human Influences on Evolution

CHAPTER SUMMARY

Any attempt to influence our own evolution has to confront the potential incompatibility of cultural and phenotypic fitness. For example, societal and medical advances maintain alleles in particular groups, or in our species as a whole, that would be deleterious in other organisms. We discuss two approaches to influencing our own evolution—eugenics and gene therapy—as well as human efforts to engineer the genomes of other species, all of which raise ethical concerns. Movements to influence evolution through selective breeding (eugenics) arose as one aspect of the application of Darwinian principles to society (Social Darwinism). Although often not recognized as such, eugenics continues to be practiced by individuals in their use of assisted methods of reproduction. Humans have been engineering the gene pools of other organisms since the origins of agriculture and the domestication of plants and animals. Today genetically modified organisms (GMOs) are commonplace. So prevailing is human influence that, when humans are present, altered evolution of other organisms is not the exception but the expectation.

INTRODUCTION

Longer human life spans are generally thought desirable, but in fact natural selection rarely influences the features of post-reproductive individuals, including features associated with longevity. Given the large size of the human population and the limits on resources, lowering fertility might be advantageous, but biologically, high fertility has selective value. A large proportion of the human population is, or will be, affected by alleles that persist because of forces that maintain genetic variation, or because the intervention of modern medicine allows individuals with what might otherwise be deleterious alleles to live longer and reproduce. These are the topics of the first and second parts of the chapter.

HUMAN INFLUENCES OVER HUMAN EVOLUTION

Humans now have the technology to change themselves and the world in which they live (Palumbi, 2001a,b; Mindell, 2006). Although cultural considerations can transcend many biological constraints, improvements in human lifestyles bring us face to face with biological limitations. Our biological heritage stresses early reproductive success, after which physical deterioration sets in. Increased life expectancy, reproduction into decades beyond our 30s, and our ability to contribute to society into our 70s and 80s all stand in stark contrast to this biological heritage. Human cultural development requires continued plasticity, adaptability, and longevity but also is limited by biological decline as we age.

INCREASE IN LONGEVITY AND NATURAL SELECTION

For a table of rankings of LIFE EXPECTANCY BY COUNTRY, *see* The World Factbook, *available online at* https://www.cia.gov/library/ publications/the-world-factbook/rankorder/2102rank .html. Chad, with an estimated 2012 life expectancy of 48.69, occupied the lowest ranking of any country. Monaco, with a life expectancy of 89.68, has the highest.

In most organisms that attain reproductive maturity relatively early in their potential life span, survival after reproducing tends to be short-lived or non-existent. In contrast, individuals in many modern human societies enjoy a long period of **post-reproductive longevity**. Early in our own evolution, only about half the human population passed the age of 20 and not more than one in 10 lived beyond 40 (**FIGURE 27.1a**). Life expectancy remained between 20 and 30 years until the Middle Ages (fifth to fifteenth centuries C.E.), then rose somewhat beginning with the Renaissance period in Europe. Life expectancy has risen sharply among Europeans, Americans, Japanese, and some others in the last century and a half, from about 40 years in 1850 to the present high 70s or low 80s (**FIGURE 27.1b**). Less developed countries also have seen dramatic increases, the major exception being those countries in Africa where the AIDS epidemic, poverty and/or war have reduced already low life expectancies to as little as the low 30s.

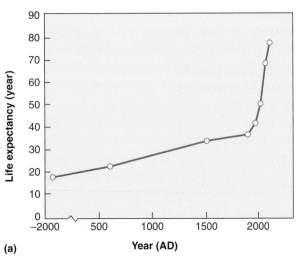

(a)

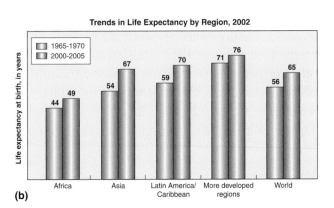

(b)

FIGURE 27.1 **(a)** Human life expectancy over 4,000 years of human evolution (from 2000 BCE to 2000 AD). **(b)** Human life expectancy varies in different human populations.

([a] Data from: National Academy of Engineering; [b] Data from: CIA World Factbook 2009.)

These changes in longevity are important. They indicate that, as a result of improvements in sanitation, diet, and medical practice in developed countries over the last century and a half, natural selection on *individuals in those societies* has little influence on "fitness"—that is, on the number of offspring produced, which more and more is a matter of personal choice. Further, many of the ills resulting from our changed lives—diabetes, heart disease, and cancer, discussed later in the chapter—affect us mostly in our later, post-reproductive years.[1] Consequently, there is little or no chance for natural selection to weed out susceptibility to these ills. In biological terms, an individual who has produced three children and, at the age of 50, develops cancer or other diseases with genetic components is no less reproductively successful than an individual of the same age who has produced three children but does not suffer from such diseases. Our increasing longevity has had an even more dramatic impact on global population, with major implications for humanity *and* for the world as a whole (**BOX 27.1**).

This discussion of longevity, however, should **not** leave you with the impression that humans are no longer subject to natural selection. Natural selection can readily be detected in populations in which technological change has been slow. Humans continue to adapt to changing environments, and those changes can be measured in rates of genetic change. In a study that looks at rates of genetic change in the historical context of migration out of Africa, Hawkes et al. (2007) documented that seven percent or more of the human genome has changed over the past 40,000 years, and that the advent of agriculture correlates with a hundred-fold increase in the rate of change in the human genome.

The Tibetan plateau lies at an average elevation of some 4,600 meters (15,000 ft) above sea level. Settlement of this plateau by humans several thousand years ago required adaptation to the low levels of oxygen at such altitude. The partial pressure of O_2 at sea level is 13.3 kPa; at 4,600 m it is 4.7 kPa. Most people can cope with oxygen levels found up to around 2,500 meters (8200 ft), at which the partial pressure of O_2 is 8.2 kPa, but at higher altitudes, humans can succumb to altitude sickness, the inability to cope with extremely low oxygen levels at higher elevations.[2] Tibetans living on the plateau have adapted to these low oxygen levels. Part of that adaptation involves the widespread presence of two genes that are found only rarely in other populations, consistent with positive selection for the genes in humans living at high altitude. One gene, *EGLN1* (*egl nine homolog 1, C. elegans*), produces a protein involved in the formation of hypoxia-inducible factor alpha proteins, transcriptional complexes, and major sensors of oxygen levels in cells that influence hemoglobin levels. The second gene, PPAR-α (*peroxisome proliferator-activated receptor α*), produces a nuclear receptor involved in the oxidation of fatty acids and energy metabolism, both of which are enhanced in Tibetans (Simonson et al., 2010).

GENETIC BIRTH DEFECTS

Despite improvements in medical care, alleles persist in human populations that, in other organisms, would be classified as deleterious because of their negative effects on adaptiveness to the environment or on fitness. Here we discuss how such alleles are maintained in human populations.

The number of children born with marked physical and/or mental abnormalities is conservatively estimated at about 20 to 25 in every 1,000 births in the United States. The mortality rate ascribed to congenital malformations accounts for around 15 percent of all

---

These **INFANT MORTALITY RATES** *are taken from estimates in the 2006 World Factbook. A recent analysis showed that a significant amount of the variation in human birth weight from region to region reflects adaptation to local selection pressures, especially parasitism.*[3]

[3] F. Thomas et al., 2004. Human birthweight evolution across contrasting environments. *J. Evol. Biol.*, **17**, 542–553.

ECOLOGICAL FOOTPRINT *is a measure defined as the amount of land and water required to produce the resources consumed and to absorb the waste of an individual, population, or lifestyle.*

---

[1] The numbers of individuals with obesity and diabetes is so high in some populations that the *average* longevity of these populations may be in decline, reversing decades-long trends.
[2] Base Camp at Mt. Everest is at 5,200 m (17,050 ft), where the partial pressure of O_2 is 3.8 kPa. Climbers normally stay at Base Camp for 4 to 8 weeks to acclimate. The Tibetan Plateau is only 600 m (just under 2,000 ft) lower than Base Camp. The summit of Everest at 8,848 m (29,029 ft) has a partial pressure of O_2 of zero.

BOX 27.1 **Population Explosion**

In most countries, infant mortality rates have declined markedly over the past century (**FIGURE B1.1**). As a result of improved sanitation, nutrition, and medical care, including control of infectious disease, infant mortality rates have fallen in more than 70 countries from 15 percent or more of all births in 1900, to less than one percent. Lower infant mortality comes at a cost, however. Syndromes such as cerebral palsy associated with premature birth cannot be brought down at the same rate as infant mortality and so has increased in incidence.

In most countries, however, it was (and still is) usual for birth rates to remain high for many years, even decades, following such a decrease in infant mortality. When combined with increased longevity, the result has been an exponential growth of the human population, an explosion that, at its height in the 1960s, saw a doubling every 35 years, a rate many thousands of times greater than that experienced in Paleolithic societies. Since 1900, the number of people has risen from a billion to the current 7.02 billion.[a] Although the overall growth rate has dropped (the doubling time is now 61 years), the absolute numbers continue to climb. Today about 203,800 people are added to the world population every 24 hours, by far the majority of them in developing nations. The U.S. Census Bureau predicts a world population of between 9 and 9.4 billion by the year 2050.[b]

This **population explosion** has major implications for the planet and all its inhabitants. The more people there are, the more food, water, space, energy, and other resources they need. The more resources consumed, the more waste produced. Our total impact on the environment depends largely on how many resources each of us uses, and on how much waste each of us produces—our per capita **ecological footprint**—which varies immensely between countries. A person living in the United States today consumes far more resources, produces far more garbage, and emits far more carbon dioxide and other pollutants than does an inhabitant of, say, Nepal. Furthermore, per capita resource use in industrialized countries, which already far exceeds sustainable levels, continues to rise. At the same time, hundreds of millions of people in developing countries are responding to improved economic conditions by adopting the patterns of resource use characteristic of wealthier regions.

FOOD SUPPLY

Here we discuss the complexities of one specific concern associated with population growth: **access to food**. Current production, though rising, appears barely able to keep up with

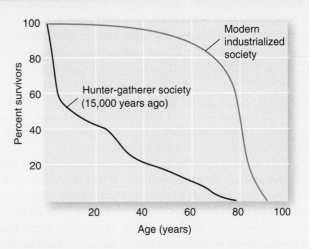

FIGURE B1.1 Survival curves for a population of hunter-gatherers who lived 15,000 years ago on the Mediterranean coast (based on skeletal remains) compared to a present-day population living in an industrialized society.

[Adapted from May, R. M., More evolution of cooperation. 1987. *Nature*, **327**, 15–17.]

present requirements. What then of the future? This turns out to be a topic with many facets, some of them political. The list below is far from complete.

- The amount of arable land is finite, and the amount cultivated is at or approaching capacity in most countries.
- The average amount of arable land per person worldwide continues to shrink as populations expand and as land is taken out of production as a result of environmental problems and urban sprawl. A decline of more than 50 percent between 1950 and 2005 has been estimated.
- Increasing amounts of land are used to grow food for livestock, a practice that ultimately results in production of far fewer food calories for humans per unit of land.
- Substantial amounts of food are lost through faulty distribution and storage systems, pests and disease, and graft and corruption.
- Extreme poverty means that a sizable fraction of the world's population cannot afford to buy nutritious food. At the same time, large surpluses exist elsewhere.
- Unwise subsidies and policies, as well as increasing globalization and domination of agriculture by multinational corporations, distort production and markets almost everywhere.

In terms of future food production, considerable amounts of arable land remain unused in several countries. There is great potential to increase both yields and the nutritional value of crops. As is evident from the list above, unless the political, environmental, and other problems that plague

---

[a]U.S. Census Bureau data are updated daily (U.S. & World Population Clocks. Retrieved July 18, 2012 from http://www.census .gov/main/www/popclock.html).
[b]The world population data are from estimates in the *World Factbook*, updated weekly. Retrieved July 18, 2012 from (https://www .cia.gov/library/publications/the-world-factbook/index.html).

BOX 27.1   Population Explosion (Cont...)

agriculture can be overcome, it is unlikely that we will be able to feed the billions more who will be added to the world's population in the years to come.

Other threats to the food supply locally and globally include soil erosion, fertilizer and pesticide contamination, water supply depletion, global climate change, and extreme climatic events. In the meantime, famine remains a recurrent problem in some developing countries, especially in sub-Saharan Africa. Indeed, United Nations statistics show that 925 million people were suffering from chronic hunger in 2010, a billion in 2011. Improving the quality of life for an increasing population while ensuring that no one lives in abject poverty are surely among the most critical issues faced by the human population.

infant deaths. Other abnormalities not immediately noted at birth may become apparent during childhood years; such defects are more common than usually realized. In various studies, about 30 percent or more of hospital admissions for children and 50 percent of all childhood deaths are ascribed to birth defects or to complications consequent to such defects. Although not all birth defects are genetic, the proportion of those that are is high.[4]

The incidence of each individual genetic disorder is low. Cystic fibrosis is 0.05% of live births. Tay-Sachs disease is 0.001%. Trisomy 21 is 0.13% and congenital heart defects 0.4%. Summing the incidence of genetic disorders in humans, whether autosomal, X-linked, or chromosomal, gives a total frequency of around 2.6% of live births. If genetic defects, such as muscular dystrophy, that appear later in life are included, the frequency doubles to around 6%. If, in addition, we include less obvious conditions that nevertheless have strong genetic components—impaired resistance to stress and infection—the effects of harmful genes touch a significant proportion of the human population.

As of July 31, 2012, the Human Gene Mutation Public Database listed 127,231 MUTATIONS IN 4,631 genes. Most of these mutations are exceedingly rare.

(The Human Mutation Database at the Institute of Medical Genetics in Cardiff, © 2010 Cardiff University. Available at http://www.hgmd.cf.ac.uk.)

MAINTENANCE OF "DELETERIOUS" ALLELES

Phenotypically normal human offspring carry a genetic load of one to eight deleterious lethal alleles, which, if homozygous, would result in early death.[5] Two important questions arise.

- What accounts for the prevalence of these harmful alleles?
- What, if anything, can be done to rid ourselves of them?

The reasons for the high frequencies of deleterious alleles are not fully agreed upon, although there is little question that they originally arose through mutation. In one hypothesis, held by the late Theodosius Dobzhansky and others, such alleles may offer considerable advantage to their heterozygous carriers by producing some form of hybrid vigor. According to this hypothesis, a gene will be maintained in the population even if the homozygote is relatively inferior in fitness.

Another school, formerly headed by the late American geneticist Hermann Muller (1890–1967), was based on the hypothesis that such genes produce no advantage of any kind; their frequency is high because the usual effect of natural selection has been artificially reduced. According to Muller (1950), genotypes that were defective and would have been eliminated in earlier times are now retained in the population by medical techniques that enable these alleles to be passed on to offspring.

HERMANN MULLER received the 1946 Nobel Prize in Physiology or Medicine for the discovery of the production of mutations by x-ray irradiation.

---

[4] For U.S.-wide statistics on birth defects see M. A. Cranfield et al., 2006. National estimates and race/ethnic specific variation of selected birth defects in the United States, 1999–2001. *Birth Def. Res. (Part A)*, **76**, 747–756.

[5] A. Eyre-Walker, and P. D. Keightley, 1999. High genomic deleterious mutation rates in hominids. *Nature*, **397**, 344–347.

If not eliminated by selection—and selection pressures themselves change over time—deleterious alleles will gradually increase in frequency in accordance with their mutation rates. Because mutation rates are usually low, the frequency of any particular allele will increase rather slowly, but as there are many possible deleterious alleles, the allelic load will increase significantly. Nearsightedness is a trait whose frequency has most likely increased in recent periods, but can be corrected quite simply by wearing glasses. Nevertheless, and despite (or because of) advanced medical care, a great many alleles that would affect fitness in other organisms are retained in human populations (Lynch, 2010).[6]

EUGENICS

In his dialogue, *The Republic*, Plato suggested that humans could be improved through **selective breeding**. In his ideal philosopher-state only the most physically and mentally fit individuals would reproduce. Their offspring would be raised by the state and the governing class would be selected only "from the most superior." In its nineteenth century form, suggestions for improving the human gene pool come under the name **eugenics**, a term coined by Francis Galton[7] in 1883 from the Greek words *eu* (good) and *gen* (birth).

Galton was fascinated by the inheritance of mental ability. His first book, *Hereditary Genius*, published in 1869, was an in-depth and statistical analysis of the inheritance of mental ability in the British upper class. Galton introduced the bell curve in his statistical analyses, coined the phrase "nature versus nurture," and wrote the first book on eugenics (Galton, 1883).[8] From 1865 on, Galton promoted the idea that the evolution of human traits through natural selection could be substituted by their evolution through selective breeding: "What nature does blindly, slowly, and ruthlessly, man may do providently, quickly, and kindly. As it lies within his power, so it becomes his duty to work in that direction."[9]

Coming from a brilliant family himself, Galton was impressed by the way in which intellectual and personality traits tend to run in families. Convinced that such traits were inherited, and drawing on his knowledge of animal breeding, Galton concluded that "judicious marriages over several generations" could "produce a highly gifted race of men," and thus thwart the "reversion to mediocrity" he believed to be threatening society as a result of excessive breeding by those who were not superior.[10]

Galton's ideas were taken up and developed by many prominent thinkers. Like Social Darwinists, early eugenicists reflected their personal and cultural biases in which traits they deemed undesirable. Strong racist and class-based biases played a part from the beginning. Charles B. Davenport (1866–1944), an influential leader of the early twentieth century eugenics movement in the United States, exemplified this racist approach by using New Englanders as the standard of comparison for all American social groups, irrespective of their country of origin. According to Davenport and other members of his Eugenics Record Office—which from 1910 to 1944 collected an enormous database of information about American families—particular groups could be characterized by social traits with identifiable hereditary components, "Italian violence and Jewish mercantilism" being two examples he used.

---

[6] A popular account of this topic is "*Survival of the Sickest*" (Moalem and Prince, 2007). Although not all the explanations provided are equally supported, it is an excellent source of information on the evolutionary bases for the continued presence of diabetes, cystic fibrosis, and other diseases.

[7] Galton is also known for his disagreements with his first cousin, Charles Darwin, over Darwin's hypothesis of pangenesis.

[8] See Castle (1932) and Kevles (1995) for analyses of eugenics at two periods in its history.

[9] F. Galton, 1904. Eugenics: its definition, scope, and aims. *Am. J. Sociol.*, **10**, 1–25.

[10] Although intelligence does have a large genetic component, much of it is remarkably plastic and influenced by prenatal diet, cultural, socio-economic, and maternal uterine environments (B. Devlin et al., 1997. The heritability of IQ. *Nature*, **388**, 468–471).

FIGURE 27.2 Selective breeding has produced one-ton Belgian blue super cows.

No evidence supports the biological superiority of any particular human group in respect to social characteristics or intelligence. Many characteristics considered desirable—high intelligence, esthetic sensitivity, longevity, and good physical health—are not associated with single genes that are easily identified, but by complexes of many genes acting together and responding to different environments. As learned from domesticating animals (FIGURE 27.2), selective breeding to develop beneficial gene complexes is fraught with difficulty. Selective breeding involves complicated schemes based on selection of parents and testing of progeny under controlled environmental conditions. Selective breeding enhances the preferred characters but at the expense of the deterioration of other characters.

Eugenicists believed in active intervention in breeding. Two avenues of action were promoted, and in some instances followed:

- **positive eugenics**, which aimed to increase the frequency of beneficial genes, and
- **negative eugenics**, which aimed to decrease the frequency of harmful genes.

Francis Galton's approach of encouraging particular people to marry is an example of positive eugenics. Singapore's campaign in the 1990s, in which young graduates were offered inducements to produce children, is another. Unfortunately, however, negative policies dominated the vast majority of government programs instituted, roughly from the 1890s to the 1940s (see below). So strong was the influence of the eugenics movement that almost every non-Catholic Western country was affected. Other areas included the southernmost Latin American states (where policies favored whiter complexions) and, since the early 1990s, China. Viewed today through the prism of human rights, we are appalled at the legislation—some of it draconian—enacted and enforced in the name of eugenics. Methods employed included:

- restrictions on immigration and marriage;
- racial segregation, including bans in the United States on marriage between whites and African Americans, overturned by the Supreme Court only in 1967;
- compulsory sterilization of the "feebleminded," certain criminals, and others deemed unfit;
- forced abortions; and, finally,
- in Germany under the Nazis, genocide of those (especially Jews) regarded as racially inferior and thus a threat to the "purity" of the Aryan race.

In the United States, home to the second largest eugenics movement (after Germany), marriage prohibitions were enacted in many states during the early decades of the twentieth century, and tens of thousands of individuals were sterilized, the last of them in the early 1960s (Kevles, 1995). Backing the eugenics movement was a large body of research. Fortunately, even from the early years, some geneticists and members of the general public were sharply critical of the methods employed and of conclusions drawn from the findings.

Modification of human reproduction continues in quite a different guise—modern **reproductive technology**.

REPRODUCTIVE TECHNOLOGY

The search for healthy children by today's parents has many eugenic overtones. Genetic counseling before conception, artificial insemination, testing the fetus for inherited disorders, selective abortion of fetuses deemed defective, ultrasound examination, and *in vitro* fertilization, including pre-implantation screening of embryos, are all in effect eugenic practices, albeit voluntary and on an individual basis. Eugenic practices and reproductive technologies both change gene frequencies in populations. In countries where reproductive technologies are widely used, the incidence of some congenital disorders has declined sharply in recent years. In Taiwan, for instance, the incidence of thalassemia amongst newborns dropped from 5.6 to 1.21 per 100,000 over eight years.

Because prenatal testing is expensive and often invasive, ultrasound is the only technique carried out as a matter of routine (**FIGURE 27.3**); other tests are usually performed only when the family history (or mother's age, in the case of Down syndrome) indicates a clear risk that a particular mutation or chromosomal abnormality could be present (Leroi, 2006). Armed with knowledge gained from sequencing the human genome, it may not be long before screening tools can be designed to test for all known disease-causing mutations. Given the financial cost and risk to the fetus of prenatal screening, however, the most likely use of such tools will be in pre-implantation testing of embryos conceived via *in vitro* fertilization (**FIGURE 27.4**).

Reproductive technologies, including genetic screening, evoke a host of ethical and social concerns, including: What constitutes a disease? How do potential parents make an informed choice among the options? Who has the right to know about an individual's genetic makeup—schools, employers, insurance companies, governments? Who owns this information? Despite the many ethical concerns, it is likely that the use of modern reproductive methods will continue to increase, and that over time they will have a larger and larger impact on our gene pool, at least in wealthy countries and populations.

For now, however, and into the future, a far greater influence is likely to be modern medicine in general. Before open-heart surgery, few if any individuals with severe congenital heart disease survived childhood (Marelli et al., 2007). Less benignly, and despite laws banning the practice, the widespread use in India and China of ultrasound followed by abortion to ensure the birth of a male child is leading to significant imbalances in the male/female ratio of those societies. The argument could be made that, because only a

In THALASSEMIA, α- *or* β-*hemoglobin chains are either not produced or are produced in reduced numbers.*

For a longer LIST OF CONCERNS, *see the Human Genome website, under Ethical Issues: http://www.ornl.gov/sci/techresources/Human_Genome/elsi/elsi.shtml.*

FIGURE 27.3 An ultrasound image of a human fetus from the left-hand side. The head is on the right.

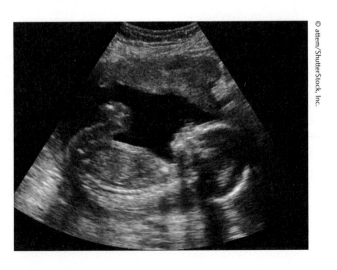

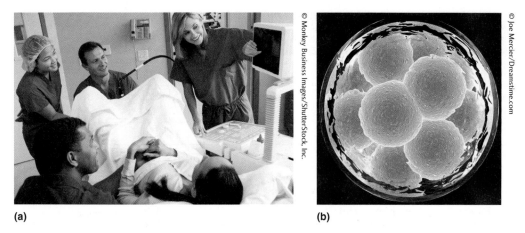

(a) © Monkey Business Images/ShutterStock, Inc.

(b) © Joe Mercier/Dreamstime.com

FIGURE 27.4 *In vitro* fertilization. **(a)** A doctor, using ultrasound imaging, retrieves a woman's egg to be inseminated *in vitro* (outside the body). **(b)** A 12-cell stage intact embryo ready to be implanted into the uterus.

handful of children now die of infectious disease in countries with robust healthcare systems, the door to a population with weaker average immune systems is being opened.

GENE THERAPY

An area of research evoking great interest is **gene therapy**, the introduction of genetic material to treat or cure a medical condition. Human trials, which began in 1990, have targeted several diseases using a variety of methods and vectors. Despite hundreds of trials, a great many problems remain.

In 1999 an individual in an American gene therapy trial died, most likely as a result of a severe immune response to the viral vector. A further major setback occurred in French trials carried out on young children with X-linked severe combined immune deficiency (X-SCID, also known as "bubble boy" syndrome). Although nine of the ten children in this trial were successfully treated, three subsequently developed cancer (T-cell leukemia). In a study conducted in 2006, one-third of mice administered the same gene as was used in the X-SCID trial developed lymphoma later in life. As of this writing, no gene therapy product had been approved for sale in the United States. In May 2007, Epeius Biotechnologies Corporation's Rexin-G™ was approved by the United States Federal Drug Administration for consideration for phase I clinical trials as an orphan drug to treat pancreatic cancer. In 2011, with phase I and II clinical trials complete, Rexin-G™ remains the first and only targeted gene delivery system designed to treat metastatic cancer.

ORPHAN DRUG *is the term for the 7-year right given by the U.S. FDA to a drug company to exclusively market a drug developed to treat a rare human disease, one affecting fewer than 200,000 individuals.*

THE RISE OF AGRICULTURE

The transformation of human societies produced by cultural change over the last 10,000 years has been dramatic. This is readily seen in the origin and evolution of **agriculture**. The evolution of agriculture is not limited to human societies, however. Cultivation of land, crops, or farming other species (agriculture) has evolved independently in four groups of animals: humans, termites, bark beetles, and ants. Ant agriculture arose when a lineage of ants began to cultivate a range of species of fungi some 50 Mya. Three novel systems of ant fungal "gardening" evolved in the last 30 My, each involving the cultivation of a separate species of fungus. A further system, cultivation of a single species of fungus by leaf-cutter ants that evolved 8 to 12 Mya, is now the dominant form of ant agriculture in the New World tropics (De Fine Licht et al., 2010).

Somewhere during the Neolithic Age, the long-prevailing hominin lifestyle of hunting–gathering–fishing began to give way to the cultivation of food using

TABLE 27.1	Major Human Agricultural Expansions from the Neolithic Age Onward			
Center of Origin	**Area of Expansion**	**Time (Years Ago)**	**Technologies**	**Crop or Product**
Middle East	Europe, North Africa, and Southwest Asia	10,000 to 5,000	Farming/Domestication	Wheat, barley, goats, sheep, cattle
North China	North China	9,000 to 2,000	Farming/Domestication	Millet, pigs
South China	Southeast Asia	8,000 to 3,000	Farming/Domestication	Rice, pigs, water buffalo
Central America and North Andes	Americas	9,000 to 2,000	Farming	Corn, squash, beans
West Africa	Sub-Saharan Africa	4,000 to 300	Farming	Millet, sorghum, gourd
Eurasian steppes	Eurasia	5,000 to 300	Pastoral nomadism	Horses

[*Source:* Cavalli-Sforza, L. L., P. Menozzi, and A. Piazza, 1993. Demic expansions and human evolution. *Science,* **259**, 639–646.]

domesticated plants and animals. Energies formerly expended in finding food were directed into methods of agriculture. Although originally developed in the Middle East, China, and Central America, such changes spread rapidly. Within 1,000 years or so, some form of agriculture had begun in many contiguous areas. Within 5,000 years agriculture and the technologies it stimulated extended widely (**TABLE 27.1**).

The most immediate and most far-reaching effect of agriculture was to increase food supply many-fold, a change permitting larger populations and a greater population density in agricultural communities. What matters here is the change in human lifestyle. Although there are uncertainties concerning whether sedentary communities preceded or followed agriculture, and whether agriculture was stimulated by climatic changes or by increased Neolithic social complexities, it did not take many generations for humans to settle arable areas on a permanent basis.

DOMESTICATION

Humans have been changing the genetic makeup of organisms since the origins of agriculture (Molina et al., 2011). **Domestication** of plants and animals involved selection to promote some traits and eliminate others (Clutton-Brock, 1999). In some species, the results have been dramatic, as in the evolution of wheat.

In the thousands of years since domestication began, humans have produced a world of agriculture, horticulture, and plantation forestry by manipulating genes and genomes, most commonly through selection. Synthesis of information gained from genetics and archaeology is a promising approach to further our understanding of how selection acted during domestication of plants (Purugganan and Fuller, 2009; Heerwaarden et al., 2011; Molina et al., 2011). From such studies we know that:

- Humans at the hunter-gather stage of their evolution cultivated food plants in 24 different locations independently.
- Grasses (from which cereal crops arose) were cultivated in half these locations.
- Loss of seed shattering (dispersal before harvest) and an increase in seed size were early and adaptive responses of plants to selection associated with domestication.
- Adaptive changes in seed size were rapid, as little as 500 years for rye.
- Small numbers of genes are sufficient to reduce seed shattering; that is, selection on a few regions of the genome is sufficient to drive domestication.[11]

[11] The genes are *Sh4* (transcription factor) and the QTL qSH1 (including a homeobox protein) in rice, and the gene *Q* in wheat (Purugganan and Fuller, 2009).

- Phenotypic variation in domesticated strains can exceed the amount of variation in the undomesticated wild ancestors.

In agriculture, it has been standard practice for millennia to develop new varieties through hybridization between different species. At first, the species concerned were closely related. By the middle of the twentieth century, however, methods had been developed that allowed crosses between more distantly related species. For instance, triticale, a widely grown cereal, is a hybrid between wheat (*Triticum*) and rye (*Secale*). The *green revolution*, a series of research, development, and technology programs aimed at increasing global agricultural production, was a major achievement of this type of breeding. Yet although intended to alleviate starvation in poverty-stricken regions, the green revolution drew a great deal of protest for its environmental and social consequences. The ethics and safety of gene transfer between species were not, however, raised as areas of concern as they have been with **genetic engineering**, the subject of the next section. Developed after the green revolution, the genes transferred in this relatively new technology often come from a different phylum or even kingdom, which was not the case with earlier breeding practices (Avise, 2004; Dyson, 2007).

GENETIC ENGINEERING OF FOOD CROPS AND ANIMALS

The early 1970s marked the beginning of the era of genetic engineering, when it was discovered that DNA fragments from different sources could be spliced together to form entirely new **recombinant DNA**, which could be used to introduce new traits to an organism (**TABLE 27.2**). A basic technique of genetic engineering is to insert a new or modified nucleic acid sequence into a virus or plasmid that can carry these novel sequences into host cells and thence host genomes where they can be incorporated and amplified many times over (**FIGURE 27.5**).[12] This technology allows active intervention into the genetic material of any organism. Genomes of many organisms have been manipulated to affect such traits as protein production, resistance to infective agents, agricultural yield, nutritional value, toxic susceptibility, environmental stamina, tumor resistance, and so forth (Palumbi, 2001a,b; Mindell, 2006). Such manipulation of genetic traits allows us to move organismal evolution from the age-old province of random mutation and natural selection to **human-directed evolution**. Genetic engineering is now big business, especially when it comes to agriculture.

The first genetically modified organism (GMO), the FlavrSavr tomato, was put on the United States market in 1994 (Table 27.2). Although the FlavrSavr failed on the market, and production was ceased in 1997, little more than a decade later, the area covered by just the four major genetically modified crops in the United States (corn, soybeans, cotton, and canola) had grown to 519,000 km² according to 2006 U.S. Department of Agriculture statistics.[13] A large proportion of food on North American supermarket shelves now contains at least some genetically modified content. The area under cultivation of GM crops continues to escalate. By 2009 it had grown to encompass 34 million km² worldwide. A major exception to this trend is Europe, where a huge public outcry against the technology on ethical, environmental, and food safety grounds continues to have an impact on government policy.

With their early emphasis on increased resistance to insect pests, diseases, and herbicides, the most common GM crops in use today were largely developed with financial profit in mind, but the potential is enormous. Improved nutrition (golden rice, for example, is rich in β carotene, provitamin A) is one such promised benefit. Other benefits

---

[12] In nature genes are transferred horizontally by hitchhiking on plasmids.

[13] By 2011, 395 million acres of land were planted with biotech crops worldwide; see http://www.usatoday.com/money/industries/food/story/2012-02-06/biotech-crops/53005000/1.

TABLE 27.2	Major Discoveries Leading to Genetic Engineering

Year(s)	Advance
1970s	Isolation of a restriction enzyme that cuts DNA at specific sites Development of method to recombine the fragments, launching recombinant DNA technology (genetic engineering).
1972	First recombinant DNA molecules produced.
1973	Introduction of recombinant plasmid vectors into bacteria for cloning.
1975	DNA sequencing technology invented.
1976	First clinical use of recombinant DNA for prenatal diagnosis of α-thalassemia.
1977	Laboratory techniques for sequencing DNA invented.
1978	Synthesis of the proinsulin peptide using recombinant DNA.
1983	Discovery of polymerase chain reaction (PCR).
1985	Polymerase chain reaction (PCR) used to amplify nucleotide sequences.
1987	Production of a vaccine (anti-hepatitis B) based on recombinant DNA technology.
1988	Incorporation and expression of genes inserted into mouse cells using retroviruses.
1989	First human gene sequenced, and a defect in the gene product shown to "cause" cystic fibrosis.
1990	Gene for defective enzyme (adenine deaminase) required for nucleic acid synthesis replaced in affected humans.
1991	Craig Venter discovers expressed sequence tags (ESTs), allowing a much faster search for specific genes.
1993	Marketing in the United States of FlavrSavr tomatoes, genetically engineered for longer shelf life.
1996	Birth of a lamb named Dolly, first animal to result from cloning an adult somatic cell.
1999	Sequencing of a human chromosome and of complete fruit fly (*Drosophila melanogaster*) genome.
2000	Three babies in France cured of genetic disorders using human gene therapy (HGT).
2001	First draft sequences of human genome completed independently by the Human Genome Project and Celera Genomics.
2002	Mouse genome sequenced; birth of first genetically modified cat.
2006	Sequencing of the human genome completed.
2008	Introgen's Advexin is the first gene therapy for cancers associated with Li-Fraumeni syndrome.

may exist as well; as documented over more than decade, transgenic corn can reduce pest damage on non-genetically engineered corn or maize in adjacent fields (Hutchinson et al., 2010; Tabashnik, 2010).[14] GM technology has been applied to many other organisms besides plants, including fish for enhanced growth, cold tolerance, and human insulin production; cows to produce human growth hormone and casein-enriched milk; pigs for lower fat content, and mice to produce fish oils (Palumbi, 2001a,b).

Although most of this work has yet to be approved for commercial application, in 2006 approval for release of Atryn®, the first pharmaceutical drug generated in a mammal (a goat), was granted in Europe. Indeed, the pharmaceutical industry is investing heavily in research into this method of drug production. Microorganisms have been used to produce a wide array of human and other proteins, for example, human growth hormone and an essential blood-clotting factor (factor VIII) for hemophilia sufferers (Dyson, 2007).

Atryn®, an anti-clotting agent, is a recombinant form of HUMAN ANTITHROMBIN, *designed to be used during surgery on people with antithrombin deficiency.*

---

[14]Corn in the United States engineered with the bacterium *Bacillus thuringiensis* (*Bt*) to make anti-insecticidal proteins gains protection against the European corn borer as does non-engineered corn that benefits from the reduced numbers of corn borers in the neighborhood.

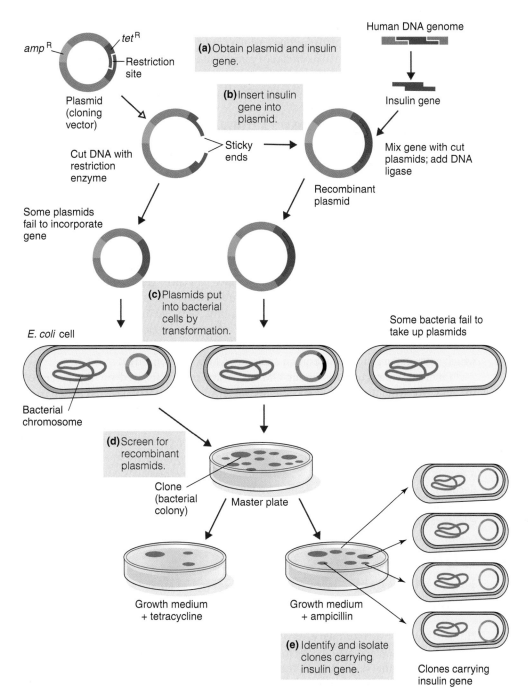

FIGURE 27.5 A general scheme for constructing a clone of recombinant DNA molecules using plasmids and restriction endonuclease enzymes. **(a)** A foreign DNA molecule (human insulin in this example) and a plasmid vector are selected, both carrying recognition sites that can be cleaved by a restriction enzyme. **(b)** Cleavage produces one or more fragments of the foreign DNA and opens the plasmid vector allowing the DNA to be inserted. **(c)** The plasmid carrying the foreign DNA is inserted into a host cell, in this example, into the bacterium *E. coli*. **(d)** Because some bacteria fail to take up the plasmid, the cells are screened for the presence of recombinant plasmids, a process that may involve growing the cells in the presence of specific molecules—tetracycline or ampicillin in this example. **(e)** Clones carrying the gene of interest are further replicated to obtain large quantities of the gene.

[Adapted from Strickberger, M. W., 1985. *Genetics, Third Edition* Macmillan, New York.]

In summary, the major benefits conferred by GM technology are that, in contrast to more traditional breeding techniques, which are time-consuming and inexact, GM is more efficient, more precise in its choice of genes, and allows genes from a far wider range of organisms to be introduced into the recipient organism's genome. When combined with cloning, its scope may be further enhanced.

CONCERNS ABOUT GM TECHNOLOGY

The physical/biological environment considered "natural" is not unchanging. Organisms have regularly affected and changed the environment, and continue to do so, while humans have long engaged in a succession of interventions into nature that have changed its form, function, and substance in most localities and for many organisms.

An important feature of GM technology is that the genes employed usually come from a species other than the receiving organism. To many, this poses a substantial ethical problem. But gene flow via horizontal gene transfer between organisms is a common event across the whole Tree of Life. Species barriers are continually being crossed by viruses, transposons, and plasmids, all of which can carry genes from one species to another. Humans readily incorporate genes introduced by viruses, though but not by bacteria (the latter was, however, controversial for a time; see Genereux and Logsdon, 2003). Human genomes contain many hundreds of transposable elements that affect gene expression, genetic recombination, and unequal crossing over of chromosomes, some of which entered the genomes of human ancestors 4 Mya (Britten, 2010).

Environmental concerns about GM crops are mostly focused on the perceived threats to local biodiversity and ecosystems, in particular the possibility that pollen from GM plants will fertilize closely related weeds (a problem if the crop is modified to resist pests or herbicides), closely related native plants (whose very existence might then be threatened), or non-GM crops of the same species. The long-term impact of such crosses will depend, of course, on the relative fitness of the gene(s) in question. Again, this is not a new thing; gene flow from crops altered by non-GM methods has occurred for many decades. Nevertheless, great care must be taken before introducing any new GM crop. After all, our experience with the technology is still relatively limited, and long-term effects can be difficult to envisage.

OUR INFLUENCE ON THE EVOLUTION OF OTHER SPECIES AND THEIR ENVIRONMENTS

Humans have had extraordinary impacts on the distribution and gene pools of organisms they have domesticated, and on organisms where the impact was more indirect. As aptly summarized at the beginning of the twenty-first century, "technological impact has increased so markedly over the past few decades that humans may be the world's dominant evolutionary force" (Palumbi, 2001a, p. 1786). In the same paper, Palumbi estimated the annual economic cost to the United States of human-induced evolution of insect pests and new drugs against drug-resistant *Staphylococcus aureus* and HIV as $50 billion. This figure doubles when other organisms are taken into account.

Starting in the Paleolithic, when human actions such as hunting or the widespread use of fire may have led to the extinction of the megafaunas of Australia, the Americas, and other parts of the world, humans have altered both the species mix and the environment of almost the entire globe (Crosby, 1986; Flannery, 1994, 2001). A notable example today is the plight of the ocean's fish stocks, all of which are under threat from over-fishing. Despite a moratorium on fishing and other conservation measures, populations of the northern cod off Newfoundland and Labrador have not recovered at all from a disastrous collapse in the early 1990s.

Wherever humans have moved, especially since the development of agriculture, they have brought with them a variety of other organisms, both domesticated (cattle, goats, etc.) and opportunistic (rats, weeds, disease-causing microbes). Changes wrought on local environments took a quantum leap when Europeans began venturing across oceans. Impacts on native humans in newly discovered regions were calamitous, a story told by Jared Diamond in *Guns, Germs, and Steel* (1997).

In the decades following the arrival of Europeans in the Americas and the Pacific, infectious diseases such as smallpox, measles, and tuberculosis decimated indigenous human populations that, up until then, had been isolated. A similar fate befell a variety of animal and plant species, especially trees (for example, the American chestnut, which was devastated by a fungal blight apparently imported from China or Japan in the early 1900s). Among the effects have been mass extinctions, especially on islands (Hawaii is a prime example) and displacement of native species by invasive weedy ones (the list is long). Fueled by globalization, such impacts continue today. Agents of destruction can arrive from any quarter—invasive species carried in the ballast water of ships or insect pests carried on lumber are two examples. Entire ecosystems are threatened or in some cases have disappeared (**FIGURE 27.6**).

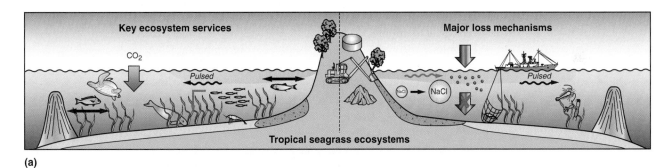

(a)

(b)

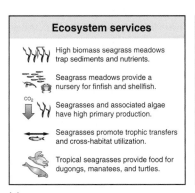

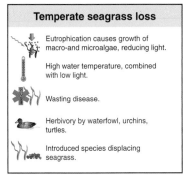

(c)

FIGURE 27.6 Tropical **(a)** and temperate **(b)** latitude seagrass ecosystems and the influences that can lead to their loss **(c)**.

*A DVD entitled,
"EVOLUTION—WHY
BOTHER?" produced by the
American Institute of Biological
Sciences and the U.S.
Biological Sciences Curriculum
Study (available at http://www
.aibs.org/bookstore), contains
an up-to-date evaluation of
how evolution impacts on
agriculture, human health,
development of new drugs,
and other issues of technology
application.*

These impacts on other species and ecosystems threaten humans themselves. In another book, *Collapse: How Societies Choose to Fail or Succeed*, Jared Diamond details the consequences of exceeding our resource base (Diamond, 2005). Using a number of case studies (the Maya of Central America, the Norse settlement in Greenland, and more), he shows how human societies in different times and places have collapsed as a result of changes in their environment. Such changes were in large part precipitated through their own actions, including faulty agricultural practices, deforestation, and water depletion, to name a few. Today our resource base is the whole world, and our impacts are accordingly global, extending even to climate change. The exploding human population has the potential to affect all organisms on Earth (Box 27.1), especially with the rapidly growing ability of humans to engineer the gene pool of other organisms to create genetically modified organisms.

■ KEY TERMS

agriculture	human-directed evolution
birth defects	negative eugenics
deleterious alleles	population explosion
domestication	positive eugenics
ecological footprint	post-reproductive longevity
eugenics	recombinant DNA
gene therapy	reproductive technology
genetically modified organism (GMO)	selective breeding
genetic engineering	

■ EVOLUTION ON THE WEB

Explore evolution on the Internet! Visit the accompanying website for *Strickberger's Evolution, Fifth Edition,* at **go.jblearning.com/Evolution5eCW** for exercises and links relating to topics covered in this chapter.

■ REFERENCES

Avise, J. C., 2004. *The Hope, Hype & Reality of Genetic Engineering.* Oxford University Press, Oxford, UK.

Britten, R. J., 2010. Transposable elements insertions have strongly affected human evolution. *Proc. Natl Acad. Sci. USA,* **107,** 19945–19948.

Castle, W. E., 1932. *Genetics and Eugenics.* 4th ed. Harvard University Press, Cambridge, MA.

Cavalli-Sforza, L. L., P. Menozzi, and A. Piazza, 1993. Demic expansions and human evolution. *Science,* **259,** 639–646.

Clutton-Brock, T. H., 1999. *A Natural History of Domesticated Mammals.* 2nd ed. Cambridge University Press, Cambridge, UK.

Crosby, A., 1986. *Ecological Imperialism: The Biological Expansion of Europe, 900–1900.* Cambridge University Press, Cambridge, UK.

De Fine Licht, H. H., M. Schiøtt, U. G. Mueller, and J. J. Boomsma, 2010. Evolutionary transitions in enzyme activity of ant fungus gardens. *Evolution,* **64,** 2055–2069.

Diamond, J. M., 1997. *Guns, Germs, and Steel: The Fates of Human Societies.* W.W.Norton, New York.

Diamond, J. M., 2005. *Collapse: How Societies Choose to Fail or Succeed.* Viking Books, New York.

Dyson, F., 2007. Our biotech future. *N. Y. Rev. Books,* **54**(12), 4–8.

Flannery, T., 1994. *The Future Eaters: An Ecological History of the Australasian Lands and People.* Grove Press, New York.

Flannery, T., 2001. *The Eternal Frontier: An Ecological History of North America and its Peoples.* Atlantic Monthly Press, New York.

Galton, F., 1869. *Hereditary Genius.* Macmillan, London.

Galton, F., 1883. *Inquiries into Human Faculty and Its Development.* Macmillan, London.

Genereux, D. P., and J. M. Logsdon, Jr., 2003. Much ado about bacteria-to-vertebrate lateral gene transfer. *Trends Genet.*, **19**, 191–195.

Hawks, J., E. T. Wang, G. M. Cochran, et al., 2007. Recent acceleration of human adaptive evolution. *Proc. Natl Acad. Sci. USA*, **104**, 20753–20758.

Heerwaarden, J. van, J. Doebley, W. H. Briggs, et al., 2011. Genetic signals of origin, spread, and introgression in a large sample of maize landraces. *Proc. Natl Acad. Sci. USA*, **108**, 1088–1092.

Hutchinson, W. D., E. C. Burkness, P. D. Mitchell, et al., 2010. Area wide suppression of European corn borers with Bt Maize reaps savings to non-Bt maize growers. *Science*, **330**, 220–224.

Kevles, D. J., 1995. *In the Name of Eugenics: Genetics and the Uses of Human Heredity.* Harvard University Press, Cambridge, MA.

Leroi, A. M., 2006. The future of neo-eugenics. Now that many people approve the elimination of certain genetically defective fetuses, is society closer to screening all fetuses for all known mutations? *EMBO Rep*, **7**, 1184–1187.

Lynch, M., 2010. Rate and molecular spectrum, and consequences of human mutation. *Proc. Natl Acad. Sci. USA*, **105**, 9272–9277.

Marelli, A. J., A. A. Mackie, R. Ionescu-Ittu, E. Rahme, and L. Pilote, 2007. Congenital heart disease in the general population: changing prevalence and age distribution. *Circulation*, **115**, 163–172.

May, R. M., 1987. More evolution of cooperation. *Nature*, **327**, 15–17.

Mindell, D. P., 2006. *The Evolving World: Evolution in Everyday Life.* Harvard University Press, Cambridge, MA.

Moalem, S., and J. Prince, 2007. *Survival of the Sickest: A Medical Maverick Discovers Why We Need Disease.* William Morrow, New York.

Molina, J., M. Sikora, N. Garud, et al., 2011. Molecular evidence for a single evolutionary origin of domesticated rice. *Proc. Natl Acad. Sci. USA*, **108**, 8351–8356.

Muller, H. J., 1950. Our load of mutations. *Amer. J. Hum. Genet.*, **2**, 111–176.

Palumbi, S. R., 2001a. Humans as the world's greatest evolutionary force. *Science*, **293**, 1786–1790.

Palumbi, S. R., 2001b. *The Evolution Explosion: How Humans Cause Rapid Evolutionary Change.* W. W. Norton, New York.

Purugganan, M. D., and D. Q. Fuller, 2009. The nature of selection during plant domestication. *Nature*, **457**, 843–848.

Simonson, T. S., Y. Yang, C. D. Huff, et al., 2010. Genetic evidence for high-altitude adaptation in Tibet. *Science*, **329**, 72–75.

Tabashnik, B. E., 2010. Communal benefits of transgenic corn. *Science*, **330**, 189–190.

CHAPTER
28

Culture, Religion, and Evolution

CHAPTER SUMMARY

An important part of human culture is a system of beliefs by which individuals order their lives. Many human cultures have developed belief systems in which the universe follows a designed order established by an intelligent deity. For some, the discovery that Earth is not the center of the universe made them question the necessity of a deity. Discovery of physical mechanisms that governed the universe made others question the necessity of a creator or designer. Darwin's theory of evolution provided a natural explanation and testable mechanisms for organismal origins, relationships, and change. The realization that organisms with a heritable system could adapt their form and function to an environment, coupled with their ability to respond to changing circumstances, provided an explanation beyond the designs of a creator. Replacement of a creator/designer by a process of chance and necessity produced tensions within individuals, established religions, and societies. Nevertheless, religious beliefs and scientific rationality coexist in many individuals and in many sections of societies. That said, it is important to remember that religious arguments have explanatory power with respect to belief systems, but they are not scientific explanations and should not be confused with, or regarded as, scientific explanations.

BELIEF, RELIGION, AND SPIRITUALITY

Interactions between SCIENCE AND RELIGION *vary enormously from one religious/ cultural tradition to another. Discussion of evolution and religion in this chapter is illustrated primarily using the Judeo-Christian religion.*

We begin this chapter begins by exploring science in relation to culture and how the theory of evolution impacted on a very specific aspect of human culture, **systems of belief**. We examine the roots of religion in a historical context, how the theory of evolution impacted on religion, and how creationists view evolution.[1] The latter presents especially difficult problems: creation operates from the top down and requires faith in the existence of a superior being(s); natural selection operates from the bottom up following processes that can be analyzed scientifically.

We distinguish religious belief from **spirituality**, by which we mean aspects of human lifestyle and experience such as meditation, love of music, or appreciation of beauty that go beyond a material view of the Universe. There is no necessary connection between spirituality and religion or religious belief.

SCIENCE AND SOCIETY

As is true of any human activity, science is not independent of the society in which it is carried out. The social context of science is amply illustrated when we consider that many regard the "invention" of gunpowder, the compass, and the printing press[2] (**FIGURES 28.1** to **28.3**) in Europe between 900 and 1500 AD as setting up the transformation of Europe during the scientific revolution in the seventeenth century. Francis Bacon (1561–1626), the philosopher, statesman, and essayist—"knowledge is power" comes from one of his essays—whose promotion of the scientific method set the stage for the scientific revolution, saw the scientific revolution as based on:

> Printing, gunpowder, and the compass: These three have changed the whole face and state of things throughout the world; the first in literature, the second in warfare, the third

FIGURE 28.1 Fireworks over palaces in China celebrate New Year's Eve.

© iBird/ShutterStock, Inc.

---

[1] See Appleman (2001), Cracraft and Bybee (2005), and Grinnell (2009) for three complimentary approaches to science and society.

[2] Printing, gunpowder, and the compass were actually invented in China between 400 and 800 BC, one to two thousand years before their "invention" in Europe. China developed an advanced system of social development through its religious belief system, reverence for nature, thirst for discovery, and understanding of the natural world, coupled with a quest to harness that understanding to benefit the Chinese people through application—that is, through technology. Science and cultural development go hand in hand.

© sgame/ShutterStock, Inc.

FIGURE 28.2 An early Chinese magnetic compass.

in navigation; whence have followed innumerable changes, in so much that no empire, no sect, no star seems to have exerted greater power and influence in human affairs than these mechanical discoveries. (*Novum Organum* [*New Instrument*], 1620).

From a social point of view, the development of evolutionary theory in Europe in the late nineteenth century coincided with an all-pervasive political and economic revolution in social behavior. Economic change posed by capitalism and its new wealthy classes allowed, even encouraged, European science to flourish by presenting ideological challenges to the prevailing religious and philosophical systems that had supported the old social order. Two of those challenges were **Social Darwinism** and the rise of **eugenics**.

The impact of Darwin's theory on all aspects of Victorian and subsequent societies was profound. By proposing that the form and function of living organisms did not arise

© Universal History Arc/Age Fotostock

FIGURE 28.3 The frontispiece of the earliest known printed book, the *Diamond Sutra*, a scroll some 5 meters long and printed in 868 AD.

Philosophers (Herbert Spencer), revolutionaries (Karl Marx), psychologists (Sigmund Freud), and writers, playwrights, and poets (Joseph Conrad, Thomas Hardy, Alfred Tennyson, George Eliot, George Bernard Shaw) are a few of those who INCORPORATED EVOLUTION *into their studies, writings, politics, and worldviews.*

by creation but rather by natural processes, Darwin made it clear that species were not fixed and unchangeable. These radical ideas, which revolutionized biology, also affected sociology, anthropology, economics, politics, women's rights, fiction, poetry, linguistics, philosophy, and psychology.[3] To quote the philosopher and education reformer John Dewey (1859–1952):

> In laying hands upon the sacred ark of absolute permanency, in treating the forms that had been regarded as types of fixity and perfection as originating and passing away, the "Origin of Species" introduced a mode of thinking that in the end was bound to transform the logic of knowledge, and hence the treatment of morals, politics, and religion. (Dewey, 1910, pp. 1–2)

LAWS OF NATURE

For many centuries, the traditional Western religious rationale for social, cultural, and biological systems was that the universe followed a designed order established by an intelligent deity. Along with belief in a god who created and maintained the universe, religion provided a set of ethical and moral values upon which to base social systems (Kitcher, 2006). Until Copernicus and Galileo in the sixteenth century, no one had seriously challenged the geocentric universe championed by the church. In a new worldview accepted by some after the discoveries of Copernicus, Galileo, and others, this god was an initial creator who established the laws of nature governing the running of the universe.

The advent of Darwinism posed threats to Western religion by suggesting that humans had no special place in the universe and that their origin, features, and relationships to the great Apes could be explained by natural selection without the intervention of a god. The Darwinian view that evolution is a historical process—extant organisms were not created spontaneously but formed in a succession of past events as organisms adapted to changing environments—contradicted the common religious view of design by an intelligent creator. Given the great age of Earth and the power of natural selection, complex designs that seem unlikely as singular spontaneous events (and previously only explained as the intent of an intelligent maker) become evolutionarily probable events. Consequently, for many, the reality of evolution challenged belief in a god or gods.

THE QUESTION OF DESIGN

The first significant cracks in the theological armor of divine intervention in nature arose through the discoveries by Copernicus, Galileo, and Kepler of natural laws regulating the motion of the solar system (**FIGURE 28.4**). Their genius was that all three provided simpler and more universal explanations of the workings of the physical universe than previously available. The elegance of their explanations made inevitable the rise of science and of the scientific method.

> A momentous change had come about when what scientists did came to be taken for granted, even by those who understood little or nothing of it. The crucial change in the making of the modern mind was the widespread acceptance of the idea that the world is essentially rational and explicable, though very wonderful and complicated.[4]

These openings were widened considerably by the mechanistic explanations proposed by Isaac Newton (1642–1727) to describe the solar system's motion through the force of gravity. Later, geologists such as Charles Lyell extended this mechanistic approach by proposing how natural forces could mold Earth's surface. Although these scholars

---

[3] For Darwin's influence on novelists, especially those writing in Britain, see the essays reprinted in Part IX (pp. 631–682) of Appleman, 2001.

[4] J. M. Roberts, 1985. *The Triumph of the West.* British Broadcasting Corporation, London (p. 242).

(a)

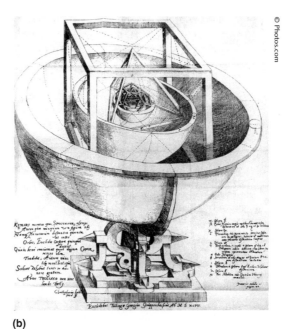

(b)

FIGURE 28.4 Johannes Kepler **(a)** (1571–1630), a German astronomer and mathematician whose improved reflecting microscope and insights published in *Astronomia Nova* (1600) and other works demonstrated that the planets orbit the sun, as shown in the historic illustration of Kepler's planetary system model **(b)**.

were not atheists, their findings about natural processes made through sciences such as mechanics, optics, and chemistry, helped relegate God from a continually active, intervening agent to a prime force—a master artisan who designed logical, self-functioning machines—and so led to a closer examination of the nature of God.

WILLIAM PALEY AND DESIGN

The essential belief that each design has a creative purpose is reflected in the doctrine expounded by William Paley (1743–1805) (**FIGURE 28.5a**), an English theologian, philosopher, and advocate of Christian ethics.[5] Paley's philosophy that Earth and its creatures were created, designed, and maintained by God is elaborated in considerable detail in his last book, *Natural Theology; or, Evidences of the Existence and Attributes of the Deity* (Paley, 1802). According to Paley,

> There cannot be design without a designer; contrivance without a contriver; order without choice; arrangement without anything capable of arranging. . . Arrangement, disposition of parts, subservience of means to an end, relation of instruments to a use, imply the presence of intelligence and mind. (Paley, 1802, p. 11)

Paley's knowledge of natural history was profound. A large number of diverse and hard to understand examples of organs are discussed in depth in his book, from air bladders in fish to the elephant's trunk and spider's webs. This treatise formed an essential part of the teaching of natural theology at Oxford and Cambridge until the mid 1850s. Darwin studied Paley while at Cambridge, was much influenced by *Natural Theology*, and credited his knowledge of Paley's writing in helping him do well in his B. A. examinations.[6] Here is what Darwin had to say:

> In order to pass the B.A. examination, it was, also, necessary to get up Paley's Evidences of Christianity, and his Moral Philosophy. This was done in a thorough manner, and I am convinced that I could have written out the whole of the Evidences with perfect correctness, but not of course in the clear language of Paley. The logic of this book and as I may add of his Natural Theology gave me as much delight as did Euclid. The careful study of these works, without

---

[5] McPherson (1972) provides a classic and accessible introduction to design, and Gliboff (2008) and Reiss (2009) excellent evaluations of William Paley and design in evolution.

[6] Darwin's comments are in his autobiography (Darwin, 1993), which was edited by his granddaughter, Nora Barlow.

(a) (b)

FIGURE 28.5 **(a)** William Paley. **(b)** Watchmaker, tinkering.

attempting to learn any part by rote, was the only part of the Academical Course which, as I then felt and as I still believe, was of the least use to me in the education of my mind. I did not at that time trouble myself about Paley's premises; and taking these on trust I was charmed and convinced by the long line of argumentation. By answering well the examination questions in Paley, by doing Euclid well, and by not failing miserably in Classics, I gained a good place among the . . . crowd of men who do not go in for honours. Oddly enough I cannot remember how high I stood, and my memory fluctuates between the fifth, tenth, or twelfth name on the list. [Darwin stood tenth.] (Darwin, 1993, p. 59).

Simple common sense seemed to support supernatural design. Organismal design presupposes a designer, who by definition (but not based on any scientific evidence) is an intelligent supernatural being, a chain of reasoning that leads back to creation and a creator.

Paley began his book with the most famous example and question concerning design and a designer: can there be a watch without a watchmaker (**FIGURE 28.5b**), and by extension, can there be a person without a person-maker, laws without a lawmaker?

In crossing a heath, suppose I pitched my foot against a stone, and were asked how the stone came to be there; I might possibly answer, that, for anything I knew to the contrary, it had lain there forever: nor would it perhaps be very easy to show the absurdity of this answer. But suppose I had found a watch upon the ground, and it should be inquired how the watch happened to be in that place; I should hardly think of the answer I had before given, that for anything I knew, the watch might have always been there. (Paley, 1802, p. 1)

Paley used this analogy to illustrate his conviction that humans could not exist without a creator. Indeed, it is no surprise to find a creation story in the history of many cultures and societies (**BOX 28.1**).

To evolutionary theory, the essential challenge that religion poses has always been, "How from the disorder of random variability can nature achieve the beauty of adaptation

BOX 28.1	Accounts of the Origin of the World from Ten Different Cultures

Egypt	God arose from the depths of the ocean, created dry land, and then created all creatures on the hill at Eliopoulos at the center of the Universe.
Mesopotamia	God made sky and Earth by splitting the powers of evil in half, and then produced humans for purposes of worship.
Iran	God created all that is good and struggled with an evil being that creates all that is bad. Each struggle lasts about 3,000 years and will continue until evil is vanquished, at which time creation will be complete and perfect.
Greece	God, a female, divided the sky from the sea, and produced a serpent with whom She copulated. She then laid a giant egg, out of which came Earth, its creatures, and all the heavenly bodies, as well as the subsidiary powers to rule these various entities.
India	God created himself from a golden egg, and from the various parts of his body everything was born. After a time, life is destroyed and the cycle begins again.
Israel	God created the universe in 6 days* ending with the creation of humans, according to Genesis 1; or, God first created Adam in the Garden of Eden and then created animals and birds and eventually Eve, according to Genesis 2.
Benin, Africa	God was a woman who produced twins: the Sun and the Moon. During various eclipses, the twins came together to create the various gods and spirits of Earth and sky that rule over humans.
Yucatán (Mexico)	God created the world in four distinct periods, each separated by a flood.
Crow Indians (United States)	God created the Earth and its creatures from mud gathered in the webbed feet of ducks that swam on a primeval ocean.
China	The Universe was originally in the shape of a hen's egg, out of which God emerged and chiseled its main physical features. After 18,000 years God died and the remainder of the world was derived from his body: the dome of the sky from his skull, rocks from his bones, soil from his flesh, rain from his sweat, plant life from his hair, and humans from his fleas.

*"Days" is the literal translation of the Hebrew work *yom,* which can refer to indeterminate periods of time.

without intelligent intervention?" Darwin's fundamental contribution was to propose a mechanism, natural selection, that no one had thoroughly explored before (Ayala, 2007). In providing his theory of evolution by natural selection, Darwin was explicit on the failure of Paley's arguments from design "now that the law of natural selection has been discovered." For Darwin there was "no more design in the variability of organic beings and in the action of natural selection than in the course which the wind blows." As Darwin stated in his *Autobiography*,

> The old argument of design in nature, as given by Paley, which formerly seemed to me so conclusive, falls, now that the law of natural selection has been discovered. We can no longer argue that, for instance, the beautiful hinge of a bivalve shell must have been made by an intelligent being, like the hinge of a door by a man.[7]

[7] N. Barlow, 1958. *The Autobiography of Charles Darwin.* Collins, London (p. 87).

Paley devoted one chapter of *Natural Theology* to the design of the human body and another to the human eye. Darwin saw the evolution of eyes in animals as a challenge to his mechanisms of chance, necessity, and evolution by natural selection. Indeed, as outlined in **BOX 28.2**, animal eyes can be extraordinarily complex. Nevertheless, given the understanding of the mechanisms of evolution outlined in the present book, the utility of simple light-gathering organs can be demonstrated and evolutionary sequences can be determined (Su et al., 2007). An example from mollusk eyes, a topic discussed by Paley and Darwin, is outlined in **FIGURES B2.1** and **B2.2**.

Darwin's theory had a profound impact on biology, religion, and virtually all spheres of human society and culture. Acceptance of religious explanations eroded as more and more natural explanations for the origin and modification of Earth and its inhabitants were discovered, as it was recognized that ethics and morality can vary between different human societies, and as it was understood that changes in such values need not depend on religious beliefs (Dewey, 1910; Ayala, 2010).

BOX 28.2	Eye Evolution

Animal eyes can be extraordinarily complex (Arendt, 2003; Gehring, 2005). Examples from lineages of mollusks in **FIGURE B2.1** show a wide range of light-gathering organs ("eyes").

The most basic form of light-gathering organ in the animal kingdom is a pigment spot with neural connec-tions stimulated by light (**a** in Figure B2.1). The light spot could consist only of a small number of cells, theoretically a single cell, as seen in species of planarians. Folding of pigment cells concentrates their light-gathering activity, providing improved light detection (**b** in Figure B2.1)., A partly closed, water-filled cavity surrounded by pigment

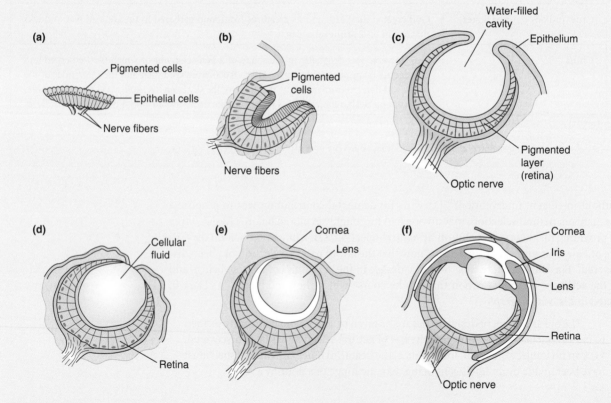

FIGURE B2.1 Light-gathering organs in mollusks from a simple pigment spot with connecting nerve fibers **(a)** to the complex eye of the squid **(f)**.

BOX 28.2 | **Eye Evolution (Cont...)**

cells allows images to form on the pigmented layer as in a pinhole camera (Figure B2.1c). In the eyes of other invertebrates, a transparent fluid secreted by the cells—a fluid extracellular matrix—rather than water, forms a barrier that protects the pigmented retina from injury (Figure B2.1d). A thin film or transparent "skin" may cover the entire eye apparatus, adding further protection. Some of the fluid extracellular matrix within the eye hardens into a convex lens that improves the focusing of light on the retina (Figure B2.1e). Highly complex eyes, as seen in species of squid and in vertebrates have an adjustable iris diaphragm and a focusing lens (Figure B2.1f).

The series of images in **FIGURE B2.2** shows stages in eye evolution, as depicted by a computerized model in which random changes in eye structure are followed by selection for visual acuity. Beginning with a light-sensitive middle layer of skin backed by pigment (**a**) successive selective steps for improved optical properties lead to a concave buckling that enhances light gathering (**b-e**), a focusing lens (**f, g**), and an eye with a flattened iris in which the focal length of the lens equals the distance between lens and retina (**h**).

Even such complex organs as eyes may not have evolved only once; multiple examples of parallel or convergent evolution have been introduced in the text. Presence of similar complex eyes in mollusks (Figure B2.1f) and vertebrates is indicative of convergent evolution as similar selective pressures led to organs that enhance visual acuity. A specialized layer that provides high spatial acuity evolved at least twice by convergence in the eyes of predatory spiders (Su et al., 2007). Genetic studies indicate that an orthologous gene *Pax-6* regulates the development of anterior sense organ patterns in invertebrates and vertebrates (which share a deep metazoan ancestry), and regulates the ability to form light-gathering cells or organs.[*] Reflecting their independent evolutionary histories, specific genetic and cellular pathways in embryonic eye development differ between squid and vertebrates.

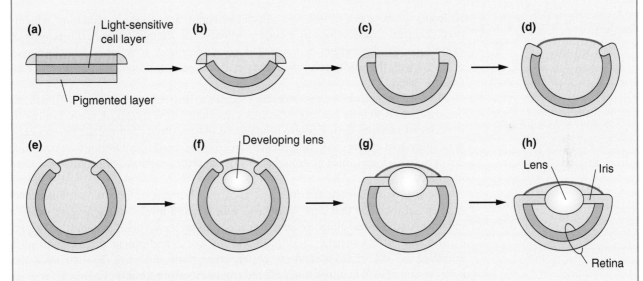

FIGURE B2.2 Proposed stages in the evolution of the eye from light-sensitive skin underlain by pigment **(a)**, through infolding **(b–e)**, evolution of a focusing lens **(f–h)**.

[Adapted from Nilsson, D. E. and S. Pelger, 1994. *Proc. R. Soc. Lond.* **256**, 53–58.]

[*] S. I. Tomarev et al., 1997. Squid *Pax-6* and eye development. *Proc. Natl. Acad. Sci, USA*, **94**, 2421–2426.

FIGURE 28.6 Plagues of locusts and other insects have devastated crops for millennia. Here, desert locusts swarm in Aleg, Mauritania.

BASES AND EVOLUTION OF RELIGIOUS BELIEFS

Religious development in different cultures provides clues to the evolution of religion itself. Religion first develops in a culture when, in an attempt to deal with aspects of experience that can neither be controlled nor understood, societies endow the forces of nature with the spirits of animals and/or with supernatural powers (La Barre, 1970; Culotta, 2009).

Ritual develops when ceremonies are repeated to help ensure their efficacy. Ritualized behavior seems to have become especially important in the transition from hunting to agricultural societies. Crops have to be planted and harvested at appropriate seasons each year. Individual efforts could be either rewarded or damned by mysterious external and uncontrollable forces, such as droughts, floods, volcanic eruptions, or plagues of insects (FIGURE 28.6).

Because societies saw the forces of nature as humanlike, religion sustained and encouraged the hope that gifts, sacrifice, obedience, and loyalty could appease nature's judgment and recrimination. Religion provided and preserved belief systems that helped explain and guide a societies' relationship to the world, and provided answers to such questions as, "Where did we and our society come from, and why?" To these ends, the belief system of each culture usually offered comprehensive accounts of how and for what purpose the world was created (FIGURE 28.7). Accounts from 10 different cultures are outlined in Box 28.1.

In support of a supernatural view of events, religion has relied on two basic concepts that arose early in human history, the *soul* and *God*. According to hypotheses proposed by psychologists and anthropologists, the idea that the soul is separate from the body originated in the separation of mind from reality in dreams. Some theologians sought a "scientific" argument for the nonmaterial nature of the soul or the personality, claiming that intellectual processes cannot have a material origin. Today, such concerns are reflected in research whose goal is to understand **consciousness**, which embodies the awareness of a self that is aware of itself, looks out at a world that it perceives, and responds to it.[8]

See Ruse (2005) and Dennett (2006) for how a philosopher of biology and a psychologist approach the issues of RELIGION, SCIENCE, NATURAL PHENOMENA AND EXPERIENCE, *and see Lewontin (2005) and Coyne (2012) for the approaches of two prominent evolutionary biologists.*

---

[8] For readings on these issues from various perspectives, see Richards (1987), Ruse (2001), Koch (2004), Carruthers (2005), and Dennett (2006).

FIGURE 28.7 A representation of how the world may have arisen from the Big Bang (top left) to DNA (bottom right)

An extensive Internet source is the Talk Origins Archive at http://www.talkorigins .org. Two further sites are http://www.nap.edu/ catalog/6024.html#toc (Science and Creationism: a View from the National Academy of Sciences, 2nd ed. The National Academies Press, Washington, DC) and http://www.hhmi.org/ biointeractive/evolution/ lectures.html#evodiscussion, BioInteractive, a website maintained by the Howard Hughes Medical Institute with discussions of RELIGION AND SCIENCE.

EVOLUTION AND RELIGION

Darwinism had an especially dramatic impact on religion. To many of Darwin's religious contemporaries and to others since, *On the Origin of Species* and *The Descent of Man* raised issues of momentous importance; to the general public, Darwinism was at least as much a religious as a scientific concern. Darwin was painfully aware of this and pointed out at least two discoveries that could refute his theory:

- an inversion of the evolutionary sequence such as evidence of humans in the Paleozoic or Mesozoic Eras; or
- finding the same species in two separated geographical locations when their presence did not result from migration between these areas.

GORILLAS *only became known after 1854 when the first bones were shipped from Africa, and in the early 1860s when a stuffed gorilla skin was taken along as a prop on lecture tours undertaken throughout England by an African explorer. "Victorians were horrified to think that these reputedly violent animals—distorted men in shape and size, representing the brutish, dark side of humanity—were possible ancestors" (Browne, 2006, p. 98).*

No such evidence has been found in the century and a half since Darwin's theory was published.

Understanding the role that evolution plays in modern life requires understanding how religion and evolution interacted (and still interact). The two views were the basis for a confrontation at a debate on November 5, 1860 at Oxford University, less than a year after the publication of *The Origin of Species*. As part of the ongoing public discussions generated by the book, Bishop Samuel Wilberforce (1805–1873) of the Church of England attacked Darwinian theory as incompatible with the Bible. Coached by the religious comparative anatomist Richard Owen, the Bishop attempted to destroy Darwin's theory through scientific arguments. Wilberforce aimed his final point directly at Thomas Huxley when he asked whether it was through his grandfather or grandmother that Huxley claimed descent from a monkey. The wit of Huxley's response, recounted a few months later in a letter to his friend Frederick Dyster, has often been quoted.

> If, then, said I, the question is put to me would I rather have a miserable ape for a grandfather, or a man highly endowed by nature and possessed of great means and influence, and yet who employs these faculties and that influence for the mere purpose of introducing ridicule into a grave scientific discussion, I unhesitatingly affirm my preference for the ape.

Such religious attacks were worldwide, frequent, harsh, and almost always focused on the same points: religious opponents accused Darwinists of seeking "to do away with all ideas of God," "to produce in their readers a disbelief of the Bible," "to displace God by the unerring action of vagary," and to "destroy humanity's unique status."

In the most sensitive area of all—life itself—Darwinian evolution offered different answers to religion's claims of why life's important events occur. Darwin's works made clear that society no longer needed to believe that only the actions of a supernatural creator could explain biological relationships. Darwin presented a nature of continual change, unpredictable events, and unrelenting competition for resources among living creatures with no obvious guidance. To religious believers, natural selection substituted waste for economy by treating life as a continually expendable commodity.

For many, it was a difficult transition from regarding the operation of nature in understandable human terms to its operation in terms of the evolutionary opportunism expressed by competition and reproductive success. The philosopher Emanuel Kant (1724–1804), who took an evolutionary approach to cosmology and to the origin of the solar system, found it at times abhorrent to admit that species could evolve, describing such notions as "ideas so monstrous that the reason shrinks before them." Darwin is reported to have felt quite uncomfortable about his role in proposing the evolution of species. In a letter written to Joseph Hooker in 1884, Darwin wrote that, "it is like confessing a murder." One source of ambivalence was the established belief that humans were created in the image of God and endowed to rule over other biological and social groups. Kinship to those "below," whether ape or servant, was a repugnant idea. An oft-reported example of this repugnance is the response of the wife of the Bishop of Worcester when informed that Huxley had announced that man was descended from apes: "Descended from apes! My dear let us hope that it is not true, but if it is, let us pray that it will not become generally known."

Essential to the preservation of religion alongside Darwin's theory was the place served by religion as the main repository for the ethics and morals of society (what is right and wrong). Religion served to maintain confidence in the social order and to unify nationalistic and provincial sentiments ("God is on our side"). In 1975, E. O. Wilson claimed in an outline of sociobiology that religion serves an important role; religion helps provide a common social identity to groups of individuals, and by increasing group power, confers "biological advantage" (Wilson, 1975). More recent analyses have reinforced the adaptive role of religious ritual over more than 100,000 years. Religious rituals emphasize, maintain, and promote cooperation and communication.[9]

Society has held on to both evolution and religion because they serve different needs. With the exception of some fundamentalists, whose views are discussed in the following section, religion essentially withdrew from the domain of biological evolution, leaving both the origin of species and the origin of humans to evolutionary biologists and anthropologists.

CREATIONISM AND INTELLIGENT DESIGN

Fundamentalist religious groups have generally not accepted the uneasy relationship between evolution and religious institutions that exist in the Western world. Some Judeo-Christian groups completely reject evolutionary explanations for biological events. That said, a wide spectrum of views is found across religions groups, from special creation and the denial of evolution at one end of the spectrum to acceptance of evolution at the other; Scott (1997, 2005) usefully discusses these as extending from "young Earth" to "old Earth" creationism.

In the United States, many individuals and groups who believe in the Judeo-Christian account of creation in Genesis have formed political pressure groups to insist that their beliefs be treated as a scientific alternative to evolution in public education. Especially in the United States, and increasingly in Europe, fundamentalist religious groups opposed to evolution have attempted to prevent its teaching and to reduce or eliminate the subject of evolution in many biology textbooks. The "creation science" movement (which, despite its name, does not use the scientific method) and more recently the "intelligent design" movement are the latest attempts to deny the reality of evolution.[10]

The origins of such groups within society date back at least to the early 1900s with strong roots among economically threatened tenant farmers and small landholders, especially in the United States South and Southwest. Sociologists have suggested that believing in the literal truth of the Bible, revivalism, and other aspects of fundamentalist religion, helped many of these rural groups defend their way of life against domination and control by the more intellectual but exploitative northern and eastern social and economic establishments.

MONKEY TRIALS

Whatever their initial motivations, fundamentalist groups were successful in pursuing antievolution goals in the South and Southwest during the first few decades of the twentieth century. By the end of the 1920s, fundamentalists had introduced antievolution bills into a majority of United States state legislatures, and had passed some in various southern states. The most famous confrontation between evolution and biblical creationism during that period was the 1925 trial of a schoolteacher, John Thomas Scopes (1900–1970), convicted of ignoring the ban against teaching evolution in Tennessee schools.

Scopes, who had earned a law degree from the University of Kentucky in 1924, moved to Dayton, Tennessee as football coach at the Rhea County High School (**FIGURE 28.8**). Occasionally he filled in as a substitute for regular teachers, a role that placed him at the center of what became known as the "Scopes

According to a 1982 GALLUP POLL, 44 percent of Americans agreed with the statement that, "God created humankind in its present form almost 10,000 years ago." In 1999, the number had risen to 47 percent. A poll taken by the New York Times in November 2004 produced a statistic of 55 percent, and a 2012 GALLUP POLL showed a statistic of 47 percent.[11]

[11]http://www.gallup.com/poll/155003/Hold-Creationist-View-Human-Origins.aspx.

Courtesy of Smithsonian Institution Archives

FIGURE 28.8 John T. Scopes, who stated in his memoirs that "I furnished the body that was needed to sit in the defendant's chair"[12] in the Scopes Monkey trial, July 10–25, 1925.

[12]J. Presley and J. T. Scopes, 1967. *Center of the Storm*. Holt, Rinehart and Winston, New York. The quotation is from p. 60.

---

[9]R. Rappaport, 1999, *Ritual and Religion in the Making of Humanity*. Cambridge University Press, Cambridge, UK; R. Sosis, 2004, The adaptive value of religious ritual. *Am. Sci.*, **92** (2), 166–172. Coyne (2012) sets out the arguments for religion and the dysfunctionality of American society.
[10]For analyses of why intelligent design is not science or based on the scientific method, see Scott (2005), Bleckmann (2006), Brockman (2006), Hallgrímsson (2008), and Miller (2009).

FIGURE 28.9 William Jennings Bryan (in the bow tie) being interrogated by the lawyer Clarence Seward Darrow, during the trial of the *State of Tennessee vs. John Thomas Scopes,* July 20, 1925. The trial had moved outdoors; the temperature in the courtroom exceeded 100°F.

© Smithsonian Institution Archives, SIA2007-0124

See http://www.hhmi.org/ biointeractive/evolution/ lectures.html#evodiscussion, a website maintained by the Howard Hughes Medical Institute for discussions of RELIGION AND SCIENCE.

Monkey Trial" (Larson, 1997) when, while substituting in a biology class, he taught evolution. The case for the State of Tennessee was placed in the hands of William Jennings Bryan (1860–1925), a congressman from Nebraska, born and raised in Salem, Illinois (as indeed was Scopes). A formidable advocate, Bryan had been the Democratic Party nominee for President in the federal elections of 1896, 1900, and 1908. Scopes was defended by Clarence Seward Darrow (1857–1938), a leading member of the American Civil Liberties Union that financed Scopes' defense (FIGURE 28.9). The trial, which lasted through 15 days of summer heat, ended on July 21, 1925. Scopes was found guilty and fined $100, though the conviction was later overturned on a technicality.

Although many biologists felt that the Scopes trial essentially defeated the intellectual validity of the creationist position, creationists apparently lost little ground in these regions and had an impact on public education far beyond the South and Southwest. By influencing textbook adoption procedures in various local and state school boards in the United States, creationists successfully minimized evolutionary explanations in secondary school textbooks for a long time; by 1942, more than 50 percent of high school biology teachers throughout the United States excluded any discussion of evolution from their courses.

In the late 1950s and early 1960s, the United States recognized that their science education was lagging behind that of other countries, specifically the Soviet Union, which in 1957 had launched the first space satellite, *Sputnik*. This awareness provided the impetus to reform the science curriculum in United States secondary schools. Among the resulting innovations were new high school textbooks in both biological and social sciences (*Biological Science Curriculum Study; Man as a Course of Study*) that discussed evolution and analyzed changes in human social relationships. By the end of the 1960s, antievolution laws were either repealed or declared unconstitutional. Nevertheless, a survey conducted in 2007 found that high school biology teachers in the United States only devoted an average of 13.7 hours to the teaching of evolution (Berkman et al., 2008).

Furthermore, within the last decade, a number of societies and institutes established by fundamentalists to promulgate creation science/intelligent design have entered the fray with the aim of including creationism in the science curriculum. Despite the name, there is no recognizable science in creation science, which does not employ the scientific method. Although considerable many books and papers deal with creationist attacks on evolution,[13] the refusal of fundamentalist creationists to accept scientific evidence shows

[13] For the contrast between science and religion, see Pigliucci (2002), Scott (1997, 2005), Brockman (2006), and Kitcher (2006).

no promise of being resolved. By contrast, both conforming and dissenting views in evolution are evaluated, as in all science, by evidence and not by faith or unverifiable decree.

Despite the overwhelming scientific evidence for evolution as a natural process, some religious groups adhering to creation have developed intelligent design as a purported scientific alternative to evolution. "Intelligent design" is latter-day creationism. Because it relies on supernatural beliefs rather than scientific explanations intelligent design is not science. Religious arguments have explanatory power with respect to belief systems, but they are not scientific explanations and should not be confused with, or regarded as, scientific explanations. Indeed, in a landmark December 2005 decision in a United States federal district court trial (*Kitzmiller v. Dover* in Dover, Pennsylvania), intelligent design was ruled a form of religion and not science, and as such "cannot be taught alongside evolution in sciences classes in U.S. public schools."[14] Evidence presented throughout the text counters unfounded intelligent design claims that evolution has not, does not, and could not occur.

LAST WORDS

Many intellectual threads led to the modern theory of evolution, a theory that recognizes that Earth is ancient, that there is a common inheritance, and a universal tree of life, and that natural events can be explained by discoverable natural laws. Essential to our understanding of evolution is that:

- Groups of organisms are bound together by their common inheritance.
- The past has been long enough for inherited changes to accumulate.

And, arguably, most essential of all, that

- Discoverable biological processes and natural relationships among organisms provide the evidence for the reality of evolution.

It took a long time to weave these threads into an evolutionary tapestry, the richness and beauty of which we hope this presentation of the evidence for, and the fact of, evolution will enable you to appreciate more clearly.

■ KEY TERMS

Clarence Seward Darrow
consciousness
creationism
design
eugenics
eye evolution
John Thomas Scopes
Laws of Nature
monkey trials

origin of the world
religious belief
science and society
Social Darwinism
spirituality
systems of belief
William Jennings Bryan
William Paley

■ EVOLUTION ON THE WEB

Explore evolution on the Internet! Visit the accompanying website for *Strickberger's Evolution, Fifth Edition,* at **go.jblearning.com/Evolution5eCW** for exercises and links relating to topics covered in this chapter.

---

[14] For the website containing the full decision in *Kitzmiller v. Dover* see: http://en.wikipedia.org/wiki/Kitzmiller_v._Dover_Area_School_District. Accessed December 6, 2010. For the consequences of the ruling, see Scott (2006) and Padian (2007).

■ REFERENCES

Appleman, P. (ed.), 2001. *A Norton Critical Edition.* Darwin: Texts, Commentary, 3rd ed. W. W. Norton & Company, New York.

Arendt, D., 2003. Evolution of eyes and photoreceptor cell types. *Int. J. Dev. Biol.*, **47**, 563–571.

Ayala, F. J., 2007. Darwin's greatest discovery: design without designer. *Proc. Natl Acad. Sci. USA*, **104**, 8567–8573.

Ayala, F. J., 2010. The difference of being human: morality. *Proc. Natl Acad. Sci. USA*, **107**, 9015–9022.

Berkman, M. B., J. S. Pacheco, and E. Plutzer, 2008. Evolution and creationism in America's classrooms: a national portrait. *PLoS Biol.*, **6**, e124. doi:10.1371/journal.pbio.0060124.

Bleckmann, C. A., 2006. Evolution and creationism in science: 1880–2000. *BioScience*, **56**, 151–158.

Brockman, J. (ed.), 2006. *Intelligent Thought: Science Versus the Intelligent Design Movement.* Vintage Books, New York.

Browne, J., 2006. *Darwin's Origin of Species. A Biography.* Douglas & McIntyre, Vancouver, Canada.

Carruthers, P., 2005. *Consciousness: Essays from a Higher-Order Perspective.* Oxford University Press, Oxford, UK.

Coyne, J. A. 2012. Science, religion, and society: the problem of evolution in America. *Evolution*, **66**, 2654–2663.

Cracraft, J., and R. W. Bybee (eds.), 2005. *Evolutionary Science and Society: Educating a New Generation.* Biological Sciences Curriculum Study, Washington, DC.

Culotta, E., 2009. On the origin of religion. *Science*, **326**, 784–787.

Darwin, C., 1993. *The Autobiography of Charles Darwin.* Nora Barlow (ed.), Norton, New York.

Dennett, D. C., 2006. *Breaking the Spell: Religion as a Natural Phenomenon.* Allen Lane, London.

Dewey, J., 1910. *The Influence of Darwin on Philosophy And Other Essays in Contemporary Thought.* Henry Holt & Co. (Reprinted 1965 by Indiana University Press, Bloomington, IN.)

Gehring, W. J., 2005. New perspectives on eye development and the evolution of eyes and photoreceptors. *J. Hered.* **96**, 171–184.

Gliboff, S., 2008. *H. G. Bronn, Ernst Haeckel, and the Origins of German Darwinism.* MIT Press, Cambridge, MA.

Grinnell, F., 2009. *Everyday Practice of Science: Where Intuition and Passion Meet Objectivity and Logic.* Oxford University Press, Oxford, UK.

Hallgrímsson, B., 2008. The false dichotomy of evolution versus intelligent design. *Evol. Biol.*, **35**, 1–3.

Kitcher, P., 2006. *Living with Darwin: Evolution, Design, and The Future of Faith.* Oxford University Press, Oxford, UK.

Koch, C., 2004. *The Quest for Consciousness: A Neurobiological Approach.* Roberts & Co., Englewood, CO.

La Barre, W., 1970. *The Ghost Dance: Origins of Religion.* Doubleday, New York.

Larson, E. J., 1997. *Summer for the Gods: The Scopes Trial and America's Continuing Debate over Science and Religion.* Basic Books, New York.

Lewontin, R., 2005. The war over evolution. *N. Y. Rev. Books*, **52**(16), 51–54.

McPherson, T., 1972. *The Argument from Design.* Macmillan, London.

Miller, K. R., 2009. *Only a Theory: Evolution and the Battle for America's Soul.* Viking/Penguin Press, New York.

Padian, K., 2007. The case of creation: last years' Dover trial resulted in intelligent design being removed from the science curriculum. *Nature*, **448**, 253–254.

Paley, W., 1802. *Natural Theology; or, Evidences of the Existence and Attributes of the Deity Collected from the Appearances of Nature.* Bridgewater Treatises, Faulder. (Reissued many times, most recently by Cambridge University Press, 2009.)

Pigliucci, M., 2002. *Denying Evolution, Creationism, Science, and Nature of Science.* Sinauer Associates, Sunderland, MA.

Reiss, J., 2009. *Not By Design: Retiring Darwin's Watchmaker.* University of California Press, Berkeley, CA.

Richards, R. J., 1987. *Darwin and the Emergence of Evolutionary Theories of Mind and Behavior.* The University of Chicago Press, Chicago.

Ruse, M., 2001. *The Evolution Wars: A Guide to the Debates.* Rutgers University Press, New Jersey and London.

Ruse, M., 2005. *The Evolution-Creation Struggle.* Harvard University Press, Cambridge, MA.

Scott, E. C., 1997. Antievolutionism and creationism in the United States. *Ann. Rev. Anthropol.*, **26**, 263–289.

Scott, E. C., 2005. *Evolution vs. Creationism: An Introduction.* University of California Press, Berkeley, CA.

Scott, E. C., 2006. Creationism and evolution: it's the American way. *Cell*, **124**, 449–451.

Su, K. F., R. Meier, R. R. Jackson, et al., 2007. Convergent evolution of eye ultrastructure and divergent evolution of vision-mediated predatory behaviour in jumping spiders. *J. Evol. Biol.*, **20**, 1478–1489.

Wilson, E. O., 1975. *Sociobiology: The New Synthesis.* Harvard University Press, Cambridge, MA.

GLOSSARY

The glossary of 900 terms serves the traditional function of providing definitions of the key terms and processes discussed in the text. Along with the index it also serves as a means to access similar or related terms and concepts (for example, **Adaptive radiation**, **Biogeography**, **Innovation**, **Speciation**, and **Zoogeography** are cross-referenced as related topics) or to draw attention to sets of contrasting terms or concepts such as **Homology** and **Analogy** (and many others), **Ontogeny** and **Phylogeny** or **Individuals** and **Populations**.

A

Abiotic   Nonbiological in origin; environments characterized by the absence of organisms. (*See also* **Biotic, Life.**)

Acheulean tools   Large stone hand axes made and used by *Homo ergaster* and *Homo erectus* 1.5 Mya and by *Homo heidelbergensis* 500,000 years ago. (*See also* **Mousterian tools, Oldowan tools, Stone tools.**)

Acidic   A compound that produces an excess of hydrogen ions (H^+) when dissolved in water and so has a pH value less than 7.0. (*See also* **Base (alkaline), pH scale.**)

Acoelomates   Animals lacking a coelom (internal body cavity).

Acrocentric   Chromosomes with centromeres near one end between metacentric and telocentric locations.

Active transport   Biochemical transport requiring the input of energy, for example, hydrolysis of adenosine-5′-triphosphate (ATP). (*See also* **Adenosine-5′-triphosphate.**)

Adaptation   The relationship between structure and/or function and an organism's environment that makes the structure or function suitable for ("adapted" to) life in that environment. (*See also* **Environment, Exaptation, Inheritance of acquired characters, Preadaptation.**)

Adaptive landscape   A model originally devised by Sewall Wright that describes a topography in which populations with high fitness occupy peaks and those with low fitness occupy valleys. (*See also* **Fitness, Genetic load, Heterozygote advantage (superiority), Hybrid breakdown (inferiority, inviability, sterility), Inclusive fitness, Natural selection, Population genetics, Relative fitness, Selection, Selection coefficient (s).**)

Adaptive peak   *See* **Adaptive landscape**

Adaptive radiation   Diversification of a single species or group of related species into new ecological or geographical zones resulting in formation of additional species. (*See also* **Biogeography, Innovation, Speciation, Zoogeography.**)

Adaptive valley   *See* **Adaptive landscape**

Adenosine triphosphate (ATP)   An organic compound commonly involved in the transfer of phosphate bond energy, composed of adenosine (an adenine base + a D-ribose sugar) and three phosphate groups. (*See also* **Active transport.**)

Aerobic metabolism   The use of oxygen for reactions that provide energy for metabolism from the oxidative breakdown of food molecules; living in the presence of oxygen. (*See also* **Anaerobic metabolism, Metabolism**)

Aerobic respiration   An electron transport system in which oxygen serves as the terminal electron acceptor.

Agriculture   The cultivation of other organisms (animals, plant, fungi) by humans and by social animals such as ants. (*See also* **Genetically modified organisms (GMOs)**)

Algae   Photosynthetic multicellular organisms now divided among several super kingdoms of life. (*See also* **Archaeplastida, Chromalveolata, Cyanobacteria, Photosynthesis, Protista.**)

Allantois   An extraembryonic membrane that functions in gas exchange and excretion of waste products in amniote embryos. (*See also* **Chorioallantoic membrane.**)

Allele   One of the alternative forms of a single gene; a nucleotide sequence occurring at a given locus on a chromosome. (*See also* **Allele frequency, Dominant allele, Gene, Gene pool, Genetic polymorphism, Heterozygote, Homozygote, Independent assortment, Linkage, Linkage equilibrium, Neutral theory of molecular evolution, Population genetics, Recessive allele, Recessive lethal, Relative fitness, Segregation.**)

Allele frequency   The proportion of different alleles in a population. (*See also* **Allele, Dominant allele, Gene,**

Gene pool, Genetic polymorphism, Heterozygote, Homozygote, Independent assortment, Linkage, Linkage equilibrium, Neutral theory of molecular evolution, Population genetics, Recessive allele, Recessive lethal, Relative fitness, Segregation.)

Allen's rule The generalization that mammals tend to have shorter extremities in colder than in warmer climates.

Allometry The proportionate relationship between body parts and the size of the organism.

Allopatric Species or populations in separate geographical locations. (*See also* **Allopatric speciation**.)

Allopatric speciation Evolutionary change between populations initiated by geographical separation or because intervening geographical populations become extinct (*See also* **Allopatric, Parapatric speciation, Peripatric speciation, Speciation, Sympatric speciation**)

Allopolyploid An organism or species that has more than two sets of chromosomes because of hybridization and chromosome doubling.

Allotetraploid An individual or species with double the chromosome number because of hybridization.

Allozyme The particular form (amino acid sequence) of an enzyme produced by a particular allele at a gene locus when there are different possible forms of the enzyme, each produced by a different allele. (*See also* **Enzyme**.)

Alternate splicing The mechanism by which single genes produce different functional proteins by rearrangement of exons. (*See also* **Exon**.)

Alternation of generations A life cycle (typically of a plant or fern) in which a multicellular haploid stage alternates with a multicellular diploid stage. (*See also* **Alternation of generations, Diploid, Haploid, Meiosis**.)

Altricial Born or hatched in a state requiring parental care.

Altruism Behavior that benefits the reproductive success of other individuals because of an actual or potential sacrifice of reproductive success by an individual known as an altruist. (*See also* **Evolutionary stable strategies, Mutualism, Reciprocal altruism, "Tit for tat"**.)

Alveolates A supergroup of Eukaryotes that share a honeycomb (alveolar) system of flattened membrane-bound sacs beneath the cell membrane. (*See also* **Archaeplastida, Chromalveolata, Classification, Excavata, Kingdom, Rhizaria, Supergroups, Unikonta**.)

Amino acids Organic molecules of the general formula R—CH (NH_2) COOH, possessing both basic (NH_2) and acidic (COOH) groups, as well as a side group (R) specific for each type of amino acid. (*See also* **Life, Organic, Peptide (polypeptide)**.)

Amino group An $-NH_2$ group.

Amniota (amniotes) The clade including all extant and extinct vertebrates with embryos encased in extraembryonic membranes. (*See also* **Amniotic egg**.)

Amniotic egg The type of egg produced by reptiles, birds, and mammals (Amniota), in which the embryo is enveloped in a series of membranes (amnion, allantois, chorion) that help sustain its development, usually outside the body of the female. (*See also* **Amniota**.)

Amoebozoans Organisms including slime molds and many types of amoebae and classified in the supergroup Unikonta. (*See also* **Unikonta**.)

Anaerobic glycolysis See **Glycolysis**.

Anaerobic metabolism Energy for metabolic processes obtained from the oxidative breakdown of food molecules) in the absence of molecular oxygen; living in the absence of oxygen. (*See also* **Aerobic metabolism, Metabolism**)

Anaerobic respiration An electron transport system in which substances other than oxygen (for example, sulfates, nitrates, methane), serve as the terminal electron acceptor.

Anagenesis Evolution within a single lineage (usually a species) as opposed to cladogenesis where a group (species) diverges into two or more branches. (*See also* **Cladogenesis, Phyletic evolution**.)

Analogy The possession of a similar character by two or more species caused by factors other than common genetic ancestry. (*See also* **Convergence, Homology**.)

Anatomically modern humans The name used for extant and fossil individuals of the species *Homo sapiens*. (*See also* **Fossils, Hominid, Mousterian tools, Upper Paleolithic tools**,

Ancestral state The characteristic(s) of an organism in the stem or ancestral state. (*See also* **Derived character, Stem group**.)

Ancient DNA The nucleic acid DNA extracted from dead of fossilized organisms. (*See also* **Fossils, Molecular fossils, Paleontology, Period (geological)**.)

Aneuploidy The gain or loss of chromosomes leading to a number that is not an exact multiple of the basic haploid chromosome set. (*See also* **Diploid, Haploid, Haplodiploidy, Polyploidy**.)

Angiosperms Flowering plants that possess floral reproductive structures and encapsulated seeds. (*See also* **Biota, Dicotyledons, Double fertilization, Flora, Flower, Gymnosperms, Mesozoic Era, Monocotyledons, Pollen, Seeds, Telome hypothesis, Vascular plants**.)

Antennapedia complex A cluster of six *Antennapedia*-linked genes first discovered in fruit flies (*Drosophila* spp.) and which determine posterior body structures. (*See also* **Bithorax complex, Homeotic mutations**.)

Antibody   A protein produced by the immune system that binds to a substance (antigen) typically foreign to the organism. (*See also* **Antigen.**)

Anticodon   A sequence of three nucleotides (a triplet) on transfer RNA that is complementary to the codon on messenger RNA, and that specifies placement of a particular amino acid in a polypeptide during translation. (*See also* **Codon.**)

Antigen   A substance, typically foreign to an organism, that initiates antibody formation and is bound by the activated antibody. (*See also* **Antibody.**)

Anterior–posterior (A-P) axis   The major axis of symmetry of bilateral animals where the head is anterior (A) and the tail posterior (P). (*See also* **Axes of symmetry, Dorso-ventral axis, P-D axis.**)

Apomixis (apomictic)   In plants, asexual reproduction from seeds. (*See also* **Parthenogenesis.**)

Apomorphic character   A feature recently derived in evolution in contrast to an ancestral (plesiomorphic) character. (*See also* **Character, Character state, Convergence, Homology, Homoplasy, Parallel evolution, Phenotype, Plesiomorphy, Symplesiomorphy, Synapomorphy.**)

Apoptosis   Cell death that is genetically programmed.

Aposematic coloration   Color or markings on an animal or plant signalling warning harm to a predator if consumed. (*See also* **Batesian mimicry, Mimicry, Müllerian mimicry.**)

Archaea   One of the three domains of organisms proposed by Cark Woese in 1990, the other two being Eubacteria and Eukarya. (*See also* **Domains, Eubacteria, Eukarya.**)

Archaebacteria   Sulfur-dependent and methane-producing organisms with cell walls and a single chromosome. Originated early in the history of life. (*See also* **Greenhouse gas, Methanogens.**)

Archaeopteryx   A basal but not the earliest fossil bird from the Jurassic Era (150–155 Mya).

Archaeplastida (also known as **Plantae**)   A supergroup of eukaryotic organisms including red and green algae, land plants and Charophyta (the latter the ancestors of plants). (*See also* **Alveolates, Chromalveolata, Classification, Excavata, Kingdom, Rhizaria, Supergroups, Unikonta.**)

Archean Eon   A Precambrian geological Eon that lasted from 3.8 to 2.5 Bya. (*See also* **Eon, Geological time scale, Phanerozoic Eon, Precambrian Eon.**)

Archenteron   The cavity that forms during gastrulation of animal embryos and that later forms the primitive gut. (*See also* **Endoderm, Gastrula**)

Archetype   A body plan ("*Bauplan*") characteristic of clades of organisms, usually phyla, but can characterize lower taxonomic levels.

Arms race   In biology, used to describe the situation where two different organisms, for example a parasite and its host, or an animal and a bacterium, evolve to counter changes in the other organisms. (*See also* **Coevolution, Parasitism.**)

Arthropods   The clade of animals (Phylum Arthropoda) that includes insects, spiders and crustaceans with hard exoskeletons, a segmented body plan, and jointed appendages. (*See also* **Phylum.**)

Artificial selection   The process of selective breeding of organisms by humans to change one or more characters over time. (*See also* **Canalizing selection, Constraint, Darwinism, Directional selection, Disruptive selection, Frequency dependent selection, Group selection, Kin selection, Natural selection, Selection, Selection coefficient (s), Sexual selection, Stabilizing selection.**)

Asexual reproduction   Offspring produced by one parent in the absence of sexual fertilization or in the absence of gamete formation (*See also* **Clone.**)

Assortative mating   Reproduction between individuals on the basis of their phenotypic or genotypic similarities (positive assortative mating) or differences (negative assortative mating).

Astronomy   The study of matter in outer space.

Atlantis   A legendary island in the Atlantic Ocean west of Gibraltar, said by Plato to have sunk beneath the sea following an earthquake.

Atmosphere   The gaseous envelope surrounding a celestial body such as Earth. (*See also* **Ozone layer, Primary atmosphere, Secondary atmosphere.**)

Australopithecines   Extinct primates (hominins) of the genus *Australopithecus*. (*See also* **Hominin, Primates.**)

Autocatalytic reaction   Chemical reactions in which the agent that promotes (catalyzes) the reaction is a product of the reaction.

Autosomes   Chromosomes other than sex chromosomes. (*See also* **Chromosomes, Polytene chromosomes.**)

Autotroph   An organism capable of synthesizing complex organic compounds needed for growth from simple inorganic environmental substrates.

Axes of symmetry   For organisms the major alignment of the body of animals or the stem of plants. Axes may be anterior-posterior, left-right, stem-root. (*See also* Anterior-posterior axis, **A–P axis, Bilateral symmetry, Dorso-ventral axis, P–D axis.**)

B

Bacteriophage   A virus (phage) that parasitizes eubacteria.

Balanced genetic load   The decrease in overall fitness of a population caused by alleles (for example, homozygotes for deleterious recessives) that persist in the population because they confer selective advantages in other genotypic combinations (for example, heterozygote advantage).

Balanced polymorphism The persistence of two or more different genetic forms through selection. (*See also* **Genetic polymorphism**.)

Banded iron formations Sedimentary rocks, some as old as 3.76Mya, comprised of thin layers of iron oxides alternating with thin iron-poor layers of chert and shale. The iron oxide layers are believed to result from the photosynthetic activity of cyanobacteria.

Baryon (Greek, *heavy*) The collective name for those subatomic particles composed of three quarks; includes protons and neutrons, which together form atomic nuclei. (*See also* **Baryonic matter**.)

Baryonic matter Material composed largely of baryons, including all atoms and thus almost all matter we experience in everyday life. (*See also* **Cosmic microwave background**, **Dark ages** (Cosmology), **Dark matter**.)

Basalt A fine-grained igneous rock found in oceanic crust and produced in lava flows. (*See also* **Igneous rock**, **Rock cycle**.)

Base (chemistry) The nitrogen-containing component of the nucleotide unit in nucleic acid consisting of either a purine (adenine, A, or guanine, G) or pyrimidine (thymine, T, or cytosine, C, in DNA; uracil, U, or cytosine, C, in RNA). (*See also* **Base**, **Purine**, **Pyrimidine**.)

Base pairs *See* **Base**, **Base substitutions**, **Complementary base pairs**, **DNA (deoxyribonucleic acid)**, **Homeobox**, **Homeotic genes**, **Nucleic acid**, **Nucleotide**, **RNA (ribonucleic acid)**.)

Base substitutions The substitution of one base for another in a nucleotide of DNA, constituting perhaps the simplest type of mutation. (*See also* **Base**, **Base pairs**, **Complementary base pairs**.)

Basic (alkaline) A compound that produces an excess of hydroxyl (OH^-) ions when dissolved in water and has a pH value greater than 7.0. (*See also* **Acidic**, **pH scale**.)

Batesian mimicry The similarity in appearance of a harmless species (the mimic) to a species that is harmful or distasteful to predators (the model), maintained because of selective advantage to the relatively rare mimic. (*See also* **Mimicry**, **Mullerian mimicry**.)

Bauplan (pl. **Baupläne**) The structural body plan that characterizes a group, usually applied at the level of a phylum. (*See also* **Archetype**, **Phylum**.)

Bayesian inference A method of likelihood analysis based on a given probability.

Behavior *See* **Altruism**, **Dominance (social)**, **Homology**, **Imitative learning**, **Imprinting**, **Instinct**, **Learning**, **Phenotype**, **Reciprocal altruism**, **Sociobiology**, **Superorganism**.)

Benthic Refers to the floor of a body of water and to organisms that live in on or near it.

Bergmann's rule The generalization that animals living in colder climates tend to be larger than those of the same group living in warmer climates.

Big Bang theory A theory that the universe (time and space) expanded rapidly from an extremely hot, dense state about 13.7 Bya, and that in the years since it has continued to cool and expand. (*See also* **Evolutionary universe**.)

Bilateral symmetry The condition in which the left and right halves of an organism are equivalent so that the organism can be divided in half along a single plane. (*See also* **Axes of symmetry**.)

Binary fission Replication of an organism by division into two parts, which is the common form of asexual reproduction in prokaryotic and protistan eukaryotic cells.

Binomial nomenclature Part of the system of classification established by Carl Linnaeus in which each species name defines its membership in a genus and provides it with a unique species name and identity, for example, *Homo sapiens*, modern humans. (*See also* **Classification**, **Genus**, **Species**.)

Biogenetic law A theory proposed by Ernst Haeckel that stages in the development of an individual (ontogeny) recapitulate the evolutionary history (phylogeny) of the clades to which the organism belongs. (*See also* **Development**, **Evolutionary developmental biology**, **Heterochrony**, **Ontogeny**, **Paedomorphosis**, **Parthenogenesis**, **Phylogeny**.)

Biogeography The study of the spatial (geographical) distributions of organisms. *See also* **Endemic**, **Phylogenetic evolution**, **Phylogeography**, **Zoogeography**.)

Bioinformatics Acquisition, management, storage and analysis of large data sets, such as entire genome sequences.

Biological species concept The thesis that the primary criterion for separating one species from another is reproductive isolation. (*See also* **Evolutionary species concept**, **Fixity of species**, **Genetic species concept**, **Morphological species concept**, **Polytypic species concept**, **Ring species**, **Species**.)

Biosphere The part of Earth in which organisms live.

Biota All the fauna and flora of a given region or time. (*See also* **Abiotic**, **Angiosperms**, **Biotic**, **Life**, **Burgess shale fossils**, **Chengjiang fauna**, **Ediacaran biota**, **Fauna**, **Organism**, **Rangeomorphs**, **Vendozoa (Vendian biota)**.)

Biotic Relating to or produced by biological organisms. (*See also* **Abiotic**, **Biotic**, **Life**, **Organism**)

Bipedalism The exclusive use of the hind limbs for locomotion, as seen in humans. (*See also* **Brachiation**.)

Bithorax complex. A cluster of three genes first discovered in fruit flies (*Drosophila* spp.) and that regulates development of posterior body structures. (*See also* **Antennapedia complex**, **Homeotic mutations**.)

Blastocoele The cavity that forms within the blastula during animal embryonic development. (*See also* **Blastopore**, **Blastula**.)

Blastopore   The opening formed by the invagination of cells in the embryonic gastrula, connecting its cavity (archenteron) to the outside. In protostome phyla the blastopore is the site of the future mouth. In deuterostomes the blastopore becomes the anus and the mouth forms elsewhere. (*See also* **Blastocoele, Blastula, Gastrula.**)

Blastula   A hollow sphere enclosed within a single layer of cells, occurring at an early stage of animal development animals. (*See also* **Gastrula.**)

Blending inheritance   A now-abandoned theory that offspring inherit a dilution, or blend of parental traits, rather than particles (genes) that determine those traits. (*See also* **Extranuclear inheritance, Inheritance, Inheritance of acquired characters, Lamarckian inheritance, Uniparental inheritance.**)

Body plan   *See Bauplan*

Bottleneck effect   A form of genetic drift that occurs when a population is reduced in size (population crash) and later expands in numbers (population flush). The enlarged population that results may have gene frequencies distinctly different than before the bottleneck. (*See also* **Founder effect, Population, Peripatric speciation, Population genetics.**)

Box (Boxes)   In molecular genetics, a DNA sequence specifying a protein that binds to DNA to act as a transcription factor activating or repressing other sequences of DNA. Families of such transcription factor genes share the same box. (*See also* **Homeobox, Homeodomain, Homeotic Genes, Hox gene.**)

Brachiation   A mode of locomotion in which an animal suspends itself from tree branches and moves about the canopy by swinging between branches or along branches (*See also* **Bipedalism.**)

Bricolage   *See* **Tinkering**

Broca's area   An area in the frontal lobe of the left cerebral hemisphere of the human brain associated with the motor control of speech. (*See also* **Speech, Wernicke's area**)

Bryophytes   A paraphyletic group of terrestrial nonvascular plants comprised of mosses, liverworts and hornworts.

Burgess shale fossils   A 545 My-old assemblage of superbly preserved fossils located in British Columbia, Canada, and which record the Cambrian Explosion of animal body forms. (*See also* **Biota, Cambrian explosion, Cambrian Period, Chengjiang fauna, Ediacaran biota, Fauna, Fossils, Macroevolution, Paleontology, Rangeomorphs.**)

C

Calorie   The amount of heat necessary to raise the temperature of 1 gram of water by 1 degree centigrade at a pressure of 1 atmosphere.

Calvin cycle   A cyclic series of light-independent reactions that accompany photosynthesis to reduce carbon dioxide to carbohydrate in the presence of water. (*See also* **Photosynthesis.**)

Cambium   The tissue of green land plants between phloem and xylem responsible fort producing new cells and for secondary growth. **Cambium, Vascular plants, Xylem.**)

Cambrian explosion   The appearance of fossils of many different forms of animals in the Cambrian Period, interpreted as a rapid diversification (explosion) of body plans. (*See also* **Burgess shale fossils, Cambrian Period, Chengjiang fauna.**)

Cambrian Period   The interval between about 545 and 495 My before the present, marking the appearance of many fossilized organisms (the Cambrian explosion). It is considered the beginning of the Phanerozoic time scale (Eon) and is the first Period in the Paleozoic Era. (*See also* **Burgess shale fossils, Cambrian explosion, Chengjiang fauna, Geological strata, Geological time scale, Period (geological).**)

Canalization   Restriction of phenotypic variation by intrinsic genetic or developmental mechanisms. (*See also* **Canalizing selection, Constraint, Variation, Developmental constraint.**)

Canalizing selection   Artificial or natural selection that results in reduced variation in expression of a phenotypic trait. (*See also* **Artificial selection, Canalization, Constraint, Darwinism, Directional selection, Disruptive selection, Fitness, Frequency dependent selection, Group selection, Inclusive fitness, Kin selection, Natural selection, Neo-Darwinism, Phylogenetic constraint, Population genetics, Relative fitness, Selection, Selection coefficient (s), Sexual selection, Stabilizing selection, Survival of the fittest.**)

Carbohydrate   A compound in which the hydrogen and oxygen atoms bonded to carbons are commonly in a ratio of 2/1 (for example, glucose ($C_6H_{12}O_6$), starch ($C_6H_{12}O_6)_n$, and cellulose, ($C_6H_{10}O_5)_n$).

Carbonaceous   Composed of or containing carbon.

Carbonaceous meteorites   Extraterrestrial objects containing carbon compounds; also known as chondrites.

Carnivores   Organisms (almost entirely animal, but some plants) that feed on animals.

Carrying capacity (K)   The number of individuals in a population that a given environment can sustain. (*See also* **Ecological footprint, Environment, Intrinsic rate of natural increase, Logistic growth curve, Niche, Population, Population explosion, Range.**)

Catastrophism (earth sciences)   A eighteenth- and nineteenth-century theory that fossilized organisms and changes in geological strata were produced by periodic, violent, and widespread catastrophic events rather than by naturally explainable events based on laws that act uniformly through time. (*See also* **Geological dating, Geological strata, Geological strata, Law of superposition, Paleontology, Sedimentary rock, Snowball, Earth, Uniformitarianism**.)

Cell wall A rigid or semi rigid extracellular envelope (outside the plasma membrane) that gives shape to plant, algal, fungal, and eubacterial cells.

Celsius scale (°C) A scale of temperature in which the melting point of ice is taken as 0° and the boiling point of water as 100°, measured at 1 atmosphere of pressure.

Cenozoic (Caenozoic) Era The time from 65 Mya to the present, marked by reduction in dinosaur diversity and radiation of mammals. The third and most recent era of the Phanerozoic Eon, divided into two major Periods, the Tertiary and Quaternary. (*See also* **Dinosaurs, Era, Mammals, Mesozoic Era, Paleozoic (Palaeozoic) Era, Phanerozoic Eon**.)

Central dogma (molecular biology) The conclusion that the direction of flow of genetic information is from DNA → RNA → proteins.

Centromere The chromosome region in eukaryotic nuclei to which spindle fibers attach during cell division.

Centrosome A cellular organelle in eukaryotic cells that functions as the place of origin of the microtubules for the mitotic spindle during cell division.

Character A feature, trait, or property of an organism or population. (*See also* **Analogy, Apomorphic character, Character, Character state, Convergence, Homology, Homoplasy, Inheritance of acquired characters, Morphology, Parallel evolution, Phenotype, Pleiotropy, Plesiomorphy, Quantitative character, Symplesiomorphy, Synapomorphy**.)

Character displacement An evolutionary response to competition in which phenotypic differences accompany resource partitioning among coexisting groups, differences in bill sizes enabling each species of Darwin's finches to feed on differently sized seeds being a classic example. (*See also* **Character state, Competition, Niche**.)

Character state The designation of a character into (usually) binary states used to code characters in phylogenetic analysis. (*See also* **Analogy, Apomorphic character, Character, Convergence, Homology, Homoplasy, Parallel evolution, Phenotype, Pleiotropy, Plesiomorphy, Quantitative character, Symplesiomorphy, Synapomorphy**.)

Chengjiang fauna A 555 My-old assemblage of superbly preserved fossils located in Chengjiang, China, which record the Cambrian Explosion of animal body forms. (*See also* **Biota, Burgess shale fossils, Cambrian explosion, Cambrian Period, Ediacaran biota, Fauna, Fossils, Paleontology, Rangeomorphs**.)

Chert A sedimentary rock composed largely of tiny quartz crystals (SiO_2) precipitated from aqueous solutions.

Chiasma The place where two homologous chromosomes establish close contact and where exchange of homologous parts takes place between non-sister chromatids during meiosis.

Chimera An organism (or part of an organism) made from (composed of cells from) two different organisms, either from the same or from different species.

Chlorophyll A green pigment found in the chloroplasts of plants that collects light used in photosynthesis. (*See also* **Chloroplast, Cyanobacteria, Photosynthesis**.)

Chloroplast A chlorophyll-containing, membrane-bound organelle that is the site of photosynthesis in the cells of plants and some protistans. Chloroplasts contain their own genetic material (circular DNA without histones) and are descendants of cyanobacteria that entered eukaryotic cells via endosymbiosis. (*See also* **Algae, Chlorophyll, Cyanobacteria, Endosymbiosis, Photosynthesis, Protista, Thylakoids**.)

Choanoflagellates Single-celled unikonts, the earliest forms of which may have been the ancestors of animals (or they and animals descended from a common ancestor). (*See also* **Opisthokonts, Unikonta**.)

Chondrites *See* **Carbonaceous meteorites**.

Chordates Members of the Phylum Chordata (which includes the vertebrates), united by the presence of a notochord and dorsal nerve cord. (*See also* **Notochord, Phylum**.)

Chorioallantoic membrane An extraembryonic membrane that forms from the fusion of the chorion and the allantois to form the outer extraembryonic membrane of amniotes. (*See also* **Allantois**.)

Chromalveolata A supergroup of organisms including various types of algae (kelp, dinoflagellates, diatoms), non-photosynthetic ciliates, and parasites such as *Plasmodium falciparum*, which causes malaria. ((*See also* **Algae, Alveolates, Archaeplastida, Classification, Excavata, Kingdom, Rhizaria, Supergroups, Unikonta**.)

Chromatid One of the two sister products of chromosome replication in eukaryotic cells, marked by an attachment between sister chromatids at the centromere.

Chromosomes Elongate, threadlike strand of DNA and proteins in the nuclei of eukaryotic cells housing the nuclear genes. (*See also* **Autosomes, Deletion, Gene locus, Heterozygote, Homozygote, Independent assortment, Locus, Quantitative trait loci (QTLs), Polytene chromosomes**.)

cis-**regulation** The process by which an element of DNA that is adjacent to a gene interacts with the promoter of that gene to control binding sites for transcriptional activators or repressors. (*See also* **Developmental control gene, Enhancer, Messenger RNA, Promoter, Repressor, Transcription, *trans*-regulation**.)

Citric acid cycle *See* **Krebs cycle**.

Clade A natural grouping of taxa (all the descendants) derived from a single common ancestor. (*See also* **Crown Group, Missing link.**)

Cladistics A mode of classification in which taxa are principally grouped on the basis of their shared possession of similar ("derived") characters that differ from the ancestral condition. (*See also* **Cladogram, Classification, Phylogenetic method.**)

Cladogenesis Branching evolution involving the splitting and divergence of a lineage into two or more lineages. (*See also* **Lineages, Monophyletic, Paraphyletic, Phyletic evolution, Phylogenetic evolution, Principle of divergence.**)

Cladogram A tree diagram representing phylogenetic relationships among taxa. (*See also* **Cladogenesis, Phyletic evolution, Phylogenetic evolution, Principle of divergence.**)

Class A taxonomic rank that stands between phylum and order. A phylum may include one or more classes, and a class may include one or more orders. (*See also* **Classification, Family, Order, Phylum, Systematics, Taxon, Taxonomy.**)

Classification The grouping of organisms into a hierarchy of categories commonly ranging from species through genera, families, orders, classes, phyla, and kingdom, each category reflecting one or more significant characters. (*See also* **Binomial nomenclature, Cladistics, Class, Family, Kingdom, Numerical Taxonomy, Order, Phylum, Supergroups, Systematics, Taxon, Taxonomy.**)

Cleavage The process of division by which a single-celled zygote becomes a multicellular organism. (*See also* **Life cycle, Sexual reproduction, Zygote.**)

Cline A gradient of phenotypic or genotypic change in a population or species correlated with the direction or orientation of some environmental feature, such as a river, mountain range, north–south transect, or altitude.

Clone A group of organisms derived by asexual reproduction from a single individual; a group of cells that are identical because they arose from a single cell. (*See also* **Asexual reproduction, Cloning.**)

Cloning (genetics) Techniques for producing identical copies of a section of genetic material by inserting a DNA sequence into a cell, such as a bacterium, where it can be replicated. (*See also* **Clone.**)

Cnidaria Animals commonly known as corals, jellyfish, and hydrozoans.

Coacervate An aggregation of colloidal particles in liquid phase that persists for a short time as suspended membranous droplets. (*See also* **Microspheres, Protocells.**)

Coadaptation The action of selection in producing adaptive combinations of alleles at two or more different gene loci.

Codominance The independent phenotypic expression of two different alleles in a heterozygote.

Codon The triplet of adjacent nucleotides in messenger RNA that codes for a specific amino acid carried by a specific transfer RNA or that codes for termination of translation (STOP codons). Placement of the amino acid is based on complementary pairing between the anticodon on tRNA and the codon on mRNA. (*See also* **Anticodon, Degenerate (redundant) code, Genetic code, Messenger RNA (mRNA), Peptide (polypeptide), Stop codon, Translation, Universal genetic code.**)

Coelom (Coelome) The internal body cavity of eucoelomate animals (true coelomates) formed within mesoderm. (*See also* **Germ layers, Haemocoel Mesoderm, Triploblastic.**)

Coenzymes Non-protein enzyme-associated organic molecules (for example, NAD, FAD, and coenzyme A) that participate in enzymatic reactions by acting as intermediate carriers of electrons, atoms, or groups of atoms. (*See also* **Enzymes.**)

Coevolution Evolutionary changes in one or more genes, developmental processes, organs or species in response to changes in genes, developmental processes, organs or species. (*See also* **Arms race, Parasitism.**)

Coextinction When two or more species, lineages or groups become extinct at the same time. (*See also* **Extinction, Lineages.**)

Cofactor A small molecule, which may be organic (that is, a coenzyme) or inorganic (that is, a metal ion), required by an enzyme in order to function.

Cohort Individuals of a population all of the same age.

Colinearity Applied to Hox genes to signify a correspondence between the positions of the genes in relation to one another on a chromosome and the antero-posterior parts of the body whose position they determine. (*See also* **Hox genes.**)

Collision theory Proposed in 1749 by Georges Louis Leclerc, Comte de Buffon that a comet (or star) struck the sun and broke off fragments that formed the planets. (*See also* **Condensation theory, Cosmology, Dwarf planets, Gravitation, Nebular hypothesis, Planetesimals, Planets, Protoplanet, White dwarf.**)

Commensalism An association between organisms of different species in which one species benefits from the relationship and the other species is not affected significantly. (*See also* **Parasitism, RNA virus, Virus.**)

Compartments (Developmental biology) Separate cell populations in embryos of animals separated by boundaries from adjacent populations.

Competition Relationship between organismal units (for example, individuals, groups, species) attempting to exploit a limited common resource in which each unit inhibits, to varying degrees, the survival or reproduction of another unit by means other than predation. (*See also* **Competitive exclusion, Environment, Group selection, Niche, Principle of divergence, Resource partitioning, Sexual selection, Species sorting**.)

Competitive exclusion The principle that two species cannot continue to coexist in the same environment (niche) if they use it in the same way. (*See also* **Competition, Environment, Group selection, Niche, Principle of divergence, Resource partitioning, Species sorting, Sexual selection**.)

Complementary base pairs Nucleotides on one strand of a nucleic acid that hydrogen bond with nucleotides on another strand according to the rule that pairing between purine and pyrimidine bases is restricted to certain combinations. A pairs with T in DNA, A pairs with U in RNA, and G pairs with C in both DNA and RNA. (*See also* **Base, Base pairs, DNA (deoxyribonucleic acid), Homeobox, Homeotic genes, RNA (ribonucleic acid)**.)

Complexity A state of intricate organization caused by arrangement or interaction among different component parts or processes: the greater the number of interacting parts, the greater the complexity. (*See also* **Design, Hierarchy, Progress, Reductionism**.)

Condensation theory The theory that the planets arose by the collision and accretion of planetesimals. (*See also* **Collision theory, Cosmology, Dwarf planets, Gravitation, Nebular hypothesis, Planetesimals, Planets, Protoplanet, White dwarf**.)

Consciousness The ability of an organisms to have an awareness of itself, to perceive the world around it and to respond to it.

Constraint Factors that influence character variation or evolutionary direction. (*See also* **Canalization, Canalizing selection, Directional selection, Discontinuous variation, Phenotypic plasticity, Phylogenetic constraint, Variation**.)

Continent Seven large bodies of land (North America, South America, Asia, Europe, Africa, Antarctica, Australia) surrounded by water. (*See also* **Continental drift, Crust, Gondwana, Laurasia, Pangaea, Plate tectonics, Tectonic plates**.)

Continental drift Hypothesis proposed in 1912 by Alfred Wegener, according to which large landmasses move relative to each other across Earth's surface. (*See also* **Continents, Crust, Gondwana, Laurasia, Paleomagnetism, Pangaea, Plate tectonics, Seafloor spreading, Tectonic plates**.)

Continuous variation Character variations (such as height in humans), the distribution of which follows a series of small, nondiscrete quantitative steps. (*See also* **Canalization, Character, Character state, Darwinism, Quantitative character, Variation**.)

Convergence (Convergent evolution) The evolution of similar characters (genes or morphologies) in genetically unrelated or distantly related species (homoplasy), mostly because they have been subjected to similar environmental selective pressures. (*See also* **Analogy, Apomorphic character, Character, Character state, Homology, Homoplasy, Parallel evolution, Phenotype, Pleiotropy, Plesiomorphy, Symplesiomorphy, Synapomorphy**.)

Co-option (Evolution) Acquisition of a new function by an existing character in response to a new environment or set of conditions. (*See also* **Adaptation, Co-option (Genetics), Exaptation**.)

Co-option (Genetics) Acquisition of a gene or assembly of genes by one part of an organism from another part of the same organism, as between germ layers. (*See also* **Co-option (Evolution), Exaptation, Germ layers**.)

Cope's rule The generalization (not always confirmed) that body size tends to increase in an animal lineage during its evolution.

Core (Earth sciences) The innermost layer of Earth, composed mostly of iron with some nickel, and consisting of a solid inner core and an outer, liquid core. (*See also* **Crust, Earth, Lithosphere, Mantle (Earth sciences)**.)

Correlation The degree to which two measured characters tend to vary in the same quantitative direction (positive correlation) or in opposite directions (negative correlation).

Cosmic microwave background (CMB) A form of electromagnetic radiation that fills the entire universe and is a relic of the Big Bang. Tiny differences in the temperature of the CMB, which averages 2.7 K, reflect minute fluctuations in the matter of the universe and provide a snapshot of the early universe. (*See also* **Baryonic matter, (CMB), Dark ages (Cosmology), Dark matter**.)

Cosmological constant The energy density of the smooth vacuum, introduced by Albert Einstein in his theory of general relativity to represent the inbuilt tendency of space-time to expand; a constant required to explain the universe as stationary. (*See also* **Expanding universe, Space-time**.)

Cosmological redshift Reflects the fact that light (photons) from distant galaxies is proportional to their distance from the observer and has been stretched so that the wavelength moves towards the red end of the spectrum; hence redshifted. (*See also* **Expanding universe, Galaxy, Photons, Redshift, Space-time**.)

Cosmology The study of the structure and evolution of the universe. (*See also* **Collision theory, Condensation theory, Dwarf planets, Expanding universe, Gravitation, Nebular hypothesis, Planetesimals, Planets, Protoplanet, Solar, nebula, Space-time, White dwarf**.)

Covalent bond A strong chemical bond that results from the sharing of electrons between two atoms.

Covariation   Correlated variation in two or more variables such as genes of features of the phenotype.

Creationism   The belief that each different kind of organism was individually created by a supernatural force. (*See also* **Evolution**.)

Critical density (Cosmology)   The density of atoms (10^{-29}g/cm^3 or six atoms of hydrogen/m^3) that determines the geometry of the universe as flat. A higher density of atoms and the universe would collapse on itself; a lower density and the universe would continue to expand forever. (*See also* **Cosmology**.)

Crossing Over   The process in which the chromatids of two homologous chromosomes exchange genetic material. (*See also* **Recombination, Unequal crossing over**.)

Crown Group   The last common ancestor of the extant members of a clade and all its descendants. (*See also* **Classification, Stem group, Taxonomy**.)

Crust (Earth sciences)   The exterior portion of the earth floating on the mantle, divided into continental (older) and oceanic (younger) crust, and composed of igneous, sedimentary, and metamorphic rocks. (*See also* **Continental drift, Continents, Core, Earth, Igneous rock, Lithosphere, Mantle** (Earth sciences), **Metamorphic rock, Rock cycle, Sedimentary rock**.)

Cryptic   A feature that resembles (mimics) the background and so is less visible

Cryptic (Sibling) species   Two or more species that are so similar that they cannot be distinguished morphologically. (*See also* **Sibling species**.)

Ctenophora   Animals commonly known as comb jellies.

Culture   The learned behaviors and practices common to a social group. (*See also* **Altruism, Dominance (social), Homology, Imprinting** (behavior), **Instinct, Learning, Phenotype, Reciprocal altruism, Sociobiology, Superorganism**.)

Cyanobacteria   Photosynthetic organisms possessing chlorophyll a but not chlorophyll b. Previously known as blue-green algae, their color is caused by a bluish pigment masking the chlorophyll. Many are photosynthetic aerobes (requiring oxygen) some are anaerobes (not requiring oxygen). (*See also* **Algae, Chlorophyll, Chloroplast, Photosynthesis**.)

Cyclomorphosis   Present of polymorphism in planktonic and terrestrial organisms often initiated in response to seasonal changes or to factors dependent on seasonality. (*See also* **Morph, Phenotypic plasticity, Polymorphism, Seasonal polymorphism**.)

Cytology   The study of cells—their structures, functions, components, and life histories.

Cytochromes   Proteins containing iron–porphyrin (heme) complexes that function in respiration and photosynthesis as hydrogen or electron carriers. (*See also* **Photosynthesis**.)

Cytoplasm   All cellular material within the plasma membrane, excluding the nucleus and nuclear membrane. (*See also* **Eukaryotic cells, Extranuclear inheritance, Fertilization, Histones, Maternal inheritance, Nucleus, Organelles, Prokaryotic cells, Protoplasm, Uniparental inheritance**.)

Cytoplasmic inheritance   *See* **Extranuclear inheritance, Inheritance, Maternal inheritance, Uniparental inheritance**.)

Cytoskeleton   Tubular or membrane-bound elements in the cytoplasm including microtubules and microfilaments.

D

Dark ages (Cosmology)   The time during the early life of the universe before the evolution of stars, when ordinary matter in the universe consisted of neutral hydrogen and helium, with a little lithium, and the universe was extremely dense, hot, and dark. (*See also* **Baryonic matter, Collision theory, Cosmic microwave background (CMB), Dark matter, Galaxy, Milky Way, Neutron stars (pulsars), Red giant** (Cosmology).)

Dark energy   A force that does not interact with light but is detectable through its gravitational effects, which accounts for 74% of the density of the universe, and which is thought to cause the acceleration of the expansion of the universe. (*See also* **Baryonic matter, Dark matter**.)

Dark matter   That portion of the universe (22 percent) that neither emits nor absorbs light, which is known only from its gravitational effects on itself and on baryons, and of unknown composition. (*See also* **Cosmic microwave background (CMB), Dark ages** (Cosmology), **Dark energy**.)

Darwinism   The theory, proposed by Charles Darwin, that biological evolution has led to the many different highly adapted species through natural selection acting on hereditary variations in populations to give rise to descent with modification. (*See also* **Continuous variation, Discontinuous variation, Domestication, Homology, Natural selection, Neo-Darwinism, Population genetics, Selection, Sexual selection, Social Darwinism, Survival of the fittest, Tinkering, Variation**.)

Deccan traps   Staircase-like basaltic lava formations in the Deccan region of western India.

Degenerate (redundant) code   Part of the genetic code for which there is more than one triplet codon for a particular amino acid but where a specific codon cannot code for more than one amino acid. (*See also* **Codon, Genetic code, Messenger RNA (mRNA), Peptide (polypeptide), Stop codon, Translation, Triplet code, Universal genetic code**.)

Deleterious allele   An allele that reduces the adaptive value of its carrier, either when present in homozygous condition (recessive allele) or in heterozygous condition (dominant or partially dominant allele). (*See also* **Allele, Allele frequency, Heterozygote, Homozygote, Recessive lethal, Relative fitness, Segregation.**)

Deletion (Genetics)   A genetic change in which a section of DNA or chromosome has been lost. (*See also* **Chromosome, Mutation, RNA editing.**)

Deme   A local population of a species; in sexual forms, a local interbreeding group. (*See also* **Gene flow, Group, Isolating mechanisms, Mendelian population, Migration (genetics), Parapatric, Population, Species, Subspecies.**)

Density dependent   The dependence of population growth and size on factors directly related to the numbers of individuals in a particular locality (for example, competition for food, accumulation of waste products).

Density independent   The dependence of population growth on factors (climatic changes, meteorite impacts, and so on) unrelated to the numbers of individuals in a particular locality.

Derived character   A feature of an organisms that differs from the ancestral character. (*See also* **Basal state.**)

Descent with modification   Darwin's phrase for biological evolution as change from generation to generation. (*See also* **Darwinism, Domestication, Homology, Tinkering.**)

Design   The concept that the complexity of organisms requires belief in a supernatural power (designer) to explain the diversity of life. (*See also* **Complexity, Fixity of species, Hierarchy, Great Chain of Being, Ladder of Nature, Progress, Reductionism.**)

Development   *See* **Embryonic development.**

Developmental constraint   A feature of embryonic development that influences the range of variation in a character. (*See also* **Canalization, Directional selection, Discontinuous variation, Phenotypic plasticity, Phylogenetic constraint, Variation.**)

Developmental control gene   A regulatory gene in a gene network that controls other (often many other) genes. (*See also* ***cis*-regulation, Downstream gene, Enhancer, Gene regulation, Genetic assimilation, Heat shock proteins, Hierarchy, Homeotic mutations (homeosis), Modularity, Promoter, Regulatory gene, Repressor, Signal transduction, Transcription, Upstream gene.**)

Diapsida (diapsids)   The clade of all extant and extinct amniotes with a post-orbital double opening of the skull.

Dicotyledons   Flowering plants (angiosperms) with embryos containing two seed leaves (cotyledons). (*See also* **Angiosperms, Double fertilization, Flower, Mesozoic Era, Monocotyledons, Pollen, Seeds.**)

Dimorphism   Presence in a population or species of two morphologically distinctive types of individuals, for example, differences between males and females, pigmented and nonpigmented forms.

Dinosaurs   Extinct terrestrial carnivorous or herbivorous reptiles that existed during the Mesozoic Era; members of the clades Saurischia and Ornithischia. (*See also* **K–T extinction, Mammals, Mesozoic Era, Phanerozoic Eon.**)

Dioecious   Organisms in which the male and female sex organs are in separate individuals.

Diploblastic   An animal with embryos comprised only of two major germ layers: ectoderm and endoderm (for example, Cnidaria). (*See also* **Ectoderm, Endoderm, Germ layers, Triploblastic.**)

Diploid   An organism with somatic cell nuclei containing two sets of chromosomes (2n), usually one from the male and one from the female parent, providing two different (heterozygous) or similar (homozygous) alleles for each gene. (*See also* **Alternation of generations, Aneuploidy, Double fertilization, Gamete, Haploid, Haplodiploidy, Meiosis, Polyploidy, Seed, Sporophyte.**)

Directional selection   A form of selection resulting in change in a character or in the phenotype of a character in one direction. (*See also* **Artificial selection, Canalizing selection, Constraint, Darwinism, Developmental constraint, Disruptive selection, Fitness, Frequency dependent selection, Group selection, Inclusive fitness, Kin selection, Natural selection, Neo-Darwinism, Population genetics, Relative fitness, Selection, Selection coefficient (*s*), Sexual selection, Stabilizing selection, Survival of the fittest.**)

Discontinuous variation   Character variations that are sufficiently different from each other that they fall into non-overlapping classes. (*See also* **Canalization, Constraint, Continuous variation, Developmental constraint, Quantitative character, Variation.**)

Disruptive (diversifying) selection   Change that favors the survival of organisms in a population that are at opposite phenotypic extremes for a particular character and eliminates individuals with intermediate values. (*See also* **Artificial selection, Canalizing selection, Constraint, Darwinism, Directional selection, Disruptive selection, Fitness, Frequency dependent selection, Group selection, Inclusive fitness, Kin selection, Natural selection, Neo-Darwinism, Population genetics, Relative fitness, Selection, Selection coefficient (*s*), Sexual selection, Stabilizing selection, Survival of the fittest.**)

Divergent evolution   Change leading to differences between lineages.

Division of labor   *See* **Social Organization.**

DNA (deoxyribonucleic acid)   A nucleic acid that serves as the genetic material of all cells and many viruses; composed of nucleotides that are usually polymerized into long double-stranded chains, each nucleotide characterized by the presence of a deoxyribose sugar. (*See also* **Base, Base Pairs, Base substitutions, Complementary base pairs, Nucleic acid, Nucleotide, RNA (ribonucleic acid).**)

DNA repair mechanisms   Excision of bases, excision of nucleotides and repair of mismatches bases are three major ways in which damage to DNA is corrected. (*See also* **Gene therapy.**)

DNA virus   A small intracellular infectious agent, often parasitic, composed of DNA in a protein coat. (*See also* **Mimivirus, RNA virus, Virus.**)

DNA world   The hypothesis that the first chemical of life was DNA. (*See also* **Protein world, RNA world.**)

Domain   In molecular biology, an amino acid sequence within a polypeptide chain that performs a particular function. The term has also been used in systematics for the tripartite division of organisms—Archaea, Eubacteria, Eucarya—as a substitute for the rank of superkingdom, which commonly designates prokaryotes and eukaryotes. (*See also* **Archaea, Eubacteria, Eukarya.**)

Domestication   The modification of plants or animals to suite human needs, often effected through artificial selection. (*See also* **Artificial selection, Darwinism.**)

Dominance (social)   The result of behavioral interactions between individuals in a group in which one or more individuals, sustained by aggression or other behaviors, rank higher than others in controlling the conduct of group members. (*See also* **Altruism, Culture, Imprinting** (behavior), **Instinct, Learning, Reciprocal altruism, Sociobiology.**)

Dominant allele   The allele that determines the phenotype in heterozygotes. (*See also* **Allele, Allele frequency, Gene, Gene pool, Genetic polymorphism, Heterozygote, Homozygote, Independent assortment, Linkage, Linkage equilibrium, Neutral theory of molecular evolution, Population genetics, Recessive allele, Recessive lethal, Relative fitness, Segregation.**)

Dorsal   The back side or upper surface of an animal; opposite of ventral. (In vertebrates, the surface defined by the location of the spinal column.)

Dorso–ventral (D–V) axis   A major axis of symmetry of bilateral animals where the nervous system is either dorsal (D, chordates) or ventral (V, invertebrates). (*See also* **Anterior-posterior axis, Axes of symmetry, P-D axis.**)

Dosage compensation   A mechanism that compensates for difference in the number of X (or Z) chromosomes between males and females, equalizing the effects of their X-linked genes. (*See also* **Haldane's rule, Homogametic sex, Sex chromosomes, Sex-linked genes, X chromosome, Y chromosome.**)

Double fertilization   A distinctive feature of angiosperm plants in which two nuclei from a male pollen tube fertilize the female gametophyte, one producing a diploid embryo and the other producing polyploid (usually triploid) nutritional endosperm. (*See also* **Angiosperms, Dicotyledons, Diploid, Flower, Mesozoic Era, Monocotyledons, Nucleus, Pollen, Pollination, Polyploidy, Seeds, Triploid.**)

Downstream gene   A gene(s) controlled by (downstream of) another gene, usually a regulatory gene. (*See also* **cis-regulation, Developmental control gene, Gene regulation, Heat shock proteins, Hierarchy, Homeotic mutations (homeosis), Modularity, Upstream gene.**)

Drift   *See* **Genetic drift.**

Duplication (Genetics)   When a section of DNA, chromosome segment, chromosome, or entire genome has been doubled. (*See also* **Gene family, Genome, Mutation, Neo fictionalization, Paralogous genes, Repetitive DNA, Selfish DNA, Serial ("iterative") homology.**)

Dwarf planets   A class of planets in our solar system based on size, established in 2006 to accommodate Pluto, Ceres, and UB3Unlike planets, dwarf planets do not dominate their neighborhoods. (*See also* **Condensation theory, Cosmology, Gravitation, Planetesimals, Planets, Protoplanet, White dwarf.**)

E

Ear ossicles   *See* **Middle ear ossicles**

Ecdysoza   One of two groups (the other being Lophotrochozoa) of protostomes, Ecdysoza being animals such as arthropods and nematodes with a molt in the life cycle. (*See also* **Arthropods, Lophotrochoza.**)

Ecological footprint   The amount of land and water required to produce the resources consumed and to absorb the waste of an individual, population, or lifestyle. (*See also* **Carrying Capacity, Ecological footprint, Environment, Intrinsic rate of natural increase, Logistic growth curve, Niche, Population, Population explosion, Range.**)

Ecological niche   *See* **Niche.**

Ecology   The study of the relations between organisms and their environment, in terms of their numbers, distributions, and life cycles. (*See also* **Carrying Capacity, Ecological footprint, Environment, Group, Intrinsic rate of natural increase, Logistic growth curve, Migration (ecology), Niche, Population, Population explosion, Range.**)

Ecomorph   A phenotypic subset of a population or a species associated with a particular environment. (*See also* **Morph, Phenotypic plasticity, Polymorphism.**)

Ectoderm   The outermost layer of cells that covers early animal embryos, from which nerve tissues and outermost epidermal tissues are derived. (*See also* **Diploblastic, Endoderm, Germ layers, Mesoderm, Triploblastic.**)

Ectotherm (Ectothermic) An organism having a variable body temperature primarily determined by the ambient (environmental) temperature.

Ediacaran biota Soft-bodied fossils found in South Australia and other places, dating to the Ediacaran Precambrian Period lasting 60 or more My and with unknown relationships to other organisms. (*See also* **Biota, Burgess shale fossils, Chengjiang fauna, Ediacaran Period, Fauna, Fossils, Macroevolution, Paleontology, Precambrian Eon, Rangeomorphs, Vendozoa (Vendian biota).**)

Ediacaran Period The last geological Period of the Neoproterozoic Era from 600 to 542 Million years. Also known as the Vendian Period and characterized by the presence of the Ediacaran biota. (*See also* **Ediacaran biota, Period** (geological), **Epoch, Era.**)

Electron carrier In oxidation–reduction reactions, a molecule that acts alternatively as an electron donor (becomes oxidized) and as an electron acceptor (becomes reduced).

Embden-Meyerhof glycolytic pathway The biochemical pathway leading to pyruvic acid and providing a net yield of two high-energy phosphate bonds in ATP used as chemical energy by cells.

Embryonic Development The progression from egg to adult in multicellular organisms. (*See also* **Biogenetic law, Developmental control gene, Embryology, Fertilization, Epigenesis, Evolutionary developmental biology, Heterochrony, Homeotic mutations (homeosis), Morphogenesis, Ontogeny, Paedomorphosis, Parthenogenesis, Preformationism, Regulation.**)

Embryology The study of the development of organisms from their inception to birth/hatching, and often into later life history stages such as larvae. (*See also* **Biogenetic law, Development, Developmental control gene, Fertilization, Epigenesis, Evolutionary developmental biology, Heterochrony, Homeotic mutations (homeosis), Larva, Life history, Morphogenesis, Ontogeny, Paedomorphosis, Parthenogenesis, Preformationism, Regulation.**)

Enation hypothesis The earliest leaves to evolve, thin microphylls, evolved from extensions of tissues along the stem and not from small branches. (*See also* **Dicotyledons, Monocotyledons, Telome hypothesis.**)

Encephalization quotient (EQ) The relative brain size as the ratio of actual brain weight to the brain weight expected for an animal of the same body size.

Endemic A species or population that is specific (indigenous) to a particular geographic region. (*See also* **Biogeography, Phylogenetic evolution, Phylogeography, Zoogeography.**)

Endocytosis Cellular engulfment of outside material, followed by its transfer into the cellular interior encapsulated in a membrane. (*See also* **Endosymbiosis, Primary endocytosis, Secondary endocytosis.**)

Endoderm The layer of cells that lines the embryonic gut (archenteron) during the early stages of development in animals, and which later forms the epithelial lining of the intestinal tract and internal organs such as the liver, lung, and urinary bladder. (*See also* **Archenteron, Diploblastic, Ectoderm, Germ layers, Mesoderm, Neural crest, Triploblastic.**)

Endosymbiosis A relationship between two different organisms, in which one (the endosymbiont) lives within the tissues or cell of the other, benefiting either or both. Eukaryotic organelles, such as mitochondria and chloroplasts, had an endosymbiotic prokaryotic origin. (*See also* **Chloroplasts, Endocytosis, Mitochondria, Primary endosymbiosis, Secondary endosymbiosis, Symbiosis.**)

Endothermic A body temperature maintained by internal physiological mechanisms at a level independent of the ambient (environmental) temperature.

Enhancer A region of the DNA (nucleotide sequence) of a gene that binds to another nucleotide sequence within the gene so that the promoter sequence can be transcribed. Individual genes may contain more than one enhancer. (*See also* ***cis*-regulation, Gene regulation, Messenger RNA, Promoter, Regulatory gene, Repressor, Signal transduction, Transcription, *trans*-regulation.**)

Entropy The measure of disorder of a physical system. In a closed system, to which energy is not added, the second law of thermodynamics states that entropy, or energy unavailable for work, will remain constant or increase but never decrease. Living systems, however, are open systems, to which energy is added from sunlight and other sources, and order can therefore arise from disorder in such systems, that is, energy available for work can increase and entropy can decrease.

Environment The complex of external conditions, abiotic and biotic, that affects and interacts with organisms or populations to provide the facilities and resources that enable hereditary data (genotypes) to produce organismal characters (phenotypes). (*See also* **Adaptation, Competitive exclusion, Genetic assimilation, Group selection, Niche, Preadaptation, Principle of divergence, Resource partitioning, Sexual selection.**)

Enzyme A protein that catalyzes chemical reactions. (*See also* **Allozyme, Coenzymes, Metabolic pathway, Metabolism, Proteinoids, Restriction enzymes, Ribozymes, Transposons (transposable elements)**

Eon A major division of the geological time scale, often divided into two eons beginning from the origin of Earth 4.5 Bya: the Precambrian or Cryptozoic (rarity of life forms) and the Phanerozoic (abundance of life forms). (*See also* **Archean Eon, Eon, Era, Geological time scale, Precambrian Eon, Phanerozoic Eon, Precambrian Eon.**)

Epigenesis The theory that tissues and organs are formed by interaction between cells and substances that appear during development, rather than being preformed in the zygote. (*See also* **Biogenetic law, Development, Developmental control gene, Embryology, Fertilization, Evolutionary developmental biology, Heterochrony, Homeotic mutations (homeosis), Morphogenesis, Ontogeny, Paedomorphosis, Parthenogenesis, Preformationism, Regulation.**)

Epigenetic The sum of the genetic and non-genetic factors that influence gene action.

Epistasis Interaction between non-allelic genes.

Epithelium (pl. **epithelia**) One of the two fundamental types of cellular organization in animals, consisting of a sheet of laterally connected and interacting cells resting on an extracellular matrix—the basement membrane—that is produced by the epithelium. (*See also* **Mesenchyme.**)

Epoch One of the categories into which geological time is divided. Periods are often divided into three epochs: Early, Middle, and Late; for example, Early Cambrian. (*See also* **Geological time scale, Period (geological).**)

Equilibrium (Genetics) The persistence of the same allelic frequencies over a series of generations.

Era A division of geological time that stands between the Eon and the Period. The Phanerozoic Eon is divided into Paleozoic, Mesozoic, and Cenozoic Eras; and each era is divided into two or more Periods. (*See also* **Cenozoic (Caenozoic) Era, Geological time scale, Paleozoic (Palaeozoic) Era, Period (geological), Phanerozoic Eon.**)

Eubacteria One of the three domains of organisms proposed by Cark Woese in 1990 for organisms previously known as Prokaryotes. The other two domains are Archaea and Eukarya. (*See also* **Archaea, Domains, Eukarya.**)

Euchromatin Normally staining chromosomal regions that possesses most of the active genes. (*See also* **Heterochromatin.**)

Eugenics The proposal that humanity can be improved by selective breeding. (*See also* **Gene therapy, Genetic engineering, Negative eugenics, Positive eugenics, Reproductive technology.**)

Eukarya One of the three domains of organisms proposed by Cark Woese in 1990 for organisms with a nucleus and nuclear membrane (eukaryotic cells). The other two domains are Eubacteria and Eukarya. (*See also* **Archaea, Domains, Eubacteria, Eukaryotic cells.**)

Eukaryotic cells Nuclear membranes, mitochondrial organelles, and other characteristics distinguish these cells from prokaryotic cells. (*See also* **Chloroplasts, Cytoplasm, Eukarya, Histones, Mitochondria, Nucleus, Organelles, Prokaryotic cells, Protoplasm, Thyalkoids.**)

Eusociality Organisms that display division of labor (workers, soldiers) and that live in cooperative groups caring for the offspring of a single female and several males. (*See also* **Social organization, Superorganism**)

Evo-devo *See* **Evolutionary developmental biology.**

Evolution Heritable changes in genes, features, organisms, populations and species through time

Evolutionary developmental biology The field in biology in which developmental processes are studied for their insights into evolutionary processes. (*See also* **Biogenetic law, Development, Developmental control gene, Embryology, Epigenesis, Heterochrony, Homeotic mutations (homeosis), Morphogenesis, Ontogeny, Paedomorphosis, Parthenogenesis.**)

Evolutionary plasticity *See* **Constraint, Developmental constraint, Phenotypic plasticity, Phylogenetic constraint, Reaction norm.**

Evolutionarily stable strategies The result of a balance between cooperation and individual action among interacting individuals in a social group. (*See also* **Altruism, Mutualism, Reciprocal altruism, "Tit for tat".**)

Evolutionary species concept A definition of a species based on isolation from other species, often identified from ancestor–descendant populations. (*See also* **Biological species concept, Fixity of species,** Genetic species concept, **Morphological species concept, Phylogenetic species concept, Polytypic species concept, Species.**)

Evolutionary trees Arrangements of the evolutionary history of one or more groups of organisms presented in a tree-like branching pattern. (*See also* **Cladogenesis, Phylogenetic tree, Tree of life, Universal Tree of life.**)

Evolutionary universe (Cosmology) A more formal title for the Big Bang hypothesis to describe the origin and evolution of the universe. (*See also* Big **Bang theory, Inflation**)

Evolvability The concept that the genotypes of some organisms evolve more rapidly/readily than do the genotypes of other organisms.

Exaptation A character that was adaptive under a prior set of conditions and later provides the initial stage (is "co-opted") for the evolution of a new adaptation under a different set of conditions. (*See also* **Adaptation, Co-option, Environment, Inheritance of acquired characters, Preadaptation.**)

Excavata A supergroup of organisms, previously members of the Protista, and including *Giardia*, which causes the intestinal illness giardiasis. *Trypanosoma brucei*, which causes sleeping sickness, and *Trichomonas*, which causes trichomoniasis. (*See also* **Alveolates, Archaeplastida, Chromalveolata, Classification, Kingdom, Protista, Rhizaria, Supergroups, Unikonta.**)

Exon   An expressed nucleotide sequence in a gene that is transcribed into messenger RNA and spliced together with the transcribed sequences of other exons from the same gene. Exons are separated from one another by intervening nontranslated sequences (*See also* **Alternate splicing, Intron, Nucleotide, RNA splicing, Split gene, Transcription, Translation.**)

Exon shuffling   The recombination or exclusion of exons such that they remain active as sources of genetic information. (*See also* **Exon, Independent assortment, Recombination.**)

Expanding universe   The theory that the space-time between galaxies is expanding so that universe is continuing to increase. (*See also* **Cosmological constant, Galaxy, Space-time.**)

Extinction   The disappearance of a species or higher taxon. (*See also* **Coextinction.**)

Extranuclear inheritance   Patterns of heredity in which the transmission is not via nuclear genes, for example transmission by mitochondrial genes. (*See also* **Inheritance, Maternal inheritance, Pangenesis, Uniparental inheritance.**)

Extremophiles   Organisms that live under and are adapted to extreme conditions of temperature, pressure, acidity or salinity. (*See also* **Thermophilic organisms.**)

F

F (inbreeding coefficient)   *See* **Inbreeding coefficient.**

Family   A taxonomic category that stands between order and genus; an order may comprise a number of families, each of which contains a number of genera. (*See also* **Class, Classification, Genus, Order, Systematics, Taxon, Taxonomy.**)

Fauna   All animals of a particular region or time. (*See also* **Biota, Burgess shale fossils, Chengjiang fauna, Ediacaran biota, Rangeomorphs, Vendozoa (Vendian biota).**)

Fecundity   A measure of the potential production of offspring. (*See also* **Reproductive success, Sexual reproduction.**)

Ferns (Pterophyta)   Spore-bearing plants that carry their sporangia on the fronds. (*See also* **Sporophyte, Telome hypothesis.**)

Fertility   A trait of organisms measured by the number of viable offspring produced. (*See also* **Asexual reproduction, Hybridization (Genetics), Inbreeding, Inbreeding depression, Parthenogenesis Reproductive success, Sexual reproduction.**)

Fertilization   In animals, the fusion of a sperm with an egg that allows sperm and egg nucleus to fuse as a zygote nucleus and development to be initiated. (*See also* **Development, Fertility, Gamete, Life cycle, Nucleus, Ontogeny, Reproductive Incompatibility, Sexual reproduction, Zygote.**)

Fitness   Central to evolutionary theory evaluating genotypes and populations, fitness has had many definitions, ranging from comparing growth rates to comparing long-term survival rates. The basic fitness concept that population geneticists commonly use is relative reproductive success, as governed by selection in a particular environment. (*See also* **Adaptive landscape, Genetic load, Heterozygote advantage (superiority), Hybrid breakdown (inferiority, inviability, sterility), Inclusive fitness, Natural selection, Neutral mutation, Neutral theory of molecular evolution, Population genetics, Relative fitness, Reproductive success, Selection, Selection coefficient (s).**)

Fixity of species   A theory held by Linnaeus and others that members of a species could only produce progeny like themselves, and therefore each species was fixed in its particular form(s) at the time of its creation. (*See also* **Biological species concept, Design, Great Chain of Being, Ladder of Nature, Species.**)

Flora   All plants of a particular region or time. (*See also* **Angiosperms, Biota.**)

Flower (Botany)   The reproductive structure of many seed-producing plants; often brightly colored and elaborate in form. (*See also* **Angiosperms, Dicotyledons, Double fertilization, Mesozoic Era, Monocotyledons, Pollen, Pollination, Seeds, Self incompatibility.**)

Fossils   The geological remains, impressions, or traces of organisms that existed in the past. (*See also* **Ancient DNA, *Archaeopteryx*, Burgess shale fossils, Chengjiang fauna, Ediacaran biota, Hominid. Living fossil, Macroevolution, Molecular fossils, Paleontology, Period (geological), Phanerozoic Eon, Piltdown man, Rangeomorphs, Sedimentary rock. Stromatolites, Vendozoa (Vendian biota).**)

Founder effect   The effect caused when a few individuals ("founders") derived from a large population begin a new colony. Since these founders carry only a small fraction of the parental population's genetic variability, different gene frequencies can become established in the new colony. (*See also* **Bottleneck effect, Peripatric speciation, Population genetics.**)

Fractal   A geometric shape that is repeated over a wide range of scales to produce a regular shape as seen in the Rangeomorphs of the Ediacaran biota. (*See also* **Ediacaran biota, Rangeomorphs.**)

Frequency dependent selection   Instances where the effect of selection on a phenotype or genotype depends on its frequency (for example, a genotype that is rare may have a higher adaptive value than when it is common). (*See also* **Artificial selection, Canalizing selection, Constraint, Darwinism, Directional selection, Disruptive selection, Fitness, Group selection, Inclusive fitness, Kin selection, Natural selection, Neo-Darwinism, Population genetics, Relative fitness, Selection, Selection coefficient (*s*), Sexual selection, Stabilizing selection, Survival of the fittest.**)

Frozen accident The hypothesis that an accidental event in the distant past was responsible for the presence of a universal feature in living organisms. Such events may include an accident in which the present genetic code was used by a group of early organisms that managed to survive some population bottleneck, thereby conferring this particular code on later organisms. (*See also* **Codon, Genetic code, Universal genetic code.**)

Fungi More closely related to animals than to plants, fungi are a lineage of multicellular organisms that lack chlorophyll, obtaining nutrition from parasitizing live hosts or digesting dead organic matter. (*See also* **Cell wall, Heterotroph, Metabolism, Opisthokonts, Organic, Saprophyte, Parasitism, Unikonta.**)

G

Galaxy A system of numerous stars, held together by mutual gravitational effects, and often spiral or elliptical in shape. (*See also* **Milky Way, Protogalaxy, Neutron stars (pulsars), Red giant** (Cosmology).)

Gamete A germ cell (eggs in females, sperm in males) that is usually haploid and that fuses with a germ cell of the opposite sex to form a zygote (usually diploid) at fertilization. (*See also* **Diploid, Fertilization, Gametophyte, Haploid, Life cycle, Meiosis, Ontogeny, Reproductive Incompatibility, Seed, Sexual reproduction, Zygote.**)

Gametophyte The haploid life cycle phase in plants. (*See also* **Alternation of generations, Gamete, Haploid, Meiosis, Polyploidy, Seed, Sporophyte.**)

Gastrula A (typically) cuplike embryonic stage in animal development that follows the blastula stage. Its hollow cavity (archenteron) is lined with endoderm and opens to the outside through a blastopore. (*See also* **Archenteron, Development, Embryology, Haeckel's gastrula hypothesis, Morphogenesis.**)

Gene A unit of hereditary genetic material composed of a segment of DNA (sequence of nucleotides), usually with a specific function—coding for a protein or sometimes coding for RNA. (*See also* **Allele, Allele frequency, Dominant allele, Gene pool, Genetic polymorphism, Heterozygote, Homozygote, Independent assortment, Linkage, Linkage equilibrium, Nucleotide, Neutral theory of molecular**

evolution, **Population genetics, Pseudogene, Recessive allele, Recessive lethal, Relative fitness, Segregation.**)

Genealogy A record of familial ties and ancestral connections among members of a group.

Gene duplication *See* **Duplication** (Genetics), **Gene family, Neofunctionalization**

Gene family Two or more gene loci in an organism sufficiently similar in nucleotide sequences to indicate they arose by duplication from a common ancestral gene. (*See also* **Duplication** (Genetics), **Mutation, Neofunctionalization, Paralogous genes, Serial ("iterative") homology.**)

Gene flow The migration of genes into a population from other populations by interbreeding. (*See also* **Deme, Group, Isolating mechanisms, Mendelian population, Migration (ecology), Migration (genetics), Parapatric, Population, Species, Subspecies.**)

Gene frequency The proportion of a particular allele among all alleles at a gene locus (also called allele or allelic frequency). (*See also* **Allele, Allele frequency, Chromosomes, Dominant allele, Gene, Gene locus, Genetic polymorphism, Heterozygote, Homozygote, Independent assortment, Linkage, Linkage equilibrium, Linkage map, Locus, Microsatellites, Population genetics, Quantitative trait loci (QTLs), Relative fitness, Segregation.**)

Gene locus The chromosomal position (nucleotide sequence) occupied by a particular gene. (*See also* **Chromosomes, Gene frequency, Genetic polymorphism, Heterozygote, Homozygote, Independent assortment, Linkage, Linkage equilibrium, Linkage map, Locus, Microsatellites, Quantitative trait loci (QTLs).**)

Gene migration *See* **Gene flow.**

Gene mutation *See* **Mutation.**

Gene pool All the genes present in the gametes of individuals in a sexually reproducing population. (*See also* **Allele, Allele frequency, Gene, Genetic polymorphism, Heterozygote, Homozygote, Independent assortment, Linkage, Linkage equilibrium, Neutral theory of molecular evolution, Population genetics, Recessive allele, Recessive lethal, Relative fitness, Segregation.**)

Gene regulation The processes by which a gene is turned on or off or by which the level of activity of a gene is controlled. (*See also* ***cis*-regulation, Developmental control gene, Downstream gene, Enhancer, Genetic assimilation, Heat shock proteins, Hierarchy, Homeotic mutations (homeosis), Modularity, Promoter, Regulatory gene, Repressor, Signal transduction, Transcription, *trans*-regulation, Upstream gene.**)

Gene therapy Human-directed repair or replacement of genes that cause inherited diseases. When confined to somatic (body) cells rather than to sex cells (sperm or eggs), such gene repairs are not passed on to future generations. (*See also* **DNA repair mechanisms, Eugenics, Genetic engineering, Reproductive technology, Transgenic.**)

Genetically modified organisms (GMOs) Animals, plants or single-celled organisms whose genetic characteristics have been altered by human intervention usually by the insertion or modification of a gene using the techniques of genetic engineering. Also, the alteration of an organism's genes following horizontal gene transfer or other processes of natural gene transfer (*See also* **Agriculture, Horizontal (Lateral) gene transfer, Transposons**)

Genetic assimilation The processes by which a trait elicited by an environmental stimulus can be genetically incorporated and appear developmentally in later generations in the absence of the environmental stimulus. (*See also* **Developmental control gene, Environment, Gene regulation, Genetic assimilation, Heat shock proteins.**)

Genetic code The sequences of nucleotide triplets (codons) on messenger RNA that specify each of the different kinds of amino acids positioned on polypeptides during the translation process. (*See also* **Amino acids, Codon, Degenerate (redundant) code, Frozen accident, Messenger RNA (mRNA), Peptide (polypeptide), Stop codon, Translation, Universal genetic code.**)

Genetic distance A measure of the divergence among populations based on their differences in frequencies of given alleles. (*See also* **Allele, Allele frequency, Genetic polymorphism, Linkage, Linkage equilibrium, Linkage map, Population genetics, Relative fitness, Segregation.**)

Genetic drift The random change in frequency of alleles in a population. (*See also* **Allele, Allele frequency, Genetic polymorphism, Independent assortment, Linkage equilibrium, Population genetics, Recessive allele, Relative fitness, Segregation.**)

Genetic engineering Manipulation of genetic material from different sources to produce new combinations that are then introduced into organisms in which such genetic material does not normally occur. (*See also* **Eugenics, Gene therapy, Reproductive technology, Transgenic.**)

Genetic load The loss in average fitness of individuals in a population because the population carries deleterious alleles or genotypes. (*See also* **Allele, Allele frequency, Balanced genetic load, Fitness, Genetic polymorphism, Independent assortment, Linkage, Linkage equilibrium, Muller's ratchet, Mutation, Mutational load, Population genetics, Recessive allele, Recessive lethal, Relative fitness, Segregation.**)

Genetic polymorphism The presence of two or more alleles at a gene locus over a succession of generations. (Called balanced polymorphism when the persistence of the different alleles cannot be accounted for by mutation alone. (*See also* **Allele, Allele frequency, Balanced polymorphism, Gene, Gene frequency, Gene locus, Heterozygote, Homozygote, Independent assortment, Linkage, Linkage equilibrium, Linkage map, Locus, Neutral theory of molecular evolution, Population genetics, Quantitative trait loci (QTLs), Recessive allele, Recessive lethal, Relative fitness, Segregation.**)

Genetic species concept Identification of species on the basis of shared genetic similarity. (*See also* **Biological species concept, Evolutionary species concept, Fixity of species, Morphology, Phylogenetic species concept, Polytypic species concept, Species.**)

Genetic variation *See* **Continuous variation, Covariation, Heritability, Variation, Wild type**.

Genome The complete genetic constitution of a cell or individual. (*See also* **Duplication** (Genetics), **Genome, Genotype, Repetitive DNA, Selfish DNA.**)

Genome duplication *See* **Duplication** (Genetics)

Genomics The science of the study of genes and genomes by analysis of their origin, structure and function; mapping and sequencing genes; analysis of the interactions between genes/genomes and their environments.

Genotype The genetic constitution of an individual. (*See also* **Environment, Evolvability, Fitness, Genetic load, Genome, Hardy–Weinberg equilibrium (principle), Heterozygote, Homozygote, Inclusive fitness, Phenotype, Random mating, Relative fitness, Reproductive success, Selection coefficient (s), Variability, Wild type.**)

Genus (pl. genera) A taxonomic category that stands between family and species. In taxonomic binomial nomenclature, the genus is used as the first of two words in naming a species; for example, *Homo* (genus) *sapiens* (species). (*See also* **Binomial nomenclature, Species**)

Geographical isolation The separation between populations caused by geographic distance or geographical barriers. (*See also* **Isolating mechanisms, Postzygotic mating barrier, Prezygotic mating barrier, Reproductive barriers, Reproductive isolation.**)

Geographic speciation *See* **Allopatric speciation**.

Geological dating The determination of the ages of rocks and strata using a variety of methods. (*See also* **Geological strata, Half-life (Radioactivity), Isotope, Radiometric dating** (Earth sciences).)

Geological strata A series of discrete layers laid down on top of each other as rock by natural forces such as lava flows, siltation, infilling of a marsh, and so on. (*See also* **Catastrophism** (earth sciences), **Geological dating, Igneous rock, Law of superposition, Metamorphic rock, Paleontology, Rock cycle, Sedimentary rock.**)

Geological time scale The correlation between rocks (or the fossils contained in them) and time in the past. (*See also* **Archean Eon, Cambrian Period, Catastrophism** (earth sciences), **Eon, Epoch, Era, Geological strata, Molecular clock, Period (geological), Phanerozoic Eon, Phyletic evolution, Precambrian Eon, Speciation, Stasis.**)

Germ layers The primary layers from which multicellular embryos develop. (*See also* **Diploblastic, Ectoderm, Endoderm, Mesoderm, Neural crest, Triploblastic.**)

Germ line *See* **Germ Plasm.**

Germ plasm Cells in animals that are exclusively devoted to transmitting hereditary information to offspring, in contrast to somatic cells, which comprise all other tissues of the body. (*See also* **Gamete, Germ plasm theory, Imprinting** (Genetics), **Somatic cells (or tissues).**)

Germ plasm theory The theory that, in some animals, gametes form from a special part of the egg (the germ plasm) and so cannot be influenced by environmental influences acting on the somatic tissues. (*See also* **Gamete, Germ plasm, Environment, Somatic cells (or tissues).**)

Glaciation The influence on Earth surface structure (landform) by the slow movement of masses of ice (glaciers) that have built up over geological time.

Globin genes A family of genes that codes for globin and related proteins (hemoglobin, myoglobin) proteins involved in oxygen transfer in animals. (*See also* **Hemoglobin.**)

Glycolysis The energy producing conversion of glucose to pyruvate under anaerobic conditions (fermentation). Subsequent steps may yield lactic acid or ethanol.

Gondwana The supercontinent in the Southern Hemisphere formed from the breakup of the larger Pangaea landmass about 180 Mya. Gondwana was composed of what are now South America, Africa, Antarctica, Australia, and India. (*See also* **Continental drift, Continents, Laurasia, Pangaea, Tectonic plates.**)

Granite A coarse-grained igneous rock commonly intruded into continental crust. (*See also* **Igneous rock, Rock cycle,**)

Gravitation The universal law discovered by Isaac Newton explaining how planets remain in their orbits because of a force (gravity). (*See also* **Collision theory, Condensation theory, Cosmology, Dwarf planets, Galaxy, Milky Way, Nebular hypothesis, Planetesimals, Planets, Protogalaxy, Protoplanet, White dwarf.**)

Great Chain of Being The eighteenth-century hypothesis that instead of a static universe, there is a continuous progression of stages leading to a superior supernatural being; the transformation of the "Ladder of Nature" into a succession of moving platforms. (*See also* **Biological species concept, Design, Fixity of species, Ladder of Nature.**)

Greenhouse gas A gas such as carbon dioxide (CO_2) and methane (CH_4) that absorbs infrared radiation, reduces the loss of heat from Earth's surface and so raises the global temperature. (*See also* **Archaebacteria, Methanogens, Snowball Earth.**)

Group A population(s) in a species that shares a geographically and/or ecologically identifiable origin and has gene frequencies and phenotypic characters that distinguishes it from other groups. (*See also* **Deme, Gene flow, Isolating mechanisms, Mendelian population, Migration (ecology), Migration (genetics), Parapatric, Population, Species, Subspecies.**)

Group selection A form of selection acting on the attributes of a group of related individuals in competition with other groups rather than only on the attributes of an individual in competition with other individuals. (*See also* **Competition, Competitive exclusion, Environment, Niche, Principle of divergence, Resource partitioning, Sexual selection.**)

Gymnosperms A group of vascular plants with seeds unenclosed in an ovary (naked); mainly cone-bearing trees. (*See also* **Angiosperms, Mesozoic Era, Seed, Telome hypothesis, Seed, Vascular plants.**)

H

Habitat The place and conditions in which an organism normally lives. (*See also* **Environment, Niche, Reproductive Incompatibility, Resource partitioning.**)

Haeckel's Gastraea theory The theory that animals developed from swimming hollow-balled colonies of flagellated protozoans.

Haldane's rule If one sex is absent, rare, or sterile in a cross between two species with sex-determining chromosomes, that sex is the heterogametic one. (*See also* **Dosage compensation, Homogametic sex, Sex chromosomes, Sex determination, Sex-linked genes, X chromosome, Y chromosome.**)

Half-life (radioactivity) The time required for the decay of one-half the original amount of a radioactive isotope. Each radioactive isotope has a distinctive and constant half-life. (*See also* **Geological dating, Isotope, Radiometric dating** (Earth sciences).)

Haplodiploidy A reproductive system found in some animals, such as bees and wasps, in which males develop from unfertilized eggs and are therefore haploid, while females develop from fertilized eggs and are therefore diploid. (*See also* **Diploid, Haploid.**)

Haploid Cells or organisms that have only one set (1n) of chromosomes. (*See also* **Alternation of generations, Aneuploidy, Diploid, Double fertilization, Gamete, Haplodiploidy.**)

Haplorhines The 'dry nosed primates' comprised of New World monkeys, Old World monkeys, and apes and human. (*See also* **Haplorhines, Hominoids, Primates, Strepsirrhines.**)

Hardy–Weinberg equilibrium (principle) The conservation of gene (allelic) and genotype frequencies in large populations under conditions of random mating and in the absence of evolutionary forces, such as selection, migration, and genetic drift, which act to change gene frequencies. (*See also* **Environment, Evolvability, Fitness, Genetic drift, Genetic load, Genotype, Heterozygote, Homozygote, Inclusive fitness, Migration, Phenotype, Random mating, Relative fitness, Reproductive success, Selection, Selection coefficient (*s*), Variability, Wild type.**)

Heat shock proteins Molecular chaperones that help other proteins maintain their 3-D conformation, prevent them from degrading, and mask changes mutations would otherwise have on the phenotype. An environmental shock can unmask these gene products creating new patterns and combinations of expressed proteins. (*See also* ***cis*-regulation, Developmental control gene, Downstream gene, Gene regulation, Genetic assimilation, Hierarchy, Homeotic mutations (homeosis), Modularity, Upstream gene.**)

Hemoglobin A protein with as basic unit of an iron-containing porphyrin (heme) that reversibly binds oxygen attached to a globin polypeptide chain. (*See also* **Globin genes.**)

Herbivores Animals that feed mainly on plants.

Heritability The degree to which variations in the phenotype of a character are caused by genetic differences. Traits with high heritabilities can be more easily modified by selection than traits with low heritabilities. A measure of an organism's potential to respond to selection. (*See also* **Inheritance, Lamarckian inheritance.**)

Hermaphrodite An individual possessing both male and female sexual reproductive systems. (*See also* **Self-fertilization, Sexual reproduction.**)

Heterochromatin A region of the eukaryotic chromosome that stains differently from euchromatin because of its tightly compacted structure. Compared to euchromatin, heterochromatin is characterized by few active genes and many more repetitive DNA sequences. (*See also* **Euchromatin.**)

Heterochrony A term Haeckel proposed to describe changes in timing of an organ's development during evolution. Its present usage varies but still hinges on a phylogenetic change in developmental timing, whether of one organ relative to other organs, or of one organ relative to the same ancestral organ. (*See also* **Biogenetic law, Development, Developmental control gene, Evolutionary developmental biology, Heterotopy, Morphogenesis, Ontogeny, Paedomorphosis, Preformationism.**)

Heterogametic sex The sex that produces two kinds of gametes for sex determination in offspring, one kind for males and the other for females. The heterogametic sex is the male in mammals and the female in birds. (*See also* **Sex chromosomes.**)

Heterosis (Hybrid vigor) The increase in vigor and performance that can result when two different, often inbred strains are crossed. (*See also* **Heterozygote advantage (superiority), Hybrid breakdown (inferiority, inviability, sterility), Hybrids, Hybridization (Genetics), Inbreeding, Inbreeding depression, Relative fitness, Reproductive Incompatibility.**)

Heterotopy A term Haeckel proposed to describe changes in the position of an organ during evolution. (*See also* **Evolutionary developmental biology, Heterochrony, Ontogeny.**)

Heterotroph An organism that cannot use inorganic materials to synthesize the organic compounds needed for growth but obtains them by feeding on other organisms or their products, such as a carnivore, herbivore, parasite, scavenger, or saprophyte. (*See also* **Fungi, Metabolism, Organic, Saprophyte.**)

Heterozygote The situation in a genotype or an individual in which the two copies of a gene are different; having different alleles at a particular gene locus on homologous chromosomes (for example, *Aa* in a diploid). (*See also* **Allele, Allele frequency, Chromosomes, Gene, Gene frequency, Genetic polymorphism, Homozygote, Independent assortment, Linkage, Linkage equilibrium, Linkage map, Locus, Neutral theory of molecular evolution, Population genetics, Recessive allele, Recessive lethal, Relative fitness, Segregation.**)

Heterozygote advantage (superiority) The superior fitness of some heterozygotes relative to homozygotes. (*See also* **Adaptive landscape, Fitness, Genetic load, Heterosis (Hybrid vigor), Hybrid breakdown (inferiority, inviability, sterility), Inclusive fitness, Population genetics, Relative fitness, Reproductive success, Selection, Selection coefficient (*s*).**)

Hierarchy A term used to designate an ordered grouping of the items within a system, often associated with increasing levels of complexity or organization. (*See also* **Complexity, Design, Developmental control gene, Homeotic mutations (homeosis), Progress, Reductionism.**)

Histones A family of small acid-soluble (basic) proteins that are tightly bound to eukaryotic nuclear DNA molecules and help fold DNA into thick chromosome filaments. (*See also* **Eukaryotic cells, Mitochondria, Nucleus, Prokaryotic cells.**)

Hitchhiking When a gene persists in a population, not because of selection, but because of close linkage to one or more selected genes. (*See also* **Gene frequency, Gene locus, Genetic polymorphism, Independent assortment, Linkage, Linkage equilibrium, Linkage map, Locus, Quantitative trait loci (QTLs).**)

Homeobox A transcriptional sequence of 180 base pairs that unites genes known as homeotic, homeobox, or Hox genes. The overall conservation of homeobox sequences, of the genes containing them, and of their linkage orders, indicate common developmental functions in different phyla preserved for many hundreds of millions of years, extending back to Precambrian times. (*See also* **Base pairs, Box (Boxes), Development, Developmental control gene, Homeobox, Homeodomain, Homeotic genes, Homeotic mutations.**)

Homeodomain A specific protein domain that is shared among a family of transcription factors (homeotic genes) and that is transcribed from a conserved DNA sequence known as the homeobox. (*See also* **Box (Boxes), Homeobox, Homeotic Genes, Homeotic mutations, Hox genes.**)

Homeosis *See* **Homeotic mutations**

Homeostasis The tendency of a physiological system to react to an external disturbance so that the system is not displaced from normal values.

Homeotic genes A class of genes that share the homeobox of 180 base pairs. (*See also* **Base pairs, Box (Boxes), Homeobox, Homeodomain, Hox gene.**)

Homeotic mutations (homeosis) Homeosis was defined by William Bateson as "something [that] has been changed into the likeness of something else." In modern genetic usage, homeotic mutations cause the development of tissue in an inappropriate position; for example, the *bithorax* mutations in *Drosophila* produce an extra set of wings. (*See also* **Development, Developmental control gene, Evolutionary developmental biology, Gene regulation, Heterochrony, Hierarchy, Homeotic genes, Morphogenesis, Ontogeny, Paedomorphosis, Regulation.**)

Hominid A member of the family Hominidae, which includes humans. Previously used exclusively for humans and the fossil species most closely related to humans. (*See also* **Anatomically modern humans, Fossils, Hominin, Hominoids, Macroevolution, Paleontology, Piltdown man.**)

Hominin All taxa on the human lineage after separation from the common ancestor with chimpanzee; members of the tribe *hominini*. In recent publications hominin has the same meaning as hominid in older literature. (*See also* **Anatomically modern humans,** **Australopithecines, Fossils, Hominid, Hominoids, Macroevolution, Paleontology, Piltdown man.**)

Hominoids A group (superfamily Hominoidea) that includes hominids (Hominidae), gibbons (Hylobatidae), and apes (Pongidae). (*See also* **Anatomically modern humans, Fossils, Haplorhines, Hominid, Hominin, Hominoids, Macroevolution, Paleontology, Piltdown man, Strepsirrhines.**)

Homogametic sex The sex that produces only one kind of gamete for sex determination in offspring, thus causing sex differences among offspring to depend on the kind of gamete contributed by the heterogametic sex. The homogametic sex is the female in mammals and the male in birds. (*See also* **Dosage compensation, Haldane's rule, Sex chromosomes, Sex-linked genes, X chromosome, Y chromosome**

Homologous *See* **Homology**

Homologous chromosomes The chromosomes that pair during meiosis, each pair usually possessing a similar sequence of genes. (*See also* **Chromosomes, Homozygote, Independent assortment, Meiosis, Segregation, Unequal crossing over.**)

Homology The similarity of characters (genes, structures, behaviors) in different species or groups because of their descent from a common ancestor. (*See also* **Analogy, Apomorphic character, Character, Character state, Convergence, Homoplasy, Novelty, Orthologous genes, Parallel evolution, Plesiomorphy, Symplesiomorphy, Synapomorphy.**)

Homoplasy Character similarity that arose independently in different groups through parallelism or convergence. (*See also* **Analogy, Character, Character state, Convergence, Homology, Parallel evolution, Phenotype, Plesiomorphy, Symplesiomorphy, Synapomorphy.**)

Homozygote The situation in a genotype or an individual in which the two copies of a gene are the same; having the same alleles at a particular gene locus on homologous chromosomes (for example, *AA*). (*See also* **Allele, Allele frequency, Chromosomes, Gene, Gene frequency, Gene locus, Genetic polymorphism, Heterozygote, Independent assortment, Linkage, Linkage equilibrium, Linkage map, Locus, Neutral theory of molecular evolution, Population genetics, Recessive allele, Recessive lethal, Relative fitness, Segregation.**)

Horizontal (Lateral) gene transfer Transmission of genes from one organism to another without reproduction. (*See also* **Gene therapy, Genetic engineering, Genetically modified organisms (GMOs), Transgenic, Transposons (transposable elements).**)

Host-plant specificity A term for the species-specific interactions between a pollinator (usually an insect) and a species of plant. (*See also* **Pollination.**)

Hox gene   A homeobox gene in vertebrates, taking the gene name from the first two letters of the orthologous gene in *Drosophila* and adding an x. (*See also* **Box (Boxes), Colinearity, Homeobox, Homeodomain, Homeotic Genes, Orthologous genes.**)

Hunter-gatherer societies   A form of human subsistence based on hunting animals for meat as well as foraging for other foods such as plants, insects, and scavenged meat.

Hybrid breakdown (inferiority, inviability, sterility)   Hybrids that suffer from loss of fitness and reproductive failure. (*See also* **Fitness, Genetic load, Heterosis (Hybrid vigor), Heterozygote advantage (superiority), Inbreeding, Inclusive fitness, Natural selection, Relative fitness, Reproductive Incompatibility Reproductive success, Selection, Selection coefficient (*s*).**)

Hybrid inferiority   *See* **Hybrid breakdown (inferiority, inviability, sterility.**)

Hybridization (Genetics)   Reproduction between individuals from two species to form a fertile offspring (the hybrid). (*See also* **Asexual reproduction, Fertility, Inbreeding, Inbreeding depression, Parthenogenesis Reproductive success, Sexual reproduction.**)

Hybridization (Molecular biology)   The formation of double-stranded molecules of DNA or RNA from a complementary single strand.

Hybrids (Genetics)   Offspring of a cross between genetically different parents or groups. (*See also* **Heterosis (Hybrid vigor), Hybridization (Genetics), Inbreeding, Reproductive Incompatibility.**)

Hybrid vigor   *See* **Heterosis (Hybrid vigor)**

Hydrogenosome   Double-membrane based organelle that generates hydrogen during the production of ATP. Found in trichomonads, fungi and other organisms and evolutionary related to the mitochondrion. (*See also* **Mitochondria, organelles.**)

Hydrothermal vents   Openings in the ocean floor from which water flows at temperatures up to 350°C and in and around which thermophilic organisms live. (*See also* **Thermophilic organisms.**)

I

Idealism   The philosophy that all objects, including the universe, have no independent existence apart from the minds of those perceiving them; perceiving objects as ideal forms.

Igneous rock   A rock such as basalt (fine-grained) and granite (coarse-grained), formed by the cooling of molten material from Earth's interior. (*See also* **Basalt, Crust** (Earth sciences), **Geological strata, Granite, Metamorphic rock, Rock cycle, Sedimentary rock.**)

Imaginal disks   Clusters of cells set aside in the larvae of insects such as *Drosophila* from which the structures of the adult fly are formed. (*See also* **Embryology, Larva, Life history, Metamorphosis** (zoology))

Imitative learning   Acquisition by an animal of a behavior through copying (imitating) the behavior of other individuals (*See also* **Behavior, Learning, Social learning.**)

Imprinting (behavior)   The learning process by which newborn/hatched organisms associate with an object (usually a parent or an adult of the same species, or an environment) and orient their behavior toward that object. Fish imprinting on a home river, newly hatched chicks imprinting on a parent are two well-studied examples. (*See also* **Culture, Dominance (social), Instinct, Learning, Sociobiology, Altruism.**)

Imprinting (Genetics)   The silencing of genes in germ cells by methylation of DNA. As a state, imprinting is usually sex-specific and is removed after development of a new individual begins from the imprinted gamete(s). (*See also* **Gamete, Germ plasm, Germ plasm theory.**)

Inbreeding   Mating between genetically related individuals, often resulting in increased homozygosity in their offspring. (*See also* **Fertility, Heterosis (Hybrid vigor), Hybridization** (Genetics), **Inbreeding depression, Relative fitness, Reproductive Incompatibility, Reproductive success, Self-fertilization, Self-incompatibility, Sexual reproduction.**)

Inbreeding coefficient (*F*)   The probability that the two alleles of a gene in a diploid organism are identical because they originated from a single allele in a common ancestor.

Inbreeding depression   Decrease in the average value of a character, or in growth, vigor, fertility, and survival, as a result of inbreeding. (*See also* **Fertility, Hybridization** (Genetics), **Inbreeding, Reproductive success, Sexual reproduction.**)

Inclusive fitness   The fitness of an allele or genotype measured not only by its effect on an individual but also by its effect on related individuals that also possess it (kin selection). (*See also* **Adaptive landscape, Allele, Allele frequency, Fitness, Gene, Genetic polymorphism, Genotype, Inclusive fitness, Independent assortment, Linkage, Linkage equilibrium, Population genetics, Relative fitness, Recessive allele, Recessive lethal, Relative fitness, Segregation, Selection, Selection coefficient (*s*).**)

Incomplete dominance   Instances where two different alleles of a gene in a heterozygote produce a phenotypic effect intermediate between the effects produced by the two homozygotes.

Independent assortment   A basic principle of Mendelian genetics—that a gamete will contain a random assortment of alleles from different chromosomes because chromosome pairs orient randomly toward opposite poles during meiosis. (*See also* **Allele, Allele frequency, Chromosomes, Gene, Gene frequency, Gene locus, Genetic polymorphism, Hitchhiking, Heterozygote, Homologous chromosomes, Homozygote, Independent assortment, Linkage, Linkage equilibrium, Linkage map, Locus, Meiosis, Mendelian genetics, Neutral theory of molecular evolution, Population genetics, Quantitative trait loci (QTLs), Recessive allele, Recessive lethal, Recombination, Relative fitness, Segregation.**)

Individual A single thing such as a person or plant that has a life with a beginning and an end. (*See also* **Populations.**)

Industrial melanism The effect of soot and other dark-colored pollution in industrial areas in increasing the frequency of darkly pigmented (melanic) forms perhaps because of selection by predators against non-pigmented or lightly pigmented forms.

Inflation (Cosmology) The theory that the universe expanded extremely rapidly after the big bang, and continues to expand. (*See also* **Big Bang theory, Evolutionary universe.**)

Ingroup Taxa more closely related to one another (monophyletic) than they are to any other group.

Inheritance The transmission of information from generation to generation. (*See also* **Blending inheritance, Extranuclear inheritance, Heritability, Inheritance, Inheritance of acquired characters, Lamarckian inheritance, Maternal inheritance, Pangenesis, Uniparental inheritance, Vertical transmission.**)

Inheritance of acquired characters The theory used by Lamarck to explain evolutionary adaptations—that phenotypic characters acquired by interaction with the environment during the lifetime of an individual are transmitted to its offspring. (*See also* **Blending inheritance, Inheritance, Inheritance of acquired characters, Lamarckian inheritance, Maternal inheritance, Preadaptation.**)

Innate Behavior *See* **Instinct.**

Innovation The evolution of a new character that facilitates adaptive radiation. (*See also* **Adaptive radiation, Novelty.**)

Instinct An inherited (innate), relatively inflexible behavior pattern that is often activated by one or several environmental factors (releasers). (*See also* **Altruism, Culture, Dominance (social), Homology, Imprinting** (behavior), **Learning, Phenotype, Sociobiology.**)

Insular dwarfism The process of the reduction of body size in organisms isolated on islands.

Intrinsic rate of natural increase (*r*) The potential rate at which a population can increase in an environment free of limiting factors. (*See also* **Carrying Capacity, Environment, Logistic growth curve, Niche, Population, Population explosion, Range.**)

Introgressive hybridization The incorporation of genes from one species into the gene pool of another because some fertile hybrids are produced from crosses between the two species.

Intron A nucleotide sequence (region of DNA) within a gene that is transcribed to produce mRNA but the mRNA is not translated into protein. (*See also* **Exon, Nucleotide, RNA splicing, Split gene, Transcription, Translation.**)

Intrusion (geology) Igneous rock inserted within or between geological strata rather than being deposited on Earth's surface.

Inversion (Genetics) An aberration in which a section of DNA or chromosome has been inverted 180 degrees, so that the sequence of nucleotides or genes within the inversion is now reversed with respect to its original order in the DNA or chromosome. (*See also* **Mutation.**)

Iridium (Ir) A hard corrosion-resistant metal in higher concentration in meteorites than on Earth and used to identify the K–T extinction boundary (*See also* **K–T extinction**).

Isolating mechanisms Biological mechanisms that act as barriers to gene exchange or interbreeding between populations. (*See also* **Gene flow, Geographical isolation, Mendelian population, Migration (genetics), Postzygotic mating barrier, Parapatric, Prezygotic mating barrier, Reproductive barriers, Reproductive isolation, Speciation.**)

Isotope One of several forms of an element, with a distinctive mass based on the number of neutrons in the atomic nucleus. Radioactive isotopes decay at a rate that is constant for each isotope and release ionizing radiation as they decay. (*See also* **Geological dating, Half-life** (Radioactivity), **Radiometric dating** (Earth sciences).)

J

"Junk" DNA" *See* **Selfish DNA.**

K

Karyotype The characteristic chromosome complement of a cell, individual, or species.

Kelvin scale (K) A temperature scale in which absolute zero (the point at which molecules oscillate at their lowest possible frequency, 273°C) is designated as 0K, and the boiling point of water as 373K.

Key Innovation *See* **innovation.**

Kingdom The highest inclusive category of taxonomic classification. Each kingdom includes phyla or subkingdoms. (*See also* **Binomial nomenclature, Classification, Domain, Numerical Taxonomy, Supergroups, Systematics, Taxon, Taxonomy.**)

Kin selection Processes that influence the survival and reproductive success of genetically related individuals (kin). This contrasts with selection confined solely to an individual and its own offspring. (*See also* **Altruism, Evolutionary stable strategies, Inclusive fitness.**)

Knuckle-walking Quadrupedal gait of chimpanzees and gorillas, performed by curling the fingers toward the palm of the hand and using the backs (dorsal surfaces) of the knuckles to support the weight of the front part of the body.

Krebs cycle The cyclic series of reactions in the mitochondrion in which pyruvate is degraded to carbon dioxide and hydrogen protons and electrons. The latter are then passed into the oxidative phosphorylation pathway to generate ATP.

K-selection Selection based on a population being maintained at or near the limit of its carrying capacity; selection is theoretically for improved competitive ability rather than for rapid numerical increase. (*See also* **r-selection**.)

K–T extinction The disappearance of many species, including all non-avian dinosaurs and pterosaurs at a time that corresponds to the Cretaceous (*Kreidezeit* in German)–Tertiary (K-T). (*See also* **Extinction, Coextinction, Iridium**.)

K value *See* **Carrying capacity** (K)

L

Ladder of Nature A concept based on Aristotle's view (the Scale of Nature) that nature can be represented as a succession of stages or ranks that leads from inanimate matter through plants, lower animals, higher animals, and finally to the level of humans. (*See also* **Biological species concept, Design, Fixity of species, Great Chain of Being**.)

Lamarckian inheritance The concept that the phenotype of an organism is itself hereditary: that characters acquired or lost during life experience, as well as characters that organisms acquire in order to meet environmental needs, can be transmitted to offspring. (*See also* **Inheritance of acquired characters, Inheritance, Use and disuse**.)

Language A structured system of communication among individuals using vocal, visual, or tactile signs to describe thoughts, feelings, concepts, and observations. (*See also* **Culture, Speech**.)

Larva (pl. **larvae**) A sexually immature stage in various animal groups, often with a form and diet distinct from those of the adult. (*See also* **Embryology, Imaginal disks, Life history, Metamorphosis** (zoology)

Lateral transmission *See* **Horizontal (lateral) gene transfer**.

Laurasia The supercontinent in the Northern Hemisphere (comprising what is now North America, Greenland, Europe, and parts of Asia) formed from the breakup of Pangaea about 180 Mya. (*See also* **Continental drift, Continents, Gondwana, Pangaea, Tectonic plates**.)

Law of superposition For any given series of sedimentary rocks, the oldest layers (strata) lie at the bottom and the youngest layers at the top. (*See also* **Catastrophism** (earth sciences), **Geological dating, Geological strata, Paleontology, Sedimentary rock, Uniformationarism**.)

Learned behavior *See* **Culture, Imitative behavior.**

Learning Acquisition by an animal of a behavior through experience, imitation and/or social interaction (*See also* **Culture, Dominance (social), Homology, Imitative learning, Imprinting** (behavior), **Instinct, Language, Learning, Phenotype, Reciprocal altruism, Sociobiology**.)

Life The capability of performing various organismal functions such as metabolism, growth, and reproduction of genetic material. (*See also* **Abiotic, Biotic, Metabolism, Organic, Organism, Saprophyte, Vitalism**.)

Life cycle The series of stages that takes place between the formation of zygotes in one generation of a species and the formation of zygotes in the next generation. (*See also* **Fertilization, Gamete, Ontogeny, Reproductive Incompatibility, Sexual reproduction, Zygote**.)

Life history All the stages of an individual life from its beginning to death. (*See also* **Embryology**, Larva, **Life cycle, Life-history traits, Metamorphosis** (zoology).)

Life-history trait Characters associated with the survival and reproduction of an individual such as number of offspring produced, age at reproduction, growth rate. (*See also* **Character, Life history, Life-history trait**.)

Light-year The distance traveled by light, moving at 186,000 miles a second, in a solar year; approximately 6×10^{12} miles or 9.5×10^{12} kilometers.

Lineage An evolutionary sequence, arranged in linear order from an ancestral (stem) group or species to a descendant (crown) group or species (or *vice versa*). (*See also* **Cladogenesis, Coextinction, Monophyletic, Parallel evolution, Phyletic evolution, Phylogenetic evolution, Phylogenetic tree, Phylogeography, Polyphyletic, Punctuated equilibrium, Species, Synapomorphy**.)

Linkage (genetic) The occurrence of two or more gene loci on the same chromosome. (*See also* **Chromosomes, Gene frequency, Gene locus, Hitchhiking, Independent assortment, Linkage equilibrium, Linkage map, Locus**.)

Linkage disequilibrium The absence of linkage equilibrium (that is, the presence of nonrandom associations between alleles at different loci). (*See also* **Hitchhiking, Linkage equilibrium**.)

Linkage equilibrium The attainment of genotypic frequencies in a population indicating that recombination between two or more gene loci has reached the point at which their alleles are now found in random genotypic combinations. (*See also* **Allele, Allele frequency, Gene, Genetic polymorphism, Hitchhiking, Heterozygote, Homozygote, Independent assortment, Linkage, Linkage map, Neutral theory of molecular evolution, Population genetics, Recessive allele, Recessive lethal, Recombination, Relative fitness, Segregation.**)

Linkage map The linear sequence of known genes on a chromosome obtained from recombination data. (*See also* **Chromosomes, Gene frequency, Gene locus, Independent assortment, Linkage, Linkage equilibrium, Locus, Restriction fragment length polymorphisms (RFLPs).**)

Lissamphibia The clade that includes all recent amphibians (frogs, toads, newts, salamanders, caecilians).

Lithosphere (Earth sciences) The term for the crust plus the uppermost portion of the mantle of Earth. (*See also* **Core, Crust, Earth, Lithosphere, Mantle** (Earth sciences).)

Living *See* **Life**

Living fossil An existing species whose similarity to a fossil taxon indicates that very few morphological changes have occurred over a long geological time. (*See also* **Fossils, Paleontology.**)

Locus (pl. **loci**) The site (nucleotide sequence) on a chromosome occupied by a specific gene. Some researchers use it more broadly as a synonym of gene. (*See also* **Allele, Chromosomes, Gene, Gene frequency, Gene locus, Independent assortment, Linkage, Linkage equilibrium, Population genetics, Segregation.**)

Logistic growth curve Population growth that follows a sigmoid (S-shaped) curve in which numbers increase slowly at first, then rapidly, and finally level off as the population reaches its maximum size or carrying capacity for a particular environment. (*See also* **Carrying Capacity, Environment, Intrinsic rate of natural increase, Niche, Population, Population explosion, Range.**)

Longevity The average life span of individuals in a population.

Lophotrochozoa One of two groups (the other being Ecdysoza) of protostomes, Lophotrochozoa being animals such as mollusks, annelids, brachiopods with tentacles. (*See also* **Ecdysoza.**)

M

Macroevolution The pattern of evolution at and above the level of the species. Fossils are the chief evidence for macroevolution. (*See also* **Fossils, Microevolution, Paleontology, Punctuated equilibrium.**)

Macromutation A hypothesis for the origin of a new species or higher taxonomic category by a single large mutation rather than by selection acting on many mutations.

Magnetosphere A magnetic "bubble" that envelops Earth, shielding it from incoming radiation, cosmic rays, ionized particles, and from particles carried by solar winds.

Mammals Homeothermic, vertebrates that suckle their offspring with milk produced in the mammary glands, have hair, three middle ear ossicles and a neocortex region in the brain. (*See also* **Lactation, Mammary glands, Marsupials, Mesozoic Era, Monotremes, Placentals, Tetrapods, Cenozoic (Caenozoic) Era.**)

Mammary glands One or more pairs of ventrally placed glands used by mammalian females for nursing offspring. (*See also* **Lactation.**)

Mantle (Earth sciences) A layer of partly plastic or ductile rock 2,900 km thick, that has experienced repeated melting and crystallization, and which constituting four-fifths of Earth's volume. (*See also* **Core, Crust, Earth, Lithosphere.**)

Marsupials Mammals of the infraclass Metatheria possessing, among other characters, a reproductive process in which tiny live young at an early stage of development are born and then nursed in a female pouch. (*See also* **Lactation, Mammals, Mammary glands, Mesozoic Era, Monotremes, Placentals, Tetrapods, Cenozoic (Caenozoic) Era.**)

Maternal inheritance Transmission of heredity information from the female parent by deposition of gene products into the egg cytoplasm. (*See also* **Blending inheritance, Extranuclear inheritance, Inheritance, Pangenesis, Protoplasm, Uniparental inheritance.**)

Maximum likelihood estimates A statistical method based on probability distribution that finds the maximum estimate of the likely distribution for a set or sets of data.

Meiosis The two eukaryotic cell (maturation) divisions that produce haploid gametes (animals) or spores (plants) from a diploid cell. One is a reduction division that ensures that each gamete or spore contains one representative of each pair of homologous chromosomes in the parental cell. (*See also* **Chromosomes, Diploid, Gamete, Gametophyte, Haploid, Homologous chromosomes, Homozygote, Independent assortment, Segregation.**)

Mendelian genetics **The science of the study of inheritance based on the law's proposed by Gregor Mendel.** (*See* **Independent assortment, Segregation.**)

Mendelian population A group of interbreeding, diploid individuals that exchange genes through sexually reproduction. (*See also* **Deme, Gene flow, Group, Isolating mechanisms, Migration (genetics), Parapatric, Population, Species, Subspecies.**)

Mendel's laws *See* **Independent assortment, Segregation.**

Mesenchyme One of the two fundamental cell types in animals, consisting of loosely connected cells in an extracellular matrix that may be "solid" as in bone or fluid as in blood. (*See also* **Epithelium.**)

Mesoderm The embryonic tissue layer between ectoderm and endoderm in triploblastic animals that gives rise to muscle tissue, kidneys, blood, internal cavity linings, and so on. (*See also* **Coelom (Coelome), Diploblastic, Ectoderm, Endoderm, Germ layers, Neural crest, Triploblastic.**)

Mesozoa A clade (phylum) of some 50 species, all parasitic on marine invertebrates, considered by some to be degenerate, but usually regarded as a separate phylum. (*See also* **Placozoa.**)

Mesozoic era The middle era of the Phanerozoic Eon, covering an approximately 220-million-year interval between the Paleozoic (ending about 248 Mya) and the Cenozoic (beginning about 65 Mya). It is marked by the origin of mammals in the earliest Period of the era (Triassic), the dominance of dinosaurs throughout the last two Periods of the era (Jurassic and Cretaceous), and the origin of angiosperms. (*See also* **Angiosperms, Cenozoic (Caenozoic) Era, Dicotyledons, Dinosaurs, Double fertilization, Era, Flower, Gymnosperms, Mammals, Monocotyledons, Paleozoic (Palaeozoic) Era, Phanerozoic Eon, Pollen, Seeds, Vascular plants.**)

Messenger RNA (mRNA) An RNA molecule produced by transcription from a DNA template, bearing a sequence of triplet codons used to specify the sequence of amino acids in a polypeptide. (*See also* **Promoter, Repressor, Transcription.**) (*See also* *cis*-regulation, **Codon, Enhancer, Micro RNA (miRNA), Nucleotide, Peptide (polypeptide), Promoter, Repressor, Ribosomal RNA (rRNA), RNA (ribonucleic acid), RNA interference (RNAi), Small interference RNA (siRNA), Transfer RNA (tRNA).**)

Metabolic pathway A sequence of enzyme-catalyzed reactions that convert a precursor substance to one or more end products. (*See also* **Enzyme, Metabolism, Proteinoids, Restriction enzymes, Transposons (transposable elements)**

Metabolism A network of enzyme-catalyzed reactions used by living organisms to maintain themselves. (*See also* **Aerobic metabolism, Anaerobic metabolism, Enzyme, Life, Metabolic pathway, Organic, Organism, Proteinoids, Restriction enzymes, Saprophyte, Transposons (transposable elements), Vitalism.**)

Metamorphic rock The class of rock that has been subjected to high but non-melting temperatures and pressures, causing chemical and physical changes. (*See also* **Crust** (Earth sciences), **Geological strata, Igneous rock, Rock cycle, Sedimentary rock, Stromatolites.**)

Metamorphosis (zoology) The transition from one form into another during the life cycle, for example, a larva into an adult. (*See also* **Imaginal disks, Larva, Life history.**)

Metaphyta The Kingdom containing the green plants; also known as Embryophyta, Plantae).

Metatheria *See* **Marsupials.**

Metazoa Multicellular animals.

Methanogens Single celled methane-generating organisms levels that only survive in the absence of oxygen and that are known to have existed at least 3.8 Bya. (*See also* **Archaebacteria, Greenhouse gas.**)

Microarray analysis The use of a gene chip to simultaneously obtain information about levels of mRNA expression in a very large number of genes.

Microevolution Evolutionary changes within populations of a species. (*See also* **Macroevolution.**)

Micro RNA (miRNA) Short (18 to 25 nucleotide) sequences of non-coding single-stranded RNA that regulate the translation of proteins in plants and animals by binding to matching target mRNAs leading to destruction of the mRNA. (*See also* **Messenger RNA (mRNA), Nucleotide, Ribosomal RNA (rRNA), RNA (ribonucleic acid), RNA interference (RNAi), Small interference RNA (siRNA), Transfer RNA (tRNA).**)

Microsatellites Tandem repeats of short di-, tri-, and tetra-nucleotide sequences, Such loci are abundant and mutate at a relatively high rate. (*See also* **Chromosomes, Gene frequency, Gene locus, Locus.**)

Microspheres Microscopic membrane-bound spheres formed when proteinoids are boiled in water and allowed to cool. Some cell-like properties, such as osmosis, growth in size, and selective absorption of chemicals, have been ascribed to them. (*See also* **Coacervates, Enzyme, Metabolic pathway, Metabolism, Microspheres, Proteinoids, Protocells.**)

Middle ear ossicles Small bones in the middle ear of tetrapods that transmit sound from the eardrum to the oval window of the inner ear.

Mid-oceanic ridges An 84,000 km-long series of mountain ranges on floors of the North and South Atlantic, Indian, and South Pacific Oceans. (*See also* **Continental drift, Continents, Crust, Subduction** (Earth sciences), **Plate tectonics, Sea-floor spreading, Tectonic plates.**)

Migration (Ecology) Movement of a population to a different geographical area or its periodic passage from one region to another. (*See also* **Ecology, Gene flow, Group, Population.**)

Migration (Genetics) The transfer of genes from one population into another by interbreeding (gene flow). (*See also* **Deme, Gene flow, Group, Isolating mechanisms, Mendelian population, Parapatric, Population, Species, Subspecies.**)

Milky Way (Cosmology)   A large spiral galaxy containing more than 200 billion stars and their planets, including our own solar system, and surrounded by more than a dozen much smaller galaxies. (*See also* **Collision theory, Dark ages (Cosmology), Galaxy, Gravitation, Neutron stars (pulsars), Planets, Red giant** (Cosmology).)

Mimicry   Resemblance of individuals in one species (mimics) to individuals in another (models) as a result of selection. (*See also* **Batesian mimicry, Mullerian mimicry.**)

Mimivirus   A genus of the largest viruses known (0.0005 mm diam.) that infects other viruses. (*See also* **DNA virus, RNA virus.**)

Missing link   A specimen or species of an extinct organism thought to be intermediate between two clades and so to provide a link between them.

Mitochondria   Organelle in eukaryotic cells that use an oxygen-requiring electron transport system to transfer chemical energy. Mitochondria have their own genetic material (circular mtDNA without histones) and generate some mitochondrial proteins by using their own protein-synthesizing apparatus. (*See also* **Chloroplasts, Endosymbiosis, Eukaryotic cells, Histones, Hydrogenosome**)

Mitochondrial DNA (mtDNA)   The DNA found within the intracellular organelles known as mitochondria. (*See also* **Mitochondria, Mitochondrial Eve.**)

Mitochondrial Eve   The name given to the common mitochondrial DNA human ancestor estimated from mtDNA sequences to have existed 140,000 years ago. (*See also* **Mitochondrial DNA, Y-chromosome Adam.**)

Mitosis   The mode of eukaryotic cell division that produces two daughter cells possessing the same chromosome complement as the parent cell. (*See also* **Chromosomes, Homologous chromosomes, Independent assortment, Meiosis, Segregation.**)

Modern synthesis (Evolutionary theory)   *See* **Neo-Darwinism.**

Modularity   The concept that units of life, such as gene networks, aggregations of cells and organ primordia, develop and evolve as units (modules) that interact with other modules. (*See also* **Developmental control gene, Gene regulation, Mosaic evolution.**)

Molecular clock   The rate at which nucleotides are substituted over evolutionary time. (*See also* **Geological time scale, Speciation, Stasis.**)

Molecular fossils   Traces of molecules produced by organisms found in rocks from the past. (*See also* **Ancient DNA, Fossils.**)

Monocotyledons   Flowering plants (angiosperms) in which the embryo bears one seed leaf (cotyledon). (*See also* **Angiosperms, Dicotyledons, Double fertilization, Flower, Mesozoic Era, Pollen, Seeds.**)

Monophyletic   A taxonomic group united by having arisen from a single ancestral lineage. (*See also* **Cladogenesis, Coextinction, Lineage, Parallel evolution, Phyletic evolution, Phylogenetic evolution, Phylogenetic tree, Phylogeography, Polyphyletic, Punctuated equilibrium, Species, Synapomorphy.**)

Monotremes   Egg-laying mammals, presently restricted to Australasia; echidnas (*Tachyglossus, Zaglossus*) and the platypus (*Ornithorhynchus*). (*See also* **Lactation, Mammals, Mammary glands, Mesozoic Era, Monotremes, Placentals, Tetrapods, Cenozoic (Caenozoic) Era.**)

Morganucodonts   The earliest mammals with a dentary-squamosal joint, which is a mammalian synapomorphy.

Morph   A subset of a population or a species with a phenotype that differs from the majority of the individuals (*See also* **Ecomorph, Phenotypic plasticity, Polymorphism.**)

Morphogenesis   Development of the form (morphology) of an organism or part of an organism. (*See also* **Development, Embryology, Epigenesis, Evolutionary developmental biology, Homeotic mutations (homeosis), Morphology, Ontogeny, Regulation.**)

Morphological species concept (Morphospecies)   Assemblages of individuals with shared structural characters that allow them to be separated from other assemblages. (*See also* **Biological species concept, Evolutionary species concept, Fixity of species, Genetic species concept, Morphology, Palaeontological species concept, Polytypic species concept, Species.**)

Morphology   The study (science) of the anatomical form and structure of organisms. (*See also* **Character, Cyclomorphosis, Ecomorph, Morph, Morphogenesis, Morphological species concept, Morphospace, Neoteny, Paedomorphosis, Phenotypic plasticity, Seasonal polymorphism, Sibling species.**)

Morphospace   A three-dimensional representation of the morphological characters of an organisms or groups of organisms used to show how much of the possible range of morphologies is expressed. (*See also* **Character, Morphogenesis, Morphology.**)

Mosaic evolution   The independent evolution of different parts of an organism. (*See also* **Modularity.**)

Mousterian tools   Stone tools, including spear points and scrapers made and used by *Homo neanderthalensis* and *Homo sapiens* 200,000 (*See also* **Acheulean tools, Anatomically modern humans, Oldowan tools, Stone tools, Upper Paleolithic tools.**)

Müllerian mimicry   Sharing of a common warning coloration or pattern among a number of species that are all dangerous or toxic to predators; resemblances maintained because of common selective advantage. (*See also* **Aposematic coloring, Batesian mimicry, Mimicry.**)

Muller's ratchet   The generalization that because of sampling errors, populations more easily lose that class of individuals bearing the fewest harmful mutations, so that classes with increasing numbers of such mutations tend to increase with time. (*See also* **Genetic load, Mutation, Mutational load, Population.**)

Multicellularity   Organisms composed of many cells and with specialization of cell function. (*See also* **Pluricellularity.**)

Multigene family   *See* **Gene family.**

Mutagenesis   Production of mutations by chemical treatment or radiation. (*See also* **Mutation, Radioactivity.**)

Mutation   A change in gene structure and often function; change in the nucleotide sequence of genetic material whether by substitution, duplication, insertion, deletion, or inversion. (*See also* **Chromosome, Deletion, Duplication** (Genetics), **Inversion, Mutagenesis, RNA editing.**)

Mutational load   That portion of the genetic load caused by production of deleterious genes through recurrent mutation. (*See also* **Genetic load, Muller's ratchet, Mutation, Population.**)

Mutualism   A relationship among different species in which the participants benefit. (*See also* **Altruism, Evolutionary stable strategies, Reciprocal altruism, Symbiont, Symbiosis, "Tit for tat."**)

N

Natural scientist (natural historian)   The term used before the twentieth century when referring to individuals we would now refer to as biologists, botanists, zoologists, physicists and so forth.

Natural selection   Differential reproduction or survival of replicating organisms caused by agencies other than humans. Since such differential selective effects are widely prevalent, and often act on hereditary (genetic) variations, natural selection is a common major cause for a change in the gene frequencies of a population that leads to a new distinctive genetic constitution (evolution). (*See also* **Adaptation, Adaptive landscape, Artificial selection, Canalizing selection, Continuous variation, Darwinism, Fitness, Frequency dependent selection, Group selection, Inclusive fitness, Kin selection, Neo-Darwinism, Population genetics, Relative fitness, Reproductive success, Selection, Selection coefficient (*s*), Sexual selection, Stabilizing selection, Survival of the fittest, Variation.**)

Nebular hypothesis   Originally proposed by Emanuel Kant and Laplace in the eighteenth century to explain the origin of the universe; a cloud of gas and dust (a nebula) collapses as gravitational force overcome the pressure of the gas. (*See also* **Collision theory,** **Condensation theory, Cosmology, Gravitation, Planetesimals, Planets, Solar Nebula.**)

Negative eugenics   Proposals to eliminate deleterious genes from the human gene pool by identifying their carriers and restraining or discouraging their reproduction. (*See also* **Eugenics, Positive eugenics.**)

Neo-Darwinism   The theory (also called the Modern synthesis) of evolution as a change in the frequencies of genes introduced by mutation, with natural selection as the most important, although not the only, cause for such changes. (*See also* **Fitness, Natural selection, Population genetics.**)

Neofunctionalization (Genetics)   Acquisition of a new role after genetic duplication, usually of a single gene but can be of a gene network or gene family. (*See also* **Duplication** (Genetics), **Gene family.**)

Neoteny   The retention of juvenile morphological traits in the sexually mature adult. (*See also* **Heterochrony, Morphology, Paedomorphosis.**)

Neural crest   A fourth germ layer found in vertebrates and from which arise skeletal tissues, cells that form the dentine of teeth, pigment cells, peripheral nerves and ganglia, some hormone-synthesizing cells, valves and septa of the heart, and other cell types. (*See also* **Ectoderm, Germ layers, Triploblastic.**)

Neutral mutation   A mutation that does not affect the fitness of an organism in a particular environment. (*See also* **Fitness, Inclusive fitness, Natural selection, Neutral theory of molecular evolution, Population genetics, Relative fitness, Selection, Selection coefficient (*s*).**)

Neutral theory of molecular evolution   The theory that most mutations that contribute to genetic variability (genetic polymorphism on the molecular level) consist of alleles that are neutral in respect to the fitness of the organism and that their frequencies can be explained in terms of mutation rate and genetic drift. (*See also* **Allele, Allele frequency, Fitness, Gene, Genetic polymorphism, Heterozygote, Homozygote, Independent assortment, Linkage equilibrium, Neutral mutation, Population genetics, Recessive allele, Recessive lethal, Relative fitness, Segregation.**)

Neutron stars (pulsars)   The collapsed core of an exploding star or supernova Type II. (*See also* **Collision theory, Dark ages (Cosmology), Galaxy, Milky Way, Red giant** (Cosmology).)

Niche   The environmental habitat of a population or species, including the resources it uses and its interactions with other organisms. (*See also* **Carrying Capacity, Character displacement, Competition, Competitive exclusion, Environment, Habitat, Intrinsic rate of natural increase, Logistic growth curve, Population, Population explosion, Range, Resource partitioning, Species sorting.**)

Normalizing selection   *See* **Stabilizing selection.**

Notochord   The dorsal axial supporting rod found in all chordates. (*See also* **Chordates.**)

Novelty   A feature in an organisms that has no homologue in the lineage. (*See also* **Homology, Innovation.**)

Nucleic acid   An organic acid polymer, such as DNA or RNA, composed of a sequence of nucleotides. (*See also* **Base, Base Pairs, Base substitutions, Complementary base pairs, DNA (deoxyribonucleic acid), Nucleotide, RNA (ribonucleic acid).**)

Nucleotide   A molecular unit consisting of a purine or pyrimidine base, a ribose (RNA) or deoxyribose (DNA) sugar, and one or more phosphate groups. (*See also* **Base, Exon, Gene, Intron, RNA splicing, Split gene, Transcription, Translation.**)

Nucleus   A membrane-enclosed eukaryotic organelle that contains all the histone-bound DNA in the cell (that is, practically all the cellular genetic material). (*See also* **Cytoplasm, Double fertilization, Eukaryotic cells, Fertilization, Histones, Organelles, Prokaryotic cells, Protoplasm.**)

Numerical taxonomy   A statistical method for classifying organisms by comparing them on the basis of measurable phenotypic characters and giving each character equal weight. The degree of overall similarity between individuals or groups is then calculated, and a decision is made as to their classification. (*See also* **Binomial nomenclature, Cladistics, Classification, Kingdom, Phylum, Supergroups, Systematics, Taxon, Taxonomy.**)

O

Oldowan tools   Stone tools made with a single blow of one stone by another, made and used by Homo habilis between 2.5 and 1 Mya. (*See also* **Acheulean tools, Mousterian tools, Stone tools, Upper Paleolithic tools.**)

Omnivores   Animals that feed on both plants and animals.

Ontogeny   The development of an individual from zygote to maturity. (*See also* **Biogenetic law, Development, Developmental control gene, Embryology, Fertilization, Epigenesis, Evolutionary developmental biology, Heterochrony, Homeotic mutations (homeosis), Life cycle, Morphogenesis, Paedomorphosis, Phylogeny, Preformationism, Zygote.**)

Operon   A cluster of coordinately regulated structural genes. (*See also* **Gene regulation, Structural gene, Transcription.**)

Opisthokonts   A subgroup of unikonts containing Unikonta some parasitic protists, choanoflagellates, fungi and animals. (*See also* **Fungi, Protista, Unikonta.**)

Order   A taxonomic category between class and family. A class may contain a number of orders, each of which contains a number of families. (*See also* **Class, Classification, Family, Systematics, Taxon, Taxonomy.**)

Organelles   Functional intracellular membrane enclosed bodies such as nuclei, mitochondria, and chloroplasts in eukaryotic cells. (*See also* **Chloroplasts, Cytoplasm, Eukaryotic cells, Hydrogenosome, Mitochondria, Nucleus, Prokaryotic cells, Protoplasm, Ribosomes, Thyalkoids**)

Organic   Carbon-containing compounds. Also refers to features or products characteristic of organisms or life. (*See also* **Amino acids, Life, Metabolism, Peptide (polypeptide), Photosynthesis, Saprophyte.**)

Organism   A living entity. (*See* **Biotic, Life, Metabolism, Vitalism.**)

Origination   In evolution, the first appearance or origin of a character, feature or organism.

Orthogenesis   The concept that evolution proceeds in a particular direction because of internal or vitalistic causes. (*See also* **Progress, Vitalism.**)

Orthologous genes (Orthologues)   Gene loci in different species that are sufficiently similar in their nucleotide sequences (or amino acid sequences of their protein products) to suggest they originated from a common ancestral gene. (*See also* **Homology, Hox gene, Paralogous genes.**)

Outgroup.   A taxon that diverged from an ingroup before the members of the ingroup diverged from one another. (*See also* **Ingroup, Taxon.**)

Oxidation–reduction   Reactions in which electrons are transferred from one atom or molecule (the reducing agent, which is oxidized by losing electrons) to another (the oxidizing agent, which is reduced by gaining electrons).

Ozone (O_3)   A molecule consisting of three atoms of oxygen. (*See also* **Ozone.**)

Ozone layer   That part of Earth's atmosphere between 15 and 30 km in altitude and with a high concentration of ozone. (*See also* **Atmosphere, Ozone, Secondary atmosphere**)

P

Paedomorphosis   The incorporation of adult sexual characters into immature developmental stages. (*See also* **Development, Embryology, Evolutionary developmental biology, Heterochrony, Heterotopy, Ontogeny, Preformationism.**)

Paleobiology   The study of the biology of extinct organisms and their ecosystems.

Paleolithic stone age   The time (Paleolithic era) in human prehistory between 2.5 million and 20,000 years ago characterized by the presence of tools made from rocks. (*See* **Stone tools, Upper Paleolithic tools.**)

Paleomagnetism   The magnetic fields of ferrous (iron-containing) materials in ancient rocks. (*See also* **Continental drift.**)

Paleontological species concept Application of the morphological species concept to identify fossils as species. (*See also* **Morphological species concept.**)

Paleontology The study of extinct fossil organisms or traces of organisms. (*See also* **Ancient DNA,** *Archaeopteryx,* **Burgess shale fossils, Chengjiang fauna, Ediacaran biota, Fossils, Hominid. Living fossil, Macroevolution, Period (geological), Phanerozoic Eon, Piltdown man, Rangeomorphs, Sedimentary rock, Vendozoa (Vendian biota).**)

Paleozoic (Palaeozoic) era The first era of the Phanerozoic Eon, extending from 545 to about 248 Mya. (*See also* **Cenozoic (Caenozoic) Era, Dinosaurs, Era, Mesozoic Era, Phanerozoic Eon.**)

Pangaea A very large supercontinent formed about 250 Mya comprising most or all of the present continental landmasses. (*See also* **Continental drift, Continents, Gondwana, Laurasia, Tectonic plates.**)

Pangenesis The theory of heredity, held by Darwin and others, that small, particulate "gemmules," or "pangenes" are produced by each of the various tissues of an organism and sent to the gonads where they are incorporated into gametes. (*See also* **Extranuclear inheritance, Inheritance, Maternal inheritance, Uniparental inheritance Vertical transmission.**)

Parallel evolution The evolution of similar characters in related lineages that do not share a recent common ancestor. (*See also* **Analogy, Character, Character state, Convergence, Homology, Homoplasy, Lineage, Phenotype, Plesiomorphy, Symplesiomorphy, Synapomorphy.**)

Paralogous genes (paralogues) Two or more different gene loci in the same organism that are sufficiently similar in their nucleotide sequences (or in the amino acid sequences of their protein products) to indicate they originated from one or more duplications of a common ancestral gene. (*See also* **Duplication (Genetics), Gene family, Mutation, Orthologous genes, Serial ("iterative") homology.**)

Parapatric Geographically adjacent, non-overlapping species or populations that at the zone of contact do not interbreed. (*See also* **Deme, Gene flow, Group, Isolating mechanisms, Mendelian population, Migration (genetics), Population, Species, Subspecies.**)

Parapatric speciation A population at the periphery of a species adapts to different environments but remains contiguous with its parent so that gene flow is possible between them. (*See also* **Allopatric speciation, Peripatric speciation, Sympatric speciation.**)

Paraphyletic A taxonomic grouping which includes some descendants of a single common ancestor, but not all. (*See also* **Cladogenesis, Lineage, Monophyletic, Phyletic evolution, Polyphyletic.**)

Parasitism An association between species in which individuals of one (the parasite) obtain their nutrients by living on or in the tissues of the other species (the host), often with harmful effects to the host. (*See also* **Arms race, Coevolution, Commensalism.**)

Parental investment Parental provision of resources to offspring with the effect of increasing the offspring's reproductive success.

Parsimony method Choice of a phylogenetic tree that minimizes the number of evolutionary changes necessary to explain species divergence. (*See also* **Cladistics, Phylogenetic tree.**)

Parthenogenesis Development of an egg without fertilization. (*See also* **Apomixis (apomictic), Asexual reproduction, Development, Embryology, Fertility, Fertilization, Epigenesis, Evolutionary developmental biology, Ontogeny, Preformationism, Sexual reproduction.**)

P–D axis A major axis of symmetry of bilateral animals where distal (D) is furthest from the midline and proximal (P) is closest to the midline. (*See also* Anterior-posterior axis, **A–P axis, Dorso-ventral axis, Axes of symmetry.**)

Pentadactyly (Pentadactyl) Literally, "five fingers," referring to the theory that the digits of tetrapods are build on an ancient five-digit plan, although we now know that the first tetrapods had more than five digits.

Peptide (polypeptide) An organic molecule composed of a sequence of amino acids covalently linked by peptide bonds (a bond formed between the amino group of one amino acid and the carboxyl group of another through the elimination of a water molecule). (*See also* **Amino acids, Organic.**)

Period (geological) A major subdivision of an era of geological time distinguished by a particular system of rocks and associated fossils. (*See also* **Ediacaran Period, Epoch, Era, Fossils, Geological strata, Geological time scale, Macroevolution, Paleontology.**)

Peripatric speciation When a population divides and becomes genetically isolated because of the budding off of a small completely isolated founder colony from a larger population. (*See also* **Allopatric speciation, Bottleneck effect, Founder effect, Parapatric speciation, Population genetics, Sympatric speciation.**)

Phagocytic Cellular engulfment of external material. (*See also* **Endocytosis.**)

Phanerozoic Eon A major division of the geological time scale marked by the relatively abundant appearance of fossilized skeletons of multicellular organisms, dating from about 545 Mya to the present. (*See also* **Archean Eon, Cenozoic (Caenozoic) Era, Dinosaurs, Eon, Geological time scale, Mammals, Mesozoic Era, Paleozoic (Palaeozoic) Era, Paleontology, Precambrian Eon.**)

Pharyngeal jaws An additional set of jaws that develop from posterior pharyngeal arches in some teleost fish.

Phenetics *See* **Cladistics**.

Phenotype The characters that constitute the structural, functional and behavioral; properties of an organism. (*See also* **Analogy, Character, Character state, Convergence, Genotype, Homology, Homoplasy, Parallel evolution, Plesiomorphy, Quantitative character, Symplesiomorphy, Synapomorphy**.)

Phenotypic plasticity Variation in the phenotype expressed in response to environmental changes and indicative of underlying genotypic plasticity. (*See also* **Canalization, Constraint, Developmental constraint, Discontinuous variation, Ecomorph, Genetic assimilation, Morph, Natural selection, Reaction norm, Variation**.)

Phloem The conducting tissue of green land plants that takes nutrient to all parts of the plant. (*See also* **Cambium, Vascular plants, Xylem**.)

Phospholipid Organic compounds composed of fatty acids, a phosphate group, and a nitrogenous base; major lipids of cell membranes.

Photons The quantum of electromagnetic energy lacking mass or electric charge. (*See also* **Cosmological red-shift, Red-shift**.)

Photosynthesis The synthesis of organic compounds from carbon dioxide and water through a process that begins with the capture of light energy by chlorophyll. (*See also* **Algae, Calvin cycle, Chlorophyll, Chloroplast, Cyanobacteria, Cytochromes, Metabolism, Thyalkoids**.)

pH scale [The negative logarithm of the hydrogen ion (H^-) concentration in an aqueous solution.] A scale used for measuring acidity (pH less than 7) and alkalinity (pH greater than 7), given that pure water has a neutral pH of 7.

Phyletic evolution Evolutionary changes within a single nonbranching lineage. Although new species are produced by this lineage over time (chronospecies) there is no increase in the number of species existing at any one time. (*See also* **Cladogenesis, Geological time scale, Lineage, Monophyletic, Parallel evolution, Paraphyletic, Phylogenetic evolution, Phylogenetic tree, Phylogeography, Polyphyletic, Principle of divergence, Punctuated equilibrium, Principle of divergence, Speciation**.)

Phylogenetic branching *See* **Cladogenesis, Phylogenetic evolution**.

Phylogenetic constraint An aspect of past evolutionary history that influences the range of variation in a character. (*See also* **Canalization, Developmental constraint, Directional selection, Discontinuous variation, Phenotypic plasticity, Variation**.)

Phylogenetic evolution (Also called branching evolution.) Evolutionary changes producing two or more lineages that diverged from a single ancestral lineage. (*See also* **Cladogenesis, Lineage, Monophyletic, Phyletic evolution, Principle of divergence**.)

Phylogenetic species concept A group of related organisms at the tip of a phylogeny that share a common ancestor. (*See also* **Biological species concept, Evolutionary species concept, Fixity of species, Genetic species concept, Morphology, Polytypic species concept, Species**.)

Phylogenetic systematics *See* **Cladistics**

Phylogenetic tree A branching diagram showing the relationships and evolutionary lineages of one or more groups of organisms. (*See also* **Evolutionary trees, Lineages, Parsimony, Phylogeny, Tree of life, Universal Tree of life**.)

Phylogeny The evolutionary history of a species or group of species in terms of their derivations and connections. A phylogenetic tree is a schematic diagram designed to represent that evolution—ideally, a portrait of genetic relationships. (*See also* **Biogenetic law, Ontogeny, Phylogenetic tree**.)

Phylogeography The study of the evolutionary processes regulating the geographic distributions of lineages/groups by reconstructing genealogies of individual genes, groups of genes or populations. (*See also* **Biogeography, Endemic, Lineage, Phylogenetic evolution, Zoogeography**.)

Phylum (pl. **phyla**) The major taxonomic category below the level of kingdom, used to include classes of organisms that may be phenotypically quite different but share some general characters or body plan. (*See also* **Arthropods,** *Bauplan* (pl. **Baupläne**), **Chordates, Class, Classification, Kingdom, Numerical Taxonomy, Systematics, Taxon, Taxonomy**.)

Piltdown man A fossil thought to be of an early human and discovered in Piltdown England in 19Subsequently shown to be a human cranium and the lower jaw of a female orangutan crafted to appear to be a fossil. (*See also* **Fossils, Hominid, Paleontology**.)

Piwi-interacting RNA (piRNA) Families of short molecules of 26 to 31 nucleotides found throughout the genome and active in silencing transposons. (*See also* **Transposons (transposable elements)**.)

Placentals Mammals of the infraclass Eutheria, possessing, among other characters, a reproductive process that uses a placenta to nourish their young until a relatively advanced stage of development compared to other mammalian groups (monotremes and marsupials). (*See also* **Lactation, Mammals, Mammary glands, Marsupials, Mesozoic Era, Monotremes, Tetrapods, Cenozoic (Caenozoic) Era**.)

Placozoa A clade (phylum) comprised of the single species *Trichoplax adhaerens*, with uncertain affinities to animals. (*See also* **Mesozoa**.)

Planetesimals Aggregations of dust and gas that form once aggregates reach one kilometer in diameter, allowing gravity to influence their formation. As they expand planetesimals can become protoplanets and finally planets. (*See also* **Condensation theory, Cosmology, Dwarf planets, Gravitation, Nebular hypothesis, Planets, Protoplanet, White dwarf.**)

Planets   Celestial bodies larger than asteroids or comets and illuminated by light from a star, which they orbit. (*See also* **Dwarf planets, Planetesimals, Protoplanet, White dwarf.**)

Plasmid   A self-replicating, circular DNA element that can exist outside the host chromosome. There are various kinds, some maintaining more than one copy per cell.

Plasticity.   *See* **Phenotypic plasticity**.

Plate tectonics   The geological theory that Earth's crust is comprised of moving plates. (*See also* **Continental drift, Crust, Earth, Sea-floor spreading, Subduction (Earth sciences), Tectonic plates.**)

Pleiotropy   Phenotypic effects of a single gene on more than one character. (*See also* **Character, Convergence, Gene.**)

Plesiomorphy   Instances when a species character is similar to that character in an ancestral species. (*See also* **Apomorphic character, Character, Character state, Convergence, Homology, Homoplasy, Parallel evolution, Symplesiomorphy, Synapomorphy.**)

Pluricellularity   The aggregation of cells or single-celled organisms into a many celled unit but without the cellular specialization associated with multicellularity. (*See also* **Multicellularity.**)

Pollen (Botany)   Small grains composed of protein, produced by the male organs (anthers) of seed plants (flowers, trees, grasses, weeds) and containing the male DNA. (*See also* **Angiosperms, Dicotyledons, Double fertilization, Flower, Gamete, Mesozoic Era, Monocotyledons, Pollination, Seeds.**)

Pollination   The process of the transfer of male sex cells (pollen) from the anther to the female reproductive organ (stigma) in flowering plants, usually accomplished by a pollinator such as an insect, by wind or by transfer via other animals. (*See also* **Host-plant specificity, Pollen, Self-fertilization, Self-incompatibility.**)

Polygene   A gene that interacts with other genes to produce an aspect of the phenotype.

Polymerase chain reaction (PCR)   A laboratory technique that can replicate a sequence of DNA nucleotides into millions of copies in a very short time.

Polymorphism   The presence of two or more genetic or phenotypic variants in a population. (*See also* **Balanced polymorphism, Cyclomorphosis, Ecomorph, Genetic polymorphism, Morph, seasonal polymorphism**)

Polypeptide   *See* **Peptide**.

Polyphyletic   The presumed derivation of a single taxonomic group from two or more different ancestral lineages through convergent or parallel evolution.

(*See also* **Cladogenesis, Convergence, Lineage, Monophyletic, Parallel evolution, Paraphyletic, Phyletic evolution, Phylogenetic evolution.**)

Polyploidy   When the number of chromosome sets (n) is greater than the diploid number (2n). (*See also* **Aneuploidy, Diploid, Haplodiploidy, Haploid, Meiosis.**)

Polytene chromosomes   Giant chromosomes that replicate many times without separating and show detailed banding patterns when stained. Especially studied in salivary glands of species of *Drosophila* and other similar insects. (*See also* **Autosomes, Chromosomes.**)

Polytypic species concept   A species consisting of individuals of two or more forms which may be varieties, races or subspecies. (*See also* **Biological species concept, Evolutionary species concept, Fixity of species, Genetic species concept, Morphological species concept, Species.**)

Population   A group of conspecific organisms occupying a more or less well-defined geographical region and exhibiting reproductive continuity from generation to generation. (*See also* **Bottleneck effect, Founder effect, Carrying Capacity, Deme, Environment, Gene flow, Group, Individual, Intrinsic rate of natural increase, Isolating mechanisms, Logistic growth curve, Mendelian population, Migration (ecology), Migration (genetics), Niche, Population, explosion, Range.**)

Population genetics   The study of evolution as represented by change in gene (now allele) frequencies because of the action of mutation, selection and genetic drift, a view of evolution that became known as the neo-Darwinian theory. (*See also* **Adaptive landscape, Allele, Allele frequency, Bottleneck effect, Fitness, Founder effect, Gene, Genetic load, Genetic polymorphism, Heterozygote, Heterozygote advantage (superiority), Homozygote, Inclusive fitness, Independent assortment, Linkage, Linkage equilibrium, Mutation, Mutational load, Natural selection, Neo-Darwinism, Neutral mutation, Neutral theory of molecular evolution, Recessive allele, Recessive lethal, Relative fitness, Reproductive success, Segregation, Selection, Selection coefficient (s).**)

Population explosion   A term used, usually for human populations, for rapid increases in numbers of individuals, with the implication that the rate of increase is unsustainable. (*See also* **Carrying Capacity, Environment, Intrinsic rate of natural increase, Logistic growth curve, Population, Range, Rate of increase (r)**)

Population growth   See **Carrying capacity (K), Density dependent, Density independent, Intrinsic rate of natural increase (r), Logistic growth curve**.

Positive eugenics   Proposals to increase the frequency of beneficial genes in the human gene pool by identifying their carriers and encouraging/permitting their reproduction. (*See also* **Eugenics, Negative eugenics.**)

Postzygotic mating barrier (isolating mechanism) The situation in which mating occurs between individuals of two species but hybridization is prevented because embryos fail to develop or if they begin to develop fail to survive. (*See also* **Geographical isolation, Isolating mechanisms, Prezygotic mating barrier, Reproductive barriers, Reproductive isolation.**)

Preadaptation A character that was adaptive under a prior set of conditions and later provides the initial stage (is "co-opted") for the evolution of a new adaptation under a different set of conditions. (*See also* **Adaptation, Co-option, Environment, Exaptation.**)

Precambrian Eon A major division of the geological time scale that includes all eras from Earth's origin about 4.5 Bya to the beginning of the Phanerozoic Eon, about 545 Mya. The Precambrian (also known as the Cryptozoic) is marked biologically by the appearance of prokaryotic cells about 3.5 Bya and small, non-skeletonized multicellular organisms in the Ediacarian Period about 50 or 60 My before the Phanerozoic. (*See also* **Archean Eon, Ediacaran biota, Eon, Geological time scale, Phanerozoic Eon, Precambrian Eon, Rangeomorphs, Vendozoa (Vendian biota).**)

Predation The killing and consumption of one living organism—the prey—by another, the predator.

Preformationism The theory that an organism is preformed at conception in the form of a miniature adult and that embryonic development consists of enlargement of preformed structures. (*See also* **Development, Embryology, Fertilization, Epigenesis, Evolutionary developmental biology, Morphogenesis, Ontogeny, Paedomorphosis.**)

Prezygotic mating barrier (isolating mechanism) The situation in which hybridization fails to occur because gametes fail to form if individuals from two species are crossed. (*See also* **Geographical isolation, Isolating mechanisms, Postzygotic mating barrier, Reproductive barriers, Reproductive isolation.**)

Primary atmosphere The first gaseous envelope surrounding Earth, which arose between 4.6 and 4.2 Bya, was composed of hydrogen and helium but not oxygen. (*See also* **Atmosphere, Secondary atmosphere.**)

Primary Endosymbiosis The formation of mitochondria and chloroplasts through the engulfment by a unicells of an aerobic bacterium. *See also* **Chloroplasts, Endocytosis, Endosymbiosis, Mitochondria, Secondary endosymbiosis, Symbiosis.**)

Primates Mammals of the Order Primates. (*See also* **Australopithecines, Haplorhines, Strepsirrhines.**)

Principle of divergence Hypothesis developed by Charles Darwin that competition between subpopulations favors specialization and separation of the popula-

tions to the point of speciation. (*See also* **Cladogenesis, Competition, Phyletic evolution, Principle of divergence.**)

Prion disease A state resulting from a change in the 3D structure of a prior. (*See also* **Prions.**)

Prions The smallest agent of infection, composed of a hydrophobic protein but neither DNA nor RNA. (*See also* **Prion disease.**)

Progenote The hypothetical ancestral cellular form that gave rise to archaebacteria, eubacteria and eukaryotes.

Progress (Evolution) A controversial concept in evolutionary biology that evolution has been accompanied by change reflected in increasing complexity (*See also* **Complexity, Design, Hierarchy, Orthogenesis, Reductionism, Vitalism.**)

Prokaryotic cells Single cells that lack histone-bound DNA, endoplasmic reticulum, a membrane-enclosed nucleus, and other cellular organelles found in eukaryotic cells. (*See also* **Cytoplasm, Eubacteria, Eukaryotic cells, Histones, Mitochondria, Nucleus, Protoplasm.**)

Promoter A DNA nucleotide sequence that enables transcription (RNA synthesis) by serving as the starting point for transcription and as the site for binding the enzyme RNA polymerase. (*See also* ***cis*-regulation, Enhancer, Gene regulation, Messenger RNA, Repressor, Signal transduction, Transcription.**)

Protein A macromolecule composed of one or more polypeptide chains of amino acids, coiled and folded into specific shapes based on its amino acid sequences. (*See also* **Amino acids.**)

Protein world The hypothesis that the first chemicals of life to evolve were DNA. (*See also* **DNA world, RNA world.**)

Proteinoids Synthetic polymers produced by heating a mixture of amino acids. Some show protein-like properties in respect to enzyme activity, color test reactions, hormonal activity, and so on. (*See also* **Enzyme, Metabolic pathway, Metabolism, Microspheres.**)

Protista One of the four eukaryotic kingdoms; includes protozoa, algae, slime molds, and some other groups. (Called "Protoctista" by some biologists.). (*See also* **Chloroplast, Excavata, Opisthokonts, Unikonta.**)

Protocells A membrane-bounded system containing molecules considered an early step in the origin of cells. (*See also* **Coacervates, Microspheres.**)

Protogalaxy (Cosmology) A galaxy in the process of formation. (*See also* **Galaxy, Gravitation, Milky Way.**)

Protoplanet (Cosmology) A forming planet that arises from the aggregation of planetesimals in a protoplanetary disc. (*See also* **Condensation theory, Cosmology, Dwarf planets, Gravitation, Planetesimals, Planets, White dwarf.**)

Protoplasm Cellular material within the plasma membrane but outside the nucleus. (*See also* **Cytoplasm,**

Eukaryotic cells, Nucleus, Organelles, Prokaryotic cells.)

Prototheria *See* **Monotremes**

Pseudogene A segment of DNA resembling a gene in structure but unable to be transcribed, often because of the presence of stop codons of nucleotide frame shifts. (*See also* **Gene, Repressor, Transcription (Genetics).**)

Pulsars *See* **Neutron stars**

Punctuated equilibrium The view that evolution of a lineage follows a pattern of long intervals in which there is relatively little change (stasis or equilibrium), punctuated by short bursts of speciation during which new taxa arise. (*See also* **Lineage, Phyletic evolution, Phylogenetic evolution, Phylogenetic tree, Species.**)

Purine A nitrogenous base composed of two joined ring structures, one five-member and one six-member, commonly present in nucleotides as adenine (A) or guanine (G). (*See also* **Base, Pyrimidine.**)

Pyrimidine A nitrogenous base composed of a single six-member ring, commonly present in nucleotides as thymine (T), cytosine (C), or uracil (U). (*See also* **Base, Purine.**)

Q

QTL effect The sum of the actions of quantitative trait loci on the variation of a feature of the phenotype. (*See also* **Quantitative character, Quantitative trait loci (QTLs).**)

Quantitative character A feature of the phenotype that can be numerically measured or evaluated; a feature displaying continuous variation. (*See also* **Character, Character state, Quantitative trait loci (QTLs), Variation.**)

Quantitative trait loci (QTLs) Regions of a chromosome containing genes (alleles) that influence a quantitative trait such as height or weight. (*See also* **Allele, Chromosomes, Gene, Gene frequency, Gene locus, Independent assortment, Linkage, Linkage equilibrium, Locus, Population genetics, QTL effect, Segregation.**)

Quantum evolution A rapid increase in the rate of evolution over a relatively short time.

R

Radiation (phylogenetic) *See* **Adaptive radiation.**

Radioactivity Emission of radiation by certain elements as their atomic nuclei undergo changes. (*See also* **Mutagenesis.**)

Radiometric dating (Earth sciences) The dating of rocks by measuring the proportions present of a ra-

dioactive isotope and the stable products of its decay. (*See also* **Geological dating, Half-life** (Radioactivity), **Isotope.**)

Random genetic drift *See* **Genetic drift.**

Random mating Sexual reproduction within a population regardless of the phenotype or genotype of the sexual partner (panmixis). (*See also* **Genotype, Hardy–Weinberg equilibrium (principle), Inclusive fitness, Phenotype, Relative fitness, Reproductive success, Selection coefficient (s), Variability.**)

Range The geographical limits of the region habitually traversed by an individual or occupied by a population or species. (*See also* **Carrying Capacity, Environment, Intrinsic rate of natural increase, Logistic growth curve, Population, Population explosion.**)

Rangeomorphs A group of organisms with a frond-like fractal organization that existed in complex ecological communities within the Ediacaran biota/fauna 570 to 575 Mya. One species, *Charnia wardi*, grew to heights of two meters. (*See also* **Biota, Burgess shale fossils, Chengjiang fauna, Ediacaran biota, Fauna, Fossils, Fractal, Macroevolution, Paleontology, Precambrian Eon, Vendozoa (Vendian biota).**)

Rapoport's rule Species adapted to cooler climates are distributed along a wider range of latitudes than are species of the same group adapted to warmer climates.

Rate of increase (r) *See* **Intrinsic rate of natural increase (r).**

Reaction norm A measure of the responsiveness of a character of the phenotype to a range of levels or concentrations of an environmental factor, temperature or predator abundance for example. A reaction norm is plotted as the phenotypic response (size, shape, number of elements and so forth) against the environmental parameter (temperatures, predator abundance and so forth). (*See also* **Genetic assimilation, Phenotypic plasticity, Variation.**)

Receptor A protein in the cell membrane that binds to a ligand to allow gene action to continue. (*See also* **Signal transduction.**)

Recessive allele An allele without phenotypic effect in a heterozygote. (*See also* **Allele, Allele frequency, Gene, Genetic polymorphism, Heterozygote, Homozygote, Independent assortment, Linkage, Linkage equilibrium, Neutral theory of molecular evolution, Population genetics, Relative fitness, Segregation.**)

Recessive lethal An allele which when homozygous causes lethality. (*See also* **Allele, Allele frequency, Dominant allele, Gene, Genetic polymorphism, Homozygote, Independent assortment, Lethal allele, Recessive allele, Relative fitness, Segregation.**)

Reciprocal altruism A mutually beneficial exchange of altruistic behavioral acts between individuals. (*See* *also* **Altruism, Evolutionary stable strategies, "Tit for tat".)**

Recombinant DNA   A DNA molecule composed of nucleotide sequences from different sources.

Recombination (Genetics)   A chromosomal exchange process that produces offspring that have gene combinations different from those of their parents. (Also used by some geneticists to describe the results of independent assortment.) (*See also* **Crossing over, Exon shuffling, Linkage equilibrium, Independent assortment.**)

Red giant (Cosmology)   A star that has exhausted its core hydrogen fuel so that only the center is hot enough to burn and in which hydrogen fusion continues in a shell around the core. (*See also* **Collision theory, Dark ages (Cosmology), Galaxy, Milky Way, Neutron stars (pulsars).**)

Red Queen hypothesis   The hypothesis that adaptive evolution in one species of a community causes a deterioration of the environment of other species. As a consequence, each species must evolve as fast as it can in order "to stay in the same place" (to survive).

Redshift (Cosmology)   The degree to which the photons reaching Earth have been stretched so that their wavelength moves towards the red end of the spectrum. (*See also* **Cosmological redshift, Photons**)

Reductionism   The theory that explanations for events at one level of complexity can or should be reduced to explanations at a more basic level. For example, that all biological events should be explained in the form of chemical reactions. (*See also* **Complexity, Design, Hierarchy, Progress.**)

Regulation (developmental biology)   The ability of an embryo or part of a developing embryo to compensate for the loss of parts. (*See also* **Development, Developmental control gene, Embryology, Epigenesis, Evolutionary developmental biology, Morphogenesis, Ontogeny, Preformationism.**)

Regulatory gene   A gene that controls other gene, either by turning them on or off or by regulating their rate. (*See also* **Developmental control gene, Enhancer, Gene regulation, Promoter, Repressor, Signal transduction, Transcription, Upstream gene**.)

Relative fitness   The relative reproductive success of an allele or genotype as compared to other alleles or genotypes. (*See also* **Adaptive landscape, Allele, Allele frequency, Fitness, Gene, Genetic load, Genetic polymorphism, Genotype, Heterozygote, Heterozygote advantage (superiority), Homozygote, Hybrid breakdown (inferiority, inviability, sterility), Inclusive fitness, Independent assortment, Linkage, Linkage equilibrium, Natural selection, Neutral theory of molecular evolution, Population genetics, Recessive allele, Recessive lethal, Reproductive success, Segregation, Selection, Selection coefficient (s).**)

Repetitive DNA   Nucleotide sequences of DNA that are repeated many times in the genome. (*See also* **Duplication** (Genetics), **Genome, Selfish DNA.**)

Replication   Doubling of DNA.

Repressor   A regulatory gene that produces a repressor (usually a protein) that binds to a particular nucleotide sequence and prevents transcription. (*See also* **Enhancer, Promoter, Regulatory gene, Repressor, Signal transduction, Transcription.**)

Reproduction   *See* **Apomixis (apomictic), Asexual reproduction, Assortative mating, Binary fission, Haplodiploidy, Hybrids, Inbreeding, Parental investment, Parthenogenesis, Pollination, Random mating, Reproductive Incompatibility, Reproductive technology, Sexual reproduction.**

Reproductive barriers   Any mechanisms that prevents individuals from two populations or species from breeding. (*See also* **Geographical isolation, Isolating mechanisms, Postzygotic mating barrier, Prezygotic mating barrier, Reproductive isolation.**)

Reproductive Incompatibility   Failure of two individuals to reproduce because of isolation by season, habitat or behavior; death of gametes, zygote or early embryo; formation of a hybrid with low viability; or hybrid sterility. (*See also* **Geographical isolation, Habitat, Isolating mechanisms, Niche, Postzygotic mating barrier, Prezygotic mating barrier, Reproductive barriers, Reproductive isolation.**)

Reproductive isolation   The absence of gene exchange between populations. (*See also* **Geographical isolation, Isolating mechanisms, Postzygotic mating barrier, Prezygotic mating barrier, Reproductive barriers, Speciation.**)

Reproductive success   The proportion of reproductively fertile offspring produced by a genotype relative to other genotypes. (*See also* **Fecundity, Fertility, Fitness, Heterozygote advantage (superiority), Genotype, Hybrid breakdown (inferiority, inviability, sterility), Hybridization** (Genetics), **Inbreeding, Inbreeding depression, Population genetics, Random mating, Relative fitness, Selection, Selection coefficient (s), Sexual reproduction.**)

Reproductive technology   Improvement of the human condition through the use of such medical advances as genetic screening, in *vitro* fertilization, surrogate pregnancies. (*See also* **Eugenics, Gene therapy, Genetic engineering, Transgenic.**)

Resource partitioning   The situation in which competing groups of organisms minimize the harmful effects of direct competition by using different aspects of their common environmental resources. (*See also* **Competition, Competitive exclusion, Environment, Group selection, Niche, Principle of divergence, Sexual selection, Species sorting.**)

Restriction enzymes Proteins that recognize particular nucleotide sequences and cut DNA molecules at or near those sequences. (*See also* **Enzyme, Restriction fragment length polymorphisms (RFLPs)**.)

Restriction fragment length polymorphisms (RFLPs) Differences between individuals in the size of DNA fragments for a particular DNA section cut by restriction enzymes. These are inherited in Mendelian fashion, and furnish a basis for estimating genetic variation. They also provide linkage markers used to track mutant genes between generations. (*See also* **Gene locus, Genetic polymorphism, Linkage, Linkage equilibrium, Linkage map, Locus, Microsatellites, Quantitative trait loci (QTLs)**.)

Rhizaria A supergroup of more than 4000 species of eukaryotic organisms including foraminiferans and radiolarians. (*See also* **Alveolates, Archaeplastida, Chromalveolata, Classification, Excavata, Kingdom, Supergroups, Unikonta**.)

Ribosomal RNA (rRNA) RNA sequences that are incorporated into the structure of ribosomes. (*See also* **Messenger RNA (mRNA), Micro RNA (miRNA), Nucleotide, ribosomes, RNA (ribonucleic acid), RNA interference (RNAi), Small interference RNA (siRNA), Transfer RNA (tRNA)**.)

Ribosomes Intracellular particles composed of ribosomal RNA and proteins that furnish the site at which messenger RNA molecules are translated into polypeptides. (*See also* **Messenger RNA (mRNA), Organelles, Peptide (polypeptide), Ribosomal RNA (rRNA), Transfer RNA (tRNA), Translation**.)

Ribozymes Sequences of RNA nucleotides that can perform catalytic roles. (*See also* **Enzymes**.)

Ring species Biological species in which subgroups (may be subspecies) overlap and interbreed except for those subgroups at the ends of the distribution, which often forms a U or ring. (*See also* **Biological species concept, Speciation, Species, Sympatric, Sympatric speciation**.)

RNA (ribonucleic acid) A typically single-strand nucleic acid, characterized by the presence of a ribose sugar in each nucleotide, and which serves as messenger, ribosomal, or transfer RNA in cells or as genetic material in some viruses. (*See also* **Base, Base Pairs, Base substitutions, Complementary base pairs, DNA (deoxyribonucleic acid), Nucleic acid, Nucleotide, RNA virus, Tobacco mosaic virus**.)

RNA editing Information changes in RNA molecules by the addition, deletion, or transformation of ribonucleotide bases after these molecules have been transcribed from their DNA templates. (*See also* **Chromosome, Deletion, Mutation**.)

RNA interference (RNAi) The process of regulating gene transcription using small interference RNA (siRNA). A class of short (18 to 25 nucleotide) sequences of non-coding RNA that repress the translation of proteins by binding to matching target mRNAs leading to degradation of the mRNA. (*See also* **Messenger RNA (mRNA), Micro RNA (miRNA), Nucleotide, Ribosomal RNA (rRNA), RNA (ribonucleic acid), Small interference RNA (siRNA), Transcription, Transfer RNA (tRNA)**.)

RNA splicing The joining of exons by the excision of introns. (*See also* **Exon, Exon splicing, Intron, Nucleotide, Split gene, Transcription, Translation**.)

RNA virus A small intracellular infectious agent, often parasitic, composed of RNA in a protein coat. (*See also* **DNA virus, RNA (ribonucleic acid), Tobacco mosaic virus, Virus**.)

RNA world The hypothesis that the first chemicals of life to evolve were DNA. (*See also* **Protein world, RNA world**.)

Rock cycle The natural sequence of events in which rocks transform from one type to another. (*See also* **Igneous rock, Metamorphic rock, Sedimentary rock**.)

r-selection Selection in populations subject to rapidly changing environments with highly fluctuating food resources. Theoretically, selection in such populations emphasizes adaptations for rapid population growth rather than for the competitive ability experienced in K-selected populations. (*See also* **K-selection**.)

S

Saltation The hypothesis that new species or higher taxa originate abruptly.

Saprophyte An organism that feeds on decomposing organic material. (*See also* **Fungi, Heterotroph, Inorganic, Life, Metabolism, Organic**.)

Sauropsids (Sauropods) A clade of amniotes that includes all recent and all or almost all extinct reptiles (excluding the Synapsida), and birds.

Scientific method A universal means of proposing a hypothesis, designing experiments, collecting data to test the hypothesis, interpreting the data in the context of past knowledge, and accepting or rejecting the hypothesis.

Sea-floor spreading Expansion of oceanic crust through the deposition of mantle material along oceanic ridges. (*See also* **Continental drift, Continents, Crust, Subduction** (Earth sciences), **mid-Oceanic ridges, Plate tectonics, Tectonic plates**.)

Seasonal isolation See **Reproductive incompatibility**.

Seasonal polymorphism The presence of different morphological types of a species in different seasons. (*See also* **Cyclomorphosis, Morphology, Polymorphism**.)

Secondary atmosphere The second gaseous envelope surrounding Earth, which arose between 4.2 and 3.5 Bya, was composed of water vapor and carbon dioxide but not oxygen. (*See also* **Atmosphere, Ozone layer, Primary atmosphere**.)

Secondary endosymbiosis. The acquisition of chloroplasts through engulfment by eukaryotic cells of another eukaryotic cell. (*See also* **Endocytosis, Endosymbiosis, Primary endosymbiosis, Symbiosis**.)

Sedimentary rock Formed by the hardening of accumulated particles (sediments) that had been transported by agents such as wind and water. Sedimentary rocks are the prime source of fossils. (*See also* **Catastrophism** (earth sciences), **Crust** (Earth sciences), **Geological dating, Geological strata, Igneous rock, Law of superposition, Paleontology, Metamorphic rock, Rock cycle, Uniformationarism**.)

Seed (Botany) A complex structure of plants containing the embryo along with parental diploid and haploid tissues. (*See also* **Angiosperms, Dicotyledons, Diploid, Double fertilization, Flower, Gamete, Gametophyte, Haploid, Meiosis, Mesozoic Era, Monocotyledons, Pollen, Seeds**.)

Segmentation The repetition of body structures along an animal's anterior–posterior axis, as found generally in annelids, arthropods, and chordates. (*See also* **Metamerism**.)

Segregation The Mendelian principle that the two different alleles of a gene pair in a heterozygote segregate from each other during meiosis to produce two kinds of gametes in equal ratios, each with a different allele. (*See also* **Allele, Allele frequency, Chromosomes, Gene, Genetic polymorphism, Heterozygote, Homologous chromosomes, Homozygote, Independent assortment, Linkage, Linkage equilibrium, Meiosis, Mendelian genetics, Neutral theory of molecular evolution, Population genetics, Recessive allele, Recessive lethal, Relative fitness**.)

Selection A composite of all the forces that cause differential survival and differential reproduction among genetic variants. When the selective agencies are primarily those of human choice, the process is called artificial selection; when the selective agencies are not those of human choice, it is called natural selection. (*See also* **Adaptive landscape, Artificial selection, Canalizing selection, Constraint, Darwinism, Directional selection, Disruptive selection, Fitness, Frequency dependent selection, Genetic load, Group selection, Heterozygote advantage (superiority), Hybrid breakdown (inferiority, inviability, sterility), Inclusive fitness, Kin selection, Natural selection, Neo-Darwinism, Neutral mutation, Neutral theory of molecular evolution, Population genetics, Relative fitness, Reproductive success, Selection coefficient (s), Sexual selection, Stabilizing selection, Survival of the fittest**.)

Selection coefficient (s) A relative measure of the effect of selection, usually in terms of the loss of fitness endured by a genotype, given that the genotype with greatest fitness has a value of 1. (*See also* **Adaptive landscape, Fitness, Genetic load, Genotype, Heterozygote advantage (superiority), Hybrid breakdown (inferiority, inviability, sterility), Inclusive fitness, Natural selection, Population genetics, Relative fitness, Reproductive success, Selection**

Self-assembly The spontaneous aggregation of macromolecules into biological configurations that can have functional value.

Self-fertilization Reproduction initiation by fertilization of a female gamete by a male gamete produced by the same individual. (*See also* **Hermaphrodite, Inbreeding, Pollination, Self-incompatibility, Sexual reproduction**.)

Self-incompatibility Mechanisms by which plants avoid inbreeding, usually by preventing their own pollen from reaching the ovules. (*See also* **Inbreeding, Pollination**.)

Selfish DNA The hypothesis that the persistence of DNA sequences with no discernible cellular function (for example, various repetitive DNA sequences) arises from the likelihood that, once present in the genome, they are impossible to remove without the death of the organism—that is, they act as "selfish," or "junk" DNA, which the cell has no choice but to replicate along with functional DNA. (*See also* **Duplication** (Genetics), **Genome, Repetitive DNA**.)

Serial ("iterative") homology Similarities between parts of the same organism, such as the vertebrae of a vertebrate or the different kinds of hemoglobin molecules produced by a mammal. The genetic basis for such homology can often be ascribed to gene duplications that have diverged over time but still produce somewhat similar effects. (*See also* **Duplication** (Genetics), **Gene family, Mutation, Paralogous gene**s.)

Sex chromosomes Those gene-carrying structures associated with determining the difference in sex. These chromosomes are alike in the homogametic sex (for example, XX) but differ in the heterogametic sex (for example, XY). (*See also* **Autosomes, Dosage compensation, Haldane's rule, Homogametic sex, Sex-linked genes, X chromosome, Y chromosome**.)

Sex determination The process by which the gender of an individual is determined, either by genes on sex chromosomes or environmentally. (*See also* **Sex chromosomes**.)

Sex linkage Genes linked on a sex chromosome. (*See also* **Chromosomes, Gene locus, Independent assortment, Linkage, Linkage equilibrium, Linkage map, Locus**.)

Sex ratio   The relative proportions of males and females in a population.

Sexual dimorphism   When males and females of a species have distinctive phenotypes.

Sexual reproduction   Zygotes produced by the union of genetic material from different sexes through gametic fertilization. (*See also* **Fertilization, Gamete, Hermaphrodite, Life cycle, Ontogeny, Reproductive Incompatibility, Self-fertilization, Zygote.**)

Sexual selection   The form of selection that acts directly on mating success through direct competition between members of one sex for mates (intrasexual selection), or through choices made between them by the opposite sex (epigamic selection), or through a combination of both selective modes. In any of these cases, sexual selection may cause exaggerated phenotypes to appear in the sex on which it is acting (large antlers, striking colors, and so on). (*See also* **Competition, Environment, Principle of divergence, Resource partitioning.**)

Short tandem repeat polymorphisms (STRPs)   Sequences or numbers of repeating nucleotide strings distributed throughout the genome.

Sibling species   A form of species so similar to each other morphologically that they are difficult to distinguish but that are reproductively isolated. (*See also* **Cryptic (Sibling) species, Morphology, Reproductive isolation, Speciation.**)

Sickle cell allele   A locus at the sickle cell gene (H^s) associated with hemolytic (sickle cell) anemia in humans. (*See also* **Sickle cell anemia.**)

Sickle cell anemia   Destruction of red blood cells in humans homozygous for the autosomal sickle cell gene, H^s; clinically presents as a usually fatal form of hemolytic anemia. (*See also* **Sickle cell allele.**)

Signal transduction   The process whereby a receptor is activated inside a cell to regulate gene activity. (*See also* **Developmental control gene, Enhancer, Gene regulation, Promoter, Regulatory gene, Repressor, Transcription, Upstream gene.**)

Sister group (sister taxon)   A group (taxon) that is the closest relative of another group (taxon). Derives from the concept that each significant evolutionary step marks a dichotomous split that produces two sister taxa equal to each other in rank. (*See also* **Taxon.**)

Small interference RNA (siRNA)   Twenty to twenty five nucleotide RNA sequences that assemble into RNA-induced silencing complexes (RISCs) that they guide to complementary RNA sequences, which they cleave and destroy. (*See also* **Codon, Messenger RNA (mRNA), Micro RNA (miRNA), Nucleotide, Ribosomal RNA (rRNA), RNA (ribonucleic acid), RNA interference (RNAi), Transfer RNA (tRNA), Translation.**)

Snowball Earth   The theory that for 10 My about 600 Mya when the average temperature was around −40°C, Earth was enveloped in a blanket of ice as much as 1 km thick, and that four such episodes may have occurred between 750 and 600 Mya. (*See also* **Catastrophism** (earth sciences).)

Social Darwinism   The theory that social and cultural differences in human societies (political, economic, military, religious, and so on) arise through processes of natural selection, similar to those that account for biological differences among populations and species. (*See also* **Neo-Darwinism, Population genetics.**)

Social learning   Acquisition by an animal of a behavior through interaction with other individuals interaction. (*See also* **Imitative learning, Learning**)

Social organization   The situation when there is a division of labor among different members of a species as seen in termites and many bees, ants and wasps. (*See also* **Eusociality, Superorganism**)

Sociobiology   The study of the biological basis of social behavior. (*See also* **Altruism, Culture, Dominance (social), Eusociality, Homology, Imprinting (behavior), Instinct, Learning, Phenotype, Reciprocal altruism.**)

Solar nebula (Cosmology)   The cloud of gas and dust from which the Solar System is proposed to have formed. (*See also* **Cosmology, Nebular hypothesis.**)

Solvent   A liquid that can dissolve molecules or compounds to form a solution. Water is regarded as a universal solvent.

Somatic cells (or tissues)   All body cells (also known as soma) other than those that produce sperm or eggs. (*See also* **Gamete, Germ plasm, Germ plasm theory, Imprinting** (Genetics).)

Space-time (Cosmology)   The four-dimensional construct proposed to contain all celestial bodies (stars, planets, galaxies, black holes). Because it continues to inflate, space-time is responsible for the ongoing expansion of the universe. (*See also* **Cosmological constant, Cosmology, Expanding universe, Galaxy, Gravitation, Planets.**)

Speciation   The splitting of one species into two or more new species (*See* **Cladogenesis, Phylogenetic evolution**) or the transformation of one species into a new species over time. (*See also* **Adaptive radiation, Anagenesis, Allopatric speciation, Principle of divergence, Parapatric speciation, Peripatric speciation, Reproductive isolation, Phyletic evolution, Sibling species, Stasis, Sympatric speciation**)

Species   A basic taxonomic category for which there are various definitions. Among these are an interbreeding or potentially interbreeding group of populations reproductively isolated from other groups (the biological species concept) and a lineage evolving separately from others with its own unitary evolutionary role and tendencies (the evolutionary species concept). (*See also* **Binomial nomenclature, Deme, Gene flow, Genus, Group, Isolating mechanisms, Lineages, Mendelian population, Migration (genetics), Parapatric, Polytypic species concept, Population, Sibling species, Species, Subspecies.**)

Species sorting   The hypothesis that two species cannot occupy the same niche indefinitely and that competition will result in the development of separate niches through resource partitioning. (*See also* **Competition, Competitive exclusion, Niche, Resource partitioning, Speciation, Species, Sympatric.**)

Speech   The ability to express thoughts, feelings, or perceptions and to transfer that information to others by the articulation of words and sentences. (*See also* **Broca's area, Language, Wernicke's area**)

Split gene   A gene with a nucleotide sequence consisting of exons and introns. (*See also* **Exon, Intron, Nucleotide, RNA splicing, Transcription, Translation.**)

Spontaneous generation   The theory that complex organisms can appear spontaneously from inert materials without biological parentage; life without parents.

Sporangia   The organ in plants and fungi that produces or contains spores.

Sporophyte   The diploid spore-producing stage of plants. (*See also* **Alternation of generations, Diploid, Meiosis, Seed.**)

Stabilizing selection   The form of selection that favors the survival of organisms in a population that are at an intermediate phenotypic value for a particular character, thus eliminating extreme phenotypes. (Also called normalizing selection.) (*See also* **Artificial selection, Canalizing selection, Directional selection, Disruptive selection, Fitness, Frequency dependent selection, Group selection, Natural selection, Neo-Darwinism, Population genetics, Relative fitness, Selection, Selection coefficient (s), Sexual selection, Survival of the fittest.**)

Stasis   A time (usually geological) without evident evolutionary change. (*See also* **Geological time scale, Molecular clock, Stabilizing selection.**)

Stem cells   Found in multicellular organisms and capable of producing more than one different cell type.

Stem group   The group of extinct organisms considered closest to and more basal than the most basal members of a clade. (*See also* **Ancestral state, Clade, Crown group.**)

Stone age   *See* **Paleolithic stone age, Stone tools, Upper Paleolithic tools.**

Stone tools   Stones that have been modified (usually by humans) for use in food capture/preparation or defense. (*See also* **Acheulean tools, Mousterian tools, Oldowan tools, Paleolithic stone age Stone tools, Upper Paleolithic tools.**)

Stop codon   One of the three messenger RNA codons (UAA, UAG, UGA) that terminates the translation of a polypeptide. (Also called chain-termination codon or nonsense codon.) (*See also* **Codon, Degenerate (redundant) code, Genetic code, messenger RNA (mRNA), Peptide (polypeptide), Translation, Universal genetic code.**)

Strata   *See* **Geological strata.**

Strepsirrhines   The `wet-nosed primates comprised of lemurs and lorises. (*See also* **Haplorhines, Primates.**)

Stromatolites   Laminated rocks produced by layered accretions of benthic microorganisms (mainly filamentous cyanobacteria) that trap or precipitate sediments. (*See also* **Molecular fossils, Fossils.**)

Structural gene   A DNA nucleotide sequence that codes for RNA or protein. Some definitions restrict this term to a protein-coding gene. (*See also* **Operon.**)

Subduction   (Earth sciences)   The process by which a tectonic plate descends (is subducted) beneath the edge of another plate into the mantle, often giving rise to earthquakes and/or volcanic activity. (*See also* **Continental drift, Plate tectonics, Sea-floor spreading, Tectonic plates.**)

Subspecies   A taxonomic subdivision of a species often distinguished by special phenotypic characters and by its origin or localization in a given geographical region. Like other species subdivisions (*see* **Group**), a subspecies can still interbreed with the remainder of the species. (*See also* **Deme, Gene flow, Group, Isolating mechanisms, Mendelian population, Migration (genetics), Parapatric, Population, Species, Subspecies.**)

Supergroups   Major groupings of eukaryotes at levels that replace kingdoms in some classifications. (*See also* **Alveolates, Archaeplastida, Binomial nomenclature, Chromalveolata, Cladistics, Classification, Excavata, Kingdom, Numerical Taxonomy, Rhizaria, Systematics, Taxon, Taxonomy, Unikonta**)

Supernova   *See* **Type Ia supernova, Type IIa supernova.**

Superorganism   A group of organisms such as an ant colony that functions as a single social unit; by analogy to groups of cells functioning as a single organisms. (*See also* **Behavior, culture, Eusociality, Social organization.**)

Survival of the fittest   The phrase coined by Herbert Spencer to describe the result of the operation of natural selection. (*See also* **Artificial selection, Canalizing selection, Constraint, Darwinism, Directional selection, Disruptive selection, Fitness, Frequency dependent selection, Group selection, Inclusive fitness, Kin selection, Natural selection, Neo-Darwinism, Population genetics, Relative fitness, Selection, Selection coefficient (s), Sexual selection, Stabilizing selection.**)

Survivorship The proportion of individuals born at a given time (cohort) who survive to a given age.

Symbiont A participant in the interactive association (symbiosis) between two individuals or two species. This term is often restricted to mutually beneficial associations. *See also* **Altruism, Evolutionary stable strategies, Mutualism, Reciprocal altruism, Symbiosis, "Tit for tat".**)

Symbiosis A close, often life-long, association between individuals of two species that usually benefits both individuals/species. (*See also* **Commensalism, Endosymbiosis, Mutualism, Symbiont**).

Sympatric Species or populations with coinciding or overlapping geographical distributions. (*See also* **Allopatric speciation, Parapatric speciation, Peripatric speciation, Ring species, Speciation, Sympatric speciation.**)

Sympatric speciation The form of speciation that occurs between populations occupying the same geographic range. (*See also* **Allopatric speciation, Parapatric speciation, Peripatric speciation, Speciation, Sympatric.**)

Symplesiomorphy A trait (feature, character) shared between two or more taxa and also shared with other taxa with which those two or more taxa share a last common ancestor; a shared or ancestral state of a character. (*See also* **Apomorphic character, Character, Character state, Homology, Plesiomorphy, Synapomorphy, Plesiomorphy.**)

Synapomorphy The possession by two or more related lineages of the same phenotypic character derived from a different but homologous character in the ancestral lineage. (*See also* **Apomorphic character, Character, Character state, Homology, Lineages, Plesiomorphy, Symplesiomorphy.**)

Synthetic theory of evolution *See* **Neo-Darwinism.**

Systematics Although defined by G. G. Simpson as the study of the diversity of organisms and all their comparative and evolutionary relationships, it is often used interchangeably with the terms classification and taxonomy. (*See also* **Binomial nomenclature, Cladistics, Class, Classification, Family, Kingdom, Numerical Taxonomy, Order, Phylum, Supergroups, Taxon, Taxonomy.**)

T

Taxon (pl. taxa) A taxonomic unit at any level of classification. (*See also* **Binomial nomenclature, Cladistics, Class, Classification, Family, Kingdom, Numerical Taxonomy, Order, Phylum, Sister group (sister taxon), Supergroups, Systematics, Taxonomy.**)

Taxonomy The principles and procedures used in classifying organisms. (*See also* **Binomial nomen-**clature, **Cladistics, Class, Classification, Family, Kingdom, Numerical Taxonomy, Order, Phylum, Supergroups, Systematics, Taxon, Type specimen.**)

Tectonic plates The fairly rigid plates composing Earth's crust with boundaries consisting of earthquake belts and volcanic chains. Continental masses ride on some of these plates, accounting for continental drift and such processes as the mountain building that occurs when these plates collide. (*See also* **Continents, Continental drift, Continents, Crust, Gondwana, Laurasia, Pangaea, Plate tectonics, Subduction.**)

Teleology The theory that natural processes such as development or evolution are guided by their final stage (*telos*) or for some particular purpose, for example, "the reason plants engage in photosynthesis and animals seek food is for survival, and the ultimate purpose of survival is for reproductive success."

Telome hypothesis Thin branches (*telomes*) evolved toward greater complexity and vascularization to produce the leaves and branches of ferns and vascular plants, or regressed toward a single unbranched form, producing bryophytes. (*See also* **Angiosperms, Dicotyledons, Enation hypothesis, Ferns, Gymnosperms, Mesozoic Era, Monocotyledons, Vascular plants.**)

Tetrapods Literal meaning, "four-footed." Commonly used to specify a member of the land-evolved vertebrates: amphibians, reptiles, and mammals.

Therapsids Previously known as "mammal-like reptiles," composed mainly of fairly large herbivorous and carnivorous forms, which were dominant reptilian stocks during the Permian and Triassic Periods. Mammals are descended from the cynodont therapsids.

Thermophilic organisms (thermophiles) Organisms that thrive at temperatures as high as 350°C as found in hydrothermal vents (*See also* **Extremophiles, Hydrothermal vents.**)

Thylakoids A vesicular organelle lined with a pigmented membrane and the site of photosynthesis in the chloroplasts of in plants and algae. (*See also* **Chloroplasts, Photosynthesis.**)

Tinkering (bricolage) The hypothesis that evolutionary change consists of minor modifications of existing genes, pathways and processes. (*See also* **Darwinism, Descent with modification Domestication, Homology.**)

Tissue A group of cells all performing a similar function in a multicellular organism.

"Tit for tat" An evolutionarily stable strategies in which an individual behaves cooperatively as the first move in an interaction, and then repeats its opponent's next move. *See also* **Altruism, Evolutionary stable strategies, Mutualism, Reciprocal altruism.**)

Tobacco mosaic virus A single stranded RNA virus that causes mottling and leaf discoloration of tobacco and related plants; the first virus discovered. (*See also* **RNA (ribonucleic acid), RNA virus, Virus**).

Tools   *See* **Stone tools.**

Trait   *See* **Character.**

Transcription (Genetics)   The process by which the synthesis of an RNA molecule (for example, messenger RNA) is initiated and completed on a DNA template by RNA polymerase enzyme. (*See also* ***cis*-regulation, Enhancer, Exon, Intron, Messenger RNA(mRNA), Promoter, Pseudogene, RNA interference (RNAi), RNA splicing, Repressor, *trans*-regulation.**)

Transcription factors   Gene products that bind to specific (regulatory) sequences of other genes to control their activity. (*See also* ***cis*-regulation, Enhancer, Promoter, RNA interference (RNAi), RNA splicing, Repressor, Transcription.**)

Transfer RNA (tRNA)   Relatively small RNA molecules (about 80 nucleotides long) that carry specific amino acids to the ribosome for polypeptide synthesis. Each kind of tRNA has a unique anticodon complementary to messenger RNA codons that specify the placement of particular amino acids in the polypeptide chain. (*See also* **Amino acids, Codon, Messenger RNA (mRNA), Micro RNA (miRNA), Nucleotide, Peptide (polypeptide), Ribosomal RNA (rRNA), RNA (ribonucleic acid), RNA interference (RNAi), Small interference RNA (siRNA), Translation.**)

Transgenic   An organism (sometimes a cell) containing a gene from another organism (or cell) that has been incorporated into its genome. (*See also* **Gene therapy, Genetic engineering, Horizontal (Lateral) gene transfer, Reproductive technology, Transposons (transposable elements).**)

Translation   The protein-synthesizing process that takes place on the ribosome, linking together a particular sequence of amino acids (polypeptide) on the basis of information received from a particular sequence of codons on messenger RNA. (*See also* **Amino acid, Codon, Degenerate (redundant) code, Genetic code, Messenger RNA (mRNA), Peptide (polypeptide), Ribosomes, Stop codon, Universal genetic code.**)

Translocation   The movement during division of a sequence of nucleotides to a different position in the genome.

Transposable elements   *See* **Transposons.**

Transposons (transposable elements)   Nucleotide sequences that produce enzymes to promote their own movement from one chromosomal site to another and that may carry additional genes such as those for antibiotic resistance. (*See also* **Enzyme, Horizontal (Lateral) gene transfer, Piwi interacting RNA (piRNA), Transgenic.**)

***trans*-regulation**   The process by which an element of DNA that is distant from a gene interacts with the promoter of that gene to control binding sites for transcriptional activators or repressors. (*See also* ***cis*-regulation, Developmental control gene, Enhancer, Messenger RNA, Promoter, Repressor, Transcription.**)

***trans*-regulatory elements**   *See* ***trans*-regulation.**

Tree of Life   A phylogenetic tree of all the organisms on Earth today. Based on the proposal by Charles Darwin that the similarities and differences between organisms reflected a single, branched and hierarchical tree of nature. (*See also* **Evolutionary trees, Phylogenetic tree, Universal Tree of life.**)

Triplet code   The genetic code in which a sequence of three nucleotides specifies the sequence of an individual amino acid. (*See also* **Genetic code, Degenerate (redundant) code, Universal genetic code.**)

Triploblastic   An animal that produces all three major types of cell layers during development—ectoderm, endoderm, and mesoderm. (*See also* **Diploblastic, Ectoderm, Endoderm, Germ layers, Mesoderm, Neural crest.**)

Type specimen   An individual of as species that defines a species and is housed in a museum where it may be examined. (*See also* **Taxonomy.**)

Typology   The study of organic diversity based on the principle that all members of a taxonomic group conform to a basic plan, and variation among them is of little or no significance. (*See also* **Archetype.**)

U

Ultraviolet (UV) radiation   Electromagnetic radiation at wavelengths between about 4 and 400 nanometers, shorter than visible light but longer than X-rays. UV radiation is absorbed by purine and pyrimidine ring structures and is therefore quite damaging to nucleic acid genetic material.

Unequal crossing over   The result of improper pairing between chromatids, causing their crossover products to differ from each other in the amounts of genetic material. (*See also* **Chromosomes, Crossing over, Homologous chromosomes, Independent assortment, Meiosis, Mitosis, Segregation.**)

Uniformitarianism   The theory in earth sciences, popularized by Lyell, that none of the forces active in past Earth history were different from those active today. (*See also* **Catastrophism** (earth sciences), **Law of superposition.**)

Unikonta   A supergroup of organisms including slime molds, many types of amoebae, some parasitic protists, choanoflagellates, fungi and animals. (*See also* **Amoebozoans, Alveolates, Archaeplastida, Chromalveolata, Classification, Excavata, Fungi, Kingdom, Opisthokonts, Protists, Rhizaria, Supergroups, Unikonta**)

Uniparental inheritance Transmission of heritable characters through a single parent, for example, mitochondria and chloroplasts through the female parent. (*See also* **Extranuclear inheritance, Inheritance, Maternal inheritance.**)

Universal genetic code The use of the same genetic code in virtually all living organisms. (*See also* **Codon, Degenerate (redundant) code, Frozen accident, Genetic code, Messenger RNA (mRNA), Peptide (polypeptide), Stop codon, Translation, Triplet code**)

Universal Tree of life (UtoL) The proposal that all organisms can be represented by branches and twigs on a single phylogenetic tree. (*See also* **Evolutionary trees, Phylogenetic tree, Tree of life.**)

Upper Paleolithic tools Hooks, barbed harpoons, needles, ivory beads and carved figurines made and used by *Homo sapiens*, some as early as 90,000 years ago. (*See also* **Acheulean tools, Anatomically modern humans, Mousterian tools, Oldowan tools, Paleolithic stone age, Stone tools.**)

Upstream gene A regulatory gene(s), that controls (is upstream of) another gene or genes. (*See also* ***cis*-regulation, Developmental control gene, Downstream gene, Enhancer, Gene regulation, Homeotic mutations (homeosis), Regulatory gene, Repressor, Signal transduction, Transcription.**)

Use and disuse A theory used by Lamarck to explain evolution as resulting from the transmission of characters that became enhanced or diminished because of their use or disuse, respectively, during the life experience of individuals. (*See also* **Lamarckian inheritance.**)

V

Variability The propensity of genotypes or phenotypes to vary. (*See also* **Environment, Evolvability, Genotype, Variation.**)

Variation A term commonly used to indicate differences in the qualitative or quantitative values of a character among individual members of a population, whether molecules, cells, or organisms. (*See also* **Canalization, Constraint, Continuous variation, Darwinism, Developmental constraint, Discontinuous variation, Natural selection, Phenotypic plasticity, Quantitative character, Reaction norm, Variability.**)

Vascular plants Land plants that have special water- and food-conducting vessels and tissues (xylem and phloem). (*See also* **Angiosperms, Gymnosperms, Mesozoic Era, Phloem, Seed, Telome hypothesis, Xylem.**)

Vendian Period. *See* **Ediacaran Period, Vendozoa (Vendian biota)**

Vendozoa (Vendian biota) The name given to organisms in the Ediacaran biota to designate them as a separate Kingdom of life. (*See also* **Ediacaran biota, Kingdoms, Rangeomorphs, Vendian Period.**)

Vertical transmission Passing of heredity from parent to offspring. (*See also* **Extranuclear inheritance, Inheritance, Maternal inheritance, Pangenesis, Uniparental inheritance.**)

Vestiges *See* **Vestigial organs**.

Vestigial organs Structures that appear to be small and functionless but can be shown to be homologous with ancestral organs and structures that were larger and functional.

Virus A small intracellular infectious agent, often parasitic, often composed of DNA or RNA in a protein coat, and that depends on the host cell to replicate its genetic material and to synthesize its proteins. (*See also* **DNA virus, Mimivirus, RNA virus, Tobacco mosaic virus.**)

Vitalism The concept that the activities of living organisms cannot be explained by any underlying physical or chemical principles but arise from unknowable internal or supernatural causes. (*See also* **Life, Metabolism, Organism, Orthogenesis, Progress.**)

W

Warning coloration *See* **Aposematic coloration**.

Wernicke's area A region of the brain, located where the temporal and parietal lobes meet, that is critical to the comprehension of speech and the formulation of coherent and meaningful speech. (*See also* **Broca's area, Speech.**)

White dwarf (Cosmology) The name for the final dying stage in the life of a planet, visible as a faint object in the sky. (*See also* **Condensation theory, Cosmology, Dwarf planets, Gravitation, Planetesimals, Planets, Protoplanet.**)

Wild type The most commonly observed phenotype or genotype for a particular character. Variations from wild type are considered mutants. (*See also* **Genetic load, Genotype, Hardy–Weinberg equilibrium (principle), Mutation, Phenotype, Relative fitness, Selection Variability, Variation.**)

X

X chromosome The name given in various groups to a sex chromosome usually present twice in the homogametic sex (XX) and only once in the heterogametic sex (XY or XO). (*See also* **Dosage compensation, Haldane's rule, Homogametic sex, Sex chromosomes, Sex-linked genes, Y chromosome.**)

X-linked genes Those genes located on the X chromosome. (*See also* **Sex linkage**.)

Xylem The conducting tissue of green land plants that takes water to all parts of the plant. (*See also* **Phloem, Vascular plants**.)

Y

Y chromosome A sex chromosome present only in the heterogametic sex (XY). (*See also* **Dosage compensation, Haldane's rule, Homogametic sex, Sex chromosomes, Sex-linked genes, X chromosome**.)

Y-chromosome Adam The name given to the common Y-chromosome human ancestor estimated from Y-chromosome nucleotide sequences to have existed 150,000 years ago. (*See also* **Mitochondrial DNA, Mitochondrial Eve**.)

Z

Zoogeography The science of the study of the distribution of animals in a particular region. (*See also* **Biogeography, Endemic, Phylogenetic evolution, Phylogeography**.)

Zootype A proposed stage in development characterized by the expression of a particular set of genes that governs spatial development in animals.

Zygote The cell formed by the union of male and female gametes. (*See also* **Fertilization, Gamete, Life cycle, Ontogeny, Reproductive Incompatibility, Sexual reproduction**.)

INDEX

Note: Page numbers in the Glossary are not included in this index. n after a page
number indicates a reference to a footnote.

innate, 530–532
instinctive, 529–532
larval and sex determination, 254–255
learned, 530–532
learned in chimpanzees, 531
parental age and chick survival in herring gulls, 532
mound construction by termites, 487
tool use as learned in chimpanzees, 529, 531
transmitted as cultural evolution, 429
and reproductive isolation, 227, 435, 455, 459–461
in rooks, 406
social and sex determination, 257
tent construction by caterpillars, 487. *See also* Instincts; Sociobiology; Tool use
Belief systems, 561–562
and accounts of origin of Earth, 567
China and inventions between 400 and 800 BC, 562n
and crops, 570
spirituality, 562. *See also* Cultures; Religion
Bicoid gene and A-P patterning, 338–339
Big Bang Theory. *See* Universe
Bioinformatics, 373
Biological evolution
intellectual origins of, 3–18
outpaced by cultural evolution, 529, 534–535. *See also* Cultural evolution
Biological species. *See* Species
Bipedalism, 507–508
environment and, 508
evolution of, 499, 507–508
identification of in primates, 499
and human brain size, 499, 507–508
key innovations and, 499, 508. *See also* Dinosaurs; Humans; Primates
Birds
Aepyornis maximus (flightless Pleistocene elephant bird), 495
Archaeopteryx, 42, 57–58, 483–485, 494
Confuciusornis from Lower Cretaceous, 485
Diatryma gigantea (Early Cenozoic flightless bird), 494–495
Dinornis maximus (flightless Pleistocene giant moa), 495
as dinosaurs, 57, 483–485
Dromornis stirtoni, flightless "thunder bird" from Australia, 495n
Enantiornithines, 485
Euornithes ("true birds"), 484
evolution of, 485
extant flightless birds, 495
extinct flightless birds, 494–495
feather evolution, 471, 494–496
flightlessness, 494–496
Gansus yumenensis, 485
as non-avian reptiles, 39, 483
origin of, 4383–485
penguins fly underwater, 496
phylogenetic relationships, 483n, 484–485
song divergence following habitat change, 462n
Waimanu (oldest penguin), 485. *See also Archaeopteryx*; Darwin's finches
Biston betularia (peppered moth). *See* Industrial melanism
Bithorax gene
mutations of, 340
thorax development in *Drosophila*, 340
transformation of haltere to wings in *Drosophila*, 340
transformation of third to second segment in *Drosophila*, 340
Blending inheritance, 225, 227, 228. *See also* Heredity, Inheritance
Blyth Edward,
Darwin's response to, 198
selection to maintain species, 198
Body plan, 22, 336–340
gradient model for determination of, 338–340. *See also* Animals; Axes of symmetry; *Bauplan (Baupläne)*; *Drosophila*; Homeobox genes; Insects
Bonnet, Charles, 5–6. *See also* Preformation; Progress
Bonobos. *See* Chimpanzees
Brain volume (cm³)

Australopithecus africanus, 536
bipedalism and human, 499, 507–508
cerebrum expanded in humans, 537
chimpanzee, 535
differential change in regions in primate evolution, 537
encephalization, 535
evolution of in genus *Homo*, 536
expansion of prefrontal cortex in humans, 526–537
gorillas, 535
growth and metabolism of human, 535
Homo erectus, 536
Homo floresiensis, 521
Homo habilis, 536
Homo sapiens, 535–537
Neanderthals, 536
relationship to body size, 535. *See also* individual species of *Homo*
Buffon, Georges
extinction of species, 196
species distinctions and, 19, 21, 24, 196
Bumpus, Hermon and stabilizing selection, 271–272
Burgess Shale fauna, 349, 351, 354
Anomalocaris, 351
Canadapsis perfecta, 354
diversity of, 351, 352
Hallucigenia sparsa, 351, 354
many phyla represented, 354
Marrella splendens, 354
Opabinia, 351
origin of, 354
plate tectonics and, 356
Wiwaxia, 351. *See also* Animals; Chengjiang fauna; Ediacaran biota
Bryan, William Jennings, 574. *See also* Scopes Monkey Trial
Bryophytes. *See* Plants

C

Calamites (Carboniferous horsetail), 176
Calvin cycle. *See* Photosynthesis
Cambrian explosion. *See* Animals
Canalization, 274, 285, 289
Conrad Waddington and, 289
canalizing selection, 274, 289
response to selection and, 285
^{13}Carbon, 60, 148, 152
^{13}Carbon/^{12}Carbon ratios, 60, 142, 148
^{14}Carbon, 78
Carbon dioxide (CO_2) levels
climate change, 493
composition in atmosphere, 83
during Snowball Earth, 492–493
Earth primary atmosphere, 69, 142
extinctions and, 472, 475
greenhouse effect and, 476
Precambrian, 493. *See also* Greenhouse effect
Catastrophism, 57
Cave-dwelling axolotl, 12
Cave-dwelling fish, 12, 344–346, 359. *See also* Mexican cavefish; Modularity
Cell division, 231–234
binary in prokaryotes, 151, 185, 231
chromosomes and, 231–234
cleavage in animal embryos, 249, 344
meiosis, 231–234, 249
mitosis, 151, 174, 231–233, 249
Cell membranes, 133
Central dogma of molecular biology, 126, 129–130, 324
does not hold for viruses, 324
Francis Crick proposes, 324
Character displacement
Darwin's finches, 410–412, 461–462

in meteorites, 103
peptides and, 112
proteinoids, 112
replicating in laboratory, 112
RNA first, 130–131
volcanoes and, 109–110. *See also* Genetic Code
Origin of species
in absence of geographical isolation (sympatric), 433, 442–450
genes associated with, 462–464
by geographical isolation, 456–461
hybridization and, 464, 464n
mechanisms of reproductive isolation, 227, 435, 459–461
sexual isolation and, 435, 465–466
sympatric speciation and, 433, 442–450. *See also* Allopatric speciation; Speciation; Sympatric speciation
Orthogenesis, 475. *See also* Extinctions
Owen, Richard, 46–47, 58. *See also* Homology
Oxygen
and animal origins, 352, 356
in atmosphere, 82, 142, 143–144
Cambrian explosion, 327, 349, 355–356
and cyanobacteria, 82, 145
in Earth atmosphere today, 83
and origin of multicellular organisms, 145
photosynthesis and, 82, 142–143, 160–163. *See also* Photosynthesis
Ozone layer, 79, 152

P

Paleomagnetism. *See* Continental drift
Paley, William, 565–568. *See also* Design
Pangaea, 87, 89, 490
fragmentation of, 87, 479, 490
mammalian radiations, 95–96, 478, 490–491
Pangenesis, 225–229,
August Weismann and, 228–229
Francis Galton's experiments on, 228, 548n
gemmules, 117, 120–122
compared with germ plasm theory, 225, 226, 228–229
continued study of, 228n. *See also* Darwin, Charles; Heredity; Inheritance
Paramecium
competitive exclusion between species of, 412, 414
Parallelism, 45, 46, 53–54
anteaters, 42, 53, 491
cacti and euphorbs, 45, 54
convergence and 46, 47
definition, 46, 53
of "wolf" phenotype, 42, 45, 53–54. *See also* Convergence; Homology
Parasitism
as coevolution, 421–422
and complexity, 171–172
fungal life style as, 423
life history stages, 171–172
myxoma virus and rabbits, 422
viruses and, 421–422. *See also* Coevolution
Parsimony. *See* Cladistics
Pax-6 gene, 54n, 569
PCR. *See* Polymerase chain reaction
Pesticides
DDT, 366–368
evolution of resistance to, 366–368
Pharyngeal jaws, 447–450. *See also* Cichlid fishes
Phenotype, 11
cultural evolution outpaces in humans, 529, 534–535
environment and, 258–259
more than one from single genotype, 258–259, 327
relation to genotype, 11, 229, 258–259, 302, 327, 434–435. *See also* Genotype

Phenotypic plasticity, 259, 435–440
in aquatic organisms, 436–437
cannibalistic tadpoles and, 437–438
catkin morph of N. American emerald moth, 438–439
chemical signals and, 436–437
cichlid fishes, 444–450
climate change and, 440
concept of, 259
Daphnia pulex (water flea), 436–437
distance birds migrate and, 440
ecomorphs, 259, 289
environment and, 259, 289–290, 439–440
habitat and, 440
Mediterranean blue tit (*Cyanistes caeruleus*) on mainland and islands, 439–440
migrating European birds and, 440
morphs, 259, 435, 438–439
moths, 270, 438–439
North American emerald moth (*Nemoria arizonaria*), 438–439
predator-prey induced, 433, 436–437
rotifers, 434, 436–437
sticklebacks, 374, 437–438
sympatric speciation and, 433–451
tadpoles, and, 437–438
tannins and catkin and twig morphs, 439n
in terrestrial organisms, 438–440
twig morph of N. American emerald moth, 438–439. *See also* Canalization; Reaction norms; Seasonal polymorphisms; Sympatric speciation
Phospholipids
cell membranes, 132–133
membranous droplets and, 132
of protocells, 132–133
Photons, 71
Photosynthesis
Calvin cycle, 160, 161
carbon source 160–163
chlorophyll and, 160–163
cyanobacteria and, 82, 145
cyclic, 160
noncyclic, 163
origin of, 144, 151, 167
oxygen and, 82, 143–144, 160, 161, 162, 163
photophosphorylation, 160, 167
primary productivity and, 144
as proton pump, 160, 162
in symbiosis between protozoan and algae, 177
thylakoids, 162, 163. *See also* Aerobic metabolism; Krebs cycle
Phyla
definition of, 34
increase in cell types, 170–171
most present in Cambrian Burgess Shale fauna, 354
Phylogenetic constraints, 286–288. *See also* Developmental constraints
Phylogenetic species. *See* Species
Phylogenetic trees
of Australopithecines, 501
of birds, 484
of cichlid fishes, 446
Darwin's illustration of, 217
of dinosaurs, 484
Ernst Haeckel and, 156
great apes, 499, 501, 504
hominins, 501
hominoids, 503
humans and apes, 501
New World monkeys, 502
Old World monkeys, 501–502
parsimony and, 39
patterns of branching, 217
of placental mammals, 288
of primates, 501, 502. *See also* Characters; Cladistics